PLANT SCIENCE
GROWTH, DEVELOPMENT, AND
UTILIZATION OF CULTIVATED PLANTS

FIFTH EDITION

Margaret J. McMahon
Department of Horticulture and Crop Science
The Ohio State University

Anton M. Kofranek
Emeritus
Department of Environmental Horticulture
University of California at Davis

Vincent E. Rubatzky
Emeritus
Department of Vegetable Crops
University of California at Davis

Prentice Hall

Boston   Columbus   Indianapolis   New York   San Francisco   Upper Saddle River
Amsterdam   Cape Town   Dubai   London   Madrid   Milan   Munich   Paris   Montreal   Toronto
Delhi   Mexico City   Sao Paulo   Sydney   Hong Kong   Seoul   Singapore   Taipei   Tokyo

Editorial Director: Vernon Anthony
Sr. Acquisitions Editor: William Lawrensen
Editorial Assistant: Lara Dimmick
Director of Marketing: David Gesell
Senior Marketing Coordinator: Alicia Wozniak
Marketing Assistant: Les Roberts
Associate Managing Editor: Alexandrina Benedicto Wolf
Project Manager: Kris Roach
Senior Operations Supervisor: Pat Tonneman
Operations Specialist: Laura Weaver
Art Director: Jayne Conte
Text and Cover Designer: Bruce Kenselaar

Manager, Cover Visual Research & Permissions:
  Cathy Mazzucca
Cover Art: Pingwin/iStockphoto
Media Director: Karen Bretz
Full-Service Project Management: Kelly Keeler,
  GGS Higher Education Resources/a division of PMG
Composition: GGS Higher Education Resources/a division
  of PMG
Printer/Binder: LSC Communications
Cover Printer: LSC Communications
Text Font: Garamond

Credits and acknowledgments borrowed from other sources and reproduced, with permission, in this textbook appear on appropriate page within text.

Library of Congress Cataloging-in-Publication Data

McMahon, Margaret.
  Plant science: growth, development, and utilization of cultivated plants / Margaret J. McMahon, Anton M. Kofranek, Vincent E. Rubatzky. — 5th ed.
    p. cm.
  Rev. ed. of: Plant science/Hudson T. Hartmann, William J. Flocker, Anton M. Kofranek.
  Includes bibliographical references and index.
  ISBN 978-0-13-501407-3 (alk. paper)
  1. Plants, Cultivated. 2. Botany, Economic. I. Kofranek, Anton M. II. Rubatzky, Vincent E. III. Hartmann, Hudson Thomas, 1914- Plant science. IV. Title. V. Title: Plant science.
  SB91.P18 2011
  630—dc22

                                                                    2010002986

14 17

SUSTAINABLE FORESTRY INITIATIVE Certified Chain of Custody
Promoting Sustainable Forestry
www.sfiprogram.org
SFI-01681

Prentice Hall
is an imprint of

www.pearsonhighered.com

Paper bound ISBN 10:   0-13-501407-7
             ISBN 13: 978-0-13-501407-3
Loose leaf   ISBN 10:   0-13-506850-9
             ISBN 13: 978-0-13-506850-2

The fifth edition of *Plant Science* is dedicated to my grandfather, Harold Deeks, and my parents, Bob and Marty McMahon, who taught me to love growing plants and to appreciate the science of growing plants.

About the Authors

Margaret (Peg) J. McMahon
Associate Professor of Horticulture and Crop Science,
The Ohio State University

Peg is a fourth generation horticulturist who grew up working on the family farm and in the ornamental and vegetable greenhouses owned by her family. She earned a B.S. in Agriculture with a major in Horticulture from The Ohio State University in 1970. After graduation she worked for fifteen years as a grower and propagator for Yoder Bros., Inc. (now Aris Horticulture, Inc.), a multinational greenhouse company. She started at their Barberton, Ohio facility, transferring to operations in Salinas, California then Pendleton, South Carolina.

While in South Carolina, she earned her M.S. in Horticulture and Ph.D. in Plant Physiology from Clemson University in 1988 and 1992, respectively. Peg's masters project focused on the early detection of chilling injury in tropical and subtropical foliage plants. Her Ph.D. research was in photomorphogenesis, specifically, the use of far-red absorbing greenhouse glazing materials as a potential technology to reduce unwanted stem elongation in greenhouse crop production.

In 1994 she started a faculty position in the Department of Horticulture and Crop Science (H&CS) at The Ohio State University. She teaches several classes including both the introductory and senior capstone courses in horticulture and crop science, as well as classes in indoor gardening (interiorscapes) and greenhouse crop production. She also teaches a graduate level class to help prepare the graduate teaching assistants in H&CS to be effective teachers.

Dr. McMahon has published over 30 peer-reviewed articles as well as many trade journal articles on photomorphogenesis, teaching methods, and other subjects.

Anton M. Kofranek
Retired as Professor Emeritus from University of California
at Davis in 1987

Dr. Kofranek obtained his B.S. in Horticulture at the University of Minnesota in 1947, an M.S. in Plant Physiology at Cornell University in 1949, and his Ph.D. in Plant Physiology at Cornell University in 1950. His first academic appointment was as Instructor in the Department of Horticulture at the University of California at Los Angeles in September 1950. In 1968 when the department was transferred to the University of California at Davis and renamed as the Department of Environmental Science, he also relocated and continued his floricultural research and teaching at Davis until retirement in 1987.

His research specializations dealt with photoperiod research for numerous floricultural crops. Other interests included postharvest physiology and nutritional investigations of flower crops. Dr. Kofranek produced over 200 publications. He co-authored *Hartmann's Plant Science: Growth, Development, and Utilization of Cultivated Plants* and *The Azalea Manual.*

Throughout his career, Dr. Kofranek was highly respected and appreciated by persons in all segments of the floriculture trade, and he received numerous awards and recognition from producers, wholesalers, landscape architects, and other allied industry segments. He has been repeatedly recognized for his efforts by the California

Flower Growers. He also received considerable academic recognition, and in 1979 was installed as a Fellow of the American Society for Horticultural Sciences. Dr. Kofranek carried out sabbatical study programs in The Netherlands, Israel, England, South Africa, and at Cornell University. He was a consultant with USAID programs with Egypt and India. His extensive travels led to collaboration with scientists in several countries.

Vincent E. Rubatzky
Retired as University of California Extension Vegetable Specialist, Emeritus in 1995

Dr. Rubatzky obtained his B.S. in Vegetable Crops at Cornell University in 1956, his M.S. in Plant Physiology at Virginia Polytechnic Institute in 1958, and his Ph.D. in Plant Physiology and Horticulture at Rutgers University in 1964. His first academic position was as Extension Vegetable Specialist at the University of California at Davis in 1964, where he conducted his extension outreach and research program until retirement.

His research dealt with variety development and evaluation, harvest mechanization, crop physiological disorders, and crop scheduling and cultural practices. He also served as a liaison with several California vegetable industry organizations and groups. For several years Dr. Rubatzky taught a class on Evolution, Biology, and Systematics of Vegetables, as well as a Field Study of the California Vegetable Industry course. Sabbatical studies and consultations within Poland, France, United Kingdom, Italy, The Netherlands, Peru, and Ecuador provided many travel experiences. In 1995, he was elected as an American Society for Horticultural Sciences Fellow, Man of the Year by the Pacific Seedman's Association in 1987, and received additional recognition from other organizations. Following retirement Dr. Rubatzky continued to serve on a voluntary basis as Specialist, Emeritus, until mid-1999. Presently residing in New Jersey, he collaborates with colleagues in vegetable production research in California and New Jersey.

Dr. Rubatzky has produced many applied research and extension publications. He co-authored *Hartmann's Plant Science: Growth, Development, and Utilization of Cultivated Plants; World Vegetables: Principles, Production, and Nutritive Values;* and *Carrots and Related Vegetable Umbelliferae*, and was an editor for the *Third International Symposium for Diversification of Vegetable Crops*, the *Atlas of the Traditional Vegetables in China*, and several other publications on vegetables.

Preface

Enter the fascinating and colorful world of plant science through the fifth edition of *Plant Science*, now in color. Discover why we depend on plants and the people who know how to grow them for our survival. Find out how plants provide sustenance for our bodies and add enjoyment to our lives (food for body and soul). Learn how to grow, maintain, and utilize plants to benefit people and the environment. Whether your interests range from running the family farm to managing a tournament golf course, or directing an international business, rewarding, challenging, and fulfilling careers are open to anyone skilled in plant science.

Human survival absolutely depends on the ability of plants to capture solar energy and convert that energy to a form that can be used as food. The captured energy stored in plant tissues also provides fiber and oil for fuel, clothing, and shelter. The production of plants that meet our needs for survival is an important application of the knowledge of plant science; however, the essentials for nutrition and shelter can be provided by relatively few plant species. Life would be boring, though, if those few were the only species produced for our needs. Fortunately for those who dislike boredom, thousands of plant species can add enjoyment to life by providing a variety of flavors and textures in food and fiber. Other plants brighten our lives when used in landscaping and interior decoration. The importance of turfgrass in athletic and outdoor recreation sites is evident around the world. Animal feeds are another critical use of plants. Animals provide nutrition and variation in our diets, along with materials for clothing and shelter. They also reduce labor in many parts of the world and add pleasure to our lives through recreation or as pets.

Plants have tremendous economic impact in developed and developing nations. The career opportunities created by the need for people with an understanding of plant growth are unlimited. *Plant Science* is written for anyone with an interest in how plants are grown and utilized for maintaining and adding enjoyment to human life as well as improving and protecting the environment. The beginning chapters of the text provide the fundamentals of environmental factors, botany, and plant physiology that affect plant growth. The later chapters integrate the aforementioned topics into strategies for producing plants for food, fiber, recreation, and environmental stewardship.

The fifth edition of *Plant Science* has been updated to include the most recent statistics, production methods, and issues concerning the production and utilization of plants. This revision has been reorganized to present the topics in a more logical order and to reflect the changing information needs of those who grow plants for a living. The concept of using sustainable practices (ecological paradigm) that help preserve resources for future generations while providing for the current one permeates the book. New information has been added, and out dated information has been deleted. Some information retained from the previous edition has been moved to a different chapter. For example, the climate chapters have been combined into one chapter. The information on managing soil, water, and nutrients has been combined into one chapter because of the interdependence of these topics. A new chapter in this edition gives an overview of general production and postharvest handling and marketing procedures. It precedes the individual commodity chapters.

Unit One focuses on the human and environmental factors and issues that influence how, why, and where plants are grown. The unit includes how natural ecosystems influence plant cultivation and how the scientific method and research is applied to growing plants.

Unit Two addresses the biological basis of plant science. Plant physiology and biochemistry, genetics, an expanded section on genetic engineering, propagation, biodiversity and germplasm preservation, water relations (between soil, plant, and air), as well as mineral nutrition are the major topics of Unit Two.

Unit Three presents the general principles of growing plants through the application of the principles presented in the previous units to the production and/or use of major commodity groups.

Acknowledgments

The authors and editors would like to thank the reviewers of the fifth edition for their thoughtful comments and ideas. They are: David Berle, University of Georgia; Jack R. Fenwick, Colorado State University; Sharon Frey, Sam Houston State University; Mary Ann Gowdy, University of Missouri; Steve Hallett, Purdue University; Malcolm Manners, Florida Southern College; Don Mersinger, Wabash Valley College; Ellen T. Papparozzi, University of Nebraska - Lincoln.

The editors also thank all those who contributed to earlier editions of this book. Those contributors are Curtis Alley, A. H. Allison, Victor Ball, Robert F. Becker, Brian L. Benson, Alison Berry, Itzhak Biran, Arnold J. Bloom, James W. Boodley, Robert A. Brendler, Royce Bringhurst, Thomas G. Byrne, Jack Canny, Will Carlson, Kenneth Cockshull, Charles A. Conover, Beecher Crampton, Began Degan, Frank Dennis, Francis DeVos, Dominick Durkin, Roy Ellerbrock, James E. Ellis, Clyde Elmore, Thomas W. Embledon, Harley English, Elmer E. Ewing, Michael Farhoomand, Gene Galleta, Melvin R. George, Marvin Gerdts, Ernest M. Gifford, Victor L. Guzman, L. L. Haardman, Abraham H. Halevy, Richard W. Harris, R. J. Henning, Charles Heuser, James E. Hill, Jackson F. Hills, Karl H. Ingebretsen, Lee F. Jackson, Subodh Jain, Merle H. Jensen, Robert F. Kasmire, Thomas A. Kerby, Dale Kester, Paul F. Knowles, C. Koehler, Harry C. Kohl, Dale W. Kretchman, Harry Lagerstedt, Pierre Lamattre, Robert W. Langhans, Roy A. Larson, Robert A. H. Legro, Andrew T. Leiser, Gil Linson, Warner L. Lipton, Oscar A. Lorenz, James Lyons, Harry J. Mack, John H. Madison, Vern L. Marble, George Martin, Shimon Mayak, Keith S. Mayberry, Richard Mayer, Donald N. Maynard, Arthur McCain, Charles A. McClurg, Harry A. Mills, Franklin D. Moore III, Yoram Mor, Julia Morton, James L. Ozbun, Jack L. Paul, William S. Peavy, Nathan H. Peck, Jack Rabin, Allan R. Rees, Michael S. Reid, Charles Rick, Frank E. Robinson, J. Michael Robinson, Norman Ross, Edward J. Ryder, Kay Ryugo, Roy M. Sachs, Robert W. Scheuerman, Art Schroeder, John G. Seeley, William Sims, Donald Smith, L. Arthur Spomer, George L. Staby, W. B. Storey, Vernon T. Stoutemyer, Walter Stracke, Mervin L. Swearingin, Herman Tiessen, Edward C. Tigchelaar, Ronald Tukey, Benigno Villalon, Stephen Weinbaum, Ortho S. Wells, Bernard H. Zandstra, Naftaly Zieslin, and Frank W. Zink. The authors who wrote or revised chapters in the current edition are listed on the opening page of his or her respective chapter.

—Margaret J. McMahon

Additional Resources

Online Supplements Accompany the Text

An online Instructor's Manual, TestGen, and PowerPoint slides are available to instructors at www.pearsonhighered.com. Instructors can search for a test by author, title, ISBN, or by selecting the appropriate discipline from the pull-down menu at the top of the catalog home page. To access supplementary materials online, instructors need to request an instructor access code. Go to www.pearsonhighered.com, click the Instructor Resource Center link, and then click Register Today for an instructor access code. Within 48 hours after registering, you will receive a confirming e-mail including an instructor access code. Once you have received your code, go to the site and log on for full instruction on downloading the materials you wish to use.

A Companion Website is also available for this textbook. Please go to www. pearsonhighered.com/McMahon.

PEARSON IS GOING GREEN

Issues of sustainability and preserving our natural resources, consistently rank among the most important concerns to our customers. To help do our part, Pearson AG is implementing the following eco-friendly initiatives to our publishing program.

- This book, as well as all future Pearson AG titles, will be printed using paper fiber from managed forests certified by the Sustainable Forestry Initiative (SFI).
- Integrating the use of vegetable-based ink products that contain a minimum of 45% of renewable resource content and no more than 5% by weight of petroleum distillates.
- Offering alternative versions to traditional printed textbooks such as our "Student Value Editions" as well as e-book versions of the text in the "CourseSmart" platform.
- Electronic versions of supplemental material such as PowerPoint Presentations, Test Banks, and Instructors Manuals can be found by registering with our Instructor Resource Center on the web at www.pearsoned.com/.
- For more information regarding the Sustainable Forestry Initiative, please visit www.sfiprogram.org/.

brief contents

contents

unit one
Environmental, Cultural, and Social Factors That Influence the Cultivation and Utilization of Plants

*1

History, Trends, Issues, and Challenges in Plant Science

MICHAEL KNEE AND MARGARET MCMAHON

key learning concepts

After reading this chapter, you should be able to:

- Discuss the role that plant science has played and continues to play in the world economy and culture.

- Explain why modern plant scientists take into consideration production efficiency, economic viability, environmental compatibility, and social responsibility when researching the solution to a problem.

- Describe the importance and principles of research in plant science.

HISTORY

As citizens of the twenty-first century, we tend to pride ourselves on how we have used agriculture to shape the modern world to serve and please us. We have reason to feel that way—our agricultural practices have changed the world. But if we were to use H. G. Wells' time machine to transport us back 150 million years, we would see many plants very similar to those common in our century. We would see some of the same trees that grow in our world, along with other members of the **angiosperms**, the group of plants to which grasses, flowers, vegetables, fruits, trees, and shrubs belong. We would also see many plants that no longer exist. Some dinosaurs would be feeding on these plants. As time progresses, the dinosaurs would disappear and other animals would appear and evolve, and some would die out. Plant life would change, some as a result of changes in climate, and become most of the plants we know today. During this time, humans had no influence on changes in life forms that occurred or disappeared.

Humans as a race appeared around 3 million years ago and modern man, *Homo sapiens,* appeared about 28,000 years ago. For thousands of years, *H. sapiens* existed without doing much to change how plants grew. As hunters and gatherers, the nomadic tribes followed herds of animals and gathered plant materials along the way. The plants they probably gathered would have been some of the same nuts, grains, and fruits we eat today. Other plants known today would have also provided shelter.

Then something happened around 12,000 to 10,000 years ago (perhaps earlier according to recent archaeological finds) that had a dramatic impact not only on human lifestyle, but the entire global ecosystem. Humans began the purposeful growing or **cultivation** of plants to improve the supply of materials obtained from these plants. The science of understanding the cultivation of plants, plant science, was born. Plant cultivation is believed to have started in tropical and subtropical regions in the Middle East and Africa.

By cultivating plants, humans reduced the need to travel to follow the food supply. Those who did the traveling became traders more than gatherers. Commerce

began when goods from distant places were transported and sold or traded to a local population. Many of these products were plants or plant products. Along with trading the goods, ideas about cultivating plants also spread. By 6,000 to 5,000 years ago, crops were being grown in Europe and Mexico. However, the types of crops grown in the various areas differed for reasons that will be explained in later chapters.

At first, the developing lifestyle had little effect on global ecosystems; as trading increased, however, commercial or urban centers developed. The population of the urban centers increased along with the need to bring in more food and other plant products from rural areas to support that population. However, the citizens of these urban areas became less and less aware of how the plants they used were produced. The **rural/urban interface** developed. In many areas, this interface was a very distinct line of demarcation with a defense wall separating urban from rural areas. As populations grew, the urban areas spread out away from the central urban area. The rural areas were forced to move outward too, quite often into areas less favorable for growing plants.

As demand for plant products increased, cultivation methods were developed to help production keep up with the demand. Some of the earliest products of the industrial revolution were farm implements for crop cultivation, such as plows, and planting and harvesting equipment. The closely related disciplines of agronomy and horticulture were born and developed. **Agronomy** is the study of field-grown crops such as wheat, soybeans, corn, forages, and those used for industrial purposes. These crops require relatively low input during the growing part of their life cycle but usually require significant processing to be used for purposes other than animal feed. **Horticulture** is the study of crops that require more intense and constant care, from planting through delivery to the consumer. Examples of horticulture crops include fruits and vegetables and ornamentals. Most edible horticulture crops can be eaten with little or no processing. There are gray areas where the disciplines overlap. Tomatoes grown for processing and **turfgrass** are examples of crops that can be designated agronomic or horticultural.

TRENDS AND ISSUES AFFECTING PLANT SCIENCE

Background

Traditionally crops are commodities that are exchanged for money in the marketplace. In the case of recreation areas such as golf courses, the use of that field is exchanged for money. According to classical economics, the price of commodities (or user fees) results from the balance between the supply of commodities (recreational areas) and the demand for them. The supply should be influenced by the availability of the resources and raw materials required for production, and the demand should be influenced by the value that consumers perceive in the commodity.

Many people have drawn attention to the mismatch between ecological and money values. If excess fertilizer from a crop pollutes a river, the cost of cleaning up the pollution will not be reflected in the price of the commodity. On the other hand, many people may enjoy seeing an ornamental tree planted by someone else, and it can also help to absorb the carbon dioxide (CO_2) produced by cars, but those benefits seem "free." Economists call costs and benefits that are not included in the price externalities. Some argue that environmental degradation occurs when no one owns a resource. If land is under common ownership, it will be overcropped or overgrazed because it is in no one's interest to conserve the resource. Thus, they refer to the general problem of the tragedy of the commons. According to this view, if everything is privately owned, individuals have an interest in its preservation, those who use resources have to pay the owners, and the value of the resources is reflected in the price of the commodity.

Classical economics assumes that buyers and sellers are fully informed of the value of commodities in the market and make wise decisions that maximize their welfare. However, many people make unwise choices on both sides of the market relationship. In the 1930s, farmers in the American Southwest contributed to soil erosion on their own farms through their crop management practices. The market for tobacco products remains strong after fifty years of health warnings. Individuals may not know all of the consequences of their decisions, and it may take the input from many people to arrive at a full accounting. The tragedy of the commons is not always the case; people are capable of making communal decisions to conserve resources that do not belong to any individual. The soil conservation service was born out of the dust bowl experience. Plant science research and education grew out of a perception of a wider public good to be obtained by applying scientific principles to food production. Now, as you will learn in this book, we need to go beyond production to look at ecological consequences and the consequences for the individuals who consume the produce.

Dependence on the market to regulate crop production practices assumes that ecological values can be translated into cash value and that ecology is a subcomponent of the market. But that assumption is difficult to achieve. The problem is partly one of time scale.

Financial analysis discounts future values relative to current value. Low yields in future years because of soil degradation appear to be less important than high yields this year. The idea of **sustainability** as a guiding principle in crop production is an attempt to transcend this mind-set. Sustainable practices consider not only what is good today, but what will allow future generations to thrive, too. You will learn more about sustainability later.

Balanced energy and resource transfers among components of an ecosystem are central to stable, sustainable ecological relationships. In other words, most of the energy and resources used in the system are eventually recycled back into the system. Crop production requires energy and resource inputs at several points and provides us with energy in food or embodied in nonfood commodities. These inputs create energy and resource imbalances because they are not recycled back into the system, but leave with the harvest, making the system unsustainable.

Ecologist Howard Odum has suggested that we use energy accounting to try to correct some of the distortions introduced by cash accounting in agriculture and other industries. Energy comes in many forms, and every energy transaction is accompanied by losses. To arrive at a common basis for accounting, Odum proposed that all forms of energy should be related back to the primary source, which is usually the sun. Odum's concept of embodied energy, or **emergy**, is the amount of solar energy represented in the resource or commodity (Fig. 1–1). Emergy accounting can be related back to financial accounting by averaging costs across all forms of energy according to their solar energy equivalent. If energy converts raw materials into valuable commodities,

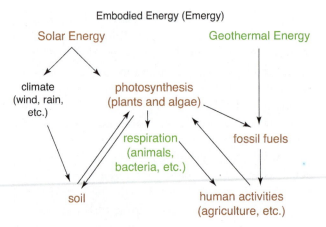

Figure 1–1

Derivation of embodied energy, or emergy, from primary energy sources. Source: Michael Knee.

then there is no reason why one form of emergy should be regarded as any more valuable than another. Emdollar accounting results in some interesting shifts of value (Table 1–1). The dollar prices of agricultural commodities appear to be much too cheap, and natural ecosystems with little or no cash value turn out to have high emdollar value in terms of services such as flood and pollution control.

Domestic Trends and Issues

In the United States, the prices of many agricultural commodities continued to fall in real terms in the last quarter of the twentieth century. Even though yields continued to increase, income per acre declined (Fig. 1–2). This is one reason that it has been impossible to break away from the system of price support for selected commodities. In 2000, the subsidies amounted

Table 1–1

EMERGY CONTENT OF COMMODITIES AND PRICES CALCULATED ON A UNIFORM COST OF $0.86 FOR 1^{12} SOLAR EMJOULES (CALCULATED AVERAGE FOR THE WHOLE US ECONOMY)

Commodity	Emergy Content (10^9 solar emjoules/g)	Calculated Price (emdollars/unit)
Corn grain	14.5	5.68/pound
Cotton	23.1	9.04/pound
Soybeans	9.9	3.87/pound
Sweetcorn	2.5	0.54/ear
Tomatoes	16.0	6.26/pound
Beef (range-fed)	48.5	19.00/pound
Eggs	107	62.00/dozen
Milk	33.7	13.20/pint
Gasoline	3.0	7.50/gallon

Source: S. Brandt-Williams, University of Florida, 2002.

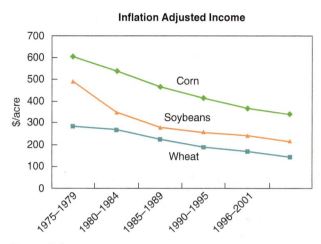

Figure 1–2

Income per acre from selected crops over the past twenty-five years, corrected for inflation by estimating on the basis of 2001 dollar value. Source: Michael Knee.

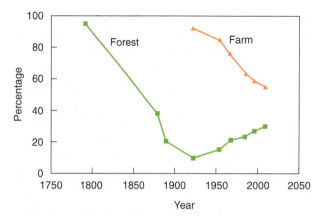

Figure 1–4

Percentage of forested land and farm area in Ohio since European settlement. Source: Michael Knee.

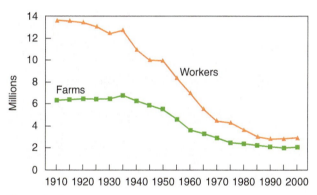

Figure 1–3

Number of farms and farm workers during the last century (1910–2000). Source: USDA National Agricultural Statistics Service Information (NASS).

Figure 1–5

Woodland developing on land in Ohio that has not been farmed for 30 years. Source: Margaret McMahon, The Ohio State University.

to half of farm income. The payments are supposed to help farms stay in business but end up as one more factor encouraging consolidation in the industry: large farms get more government assistance than small farms do. The number of farms decreased as individual holdings got larger in the second half of the century.

The number of farm workers also continued its long-term decline so that there is now a little more than one full-time worker for each farm. The numbers of farms and farm workers seem to have stabilized toward the end of the century, and these numbers may be minimum sustainable values (Fig. 1–3). Most farms are run as part-time businesses, and about 10 percent of the farms remaining in the United States account for 70 percent of production. The profitability of farming has been helped to some extent by diversifying the uses of staple crops (e.g., ethanol and syrup from corn and tofu

from soybeans) and by adopting alternative crops. The area of farmland has also declined from its high point at the beginning of the twentieth century. In contrast to many other countries, the United States has seen an overall increase in the area of woodland (Fig. 1–4). Some of the increase comes from the conversion and/or reversion of farmland to woodland (Fig. 1–5).

The increase in US population during the last century was mainly in urban areas so that rural population decreased in relative terms from 60 percent in 1900 to 25 percent in 2000. In the upper Midwest, the rural population fell in absolute terms from the middle of the century. This decline was associated with a loss of economic and cultural vitality in rural communities. Such communities were more likely to survive if there were large towns that provided an economic stimulus to the surrounding areas. This may be the reason why

rural populations persist in states such as California, Florida, New York, New Jersey, and Ohio, which have several large towns and cities.

Urban populations have increased in every US state. The increases were most marked in the coastal states, the East and West, and along the shores of the Great Lakes. Although this is classed as urban development, it is more accurate to call it suburban. Average lot sizes for new homes are about 0.15 hectare, or 0.4 acre. The spacious lifestyle of the suburbs depends on personal transportation for access to work, shops, and leisure, which accounts for much of our energy demand. Many people have criticized this and other aspects of suburban sprawl, but it leads to new opportunities. The new homes are an expanding market for landscape supplies and services. Surviving farms can market directly to the surrounding population. Families can enjoy a visit to the local farm to buy or pick their own produce. This interaction may help maintain contact and understanding between the mass of the population and the few remaining farmers.

The aesthetic and recreational use of plants is important in urban and suburban areas. The appearance and playability of fields is vitally important to many sports. Golf alone is now equivalent to about two-thirds of major crop sales and involves 12 percent of the US population. The number of golf courses has increased over threefold in the past fifty years. Residential and commercial landscaping is a business that generates billions of dollars annually in the United States alone. Even the most humble of homes is likely to have flowers for decoration at times throughout the year. It is not realistic to give a representative list here of plants used for ornamental and recreational purposes because species grown for these uses outnumber by far the species used in other commodity areas, and the number increases every year. Producing and maintaining a vast array of beautiful plants in ecologically sound ways is particularly challenging.

The rising cost of fuel to grow, maintain, harvest, and distribute plants and plant products has greatly increased costs to both the producer and consumer. Even though energy use may not have increased markedly, the increased cost has created a fundamental need for greater fuel-efficient production and delivery methods.

Global Trends and Issues

Two major trends that will affect crop production and the global environment are the increases in human population and energy use. After two centuries of exponential growth, world population shows signs of stabilizing at about 9 billion in 2050 (Fig. 1–6). Energy use, however,

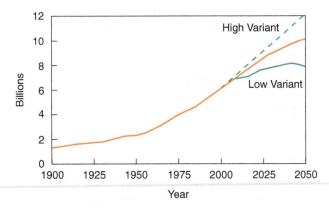

Figure 1–6

Projections of world population growth. Source: US Global Change Research Program, USGRCP Seminars, *http://www.usgcrp.gov/usgcrp/seminars/*

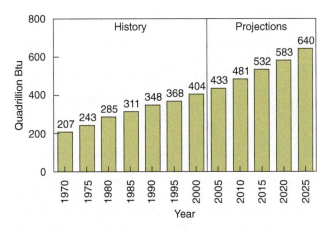

Figure 1–7

History and projection of world energy consumption, 1970–2025. Sources: **History:** Energy Information Administration (EIA), International Energy Annual 2001 DOE/EIA-0219 (2001) (Washington, DC: February 2003), www.eia.doe.gov/iea/. **Projections:** EIA, System for the Analysis of Global Energy Markets (2003).

is projected to rise about twice as fast as population because of economic development (Fig. 1–7). People have long argued about the potential for the future growth of population in relation to energy and other resource use and about the related question of the earth's carrying capacity. Estimates of carrying capacity have varied from the low billions to the trillions.

Assessing the world's food situation involves other factors besides the utilitarian one of meeting minimal food needs. When the people of a country become more affluent, they want and can afford to purchase a greater proportion of their protein requirements in the form of the more palatable animal products—steaks, chops, eggs, processed meats, and dairy products. This shift in food consumption patterns coupled with the tremendous increase in world population, especially in

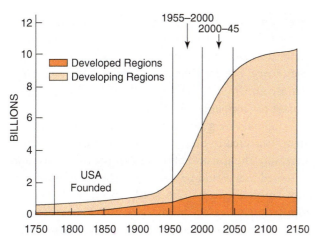

Figure 1–8

Trends and Projections (UN Medium Variant) in World Population Growth, 1750–2150. Source: World Resources Institute (WRI) in collaboration with United Nations Environment Programme (UNEP), United Nations Development Programme (UNDP), and World Bank, *World Resources 1996–1997: The Urban Environment* (Washington, DC: WRI, 1996).

developing regions (Fig. 1–8), requires continuing increases in the world's food-producing capability. Much of the world's best agricultural land is already under cultivation, although there is still unused productive land awaiting development in Argentina, Brazil, Canada, Sudan, and Australia. However, in most if not all developed countries such as the United States (see Fig. 1–3), Japan, and those in Europe, farmland is being lost forever to industrial, residential, or recreational development. In addition, in the United States, even though the number of farms is decreasing, the size of farms is increasing. The loss of the small family farm and the increase in corporate megafarms have become a serious social and political issue. In the mid-1970s various projections implied that the world was on the brink of famine or ecological disasters due to desperate food needs. But this assessment changed in the 1980s, especially in the less-developed countries, by improved production technology and greater incentives to use it. Agricultural research made new **cultivars** (cultivar = *culti*vated *vari*ety) of high-yielding wheat, rice, corn, and other crops available to highly populated developing countries. Much of this improved technology can be attributed to assistance from agricultural researchers in the United States and other developed countries working with less developed countries.

But there are still areas where people starve or are badly malnourished. A high proportion of these underfed populations are in the sub-Saharan regions of Africa, where drought, wars, and political instability are major problems, along with high human fertility rates,

low per capita income, and insufficient monetary investment in agricultural production. In addition to the above-mentioned factors, poor distribution of available food contributes to starvation in many areas, including areas of developed countries. The United States Department of Agriculture (USDA) in its 2007 Food Security Assessment, reports that the number of people consuming less than the nutritional target of 2,100 calories per day in the 70 lower income countries rose between 2006 to 2007 from 849 million to 982 million. The distribution gap which indicates access to food is projected to rise from 44 million tons in 2007 to more than 57 million tons in 2017. Improving the nutritional value of food, so that the food distributed provides better nourishment, is a challenge to current and future plant scientists.

As you will learn in Chapter 3, it is possible to feed an adult on a plant diet from about 0.2 hectares of land, and this is what will be available on average when the world population reaches 8 billion. However, the yields we achieve in industrial countries need to be achieved worldwide. These yields in turn require that inputs, at least of fuel, fertilizer, and probably pesticides, have to rise in other countries to match those in North America and Europe. While agriculture accounts for only about 2 percent of energy consumption in North America and Europe, it would amount to about 10 percent of energy demand in the rest of the world.

China alone represents a major challenge to the world's food supply. It has about 20 percent of the world's population but less than 7 percent of the world's cropland. It had about 0.2 hectares per person in 1950 and now has about 0.08. It has managed to feed itself and actually improve people's diet over the past decade. However, with its population set to increase by an additional 25 percent before stabilizing at 1.6 billion, it is doubtful that this trend can continue. Fertilizer use is high and further increases will not bring much additional yield. There are already problems of nutrient pollution and soil degradation, and the area of agricultural land is shrinking because of urban development.

The United States has about twice as much cropland as China but one-fifth of its population. It is one of a handful of countries with a relative excess of cropland in relation to population. Others include Canada, Australia, and the Russian Federation. China is undergoing rapid industrial development and now supplies other industrial nations with a wide range of the manufactured articles that they used to make. But in the future it is likely to be a major customer for US agricultural products. Other countries may try to follow China's road

to economic development. Even if they do not, it is likely that China's demand for food will raise prices of agricultural commodities, which could be good news for US producers. Of course, US consumers are likely to be unhappy as food becomes more expensive after many years of falling prices.

Grain-fed meat production is increasing in China, which increases the land required to meet its dietary needs and cannot be sustained on a global scale. To provide an adequate diet for everyone, good cropland needs to be devoted to feeding people. Animal production needs to be based on more marginal land that will support grazing but not crops. But overgrazing and desertification are a constant threat in arid lands.

A 50 percent increase in global energy use is expected over the next fifty years, and it seems likely that most of this energy will come, as it does now, from combustion of CO_2-generating fossil fuels. Although the oil industry is confident that supplies will extend beyond the forty years of known reserves at current rates of consumption, concern for negative effects on the environment from exploration for and extraction of oil has limited the exploitation of potential new sources. Natural gas reserves are a little higher: at about sixty years (but little of this is in the United States). Coal reserves are considerably higher: at over 200 years. Coal is more evenly distributed, and its consumption is rising faster than consumption for gas and oil outside the United States and Europe. However, in some areas the coal is of low grade (high polluting) or in environmentally protected areas. As we continue to increase the use of fossil fuels atmospheric carbon dioxide will continue increasing for the next fifty years. Other gaseous pollutants will also increase as coal consumption rises. The threat to the environment from the increased production of fossil fuels will also increase during that time.

Although a few individuals continue the argument about the connections between rising atmospheric CO_2, rising temperatures and other weather changes, evidence overwhelmingly supports the connections. Productivity of some crops could increase with rising CO_2, as long as rainfall patterns are not disrupted, but disruption is likely to occur. High temperatures could lead to decreases in yields of many crops, even if precipitation patterns do not change. A likely scenario is that favorable rainfall and temperatures will shift to higher latitudes where soils are less fertile. Climatic changes impose additional stress on natural plant communities, and many doubt whether forest ecosystems can adapt fast enough to survive the changes. For example,

American beech and sugar maples may become extinct over a large part of their present range, and they may not be able to spread north fast enough to take advantage of new habitat. The other atmospheric pollutants that accompany CO_2 from fossil fuels will cause more damage to crops and wild plants.

Global energy demand is only a fraction of the energy captured in photosynthesis and stored in plants. Increasing the use and efficiency of plant biomass and bio-based fuel would be a way to utilize photosynthetically captured energy. About 40 percent of the world's population relies on biomass as its primary energy source, but these same people are mostly in undeveloped countries and consume very little energy by US standards. Even in areas where biomass is the primary energy source, gathering firewood for energy consumes more and more of these people's time as the supply progressively dwindles. Developing methods of sustainable production of biofuels seems like the answer to many of the problems of energy supply. However, this approach requires massive changes in land use and a large investment in equipment for conversion of biomass to usable fuels. The major growth in fuel requirements is predicted for use in transportation, but biomass-to-fuel conversion processes (such as corn to ethanol) may not ultimately generate enough of an energy profit to support this growth.

The demand for food and fuel is leading to deforestation in many tropical countries. The loss of biodiversity and environmental quality is unfortunate for the resident population, but it also has implications for us. The tropical forests are a major **sink**, or depository, for the carbon dioxide that we generate through our fossil fuel consumption. Losing that depository contributes to increasing atmospheric CO_2 and global warming.

International trade is being progressively expanded, removing tariffs and duties that restricted imports of food and other commodities. Industrial nations are also under pressure to eliminate the subsidies that favor their own producers. This change will allow poor countries to export commodities such as sugar and grains to the rich countries. Classical economics predicts that production will become more efficient and profitable in the long run when the market is no longer distorted by the subsidies. This prediction does not seem to account for the fact that poor countries often cannot get good prices even for commodities such as bananas or coffee that are not produced by the United States and Europe. A side effect of the globalization of trade is that pests and diseases are spread around the world more easily. A recent example

is the introduction of the Emerald Ash Borer (*Agrilus planipennis*) into the Great Lakes region from wooden pallets carrying goods from China to the Port of Detroit. The borer is threatening the survival of all North American ash species, much as Dutch Elm disease decimated the elm population in the United States a century ago.

Bioterrorism is a recent, but major threat to agriculture. Terrorists may be able to introduce pests or chemicals that will destroy a good portion of crop production in certain areas. Because of this fear, quarantines and restrictions to free trade may increase. In 2003 and again in 2004, some geranium growers in the United States were ordered by the USDA to destroy much of their geranium crop because it came in from abroad infested with *Ralstonia solanacearum* race 3 biovar 2, which is listed in the Agriculture Bioterrorism Act of 2002. *Ralstonia* is a bacteria that is endemic in some parts of the world but is not found in the United States. However, it was not fear of what *Ralstonia* would do to geraniums that prompted the USDA's action. *Ralstonia* is a serious pathogen of several of our food crops. The potential of the fungus to devastate those crops is the reason it has been listed as a bioterror organism. Although the disease was imported unintentionally, its designation as a bioterror organism prompted the need to destroy the geranium crops. The negative financial impact on foreign and domestic geranium growers was devastating.

MEETING THE CHALLENGES IN PLANT SCIENCE

Many of the changes in the issues of crops and their role in the world have prompted plant scientists to change the focus of their research. For several centuries, plant scientists studied ways to improve crop productivity in a cost-effective way. They studied light, soil, water, and temperature and developed ways of managing or monitoring those factors to influence or predict plant growth. Improved understanding of plant genetics lead to breeders developing plants that would produce more reliably. A scientific approach to pests and their management reduced losses to those factors. Traditional economic analysis was used to see if the new production methods were cost-effective. Great gains were made, but as you have read, increasing evidence showed that some agricultural practices were having negative effects on the environment. Agriculture became the focus of public scrutiny, and a negative public opinion of agriculture developed. This

opinion was exacerbated by the fact that most urban dwellers today have very little understanding of their dependence on agriculture. Plant scientists have to find ways to meet our need for food, fuel, and other products and services from plants without negatively affecting the environment. When calculating the cost-effectiveness of a production procedure, costs can no longer include just material and labor costs. The cost to repair any resulting environmental damage must also be factored into cost analysis.

Although it may seem like a relatively simple calculation to factor in environmental costs, it is not. Assessing environmental impact and the cost to repair negative impact can be a challenge. It can be difficult to predict what will happen when a new production practice moves from the lab or experimental field to the real world. For example, genetically engineered plants, sometimes called **genetically modified organisms (GMOs)**, have been created that dramatically reduce the need for pesticides and field tillage. Pesticide runoff and erosion are reduced, which is beneficial for the environment. But there is concern that heritable traits from the GMOs will "escape" and become part of the wild or native plant populations if engineered plants can breed with native plants. The fear is that wild plants with these traits may have serious negative impacts on the ecosystem in which they grow. Determining the likelihood of a gene escaping requires many long term studies both in the lab and in the field. Organic farming has been proposed as a solution to many problems related to crop production and environmental impact. Organic farming does not allow the use of GMOs and certain types of chemicals for pest control and fertilization. Many organic farms have proved to be successful, thus demonstrating that the process works. However, it is not known if organic farming has the capability to produce the quantities of crop products needed in today's world. Also there is great debate regarding what constitutes organic farming. Some organic farming practices, such as erosion from soil cultivation, may have negative environmental impact.

The public often wants to see plants growing "perfectly"—a very unnatural state for plants. Generally this occurs when plants are used in leisure and recreational settings. Immaculately groomed and weed-free landscapes, flawless floral arrangements, and impeccable golf courses and athletic fields are sources of pride to those who own or use them. Ask any golf course superintendent what his greens committee would say if the fairways and greens had even a few weeds or insect and/or disease problems in them. Who

would buy a bouquet if some petals were chewed? Maintaining perfect plants almost certainly has some degree of negative environmental impact.

Plant scientists and growers can find all the current uncertainty and controversy about their fields discouraging and wonder why we should even bother to try to solve seemingly insurmountable problems. We must remember that, although most plants grow very well without human intervention, the cultivated grains and grasses, fruits, vegetables, and ornamentals have become dependent on human intervention to survive. Without cultivation, these plants would likely die out after several generations and be replaced by hardier species such as wild grasses and thistles. But the dependence is mutual—we need these plants to survive and that is why we have to solve the problems.

The estimated 6.5 billion people now living in the world depend on cultivated plants for nourishment and to provide quality to their lives. The global population cannot survive as hunters and gatherers. The need for plant scientists to increase knowledge about crop plants and their place in the ecosystem and the need for professionals who know how to use that knowledge to grow plants in environmentally sound ways will not disappear. In fact, it should increase, perhaps dramatically, if the world population increases as predicted.

Currently, enough food is produced to feed the world's population. However, malnutrition and starvation exist in both developed and undeveloped countries mostly because social and political issues prevent the distribution of food to those who need it. If the social issues cannot be resolved, it will likely become the responsibility of plant scientists and growers to find creative ways to produce food crops locally in starvation-prone areas.

The solution to the loss of small family farms will no longer lie primarily in increasing productivity of traditional crops. New uses for old crops and production strategies for new crops will have to be developed to allow the family farm to remain viable.

Recently, many of the improvements in production were developed not only to reduce labor and increase productivity and profit, but also to allow farmers to be better stewards of the environment. But these improvements come with their own issues. High-oil corn has been bred to yield a product that can be used to replace petrochemicals in some industrial uses but may replace some food crops. No- or low-till farming reduces labor costs and is less detrimental to the soil than traditional cultivation practices are. However, no- or low-till farming requires the use of herbicides to control weeds. To make herbicide use more effective and to reduce the amount of herbicide used, GMOs were developed that are resistant to a very effective herbicide, Roundup®. Weeds are susceptible to the chemical, but Roundup Ready® crops are resistant. As a result, only one or two applications of a single herbicide is required in a season to get the same or better weed control compared to multiple applications of several herbicides in nonresistant fields.

A genetically modified corn (Bt®) synthesizes a naturally occurring protein that is lethal to the larva of many species of *Lepidoptera*, such as corn borer. The production of a natural larval toxin in corn has reduced the need to spray insecticides for control of a very destructive pest. Rice has been genetically modified to produce beta-carotene in the grain (golden rice). Beta-carotene (vitamin A) provides a critical nutritional element that is predicted to save the eyesight of millions of children in areas where vitamin A deficiency causes blindness.

But the public has demonstrated a negative response to each of the aforementioned uses of GMOs. There is fear the Roundup Ready® gene will escape into the native weeds adjacent to the fields where Roundup Ready® crops are being grown, thus creating weeds that are resistant to Roundup® though they would not be resistant to other types of herbicides. The Monarch butterfly (a member of the *Lepidoptera* species) was at one time thought to be threatened by Bt® corn. Golden rice cannot be grown in many of the areas where vitamin A is needed most because of concern about the beta-carotene gene escaping into the traditional rice crop. Flavr Savr® tomatoes have disappeared from the market because chefs and diners refused to buy or eat them out of fear of what the altered gene would do to human health if consumed. It will be up to those who study and work with crops to determine if there are undue risks with the new plants. However, one approach could work.

Perhaps the best way for plant scientists to meet today's challenges is to include the ecological paradigm of agriculture in all scientific studies. A model for this concept was developed in the College of Food, Agricultural, and Environmental Sciences at The Ohio State University (OSU). As described by the college:

[T]his model is built on four areas of focus: production efficiency, economic viability, environmental compatibility, and social responsibility. A pyramid (Fig. 1–9) has been chosen to provide a visual representation of this model. Each wall of the pyramid represents one of the four dimensions. Four equal walls provide support and strength to each other and emphasize the critical need for balance and integration of the four areas.

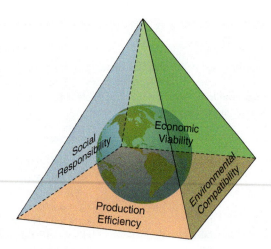

Figure 1–9

Pyramid representing the four sides of the ecological paradigm of scientific study in the agricultural sciences. Source: Adapted from College of Food, Agricultural and Environmental Sciences, The Ohio State University, Columbus, OH.

Disregarding one of the dimensions would weaken the structure significantly; operating without considering two dimensions would cause the system to collapse.

This approach is not unique to OSU. Most if not all institutions where plant science is studied have taken the same or a similar approach. This approach is also called sustainable agriculture, which is discussed in detail in Chapter 16.

SOLUTIONS THROUGH SCIENTIFIC INQUIRY

Scientific inquiry, or the scientific method, is the systematic approach to understanding and solving a problem. The steps to scientific inquiry are: identify the problem or question and make an hypothesis (educated guess) of the cause of the problem/question; test the hypothesis by doing experiments or a study; check and interpret the results; and report the results.

Hypothesis

Scientific inquiry or research begins with an objective and some idea of how the objective can be met. The objective might be to solve a problem or improve a procedure. The idea can be stated as a hypothesis, such as, "This is the source of the problem" or "This is how the process can be improved." A simple problem might be determining if a common landscape species will grow well in compacted soil. You may have observed what appears to be suitable growth in compacted soils, but maybe you would like more scientific evidence before you start using that species in compacted landscape soil.

From your observations or other information (such as claims from seed companies), you can create the hypothesis that the species will have the ability to grow acceptably well in compacted soils. From this hypothesis, a simple experiment can be developed to test the validity of the hypothesis. The two species can be grown in compacted and porous soil, and the differences in plant growth between the two soils (treatments) can be measured. From our hypothesis, we predict that compaction would not affect growth when compared to the plants growing in porous soil. If this prediction turns out to be validated, then we have confirmed our hypothesis.

If our supposed compact-tolerant species did not grow as well in compacted soil as in porous soil, perhaps our hypothesis was false or perhaps we might have achieved better results if the experiment had been designed differently. By doing more experiments, we can gain confidence that our hypothesis is correct or incorrect. Our confidence can be increased if we can explain why the growth was no different in porous soil. We can set up more experiments that look at subhypotheses; for example, maybe the root system's growth was not affected by compaction.

The example above is a relatively simple problem. At the other end of the scale are complex ecosystem problems such as determining if agricultural practices have a role in creating the dead zone of the Gulf of Mexico. The dead zone is an area off the coast of Louisiana where an unusually large number of marine life dies each summer. It is generally not wise to focus on a single hypothesis to solve a complex problem such as an ecosystem problem. It is quite likely that there is no single cause of the problem. For example, the dead zone may also be caused by industrial activity, natural causes, and/or something else. It is best to develop several hypotheses that represent as many explanations as we can imagine. We can then look for experimental evidence that will support or refute the different hypotheses. After extensive research, we can build a plausible description that is consistent with the data our experiments have generated. From this research, we may be able to suggest ways to correct the problem.

Experiments

Controlled experiments are the most certain way of obtaining reliable information on which to base decisions. Experiments are artificial situations where we try to exclude everything except our immediate area of interest, such as plant response to compacted soil or the effect of erosion sediment on marine life. In large and complex problems, a full-scale experiment is most

UNITS OF MEASURE FOR VARIOUS SYSTEMS

All scientific studies present data in scientific units. Although they may be less familiar than US units, they are often much easier to work with and they are used in scientific literature (including agronomy and horticulture). There are two versions of the metric system. The old system is being replaced by Système Internationale (SI). This table lists some corresponding units and common conversions that might help as you move through the book.

Quantity	Old Metric	SI	US	Conversion
Length	Centimeter (cm)	meter (m)	foot (ft)	1 m = 3' 3"
Area	cm^2 hectare (ha)	m^2 km^2	ft^2 acre	1 ha = 2.47 acre
Volume	cm^{3}liter (L)	m^3	gallon	1 L = 0.265 gal
Mass	g, ton (1,000 kg)	kg	pound (lb)	1 kg = 2.2 lb
Force	kg	Newton (N)	pound	1 kg = 9.81 N
Pressure	kg cm^2	Pascal (Pa)	psi	1 kPa = 0.15 psi
Energy	calorie (cal)	Joule (J)	btu	1 kJ = 0.95 btu
Power	dyne	Watt (W)	hp	1 kW = 1.34 hp
Temperature	°C	°K	°F	°C = (F − 32)* 5/9
Light	lux	mol m^{-2} s^{-1}	foot candle	

Note: Common prefixes: micro (μ)(÷1,000,000), milli (m)(÷1,000), kilo (k)(×1,000), mega (M)(×1,000,000)

likely impossible for testing our hypotheses. However, we can isolate parts of the problem and study those parts, eventually putting it all together to address the complex situation.

The key to a meaningful controlled experiment is to be sure that the only factor affecting the outcome is the factor that we set out to study. If we are testing whether the growth of a plant is affected by soil compaction or not, we need to be sure it is only compaction that is affecting the plant. For example, we would select plants of the same age to evaluate for response to compaction. If we had plants of different ages, we cannot be sure that age was not a factor in response to soil compaction.

In a typical experiment, plants of uniform characteristics are planted in compacted soil and others with the same characteristics are grown in uncompacted soil. The type of soil is called the **independent variable**. Plant growth is the **dependent variable** because its value depends on changes in the soil. Compacted soil is called the treatment. Uncompacted soil is called the control, or check. Controls allow for the comparison needed to draw conclusions about treatment effects. In practice, it is difficult or even impossible to make everything completely uniform. Replication and randomization are used to avoid unintentional bias resulting from lack of uniformity. Replication means that we base our conclusions on more than one observation on a single experimental unit (Fig. 1–10). The average of the response of experimental units to a treatment is what we base our conclusions on so there is less chance that an aberrant individual is the basis for a conclusion. Working with more than one

individual is the first step to ruling out chance variation as a factor that can affect our conclusions.

Randomization takes into account that no two plants are exactly alike. Even cuttings from the same plant may have come from different size branches or have other differences, such as mutations (Fig. 1–11). Randomly selecting the plants to be treated helps to reduce the chance that all the plants for one treatment are inherently different from those for another treatment. For example if, from a group of plants chosen for an experiment, only the short ones were used for the compact soil treatment while the tall plants grew in

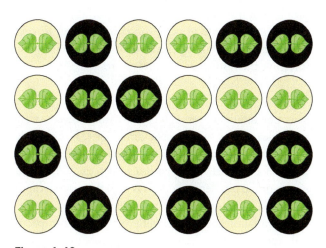

Figure 1–10

Randomized arrangement of pots containing compacted (dark-colored) and noncompacted (control) soil (light-colored). Source: Michael Knee.

noncompact soil, then the difference in height might carry all the way through the experiment and the conclusion that the plants grew best in noncompact soil would not be accurate. By mixing up the size of the plants within the treatments, changes in plant growth would more likely be the result of the treatment (Fig. 1–12).

Blocking is another way to reduce the effect of variation. An example of blocking is to arrange plants in groups by height, then subdividing those groups so that each height group receives all the treatments, not just one. Similarly, in a field trial, an experimental field can be divided into treatment plots where the treatments are replicated. These plots are randomly assigned treatments (Figs. 1–13 and 1–14). With this method, you reduce the chance that another factor, such as a wet area of the field, influences only one treatment. The more you know about the area you are using for an experiment, the more effective the blocking design can be. In greenhouse experiments, pots with compacted and noncompacted soil can be arranged randomly on a bench to eliminate the effect of drafts and drying out on those pots on the perimeter of the bench. Controlled experiments are ideal for studying one or a few variables;

however, one of their disadvantages is that treatment effects on factors other than the defined dependent variable are often not considered. Other factors may be detrimentally affected.

Although current scientific inquiry into agricultural and horticultural problems sometimes involves simple situations, such as determining how insect-resistant one new cultivar is compared to another, or what fertilizer rate is the best for flower development of a newly introduced ornamental, most research now involves some aspect of ecosystem study. We have learned the hard way that plants do not grow in isolation and what we do to help in one aspect of plant

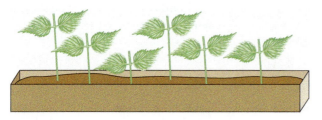

Figure 1–11
Plants showing natural variation in height from genetic and/or environmental conditions. Source: Michael Knee.

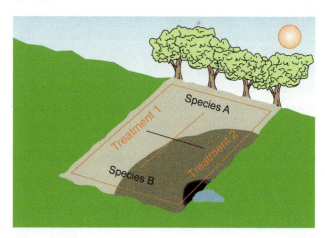

Figure 1–13
Example of a field experiment where the treatments (1 and 2) are represented only in the experiment. The darker portions of the field are wet areas that could distort or confound the data and increase the chance of making an incorrect conclusion. Source: Michael Knee.

Figure 1–12
Randomized placement of plants with different heights at time of planting into pots with compacted (dark-colored) and noncompacted (light-colored) soil. Source: Michael Knee.

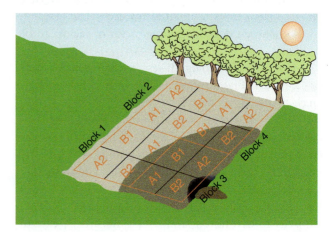

Figure 1–14
Example of a field experiment with multiple plots or blocks in which each of the two treatments (1 and 2) are randomly placed. Blocking helps to separate the effects of the wet areas from the effects of the treatment, reducing the chance of making an incorrect conclusion. Source: Michael Knee.

growth may have detrimental consequences elsewhere. These oversights created the suspicion that the public attaches to scientific study and the reassurances from the scientific community that a procedure is safe. For example, organochlorine insecticides were widely adopted after they were shown to be effective against a wide range of agricultural pests. After many years, however, the chemicals accumulated in animals at the end of the food chain. They were particularly harmful to birds of prey because they caused the birds' eggshells to be thin and brittle. Failure to hatch young led to a collapse in the populations of many birds of prey. Pesticides are now tested for effects on nontarget organisms, but no amount of research can ever prove that any chemical will be safe under all conditions of use and for all time.

The complexity of ecological research requires the use of different methods than experimental research that has one or only a few variables. Ecological research is often descriptive of a situation within an area that represents the larger ecosystem. Here, correlation analysis generates a descriptive model of the relationship between or among variables. Determining the scale of the area to be described is analogous to deciding what factor is to be the subject of an experiment. The scale has to be sufficiently large to represent the ecosystem but small enough to be effectively described. Many more variables are often recorded than in an experiment in a laboratory or greenhouse. In an experiment, variables that we do not want to influence the treatment must be controlled or held constant. In the real world, this condition is usually not possible, so the extra variables are recorded in case they are found to have an influence in the system. Dependent and independent variables are less clearly defined in descriptive research than in experimental research. The research method chosen depends on the factor or factors to be studied.

Examples of ecological research may be the study of the effects of farm runoff on the aquatic life in a single creek that feeds a larger river (Fig. 1–15). By studying the creek, inferences about the larger river's watershed can be made. Likewise, it might take fifty years to see the consequences of clear-cutting a rainforest to create farmland. However, by studying other rainforests that have been cut at different times during the past, we can look for patterns in the data that might show what happens to deforested land over time. In the case of Bt® corn, a more comprehensive study of the ecosystem of the Monarch butterfly, not just of how the butterfly larva was affected when fed only Bt® corn pollen, revealed that Bt® corn pollen was one of the least of the

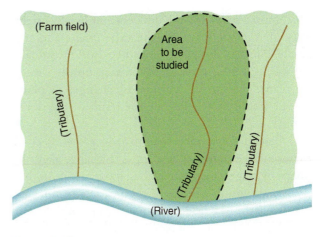

Figure 1–15

Correlative studies a small area and make correlations to a larger area. For example, a section of a field that has a small tributary stream running through it can be examined closely for effects from a farming practice. Inferences can be made regarding how other tributaries in other parts of the field are also affected by that practice.

Source: Margaret McMahon, The Ohio State University.

butterfly's threats. The conclusions from that study showed that the original public outcry stating that Bt® corn threatened the Monarch population was unfounded. However, in other cases it might be shown that a genetically engineered plant was detrimental to an ecosystem.

Data Interpretation

So how does a scientist arrive at conclusions from a set of experimental data? The data must be analyzed statistically to see if the original hypothesis is supported or not. But we can never be sure that it was our treatment and only our treatment that caused a difference in the dependent variable. We usually assign a percentage probability that something else was the cause. In most cases, if we can say that there was a 95 percent or greater chance that the treatment caused the effect, then we can claim our treatment had an effect. If the probability is less than 95 percent, we say that our results were nonsignificant.

To determine the probability, the natural variability in the population being tested has to be compared to the variation caused by the experimental treatments. In our compacted soil experiment, suppose we look at the number of roots of the intolerant plant in the pots of compacted soil and the number of roots in the pots of loose soil. We may see a wide range of numbers in both situations. If the variation in the pots is great enough, then the compacted pots may be no different from the

control, even though the two treatments may have different average root sizes. We would have to say that our treatment had no effect. For example, suppose we had four pots of porous soil and four pots of compacted soil. The average number of roots in the loose soil was fifteen, and the average number of roots in the compacted soil was ten. We would be tempted to say that the plant roots grew better in the porous soil. However, suppose we looked at the individual numbers and saw that the porous pots had twenty-two, eight, twenty-one, and nine roots and the compacted pots had fourteen, twelve, eight, and six. There may be so much variation that the differences in the two averages could have occurred without any treatment. One reason for having as much replication of experimental units as possible is to get a better estimation of natural variation.

In complex systems, correlation analysis is used to generate a relationship model. We can use the relationship model to determine whether it would be worthwhile to change a variable under our control, such as fertilizer use in a watershed, or whether changing the variable would be unlikely to have any effect.

No statistical method is completely reliable. There is always a risk of saying that there was an effect when there really was not. Conversely, we can say that results were nonsignificant when they were. That is why it is important when reporting experiments to describe in detail how they were done. Thus, future researchers can compare their results and methods to yours. It is not failure to have the results of a well-conducted experiment later contradicted by other researchers. The difference in results from different experimental situations quite often leads to a better understanding of a problem.

Reporting Findings

Once results are obtained and interpreted, the study is reported in a manuscript. First it goes for review to the scientific community. The reviewers examine the methods used and conclusions drawn and accept or reject the conclusions. If accepted, the manuscript is published and presented to the public in scientific and then trade journals.

As you can see, understanding the principles of scientific study is important to plant scientists, who must be trained in these principles. However, it is equally important that others understand the principles. Anyone who works with growing plants for a living has to be able to read and interpret the scientific and trade literature of his or her discipline. Without this ability, it is likely that decisions will not be based on sound scientific evidence.

SUMMARY AND REVIEW

Humans and cultivated plants are mutually dependent on one another. Cultivated plants require attention from humans to survive. Humans need the plants to fill basic nutritional needs and to add quality and enjoyment to their lives. Early plant scientists studied ways to improve crop productivity to meet the demands of a growing population. Plant scientists were successful at increasing production and today we are capable of feeding everyone on earth; however, problems of starvation and uneven supply and distribution of crop commodities continue to plague the world. In addition, the population continues to grow, placing higher and higher demands on our farms, which are shrinking in number.

Complicating the situation is the fact that many improvements in production were shown in time to be detrimental to the global ecosystem. Plant scientists and growers had to find ways to grow crops without creating undue risk to the global ecosystem. Because of past mistakes, however, new production practices, such as the use of GMOs to reduce pesticide use or improved shipping quality of produce, are met with suspicion because of the fear that the new methods will also have negative effects. Public opinion is proving to be a powerful force in plant science. The difficulty is exacerbated by the fact that much of the population in many countries resides in urban areas and has little understanding of how food and other plants are produced.

Today's plant scientist and grower must have a firm understanding of the relationship among production efficiency, economic viability, environmental compatibility, and social responsibility when studying ways to improve productivity. Integrating that understanding with a sound scientific approach to research in plant science will help restore public confidence by reducing the chance that a procedure will have an undesirable side effect. A sound approach begins with the formation of a hypothesis, designing the proper experiments to test that hypothesis, and finally analyzing and interpreting the data gathered correctly. Randomization and replication in experiments are important to reduce the chance of reaching a wrong conclusion.

KNOWLEDGE CHECK

1. In what group of plants do flowers, fruits, vegetables, trees, and shrubs belong?
2. What is cultivation and when did it start?
3. When did the rural/urban interface start and what problems for farmers does it cause?
4. What is agronomy, what is horticulture? Give an example crop of each and one that could fall in either category.
5. Money value and ecological value are mismatched in agriculture. What would happen to the price of food if the cost of clean up from pesticide or fertilizer run-off was added to the production cost?
6. How has our growing understanding of ecology changed plant science research and education?
7. How has the reduction in the number of farms affected the vitality of many rural communities? What has helped those communities survive in states with large metropolitan areas?
8. How was the famine predicted to be caused by diminishing farmland averted in the 1970s?
9. The ability to feed the world on 0.2 hectares of land per person (the amount of farmland predicted to be available then) when the population reaches 8 billion will require that all that land be farmed using the production techniques of what countries? How will that affect global resources, especially fossil fuels?
10. How does globalization of marketing affect the spread of pests and diseases?
11. What caused negative public opinion and close scrutiny of agriculture to develop?
12. What kind of environmental impact does the demand for perfect plants usually create?
13. What are some concerns with GMO plants and crops?
14. What are the four sides of the ecological pyramid that support sustainable agriculture?
15. How do randomization and replication improve the reliability of the results of an experiment?
16. How does research of a complex ecological problem differ from a controlled experiment?

FOOD FOR THOUGHT

1. How would you explain to someone who is complaining about food costs why food priced on emergy value has become so expensive?
2. There is great concern that the number of people dying of starvation will increase dramatically when the world's population reaches 8 billion. What do you think could be the consequences of that increase in death and how can it be prevented?
3. Sustainable farming takes into account many more factors (e.g. social issues, more environmental than just the use of chemicals) than organic farming but is not as strict in the use of chemicals. Which practice do you favor and why? How would you convince a homeowner who wants a perfect yard or a golfer who wants a perfect fairway that their demands have tremendous ecological consequences?

FURTHER EXPLORATION

DUM, H. T. 1996. *Environmental accounting: Energy and environmental decision making.* New York: John Wiley & Sons.

FOOD AND AGRICULTURE ORGANIZATION OF THE UNITED NATIONS (FAO). 2002. *The state of food and agriculture.* Rome. Available online at http://www.fao.org/DOCREP/004/y6000e/y6000e00.htm

NATIONAL RESEARCH COUNCIL. 2005. *Valuing ecosystem services: Toward better environmental decision-making.* Washington, DC: National Academy Press.

POWERS, L., AND R. MCSORLEY. 2000. *Ecological principles of agriculture.* Albany, NY: Thomson Learning.

SMITH, B. D. 1995. *The emergence of agriculture.* New York: Freeman.

UNITED STATES DEPARTMENT OF AGRICULTURE. National Agriculture Statistics Service (NASS) Publications. Available online at http://www.usda.gov/nass/

*2

Terrestrial Ecosystems and Their Relationship to Cultivating Plants

Almost all terrestrial ecosystems depend on the ability of plants to capture energy from the sun and store it in complex organic molecules. Climate determines the kinds of plants that grow naturally in an area and the rate at which they grow. All of the other organisms in the ecosystem depend on and are influenced by the nature of the vegetation and its productivity. Natural ecosystems are usually more complex than crop ecosystems; they illustrate the gamut of ecological relationships and ecosystem processes and provide a reference point from which we can evaluate crop ecosystems. Generally, the more a crop ecosystem differs from the ecosystem that would occur naturally in an area, the more difficult it is to sustain and the more resource input it requires.

ECOSYSTEM COMPONENTS

An ecosystem consists of a community of organisms in a physical environment (Fig. 2–1). The community consists of populations of individual species; different kinds of plants, animals, fungi, bacteria, and so on. A population can be defined as all of the individuals of a species that inhabit a particular environment. The way that we define the environment determines the boundaries of the population. Often we are thinking of a limited area in which the individuals share the available resources. In an isolated area of woodland, a field, or a greenhouse, the plants may be competing for light, water, or nutrients. However, ecological relationships may extend over larger areas, particularly for animals that can roam from one place to another. Even for plants, pollination may occur between individuals in physically separate environments many miles apart. It is fairly obvious that plants found naturally on different continents belong to different populations, but if there is any possibility of an ecological interaction between individuals of a species, then they can be viewed as belonging to the same population.

We tend to think of ecosystems in terms of the plants and animals that we might see on a visit to a "natural" area in a state park or wilderness location. However, anywhere organisms exist is, in a strict sense, an ecosystem. Many of the most important

key learning concepts

After reading this chapter, you should be able to:

- Describe the fundamental importance and relationship of plants and other organisms in terrestrial ecosystems.
- Describe the different biomes of the world, how they are created, and how they determine what plants grow there.
- Explain the relationship between natural ecosystems and the ecosystems we create when we grow plants.
- Explain what influences photosynthetic productivity in natural and cultivated ecosystems.
- Discuss the impact that cultivating plants has on ecosystems.

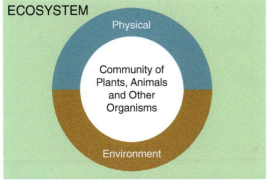

Figure 2–1

An ecosystem consists of a community of organisms in a physical environment (consisting of soil and atmosphere in the case of a terrestrial ecosystem). Source: Michael Knee.

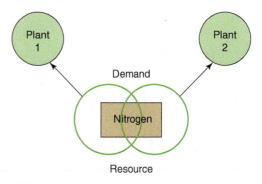

Figure 2–2

Competition occurs when plants of the same or different species attempt to use a resource that is in limited supply. Source: Michael Knee.

organisms and ecological processes in all ecosystems are invisible to us because the organisms are small or because they and the processes occur in the soil. A cornfield or an area of turfgrass may look too simple to qualify as an ecosystem, but a whole community of organisms interacts with the single species of plant, both above and below the ground. Ecology does not disappear when we grow plants in a home or a greenhouse. The physical environment, including the growing medium, may be artificial, but it will be colonized by other organisms, some harmful and some beneficial. Although books and television programs about nature may focus on the animals and birds that enliven natural ecosystems, many of the functions and ecological interactions that sustain the ecosystems are carried out by small invertebrate animals, fungi, and bacteria, and many of these organisms are in the soil. The microflora and -fauna are, relatively speaking, even more important in crop ecosystems, where we generally try to exclude macrofauna (birds and mammals) and noncrop macroflora (in other words, weeds). What distinguishes these "cultivated" ecosystems is that some of the organisms, processes, and interactions that would sustain a natural ecosystem are absent or heavily modified by human intervention. For example, inputs of nutrients are necessary to promote crop growth, or chemicals are needed to suppress disease. We can describe crop production in terms of these inputs and ignore the ecological processes and interactions. However, the goal of sustainable production requires that we minimize the inputs and maximize the contribution of ecosystem processes. By observing the full range of interactions and processes in natural ecosystems, we can learn how to manage crop ecosystems to achieve the goal of sustainability.

The organisms in an ecosystem interact according to the nature of the species and their role in the ecosystem. Each individual organism draws on the resources of the ecosystem to meet its requirements. Plants require light, water, and nutrients for growth and may require pollinators and dispersal agents for reproduction. Competition occurs when more than one organism draws on a resource that is in short supply (Fig. 2–2). The resource might be required for growth or the reproduction of the organism.

In ecosystems with a complete cover of vegetation, the most intense competition among plants is likely to be for light. A plant with a greater leaf area is likely to capture more light, particularly if the leaves are positioned above those of another plant. Trees can capture most of the light and are the dominant vegetation in many parts of the world. However, they have to invest considerable resources in the trunk and branches that support their leaves, which is one factor that limits the size of trees. Vines such as Virginia creeper, English ivy, and poison ivy can compete for light with less structural investment by using the support provided by the trees to grow tall enough to reach sunlight.

Usually a single species cannot utilize all of the resources or the whole of any one resource in the environment. For example, some light does penetrate the canopy in most forests, and herbaceous plants, such as ferns, are adapted to growth in the low-light environment of the forest floor. Many of the trees in temperate forests are deciduous, and for a period in the spring, light is available for herbaceous flowering plants such as bloodroot, *Trillium,* and *Hepatica* to grow, flower, and set seed before the trees develop their full canopy. These kinds of plants are called spring ephemerals. The shade-adapted plants and spring ephemerals occupy niches. Niches exist when a resource is partitioned so that different portions of it are accessible to only certain species (Fig. 2–3).

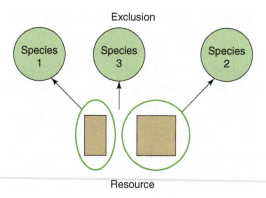

Figure 2–3

Species 1 is able to occupy a niche because species 2 does not use all of the resources. However, species 3 is subject to exclusion because it also needs the resources, but none remain.

Source: Michael Knee.

Although a critical environmental resource may be required for a species to find its niche in an ecosystem, each species has more or less strict requirements for a long list of environmental resources. In the fullest sense, the niche includes everything that is required for an organism to flourish in the environment. For a plant, this list can include light intensity, availability of water, nutrient concentrations, and the presence of the right pollinators. If two plants have identical environmental requirements and they are trying to occupy the same niche, the competition will be more severe than it is for two plants with different requirements. Competition is thus more intense between plants of the same species than it is between plants of different species. For species to coexist in an ecosystem, they must occupy different niches. It can be difficult to see how species requirements differ, for example, when looking at different species of deciduous trees that make up the forest canopy. However, species are separated by their physiological requirements as much as by their more obvious morphological differences.

In healthy, natural ecosystems, every available niche tends to be filled, and consequently resources are fully exploited. It is difficult for new species to colonize the environment because it lacks open niches or unexploited resources, and the result is called exclusion (Fig. 2–3). Thus, annual herbaceous plants cannot colonize mature woodland. If seeds blow in and germinate, the development of the foliage in the canopy during the spring will close off the light before the annual plants can flower and set seed. On the other hand, if some trees come down in a storm, annual plants may be able to colonize the open area or gap until new trees grow and the canopy closes.

Crop ecosystems have one or a few species that do not exploit all of the resources of the environment.

Thus, niches tend to be available for other plants to colonize. We call the other plants weeds, and their presence is a problem to the extent that their resource requirements overlap those of the crop species. One of the strategies in weed control is to plant additional, noncompetitive crop or noncrop plants with the main crop to occupy niches that would otherwise be occupied by weeds.

Resources captured by plants may be lost, however, to predation or parasitism. Plants are consumed by a wide range of animals, which we call herbivores. (We use the term predator for animals that consume other animals). Herbivores include mollusks (slugs and snails), arthropods (insects and mites), mammals, and birds. The diversity and large numbers of herbivores emphasize the importance of plants as a food source in terrestrial ecosystems. Many believe that herbivores do not compete with each other or come close to consuming all of the plant food resources of natural ecosystems because their numbers are kept in check by their own predators. There is a wide range of small predators among the arthropods, and there are many insect-eating birds and mammals. Large predators that feed on birds and mammals are less common because ecosystems cannot support many of them. As the number of herbivores increases, they become easy prey for the predators. Crop ecosystems are usually more vulnerable to herbivory because the number of predators is low or nonexistent. Thus, it may be necessary to use chemicals, introduced predators and parasites (biocontrols), or other means to control slugs, insects, rodents, birds, or deer.

Parasites are organisms that derive their nutrition by living in or on the tissues of another organism, often producing symptoms of disease such as swellings or discolored tissue. Plant parasites include many kinds of viruses, bacteria, and fungi. Some parasites can grow only in the host and are called obligate. Other parasites can grow outside the plant and are called facultative. All viruses are obligate parasites, whereas fungi and bacteria may be obligate parasites or facultative (as when they commonly exist outside the plant in the soil). A few parasitic plants, such as dodders and mistletoes, can infect plants in natural and crop ecosystems. Other classes of plant parasites include nematodes and some insects and mites, particularly those that form galls on leaves and stems.

Parasites differ from herbivores because they cause a disease rather than simply eating their way through the tissue. In a disease, the physiology and often the morphology of the host is altered as its resources are redirected toward the pathogen or disease-causing

organism. Parasitism is a long-term relationship between a host and a pathogen, in which the host does not usually die because this would also entail the death of the pathogen. Although the host may not die, it is usually weakened making it vulnerable to attacks by other (secondary) organisms. Parasitoids are animals, often insects, that spend the juvenile phase of their lifecycle in the tissues of another insect. When they emerge as adults, the host is killed. Parasitoids can be helpful in crop ecosystems when they infect herbivorous insects such as caterpillars.

Some associations between organisms are mutually beneficial rather than antagonistic. In natural ecosystems, the roots of most plants are colonized by fungi that draw nutrients from the soil and pass some of them on to the plant. In return, the plant passes sugar and other organic molecules to the fungus. The fungi are known as mycorrhizae and can either live on the surface of the root (ectomycorrhizae) or invade the root tissue (endomycorrhizae) (Fig. 2–4). This relationship is an example of symbiosis, which involves a permanent and close association between two organisms that are known as symbionts. Mycorrhizae are less common in crop ecosystems, particularly on herbaceous plants, than in the wild. The nutritional requirements of crops are often supplied by fertilizers and fungi are not needed to scavenge for them. However, mycorrhizae are now being supplied to some crops to reduce the need for chemical fertilizers. Lichens that are found on tree trunks, rocks, and the soil surface are formed by the symbiosis of a fungus with an algae species. Another kind of plant symbiosis involves a bacterial symbiont, *Rhizobium*, that forms the nitrogen-fixing nodules on the roots of many species in the legume family, or *Fabaceae* and a few other families. Nitrogen-fixing plants are important for maintaining the nitrogen supply in natural ecosystems. The legume family, includes many important crops such as beans, peas, and alfalfa. Rotating a nitrogen-fixing legume crop such as soybean or alfalfa with a nonlegume on a regular basis is a common farming practice to keep nitrogen levels up in a field. A more general term for a mutually beneficial association between organisms is commensalism. In addition to symbiosis, commensalism includes looser and less permanent associations such as that between a plant and its pollinating insect.

Although parasitic organisms may be the most conspicuous fungi and bacteria in crop ecosystems, most microorganisms are saprophytes that digest dead plant and animal material at or below the soil surface (Fig. 2–5). Saprophytes are essential to the survival of the ecosystem because they recycle nutrients that would otherwise be tied up in dead organisms. In natural ecosystems, a few plant species are saprophytic. Saprophytes are aided in their activity by detritovores that break up large pieces of organic matter as they consume it. Earthworms are an important part of this group, which also includes other kinds of worms, many insects, and small mammals. Detritovores and saprophytes may be less important in crop ecosystems than in nature because crops and crop residues are removed and the nutrients are replaced by synthetic fertilizers. They become more important, however, as we try to recycle crop waste and minimize fertilizer inputs to develop more sustainable production systems.

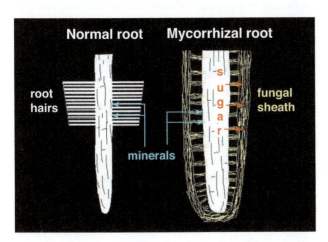

Figure 2–4
Many plant roots form a symbiosis with fungi, a relationship in which the plant supplies sugar to the fungus and the fungus draws mineral nutrients from the soil, which it passes on to the plant. Source: Michael Knee.

Figure 2–5
Recycling is essential for maintaining the supply of nutrients in natural ecosystems, and soil microorganisms are the primary agents in this process. By decomposing the leaf litter and tree trunk in this picture, saprophytes and detritovores are recycling the nutrients in those items. Source: Margaret McMahon, The Ohio State University.

Primary producers (mainly plants), herbivores, parasites, commensals, detritovores, and saprophytes exist in all terrestrial ecosystems. In a natural ecosystem, each of these roles is filled by many species of organism that exist in a complex web of relationships (Fig. 2–6). Nutrients are passed continuously from one organism to another along with the energy and carbon compounds that sustain the whole ecosystem and that originally came from plants. The system tends to be highly conserved. Many organisms compete for the resources available at any level in the food chain. If there is any gap in the utilization of resources, species are recruited to fill it. Crop ecosystems are based on one or a few plant species, which are usually intended to be harvested rather than allowing resources to be consumed by other species. While the crop is growing, however, surplus resources exist that could be exploited by competitors, herbivores, and parasites. We have attempted to manage these problems by mechanical and chemical means. By eliminating species, we attempt to simplify the ecosystem and minimize ecological interactions. More recently, however, we have begun to introduce species that help control some of the problems, for example, companion plants, predators, parasitoids, and so on.

BIOMES, CLIMATE, AND MICROCLIMATE

We often like to go to different climatic zones for vacations, but if you get the chance to travel to a similar one, say, from eastern North America to northern Europe, you may be struck by the fact that similar but not identical trees are present in the forests. You might see oak, maple, beech, and ash but not the same species that you would see in the United States. The explorer Alexander von Humboldt noticed that there was a similar progression of vegetation types as he moved north or south from the equator or ascended in altitude in different parts of the world. He realized that the same kind of vegetation tended to occur in similar climatic conditions around the globe. The vegetation appears similar because the plants are functionally similar but not necessarily the same species. The functional groups include deciduous broad-leaved trees, coniferous evergreens, succulent plants, tall and short grasses, spring-flowering bulbs, and so on. Today we call the recurrent vegetation types biomes; a **biome** is a collection of ecosystems with similar climate, soil, and plant composition.

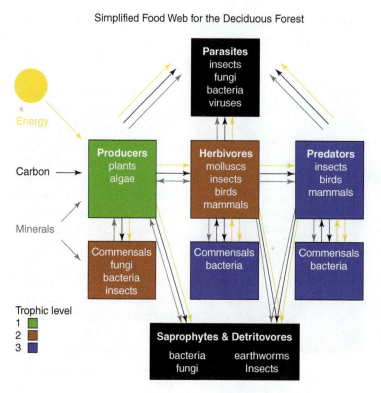

Simplified Food Web for the Deciduous Forest

Figure 2–6

Natural ecosystems operate with inputs of carbon, energy, and water and are highly conservative of nutrients because of a complex web of ecological relationships. Source: Michael Knee.

Climate is the main influence on the type of vegetation that develops. The most important climatic variables are temperature, rainfall (or, more correctly, precipitation), and any seasonal variation in both. A key component in the relationship between climate and vegetation is soil. As we will see in Chapter 5, soils are produced from parent material, such as rock, by the interaction of climate and organisms. Plants are key components of the soil ecosystem and influence all of the other organisms present. Although a world map may show the natural distribution of the different biomes across all of the continents, many of these areas have now been converted to agriculture or urban development. The soils that developed in the biomes differ greatly in their suitability for crop production. In general, the greater the difference between a crop production system and the preexisting biome, the more difficult it is to sustain that system. Sometimes it may even prove to be impossible.

Temperature is primarily influenced by latitude, and within latitudes, it is influenced by the height of the land above sea level. The other factor, moisture availability, is affected by patterns of air circulation around the globe. Air moves in one direction in the winds that we experience at ground level and then rises to the upper atmosphere to flow in the opposite direction, before returning to ground level. This circulation is rather like a donut-shaped mass of air, with the wind moving around it from the hole in the middle to the outside and back. A series of three such systems operates between the equator and each pole (Fig. 2–7).

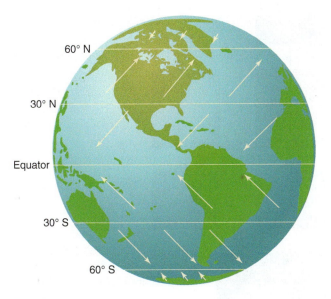

Figure 2–7
Three main belts of surface winds exist in each hemisphere. Each one determines patterns of precipitation and the vegetation that develops. Source: Margaret McMahon, The Ohio State University.

At the equator, temperatures are high throughout the year. The trade winds from the north and south carry moisture from the oceans and converge in this region. The warm, moisture-laden air rises and cools down so that the moisture condenses as rain. The warmth and readily available water are favorable for plant growth throughout the year, and all kinds of other organisms can benefit from the high productivity of the plants. The tropical rainforests of South America, Africa, India, and southern Asia developed under these conditions and make up the most productive biome, with the greatest range of biological diversity on earth.

The forest is dominated by broad-leaved and evergreen trees whose dense canopy captures most of the light. Rainforest trees include kapok, mahogany, and rosewood. Rainforest soils are shallow, and the tree roots spread outward rather than down. The trees gain stability from their shallow root systems by forming buttress roots. Because of the low-light levels, only a few shade-tolerant plants grow under the canopy; however, the trees provide support to many **lianes**. Also many herbaceous plants called epiphytes grow on the trunks and branches of the trees. The roots of these epiphytes never reach the soil; they catch their water as it runs down the surface of the tree or from the air itself with specialized aerial roots. Many orchids and bromeliads are rainforest epiphytes.

Much of the life of the forest occurs in the canopy, unseen by humans and other groundlings. Most of the nutrients in the ecosystem are also in the canopy and are cycled from one organism to another without passing through the soil. Because of high rates of growth, nutrients are rapidly taken up by plants and rapidly recycled from dead organisms, so rainforest soils contain little in the way of free nutrients. After the forest is cleared, good crops can be raised for a year or two, but the nutrients are soon exhausted. To make matters worse, nutrients are washed out by the rains, and the soils themselves are unstable in the absence of vegetation. So production of annual crops is not easily sustainable in this region.

About 30° north and south of the equator are bands of divergence in the global air circulation. Air moves away from these latitudes in the trade winds toward the equator and in the westerlies toward the temperate latitudes. This movement is powered by a downdraft of air that carries moisture away, and it is most pronounced over large land masses. It gives rise to the deserts of North and South America, Africa, Asia, and Australia, which make up the largest biome on the planet. By day, temperatures are warm, but at night the

absence of moisture in the air permits rapid loss of heat so that it can be surprisingly cold. The plants of this environment are highly specialized and grow slowly or intermittently. They can be slow-growing perennials with succulent stems or leaves that help conserve water. Cacti and yucca are an example of these kinds of plants. They can also be fast-growing annuals such as Ghost Flower (*Mohavea confertiflora*) or Desert Sunflower (*Gerea canescens*) that take advantage of occasional rains to complete their life cycles while the moisture is available. Because of slow growth, there is little accumulation of organic matter that can be recycled to build up soil fertility in these ecosystems. Because temperatures can be favorable to plant growth and deserts occupy about two-fifths of the land surface, many people have had the dream of "making the desert bloom." This requires massive amounts of water that can be economically transported or diverted to the desert region. Fertilizers must be used to compensate for the low fertility of the soils. All water sources contain small amounts of salts, even if they are so-called fresh water. The salt level increases when fertilizers are added. When the desert is irrigated, the salts are left behind after the water evaporates from the soil or through plant transpiration. Desert soils are difficult to manage in the long term because the salts tend to accumulate to levels that inhibit the growth of plants.

Moving north or south away from the deserts, the climate becomes moister but temperatures are not continuously favorable for plant growth. In latitudes between 40° and 50° and when there is more than about 75 cm of rainfall, temperate deciduous forest develops (Fig. 2–8). Ash, oak, maple, hickory, and beech are some of the hardwood species found in temperate forests. These forests develop all of the land masses in the northern hemisphere, but there is little land area in these latitudes in the southern hemisphere so temperate deciduous forest is nearly absent south of the equator. During the summer, the trees in these forests can be almost as productive as those in the tropical rainforests. However, growth nearly ceases in the winter, when most of the trees have lost their leaves. Although lianes, more commonly called vines in this biome, are present in old-growth forests of this type, the canopy is not as favorable a habitat for epiphytes as in the rainforest. However, **spring ephemerals** which are **understory** shrubs and herbaceous plants such as trillium (*Trillium grandiflorum*), dogtooth violet (*Erythronium dens-canis*), and bloodroot (*Sanguinaria canadensis*) can exploit the light that comes through the canopy in the spring before the new leaves are fully formed. (The **understory** is composed of the plants

Figure 2–8
A clear area under the canopy of a mature temperate forest where other plants cannot grow in the low light. In the spring, however, it is filled with spring ephemerals.
Source: Margaret McMahon, The Ohio State University.

that grow under the canopy of other taller plants.) The botanical diversity of the Appalachian forests of North America is a distant second to the rainforest, but it is more readily apparent to a wanderer in the sunlit carpets of spring ephemerals than it is to a ground-based visitor in the comparative (and permanent) gloom of the rainforest.

The annual leaf fall adds a certain amount of nutrients to the litter layer of the temperate deciduous forest. However, the leaves contain only a fraction of the nutrients locked up in the permanent structure of the tree, and trees manage to withdraw much of the nitrogen and phosphorus from their leaves before they are shed. So forest soils develop slowly and tend to have only a thin organic-rich layer. As with the rainforest soils, fertility can be soon exhausted if not managed properly when crops are grown after the forest is cleared.

The area between the deserts and the forests consist of grasslands and savannas (Fig. 2–9). Tree growth in grassland and savanna is limited to some extent by lack of water, but fire and grazing animals are also important in maintaining these ecosystems. Grassland vegetation survives and may even benefit from a certain level of grazing that woody plants cannot tolerate. Also dead grass is readily ignited by lightning strikes. When fires sweep through grassland, most trees are killed, but grasses and herbaceous perennials, which have their growing points at or below the soil surface, typically survive. Grasslands and savannas were favored by early people and our first efforts at land management may

Figure 2–9
The prairie in Kansas is characterized by expanses of grassland interspersed with a few trees and shrubs.
Source: Kimberly Williams, Kansas State University.

have been to set fires to provide habitat for animals that we were trying to encourage into or deliberately keep in herds. We still find prairies, meadows, and forest glades attractive but may not fully realize our own role in their ecology.

Grassland plants accumulate nutrients in the roots and stems that are renewed from year to year. In temperate climates, the old stems and roots are not fully recycled each year and organic matter builds up over time. This organic matter contains reserves of nutrients and makes up a large part of the black earths that are characteristic of moist grasslands around the world. The nutrients in such soils can sustain crop production for many years, and the former grasslands of North America, Europe, and Asia have proved to be the world's most productive agricultural land. Agriculture itself originated in such a region, the so-called "fertile crescent" that occupied river valleys from present-day Egypt to Turkey and Iran. Grassland becomes less productive as it merges with drier regions on its margin. In the United States, for example, the tall grass prairie of the central states gives way to short grass prairie to the west. Attempts to raise productivity with irrigation can meet the same problems of salinization as in the desert. Arid lands can all too easily be turned to desert by overgrazing, and soil erosion can be a devastating consequence of attempted cultivation. The extent to which human beings are responsible for the spread of deserts throughout the world is debatable. Climatic changes, such as unusual and persistent droughts, are certainly involved in the process, but questionable agricultural practices can increase the risk of degradation. This interaction contributed to the US dust bowl of the 1930s, and recovery from it required massive changes in land management. Arid lands are generally sparsely populated, but the quarter billion or so people who live in such areas around the world are among the most vulnerable to climatic change.

At about 60°N, the convergence of arctic winds and the westerlies from lower latitudes creates an upcurrent leading to precipitation, just as at the equator. Of course, it is much colder at this latitude than at the equator, and the long winter allows for only a short season of plant growth each year. Coniferous trees make up the dominant vegetation in this boreal forest, or taiga, and plant diversity is much lower than in the temperate and tropical forests. Because the conifers are mostly slow-growing evergreens, recycling and the availability of nutrients is lower than in the deciduous forest. Nutrients are also depleted as water drains through soils that are wet for most of the year. This type of vegetation is almost absent from the southern hemisphere because there is little land around 60°S. However, taiga occurs in a band from North America through northern Europe and Asia, making it the second largest biome after the desert. Of course, coniferous trees are a valuable crop for

A   B

Figure 2–10

In the taiga, farming is done on a very small scale. (A) Hay making in the taiga. (B) A small farm producing fruit and vegetables in the taiga. Source: Margaret McMahon, The Ohio State University.

lumber and paper making, but there is little prospect for large-scale production of other kinds of crops in this area (Fig. 2–10).

In the tundra, north of the taiga, the ground is permanently frozen except for a surface layer throughout the year. The short melt period in the summer allows only a low vegetation of dwarf shrubs, sedges, and grasses to develop. Similar vegetation occurs in high mountains. Some of the plants have showy flowers that make the tundra briefly more colorful than the taiga. Alpine plants are attractive for specialist collectors, but there is almost no prospect for crop production in the tundra itself.

The biomes are described in terms of the vegetation that existed before human interference, which began several thousand years ago to change the vegetation to provide food directly for humans or for animals that could be eaten by humans. At this stage, almost all the biomes have been influenced by human activity and some have been almost destroyed. Grasslands have been the most extensively modified because of their value for agriculture. In the United States, only 2 percent of the tallgrass prairie remains. Temperate forest has nearly all been harvested at some time and much was converted to farmland, although small areas have returned to woodland in many parts of the United States and Europe over the past fifty years (see Fig. 1–5 on page 5). These historic changes have been followed by clearance of the tropical forests in the twentieth century. This development is widely regarded as an ecological catastrophe, but it is not clear whether we can prevent it or have the right to act to expect it to stop.

Even when small areas of ecosystems are allowed to remain in or return to their natural state, they do not

Figure 2–11

Wetlands provide many important ecological benefits. Source: Margaret McMahon, The Ohio State University.

function in the original way. One of the most pervasive causes of change has been the widespread practice of land drainage. All of the kinds of ecosystems that made up the biomes contained wet areas or wetlands (Fig. 2–11), which were often the most productive areas because water was permanently available. Wetlands also acted as ecological buffer zones, providing refuge for animals and sources of moisture for plant life during dry seasons. Drainage proved necessary for most kinds of crop production in moist temperate areas, but this practice left remaining natural areas more vulnerable to periods of drought. Because we now understand the benefits of wetlands, the creation or restoration of wetlands has become common on farms, golf courses, housing developments, and other land use projects.

Crop production has displaced many preexisting ecosystems, and it is more realistic to describe global ecology in terms of new biomes such as farm fields, rice paddies, or pasture. Just about all of the land that can be cultivated easily is now used for the production of crops. With the pressure of world population, we need to produce more food. It is also desirable to increase the diversity of food available to many people who currently depend on one or two staple crops. Is it realistic to increase the land area available for crops? Substantial gains could be made only from biomes that have not been exploited so far. But these biomes present serious limitations to crop production. Although tomatoes can be grown in greenhouses in Alaska, the temperature limitations on crop production in the taiga and tundra cannot be overcome on a large scale. Lack of water can be overcome more easily and irrigation will continue to be important for crops in many parts of the world. Fresh water is a limited resource in all parts of the world where it could be most useful, however, and it is not realistic to contemplate massive conversion of desert areas to any kind of crop production. In addition, marginal areas are often the most vulnerable to environmental degradation. When plant growth is slow, recovery from disturbance is also slow, and other organisms that depend on plant life can easily lose their place in the ecosystem. The low productivity of the ecosystems limits the earnings of people that depend on them and can encourage overexploitation. Many argue that we should focus on increasing the productivity of existing crop areas rather than expand into currently unproductive land.

Within every biome and major climate area there are innumerable **microclimates**. These are determined by hills and valleys or even small depressions in a field, proximity to large and small bodies of water, presence of large land masses or buildings. Hills, valleys, and depressions create microclimates that can be warmer or colder than the climate in general. Bodies of water can delay spring warm-up and fall cool-down. Buildings that have been warmed by the sun can provide a source of heat, at least temporarily. Often, a microclimate is more important than the biome or climate in determining what plants can survive or thrive there. People who grow plants can take advantage of these areas to grow plants that may not do well in the regional climate or protect the plants that may be are harmed by it. For example, a landscaper can put tender plants close to a building to prolong their growing season in the autumn. The heat from the building may provide enough warmth to allow the plants to live a few weeks beyond what similar plants would live if not near the building. Farmers can avoid planting too early in parts of their fields that are known as "frost pockets," which are depressions where cold air collects and temperatures are lower than in the surrounding area.

PRODUCTIVITY OF TERRESTRIAL ECOSYSTEMS

The relative productivity of different biomes has already been discussed. As we modify more of these areas for human use, it is important to know what effect this modification has on global productivity. The productivity of ecosystems is an indication of how much we can expect to harvest from them without causing loss of species or collapse of the whole ecosystem. Some of the most obvious examples of the failure to consider sustainable harvesting practices are the exploitation of fish stocks and the wood supply in many parts of the world. On a larger scale, productivity is related to the carbon balance of the world. Plant productivity can be defined in terms of the amount of carbon taken from the atmosphere by photosynthesis. When we change ecosystems for our own use, we may also change the rate at which carbon dioxide is removed from the air, which will have an effect on global climate. We are thinking increasingly in terms of changes that we can implement to maximize atmospheric carbon removal, one way of accomplishing this objective is to maximize plant productivity.

Photosynthesis fixes carbon into organic compounds, but this process is not quite equivalent to plant productivity. The plant uses a portion of the organic compounds for its own maintenance. At the end of a growing season, less carbon is retained than was originally fixed. Net primary productivity (NPP) is carbon fixed minus the amount consumed by the plant and returned to the atmosphere through respiration.

It is quite easy to measure the carbon accumulation by a single plant, but it is difficult to measure for complex communities consisting of many species under field conditions. Two-thirds of our planet is covered by water, and it is at least as difficult to estimate the productivity of the oceans as it is for the land. Estimates of productivity have varied greatly over the years, but it is now thought that although terrestrial ecosystems occupy about 30 percent of the surface of the planet, they account for about 55 percent of global photosynthetic productivity. The world's terrestrial ecosystems can be divided into three broad groups of high, intermediate, and low photosynthetic

productivity (Fig. 2–12). Tropical forest and savanna are the most productive biomes and account for about half of terrestrial productivity, although they make up less than one-quarter of the land area. Temperate forest is almost as productive but contributes about 10 percent of total terrestrial productivity because it occupies only about 7 percent of the land surface and is photosynthetically active only part of the year. Temperate grassland, boreal forest, and cropland are about half as productive as temperate and tropical forest and tropical savanna. The contributions to global productivity from these biomes with intermediate productivity tend to match their share of land area (8 percent for temperate grassland, 6 percent for boreal forest, and 10 percent for cropland). One-third of the land surface is occupied by tundra, permanent ice, desert, and semidesert, which contribute about 4 percent of global productivity. Urban areas have about the same productivity as tundra and occupy about 1.5 percent of land surface. In all parts of the world, wetlands tend to be the most photosynthetically productive areas; on average their contribution to productivity is three times their share of land area but because their area is only about 1.8 percent, their total contribution is small. In spite of their natural productivity, we have not yet exploited the inherent productivity of wetlands, with the exception of rice and a few minor crops, probably because it is more difficult to work in flooded fields than on dry land.

In terms of carbon fixation, cropland is usually less productive than the biomes that it replaced. This is particularly true for annual crops because they do not fully occupy the land for most of the year. Of course, crop productivity is not usually the same as net primary productivity of an ecosystem. Usually crop productivity is measured in terms of a specific part of the plant that is harvested, rather than the whole plant. Part of the challenge of production is to maximize this part in relation to total biomass, or in other words to maximize the **harvest index**. In principle, annual crops are wasteful of primary productivity because most of the fixed carbon is removed at harvest and must be fixed again in the following year to make a new corn or tomato plant. However, perennial crops must allocate part of the fixed carbon to their survival structures and if this is not the part that is harvested, less fixed carbon may be accounted for in the harvest index than from an annual crop. However, some of the most productive crops are perennial plants from which the survival structures are harvested for consumption. Such crops have historically sustained people with limited land and resources, such as the Irish who grew potatoes in the nineteenth century, and the 10 percent of the world's population that today depends on cassava for its nutrition.

SUCCESSION

Descriptions of biomes can leave the impression that they consist of communities of plants and other organisms that would always be there in the absence of human interference. However, we know that organisms have evolved, landforms have changed, and climate has varied over time. The change in plant communities over time is known as **succession**. It is often described in terms of the development of vegetation from a blank-slate situation such as a rock surface, a pool of water, or bare soil (Fig. 2–13). These situations may

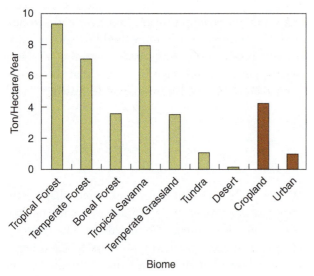

Figure 2–12
The net primary productivity of biomes represents the amount of carbon fixed in photosynthesis for a given area in a year. Source: Michael Knee.

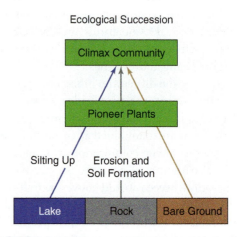

Figure 2–13
All locations in a geographic area tend to develop the same kind of vegetation over time. Source: Michael Knee.

have arisen because of natural causes (an earthquake, volcanic eruption, or the passage of a glacier) or because of human interference.

If there is no soil, succession must begin with soil formation. A rock surface can be colonized by lichens, which trap windblown dust particles and eventually decompose to form a thin layer of soil that can be colonized by mosses. Further soil accumulation allows herbaceous plants to move in. Water might be colonized by floating plants, which decompose and sink to the bottom of a pond that may also be receiving sediment from streams. As the pond becomes shallower, bottom-rooting plants with emergent leaves can colonize, followed by wetland plants. If some kind of soil is already present, fast growing herbaceous plants will exploit the space but will give way to slower-growing plants that can use resources more efficiently. The composition of the ecosystem is always evolving toward the group of organisms that can best exploit the available resources at that time. Succession begins with pioneer species that can survive under special conditions and tend to get displaced as the ecosystem develops. As niches become available in the ecosystem, they are filled by the best available candidates.

The theoretical end point of succession is known as the **climax community**. At one time, this point was thought of as a permanent assembly of species that were linked together and would always return after any kind of disturbance. Now it is clear that there is no end point to succession and that species just happen to occur together, not because of necessary associations. The climax community can be seen as a group of species within the broad structure of the biome that is appropriate to the climatic conditions of the area. Over much of the eastern United States, this would be temperate deciduous forest with a mix of forest trees, shrubs, and herbaceous plants. Although we can characterize typical associations such as oak-hickory forest, the makeup of the ecosystem varies with location and has changed over time.

In spite of the difficulties inherent in the idea of a climax community, the broad idea of succession is sound and has implications for crop production. Many crop ecosystems are maintained at an early successional stage where they invite colonization by pioneer species, which we also know as weeds. It is easy to see that without cultivation, weeds would move in, followed by a thicket of woody plants, and the process would culminate in a forest of the longer-lived tree species. The further from the climax community we try to stay, the more work and inputs are required to maintain the ecosystem. The success of the corn belt in part can be

explained by the fact that a mix of tallgrass species in the original prairie was replaced by another tallgrass that happens to be physiologically similar. In the temperate zone, we have often been lucky to be able to make drastic changes in the vegetation without suffering excessive losses of soil fertility and structure. In more difficult climates, the consequences have not been so positive, and it can be a good idea to model crop ecosystems on the broad structure of the climax vegetation.

Once a climax community is established, no other species can easily invade because all the niches in the ecosystem are filled and no surplus resources are available to exploit. This outcome is related to the idea of the climax community as a well-defined group of mutually dependent species. However, alien species do invade natural ecosystems, often with quite disastrous consequences, and this outcome is part of the evidence that no unchangeable association exists between the species in a climax community. Long-standing species may be well-adapted to secure the resources that they need from the environment, but they are also vulnerable to pathogens and herbivores, which are equally well-established in the environment. While an alien species may not be able to exploit the resources as efficiently as natives, it may be immune to the diseases and unpalatable to the herbivores. These characteristics would give it a competitive advantage, at least until the pathogens and herbivores catch up with it.

Human beings have a long history of moving plants and other organisms around the world, deliberately or by accident. Many ecosystems have been irreversibly altered by introduced species. In recent years, the movement of organisms across national borders has become much more highly regulated and actively policed. Travel and trade have also increased greatly, however, and there is no sign that the flow of potentially damaging organisms has decreased.

IMPACT OF CULTIVATING PLANTS ON ECOSYSTEMS

We have just looked at the natural forces shaping the terrestrial ecosystems of the world and the interactions that occur within them. We considered how resources are allocated and cycled within ecosystems and began to see how human beings have modified the ecosystems. In many parts of the world, human beings are now the dominant organisms shaping the environment and utilizing its primary productivity. Ten thousand years ago, at the beginning of the agricultural

revolution, early farmers could afford to burn vegetation to clear areas for crops without worrying about the ecological consequences (which were largely unknown, of course). At that time, perhaps 5 million people lived on earth. Today, with a global population of more than 6 billion people and rapidly rising, the impact on the environment is impossible to ignore. As we try to fit even more people onto the planet, we need to be more aware of our use of its resources. In this chapter, we will consider most of the ways in which we make use of plants and the ecological impact of that use. In the interest of space rather than scientific reasons, forest crops will not be included, although we will look to forests for the absorption of much of the carbon dioxide that we generate in our lives.

Human impact on natural ecosystems is both physical and biological. Worldwide we now deliberately choose or influence the organisms that inhabit about 35 percent of the land surface of the earth. As we saw in the previous chapter, much of the remaining area is desert or tundra, with low productivity and biological diversity (Fig. 2–14). Thus, we exclude or even deliberately exterminate organisms in many of the most diverse and productive regions of the earth. Some ecologists allege that we have started an era of mass extinction of species like those that have occurred in geological history from meteoric impacts or other physical causes. This proposition is somewhat controversial, and people often argue about the value of preserving

individual species or populations of species. However, we cannot escape the fact that we are increasingly presented with the choice about whether to try to preserve plants and animals from extinction or to allow them to disappear. Having admitted the existence of this debate, we will spend the rest of the chapter focusing on the physical cost of providing the crop plants that we utilize or consume. The physical cost comprises the land and energy inputs required in crop production.

Energy requirements and land use are the basis for two ways of evaluating the environmental impact of human activity. Accounting for energy inputs enables us to evaluate the efficiency of technological systems such as crop production, and we can look at energy demand in relation to availability on a national or global level. The land area required for the same activities is a measure of their contribution to the human footprint. The total **human footprint** is an estimate of the land area required by an individual, a geographic area, a sociopolitical group, or the world population taken as a whole. It accounts for the space required to provide all of the items we use in our lives and to absorb all of the wastes that we generate. Thus, crop production contributes to our footprint not just through the area required to grow crops, but also the areas required to provide inputs of fuel and chemicals, to transport the produce, and to absorb the wastes produced.

A footprint analysis often takes account of the area of land without regard for the intensity of use. Thus, a square meter of land occupied by a home is not differentiated from a square meter of forest that provides lumber or pasture that provides meat or dairy products. But the home uses the area almost entirely for human purposes, whereas the forest or pasture can support many other organisms. A more sophisticated equation is:

$$\text{Footprint} = \text{Area used} \times \text{Intensity of use}$$

This equation implies that we can reduce our footprint by reducing the area or the intensity of our activities. This calculation is particularly relevant to agriculture and horticulture, including landscape installation and maintenance. For many industries, the intensity of land use is 1, implying that they use 100 percent of the resource area. However, crop production need not completely displace various plants and animals from the space that is used. The global human footprint is said to exceed the area available by 20 percent. While it is desirable that some wilderness areas should remain comparatively free from human activity, it is unrealistic to expect that people withdraw from many of the areas of the globe that they now occupy. Although governments have generally supported the maximum intensity of crop production, they can

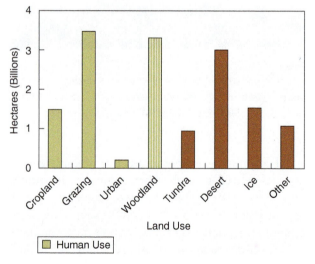

Figure 2–14

Global land use. The shaded columns indicate land used by people. Woodland is striped because it includes areas that are natural and unexploited, natural but managed or harvested, and planted purely for economic use. In a sense, all areas are used for carbon dioxide absorption, if nothing else.

Source: US Global Change Research Program, http://www.globalchange.gov/

begin to support greater biodiversity in crop production systems. This kind of policy would be consistent with attempts to rely on ecological processes rather than manufactured inputs to achieve production goals, such as maintenance of soil fertility or reduction in pests.

Much of our footprint on the planet is related to energy use. Most of our energy consumption is based on fossil fuels, and a footprint is associated with extraction, processing, and distribution of these fuels. However, a far greater footprint arises from the carbon dioxide produced as the fuel is burned to generate energy. This requires 8 hectares of forest for each person in the United States. Cropland makes a smaller contribution to carbon absorption because much of the carbon is returned to the atmosphere as the crop is consumed. The maintenance of forest for carbon fixation is a low-intensity component of our total footprint because we can allow natural ecosystems to perform this function. It is important that the most productive forests in equatorial Asia, Africa, and South America should not be converted to farmland. It has been said that we should probably pay for them to be maintained because we use their services. In fact, in 2008, the World Bank had committed $165 million dollars to offer to tropical countries for "carbon offset credits" for preserving their forests. Carbon credits can be bought by countries that emit large amounts of CO_2, essentially allowing them to buy a place to sequester the carbon. Other organizations, such as Carbon Conservation and Merrill Lynch, are also involved in "payment for preservation" programs.

Agriculture accounts for only about 2 percent of US energy use, so it can make only a minor contribution to energy conservation. However, because it accounts for less than 1 percent of gross domestic product, it is using above average amounts of energy and has to be regarded as an energy intensive industry (Fig. 2–15). Production of fertilizer accounts for nearly half of agricultural energy use in the United States. If the whole world adopted a similar mechanized and high-input agriculture, it would require about 10 percent of current global energy consumption. Most of the energy used for our food supply is consumed (for transportation, processing, and storage) after produce leaves the farm. It is difficult to account for all of the energy inputs between the farm and the dinner table, but the food sector has been estimated to account for up to 10 percent of total energy consumption. Surprisingly, less than 10 percent of this energy, or 1 percent of total energy consumption, is used in transport. Much of the energy is consumed in food marketing and preparation in the home (Fig. 2–16). If the rest of the world adopted this aspect of our lifestyle, it would use up about half of the world's energy supply.

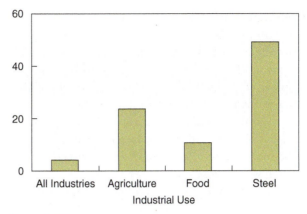

Figure 2–15
Energy use by US industries in 2000, in relation to contribution to gross domestic product. Agriculture appears to be energy intensive partly because profit margins are low. Food manufacture adds value to its raw materials and appears more energy efficient. Source: Earth Observing System Data and Information System, http://esdis.eosdis.nasa.gov/

Figure 2–16
Energy used in the food system includes production, transportation, processing, and storage.
Source: Margaret McMahon, The Ohio State University.

Footprint and energy accounting are complementary and related to each other, although no simple equivalence exists between the two. As we saw, use of fossil fuels entails land use both on the supply and consumption side. In terms of supply, these fuels minimize our present-day footprint by using the long-ago photosynthetic productivity of the earth to meet our needs now. But as you learned earlier, they also increase our footprint. A decision to use renewable energy to reduce our carbon resource footprint, often enlarges our land use footprint by increasing the need for land devoted to wind farms, solar panels, or biomass production.

SUMMARY AND REVIEW

Plants provide the energy that drives almost all terrestrial ecosystems. An ecosystem consists of populations of different species of organisms in a physical environment. Crop ecosystems require inputs because they usually lack some of the organisms and processes that sustain natural ecosystems. The organisms can interact in several ways to exploit the resources of the environment. Competition is likely to occur between individuals of the same species or closely related species. A single species cannot fully exploit all the resources in an environment. Resources are partitioned to provide niches that are occupied by different species. A natural ecosystem may not have any unoccupied niches, but crop ecosystems often provide niches that can be occupied by other plants. Small animals, fungi, and bacteria help to recycle nutrients in natural ecosystems and make them available to plants. Nutrients are often withdrawn from crop ecosystems when the crop is harvested, and they cannot generally be sustained by natural recycling.

The natural ecosystems that occur in different parts of the world can be grouped together in biomes. Each biome is a collection of ecosystems with similar climate, soil type, and vegetation, and these three elements are interrelated. Temperature and precipitation are the main climatic factors. Temperature is determined mainly by latitude and height above sea level. Precipitation patterns are related to the predominant wind patterns that occur in three bands in the northern and southern hemispheres. Temperature and precipitation are consistently high in the equatorial region, leading to continuously high rates of growth and low nutrient availability in the soil for the tropical rainforest. The divergence of wind systems north and south of the equatorial zone leads to dry conditions and the low productivity of the desert biome. In the northern hemisphere, the moist temperate zone is occupied by deciduous hardwood forest, which is highly productive in summer but dormant in winter. On the major continental land masses, the region between forest and desert was naturally occupied by grasslands. While these areas were not as productive as forests, highly fertile soils developed, and they became some of the world's most productive agricultural areas. To the north of the temperate deciduous forest, coniferous trees predominate in the taiga, or boreal forest; further north still, the subsoil is permanently frozen and trees cannot grow.

Microclimates exist everywhere and vary from the prevailing climate because of a feature such as body of water, land mass, building, and other features. The environment in a microclimate may be more favorable or less favorable to plant growth than the general climate in the area. Growers can use growing strategies that account for microclimate effects.

Many of the ecoregions that were most suitable for crop production have now been destroyed. The remaining natural areas of the planet are too dry or too cold for agricultural exploitation. Terrestrial ecosystems account for about half of global carbon fixation, but this proportion has been reduced by conversion to agriculture and urban development. Soils and ecosystems develop over time in a process known as succession to a stable climax community. Most crop ecosystems are maintained at an early, unstable successional stage that requires significant energy input to maintain.

Agricultural and other human uses now dominate the ecology of nearly all of the most photosynthetically productive terrestrial areas of the earth. Crop plants and farm animals have replaced the organisms that inhabited these areas, even to the extent that we are causing extinction of other species. We can summarize our impact on global ecology in terms of the land or the energy that is used for human purposes, including crop production. Land/energy use is summarized in the concept of the human footprint. While crop production represents a small fraction of total footprint and energy use, the intensity of land and energy use for crops is higher than many for other industries. The food supply system (processing, transportation, and storage) uses much more energy after it leaves the farm than was used in production.

KNOWLEDGE CHECK

1. Briefly describe the parts of an ecosystem.
2. What distinguishes the ecosystem created by cultivating plants from natural ecosystems?
3. What is meant by a niche in ecosystems?
4. Why is it usually easy for weeds to get established among cultivated plants?
5. Although disease parasites often do not directly kill a plant they infect, the plant will often die. Why?
6. What is the beneficial relationship that mycorrhizae have with plants? That *Rhizobium* has?
7. What role do saprophytes and detritivores have in ecosystems? When do they become most important in plant cultivation systems?

8. What is a biome?
9. How do global wind patterns affect the creation of the major biomes?
10. What types of plants are found in: tropical rainforest, desert, temperate forests, grasslands and savannas, taiga, tundra?
11. What is a microclimate?
12. Describe two ways cultivating plants has impacted biomes.
13. Of the major biomes, which are the most photosynthetically productive?

14. Are crop systems usually more or less photosynthetically productive than the native plants they replaced?
15. What is a harvest index?
16. Describe ecosystem succession. In terms of succession, what is the difficulty with maintaining many of our crops?
17. How does cultivating crops impact our carbon footprint?
18. What is the concept of "carbon credits"?
19. Give two reasons why the United States uses so much energy for its food supply.

FOOD FOR THOUGHT

1. Decide if you think it would be a good idea for countries with tropical rainforests to convert the forests to crop land or sell them for "carbon credits" instead. Then give the reasons for your decision.
2. Think about the biome in which you live. How do the plants growing in the landscape of your college or at your home compare to those that would be growing there naturally? What, if anything, has to be done to alter the environment so that the landscape plants thrive? How would these practices affect the carbon footprint?

FURTHER EXPLORATION

CHRISTOPHERSON, R. W. 2008. *Geosystems: An introduction to physical geography.* 7th ed. Upper Saddle River, NJ: Prentice Hall.

HUNTER, M., D. LINDENMAYER, and A. CALHOUN. 2007. *Saving the Earth as a career.* Hoboken, NJ: Wiley-Blackwell.

NATIONAL RESEARCH COUNCIL. 2005. *Valuing ecosystem services: Toward better environmental decision-making.* Washington, DC: National Academy Press.

SCHMITZ, O. J. 2007. *Ecology and ecosystem conservation.* Washington, DC: Island Press.

*3

Growing Plants for Human Use

Michael Knee

The previous chapter looked at ecosystems and the impact humans have on them. You learned about our energy footprint. Now let's look at why we grow plants and take a closer look at how our need for plants and the way we grow them influences ecosystems, the energy footprint, and society.

PLANTS FOR HUMAN USE

Nutrition

A healthy diet requires an energy source such as carbohydrate or fat, protein, linoleic acid, and various vitamins, minerals, and water. Apart from the minerals and water, these requirements are complex organic molecules that start with photosynthesis. Plants are essentially the only terrestrial organisms that convert inorganic carbon, oxygen, hydrogen nitrogen, and sulfur to organic forms through photosynthesis. This characteristic is why they are called **autotrophs** (meaning "self-feeder"), whereas humans and just about everything else that live on land are **heterotrophs** (or "other feeders") because they feed off autotrophs or other heterotrophs.

When we look at cultivated plants, we see that only a small percentage of all the plant species in existence feed the world's people either directly or indirectly (through animals). These plants are:

1. Cereal crops—wheat, maize (corn), rice, barley, oats, sorghum, rye, and millet. (Over half the world's food supply comes from the photosynthetic activity of these crops.)
2. Roots and tubers—potatoes, sweet potatoes, and cassavas.
3. Oil crops—soybeans, corn, peanuts, palm, coconuts, sunflowers, olive, and safflower.
4. Sugar—sugar cane and sugar beets.
5. Fruit crops—bananas, oranges, apples, pears, and many others.
6. Vegetable crops—tomatoes, lettuce, carrots, melons, asparagus, and so forth.

Fruits and vegetables add to the variety and palatability of our daily meals and supply much-needed vitamins and minerals.

key learning concepts

After reading this chapter, you should be able to:

- Discuss why plants must be cultivated for human use.
- Describe the many ways plants are needed and used by humans.
- Describe how growing plants impacts our energy use and carbon footprint.

Table 3–1 shows some of the common crops ranked in relation to the calories and proteins produced per unit of land area. Not all of the total production of food materials becomes available for human consumption. Much is lost during harvesting, transportation, and marketing, primarily from attacks by insects, diseases, birds, and rodents. Also, some of the production is saved to be used as seed for future plantings.

Adults need 2000 to 3000 kcal of energy per day, depending on their size and level of activity. This energy can be provided by carbohydrates, which are typically found in plant foods. So twenty-four to thirty-six slices of bread is one way to get our daily energy requirement. Lipids provide energy in more condensed form, and 9 to 14 oz of vegetable oil can provide our energy for the day (Fig. 3–1). Fats are more characteristic of animal foods, but many health problems are associated with consumption of animal fats partly because they tend to be saturated fats, which lack double bonds in their fatty acids. Unlike animals, plants can make polyunsaturated fatty acids (PUFAs) such as linoleic and linolenic acids. Apart from the health benefits of PUFAs, we need linoleic acid to make hormones such as prostaglandin, and plant oils are a more direct source than animal fats. We can also derive energy from protein-rich foods such as meat. Eating only meat for our energy needs is wasteful and probably unhealthy because of the need to excrete the excess nitrogenous material. Most nutritionists believe that reliance on plant foods as an energy source is

About 700 kcal

9 oz or 250 ml = about 2000kcal

Figure 3–1
Comparative amounts of bread (carbohydrate) and vegetable oil (lipid). Source: Margaret McMahon, The Ohio State University.

Table 3–1
SOME IMPORTANT FOOD CROPS RANKED ACCORDING TO CALORIE AND PROTEIN PRODUCTION PER UNIT OF LAND AREA

Rank	Calories Produced per Unit Area	Protein Produced per Unit Area
1	Sugar cane	Soybeans
2	Potato	Potato
3	Sugar beets	Corn
4	Corn	Peanuts
5	Rice	Sorghum
6	Sorghum	Peas
7	Sweet potato	Beans
8	Barley	Rice
9	Peanuts	Barley
10	Winter wheat	Winter wheat

Source: USDA, IR-1 Potato Introduction Station, Sturgeon Bay, Wisconsin.

healthy because the energy providing carbohydrates is associated with indigestible fiber, which protects us from colon disorders and other so-called diseases of affluence. Of course, these benefits come only with unrefined carbohydrates, such as whole wheat or brown, unpolished rice. Purified carbohydrates such as refined sugar or starch are not as beneficial and carry their own health risks.

An adult needs about 70 grams of protein in a day. Plant foods, especially cereal grains and pulses (peas and beans), often provide enough protein along with the energy that they supply. However a plant diet may not satisfy our protein requirement because it may be deficient in one or more of the amino acids that we require. The essential amino acids are leucine, isoleucine, valine, threonine, methionine, phenylalanine, tryptophan, and lysine; children also require

histidine. Plants make these and the rest of the twenty amino acids that make up proteins, but plant proteins often have only a low proportion of some of the essential amino acids, particularly the sulfur-containing acid, methionine, and the basic amino acid, lysine. No common plant food provides the right balance of amino acids, so we would need to overeat to meet our dietary requirements. An exception is the so-called "miracle grain," quinoa (*Chenopodium quinoa*), which provides enough of the essential amino acids along with its energy content. Animals, being more closely related to us, have a similar amino acid composition and generally provide better-quality protein. Eggs are generally regarded as providing perfect protein and are given a protein score of 100. A balanced diet can be achieved on a plant diet supplemented with a small amount of meat or other animal food. Alternatively, the deficiencies of individual plant foods can be remedied to a large extent by combining them. Grains tend to be low in methionine, whereas pulses (legumes) are low in lysine. A mixture of a grain and a pulse provides a more balanced protein. Many cultures in different parts of the world have based their diet on such mixtures: rice and soybeans in China, wheat and chick peas in the Middle East, corn and beans in South America (Fig. 3–2).

Carbohydrates, lipids (fats), and proteins are the bulk constituents of our diet. We also need small amounts of other organic molecules, called vitamins. We have some ability to make our own vitamin D, depending on the exposure of our skin to sunlight. All of the other vitamins are plant or microbial products. Most vitamins can be obtained from fruits, vegetables, or grains. Grains and grain flours are a poor source of some vitamins, particularly if they are polished or refined. Vitamin C (ascorbic acid) is found only in fresh fruits and vegetables. Some of the fruits and vegetables that are high in vitamin C are: citrus fruit, pineapples, watermelon, cantaloupe, raspberries, blueberries, cranberries, the cole crops (cabbage, broccoli, etc.), peppers, tomatoes, leafy greens (Fig. 3–3A). Vitamin A is derived from carotene, which is found in green, yellow, orange, or red fruits, vegetables, and grains (Fig. 3–3B). Plant foods do not provide vitamin B$_{12}$ (cyanocobalamin), which is manufactured by bacteria, especially in the guts of ruminant organisms, so it is available to us in dairy products and meat.

Figure 3–2

Grain and pulse crop combinations such as wheat and peas provide a balance of carbohydrates and protein.

Source: Margaret McMahon, The Ohio State University.

A

B

Figure 3–3

(A) Fruits and vegetables rich in vitamin C. (B) Green, yellow, orange, and red fruits, vegetables, and grain provide carotene, which is the precursor to vitamin A. Source: Margaret McMahon, The Ohio State University.

In addition to the organic components, we require six major inorganic nutrients (calcium, phosphorus, magnesium, sodium, potassium, and chloride) and seven micronutrients. All can be obtained from plant foods, although they are often more abundant in animal foods. We can assume most of the time that we are getting enough of the nutrients in the food that we normally eat, but calcium, phosphorus, iron, and iodine are the most likely to be deficient and are called critical nutrients.

If an adult needs 3,000 calories of energy per day and if average US yields can be achieved, it would take about 0.15 hectares of wheat or 0.06 hectares of corn to satisfy this dietary requirement for a year. To improve the protein quality of the diet, beans can be substituted for one-quarter of the grain intake. Again assuming average US yields of soybeans, this diet would take about 0.05 hectares. Yields of soybeans are about the same as those for wheat, so the total area required (0.16 hectares) would not change much on a diet of beans and wheat. However, corn yields are much higher, and a diet of corn and beans would take about 0.09 hectares. This diet would be boring, and it would lack important vitamins (particularly A and C) and minerals. The USDA recommended intakes could be supplied by about 400 grams of fruit and vegetables per day, which would require another 0.02 hectares of land (once more assuming typical US yields). So the wheat-based diet would require a total of 0.16 hectares, and the corn-based diet 0.11 hectares (Fig. 3–4). These figures are optimistic because cereal yields in most countries are lower than in the United States, and soybeans are the highest yielding of the pulse crops. The peas and beans that are mostly consumed by people have much lower yields. It is questionable whether the inputs that are used to attain US yields are physically available or environmentally desirable on a worldwide scale. The United States is exceptionally fortunate in having stable and productive soils and an excess of cropland relative to its population. Worldwide there are about 0.25 hectares (0.6 acres) of cropland for each person. With realistic inputs and yields, this land is about enough to provide everyone with a vegetarian diet.

In addition to the land required to grow crops, we should count the area needed to provide the inputs of energy and chemicals that are used in modern production systems. However, chemical inputs of fertilizer and pesticides can reduce the footprint because they increase the yield from a given area. Area is also required to absorb the wastes generated by food production. In the past, this factor tended to be ignored,

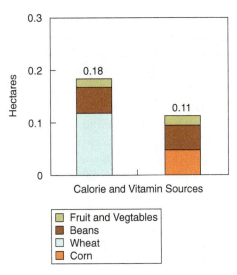

Figure 3–4

Amount of land required to provide an adult with vitamins from 400 g fruits and vegetables, 750 kcal from soybeans, and 2,250 kcal from either wheat or corn. Source: Michael Knee.

with the result that groundwater and surface water became contaminated. Now many crop production facilities large and small use fertilization and pest control practices that minimize chemicals leaching into the groundwater. Many growers provide some kind of buffer vegetation to prevent soil particles and fertilizer from running off from cropland into streams and lakes. Plant waste can often be recycled by composting or incorporation into soil or by feeding it to animals. Animal wastes are often spread on the land. Major problems still occur, however, when wastes cannot be recycled back into the production of crops.

Crop production has become concentrated and specialized in regions and countries around the world. For example, California and Florida produce more than 60 percent of the fruits and vegetables grown in the United States, which means that crops are often transported thousands of miles between production and consumption points. Although some argue from an ecological perspective that we should consume local produce, the energy costs of transport are much lower than for out-of-season production in a greenhouse (Table 3–2). On the other hand, the nutrient flows arising from food transport do present some problems. The nutrients in the produce are removed from the area where the crop was grown. For example, several countries in Africa and South America produce high-value horticultural crops for export to Europe and North America. These crops tend to have a high potassium content, so an essential plant nutrient is being exported and may not be easy to replace.

Table 3–2
ENERGY COSTS OF SUPPLYING FRUITS AND VEGETABLES OUT OF SEASON

Commodity	Provided by	Energy Input (kJ/kg)	Food Energy (kJ/kg)	Relative Energy Cost*
Strawberries	Road from Florida to Ohio	1,400	1,250	1.12
Broccoli	Road from California to Ohio	2,800	1,110	2.53
Lettuce	Greenhouse production	230,000	500	460

*The relative energy cost is estimated by dividing energy input by food energy value. A positive value indicates that the energy put into producing and shipping a crop is greater than its calorie value.

Source: Michael Knee.

The sewerage system transporting human wastes can also be viewed as part of the agricultural footprint. Like animal manure and urine, human waste contains a high level of nutritional elements. Although some municipalities have found ways of composting this waste, it cannot be recycled in food crop production because it is often contaminated with heavy metals and chemicals.

In moist temperate parts of the world, it is easy to forget how much water is needed to produce crops. It can take a ton of water to produce a kilogram of grain, and the water needed to produce our food is at least 500 times our direct intake. In dry areas, a large land surface may be needed to collect water to be used for irrigation to grow crops. The necessity for water in arid locations causes political problems in places such as the American Southwest, the Middle East, and Africa.

Each person in an industrialized country requires about 1 hectare of cropland, which is about 12 percent of their total footprint. While it seems that food consumption is not a major part of the footprint, the area is far greater than it was once thought to be. Furthermore, if everyone lived in an industrial economy, agriculture would occupy all of the world's land. Because most of the presently unused land is too cold or too dry to support crop production, agricultural use is clearly impossible. The problem arises because people are not eating beans and corn, these crops are being fed to animals, and only 10 percent of the energy that is available in food crops is being recovered in the meat or other animal products. Because of specialization, the crops are often produced on one farm and then fed to animals on another farm that may be hundreds of miles away. The crop grower needs fertilizer to maintain high yields, whereas the stock farm has the problem of disposing of animal wastes, which contain the same nutrients as the fertilizer. Currently, this method is apparently the most profitable way to raise animals, and it is cheaper to buy inorganic fertilizer than to take

animal waste back to the cornfields. However, this would most likely change if total energy costs of raising the animals are accounted for.

Forage, Fiber, and Fuel

Ruminant animals can extract energy from plant materials that are indigestible for us, so if they are fed on forage crops or allowed to graze on pasture, the energy recovery can be as high as 50 percent of what we might have obtained from crops grown for human consumption. Although soil problems and climatic limitations make it unlikely that we can increase the area of cropland substantially, much of the land that is unsuitable for crop production can provide grazing for animals. Thus, it should be possible to provide a varied diet for a world population advancing toward 10 billion if we reserve the best land for the production of crops for human consumption and confine animals to grazing on marginal land or limit their consumption to crop wastes. Grazing needs to be managed so that it does not lead to degradation of the soils and vegetation that supports it. On every continent, large land areas have been converted to unproductive scrub or near-desert through overgrazing. The difficulty of managing sustainable grazing is greatest in tropical countries, where the demand for food is often most acute. In principle, pasture can support higher levels of biodiversity than can fields from which crops are harvested. European countries are increasingly seeing the advantage of maintaining countryside that is ecologically diverse and also attractive for visitors.

Historically, farming provided many useful products in addition to food. Most of our clothing and textiles in general were made of plant and animal fibers. Animal skins were used to make most of our footwear and some of our clothing. Agriculture made some contribution to structural materials used for houses and other buildings. Straw was and still is used as a roofing material (thatch) and combined with clay to make

bricks. Forestry, however, has been a much more important source of construction materials and fuel. Agriculture provided plant and animal fats that were burned for light but was of minor importance as a source of heat. Forestry and farming compete in providing paper and cardboard; the cheaper products are usually made from wood pulp and the higher grades are made from herbaceous plant fibers such as cotton and hemp. Plants also provided dyes, perfumes, medicines, and other specialized chemical ingredients. Today most of these nonfood materials have been more or less replaced by synthetic materials, often produced from petrochemicals.

Only about 2.5 percent of cropland is devoted to fiber crops, both in the United States and worldwide. Cotton (*Gossypium spp.*) is the most important and best-known plant fiber crop. Jute (*Corchorus spp.*) is probably the next most important; it is grown in tropical countries such as Bangladesh and used for coarse string and fabrics such as burlap. Other fiber crops include flax and bamboo, both of which are used for fabrics, and hemp which is used for fabric, rope, and other products (discussed in the following paragraph). The market for natural fibers is stagnant or declining. Although cotton is and is likely to remain a major crop, it is unlikely that production and the area of fiber crops will increase in the near future.

One crop that shows promise of increasing though is hemp (*Cannabis sativa*), which has growth potential as there is a high demand for its fiber for clothing and other uses. Hemp has a deep root system, which can aid in reconditioning a compacted soil. Unfortunately, hemp grown for fiber is the same species as marijuana. Marijuana plants produce the drug tetrahydro-cannabinol (THC). As a result, the production of hemp for fiber is illegal in many areas of the world. Although closely related, the two types of *Cannabis* are very different in morphological characteristics. Fiber hemp has the long, unbranched, and sturdy stems required for fiber production. Hemp leaves are relatively small with low THC content. In contrast, the marijuana-type plant is more branched with larger leaves that have high THC levels. Morphological identification of one plant from the other is not difficult. Because the hemp plant has environmental benefit as a field crop and because the fiber from hemp produces an excellent clothing fabric, many countries, such as Canada, are once again allowing it to be grown, although under strict government control. Currently, hemp production in the United States is illegal but increasing pressure on government agencies is forcing a reevaluation of that policy.

Biofuels result from the processing of plants to produce biodiesel and ethanol for use mainly in internal combustion engines. Oil from soybeans (*Glycine max*) and other oil crops such as sunflower (*Helianthus annus*) and canola (*Brassica rapa*) is extracted and processed to be used as biodiesel fuel. The starches and sugars in corn (*Zea mays*), sugarcane (*Saccharum officinarum*), and switchgrass (*Panicum virgatum*) are fermented to produce ethanol. The advantage of these over fossil fuels is that they are renewable and they sequester CO_2 during their growing cycle thereby reducing the carbon footprint. Also, the oilseed meal from crops processed for biodiesel and the fermentation residue from ethanol production can be used as animal feed. However, because the crops used are often food crops (e.g., corn, soybeans, sugarcane), using them for fuel negatively affects the food supply for humans and domesticated animals. Even if the crops are not food (e.g., switchgrass), they may occupy land that was formerly used for food crops. Energy is required to produce the fuels so the net gain in fuel energy may be as low as 50 percent.

When the fourth edition of this book was being written in 2003, it was thought that using food crops for fuel would not greatly impact the food supply for many years. However, the use of biofuel sources increased so quickly that by 2007 some food commodities were already being negatively impacted. Corn for human and animal consumption became much more expensive. Also, land that had been used for other crops, such as wheat, was converted to grow crops for biofuels, thus reducing the supply and raising the price of the displaced food crops.

Although corn is one of the most productive ethanol crops, it is not the ideal fuel crop. It requires high-quality land and inputs of fertilizer for efficient production. Furthermore, the gasohol process uses the kernel, the most nutritious part of the plant, which is less than 50 percent of the total dry weight. It would be more advantageous if the fuel crop could be grown on low-quality land and if the fuel could be derived from the whole plant or waste after high-value crops have been harvested. Most of the whole plant or waste is cellulose and no economically viable process converts cellulose to ethanol at present, which is unfortunate because the critical need in the United States and other western countries is for fuel that can be used for transportation.

Nonfuel oil Apart from fiber and biofuel, the major nonfood use of crops is for oil-based industrial products. Tropical oil crops, such as coconut (*Cocos nucifera*) and oil palm (*Elaeis guineensis*), are the main

sources of oils used in soaps and detergents. Temperate oil crops, such as canola and soybean (*Glycine max*), are not suitable for these uses but they are used in the manufacture of lubricants and plastics. Concerns about the future availability of petroleum and the pollution produced by the petrochemical industry have led to suggestions that we should return to using plants as chemical feedstock for detergents, plastics, pharmaceuticals, and other industrial chemicals. Biotechnology could be used to increase the production of existing chemicals or introduce new chemicals into crop plants. For this approach to be economically feasible, the chemical must be sufficiently valuable and the market must be large enough to recover development and production costs.

The current annual value of the US plastics industry, at $110 billion, exceeds that of major agricultural commodities, at $90 to $100 billion. Genetically manipulated canola and soybeans could be used to produce adipic acid, which is used in the manufacture of nylon. For agriculture to capture this $2 billion market, about 5 percent of the total US crop area would be required for canola and 15 percent for soybean. Small molecules like adipic acid must be processed to produce plastic polymers such as nylon. The direct production of plastics in plants is a more attractive possibility. *Arabidopsis* plants (Fig. 3–5) have been engineered to produce small amounts of polyhydroxybutyrate, which suggests that production of plastics in

plants is feasible. Much development is required, however, and a major commitment of cropland would be necessary to make a significant contribution to the total consumption of plastics. The most rewarding applications of biotechnology may be in the production of high-value therapeutic and diagnostic proteins in plants. These proteins are impossible to produce synthetically and are required only in small quantities. So production of transgenic plants expressing proteins such as vaccines against malaria or rabies could be a profitable small-scale enterprise.

Before the industrial revolution, most of the fuels and raw materials (except for metals) used to manufacture everyday goods were of biological origin. These sources were gradually supplanted by coal and oil, but even in 1930 about 30 percent of raw materials came from plants. At that time, the dominant source was coal, whereas today over 80 percent of chemicals used in industry are petroleum products. Some suggest that this trend will be reversed so that 50 percent of chemical feedstock will be biological by 2050. However, this reversal would require a massive restructuring of the chemical industry and diversion of land from food production.

Medicine

Plants are the source of many drugs and medicines. In fact, until the last century, most medicines were derived directly from plants. From the earliest history of modern humans, the shaman and wise elders of a tribe used herbs to treat diseases and disorders among tribal members. Much of this wealth of plant lore is in danger of being lost as indigenous cultures are absorbed into modern civilizations. Ethanobotany, the study of plant usage by indigenous cultures and the preservation of that knowledge about plants, has become an important branch of plant science.

Today, about 25 percent of pharmaceuticals are based on plant products, and the rest are produced synthetically. One example of a modern-day medicine that can trace its roots to usage by Native Americans is aspirin. Salicylic acid, the pain-relieving component of aspirin, is found in the bark of willow trees (*Salix*). Native Americans chewed willow bark to relieve pain.

Many other modern medicines are directly derived from plant materials. Examples include quinine, which is used to treat malaria, and the heart medication, digitalis. Morphine, opium, codeine, and heroin are all pain-relieving extracts of the opium poppy (*Papaver somniferum*). Cocaine, another painkiller, is derived from the coca plant (*Erythroxylum coca*). However, each of these pain-relieving drugs also has addictive

Figure 3–5

In the future, plants such as Arabidopsis may be genetically engineered to produce chemicals for use in manufacturing, pharmacy, and other industries.

Source: Margaret McMahon, The Ohio State University.

properties. Addiction to these chemicals has become a major problem worldwide, as is finding alternative crops for farmers who grow illegal crops.

One of the newest plants to be recognized for its medicinal benefits is a yew (*Taxus*) that produces taxol, a chemical that shows promise in the treatment of cancer. The taxol-producing species grows wild in the Pacific Northwest and is now being evaluated as a potential crop plant for areas where it would grow well.

Pleasure, Ornamental, and Recreational Uses

After looking at the challenges of obtaining food and industrial resources from plants, it may seem frivolous to consider the ways in which plants give pleasure. However, strong reasons for consuming or using plants in these ways lead to some of the most profitable plant-based industries. In rich countries, food consumption is driven as much by hedonic (pleasure-giving) factors as nutritional requirements. We consume particular foods because of the sensory stimuli of appearance, texture, taste, and aroma that they provide. Some of the plant products that we consume provide pleasure but no nutritional benefit; they may even be harmful to our health. Tobacco and various forms of alcohol fall in this category, although they also represent some of the most profitable ways of using land. However, because of tobacco's connection to cancer and other health problems, the demand for the crop has dropped. Like opium poppy farmers, tobacco farmers are now looking for alternative crops to grow.

Tropical countries can find it much more profitable to grow coffee or tea for export than to produce staple food crops. The craving for sweets is one of the most pervasive human desires and underlies our demand for sugar. Sugar provides energy but we do not really need it in our diet. Sugarcane in the tropics and sugar beet in temperate countries tend to be more profitable than staple food crops and often supplant the staples. In the United States, the bulk of the market for sweetener has been taken over by corn syrup. About 8 percent of the corn crop is used in this way, which (like ethanol production) helps raise corn prices.

Since neolithic times, human beings have coevolved with agricultural plants and animals. People have shaped plants and animals through selection and breeding, and these organisms have influenced our physiology, psychology, and sociology. As cities developed, people were less intimately involved with plants and animals in their daily lives. It is even possible to forget that bread, meat, or milk were once part of living organisms. However, people retain an affinity for plants and animals: our desire for pets and ornamental plants

may be an aspect of a deep-seated human tendency that E. O. Wilson has called "biophilia" (Fig. 3–6). One of the most persistent, if minimal, expressions of biophilia is the desire for cut flowers or potted plants in the home or workplace. In cool-temperate areas, these plants are mostly produced in greenhouses. Because of the high market value of floral crops, greenhouse production can be more profitable than vegetables production. However, as we saw with vegetables, it is often cheaper in energy and economic terms to produce cut flowers in warmer countries and then transport them to the United States or Europe. Flowers are generally more perishable than vegetables, however, and it is usually necessary to ship them by air, so the energy advantage may not be so great, especially as fuel costs rise.

Lack of space and limited transportation encouraged intensive development of cities in Europe in the nineteenth century. There was little room for plants where people lived, but even today nearly every home has plants in small gardens, windowboxes, in pots and vases inside the home, and even on rooftops. In the twentieth century, the United States developed a new kind of city with extensive open space between houses and other buildings. While 75 percent of Americans are said to live in urban areas, it is more accurate to say that most of them live in suburbs. Cities and suburbs themselves occupy an area equivalent to about half the area of cropland in the United States. Built-up areas are expanding at a rate equivalent to 0.5 percent of crop area per year. Many argue that this expansion occurs at the expense of agriculture and will cause farm commodity prices to rise by taking land out of production and reducing the supply of agricultural products.

Figure 3–6
Biophilia is the term used to describe our desire for pets and ornamental plants. Source: Margaret McMahon, The Ohio State University.

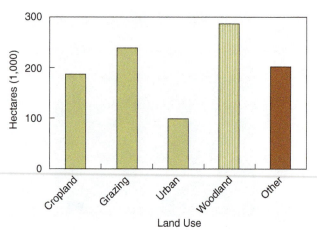

Figure 3–7

Land use in the United States. The areas of human use are shaded. Woodland is highlighted because some of it is unused (compare with Fig 2–14 on page 29). Source: Michael Knee.

Figure 3–8

The "suburban biome" consists of plants that we look at or utilize for aesthetic appeal, sports, and recreation rather than for food, fiber, or industrial purposes.
Source: Margaret McMahon, The Ohio State University.

Figure 3–9

These plants come from many different biomes and have different environmental requirements which can create challenges to maintaining the plants.
Source: Margaret McMahon, The Ohio State University.

Much of this built-up land still supports vegetation of one kind or another though (Fig. 3–7). As a result, the suburban landscape has become, in effect, a major biome. Most of the urban and suburban area that is not under buildings or blacktop is covered by turf. Some of this area is devoted to organized sports, but mostly it is maintained for informal use or merely appearance. In addition to turf, there are shade trees, shrubs, and herbaceous ornamentals. Rooftop gardens or green roofs are becoming much more popular in the United States. A new area of the construction and ornamental industries is developing around this, including architectural design, production of suitable plants, installation, and maintenance. Very little land in the urban and suburban biome is used for plants that we need for food or fiber (Fig. 3–8). However, in many cities, small "urban farms" are now being created on plots of land that are not suitable for development. These "farms" provide both food and ornamental plants and plant products to local residents.

There are thousands of ornamental species, most of which are not native to the US. The introduction of so many species occurred because early European settlers struggled not only to grow food, but to make a congenial home out of the North American forest and prairie. Most of the native flora did not seem attractive and was removed. New plants were brought deliberately or by accident to repopulate the land. The trend continues today with most of the new ornamental species being nonnative. As a result, a modern landscape may include plants from different biomes: spruce from the boreal forest, cannas from the tropics, yucca from the desert, and hostas from the temperate forest (Fig. 3–9). A large part of the economic and environmental cost of landscapes results from the attempt to maintain nonnative plants in unfavorable climates. Another downside to this humanmade diversity is that, after introduction to the United States, some ornamental plants have escaped to become agricultural weeds. Others have become invasive and displaced native species. This creates a serious disruption in the ecosystem and negatively affects the lifecycles of species dependent on the native plants. Most states have invasive ornamental plant lists and discourage or even make

it illegal to grow or use these plants. Many people have argued recently that we should use native plants to try to create landscapes with a more distinctively American style and heritage and to avoid the problems that come with nonnative plants. However, strong social and cultural reasons lead most people to continue to plant landscapes using the introduced species.

The scale of urban areas and the cultivated landscapes within them means that agriculture can no longer be regarded as the only significant source of pollution from fertilizers and pesticides. About the same rate of fertilizer application is recommended for turf as for corn or other cereals. Environmental Protection Agency (EPA) data imply that more pesticides, mostly herbicides, are applied to lawns than to agronomic crops. The productivity promoted by these chemical applications is mostly unwanted. Fruits, seeds, and even flowers are regarded as a nuisance when they drop from ornamental plants onto sidewalks. Almost every homeowner is encumbered by a surplus of grass clippings throughout the summer and leaves in the fall. The disposal of this waste has created problems in landfills and other waste collecting facilities. Several municipalities now have mulching/composting facilities and sell the mulch and compost to help pay for the cost of collecting the waste. Many homeowners now collect grass clippings for use as mulch or compost. Some chip their own wood waste for use as mulch.

Nurseries, greenhouses, and sod farms provide the plants for the urban environment (Fig. 3–10). These crops with purely aesthetic value are produced in about 6,800 hectares of greenhouses and 280,000 hectares of open area, which corresponds to about 0.15 percent of the US crop area or about 1 percent of the area used for corn. Yet the combined value of the ornamental crops is about 50 percent of the value of corn. Ornamental crops are much more labor intensive than are other crops, accounting for nearly 20 percent of US agricultural labor costs. But on the other hand, these enterprises provide jobs for many people, from day laborers to corporate executives. Ornamentals may also require more intensive use of fertilizers and pesticides than traditional agriculture. However, many greenhouses and nurseries now practice water and fertilizer recycling and use low environmental impact methods of pest control whenever possible.

Although only a few crops provide most of the material for our food and fiber needs, there are an inestimable number of crops grown on a smaller scale for many different reasons. These crops provide variety in our diets, materials for manufacturing, pharmaceuticals, health aids, perfumes, etc. In addition,

A

B

C

Figure 3–10

(A) Nurseries, (B) greenhouses, and (C) sod farms provide most of our ornamental and recreational plants. Source: A and B, Margaret McMahon, The Ohio State University; C, Dave Gardner, The Ohio State University.

many different plants are grown for ornamental and recreational use. These relatively small-scale crops can play a very important role in local and regional economies.

SUMMARY AND REVIEW

The human nutritional requirements for carbohydrate, fat, protein, vitamins, and minerals can be met from plant sources, with the important exception of vitamin B_{12}. Assuming average US yields, adequate nutrition can be obtained from the less than 0.25 hectares of cropland that is available for each of the 6 billion people on the planet. Energy and nutrients are lost when crops are fed to animals, and less people can be fed in this situation. However, meat and dairy products can be produced by animals that graze on land unsuitable for crop production. The wastes generated by the food production system and by human and animal consumption of the food are part of the human footprint. Waste management is an important consideration when growing plants.

Historically, farming provided raw materials for the manufacture of textiles, fuels, and various chemicals, including dyes, perfumes, and medicines. Today, many of these materials are produced from synthetic (usually petrochemical) sources. As oil supplies diminish, however, there is renewed interest in plant sources.

In the future, plants may be used to produce not only traditional products for industrial use but new products, especially chemicals, that result from genetically engineering the plants.

The production of sugar, alcohol, and tobacco for enjoyment can be highly profitable, but they are not a necessary part of nutrition and can even be harmful. In the case of tobacco, alternative crops are being developed. Other crops are appreciated for their visual appeal or recreational use. For the urban and suburban population, flowers, turfgrass, shrubs, and trees are an important part of the environment. These crops are at least as demanding as agriculture for energy and chemical inputs and generate significant quantities of waste. Non-native landscape plants can become invasive and displace native biodiversity. Displacing native diversity can negatively impact the environment. Practices that reduce the ecological impact of ornamental and recreational plants are as important as those practices for food, fiber, and fuel crops.

KNOWLEDGE CHECK

1. What is the advantage of plant fats over animal fats in the human diet?
2. What is the advantage of having a little meat in a human diet?
3. Why do we say that using chemical inputs of fertilizer and pesticides both raises and lowers our energy footprint?
4. What happens to the mineral nutrients (that would normally be recycled back into the soil after a plant dies) when a crop is harvested and shipped to another location? How does that influence the need to add fertilizer for the next crop?
5. What advantage is there to grazing animals on land that is not suitable for other types of crop production?
6. Why is it illegal to grow hemp in the United States and why is there an interest in making it legal?
7. What are the two main types of biofuels and what are some of the plants used for each?
8. Describe two of the issues associated with growing plants for biofuels.
9. What kinds of chemicals are likely to be the most rewarding to produce in genetically engineered plants?
10. What is biophilia and how does it relate to the kinds of plants we grow?
11. Why do we now have to consider urban and suburban landscapes as significant sources of environmental pollution?
12. What are some ways that nurseries and greenhouses are reducing their footprint on the environment?

FOOD FOR THOUGHT

1. Eighty percent of our food energy comes directly or indirectly from six major crops (wheat, corn, rice, sweet potato, white potato, and cassava or manioc). Many people in the world eat little other than one or two of these crops and lack the balanced nutrition that comes with eating a wide variety of fruits, vegetables, and grains. What are some ways you would suggest to get more variety into the diets of those people?
2. If you were asked to give your opinion on the use of biofuels to lower our reliance on petroleum-based fuels, what would you say and why would you say it?
3. If you are ever considering plants to landscape the yard where you live, what types of plants would you want to use and why would you want to use those types?

FURTHER EXPLORATION

POWERS, L. E., AND R. MCSORLEY, 2000. *Ecological principles of agriculture.* Albany, NY: Delmar

SIMPSON, B. AND M. OGORZALY. 2000. *Economic botany: Plants in our world.* 3rd ed. Columbus, OH: McGraw-Hill

THOMPSON, J. W. AND K. SORVIG, 2000. *Sustainable landscape construction.* Washington, DC: Island Press.

UNITED STATES GOVERNMENT. 2006. *Energy crops: Corn, oil seeds, wheat, rice, residue.* Washington, DC: Progressive Management. *Note:* This is a two-CD set featuring over 50,000 pages of government documents in PDF form. Its long title is *21st Century Guide to Energy Crops and Biofuels, Agricultural Residue, Corn and Wheat Stover, Rice Straw, Oil Seeds, Switchgrass, Feedstocks, Sugars, Biorefineries, Ethanol, Syngas.*

WILSON, E. O. 1996. *Biophilia.* Cambridge, MA: Harvard University Press.

*4
Climate

Michael Knee and Margaret McMahon

CLIMATE

Climate is defined as the prevailing weather conditions of an area. It includes the intensity and duration of solar radiation; the temperature, rainfall, and wind speed and direction that may be expected; and how these characteristics vary according to the season. These characteristics may be summarized as averages or ranges of expected variation to define climate. Because variation or instability in climatic variables is itself an aspect of climate, the summary values may not correspond to the actual weather experienced at a particular time. So **weather** is defined by the actual values of climatic variables and is not the same as climate.

The climate of an area is determined primarily by the input of solar radiation, which varies with latitude and season. Solar radiation determines or influences all of the other variables that make up climate. Climate is modified by local features such as altitude, the presence of land or water, and barriers to air circulation. Large-scale features such as continental land masses, mountain ranges, or oceans contribute to macroclimate. Similar but smaller effects can be caused by small-scale features such as buildings, ponds, or depressions in the ground; these features can contribute to a **microclimate**.

As we saw in Chapter 2, climate determines the kind of natural vegetation that grows in an area, but vegetation also has a modulating effect on climate, which can occur on a large or a small scale. A large area of forest can make the climate more moist, whereas a small group of trees may act as a windbreak or provide shade. Various human activities also modify climate, often inadvertently. Areas of human settlement tend to be heat islands, with higher temperatures than the surrounding countryside. Forest clearance tends to lead to drier climates. Dust and other pollutants generated by human activity reduce light intensity and can increase rainfall.

All plants are characterized by their climatic adaptation. Wild plants evolved to survive and grow optimally in the climatic conditions of their natural habitats. The requirements might apply to a whole species, but there are often **ecotypes** whose adaptation differs from other members of the same species. For example, one ecotype might have greater tolerance to cold, require a shorter growing season, or require a longer day length for flowering than

key learning concepts
After reading this chapter, you should be able to:
- Describe the factors that create climate.
- Explain the interaction between climatic variables and how they vary from location to location.
- Describe how climate factors influence plant growth and determine what plants can grow in an area.
- Discuss what can be done to modify climate factors to improve crop growth.

another. The same adaptations occur in cultivated plants, and we can improve a plant's success in an area by exploiting ecotypic variation through the selection of plants adapted to local conditions.

SOLAR RADIATION

The major energy input for the earth comes as electromagnetic radiation from the sun. The sun itself gives off a smooth and continuous spectrum of radiation from short-wave UV light to long-wavelength infrared. Atmospheric components absorb some wavelengths in this range more than others so that the spectrum has peaks and valleys. Much of the solar radiation is in the infrared wavelengths, and only about 40 percent of the energy occurs between 400 and 700 nanometers (nm), the region that supports plant photosynthesis (Fig. 4–1).

The intensity of solar radiation is highest at the equator and decreases at higher and lower latitudes for two reasons, both relating to the angle at which the radiation is received (Fig. 4–2). At the equator, the angle of the sun is always close to 90° relative to the earth's surface. The radiation travels the least distance through the atmosphere and reaches the minimum surface area. Moving toward the poles, the angle of approach decreases toward zero so that light travels further through the atmosphere and spreads over a wider surface area. The result is that the intensity at noon at 40°N is about 60 percent of the value at the equator. Clouds absorb a high proportion of incoming radiation and accentuate the difference between latitudes (Fig. 4–3).

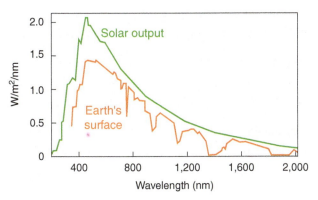

Figure 4–1

Effect of atmosphere on the spectrum of solar radiation.

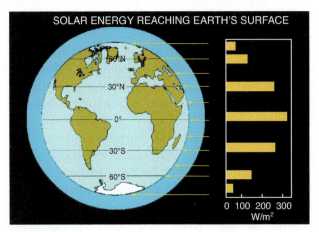

Figure 4–2

Solar energy reaching the earth's surface. Solar radiation passes through more air and is spread over a wider area at higher latitudes. Source: Michael Knee.

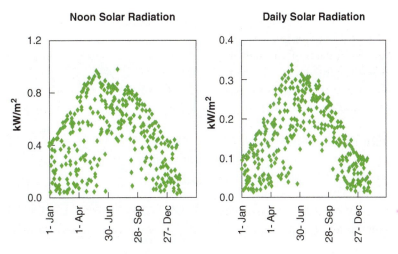

Figure 4–3

Example of variation in solar radiation over a year in Columbus, Ohio (40°N), showing peak radiation at noon and the average for the day. The upper boundary points represent clear skies. Cloudy conditions reduce radiation, particularly in the winter months.
Source: Michael Knee.

Because the earth revolves once in twenty-four hours, the average day length throughout the world is twelve hours. However, day length varies through the year because of the tilt of 14.5° in the earth's axis (Fig. 4–4). At the summer solstice, the tilt is toward the sun, and day length varies from twelve hours at the equator to twenty-four hours at the pole. At the winter solstice, the tilt is away from the sun, and day length varies from 0 at the pole to twelve hours at the equator. At the spring and autumn equinox, the axis is not tilted relative to the sun, and day length is twelve hours everywhere on earth. In Hawaii, at 20°N, day length varies between 10.8 and 13.2 hours; in Columbus, Ohio, at 40°N, it varies between 9.2 and 14.8 hours; and in Anchorage, Alaska, at 60°N, it varies between 5.6 and 18.4 hours. Variations in day length add to the effect of variation in light intensity with latitude. In higher latitudes, winter is doubly unfavorable for plant photosynthesis and growth because of short days and low light intensities. On the other hand, cool-season vegetables can grow fast and large during the long cool days of summer in Norway beyond the Arctic Circle. Greenhouses allow for the production year-round of plants, but low light in the winter can make it difficult for the plants to photosynthesize enough to produce quality plants and sufficient yield.

When light intensity is too low for adequate photosynthesis, artificial light can be used. However, even the most efficient artificial light sources convert only a fraction of the electrical energy to photosynthetically active radiation. Thus, it is expensive to achieve light intensities similar to sunlight and almost impossible to do so on a large scale. **High intensity discharge (HID)** lights are the only type of artificial light source currently available that can be used in greenhouses to promote growth of high-value crops when natural light intensity or duration is too low. HID's are most effective when used to supplement solar light and not as the sole source of light.

Day length has more subtle effects on plant growth apart from photosynthesis. Like other organisms, plants have a biological clock and can keep track of the duration of light and dark. Some plants flower and bud-break occurs in many trees when days exceed a certain critical length. Conversely, other plants flower and leaf fall occurs in trees when days are shorter than a certain critical length. Plants that flower in response to day length are **photoperiodic** and the response is called **photoperiodism**. Plants that flower when the light period exceeds a critical length are called **long day plants**. **Short day plants** are those that flower when the light period is less than a critical length. A short day with a night break of a few minutes has been shown to be equivalent to a long day. For this reason, some have argued that short- and long-day plants would be better described respectively as long-night plants and short-night plants. However, because the terms short- and long-day plants still predominate, they will be used in this book.

As with light intensity, plants are often adapted to the conditions of their natural habitat. So ten hours could count as a short day for a tropical plant but a long day for a plant from northern latitudes. That is, when the day is longer than 10 hours, the tropical plant will not flower and when the day length is less than 10 hours, the northern plant will not flower. When a plant species occurs over a wide range of latitudes, local ecotypes are often adapted to the local day length. This feature applies to flowering dogwood (*Cornus florida*) and many other ornamental trees in North America. If northern ecotypes are taken to the south, bud-break can occur too soon and flowers can be injured by late frosts. Similarly, soybeans are short day plants and varieties have different critical day lengths. A variety developed for Minnesota would be ill-suited for Kentucky.

Plants tend to adapt to the light conditions prevailing in their natural environment. Plants growing in open environments in the tropics grow well at higher light intensities than do those from higher latitudes. Of course, the presence of taller vegetation, particularly trees in forests and jungles, reduces the light intensity reaching lower vegetation. Under a continuous forest canopy, the light intensity may be 1 to 10 percent of that above the canopy. Plants of the forest floor have adapted to

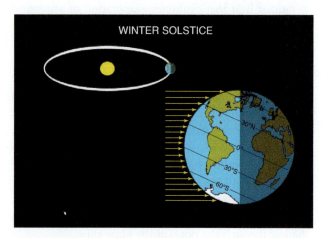

Figure 4–4
The tilt in the earth's axis causes variation in maximum intensity of solar radiation and day length throughout the year.
Source: Michael Knee.

photosynthesize and grow under these conditions and often cannot tolerate direct exposure to the sun. Indoor light levels are similar to those of the forest floor. That is why many indoor plants are understory species, such as ivies and ferns from temperate forests and *Spathiphyllum* and *Epipremnum* from tropical forests.

Plants adapted to high light often grow tall and spindly when they are shaded by other plants. This is called the **shade avoidance response**. Both photoperiodism and shade avoidance are examples of **photomorphogenesis**, which means plant shape determined by light quality. Part of the process by which photomorphogenesis occurs involves a light-absorbing protein called **phytochrome**. This protein responds differently to light in the red (centered at 660 nm) and far-red (centered at 730 nm) regions of the spectrum. The state of the protein and the response of the plant are affected by the relative energy at these wavelengths. There are naturally occurring variations in the amount of red and far-red light reaching the earth's surface and the phytochrome molecule responds to these variations. For example, the end of the day is marked by a decrease in the ratio of red to far-red (R:FR) light. This shift in ratio allows the plant to sense daylength.

The shade avoidance response occurs because leaves absorb more red light than far-red light so that plants growing in the shade of other plants experience a lower R:FR than unshaded conditions. Far-red light promotes stem elongation so the low R:FR accounts for the spindly growth. Although the spindly growth results in a weaker plant, by rapidly growing tall, the plant can become taller than its neighbors and begin to capture more light for photosynthesis and become stronger.

The sensing of light for photoperiodic and other photomorphogenic responses occur at much lower light intensities than are needed by most plants for photosynthesis. So it is quite possible to use lower intensity artificial light to promote flowering in long-day plants or prevent flowering in short-day plants. However, when shade material is used to prevent flowering in long-day plants or to promote it in short-day plants, almost all of the light must be blocked by using opaque material When using shade to control photoperiod, it is particularly important to avoid accidental light exposure during the dark period. The intensity of street lighting in urban areas is enough to cause problems of delayed leaf-fall and premature bud-break in shrubs and trees through similar effects.

When crops are grown in rows oriented north–south, the plants tend to experience more far-red light as the sun moves from east to west than do plants in rows oriented east–west. Soybean plants in north–south rows have been shown to have longer internodes and smaller leaves than soybean plants in east–west rows. Greenhouse crops often grow too tall because they are grown closely together and the shade avoidance response is induced. One way to reduce plant height is to give the plants more space. In addition to the photomorphogenic crowding response, the plants are not exposed to the mechanical stimulation of wind and weather, which inhibits stem elongation in outdoor plants. Photomorphogenic responses may be involved in differences in plant growth observed when crops are grown with different-colored mulches on the soil surface. Current research is investigating the potential of far-red-absorbing filters to control stem elongation in the greenhouse. For a more detailed description of photomorphogenic response and control see Chapters 8 and 24.

MOISTURE AVAILABILITY

Nearly all of the world's water (97.6 percent) is in the oceans and much of the remainder (2.1 percent) is frozen. Groundwater accounts for 0.3 percent of the total, while only 0.01 percent exists as fresh water in lakes and streams. A tiny but important fraction (0.001 percent) is present in the atmosphere. Plants can draw on a fraction of the water in the ground, but much of it is below the depth of root penetration. We can use fresh water to supplement the supply in the soil for the growth of crops, but the available water would soon be exhausted if there were not some way of replacing it. Groundwater and the surface fresh-water supply are replenished through the **hydrologic cycle** (Fig. 4–5), which is driven by the input of solar energy, both directly as radiant heat that promotes evaporation of water and indirectly through air currents that move water vapor from one area to another. Water vapor then can precipitate as rain or snow bringing fresh water to that area.

Water evaporates from open-water surfaces, the ground, and from plants. Evaporation from plants involves the process of moving water through the plant and evapotranspiration, often just called transpiration. Evapotranspiration cools the plant to keep it from overheating. For most plants, photosynthesis cannot occur without simultaneous transpiration. As water vapor escapes through pores in the leaf surface called stomata, the carbon dioxide needed for photosynthesis enters the leaf through the pores at the same time.

The amount of water vapor that air can absorb depends on air temperature: the cooler the air, the less water it can hold. As moisture-laden air rises or moves to

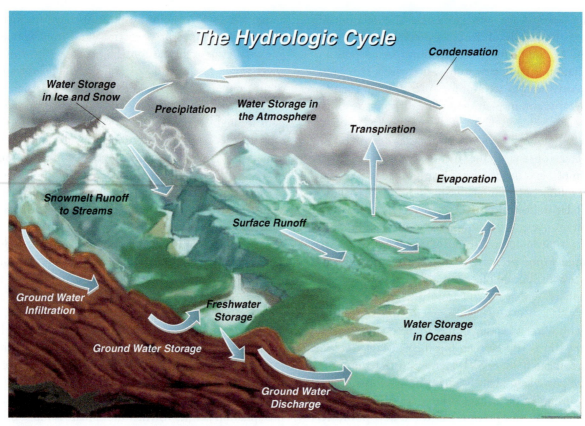

Figure 4–5

The hydrologic cycle. Source: U S Geological Survey, http://ga.water.usgs.gov/edu/watercycleprint.html

cooler regions of the earth, condensation forms fog close to the ground or clouds in the upper atmosphere. Further cooling in the clouds causes tiny ice crystals to grow and form snow. In winter, this precipitation may fall to the ground directly, but if it passes through warmer air on the way down through the atmosphere, it melts to form rain. Hail forms when the ice crystals build up layers as they bounce up and down between warm and cold regions in the cloud before falling to earth.

When rain hits the ground, it can penetrate pores in the soil through the process of infiltration or it can run off. The amount of infiltration and runoff is determined by the properties of the soil surface, the amount of slope, and the degree of saturation of the soil with water. Plants help to prevent runoff and encourage infiltration partly by capturing water on their surface, thus slowing the passage of the water to the soil. They also encourage infiltration when they die by creating a layer of loose, absorbent, organic debris at the soil surface rather than the smooth, nonabsorbent, caked surface that tends to form on bare soil. Runoff may find its way to streams, rivers, or lakes. If it is not lost by evaporation on the way, much of it ends up in the sea. The water that infiltrates the ground can evaporate from the soil surface or through plant transpiration, or

it may permeate to lower levels to meet the water table. Groundwater can accumulate in natural underground reservoirs called **aquifers.** Depending on geological conditions, water can remain in these aquifers for many years, or it may gradually flow back to the surface through springs or into surface bodies of water.

The **hydrologic cycle** is a global cycle in which evaporation and precipitation are equal over time and on a global scale. Everyday observation tells us that the balance between the two processes varies from day to day in any one area or region. Persistent imbalances also occur so that one process exceeds the other over long periods of time. The relationship between precipitation and evaporation is a major factor defining the climate of a region and its suitability for plant or crop growth. The Köppen system of classification of climate combines temperature and moisture conditions: A is tropical and rainy; B is dry; C is warm, temperate, and rainy; D is snow forest; and E is polar. People live and crops are grown primarily in zone C. Although a lot of water occurs in zones D and E, the water is mostly unavailable because it is frozen; these water zones are in effect ice deserts. Most of the earth's land surface falls in zones B, D, and E.

If we want to find out the suitability of an area for crop production or determine the day-to-day water

needs of crops, we need to know the relationship between the supply of water or precipitation and the demand by the plants in terms of evapotranspiration. Precipitation is fairly easy to measure with a rain gauge, but we may need to account for losses through runoff or drainage. Evapotranspiration is more difficult to estimate because it is influenced by several factors. In a closed system, evaporation from a water surface is determined by the water content of the air above the surface. The amount of water in the air can be expressed as a percentage by volume, percent relative humidity, or partial pressure. **Relative humidity (RH)** expresses the actual water content relative to the maximum amount that the air can hold at a given temperature, which is known as the **saturated vapor pressure (SVP).** Partial pressure indicates the contribution of a gas to the total pressure of the atmosphere, which for the atmosphere is around 102 kPa (15 lb/in.^{2}) at sea level. When air comes to equilibrium with water, meaning evaporation equals condensation, the RH is 100 percent, but the percentage volume that the air can hold, or SVP, increases with temperature. It is also possible to identify a temperature for any actual vapor pressure or percentage volume at which RH would be 100 percent; this point is called the **dew point.** RH tends to be 100 percent when it has just rained or when the temperature drops and the dew point is reached because the partial pressure becomes equal to the SVP. At other times, RH is less than 100 percent and the difference between the actual vapor pressure and the SVP is called the **vapor pressure deficit (VPD).** Evaporation and plant transpiration are influenced by VPD rather than humidity. Much more water evaporates at 70 percent humidity at 30°C than 70 percent at 10°C because the VPD is much higher at 30°C (Table 4–1).

The rate of evapotranspiration at a given VPD is increased with the input of solar radiation, temperature, and air movement. Researchers such as Penman and Monteith have developed complex equations that relate evapotranspiration to these factors. Both expected and actual values for evapotranspiration have important implications for crop selection and management. Potential evapotranspiration (Et$_0$) specifies how much water would be lost if it was available. This loss would occur from an ideal open surface of pure water in equilibrium with the environment. Standard evaporation pans located in a crop production area can be used to estimate Et$_0$, but it is more usually calculated from data provided by a weather station. Actual Et may be much lower than Et$_0$. The water use of a crop can be expressed as a crop coefficient, which is the fraction of Et$_0$ that the crop would transpire when well-supplied with water. This coefficient is typically 0.6 to 0.8 for plants in the temperate zone. Water evaporates from the soil, and for a bare surface of saturated soil, the rate may be similar to Et$_0$. However, evaporation decreases sharply as the soil dries out, and the moisture is drawn from lower and lower levels. When soil is covered by vegetation, air movement and input of solar radiation tends to be low at the soil surface. The humidity under these conditions is higher than above the crop, and evaporation from a wet soil surface may be 10 percent of Et$_0$ or less.

The balance between evapotranspiration and precipitation defines whether the climate is moist or dry and whether irrigation is likely to be necessary to produce crops. In the United States, east of about 95°W, there is a surplus of precipitation over Et$_0$, but rainfall is spread throughout the year and Et$_0$ regularly exceeds precipitation in the summer (Figs. 4–6 and 4–7). The

Table 4–1

EFFECT OF TEMPERATURE CHANGE ON VAPOR PRESSURE DEFICIT AT A RELATIVE HUMIDITY OF 70 PERCENT

	10°C	30°C
Vapor pressure (kPa)	0.86	2.97
Vapor pressure deficit (kPa)	0.37	1.27
Saturated vapor pressure (kPa)	1.23	4.24

Source: Michael Knee.

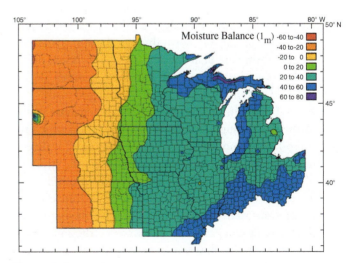

Figure 4–6

The balance between potential evapotranspiration and precipitation in the central United States. East of 95°, precipitation exceeds evapotranspiration. Precipitation is spread nearly uniformly throughout the year. However, high evapotranspiration rates in the summer often exceed precipitation during that season, causing drought conditions at times. Source: USDA.

situation is more extreme in areas such as the Pacific Northwest, where there is a **Mediterranean climate** and most of the rain falls in the winter (Fig. 4–8). (A Mediterranean climate is characterized by cool, wet winters and warm, dry summers.) Soil typically holds about 100 cm of water in the plant root zone. After the soil is saturated, the excess water runs off or drains away. It is important for spring planting that the soil should be fully charged with water, but waterlogged conditions are harmful for most crops. In many parts of the United States, it is necessary to install drainage systems to get rid of a seasonal excess of water. Successful water management involves regulating the availability of water to meet evaporative demand. Of course, if water is not present, evaporation does not occur and actual Et is much lower than Et_0. For most crops, this

situation means that in the short term, they stop growing; if the drought persists, they will die.

Providing water in order to grow crops where natural precipitation is inadequate during some or all of the growing season is one aspect of climate that can be and is modified on a large scale. Irrigation has a long history and allows us to grow many crops in areas that otherwise would be unsuitable for growing. Water was supplied to crops through open irrigation channels in the Middle East in prehistoric times. The water flowed by gravity to furrows in the fields much as it does in many parts of the world today (Fig. 4–9). Now, large areas are often irrigated with water piped under pressure to some form of overhead irrigation (Fig. 4–10). This method can result in losses because the water may

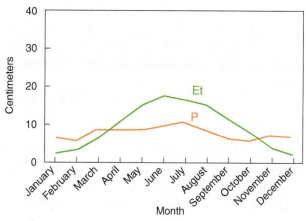

Figure 4–7
Seasonal variation in precipitation and evapotranspiration in Indianapolis, Indiana. Source: US Department of Commerce.

Figure 4–9
Irrigation channels in Iraq.
Source: Margaret McMahon, The Ohio State University.

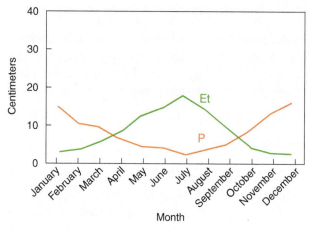

Figure 4–8
Seasonal variation in precipitation and evapotranspiration in Seattle, Washington, which has a Mediterranean climate.
Source: US Department of Commerce.

Figure 4–10
Overhead irrigation of field crop. Source: USDA Natural Resource Conservation Service, http://photogallery.nrcs.usda.gov/

fall outside the crop area and a proportion evaporates before it reaches the root zone. Smaller-scale irrigation of horticultural crops, including ornamentals, often aims to supply the water more directly to the plant roots through trickle-feed pipes or spray emitters close to the soil surface (Fig. 4–11). In a moist climate, water can be held in local ponds or lakes to supply irrigation during periods of water deficit. In a dry climate, irrigation may supply the entire water needs of the crop. The water may be drawn from a distant source or from underground aquifers.

An advantage of irrigating is that pests and diseases may be less troublesome for irrigated crops grown in dry climates than in moist areas where the crop itself might grow more readily but insect pests are more prevalent too. On the other hand, the world's freshwater supplies are limited, and it is getting increasingly difficult and expensive to maintain water supplies in the western United States and many countries. It is important to supply water efficiently so that enough reaches the plant roots and as little as possible is lost through evaporation, runoff, or drainage. Irrigation managers increasingly use estimates of evapotranspiration to decide how much water to apply and when to apply it. However, conservative use of water itself leads to problems in dry climates. Irrigation water adds small amounts of various salts to the soil. If these salts are not washed away by a sufficient volume of irrigation water or natural rainfall, they can build up to concentrations that inhibit root growth. Large areas of land have been damaged by salinization as a result of irrigation in many parts of the world.

An alternative to irrigation is to select a crop that has a low water demand. This approach is particularly relevant for landscape horticulture in the southwestern states, where water supply is at a premium. Researchers are trying to develop **xeriscapes** using locally adapted plants to replace those that are more appropriate for the eastern United States. Conversely, wetland plants are making a comeback in the midwestern states as wetland areas are restored. The restoration is replacing the wetlands of this region that were drained over the past two centuries when agriculture was introduced. The loss of wetlands led to its own problems of erosion and flooding as water ran directly off fields, carrying soil particles into drainage channels that could not cope with the sudden flow. Wetlands act as a buffer zone, temporarily holding water to spread the flow and catch sediments. Furthermore, wetland vegetation removes nutrients from excess fertilizer washing into the drainage water and helps to prevent **eutrophication** of lakes, ponds, and waterways.

As with other climatic variables, plants adapt to the water availability of the region where they originated. Sometimes plants have developed drought-avoidance strategies. For example, many plants in the Mediterranean area grow and flower during the late winter when it is wet and go dormant over the summer when it is dry. Many of our ornamental spring flowering bulbs originate from this region. Other plants tolerate drought through various physiological mechanisms. Many desert succulents minimize transpiration by absorbing carbon dioxide at night and keeping their stomata closed during the day through a process called **CAM metabolism**. Some grasses, called warm-season grasses, have a mechanism called **C4** that lets them reduce transpiration and water loss but still maintain high photosynthetic rates during prolonged hot and dry spells. Corn, sugarcane, and several turfgrasses and forage grasses fall into this group.

Figure 4–11
Trickle or drip irrigation puts water on the surface of the water near the plants, thereby reducing evaporation compared to overhead irrigation. Source: USDA Natural Resource Conservation Service (NRCS), http://photogallery.nrcs.usda.gov/

TEMPERATURE

Although sunlight and water are vital for plant life, temperature may be the most important factor in the decision about whether a crop can be grown or when to plant a crop in a given location. We generally think of sunshine as providing warmth, but wind and water move huge amounts of thermal energy around the planet.

As you learned earlier, the amount of solar radiation reaching the earth's surface decreases with increasing latitude. The energy balance of the earth is maintained by radiating energy back into space as heat. Although the total amount of energy received and radiated equal, the balance varies across the earth's surface. Radiant heat loss does not vary with latitude as much as solar radiation resulting in a net energy gain at the equator and a net loss at higher north and south latitudes. Gain and loss are equal at about 35° to 40°. These imbalances are the underlying reason for decreasing temperatures as we move north or south from the equator. Temperature variations are moderated by convection currents in the oceans and atmosphere that carry huge amounts of heat from the equator to higher and lower latitudes. The amount of energy received in this way varies greatly according to geographic position relative to ocean currents and prevailing winds. Thus, western Europe enjoys mild winters because of the northward drift of water in the Atlantic, whereas the central United States suffers from blasts of arctic air coming from Canada.

Temperature is also affected by altitude. As with other gases, the temperature of air drops as pressure decreases. The pressure of air at a given location is generated by the mass of air above that location. As elevation increases, air pressure decreases and air temperature drops. In dry air, the decrease is about 10°C for 1,000 m (or 5.5°F for 1,000 ft); in moist air, it is 6°C for 1,000 m (or 3°F for 1,000 ft). This feature means that temperate fruits and vegetables can be produced at high altitudes in tropical countries such as Kenya.

Water vapor can absorb heat radiation. Thus, cloud cover limits radiative heat loss, which can be particularly important at night. In areas with little cloud cover, temperatures can swing wildly between day and night, a feature often noticed by desert travelers. Conversely, severe freezes almost never occur in the wet winters that are characteristic of Mediterranean climates. The variation in winter weather in temperate climates is partly a matter of the degree of cloud cover.

Heat transfer by convection (movement by a carrier) is an important factor in both weather and climate. Water and air are the two carriers for convective heat transfer. Heat transfer can occur at any scale between a field or city block and a whole country or continent. Thus, northern Europe is warmed by the Gulf Stream from the south Atlantic, but the west coast of the United States is cooled by the California current from the north Pacific. Air movement called "Alberta clippers" bring sudden freezing temperatures and snow to the central United States, but warm chinook winds can cause sudden snowmelt on the east face of the Rockies.

It takes more heat input to raise the temperature of water than that of the ground, and water cools down more slowly than land. Thus, large bodies of water can moderate temperature changes in adjacent land during the day and across longer time periods. The Salinas Valley in California is more favorable for temperate vegetable production than the San Joaquin Valley because it is up to 10°C cooler in summer (Fig. 4–12). Although the valleys are only about 50 miles apart, the Salinas Valley is close enough to the Pacific Ocean to receive cool onshore wind from the ocean while the San Joaquin does not. Similarly, grape and peach production is possible around the southeastern shore of Lake Erie because average winter temperatures are up to 5°C warmer than a few kilometers inland. Although temperature generally falls with altitude, this pattern can undergo a temporary inversion, which can have local effects on crops growing on slopes or on undulating ground. This type of change tends to happen on cloudless nights when heat radiates from the ground to the sky. Cold air then flows down the slope or into hollows, leaving warm air above (Fig. 4–13). Crops at the bottom of the slope or in the hollows may be damaged or killed by the sudden cold, whereas those growing a few feet up the slope may be undamaged.

Urban development has a major effect on temperature at all times of the year. In winter, buildings radiate heat to the surrounding environment. In summer, buildings and paved surfaces absorb solar radiation and release it as so-called sensible heat over time. In the suburbs and countryside, much of the radiation is absorbed by vegetation and is used to drive transpiration which helps to cool the air. Most of the rainfall is

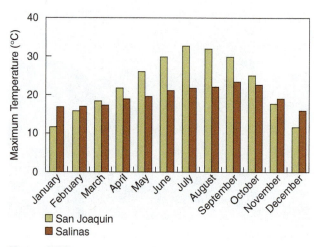

Figure 4–12

Salinas is slightly nearer to the Pacific Ocean than San Joaquin and enjoys much cooler summers. Source: Michael Knee.

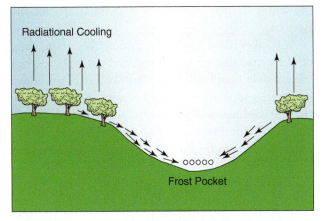

Figure 4–13

On a clear night, heat lost from the ground leads to cooling of the air, which rolls down slopes and leaves warmer air above it. Source: Michael Knee.

drained away from urban areas as soon as it falls. Because of the absence of standing water and vegetation, the city does not enjoy the evaporative cooling that lowers the temperature in rural areas (Fig. 4–14). At all times of the year, buildings provide a barrier to air movement, and carbon dioxide emissions form a blanket that reflects long-wave radiation back to earth. The cumulative result of these factors is that cities can be as much as 5°C warmer than the surrounding countryside. A dome of warm air exists over cities, forming the so-called urban heat island or dome. Warm air flows out toward the suburbs, which can be 2° to 3°C warmer than the countryside.

A defining feature of the climate of the temperate zone is that temperatures are not conducive to plant

growth throughout the year. The seasonal variation in temperature has different implications for annual and perennial plants. Winter temperatures are relatively unimportant for annuals, which survive the winter season as seeds. Perennial plants have two strategies for winter survival. Many herbaceous perennials die down to ground level but survive as underground storage structures with buds, from which new growth begins each year. These plants avoid the coldest weather because soil temperatures do not fall nearly as low as air temperatures. Woody perennials and some herbaceous plants survive above ground. Their stems become dormant and acquire a degree of cold hardiness that varies with plant species. Thus, a plant's survival depends on temperatures never falling below the minimum that it can tolerate. US plant hardiness zones are defined according to the minimum temperatures that can be expected. Over the years, growers have gathered experience about the minimum temperatures that particular plants can survive. The USDA has created—and recently updated as of 2009—a Plant Hardiness Zone Map that shows the minimum temperatures that can occur throughout the country (Fig. 4–15). We can match plants to zones and generally be confident of their survival. However, the plant hardiness zone map shows normal minimum temperatures; occasional extreme temperatures can still result in plant death.

The America Horticulture Society (AHS) has produced a heat zone map that is based on the normal high temperatures for a region. Fewer plants have had their

Figure 4–14

It feels hotter in the city than in the country because heat is absorbed and radiated by buildings and paved surfaces rather than being used to drive evapotranspiration.

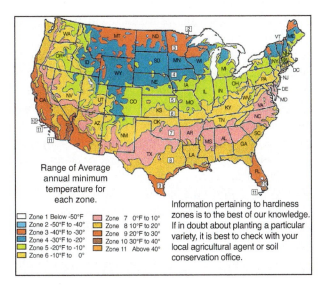

Figure 4–15

USDA plant hardiness zones are defined by usually expected minimum temperatures, not the average winter temperature or the extreme lows recorded. Source: USDA, http://www.usna.usda.gov/Hardzone/ushzmap.html

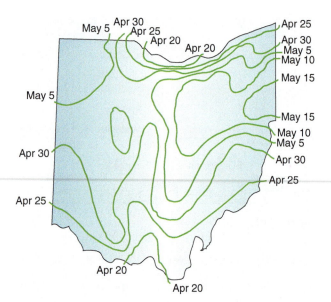

Figure 4–16
Variation in the date when there is a 50 percent chance that no more frosts will occur in Ohio. Similar data are available for other states at http://www.ncdc.noaa.gov/oa/climate/stateclimatologists.html

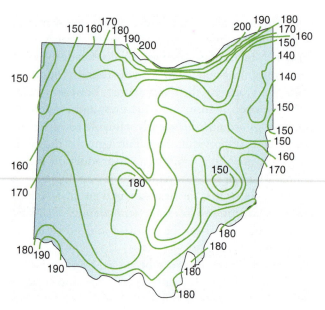

Figure 4–17
Map showing the growing seasons for the state of Ohio. The growing season is defined as the time between the average dates for the last frost in the spring and the first frost in the autumn. Similar data are available for other states at http://www.ncdc.noaa.gov/oa/climate/stateclimatologists.html

heat tolerance defined compared to cold tolerance, but that number is increasing.

Springtime temperatures are critical for both annuals and perennials. Many seeds can be planted only after the danger of frost has passed, and spring-flowering perennials, including many fruit crops, can be injured by late frosts. Maps showing expected dates of last frost can be used to decide when to sow seeds and to select areas suitable for fruit production (Fig. 4–16). Other maps can show the expected dates of first frost in the autumn. For many crops, this point marks the end of the growing season. If a mature crop cannot be harvested by this time, there is little point in growing it (Fig. 4–17). A cotton plant grows well in the summer in Ohio, but the growing season is just not long enough to allow mature cotton bolls to be harvested.

Plants generally show a parabolic relationship between growth and temperature. A species has a minimum temperature, below which growth does not occur; an optimum temperature for growth; and a maximum temperature, above which injury occurs. The optimum varies according to the geographic origin and season of growth of the plant. For most plants in the temperate zone, the minimum is between 0° and 5°C, and the maximum is around 35°C. Many tropical plants are injured below 10°C and show optimal growth around 35°C (Fig. 4–18). Spring- and early summer-flowering plants such as pansies and winter annual weeds may grow well at cool temperatures but die in the summer because of heat.

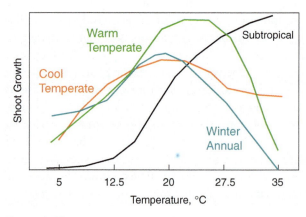

Figure 4–18
Growth of different kinds of plants in relation to temperature.

Many methods have been used to manipulate the temperature around growing crops throughout their growth or at critical times, and temperature manipulation is the major function of greenhouses. Methods can range from laying cloth or plastic-coverings on the plants to growing the plants in heated greenhouses. Various kinds of cover and wind protection have been used to promote plant growth early in the spring. Now, the most commonly used methods involve polyethylene covers, which can be applied to the soil to raise its temperature for seed germination or used in the form of a tunnel covering the whole plant, at least for part of its growth cycle. This method can be enough to prevent

injury in marginally hardy crops during the winter or to prolong the growing season for fruits, vegetables, or flowers. More sophisticated greenhouses with full temperature control may allow for year-round production of certain crops. By their very nature, however, greenhouses are poorly insulated structures, and even many floral crops are not sufficiently valuable to offset the energy costs of providing heat. Furthermore, the available light may be a limitation on production in the winter. Much of the heat requirement is at night. On a sunny day in winter, the heat gain of a greenhouse may require ventilation to lower the temperature. That heat is lost unless that it can be captured and stored to use during the cold nights.

Local temperatures can be altered by the presence of a body with a high thermal mass that absorbs solar energy during the day and radiates it over time. In natural situations, this body could be a cliff face, but horticulturalists down the ages have used walls to provide shelter and stored heat to grow tender crops. Sometimes the input of solar radiation was augmented by fireplaces and flues built into the structure of the wall.

The vulnerability of fruit blossom to spring frosts has led to several methods of short-term temperature manipulation. The danger arises mainly from radiational cooling of the ground on clear nights, leading to an inversion of cold air close to the ground with warmer air above. One way of counteracting this phenomenon is to mix the air layers with a wind machine. Various fuels have been used in heaters to provide frost protection in orchards. Water can also be sprayed over trees that are in danger of freezing. The trees become coated with ice, but as the water freezes, it gives up latent heat that may be sufficient to prevent injury because the buds freeze at a slightly lower temperature than water.

AIR MOVEMENT AND COMPOSITION

Winds are driven by differences in air pressure resulting from corresponding differences in the input of solar energy. Air movement around the earth is dominated by three belts of wind in both the northern and southern hemispheres. Although the winds move air masses between different latitudes, the rotation of the earth gives them an easterly or westerly shift via the Coriolis effect. The trade winds blow toward the equator from a northeast direction in the upper atmosphere and a southeast direction in the southern. Between about 30° and 50° latitude, winds blow in opposite directions: southwest in the northern hemisphere and northwest

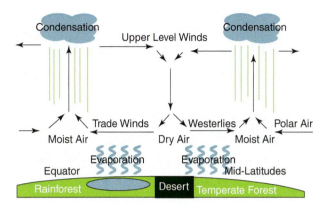

Figure 4–19

Major global air and water circulation patterns. Source: Michael Knee.

in the southern. Beyond 50°, polar winds blow predominantly from the northeast in the northern hemisphere and southeast in the southern. These surface-level winds are accompanied by upper-level winds blowing in the opposite direction. The convergence of the trade winds at the equator leads to an updraft of air, from which rain falls as it cools. The divergence of prevailing winds around 30° latitude is associated with a downdraft that carries moisture away and results in low rainfall (Fig. 4–19).

Air circulation tends to be more orderly in the southern hemisphere than in the northern, where large land masses capture the sun's heat and give rise to high pressure areas that break up the circulation patterns. The high pressure areas shift, leading to changes in the weather. In the central United States, summer weather is dominated by southeast winds bringing moisture-laden air from the Gulf of Mexico, whereas in the winter, northwest winds bring polar air (Fig. 4–20). Sudden shifts can occur in these patterns, particularly in the spring and autumn, leading to unstable weather. When a cold front meets a mass of warm, moist air, powerful air currents can lead to thunderstorms. When crosswinds blow through the rising clouds, conditions are set for tornado development.

On a more local level, bodies of water affect air circulation because they warm up more slowly than the land does in the daytime and cool more slowly at night. This characteristic leads to an onshore breeze toward the land in the morning and an offshore wind in the evening. In mountainous areas, breezes tend to blow up valleys and hillsides as they warm up during the day, but cool air flows in the opposite direction at night. Buildings tend to slow down air circulation in urban areas; however, depending on the height and distribution of buildings, areas of turbulence or convergence of winds can produce strong gusts. Air movement tends

Figure 4–20

Prevailing winds in the United States in summer and winter.

to be slowed down by vegetation. This effect is most noticeable under a continuous canopy of herbaceous or woody vegetation. Air tends to move across the top of the canopy rather than through it (Fig. 4–21). Humidity tends to be higher within the canopy than above it because of limited air exchange. Because transpiration is related to vapor pressure deficit and wind speed, water loss tends to be much lower for plants

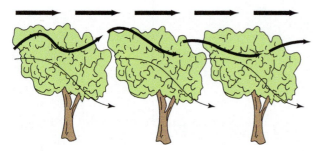

Figure 4–21

Air movement within plant canopies is generally slower than outside the canopies. The thicker the line, the faster the air movement. Source: Michael Knee.

within continuous stands than for isolated plants of the same species or for plants at the edge of the stand.

Wind can be a major cause of damage to crops, either because of direct physical damage or excessive transpiration. In exposed areas, it may be necessary to construct windbreaks or grow shelter belts of tall vegetation. However, plants are mostly adapted to grow with a certain amount of physical disturbance by wind. When this disturbance is absent, they may tend to become tall and weak-stemmed, which is a problem for woody plants that are staked in nursery production and for plants grown in a greenhouse. For greenhouse plants, it can be beneficial to provide physical stimulation by creating air movement with fans or brushing the foliage on a regular basis.

In some areas, wind has been an agent of soil formation. Loess soils are formed by the accumulation of wind-blown particles. As with water, however, wind is an agent of soil erosion, which is a problem particularly in the drier parts of the western United States and other arid parts of the world.

ATMOSPHERIC COMPOSITION

Dry air (absence of water vapor) is comprised of 78 percent (780,000 parts per million or ppm) nitrogen (N), 20.9 percent (209,000 ppm) oxygen (O), 0.037 percent (370 ppm) carbon dioxide, and the rest is all other gases.

Nearly all of the earth's nitrogen is in the atmosphere (77.5 percent) or in the lithosphere (rock and soil, 22.4 percent). The rest is mainly in water with a tiny amount in living organisms. However, life would cease if this tiny fraction could not be replenished continuously by converting atmospheric nitrogen which is not available to plants into forms that can be absorbed by plants and passed up the food chain. Nitrogen oxides are formed from free nitrogen and oxygen by lightning during thunderstorms and they fall to earth as nitrates in rainwater. Soil bacteria, both free-living and in association with plant roots, convert nitrogen to ammonia. In the last century, the same reaction was developed on an industrial scale in the Haber–Bosch process. The ammonia fertilizer from this process makes an essential contribution to the levels of crop productivity we depend on for survival. On the other hand, oxides of nitrogen are generated as by-products of gasoline combustion in vehicles and have become pollutants in the atmosphere. The oxides dissolve in rainwater and contribute to the soil and water nitrogen pool wherever they fall as acid rain. This effect is mostly undesirable in natural ecosystems and contributes to eutrophication of bodies of water, the disappearance of

nitrogen-fixing plant species (which are no longer competitive), and apparently forest decline in the United States and Europe.

The oxygen in the atmosphere that we depend on for our survival is produced by plants and algae during photosynthesis. Land plants account for about 60 percent of global photosynthesis, which keeps the balance between oxygen and carbon dioxide (CO_2) at about 600 to 1. By competing so effectively for carbon dioxide and keeping it at low levels, plants and algae modify the climate of the earth. As is now well known, carbon dioxide absorbs infrared radiation from the earth so that heat is retained in the atmosphere. This phenomenon is loosely referred to as the greenhouse effect. The level of oxygen in the atmosphere is so high (209,000 parts per million or ppm) that it is not much affected by slight changes in the balance of oxygen and carbon dioxide.

However, this balance has a much greater impact on carbon dioxide levels as a slight change in the balance can have a dramatic effect on carbon dioxide values. Currently global atmospheric carbon dioxide is at ≈370 ppm. Locally, carbon dioxide levels undergo fluctuation depending on the amount of photosynthesis occurring. Thus, concentrations can fluctuate between day and night and between winter and summer by up to 20 ppm (Fig. 4–22). The increasing combustion of fossil fuels and the clearance of natural vegetation on all continents have caused a 30 percent increase in atmospheric carbon dioxide over the past two centuries. There is now little doubt that this rise in carbon dioxide is contributing to an upward trend in temperature called global warming, particularly in higher latitudes. Whereas this trend has many negative effects, crop yields could actually increase as long as water availability and other factors did not change. It has been suggested that increased crop productivity would decrease CO_2 concentrations because the plants would be absorbing more CO_2. However, because crops are not generally as effective in removing carbon dioxide from the atmosphere as is natural vegetation, their ability, decrease global warming would be limited. Also to effectively sequester CO_2 out of the atmosphere, plants have to hold CO_2 for many years. The CO_2 in most crop plants is nearly all released shortly after consumption.

Human activity has added other gases to the atmosphere that are more obviously pollutants than is CO_2. Sulfur dioxide is another product of combustion, particularly from coal. It is directly injurious to plants. Along with nitrogen oxides, it contributes to acid rain, which has caused problems in terrestrial and aquatic habitats in areas around coal-burning power stations. Ozone (O_3) is formed from molecular oxygen by photochemical reactions with other pollutants, including oxides of nitrogen (Fig. 4–23). Plants vary in their sensitivity to ozone, which is also injurious to people. Ozone is beneficial in the upper atmosphere, however, where it is formed directly from oxygen by ultraviolet radiation and is itself a strong absorber of ultraviolet (UV) light, which would be damaging for all life if it reached the earth's surface in a large amount. Some UV light though is required for the production of flavonoids, the plant pigments that give plants some of their distinctive colors on leaves and flowers. Ultraviolet light also plays a role in how some insects detect plants.

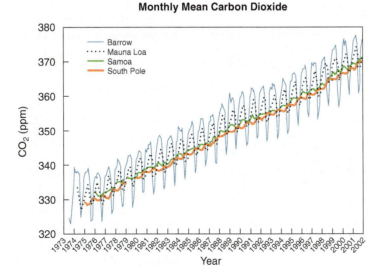

Figure 4–22

Generally, rising carbon dioxide levels show seasonal variation, depending on the photosynthetic activity of vegetation. Data shown here through May 2002. Source: http://www.cmdl.noaa.gov/

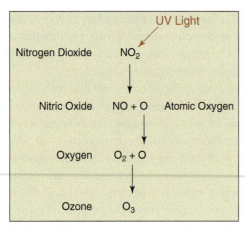

Figure 4–23

The UV component of solar radiation promotes ozone formation when the pollutant nitrogen dioxide is present in the lower atmosphere. Source: Michael Knee.

Several hydrocarbons are present in the atmosphere in trace quantities. The most abundant is methane, which arises from sewage treatment and intensive animal production as well as natural sources, such as wetlands. It is not very toxic, but like CO_2, it is a contributor to global warming. Ethylene is produced by plants and several microbes; it is one of the plant hormones and is active at about 0.1 ppm. Its effects can be benign or even beneficial, such as its role in the promotion of fruit ripening, but it also causes premature senescence (aging) of plant tissues and abnormal growth. It is a minor product of the combustion of fossil fuels and can be troublesome for plants in enclosed spaces, especially greenhouses. Although enhanced levels tend to occur in urban areas, it is not clear that this level has had any adverse effects on plant growth.

In addition to the gases, all air contains particulate matter, which can also be increased by human activity. Soil cultivation and major construction projects cause soil particles to be released into the atmosphere. In addition to causing loss of soil, the dust cloud can block sunlight, clog the stomates in plant leaves, and create a general nuisance for people. The problem is greatest in dry seasons and arid areas. Burning of forests and other natural vegetation in preparation for agriculture or other development also contributes to airborne particulates. Volcanic eruptions are natural sources of dust. Dust from local sources such as the eruption of Mount St. Helens, the burning of the Indonesian rainforests, or agricultural operations in China can have a worldwide impact, reducing light levels and temperatures. (Global warming might be worse but for all the dust we have kicked up!) A fraction of the number of the particulates consists of living organisms, plant pollen, microbial spores, and insect eggs. These "particles" also move around the world, causing pest and disease outbreaks on plants as well as allergic reactions in humans.

Attempts to modify the atmosphere focus on preventing the emission of pollutants that can be harmful to plant, animal, and human life. Carbon dioxide is sometimes added to the atmosphere in greenhouses to promote crop growth, and optimistic people suggest that rising levels will be beneficial for field crops. General interest has developed in agricultural practices that will lead to sequestration of CO_2, which means that carbon will somehow be taken out of circulation, or its reentry into the carbon cycle will be greatly delayed. Of course, when most crops are consumed, their carbon content is returned to the atmosphere as CO_2 through respiration, so the benefit of each crop may be temporary. In natural ecosystems, carbon can be sequestered in organic debris that gets incorporated into the soil or in the stems of long-lived woody plants. There is great interest in cultivation practices, such as low- or no-till cultivation, that could increase carbon sequestration in agricultural soils.

SUMMARY AND REVIEW

Climate is not the same as weather. It is defined by the prevailing weather conditions of an area. The input of solar radiation influences all other aspects of climate. Vegetation and human interference modify climate, which in turn influences natural vegetation and the possibilities for crop production in an area. Much of the solar radiation reaching the earth's surface is infrared, while about 40 percent is photosynthetically active radiation between 400 and 700 nm. Away from the equator, the intensity of solar radiation decreases and the duration of daylight varies throughout the year.

Cloud cover further decreases light intensity. Plants are adapted to the light conditions of their natural habitat. Day length influences processes such as flowering, leaf-fall, and bud-break. The spectral quality of light reflecting from or passing through foliage changes, which affects the growth of neighboring plants. Although it is uneconomic to supply high-intensity artificial light for photosynthesis on a large scale, supplementary lighting may be used to control development and flowering along with increasing photosynthesis of greenhouse crops.

Only a small proportion of the world's water is available to plants, but this amount is continuously replenished through the hydrologic cycle. Most crops are grown in the small proportion of the earth's surface with a moist temperate climate. The availability of water is determined by the supply (rainfall and irrigation) relative to the demand, which can be summarized as evapotranspiration (Et). Potential evapotranspiration (Et_0) is influenced by solar energy input, temperature, air movement, and vapor pressure deficit (VPD). VPD is one way of expressing the dryness of the air and is the difference between its actual water content and the water holding capacity, both expressed as partial pressures. Actual Et is usually less than Et_0. It is influenced by water availability and the nature of the vegetation. Seasonal estimates of rainfall and Et can be used to determine what crops to grow in a region and whether irrigation will be necessary. Variations in VPD can be used to manage irrigation supply. Water availability is the only aspect of climate that can be modified on a large scale for crop production. However, increasing problems of water availability and salinization of soils resulting from irrigation are occurring in many parts of the world. Drainage is sometimes necessary to manage seasonal excess of water, but maintenance of wetlands can help to control flooding and the pollution of lakes and waterways with nutrients and sediments. Plants differ in their requirements for water, and water management should include selection of plants with water needs that are similar to local precipitation patterns.

As with other aspects of global ecology, we have to think of crop plants as part of the hydrologic cycle. Thus, we must consider how the demand for water for crops can be met without disrupting the cycle. We also need to think about the effects of crop production on water quality. Fertilizers and pesticides applied to crops can be transported through groundwater and runoff into streams and rivers. The timing of the application of a chemical in relation to rainfall affects how much reaches its target in the plant and how much enters the hydrologic cycle.

Although it is necessary for plant growth, water can cause problems. We already mentioned the need for drainage in many growing areas. Excessive rains can still cause flooding, and plants vary in their tolerance of partial or complete submergence. Hail storms can devastate almost any crop, which is why growers may specifically insure against hail damage.

As long as the water supply is at an adequate level, temperature is the main factor that determines whether a crop can be grown in an area. Temperature is influenced by solar radiation and heat transfer by air and water. Heat input from the sun drives air movement and heat transfer. Air movement can modify temperature, and carry water vapor, soil particles, pests and diseases, and other atmospheric components to or from an area where plants are growing. Many aspects of crop growth and its management are influenced by temperature conditions. These conditions are determined by latitude, altitude, large bodies of water, land or other masses such as city buildings. Local features modify these conditions and some can be manipulated to improve crop growth.

As with other climate factors, plants have adapted to local temperature patterns. Temperate plants survive the winters by overwintering as seeds (annuals) or in state of dormancy (perennials). Some plants retain their above ground structures during dormancy while in others the above ground parts die while the underground structures are dormant. Some semi-hardy plants can be protected from the cold by covering them during the winter. Plants that cold intolerant are grown in greenhouses during the cold periods of the year.

The three major gas components of the atmosphere, aside from water vapor, are nitrogen, oxygen, and carbon dioxide. Nitrogen is present at about 780,000 ppm, oxygen at 209,000 ppm, and carbon dioxide at approximately 370 ppm. Plants require nitrogen for growth, most of which is in the atmosphere. However, atmospheric N is not available to plants and has to be converted to a usable form by such things as lightning, bacteria, or an industrial process before it can be used by plants. Plants have an important role in maintaining atmospheric composition particularly in the balance of O_2 and CO_2 and consequently play a major role in controlling global warming. Plant growth is affected by particulate and gaseous pollutants in the atmosphere.

KNOWLEDGE CHECK

1. What is the difference between climate and weather and how do they relate to each other?
2. Why is solar radiation considered to be the primary factor that determines climate?
3. What kind of features can modify the climate in an area?
4. What is a microclimate?
5. What is a plant ecotype and why are they important to cultivated plants?
6. How does latitude influence solar radiation intensity on earth? Where is radiation the highest? Where is it the lowest?

7. What are two ways the atmosphere influences how much solar radiation reaches the earth?

8. How do the rotation of the earth on its axis and the rotation of the earth around the sun influence how much solar radiation a region receives?

9. What is photoperiodism and what are two ways it can affect plant growth and development?

10. What is the shade avoidance response and what two regions of the solar spectrum influence the response?

11. What are two ways we manipulate photoperiod in a greenhouse to control flowering in photoperiodic plants?

12. What are some ways that water is moved from one area to another in the hydrologic cycle?

13. Why does vapor pressure deficit influence (VPD) water evaporation from plants more than relative humidity?

14. How does vegetation reduce evaporation from the soil?

15. What are two advantages of irrigating crops and what are two problems associated with irrigation?

16. Although the amount of solar energy received on earth equals the amount radiated back into space, what are three reasons for temperature variations among regions?

17. What is the USDA Plant Hardiness Zone map and how is it useful to anyone who wants to grow plants?

18. Why is water loss from plants less for plants on the inside of a group of plants than for those along the edge or standing alone?

19. Although high winds can be damaging to plants, why is some air movement around plants beneficial to stem development?

20. In ppm, what are the amounts of N, O, and CO_2 in the atmosphere?

21. Although crop plants may absorb more CO_2 more because of global warming, what are two reasons that they probably would not be effective at reducing CO_2 in the atmosphere?

22. What is a beneficial way ethylene affects plants; what is a negative way?

23. How does dust negatively affect plants? How may it be helping to reduce global warming?

FOOD FOR THOUGHT

1. What are some local features in your hometown that could create microclimates in that area?

2. If you wanted to control an irrigation system effectively, explain why you would prefer to use a sensor that measures VPD or one that measures RH?

3. On some farms, the wheat plants that grow on the edge of the field are harvested not for use as a grain but as stalks that are sold to local florists for flower arrangements. Explain why the stalks from the edge of the field and not ones from inside the field are probably more suitable for arrangements.

4. Suggest types of plants that could be grown to absorb carbon dioxide from the atmosphere and not release very quickly back into the atmosphere.

FURTHER EXPLORATION

ADAMS, J., AND N. ZENG. 2007. *Vegetation-climate interaction: How vegetation makes the global environment.* New York: Springer.

AGUADO, E., AND J. BURT. 2010. *Understanding weather and climate.* Upper Saddle River, NJ: Pearson Education.

CHRISTOPHERSON, R. W. 2009. *Geosystems: An introduction to physical geography.* 7th ed. Upper Saddle River, NJ: Prentice Hall.

*5

Soils

Don Eckert

Don Eckert

key learning concepts

After reading this chapter, you should be able to:

- Discuss the concept that soil ecology is a complex system made up of many living and nonliving components.
- Describe the components that make up a soil ecosystem and how they interact.
- Describe the factors that influence soil formation and give soil its physical and chemical characteristics.

People often refer to soil dismissively as dirt and probably regard it as dead stuff. However, it is an important part of the living world; along with climate, it determines the composition and function of the ecosystems that can exist in different areas. Plants live in the soil as much as they live in the air. Some of the oxygen that plants require comes from the soil and enters the roots. Many if not most of the ecological interactions between plants and other organisms occur below ground. Most of the individual organisms in an ecosystem and a large part of the biomass exist in the soil. Over time, the composition and properties of the soil are influenced by soil organisms. As you will learn later in the chapter, soil is the product of the interaction of parent material, climate, topography, and soil organisms over time. This chapter looks at how these factors interact to produce the different kinds of soils found around the world.

DEFINITION OF SOIL

Soils differ around the world, but they are basically composed of weathered rock, air, water, decomposed organic material, and various organisms, all working together in a complex ecosystem capable of supporting the growth of land plants. Not all soils will have all of these components. The type of components and their relative amounts determine what kinds of plants can grow in that soil. Soil is a very complex and dynamic system of many interacting factors and components that affect and are affected by plants. Soil consists of:

1. A solid fraction; that is, rock fragments and minerals.
2. An organic fraction; the decayed and decaying residues of plants, microbes, and soil animals.
3. A liquid fraction, including water and dissolved minerals.
4. A soil atmosphere or soil air.

The voids found between the solids are called **pore spaces**. Pore spaces vary in size and continuity and are an important physical property of soils.

The kind of soil and its ability to support plant growth are influenced by the relative amounts of each of these four components. Typically, a natural soil is approximately half solids and half pore space. The solid phase of the soil might average 40 to 45 percent mineral and 5 to 10 percent organic material by

volume, while the pore space might be filled with a mixture of water and air, the ratio depending on the wetness or dryness of the soil. In general a productive soil has 50 percent pores that will drain water quickly after a thorough watering and 50 percent that hold water through capillary forces. The water that is left after gravity has removed as much as it can is called **field capacity** of the soil. All of these relationships vary from one soil type to the next.

Kinds of Rocks

Parent rocks contain the nutrients that will be found in parent material, and later in the soil. Rocks are made up of consolidated material, unconsolidated material, or both (Fig. 5–1):

1. **Igneous rocks** (i.e., lava, magma) are formed from the hardening of various kinds of molten rock material and are composed of minerals such as quartz and feldspar.
2. **Sedimentary rocks** are generally unconsolidated and composed of rock fragments that have been transported and deposited by wind, water, or glaciers. Limestone, sandstone, and shale are examples.
3. **Metamorphic rocks** form from igneous or sedimentary rocks that have been subjected to sufficiently high pressures and temperatures to change their structure and composition. Slate, gneiss, schist, and marble are examples of metamorphic rocks.

Figure 5–1
Parent rocks from which soil can be formed: (A) igneous, (B) sedimentary, and (C) metamorphic.
Source: Margaret McMahon, The Ohio State University.

FACTORS INVOLVED IN SOIL FORMATION

Soil is derived from rocks, minerals, and decaying organic matter. The two major processes in soil formation are (1) accumulation and (2) transformation of the parent material. **Parent material** accumulates from the breakdown of rocks by weathering. This process must occur before soil can begin to form. The parent material accumulates as an unconsolidated mass that later differentiates into characteristic layers called horizons. A **horizon** is a distinct layer of soil having physical and/or chemical differences resulting from soil-forming processes as seen in a vertical cross section. Differentiation occurs by mechanical separation and/or transformation of the parent material. As the process continues, the horizons generally become more distinguishable and finally develop into a soil profile. A **soil profile** is a vertical section of soil extending through all its horizons from the surface to the parent material.

The factors responsible for soil formation are: (1) parent material, (2) climate, (3) biology, (4) topography, and (5) time.

Parent Material

The formation and accumulation of material by chemical and physical weathering of parent rocks is the first step in the development of soil.

Physical Weathering Physical weathering is the physical breakdown of large pieces of rock into smaller and smaller pieces. Changes in temperature greatly affect the rate of physical weathering. Differential rates of contraction and expansion caused by temperature changes bring about cracking and peeling of the outer layers of rocks by a process called **exfoliation**. A second process occurs due to the presence of different materials within a rock, each with its own characteristic coefficient of expansion. Because of these differences, sudden large temperature changes cause uneven expansion or contraction, cracking the rocks. A third process is the cracking of rocks caused by the expansion of water as it freezes in rock fissures.

The mechanical action of glaciers causes rocks embedded in the ice to scrape against other rocks as the glacier moves. This action grinds the rocks into increasingly smaller rock fragments. This physical process is powerful: it reached a tremendous magnitude during the Ice Ages (Pleistocene epoch). The product of glacial weathering is called **glacial till**, and it comprises rock particles ranging in size from clay to

boulders. This material is deposited beneath, beside, and at the terminus of the melting glacier.

Physical weathering is also caused by moving water, as in stream erosion, sheet erosion, rill erosion, or wave action (Fig. 5–2). The action is similar to that of glaciers. Water from rains and melting snow moves rapidly down streambeds, carrying parent rock fragments of varying sizes. As these fragments move along, they are gradually worn down to create smaller and smaller particles, eventually forming parent material. In arid regions, wind acts similarly to water. Coarse sand particles (parent material) are swept along the ground with sandblasting action wearing away other larger parent rocks.

The action of plant roots can sometimes physically break down parent rocks. For example, a tree root growing into a crack in a rock can ultimately fracture the rock. While this example is not considered weathering, it is a physical soil-forming process. In this case, some chemical weathering must occur first to provide nutrients for the plants before they can begin to grow.

Chemical Weathering   Chemical weathering entails four distinct processes.

Dissolution is the process by which the constituents of parent material dissolve in water or weak, naturally occurring acids and are leached away. Chloride, nitrate, and sulfate salts are generally very soluble in water and can be leached by rainfall alone. Other materials, such as carbonates, dissolve very slowly in water, but they dissolve much more quickly when acted upon by

organic acids produced when organic matter decomposes or the carbonic acid that forms when atmospheric carbon dioxide combines with rainwater.

Hydration adds molecular water to another compound to form a hydrated material more vulnerable to pulverization. An example is calcium sulfate ($CaSO_4$) absorbing water to form gypsum ($CaSO_4 \cdot 2\,H_2O$), a hydrated calcium sulfate:

$$CaSO_4 + 2\,H_2O \rightarrow CaSO_4 \cdot 2\,H_2O$$

Hydrolysis is the reaction between a compound and water to form a more soluble product. In the following hydrolysis reaction, potassium ions (K^+) are made more available to plants by the reaction of the slowly soluble feldspar mineral ($KAlSi_3O_8$) with water (H_2O) to form soluble potassium hydroxide (KOH):

$$KAlSi_3O_8 + H_2O \rightarrow HAlSi_3O_8 + KOH$$

Oxidation reactions form oxides of parent material by reaction with oxygen. For example, ferrous oxide (FeO) reacts with oxygen (O_2) to yield ferric oxide (Fe_2O_3), a product more oxidized than the reactant:

$$4\,FeO + O_2 \rightarrow 2\,Fe_2O_3$$

Climate

The climate affects soil formation. In areas of high rainfall, soils are often highly leached and acidic in reaction. Chemical weathering proceeds at a rapid rate, especially if high rainfall is coupled with high temperature. The fertility level of soils formed under high rainfall is generally low because many of the plant nutrients are leached from the root zone. Many of these soils are red or yellow in color, indicating a relatively high percentage of iron oxide, which remains after other elements have been removed.

On the other hand, soils developed in arid climates are not highly leached. Calcium and magnesium carbonates tend to accumulate, and chemical weathering proceeds at a much slower rate. Soils formed under arid conditions often contain excessive quantities of salts other than carbonates, and are not productive until the amounts of salts are reduced. Land can be desalinated by flooding with water and leaching the salts downward through the soil profile. Fields treated in this manner must possess excellent subsurface drainage, either natural or improved by human intervention. If heavily salted water accumulates in the soil, it can cause the soil to disperse and become waterlogged. Such degradation can render soils useless for future production unless expensive remediation practices are employed.

Figure 5–2
Massive gullies formed by severe stream erosion during periods of heavy rainfall. This gully is beyond reclamation by practicable methods. Source: USDA Natural Resources Conservation Service, http://photogallery.nrcs.usda.gov/

Biology

Just as in aboveground ecology, plants are the base of the food chain in the soil. Between 10 and 50 percent of the plant's dry matter is in its root system. Roots have three functions: they support and anchor the plant stem, they absorb water and nutrients, and they provide storage for food reserves. To keep the plant supplied with water and nutrients, roots must grow. While the root hairs in the zone behind the root tip are absorbing material from the soil, the tip is moving ahead, exploring the gaps between soil particles for fresh supplies that will be needed for future growth.

Plant roots secrete chemicals into the soil that help in the absorption of nutrients. They also secrete mucilage, which helps lubricate the passage of the tip between soil particles. The root tip is covered with a protective cap from which cells are sacrificed continually (about 20,000 a day from a single corn root). A plant's fine, absorbing roots do not grow indefinitely; old roots die off and are replaced by new ones so that more than half of the root system will be turned over in a year. So plant roots are continually adding organic matter to the soil quite apart from any that drops from above ground or is taken by root parasites or herbivores.

The soil is home to a great diversity of organisms from all of the five kingdoms: Plantae, Animalia, Fungi, Protista, and Monera. These organisms are involved in all possible types of ecological interaction, and many of them have profound effects on soil chemistry, structure, and function. Many soil organisms have not yet been classified or characterized.

Bacteria are the most abundant soil organisms; some of them are parasites of plants and other soil organisms, but many are involved in decomposition and recycling of nutrients in soil organic matter. Some have the ability to fix atmospheric nitrogen.

Fungi may also be parasites, but most of the fungi in the soil are saprophytes, breaking down soil organic matter. The roots of nearly all plants in the wild are associated with **mycorrhiza** which are fungi that assist in plant uptake of nutrients and water.

Soil Protista include algae that can carry out photosynthesis at the soil surface and protozoa that are predators of soil bacteria. Soil animals include nematodes, mollusks (slugs and snails), annelids (including earthworms), and arthropods (mites, insects, millipedes, spiders, etc.). These groups can be subdivided into herbivores, microbivores, predators, and detritovores.

The nonliving organic components of the soil are residues of plants and animals. The amount and kind of these organisms are influenced by the climate. For example, the climate of an area determines the quantity of biomass produced on a site, which in turn influences the soil-forming processes.

Vegetation aids soil formation by supplying organic matter in the form of dying and decomposing plants. Grasses decompose into a different kind of organic residue than do trees. Also, the amount of organic matter varies according to the amount and type of vegetation (Fig. 5–3). Peat soils form where reeds, sphagnum moss, and grasses grow and where decomposition of the organic matter is minimized. Soils in arid regions normally contain low amounts of organic matter because of the limited growth of desert grasses, shrubs, and cacti.

The amount of organic matter in soil is influenced by the difference between the rates of accumulation and decomposition of organic material. In cases where decomposition rates are very high, the organic fraction

Figure 5–3
The organic matter produced from the decomposition of forest litter is different from that produced from prairie grasses. In this example, the forest is not so dense that it excludes grasses, and both types of vegetation are present and produce organic matter. Source: USDA Soil Conservation Service.

accumulates very slowly. Thus, the amount of organic matter remains low. Consider the case where temperatures are high and rainfall or irrigation keeps the soil moist. Under these conditions, the rate of decomposition nearly equals the rate of accumulation, and the organic matter content changes little over time. On the other hand, in cool areas or sometimes under anaerobic conditions, decomposition is inhibited and organic matter accumulates. Many factors affect the rates of accumulation or decomposition, all of which play a role in determining the amount of organic matter present in a given soil at a particular time and location.

The type of root system produced by the different plant species also influences soil formation. Dense

Figure 5–4
Prairie grasses such as these in Kansas produce a high level of organic matter in the topsoil.
Source: Kimberly Williams, Kansas State University.

fibrous root systems of grasses often lead to soils with a deep, organic rich A horizon, such as some of the Mollisols (prairie soils) formed in the central United States (Fig. 5–4) and in some areas of eastern Europe.

Topography

Topography influences drainage and runoff. Steep slopes are subject to erosion because water flows downhill quickly, with little percolating into the soil. Resulting erosion leads to the formation of shallower soils because the soil surface is removed almost as quickly as it is formed. In depressional areas, however, deposition of eroded material leads to formation of deeper soils. Gentle slopes that are heavily covered with vegetation slow the water flow and allow more time for water to percolate into the soil, permitting the development of a well-defined profile. The same is true even of flatter areas. Rapid surface runoff causes more erosion, and if vegetation is removed or absent, deep gullies can be cut into the gently sloping land (Fig. 5–5). Soil on very flat land is more subject to wind erosion than water erosion. The presence or absence of plant material influences wind erosion in the same way as water erosion.

Topography has a marked effect on climate. High altitudes mean lower soil and air temperatures, which influence the amount and type of vegetation. As you learned in the previous chapter, In dry air, the decrease is about 10°C for 1,000 m (or 5.5°F for 1,000 ft); in moist air, it is 6°C for 1,000 m (or 3°F for 1,000 ft). Thus, an increase in elevation changes the kind and type of vegetation. Grasses and deciduous trees grow at lower elevations, with coniferous evergreens at higher

Figure 5–5
Sloping topography accelerates rill erosion, which removes topsoil. Constant removal of topsoil exposes the parent material to weathering and more rapid soil formation. Source: USDA Natural Resources Conservation Services, *http://photogallery.nrcs.usda.gov/*

Figure 5–6
At the timberline, no vegetation grows because of cold temperatures and low humidity.
Source: Margaret McMahon, The Ohio State University.

Figure 5–7
A soil profile showing different horizons. Source: USDA Image Gallery, *http://www.ars.usda.gov/is/graphics/photos/*

elevations. Above the timber line, little or no vegetation exists because of the lower temperature and humidity (Fig. 5–6). The type of vegetation present can greatly influence the characteristics of the soil at different elevations.

Time

Hard-to-decompose rocks such as granite require millions of years to form parent material; softer rocks such as limestone require less time. Interactions between biological and chemical agents reacting with parent material over long periods of time differentiate the soil into horizons. Soils without well-developed horizons are classified as young soils, even though the parent material may have been present for a great many centuries. For example, some desert soils, constantly shifted and transported by wind, are young soils because they lack profile development. Also, silt is often moved along in rivers, deposited, and then moved again, leaving little chance for profile development. Such soils remain pedologically young no matter how many years they may have been in existence.

As they develop, mature soils differentiate into well-defined profiles consisting of three principal horizons (Fig. 5–7). The surface layer, the **A horizon**, varies in depth and contains most of the plant roots. This leached zone often lacks some of the important mineral nutrients, but it does contain the largest amount of organic matter. The organic matter makes the A horizon permeable and dark-colored, and the A horizon is normally the zone of greatest biological activity in the soil profile.

Below the A horizon is the **B horizon**, the **zone of accumulation**. Plant nutrients, silts, clays, and other materials from the upper layer are leached into and accumulate in this horizon. The color is generally lighter than that of the A horizon, and less organic matter is present (although the roots of deep-rooted plants do reach into the B horizon).

The **C horizon** consists of unweathered to slightly weathered material from which the A and B horizons are formed. It can also include accumulated calcium carbonate or other salts.

The factors affecting soil formation are obviously interrelated. Parent material affects, along with other factors, the capacity of the soil to support plant life, which in turn influences the kind of vegetation. Topography and temperature also influence vegetative growth. Temperature interacts with organic matter. Topography, temperature, and time influence parent rock conversion into parent material, each interacting to yield different soil formation and soil characteristics. Thus, each factor affects and is affected by all the others. It would be difficult to say that any one is more important than another in soil formation.

PHYSICAL PROPERTIES OF SOIL

Soil Texture

An important physical property of soil is its texture. **Soil texture** is defined as the percentage of sand, silt, and clay particles in a soil. Soil particles vary in size

Table 5–1
DIMENSIONS OF SOIL PARTICLE SIZE CLASSES

	Diameter of Particles		
Millimeters	2.0	0.05	0.002 or less
Inches	0.080	0.002	0.00008 or less
Gravel, stones	Sand	Silt	Clay
Particles visible with the naked eye		Particles visible under microscope	Particles visible under electron microscope

from coarse rock fragments (>2 mm) to those so small (<0.002 mm) than an electron microscope is needed to observe them (Table 5–1). To measure soil texture, the soil particles are separated into their respective sizes and the percentage in each size category is calculated. A textural classification is then made with the aid of a soil textural triangle (Fig. 5–8).

Soil texture influences many of the soil's properties related to crop production (Table 5–2). The distribution of different particle sizes determines the ability of soils to hold and transmit water. Soil with a high percentage of sand loses water quickly, retaining little for plant use. Plants growing in these soils will experience water deficits sooner after wetting than those grown in loam or clay soils. Texture also influences soil aeration. In soils largely composed of very fine clay particles, movement of both air and water can be limited. Plant roots need oxygen for respiration, and soils with low rates of gaseous diffusion

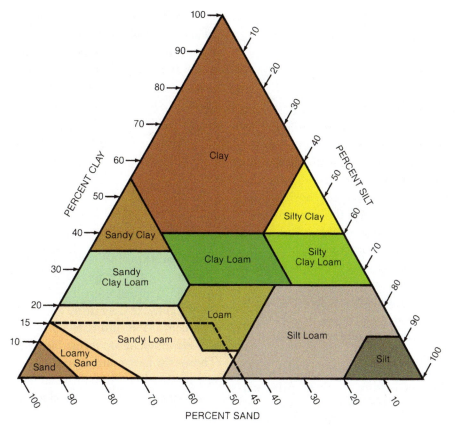

Figure 5–8
A soil textural triangle. To illustrate how the triangle works, assume that a sample of soil has been analyzed and found to contain 45 percent sand, 15 percent clay, and 40 percent silt. On the triangle, locate the 45 percent value on the sand (bottom) axis and draw a line parallel to the lines in the direction indicated by the arrow. Next, locate either the 15 percent value for clay on the clay axis (left side of triangle) or the 40 percent silt on the silt axis (right side of triangle), and draw another parallel line along either of the two axes in the direction indicated by the respective arrows. The two lines intersect in the loam area of the triangle; thus, this soil is classified as a loam. Source: USDA.

Table 5–2
SOME SOIL PROPERTIES INFLUENCED BY SOIL TEXTURE

Soil Property	Textural Class		
	Sand	*Silt Loam*	*Clay*
Aeration	Excellent	Good	Poor
Cation exchange	Low	Medium	High
Drainage	Excellent	Good	Poor
Erodibility*	Easy	Moderate	Difficult
Permeability*	Fast	Moderate	Slow
Temperature (spring)	Warms fast	Warms moderately	Warms slowly
Tillage	Easy	Moderate	Difficult
Water-holding capacity	Low	Moderate	High

*By water.

restrict respiration and plant growth. Also, many beneficial soil microorganisms require well-aerated soils.

A soil consisting of a mixture of 40 percent sand, 40 percent silt, and 20 percent clay produces a loam soil that retains sufficient water for good plant growth and permits its movement without restricting aeration. Due to their excellent water-holding properties, loams are usually excellent soils for crop production.

The ease of tillage (plowing, disking, cultivating) is influenced by soil texture. Sandy or loam soils at the proper moisture content are easier to till than clay soils. Root penetration is sometimes restricted in soils of high clay content. Other factors being equal, most crops grow better in loam soils than in either sandy or clay soils.

Soil Structure

Soil structure is defined as the arrangement of primary soil particles into secondary units, that is, the manner in which individual primary particles clump and hold together. The secondary unit (aggregate) is a clump of soil particles that acts as an individual larger particle with specific characteristics. The kind of soil structure is determined not only by the relative amounts of each primary particle but also by the manner in which these particles are arranged into aggregates. The size and form of aggregation is known as the structure of soil.

Descriptive words are used to classify soil structure, for example, prismatic, subangular blocky, blocky, columnar, platy, and granular. These words describe the shape, character, and appearance of the aggregates (Fig. 5–9). Aggregates may vary from a fraction of a centimeter to several centimeters in diameter and may be held together strongly or weakly.

CHEMICAL PROPERTIES OF SOIL

Effect of Climate

The type of parent material predominately influences the chemical characteristics of young soils. As weathering proceeds, soils tend to show the effects of climate, and the resemblance to parent material lessens or disappears. The chemical properties of the soil are determined largely by the colloid-sized (not visible with an ordinary microscope) aluminosilicate clay minerals.

In the temperate zones, chemical weathering is less intense in arid regions than it is in humid regions. Soluble salts released by weathering are not lost by leaching from soils in arid regions.

In tropical zones, with higher temperature and more rainfall than temperate regions, weathering and leaching are greater. The silicate and aluminosilicate minerals are more weathered, resulting in soils known as **Oxisols**. These soils contain high concentrations of iron and aluminum oxides, are generally red to reddish brown in color, and are low in fertility and organic matter. Even though large amounts of vegetation are produced, organic matter is low because dead plant matter is rapidly decomposed by high microbial activity.

In the cold humid regions, forest vegetation, mainly conifer trees, combine with climatic factors to produce **Spodosol** soil groups. The silica content of these soils is high in the surface layer in contrast to the iron and aluminum content in the Oxisols because the parent material of these soils is usually silica sand. These soils are highly leached and inherently low in plant nutrients. Organic compounds produced by decaying conifer needles form acid solutions that dissolve iron and aluminum oxides and basic compounds (calcium and magnesium salts). The ions are leached to

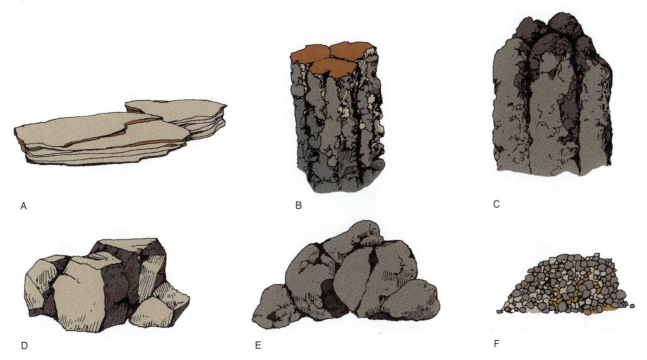

Figure 5–9

Several kinds of soil structure: (A) platy, (B) prismatic, (C) columnar, (D) angular block, (E) subangular blocky, and (F) granular.

lower depths in the soil profile and redeposited there together with some aluminum and dissolved organic matter to produce a dense layer 60 to 90 cm (2 to 3 ft) below the surface. These soils can be identified by an ashy, bleached white layer immediately above the dense layer. This layer is known as the **E horizon**, and it forms between the A and B horizons.

Soil Acidity and Alkalinity

The acidity or alkalinity of the soil is expressed as its pH. pH is defined as the negative logarithm of H^+ activity in the soil solution:

$$pH = -\log[H^+]$$

It does not include any H^+ adsorbed to the CEC or otherwise not dissolved in solution. pH is not a fixed characteristic of soil and, depending on a number of conditions, varies over time. Values for pH vary considerably among soils ranging from about 4.0 for an acid soil to 10.0 for an alkaline soil (Fig. 5–10).

Most plants do not grow well in highly acid or highly alkaline soils. In general, most plants will grow in the pH range of 5 to 7. However, most of these plants have a narrower range of optimum pH (Table 5–3).

The availability of nutrients to plants is influenced by soil pH. Some nutrients are available in greatest amounts at one pH while others are most available at a different pH (Fig. 5–11). Managing soil pH is an important strategy in fertility management as you will learn in Chapter 14.

Soils in climates with high rainfall and humidity generally tend to be acid, while those found in arid climates tend to be alkaline. In wet climates, the base elements (sodium, potassium, calcium, and

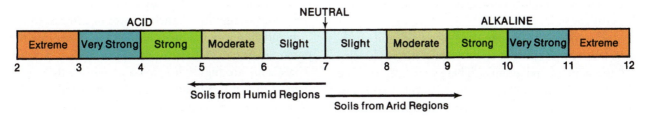

Figure 5–10

The pH for many mineral soils ranges from about 4 to 10. The pH range for most agricultural soils lies between 5 and 8.5.

Table 5-3
SOIL pH RANGE FOR OPTIMUM GROWTH OF SEVERAL CULTIVATED PLANTS

4.5 to 5.5	5.5 to 6.5	6.5 to 7.5
Azalea	Barley	Alfalfa
Bent grass	Bean (snap, lima)	Apple
Blueberry	Carrot	Asparagus
Camellia	Chrysanthemum	Beets (sugar, table)
Chicory	Corn (field, sweet)	Broccoli
Cranberry	Cucumber	Cabbage
Dandelion	Eggplant	Cauliflower
Endive	Fescue	Celery
Fennel	Garlic	Chard
Fescue	Oats	Hydrangea
Gardenia	Peas	(sepals become
Hydrangea	Pepper	pink)
(sepals become	Poinsettia	Leek
blue)	Pumpkin	Lettuce (head, Cos)
Potato	Radish	Muskmelon
Poverty grass	Rye	Onion
Red top	Squash	Parsnip
Rhododendron	Strawberry	Soybean
Rhubarb	Timothy	Spinach
Shallot	Tobacco	Sweet clover
Sorrel	Tomato	
Sweet potato	Turnip	
Watermelon	Wheat	

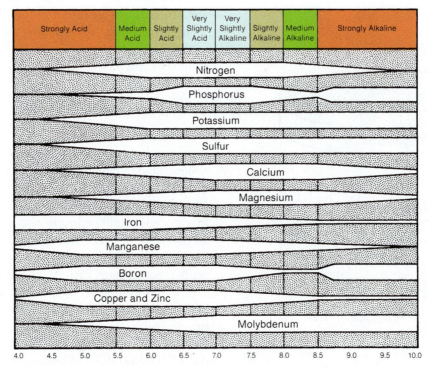

Figure 5-11
Availability of some essential nutrient elements as influenced by soil acidity or alkalinity.

magnesium) are removed from the soil by leaching as well as by the harvested crops that have absorbed them. As the base elements are lost, the exchange sites on the clay colloids become occupied with hydrogen ions, making the soil acid.

Normally, sudden and large changes in soil pH do not occur. Change is generally gradual, especially with fine-textured soils. These and other soils resist such change due to **buffering**.

Cation Exchange Capacity

An important property of clay and of the organic humus fraction of the soil is its ability to attract and hold cations—positively charged ions, some of which are essential plant nutrients (for example: NH_4^+, Ca^+, K^+). Clay colloids carry thousands of negative charges throughout the clay particle and at the broken edges of the clay's layers. Thus, a clay colloid acts as a large, highly negatively charged particle (anion) (Fig. 5–12). A soil's capacity to hold cations is called its **cation exchange capacity (CEC)**. It is called exchange capacity because cations are held very loosely by the colloids, and can be replaced, or "exchanged" with cations in the soil water (often called the **soil solution**). Ions on the CEC are held tightly enough, however, to prevent their leaching in percolating water. Organic matter has a greater net negativity than clay and has a higher CEC.

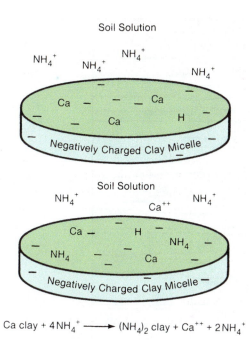

Soil Solution

Ca clay $+ 4NH_4^+$ ⟶ $(NH_4)_2$ clay $+ Ca^{++} + 2NH_4^+$

Figure 5–12

A soil colloid such as clay that is made up of numerous negatively charged micelles will attract many positively charged nutrient elements giving that soil a high cation exchange capacity (CEC).

Not all cations are attracted to or held by the CEC with equal energy. The strength of attraction for some cations when present in equivalent amounts is:

$$\text{calcium } (Ca^{++}) > \text{magnesium}$$
$$(Mg^{++}) \text{ potassium } (K^+) > \text{ammonium } (NH_4^+)$$
$$> \text{sodium}(Na^+) > \text{hydrogen}(H^+)$$

Most plant nutrients are cations and can be held by soil particles. The most notable exception is nitrate (NO_3^-), which is not held and is readily leached out of the root zone. Cation exchange capacity is influenced somewhat by soil acidity: it is greater under alkaline than under acidic conditions. The difference is slight in mineral soils, but it can be substantial in soils containing large quantities of organic matter.

Saline and Sodic Soils

Saline soils contain unusually large quantities of soluble salts. The soluble salts are typically chlorides and sulfates of calcium (Ca^{++}), magnesium (Mg^{++}), and sodium (Na^+), although other soluble cationic salts may contribute as well. Sodic soils differ from saline soils because a large percentage (over 15 percent) of the total cation exchange sites of the soil are occupied specifically by sodium ions (Table 5–4). Displacement of the sodium (Na^+) ion is the main objective in reclamation of a sodic soil. Agriculturally, saline and sodic soils are problem soils that require special handling for successful farming. Excessive amounts of soluble salts are harmful to plants and, when cations are predominantly monovalent (with a single charge), they have adverse effects on soil structure. Soils can be classified on the basis of the kind and amount of salts present, as shown in Table 5–4.

SOIL ORGANISMS

A microscopic examination of a soil sample reveals a wide variety of animal and plant life, some beneficial and essential to human well-being, and some harmful, often causing problems for people, their livestock, or crop plants. The animals include earthworms, gophers, insects, mice, millipedes, mites, moles, nematodes, slugs, snails, sowbugs, and spiders. Plants, plant-like organisms, and microorganisms in the soil include actinomycetes, algae, bacteria, and fungi.

Soil organisms act both chemically and physically on the soil. They digest plant residues and other organic matter enzymatically, and may physically move the residues from one place to another, mixing it with the soil. Earthworms and burrowing animals mix large quantities of material with the soil mass. The kind and amount of soil organisms depend on several factors,

Table 5–4
CHARACTERISTICS OF SALINE AND SODIC SOILS*

Soil	Electrical Conductivity	Exchangeable Sodium	pH
Saline	>4 mmhos/cm†	<15%	<8.5
Saline-sodic	>4 mmhos/cm	>15%	<8.5
Nonsaline-sodic	<4 mmhos/cm	>15%	>8.5

*Specific thresholds are approximate because many factors affect the measured characteristics.

†mmhos = millimhos = 1/1,000 mho. The mho is the reciprocal of ohm, a unit of electrical resistance.

including climate, vegetation, soil pH, fertility level, soil temperature, and soil moisture.

The roots of higher plants are a good source of organic matter. Their decomposition by soil organisms produces organic acids and gluelike materials that bind soil particles together to form the aggregates necessary for good soil structure. Following the decomposition of roots, open channels are left in the soil, improving drainage and aeration. Organisms also decompose stems, leaves, and other crop residues.

Certain beneficial fungi called mycorrhiza live in association with plant roots. These fungi help the plants take in water and certain mineral nutrients such as phosphorus. In return, the fungus receives carbohydrates from the plant. Such a relationship between two dissimilar organisms living together for mutual benefit is called **symbiosis**.

Nitrogen fixation, sulfur oxidation, and nitrification are processes carried on by soil bacteria essential to higher plants. Many legumes develop nodules on their roots that contain ***Rhizobium***, a nitrogen-fixing bacteria (Fig. 5–13) that has a symbiotic relationship with the plant. As the legume grows, the bacteria convert unavailable atmospheric nitrogen (N_2) into nitrogenous compounds that the legumes use while the plant furnishes energy to the bacteria. When these legumes die and decomposes the nitrogen compounds become a source of nitrogen for succeeding plants.

Elemental sulfur is not immediately available to higher plants; it must first be oxidized to the sulfate form. Autotrophic *Thiobacillus* bacteria bring about this transformation through a complicated series of reactions. Under certain conditions, autotrophic bacteria oxidize iron and manganese to compounds that are less soluble and thus less available to plants. The action of these bacteria helps prevent toxic amounts of iron and manganese from being taken up by the plants.

Figure 5–13
Nodules of Rhizobium *bacteria on a legume (soybean).*
Source: Margaret McMahon, The Ohio State University.

Not all soil organisms are beneficial. Some of the most injurious plant pests are soil borne. For example, nematodes attack and destroy plants in a wide range of species. *Phylloxera,* an aphid that attacks grape roots, devastated large vineyard areas until resistant rootstock cultivars were developed. Pathogenic soil-borne bacteria and fungi are also responsible for significant crop losses.

SOIL ORGANIC MATTER

The soil's organic matter content has a profound effect on its biological, chemical, and physical properties. Through the decomposition (composting) of organic matter, many nutrients become available to crop plants. Organic matter provides food and energy for soil organisms. Most organic matter, except for a small animal fraction, comes from plants. By weight, about

90 percent is made up of carbon, hydrogen, and oxygen. The remainder is usually nitrogen, sulfur, phosphorus, potassium, calcium, and magnesium plus a minute amount of microelements.

The speed of organic matter decomposition varies according to its chemical composition. It is rapid for simple carbohydrates and slow for fats and lignins. Essentially the decomposition reaction is the oxidation of carbon compounds to carbon dioxide, water, and energy:

$$(CH_2O)_n + O_2 \rightarrow CO_2 + H_2O + energy$$

Proteins, fats, and other complex compounds decompose in a multitude of reactions to form amino acids, ammonia, nitrates, phosphates, carbon dioxide, and other compounds. After complete decomposition, a complex, amorphous, colloidal substance called **humus** remains that is resistant to further decomposition. This is the material that helps improve soil structure, imparts the dark color to the soil mineral fraction, and increases the soil's water-holding and cation exchange capabilities. For example, the cation exchange capacity of a mineral soil ranges from about 10 to 50 cmol kg^{-1}, while the capacity of humus ranges from about 100 to 300 cmol kg^{-1}. Being colloidal in size, humus acts similarly to clay colloids in cation exchange reactions, but it is composed chiefly of carbon, hydrogen, and oxygen with small amounts of other elements, while clay colloids largely consist of aluminum, silicon, and oxygen.

Carbon:Nitrogen Ratio

Under natural conditions, there is a close relationship between the amount of carbon and the amount of nitrogen in the soil. This carbon:nitrogen (C:N) ratio is nearly constant at about 12 parts of carbon to one part nitrogen, worldwide. Variations when present seem to correlate with climate, especially temperature and rainfall. For instance, the C:N ratio tends to be smaller (less carbon, more nitrogen) in arid and warmer regions than in humid and cooler regions. For a period of time in an area that has had a high input of organic material with a high C:N, nitrogen is tied up by the organisms that are causing the decomposition (Fig. 5–14).

Eventually, the organic residues decompose to the level where the soil organisms exhaust their supply of food and begin to die. The nitrogen from the decomposing organisms is then returned to the soil and made available to plants again at about the original nitrogen level.

SOIL DEGRADATION

Natural soils are not static entities; their properties change constantly, but the characteristics of the soil are usually maintained when these changes occur in response to the slow changes that normally occur in the natural environment. All upland soils experience small episodes of erosion yearly, but the severity of that erosion is usually minimized by the protection of permanent vegetative cover. However, when that cover is removed by something such as a fire or flooding, the soil surface is exposed directly to the erosive forces of wind, rainfall, and runoff. Accelerated erosion removes soil faster than new soil can be formed, reducing the depth of the productive topsoil. If erosion continues unabated, eventually all of the rich A horizon can be lost, forcing plants to root wholly in the undesirable B horizon. Even if the entire A horizon is not lost,

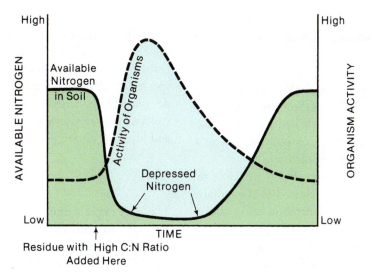

Figure 5–14

The relationship between the available nitrogen in a soil and the activity of soil microorganisms after a heavy application of organic matter with a high C:N ratio.

erosion preferentially removes smaller particles (including organic matter) first, degrading topsoil productivity long before the topsoil itself is lost completely.

Loss of soil organic matter is another facet of soil degradation. The amount of organic matter in a soil profile is dictated by the balance between additions (mainly plant residues) and losses (mainly as carbon dioxide during respiration). Additions are regulated by the amount of residue the ecosystem supplies on a continuing basis, while losses are limited by the usually low concentrations of oxygen in the soil air.

As you have just learned, soils are highly variable and many factors interact to give them their characteristics. In the cultivation of plants, growers use many techniques to manipulate soil characteristics in order to influence plant growth. These techniques and their affects on soil will be discussed in detail in Chapter 14 along with irrigation and soil fertility management.

SUMMARY AND REVIEW

Soils form a complex ecosystem composed of biological organisms, living and dead; inorganic materials; water; and air. The interaction over time of many factors including parent material, physical and chemical weathering, types of organisms present, the topography of the area, and other climate factors influence the formation and properties of soil.

Physical soil properties include texture (percentage of sand, silt, and clay particles) and structure (the arrangement of the primary particles). Soils with good structure have enough pore (air) spaces that are large enough to transmit water and air without restriction and sufficient smaller pores that retain some water against the pull of gravity. Chemical soil properties include cation exchange capacity (CEC), soil acidity and alkalinity, and salinity. Cation exchange capacity is the ability of the soil to exchange cations it already holds with cations in the soil water. This factor is very important in fertility management.

Soil acidity and alkalinity affect the availability of nutritional elements to the plant. Saline soils naturally contain large quantities of soluble salts. These salts may be harmful to plants if they are present in great enough quantities.

Soil organisms play a crucial role in determining soil characteristics. They create pore space by their tunneling activities. They add organic material when they die and decay. They often form beneficial and sometimes even necessary symbiotic relationships with the roots of plants growing in the soil. On the other hand, plant pathogen and herbivore organisms that can infest or feed on roots can also be a part of the soil ecosystem.

Soil degradation occurs when erosion removes the productive A horizon, making it difficult for plants to grow well. It also occurs when the deposition or organic material becomes less than its decomposition resulting in an overall loss of organic material.

KNOWLEDGE CHECK

1. What are the five basic components of soil?
2. How does pore space relate to a productive soil?
3. What is field capacity of a soil?
4. What is parent material?
5. What is a soil horizon? What is a soil profile?
6. What is physical weathering?
7. List the four processes of chemical weathering.
8. Why are soils that are formed in regions of very high rainfall or very low rainfall usually not productive?
9. How do plant roots add to the organic material in the soil?
10. What role do bacteria and fungi play in soil formation?
11. What are mycorrizha and how do they help plants?
12. What kind of topography promotes water erosion? Wind erosion?
13. How do plants influence erosion?
14. Even though some deserts have been in existence for thousands of years, why is the soil there considered to be very young?
15. In which soil horizon are the most roots and biological activity found?
16. What is soil texture and what are the components that contribute to a soil's texture?
17. What is soil structure?
18. What are the differences between mollisols, spodosols, and oxisols?

19. What is soil pH and what is the pH range that most plants can grow in?
20. How does cation exchange capacity (CEC) influence a soil's ability to hold the positively charged plant nutrients such?
21. What are saline and sodic soils?
22. What are the ways that soil organisms act on the soil?
23. What do plants get from the mycorrhiza relationship? What do legumes get from *Rhizobium*? What do the mycorrhizia fungi and the *Rhizobium* get in return?
24. What does the decomposition of organic matter provide for plants?
25. What is humus and how does it affect soil characteristics?
26. What is C:N in terms of a soil characteristic?
27. Along with erosion, what is another factor that contributes to soil degradation?

FOOD FOR THOUGHT

1. What would you say to someone who says we should not worry about global warming because we can just start growing our grain belt crops (growing in prairie soil) farther north in the humid regions where the conifer forests now grow in Alaska?
2. Mycorrhiza are now available for some crop plants. Do you think it would be worth it to a grower to purchase these?
3. During difficult economic times, many municipalities cut back on public services. What would you tell a city council that is considering stopping the pickup of yard waste that is used to create organic compost to improve soil in the city parks?

FURTHER EXPLORATION

ASHMAN, M. R., AND P. GEETA. 2002. *Essential soil science: A clear and concise introduction to soil science.* Oxford, UK: Blackwell Science.

COYNE, M. S., AND J. A. THOMPSON. 2005. *Fundamental soil science.* Clifton Park, NY: Thomson Delmar Learning.

unit two

Plant Structure, Chemistry, Growth, Development, Genetics, Biodiversity, and Processes

✳6

Structure of Higher Plants

key learning concepts

After reading this chapter, you should be able to:

- Define the terminology that describes plant cells, tissues, and organs.
- Explain the basic functions of plant cells, tissues, and organs.
- Explain how some of the practices we use to grow plants are directed at specific tissues and organs.

Every day we can identify familiar plants in our immediate surroundings. Some plants and their features can be identified and appreciated from their external structure, but their internal structure and function are often overlooked. The beauty of an orchid blossom is greatly admired, but just as impressive are the parts of a cell as recorded with a scanning electron microscope. The purpose of this chapter is to develop an understanding of the internal and external structures of the higher plants.

An approach that capitalizes on what is already known is to follow a plant from seed germination to full size and then to observe the formation of fruits and seeds. We can then appreciate how the plant grows and develops and, at the same time, acquire the vocabulary necessary to understand the growth processes of plants. This approach allows us to study both the external form of the plant, or its **morphology**, and its internal structure, or anatomy and **histology** (microscopic features).

Our major food, fiber, wood, and ornamental plants belong to two main classes—the **gymnosperms**, represented mainly by the narrow-leaved, evergreen trees; and the **angiosperms**, usually broad-leaved, flowering plants. In the temperate zone, angiosperms are by far the most common in everyday life. Angiosperms are divided into two subclasses: the **monocotyledons**, which have an embryo with one cotyledon, and the **dicotyledons**, which have an embryo with two cotyledons. These names are often shortened to **monocot** and **dicot**. We begin by examining the life cycle of a common monocot, corn (Fig. 6–1), and a common dicot, the bean.

THE LIFE CYCLE OF A CORN PLANT (A MONOCOT)

When a corn seed is planted in moist soil, it imbibes (absorbs) water from the soil. Germination begins with the emergence of the **radicle** (the primary root) and the **plumule** (the primary shoot). These two enlarging axes form the primary body of the plant.

The radicle grows downward through a protective sheath, the **coleorhiza**, from which the primary root develops and the secondary roots branch. A mature corn plant can develop roots 2 m (6.1 ft) long. **Adventitious roots** (roots other than those that develop from the radicle) grow from the shoot axis just at or above the soil surface (Fig. 6–2): These roots, also called anchor, brace, or prop roots, branch out in the soil to give added support to the plant.

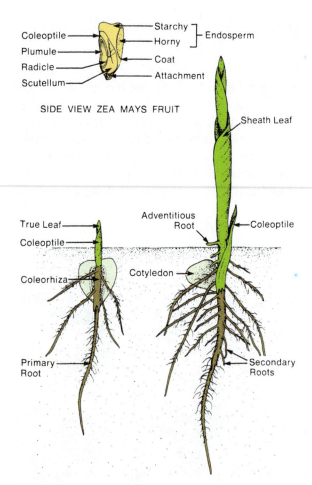

The emerging plumule is protected by a sheath-like leaf, the **coleoptile**, that envelops the main stem as it grows upward through the soil. As the true foliage leaves develop, the main stem continues to produce sheathing leaves that encircle the stems at each node.

When the corn plant has reached a given size, producing a set number of leaves, female flowers, known as **pistillate flowers** or ears, appear at the base (axil) of one or more sheath leaves. Later, the male flowers, known as **staminate flowers** or tassels, develop at the top of the plant. Fig. 6–3 shows both

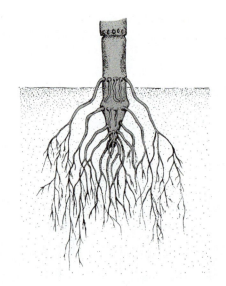

Figure 6–1
Structure of the seed and seedling of corn (Zea mays), a monocotyledonous plant.

Figure 6–2
Adventitious or anchor roots develop above the soil line on the lower stem of a corn plant. These structures add support to the plant.

Figure 6–3
(A) A pollen-bearing corn tassel (staminate flower). (B) The ear (pistillate flower), showing the "silks" that intercept the wind-blown pollen grains. (C) Each silk is attached to a single grain of corn. Source: A & B: USDA, http://www.ars.usda.gov/is/graphics/photos/;
C: Margaret McMahon, The Ohio State University.

kinds of flowers. Blown by the wind, pollen grains from the tassels fall on and pollinate the long pistillate filaments (silks) and subsequently fertilize the ovaries, which become the individual corn kernels borne on a stalk (cob). Each ovary develops into a fruit, called a **caryopsis**, which encloses the true seed. After the kernels mature and dry, the fruits (containing the seeds) are harvested and stored over the winter. The seeds can be sown when weather conditions are favorable for germination, and the life cycle repeats itself.

THE LIFE CYCLE OF A BEAN PLANT (A DICOT)

After a bean seed has been sown in moist soil, it imbibes water and swells. The seed coat bursts and the radicle emerges (Fig. 6–4). The radicle grows downward and the hook of the bean, known as the **hypocotyl**, emerges above the soil, carrying the two cotyledons with it. Between the cotyledons lies a growing point (**apical** or **shoot meristem**) flanked by two opposite primary foliage leaves. The stem region just above the cotyledons

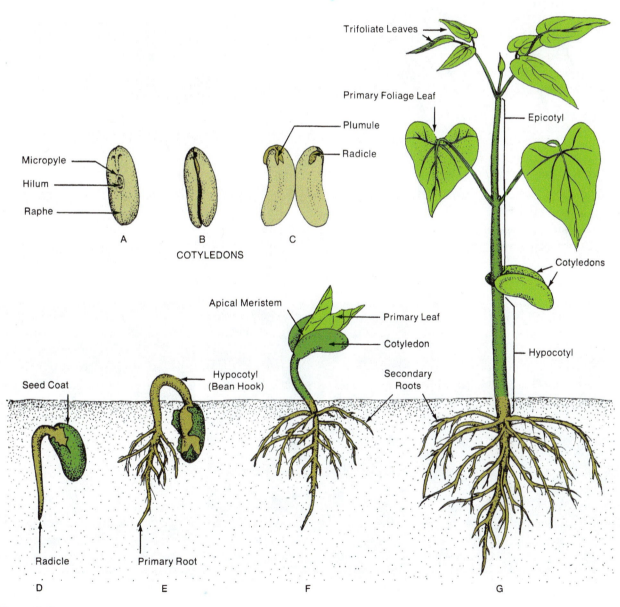

Figure 6–4
Structure of the seed and seedling—in several growth stages—of the bean (Phaseolus vulgaris), a dicotyledonous plant.

and the first trifoliate leaves is called the **epicotyl** (Fig. 6–4). Under favorable conditions the shoot apical meristem rapidly produces two trifoliate leaves opposite each other on the stem. The cotyledons, which have been supplying much of the reserve food for this initial growth, shrivel and abscise (drop off). The plant's green leaves are now capable of manufacturing food for future growth of the seedling. The bean plant produces trifoliate leaves, and flowers begin to develop in the axils of about the fourth set of leaves and in each succeeding set. These flowers are self-pollinated; thus, fruits (pods) develop as long as environmental conditions are favorable. The seeds mature and dry within the pod, and they can be sown at once to produce another generation of bean plants. The difference in emergence of the growing points of beans and corn from beneath the soil affects the tolerance of each crop to light frosts. A late frost would be more likely to severely damage a newly emerged bean seedling than a newly emerged corn seedling because the growing point of corn is below the soil and protected. The leaves of the corn plant may be damaged, but the growing point will survive to generate new leaves. However, the bean plant's growing point would be severely injured or killed and unable to generate new growth.

The life cycles of plants like the corn and bean are more or less familiar to most of us from our own observations, but to enlarge our knowledge about higher plants, we must consider the largely unfamiliar areas of plant anatomy and histology.

THE CELL

The plant **cell** is the basic structural and physiological unit of plants, in which most reactions characteristic of life occur. The tissues of the plant develop through an orderly process of cell division and differentiation. **Cytology** is the branch of biology involved in the study of the components of cells and their functions.

Cells vary greatly in size. The smallest must be measured in micrometers (1/1,000 of a millimeter), but some wood fiber cells are several centimeters long.

Early cytological studies were conducted with light microscopes, which can demonstrate general cellular features but which cannot resolve all the fine details within cells. Electron microscopes and enhanced light microscopes have revealed that living cells are not empty chambers but highly organized complexes of subcellular compartments with specialized metabolic functions. In the living cell, these complexes are distributed through a dynamic and orderly flow of materials within the **cytoplasm**.

CELL STRUCTURE

There are two types of cells. **Prokaryotic cells** have no separate subcellular units; for example, nuclear material is not enclosed in a membrane. These cells, considered primitive, are found in bacteria and blue-green algae. **Eukaryotic cells** are made up of compartments bounded by membranes, with specialized structures and functions. These units, called **organelles**, include the **nucleus**, **mitochondria**, **plastids**, microbodies, **vacuoles**, dictyosomes, and **endoplasmic reticulum** (Fig. 6–5). Plant cells are eukaryotic cells.

The Protoplast

The organelles of the plant cell are contained within a membrane-bounded **protoplast**, which in turn is encased within a cell wall. The major features of the protoplast are the outer membrane or plasma membrane, the cytoplasm, the nucleus, and the vacuole.

Plasma Membrane   The plasma membrane, also called the **plasmalemma**, is a lipid bilayer surrounding the cytoplasm. This membrane is important in maintaining a surface area for selective absorption and secretion by the cell, and plays a role in generating energy as well. Proteins embedded within the bilayer can function as enzymes; as surface receptors, both on the inside surface and in communication with the cell environment; or as channels for the uptake and efflux of ions.

Cytoplasm   The **cytoplasm** is a viscous fluid composed of matrix proteins, bounded by the semipermeable plasma membrane. The flow of organelles within the cytoplasmic matrix, called *cytoplasmic streaming*, is clearly visible in active leaf cells under a light microscope. Also within the cytoplasm is a very important network of membranes, the **endoplasmic reticulum (ER)**. **Proteins** are synthesized on the surfaces of the ER throughout the cell, on small discrete structures called **ribosomes**. Proteins may be further processed inside the ER and transported to destinations that will be sites of activity within the cell.

Plastids of several types are located within the cytoplasm. The colorless leucoplasts serve as storage bodies for oil, starch, and proteins. Chromoplasts contain the various plant pigments, including **chlorophyll**. Chromoplasts with chlorophyll are called **chloroplasts** and are responsible for photosynthesis in leaves and in some stems. Enclosed by a double membrane, most chloroplasts also contain other **pigments**, large quantities of proteins and **lipids**, and some stored **starch**. Within the chloroplast, light energy is first harvested by

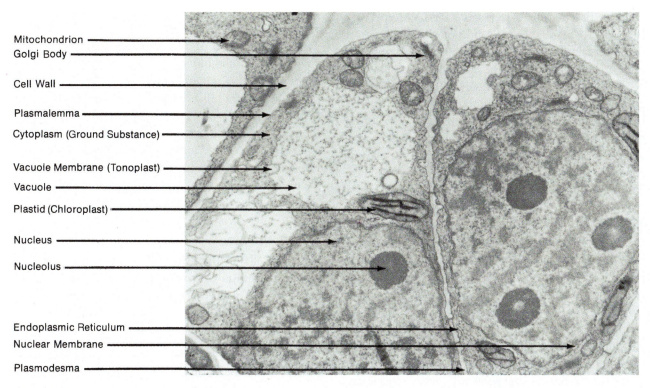

Mitochondrion
Golgi Body
Cell Wall
Plasmalemma
Cytoplasm (Ground Substance)
Vacuole Membrane (Tonoplast)
Vacuole
Plastid (Chloroplast)
Nucleus
Nucleolus
Endoplasmic Reticulum
Nuclear Membrane
Plasmodesma

Figure 6–5
Photomicrograph of a plant cell showing the various parts and organelles, × 6,000. Source: Keith Weinstock.

pigments bound to stacked membranes called grana and then converted into chemical energy in the form of sugars.

Mitochondria are cytoplasmic bodies that are smaller than plastids. Like the chloroplasts, they are surrounded by a double membrane, and contain a specialized inner membrane system. The mitochondria are sites of respiration and are also involved in protein synthesis. They produce energy-rich compounds such as **adenosine triphosphate (ATP)**.

The Nucleus

The **nucleus** is a prominent organelle within the cell, enclosed by a double membrane and containing one or more bodies called nucleoli. Within the nucleus are the **chromosomes**, long lengths of **deoxyribonucleic acid (DNA)** and associated proteins that contain the genetic information coding for all cell functions, for differentiation of the organism, and for reproduction. During cell division, the chromosomes replicate and a set of chromosomes is passed on to each of the daughter cells (see Chapter 14), thus ensuring continuity of genetic information from old to new cells. During reproduction the new cell receives half of its chromosomes from the male parent and half from the female parent, resulting in

genetic segregation. **Genetic codes** are transcribed from the DNA in the nucleus and translated into proteins on the **ribosomes**.

DNA is also found outside the nucleus in the mitochondria and in the chloroplasts, thereby giving these bodies a role in heredity independent of the nucleus. Unlike nuclear DNA, mitochondrial DNA and chloroplastic DNA are inherited only from the female parent. This has proven useful in determining the relationship between species or individuals.

Vacuoles

Vacuoles may occupy a major portion of the interior of plant cells. In actively dividing cells, vacuoles are very small, but they can account for up to 90 percent of the volume of mature cells. Vacuoles serve as a storage reserve for water and salts, as well as for toxic products. They contain a watery solution of dissolved materials, including inorganic salts, blue or red pigments (**anthocyanins**), sugars, organic acids, and various inclusions of crystals. The membrane surrounding the vacuole is called the **tonoplast**, and it serves an important role by controlling the flow of water and dissolved materials into and out of the vacuole, maintaining cell turgor, and other functions.

The Cell Wall

The **cell wall** protects the protoplast, provides an external structure, and in some tissues (e.g., bark, wood) may act as a strong support for the plant. The cell wall is nonliving, made up of **cellulose**, pectic substances, and lignins. Between cells lies an intercellular layer called the **middle lamella**, which contains many of the mucilaginous pectic compounds that hold adjacent cell walls together. Adjacent to the middle lamella is the primary wall, which is composed mostly of cellulose. This elastic but strong material is the chief constituent of most plant cell walls.

The secondary wall layer, which lies within the primary wall and is laid down only after the primary wall is complete, is usually thicker than the primary wall when fully developed. The secondary wall is also composed of cellulose, but in some cells and tissues it may contain **lignins**, **suberins**, or **cutins**. Lignins are closely associated with the cellulose and give it added strength, as is well demonstrated by wood fibers. Large quantities of water are contained and transferred in cellulosic walls, which act as wicks. In some specialized cells (for example, cork), water loss or flow is prevented by the presence of the waxy material, suberin, in the walls.

Individual cells in a tissue are connected to one another via strands of cytoplasmic material, called **plasmodesmata**, which extend through the plasma membrane. The surrounding cell wall forms channels around the plasmodesmata, called **pits**. As the wall grows thicker, these pits are preserved and can become quite long, clearly visible in the light microscope. Water and dissolved materials can move from cell to cell through these connections.

PLANT TISSUES

Large tracts of organized cells of similar structure that perform a collective function are referred to as **tissues**. Tissues of various types combine to form complex plant organs such as leaves, flowers, fruits, stems, and roots. Roots, stems, and leaves make up the vegetative parts or the plant. Flowers, fruits, and seeds are the reproductive parts.

In all plants, both young and mature, two basic kinds of tissues can be distinguished. One kind is the **meristem**, or **meristematic tissue**, which is comprised of actively dividing cells that develop and differentiate into yet other tissues and organs. Cells in the meristematic tissues have thin walls and dense protoplasts. Meristematic tissues are found in the root and shoot tips, just above the nodes (intercalary meristems) and in woody perennials, as cylinders in the shoots and roots (the **cambium layer**).

The second kind of tissue is that which develops from the meristems and has differentiated fully. This is the permanent tissue, of which there are two kinds: the simple, which includes the epidermis, parenchyma, schlerenchyma, and collenchyma; and the complex, which includes the xylem and phloem.

Meristematic Tissues

The common categories of meristematic tissues are:

Apical meristems

Shoot

Root

Subapical meristems

Intercalary meristems

Lateral meristems

Vascular cambium

Cork cambium

Apical Meristems Shoot meristems, frequently referred to as shoot apical meristems, are the termini of the above-ground portions of the plant (Fig. 6–6). They are responsible for producing new buds and leaves in a uniform pattern at the terminus of the stem and laterally along stems. The pattern of leaves and lateral buds that form from the shoot meristems vary with the species of plant. For example, in the maples (*Acer*), ashes (*Fraxinus*), and in members of the mint (*Amiaceae*) and olive (*Oleaceae*) families, the leaves and buds are opposite—at a 180° angle—to one another. On the other hand, in the oaks (*Quercus*) and walnuts (*Juglans*), for example, the leaves and buds alternate from one side of the stem to the other, and in the pines (*Pinus*), they form a spiral pattern.

The shoot apical meristem produces epidermis, cortex, primary xylem and phloem, and the central pith, tissues that form the primary structure of the stem. The shoot apex may eventually develop terminal inflorescences (floral groupings) instead of continuing to produce leaves and lateral buds as, for example, in the chrysanthemum, poinsettia, and sunflower.

Some shoot meristems always remain vegetative and continue to produce leaves and lateral buds, as in many vines such as the grape. In such cases, flowers or inflorescences are borne in the axils of lateral leaves somewhat behind the terminal growing point. Many trees also have this growth pattern; the dominant meristem (central leader) enables the plant to grow upright and gain height. In this case, individual flowers

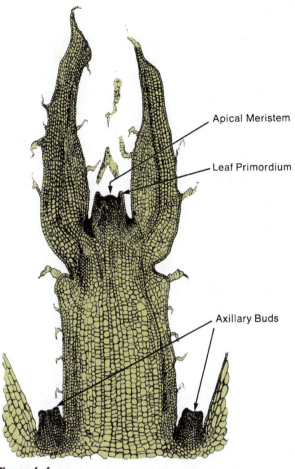

Figure 6–6

Longitudinal section of a shoot tip showing the apical meristem. Cell division in this region, along with cell elongation, is responsible for shoot growth. Buds in the axils of leaves also have meristematic regions at their tips that can develop into shoots. Should the apical meristem die or be broken off, an axillary bud can become the dominant shoot tip and continue shoot growth.

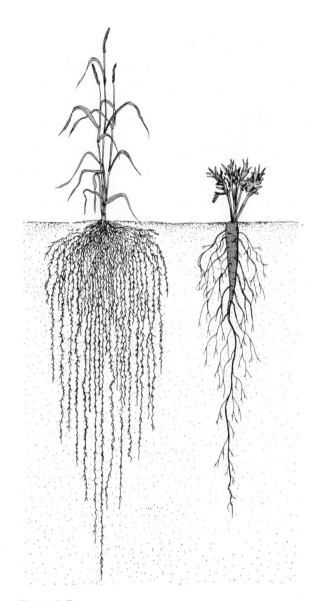

Figure 6–7

Two types of root systems. Left: Fibrous root system of a cereal plant. Right: Tap root system as developed by the carrot plant (Daucus carota).

or inflorescences are usually borne on side branches in the axils of leaves at some distance below the apex. These growth and flowering characteristics are known for most cultivated plants. Gardeners, orchardists, and foresters use this knowledge to prune trees and other plants to direct the growth in the manner that they desire.

Root meristems, located at the various termini of the roots, are the growing points for the root system. Some plants have a dominant **tap root**, which develops downward, together with limited lateral root growth (Fig. 6–7). Examples of plants with tap roots are carrots, beets, and turnips, all well-known root crops. Other species with tap roots include oaks, pecans, alfalfa, and cotton. Many plants, however, do not have a dominant tap root. Instead, the roots branch in many

directions creating a **fibrous root** system (Fig. 6–7). Examples are the grasses, grain crops, and many kinds of shallow-rooted trees.

The root meristem lies just behind the root cap, which protects the meristem as the root grows through the soil. These root cap cells are constantly being destroyed, but the apical root meristem produces more to replace them. The root meristem produces the primary tissues—such as protoderm, ground meristem, and procambium—that later become the epidermis, cortex, and vascular cylinder of the mature root.

Subapical Meristems   The **subapical meristem** produces new cells in the region a few micrometers behind an active shoot or apical meristem. The subapical meristematic region has long been thought of as a region where cells only elongate and expand. Cells do, indeed, expand in this region and thus increase internode length, thus adding to the growth in height of the plant. However, because new cells also form in this subapical region, it is a true meristem. The activity of the subapical meristem can be seen particularly in certain plants that lack tall stems when they are first producing leaves and that grow as a rosette. Examples are beets, carrots, China asters, lettuce, mustard, and turnips. These plants form rosettes of leaves on very short internodes. Later, when the shoot apical meristem initiates flowers, the stem below the flower elongates rapidly (bolts) because of the activity of the subapical meristem. During the period of fast growth, cells divide as well as elongate. The division and elongation together account for the rapid stem growth below the terminal flower buds.

Intercalary Meristems   The **intercalary meristems** are active tissues that have been separated from the apical meristem by regions of more mature or developed tissues. The separation occurs at an early stage of development, and the separated cells therefore retain their ability to divide. The best examples of intercalary meristems are found in monocots and especially in the grasses. The active meristematic cells just above the nodes in the lower region of the leaf sheath divide, and those cells develop (expand and elongate) rapidly. Grass leaves elongate from the base in a like manner which explains why grass leaf blades continue to grow after mowing even though the top has been cut off.

Lateral Meristems   The lateral meristems, which produce secondary growth, are cylinders of actively dividing cells starting somewhat below the apical or subapical meristems and continuing through the plant axis. These meristems are the **vascular cambium**—which produces new xylem (water and mineral conducting elements) and new phloem (photosynthate conducting elements)—and the **cork cambium**—which chiefly produces bark, the protective covering of old stems and roots (Fig. 6–8). Stem girth of woody perennial plants and trees increases mainly by the activity of these lateral meristems. The number of growth rings indicates the tree's age (Fig. 6–9). Measuring the width of the annual growth rings in the stems is one way to determine the rapidity of lateral growth of a tree. Long-lived trees, such as the redwood (*Sequoia*), and certain pines, such as the bristle cone pine (*Pinus aristata*), increase greatly

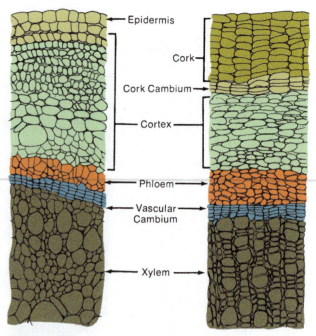

Figure 6–8
Left: Cross section of the stem of a young dicotyledonous plant showing the tissues of the meristematic region, the vascular cambium. The epidermis lies on the outside. The dividing cells of the vascular cambium layer, producing phloem to the outside and xylem to the inside, account for the thickening of the stem as it grows older. *Right:* Cross section of an older dicot stem where a cork cambium has developed. This meristematic layer produces the cork cells (bark), which protects the inner, more tender tissues.

in size by lateral growth, but short-lived summer plants (annuals) such as marigolds (*Tagetes*), tomatoes (*Lycopersicon*), and peppers (*Capsicum*) develop only a limited girth of the lower stem before the frost destroys them. Tomatoes, however, are perennials if they are grown in the frost-free tropics. Vascular and cork cambial growth turns the lower stem of the tomato into a small trunk if growth continues. In many plants, the apical meristem produces a plant hormone called auxin that suppresses the development of the dormant meristems in the leaf axils. Pinching (removing) the apical meristem is a practice growers often use to encourage lateral branch formation.

Permanent Tissues

Permanent tissues can be classified into simple and complex tissues. The simple tissues are uniform, composed of only one type of cell. Examples are epidermis, parenchyma, sclerenchyma, collenchyma, and cork. Complex tissues are mixed, containing different kinds of cells. Examples are xylem and phloem.

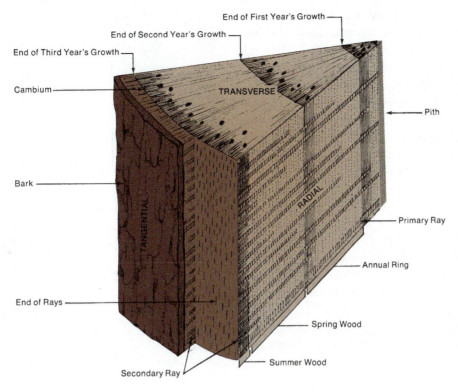

Figure 6–9

Section of a three-year-old stem of pine (Pinus) showing the annual rings by the end of the third summer. The porous, fast-growing spring wood is followed by the more dense, slower-growing summer wood.

Simple Tissues   The **epidermis** is a single exterior layer of cells that protects stems, leaves, flowers, and roots. The outside surface of epidermal cells is usually covered with a waxy substance called cutin, which reduces water loss. The epidermis of leaves is usually colorless except for the **guard cells** of the **stomata** (Fig. 6–10), which contain chlorophyll and are green. Some leaf epidermal cells are elongated into hairs and are called trichomes (Fig. 6–11). The root epidermis lacks cutin. It develops protuberances called root hairs, which actively absorb water from the soil.

Parenchyma tissue is made up of living thin-walled cells with large vacuoles and many flattened sides. This is the principal tissue of the cylindrical zone under the epidermis extending inward to the phloem in a region called the **cortex**. Parenchyma tissue, however, is not confined to stems but can be found in all plant parts. Parenchyma in leaves is active in photosynthesis. Parenchyma cells, when wounded, are capable of becoming meristematic and then proliferating to heal wounds and to regenerate other kinds of tissues.

Sclerenchyma tissue is composed of thick-walled cells found throughout the plant as fibers or **sclereids**. The protoplasts eventually die, leaving the empty hard cell walls behind. Sclerenchyma cells are common in

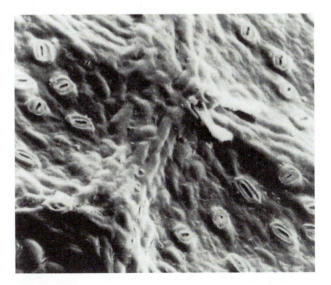

Figure 6–10

Stomates on the underside of a leaf. Source: Dr. Irving B. Sachs, Forest Products Laboratory, USDA Forest Service.

stems and bark and are also found as stone cells in pear fruits and walnut shells.

Collenchyma tissue gives support to young stems, petioles, and the veins of leaves. The walls and corners of the cells are thickened, primarily by cellulose, to provide reinforcement.

Cork tissue occurs commonly in the bark of maturing stems, the trunks of trees, and potato skins. The cell walls are waterproofed with a waxy material called suberin. Cork cells soon lose their protoplasts and die but continue to retain their structure and shape.

Complex Tissues **Xylem** is a structurally complex tissue that conducts water and dissolved minerals from the roots to all parts of the plant. The cells found in the xylem may be vessels, tracheids, fibers, and parenchyma. **Vessels** are long tubes made up of short vessel members (Fig. 6–12) that are joined end to end after the end walls of the cells have dissolved (Fig. 6–13). **Tracheids** are long, tapered, dead cells that conduct water through pits (Fig. 6–14). Tracheids contribute significant strength and support to the stems of gymnosperms. **Fibers** are thick-walled sclerenchyma cells that provide support. The parenchyma cells in xylem are arranged in vertical files (Fig. 6–15) and act as food storage sites. Not all these cell types occur in the xylem tissue of any one plant species; one or two are usually absent. Movement of water and minerals through the xylem is mostly through physical, not biological, processes.

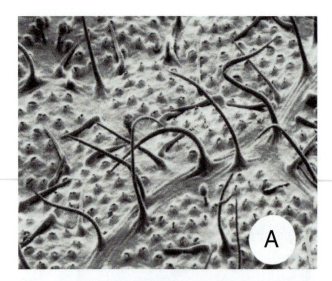

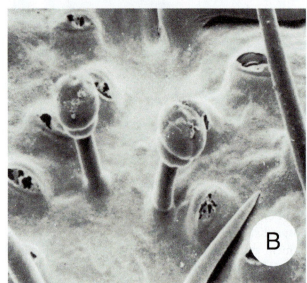

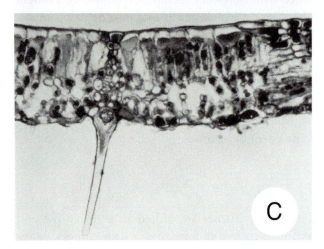

Figure 6–11
Leaf surfaces of Ulmus elata: *(A) Lower surface (× 110). (B) Lower surface stomata and trichomes (× 550). (C) Leaf lamina transection (× 230).* Source: R. E. Meyer and S. M. Meola, Morphological characteristics of leaves and stems of selected Texas woody plants. USDA Tech. Bull. 1564 (1978).

Figure 6–12
A section of the xylem of Quercus rubra, *red oak, showing the barrel-shaped vessel members. The pitting in the lateral walls allows the movement of water to adjacent cells.* Source: Scanning electron microscope (SEM) photo provided by Dr. Irving B. Sachs, Forest Products Laboratory, USDA Forest Service.

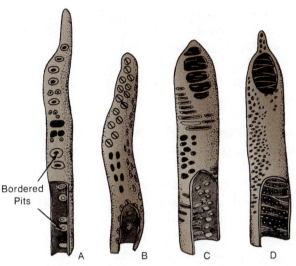

Figure 6–13

Types of tracheids and vessel elements found in xylem tissue. Left to right: (A) tracheid in pine (Pinus); (B) tracheid in oak (Quercus); (C) vessel element in magnolia (Magnolia); (D) vessel element in basswood (Tilia). Source: T. E. Weier, C. R. Stocking, M. G. Barbour, and T. L. Rost, *Botany, An Introduction to Plant Biology,* 6th ed. (New York: John Wiley & Sons/State University of New York College of Environmental Science and Forestry, 1982).

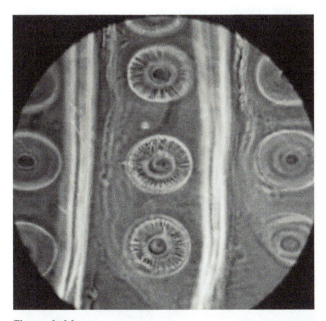

Figure 6–14

The tracheids with bordered pits of Larch (Larix lyalli) (× 800). The central part is the torus, and surrounding is the thin margo through which liquids diffuse. Source: Dr. Irving B. Sachs, Forest Products Laboratory, USDA Forest Service.

Phloem conducts food and metabolites from the leaves to the stem, flowers, roots, and storage organs. A complex tissue, phloem comprises sieve tubes, sieve tube members, companion cells, fibers, and parenchyma.

Figure 6–15

Long files of parenchyma cells surround the vessel. Pitting in lateral vessel wall in red maple (Acer rubrum). Source: Dr. Irving B. Sachs, Forest Products Laboratory, USDA Forest Service.

Sieve-tube members are long slender cells with porous ends called sieve plates (Fig. 6–16). Sieve tube members occur only in angiosperms. The equivalent cell in gymnosperms is the sieve cell, which is like the sieve-tube element except that it lacks a sieve plate. **Companion cells** aid in metabolite conduction and are closely associated with sieve-tube members (Fig. 6–16). **Phloem fibers** are thick-walled cells that provide stem support (Fig. 6–17). The parenchyma cells in the phloem serve as storage sites. Unlike xylem, which is mostly made up of hollow tubes created from dead cells, phloem is made up of living cells. Movement of food and metabolites through the phloem is by biological activity, which can be highly controlled by the plant and to some extent by a grower.

THE PLANT BODY

The various tissues are united in a structured and organized pattern to form organs such as roots, stems, leaves, flowers, fruits, and seeds. These make up the plant body. When a plant first begins to grow from seed, the original organs are the radicle and plumule. These organs form the primary plant body. As the plant continues to grow, the primary organs develop into mature organs made up of permanent tissues.

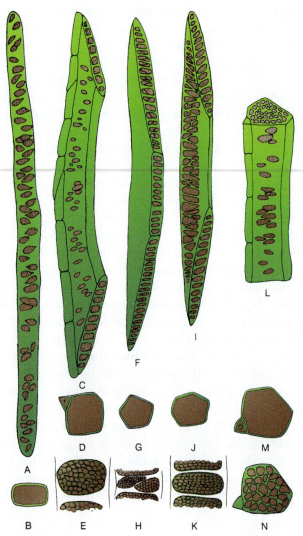

Figure 6–16

Sieve cells, sieve-tube elements, and companion cells in side view and cross section, showing detail structure of sieve plates: (A, B) from Canadian hemlock (Tsuga canadensis), only one-third of cell shown; (C, D, E) from tulip tree (Liriodendron tulipifera), (C, D) with companion cells attached, (E) detail of sieve plate; (F, G, H) from apple (Malus pumila), (H) detail of sieve plate; (I, J, K) from black walnut (Juglans nigra), (K) part of sieve plate in detail; and (L, M, N) from black locust (Robinia pseudo-acacia) with companion cells attached, (N) detail of sieve plate. Source: A. J. Eames and L. H. McDaniels. *An Introduction to Plant Anatomy* (New York: McGraw-Hill, 1947).

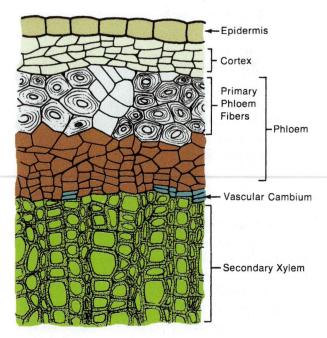

Figure 6–17

Cross section through stem of flax (Linum) showing thick-walled strengthening phloem fibers. Source: K. Esau, *Plant Anatomy*, 2nd ed. (New York: John Wiley & Sons, 1965).

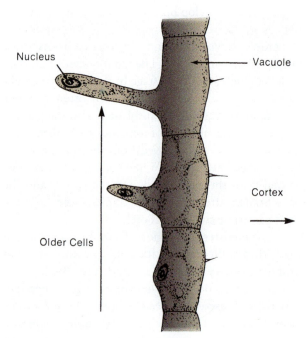

Figure 6–18

Section of epidermis of a young root showing three stages (bottom to top) in the development of root hairs.

Roots

Roots are responsible for absorbing and conducting water and mineral nutrients and for anchoring and supporting the plant. In addition, some roots, as in sugar beets and carrots, act as storage organs for photosynthesized food. The roots of some plants develop secondary xylem from cambial activity and an abundance of parenchyma cells, which are able to store photosynthates and water. Dissolved mineral nutrients and water required for growth are absorbed by the root hairs, which are extensions of the epidermal cells (Fig. 6–18).

A few kinds of trees develop aerial roots from the underside of branches. Once these roots reach the soil and penetrate it, they become functional as ground roots. Good examples of this are the strangler fig (*Ficus aurea*) and the banyan tree (*Ficus benghalensis*) of the tropics. Some of the strangest roots are those in certain tropical orchids called epiphytes (*Cattleya, Phalaenopsis, Aerides,* and *Vanda*). The roots contain chlorophyll for photosynthesizing food; they cling to rocks or tree surfaces and are fully exposed to receive light. Frequent rains and mist supply the moisture and nutrients necessary for growth.

The root system is a significant portion of the entire dry weight of any plant, about one-quarter to one-third of the total, depending on the storage or fibrous nature of the root. Measuring the total root system of a single mature rye plant showed that the plant had about 600 km (380 miles) of roots! Many of the functional roots of woody plants extend only into the top 1 m (3 ft) of soil. The depth that tree roots penetrate depends largely on the species of tree and on the structure and water status of the soil.

After the radicle or primary root emerges from the seed, it can continue to grow principally as a tap root, or it may develop branch roots and form a fibrous root system (see Fig. 6–7, page 84). The tap root usually grows downward, and the branch roots grow downward or horizontally. The tap root can be encouraged to branch at an early stage by removing or breaking the apical root meristem, which often happens when some seedlings are transplanted into the garden; for example, the tap roots of tomatoes are broken when the young plants are removed from the container, and the roots become fibrous. When young trees are transplanted from the nursery, the tap root is cut and the roots branch after the tree is transplanted into the home garden or commercial orchard.

The meristematic region of a root is composed of small, thin-walled cells with dense protoplasm that produce primary tissue at a rapid rate. Behind this active meristematic region lies the zone of elongation. Here the cells expand, especially in length; new protoplasm forms; and the size of the vacuoles increases. The apical meristem and the region of elongation take up only a few millimeters of each root. Behind the region of elongation is the region of maturation, where the enlarged cells differentiate into the tissues of the primary body. In the epidermis of this young region, the cells protrude and elongate and begin to form root hairs (Fig. 6–19). New root hairs arise in the newly developed region to replace old root hairs destroyed as the roots penetrate the soil. Because most

of the water and nutrients a plant requires enter through the root hairs, a healthy actively growing root system that is constantly producing root hairs is necessary for good plant growth and development.

The root apical meristem produces tissues different from those produced by the shoot apical meristem. The root meristem gives rise to the **root cap**, **epidermis**, **cortex**, and central vascular cylinder. The root cap is a thimble-shaped group of cells that protect the actively dividing meristem as it penetrates the soil (Fig. 6–19). These moist cells are sloughed off as the root comes in contact with sharp soil particles; the meristem forms new cells on the inner part of the cap to replace the damaged or lost cells. The meristem produces long rows of cells under the root cap. One of these becomes the protoderm that gives rise to the epidermis, or outer layer, of the root. The ground meristem is the tissue layer that gives rise to the cortex just below the epidermis. The cortical region is mainly composed of storage parenchyma cells. A single layer of inner cortical cells forms the **endodermis**, a tissue found only in the root and not the stem. Each thin-walled endodermal cell is completely encircled by a narrow, thickened band of waterproofed material known as the **Casparian strip**. The solution of water and nutrients entering the root from the soil cannot penetrate the Casparian strip. For the soil solution to enter the inner tissue (pericycle) of the root, it must pass through the permeable endodermal cell walls and the protoplast (Fig. 6–20).

The Procambium Layer

The procambium layer gives rise to various tissues of the vascular cylinder. These include the **pericycle**, which is the outermost layer of cells of the central core and lies just inside the endodermis. The pericycle develops from a single parenchyma cell layer on the outer portion of the procambium. The pericycle is a meristematic region producing lateral (branch) roots that grow outwardly through the cortex and epidermis. It may also give rise to vascular and cork cambium. The procambium layer also produces a vascular cylinder of primary phloem and xylem, vascular cambium, and, in some species, pith. The pericycle and the vascular cylinder are collectively called the **stele**. The primary xylem is a central mass of tissue that may extend as arms beyond the primary phloem (Fig. 6–19). A layer of procambium cells separates these two primary tissues; this layer is the meristematic region for any new vascular tissue that subsequently forms.

As the root grows in girth and the plant matures, a continuous ring of secondary phloem forms outside the

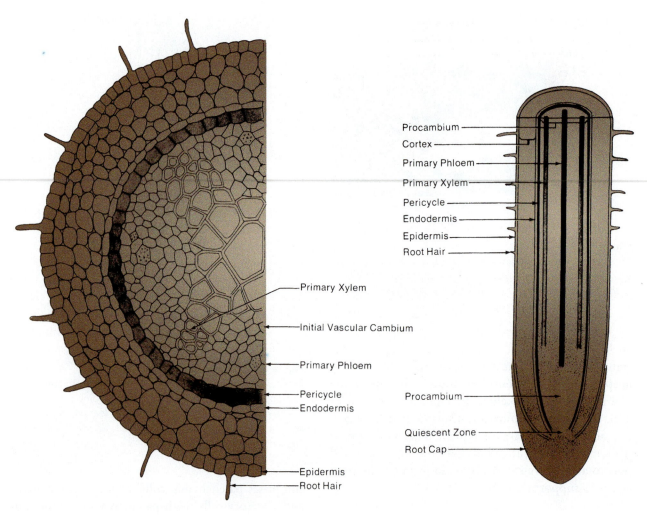

Figure 6–19
Left: *Cross section of a young root showing the parts of the primary plant body and their location.* Right: *Developmental occurrences in the root tip, showing the various components and their relative location.* Source: T. E. Weier, C. R. Stocking, M. G. Barbour, and T. L. Rost, *Botany, An Introduction to Plant Biology,* 6th ed. (New York: John Wiley & Sons/State University of New York College of Environmental Science and Forestry, 1982).

vascular cambium, and the primary phloem becomes less important than at earlier growth stages. The cambium layer develops from the procambium and from pericycle cells outside the primary xylem. The primary xylem with its extending arms remains, but it is encircled by secondary xylem formed by the adjacent vascular cambium. Annual rings develop in the secondary plant body of the root as it grows in girth, much as in stems except for the star-shaped primary xylem at the core of the root. The cork cambium formed from pericycle produces a corky layer outwardly from the vascular system.

Adventitious roots form at any place on plant tissue other than the radicle of a germinating seed and its extensions. Adventitious roots arise from meristematic cells adjacent to vascular bundles (in herbaceous dicot stems) or from cambium or young phloem cells in young stems of woody perennials. This production of adventitious roots is the basis for propagation by stem cuttings (see Chapter 10). Adventitious roots can arise from plant parts other than stems, such as from leaf petioles or leaf blades or even from old root pieces. Plant parts to be rooted are usually detached from the parent plant and placed under favorable environmental conditions for rooting. In the propagation procedure known as layering, adventitious roots are induced to form on plant parts still attached to the parent plant. Adventitious roots also develop on intact plants, as in the corn plant (see Fig. 6–2, page 79).

The ability of some plants to readily form adventitious roots allows seedlings and transplants to be planted deep in the ground if the seedling or

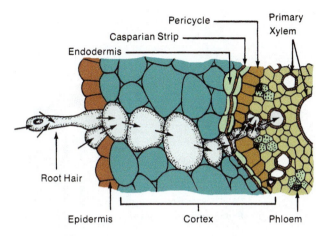

Figure 6–20

Cross section of a young wheat (Triticum) root showing the path of entrance of soil solution into the root from a root hair to the tracheary elements of the xylem. Source: K. Esau, *Plant Anatomy*, 2nd ed. (New York: John Wiley & Sons, 1965).

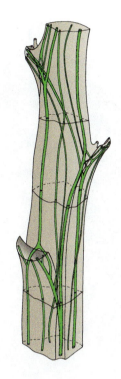

Figure 6–21

Three-dimensional skeletal view through a potato stem showing the primary vascular system extending through the stem with branches into the cutaway leaf petioles. Source: A. J. Eames and L. H. McDaniels, *An Introduction to Plant Anatomy* (New York: McGraw-Hill, 1947).

transplant has become tall and spindly. By planting deep and having adventitious roots form, the young plant is much less likely to be damaged by wind or rain. For example, tomato transplants that have gotten tall and spindly are often planted deep in the field or home garden. Easter lily bulbs are planted in tall pots to allow adventitious roots to form on the stem and help support the plant.

Stems

The main stem and its branches are the scaffold of the plant, supporting the leaves, flowers, and fruits. The leaves and herbaceous green stems manufacture food, which is transported to the roots, flowers, and fruits through the phloem. Fig. 6–21 illustrates the complexity of the primary vascular system. The greater part of the vascular system consists of xylem and phloem. The secondary xylem also serves as the major structural support in woody perennial plants.

The stem develops from three primary tissues produced by the apical meristem: the protoderm, the ground meristem, and the procambium. These give rise to the epidermis, cortex, and vascular cambium, respectively.

The epidermis, which is usually a single layer of surface cells, protects the stem. Epidermal cells are usually cutinized on their outer surface to retard desiccation. The epidermis of leaves and young stems has pores, the stomata (see Fig. 6–10, page 86) that allow for the exchange of gases.

The cortex lies just beneath the epidermis and encircles the inner core of the vascular tissue. The cortex comprises parenchyma, collenchyma, sclerenchyma, and secretory cells, with parenchyma cells the most numerous. Some parenchyma cells have chloroplasts and are called chlorenchyma. Parenchyma cells have the ability to divide and form new tissue when wounded, thus providing a protective mechanism for the stem. Collenchyma is the outer cell layer of the cortex adjacent to the epidermal layer (Fig. 6–22). These cells may be thickened at the corners, and their walls contain cellulose, hemicellulose, and pectin. This tissue therefore adds strength to the stem. Sclerenchyma cells have thick lignified walls. They can form long fibers, which are the source of strength in mature stems. Secretory cells produce resinous substances and are commonly found, for example, in the resin ducts of pine trees.

The **vascular system** of seed-bearing plants consists of the pericycle, phloem, vascular cambium, xylem, pith rays, and pith. The arrangement of these complex tissues in the vascular system differs among three broad groups of plants: (1) the gymnosperms and the woody dicotyledonous angiosperm perennials (which live for long periods), such as trees and shrubs; (2) the herbaceous dicotyledonous plants, such as potato, petunia, and phlox; and (3) the monocotyledonous

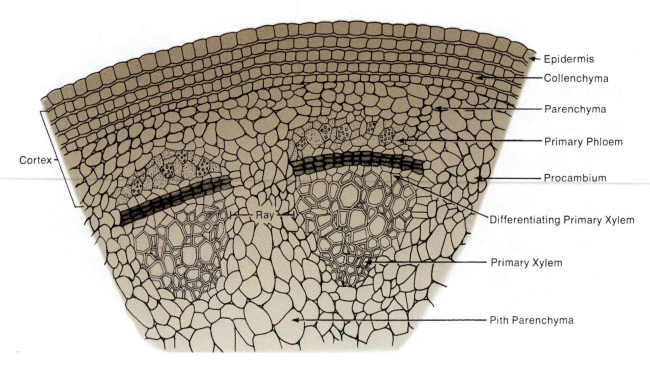

Figure 6–22

Cross section of a young woody plant stem toward the end of primary growth, showing various tissues present.

plants, such as corn and date palms. These three groups are discussed separately to distinguish among their different internal stem structures and growth patterns.

Woody Perennials (Dicotyledonous Angiosperms and the Gymnosperms)

All the cells and tissues originate from a terminal shoot meristem that forms protophloem and protoxylem (primary tissue) (Fig. 6–22). As the stem grows in length, the secondary tissues form from the vascular cambium. The secondary phloem develops toward the outside of the stem and the secondary xylem forms inwardly from the vascular cambium. These growing secondary and permanent tissues crush the primary tissues until they become difficult to see. Secondary xylem is actively produced by the vascular cambium in the early spring and less actively in late summer. This xylem tissue becomes the early (porous) and late (dense) wood that form the annual growth rings in trees (see Figs. 6–9, page 86 and 6–23). The vessels or tracheids formed during the spring flush of growth are larger than those formed during the summer (Figs. 6–21, 6–23, and 6–24). The narrow-leaved evergreen trees belonging to the gymnosperms are usually referred to as the softwoods or nonporous wood trees (Figs. 6–24 and 6–25). The xylem of gymnosperms consists mainly of tracheids (see Fig. 6–13, page 88).

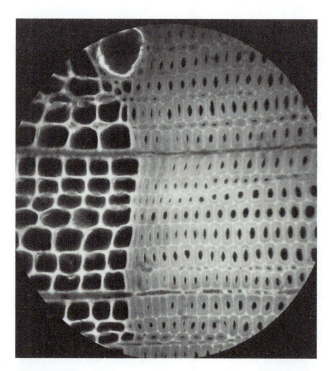

Figure 6–23

A cross section through the xylem of Douglas fir (Pseudotsuga menziesii) (× 200). The large pores to the left are the spring wood and the more dense cells were formed in the summer of the same year. The very large pore at the top is a resin duct. Source: Dr. Irving B. Sachs, Forest Products Laboratory, USDA Forest Service.

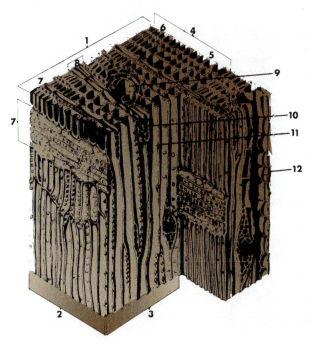

Figure 6–24
Three-dimensional view of the wood of a softwood forest species: (1) cross-sectional face; (2) radial face; (3) tangential face; (4) annual ring; (5) early wood; (6) late wood; (7) wood ray; (8) fusiform ray; (9) vertical resin duct; (10) horizontal resin duct; (11) bordered pit; (12) simple pit. Source: U.S. Forest Products Laboratory, Madison, Wis.

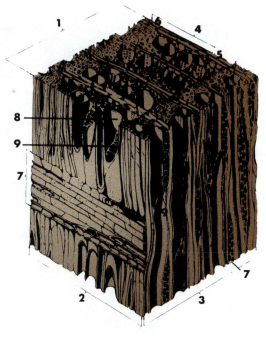

Figure 6–26
Three-dimensional view of the wood of a hardwood forest species: (1) cross-sectional face; (2) radial face; (3) tangential face; (4) annual ring; (5) early wood; (6) late wood; (7) wood ray; (8) vessel; (9) perforation plate. Source: U.S. Forest Products Laboratory, Madison, Wis.

The broad-leaved angiosperm trees are called hardwoods or porous wood trees (Fig. 6–26); the xylem tissue is made up mostly of vessel elements (see Figs. 6–12 and 6–13, pages 87 and 88).

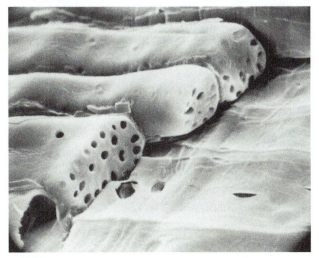

Figure 6–25
End-wall perforations in the ray cells of Douglas fir (Pseudotsuga menziesii). Source: Dr. Irving B. Sachs, Forest Products Laboratory, USDA Forest Service.

Both the gymnosperms and the woody perennial angiosperms grow in girth each year when the cells of the vascular cambium divide, forming annual rings of xylem. A stem nearing the end of its first season of growth is mostly xylem. The phloem, as it is crushed by the expanding xylem, is constantly being renewed by the vascular cambium. The phloem is a relatively thin layer of complex tissue protected by the bark or cork layer. When the trunk of a tree is girdled, the bark and phloem are removed from an area encompassing the entire trunk. The damage to the phloem stops the flow of metabolites to the plant below the girdling, resulting in a weakening of the tree and often death.

The cork cambium (**phellogen**), which is a meristematic tissue, provides cells that grow both outward and inward. The outward cells become cork cells; the inward, phelloderm. The cork cells become suberized and are, therefore, resistant to entry or loss of water. The cork cells soon die but retain their ability to resist desiccation, disease, insects, and extreme temperatures. The unusually thick bark of the cork oak (*Quercus suber*) is stripped for a multitude of commercial uses such as corks and insulation (Fig. 6–27). In the case of cork, the bark of the tree is thick enough that the phloem is not harmed if bark removal is carefully done.

Figure 6–27
The rough, thick bark of the cork oak (Quercus suber), which is commercially stripped for cork products.

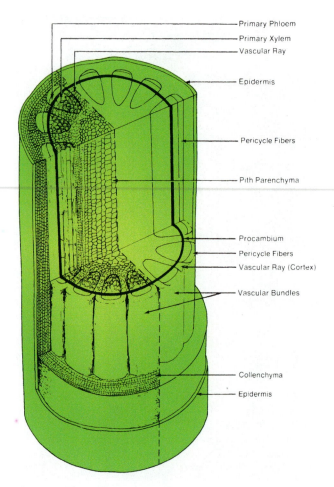

Primary Phloem
Primary Xylem
Vascular Ray
Epidermis
Pericycle Fibers
Pith Parenchyma
Procambium
Pericycle Fibers
Vascular Ray (Cortex)
Vascular Bundles
Collenchyma
Epidermis

Figure 6–29
Three-dimensional cutaway view of an herbaceous dicot stem, showing the vascular bundles.

In the young twigs and small trunks of many kinds of trees and shrubs, pore openings (**lenticels**) allow the inward and outward diffusion of gases (Fig. 6–28).

Herbaceous Dicotyledonous Plants   The early stem growth of plants in this category is much like the early growth of woody dicot stems. The vascular bundles (fascicles) of a herbaceous dicot usually remain separated and distinct; they are arranged in a single circle in the stem (Fig. 6–29). A larger proportion of the herbaceous stem is cortex and pith rather than xylem or phloem. Stem strength comes from the pericycle fibers adjacent to the phloem or from collenchyma or sclerenchyma tissue just beneath the epidermis (Fig. 6–29).

Herbaceous Monocotyledonous Plants   Stem growth originates from an apical meristem that produces vascular bundles scattered throughout the parenchyma (Fig. 6–30). The vascular bundles form most frequently near the epidermis. The sclerenchyma cells near the epidermis and thick-walled cells surrounding the bundles provide the principal support in monocot stems. Monocots (Fig. 6–30) have no continuous cambium and, therefore, lack secondary growth. Stem diameter from the base to the apex is usually more uniform in monocot stems than in dicot stems with secondary vascular growth.

Woody Perennial Monocotyledonous Plants   In trees such as date and coconut palms (*Arecaceae*), the thickness at the stem apex increases by the activity of a

Figure 6–28
The bark of a birch tree (Betula verrucosa) showing the lenticels.

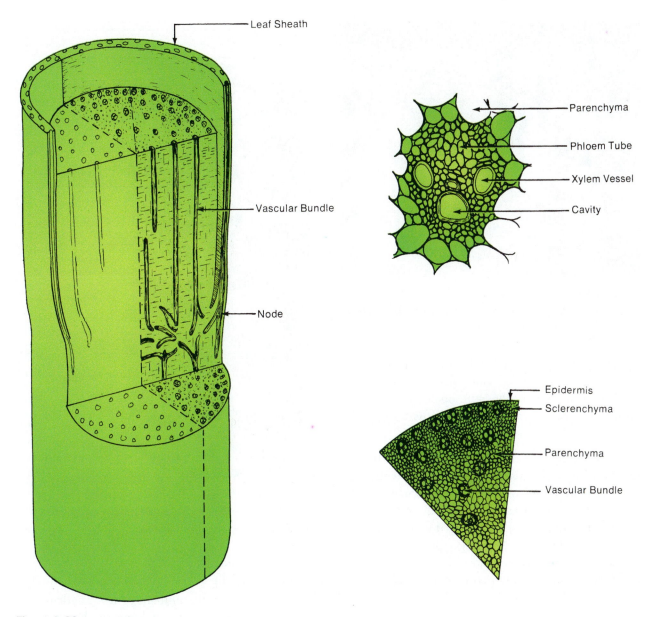

Figure 6–30
Left: *Three-dimensional cutaway view of the stem of a herbaceous monocot (solid stem), showing the scattered vascular bundles.*
Upper right: *Enlarged view of vascular bundle.* Lower right: *Enlarged cross-section area.*

primary thickening meristem. In the trunk below the terminal growing point, parenchyma cells continue to divide and enlarge, thus allowing for lateral stem enlargement. This is termed diffuse secondary growth because no actual lateral meristem is involved.

Stem Forms

When most people think of the stem of a plant, they envision the upright portion that bears branches, leaves, flowers, and fruits. Stems come in other forms, too. For example, certain fruit trees, such as apples, cherries, plums, and pears, bear flowers and fruits each spring on persistent shortened stems called spurs. Stems can also

grow horizontally, as in a pumpkin or cucumber vine. Some species of plants have underground stems; only a small portion of the stem shows above ground for a relatively short period in the spring. These are the so-called bulbous plants. The white (Irish) potato plant (*Solanum tuberosum*), ready for harvest, exemplifies two kinds of stems: the above-ground stem that bears the leaves and flowers, and an underground stem whose terminal portion swells into a tuber as it accumulates starches and sugars from photosynthesis in the leaves (Fig. 6–31). Just like other stems, the white potato tuber has buds (eyes) that sprout, when planted, to form new above-ground stems.

Figure 6–31
A potato plant at time of harvest. The tubers are stem structures growing below ground whose terminals have stored sufficient photosynthate to swell into tubers.

Figure 6–32
A gladiolus corm, which is thickened compressed stem tissue, toward the end of the growing season. The spring-planted corm (below) has shriveled with a new corm (above) forming above the old corm. New small cormels are forming at the base of the new corm.

A **rhizome** is an underground stem that grows horizontally. Examples of plants with rhizomatous stems are bananas, cannas, certain irises, certain bamboos, and some grasses, such as quack grass, Johnson grass, and Bermuda grass.

Stolons are stems that grow horizontally above ground. Sometimes called runners, stolons can develop roots in the soil at every node or at every other node (strawberry). Examples of species with stolons are ajuga, Bermuda grass, and some ferns.

Corms are thickened compressed stems that grow underground (Fig. 6–32). Buds on corms sprout to produce upright stems, which bear leaves and flowers. Gladiolus, crocus, and freesia are some examples.

Bulbs are highly compressed underground stems to which numerous storage leaves (scales) are attached. These highly developed stems provide a means for some species to survive the cold of winter and the dry soil of summer. One or more buds on the bulb sprout in the spring to produce an elongated stem with leaves and flowers. Hyacinths, lilies, onions, and tulips are examples of bulbous plants.

Stem **tubers** are the enlarged, fleshy, terminal portions of underground stems. The white potato, as discussed previously, is a good example (Fig. 6–31).

Leaves   Leaves develop in a complex series of events closely associated with stem development. Some of these developmental events are still not clearly understood. Leaves are initiated by the apical shoot meristem. Their prescribed pattern, position, and shape are influenced to some extent by their environment. For example, the leaves of cacti are adapted to growth in the desert, whereas leaves of ferns are adapted for growth in a rainforest. Most monocots, such as the grasses and palm trees, have strap-shaped leaves with parallel veins and interveinous connections between major veins. The veins contain sheaths of vascular bundles including xylem and phloem elements. **Palisade** and **spongy mesophyll parenchyma cells**, containing chlorophyll for photosynthesis, surround these veins (Figs. 6–33 and 6–34).

The leaves of dicotyledonous plants vary considerably in size and shape (Fig. 6–35). Practically all dicots leaves have veins arranged in the shape of nets. The large primary veins divide into smaller secondary veins. The veins are made up of xylem and phloem connected to all segments of the leaf. The spongy **mesophyll** parenchyma (Figs. 6–33 and 6–34) contains the intercellular spaces through which carbon dioxide, oxygen, and water pass. The outside layer or skin of the leaf is largely made up of epidermal cells. This epidermal layer contains openings or pores called **stomates**, each surrounded by two **guard cells**. There are generally more stomates in the lower epidermal layer of the leaf than in the upper epidermal layer. Waterplants such as water lilies, however, are an exception to this pattern.

The primary function of leaves is photosynthesis; a secondary function is transpiration. The guard cells, which occur in pairs on both sides of the stomata,

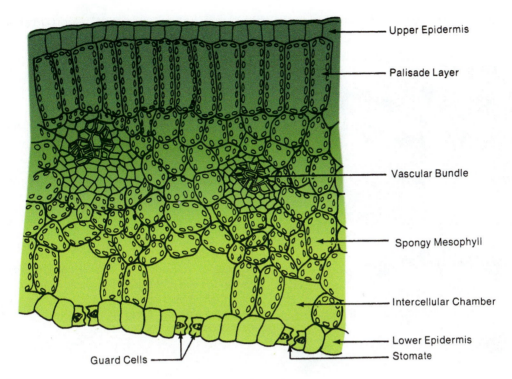

Figure 6–33

Cross section through a lily leaf showing the tissues involved in photosynthesis, transpiration, and translocation.

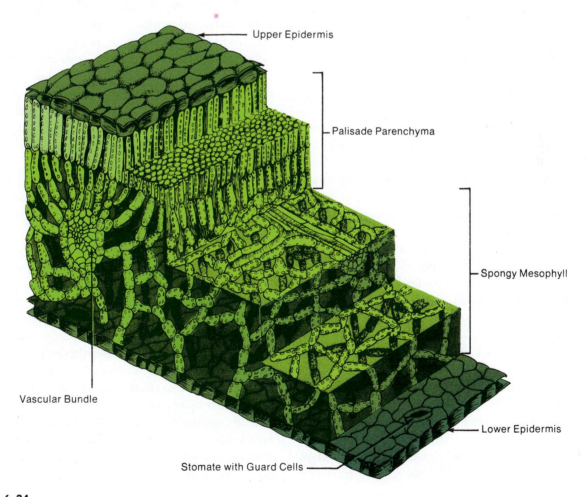

Figure 6–34

Three-dimensional cutaway view of an apple leaf showing the relation of cells in the various tissues. Source: A. J. Eames and L. H. McDaniels, *An Introduction to Plant Anatomy,* 2nd ed. (New York: McGraw-Hill, 1947)

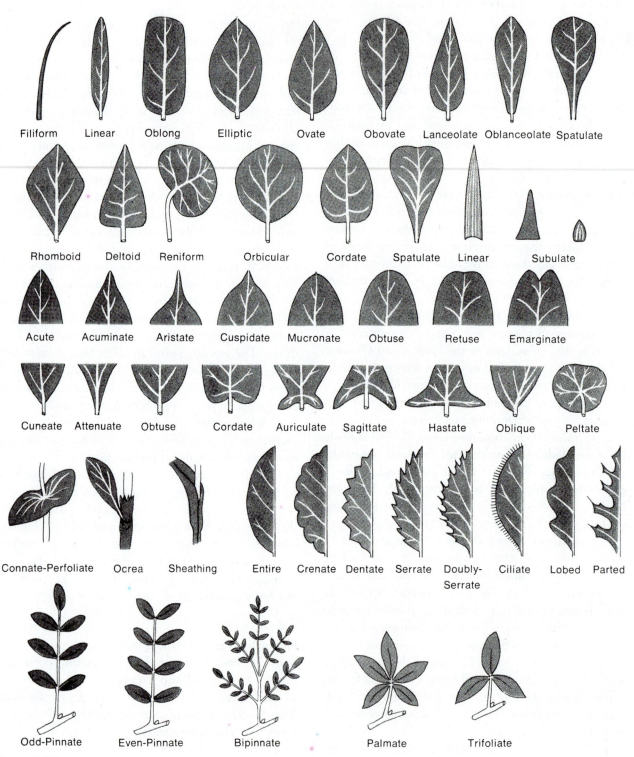

Figure 6-35

Examples of leaf patterns, with descriptive names. The top two rows are the shapes of simple leaves; the center rows are simple leaves cut to depict leaf tips or bases; one row up from the bottom illustrates examples of leaf attachments or leaf edges; the bottom row shows examples of compound leaves. These characteristics are used to describe leaves in plant identification keys.

control the opening and closing of the stomata through which carbon dioxide, one of the raw materials for photosynthesis, enters the plant, and oxygen, a product of photosynthesis, is released. Water also enters or escapes through the stomata. The loss of water from the leaf by evaporation is a process called **transpiration**. Transpiration helps regulate leaf temperature and provides the force that draws water into and through the xylem. Some plants have modified leaf surfaces that affect the rate of transpiration. The leaves of some plants, such as cabbage, have a thick waxy surface (**cuticle**) that greatly reduces water loss. In the leaves of other plants, the epidermal cells produce elongated hairs that reduce the wind velocity at the leaf surface, thus reducing the transpiration rate. Some kinds of plants, especially those native to hot dry climates, such as the olive tree, minimize water loss by having stomata sunken deep in the epidermal layer.

Plants often have leaves modified to perform functions other than photosynthesis and transpiration. For example, the leaf stipules of the black locust (*Robinia pseudoacacia*) are modified to become thorns that aid in protecting the plant. Some vining plants, like the grape, have leaves modified in the form of tendrils that help support the vine when trained on trellises.

In most dicotyledonous plants, the leaf is made up of the **blade**, the flat thin part; the stemlike **petiole**, which attaches the blade to the stem; and, in some plants, the **stipules** at the base of the petiole. Some leaf blades are attached directly to the stem and lack a petiole or stipules. These are termed **sessile** leaves.

Leaf shapes among the many plant species vary greatly, and a special morphological terminology describes the leaf shape, margin, tip, and base. Leaves are usually classified (Fig. 6–35) as **simple** (a single leaf) or **compound** (one with three or more leaflets). Distinguishing between simple and compound leaves can be difficult. The best test is to examine the base of the petiole. A true leaf has a bud in this location; a leaflet does not. A compound leaf resembling a feather is termed **pinnate**; one resembling the palm of a hand, **palmate**. A **trifoliate** compound leaf has three leaflets, as in the bean plant.

The shapes of simple leaves are described as linear, oblong, elliptical, lanceolate, deltoid, and so forth. The shapes of the tips and bases of simple leaves are also categorized as cordate (heart-shaped), sagittate (arrow-shaped), auriculate (ear-lobed), and sheathing. Leaf edges or margins range from entire (smooth), dentate (tooth like), serrate (sawlike) to lobed (rounded edges). Leaf shape is a very important feature when identifying plants.

Buds

Plant stems generally produce buds in the axils of leaves at the nodes or terminally on shoots. Buds usually do not occur on roots. A **bud** can be defined as an undeveloped shoot or flower, largely composed of meristematic tissue, and generally protected by modified leaf scales. Buds include: (1) vegetative buds, which develop into a shoot; (2) flower buds, which open to produce a flower or flowers; and (3) mixed buds, which open to produce both shoots and flowers. Cutting through a bud longitudinally reveals the miniature parts of either a stem growing point or, in a flower bud, all the miniaturized parts of a flower.

Buds are especially prominent in winter on deciduous plants when the leaves have fallen. A leaf scar, where the leaf petiole was attached, is visible just below each bud. Buds may occur opposite each other on a stem or in an alternate arrangement around the stem.

Buds are initiated by terminal growing points as shoots elongate during the growing season. Some buds continue to grow after they are formed, developing into shoots. The growing points in other buds remain dormant until the following spring. Some buds may remain latent for long periods of time and become embedded in enlarging stem tissue; these become latent buds.

Adventitious buds can develop in places where buds generally do not form, such as buds arising on root pieces when root cuttings are made.

Buds of deciduous woody species usually go into a physiological resting or quiescent state shortly after they are formed in the summer and stay that way until they are subjected to low-temperature winter chilling, which overcomes their resting condition and enables them to resume growth the following spring. Buds of tropical and subtropical plants, however, generally do not develop such a resting condition.

Flowers

Flower buds form by the differentiation of vegetative buds into flower parts. In the angiosperms, specialized floral leaves borne and arranged on the stem are adapted for sexual reproduction; these are the **flowers**. After fertilization portions of the flower develop into a **fruit**, which bears the **seed(s)**.

The flowers may be borne at the apex of a stem, as in the sunflower, rose, and poinsettia, or in the axils of the leaves lower down on the stem, as in the tomato, fuchsia, and many of the temperate zone fruit trees. Flowers or **inflorescences** vary in shape and form among the species, a fact that aids in identifying a plant's **species**, **genus**, and **family**. As with stems, botanists classify flowers in a specialized morphological terminology (Figs. 6–36 and 6–37).

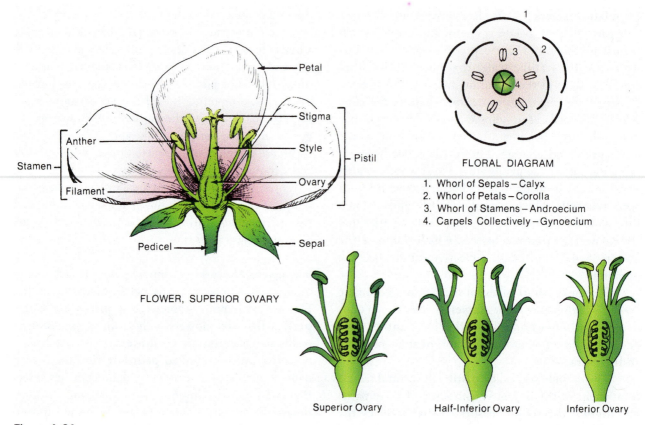

FLORAL DIAGRAM

1. Whorl of Sepals – Calyx
2. Whorl of Petals – Corolla
3. Whorl of Stamens – Androecium
4. Carpels Collectively – Gynoecium

Figure 6–36

Types of flower patterns in angiosperms, showing the relative position of the ovary in relation to accessory structures.

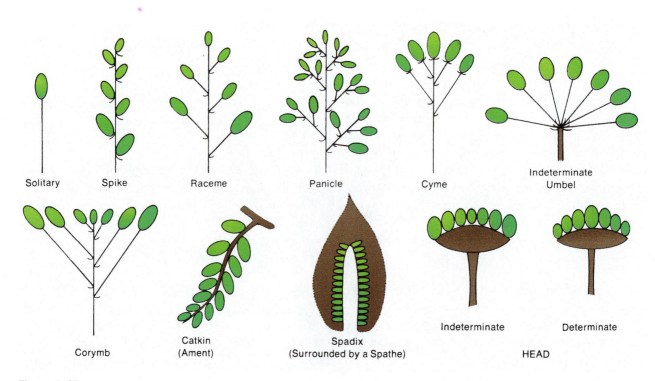

Figure 6–37

Some typical stylized inflorescences used to describe plants.

Complete Flowers   Complete flowers usually have four parts—sepals, petals, stamens, and pistil—which are usually borne on a receptacle (Fig. 6–36). The **sepals** are the leaflike scales that encircle the other flower parts, as in the carnation and rose. Most often the sepals are green, but sometimes they are the same color as the petals, as in tulips and lilies. The sepals may fold back (rose) or remain upright (carnation) as the petals grow and emerge. The sepals collectively are called the **calyx**. The **petals** are the next whorl of floral leaves inward from the sepals. The collective term for petals is **corolla**. The petals are usually brightly colored with some yellow, and they often contain **nectaries** that secrete nectar to attract insects, which pollinate the flowers. Sepals and petals collectively are called the **perianth**.

The next whorl of floral organs in a complete flower is the male part, or **stamen**. Each stamen consists of a **filament** and an **anther**; the anther produces the **pollen**. A group or whorl of stamens is the **androecium**.

The **carpel** (also called **pistil**), the central female component of the flower, is composed of three parts: the **stigma**, the receptive surface that receives the pollen; the **style**, a tube connected to the stigma; and the **ovary**, attached to the lower end of the style. The ovary contains undeveloped **ovules** that are attached to a **placenta**; the ovules develop into seeds after pollination and fertilization. An ovary can contain one or multiple ovules. The pistil can be simple (i.e., has but one carpel) or compound (i.e., has two or more fused carpels). Collectively, the carpels are known as the **gynoecium**. The apricot and the apple (Figs. 6–38 and 6–39) are examples of complete flowers with a simple and a compound pistil, respectively.

The stamens and pistils are considered the essential parts of the complete flower for sexual reproduction. The sepals (calyx) and the petals (corolla) are accessory flower parts.

Incomplete Flowers   Incomplete flowers lack one or more of the four parts: sepals, petals, stamens, or pistil. Flowers with both stamens and pistils are called **perfect flowers**. Flowers with stamens only and no pistils are called **staminate flowers**; those with pistils but no stamens are called **pistillate flowers**. Staminate or pistillate flowers are by definition **imperfect flowers**. Plants having both staminate and pistillate flowers borne on the same plant are termed **monoecious** (e.g., alder, corn, walnut) (Fig. 6–40). If the

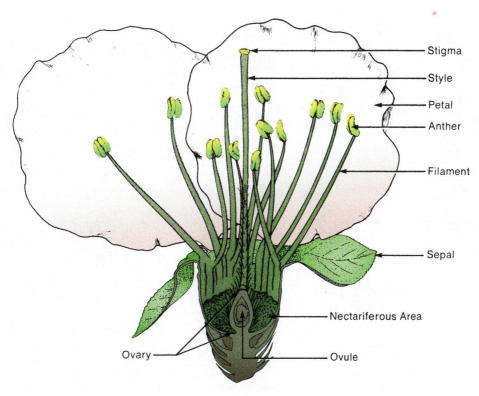

Figure 6–38
The typical complete flower of the apricot, showing a superior ovary with a simple pistil. Source: USDA.

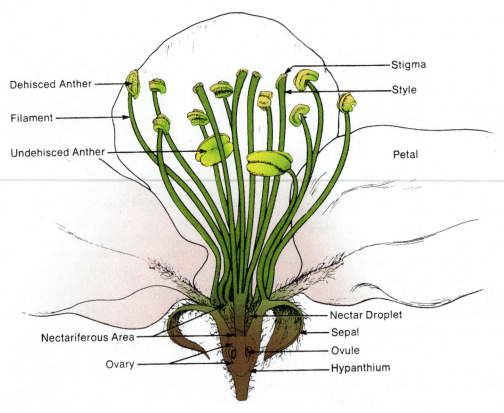

Figure 6–39

The typical complete flower of the apple, showing an inferior ovary with a compound pistil. Source: USDA.

Figure 6–40

Flowers of a monoecious species, the walnut. Left: Female flowers. Right: Male flowers, or catkins, are borne in structures separate from the female flowers on the same plant. The female flowers are wind pollinated.

pistillate and staminate flowers are borne on separate individual plants (male and female plants), the species is called **dioecious**. Examples are date palm, kiwifruit, gingko, pistachio, and asparagus.

Some flowers, like the tulip, are borne singly on a stalk and are called solitary, but others are arranged in multiples or in clusters known as **inflorescences** (Fig. 6–37). The **corymb** is a short, flat-topped flower with an indeterminate cluster that continues to produce flowers until conditions become unfavorable; the lower flowers open first. An example is the cherry. The **cyme** resembles the corymb, except that the central or topmost flower is the first to open. Examples are chickweed and strawberry. The **raceme** is a single elongated indeterminate arrangement of stalked flowers, found in the mustard and cole crops (*Brassicaceae*), for example. The **spike** is an elongated, simple, indeterminate inflorescence with sessile (no stalk) flowers, as in wheat, oats, and gladiolus. The **catkin** is a spike with only pistillate or staminate flowers exemplified by alder, poplar, walnut, and willow. The **panicle** is an indeterminate branching raceme found in many of the grasses. The **umbel** is an indeterminate, often flat-topped, cluster of

flowers that are of equal length and arise from a common point, as in carrots, dill, and onions. A **head** is a short dense spike; daisies and sunflowers have heads. A **spadix** is a complete densely flowered structure surrounded by a spathe (calla lily in the *Araceae* family). Many ornamental plants are grown for their distinctive flower forms.

Fruits

A fruit is a matured ovary plus associated parts; it is generally a seed-bearing organ, but there are parthenocarpic fruits that are seedless. The fruit protects the seed in some plants and helps disseminate it. For example, the seeds of fleshy fruits like peaches or plums occur inside their pits, which are discarded by animals or humans eating the fruit. The seeds in the discarded pits may eventually germinate and grow. One might tend to think of all fruits as fleshy organs, but there are many dry fruits such as nuts, capsules, legume pods, and grains. Many of the vegetables we eat are actually fruits. Tomatoes, peppers, squash, green beans and peas, and sweet corn botanically are all fruits though dietarily they are considered vegetables. Dietarily, fruits that are not very sweet are usually considered to be vegetables.

Fruits develop after pollination and fertilization. Flowers are self-pollinated or cross-pollinated by wind or insects. The pollen grows from a pollen grain on the stigma through the style and fertilizes the egg, causing the fruit to develop. Fruits may consist of a single carpel, as in beans and peas, or a combination of several carpels, as in apples (see Fig. 6–39) and tomatoes. The fruit matures quickly, in a matter of weeks in the case of some summer annuals and strawberries, or it requires as long as nine months, as with oranges. The ovary wall, which is called the **pericarp**, can develop into different structures. The peel of an orange is part of the pericarp. The pod of a pea, or the shell of a sunflower seed, and the skin, flesh, and pit of a peach are all derived from the ovary wall or pericarp.

Simple Fruits   **Simple fruits** have a single ovary formed from one flower. The most common classification of simple fruits categorizes them as fleshy, semifleshy, or dry by the texture of the mature pericarp.

Fleshy Fruits   The entire pericarp and accessory parts develop into succulent tissue.

> *Berry.* A pulpy fruit from one or more carpels that develops few to many seeds. Examples are bananas, dates, grapes, peppers, tomatoes, and papayas. (Fig. 6–41)

Figure 6-41
Grapes and peppers are examples of berries, which are fruits that have one or more carpels that develop few to many seeds.
Source: Margaret McMahon, The Ohio State University.

Figure 6–42
Hesperidium fruit such as oranges and limes have several carpels with pulp juice sacs enclosed in a leathery rind. The carpels, juice sacks, and leathery rind are easily seen in the cross-section. Source: Margaret McMahon, The Ohio State University.

> *Hesperidium.* A fruit with several carpels with inner pulp juice sacs or vesicles enclosed in a leathery rind; for example, orange, lemon, lime, and grapefruit. (Fig. 6–42)

> *Pepo.* A fruit formed from an **inferior ovary** that develops from multiple carpels each bearing many seeds. The pericarp is a thick and usually hard rind. Cucumbers, melons, squashes, and watermelons are examples.

Dry-Fleshy Fruits   Some parts of the pericarp become dry and the other portions remain succulent.

> *Drupe.* This is a simple fruit derived from a single carpel. The **exocarp** (the outer layer) becomes the thin skin; the **mesocarp** (the middle layer) becomes thick and fleshy; the **endocarp** (the inner

Figure 6–43
Drupes or stone fruit such as peaches, plums, almonds, and olives are a single carpel made up of an exocarp or thin skin outer layer, the thick, fleshy mesocarp, and a hard, stony endocarp all of which surround the seed. Source: Margaret McMahon, The Ohio State University.

Figure 6–44
Strawberry "fruit." The true strawberry fruits are the achenes (resembling seeds) borne on the surface of the large fleshy receptacle. Source: Margaret McMahon, The Ohio State University.

layer) becomes hard and stony and is often referred to as the pit (and often erroneously as the seed). Peaches, plums, cherries, apricots, almonds, and olives are examples of drupe fruits. (Fig. 6–43)

Pome. This is a simple fruit made up of several carpels. The outer (and edible) portion forms from an accessory structure, the hypanthium of the flower, which surrounds the multiple **carpels**. Apple (Fig. 6–39), pear, and quince are examples.

Dry Fruits (Papery or Stony)   The entire pericarp is dry at maturity.

Dehiscent fruits. These fruits dehisce (split) at maturity to expose the seeds.

Legume or pod. A fruit from a single carpel which usually dehisces along both sutures (seams) of the carpel. This fruit is typical of the pea family (*Fabaceae*), such as peas and beans.

Capsule. The fruits form from two or more carpels, each of which produces many seeds. Splitting can occur in several different ways. Iris, poppy, and jimson weed are examples.

Follicle. The fruits form from a single carpel that splits along one suture. *Delphinium* and *Helleborus* are examples.

Silique. Fruits form from two carpels with a septum between. The two halves separate longitudinally, exposing the seeds on a central membrane. Mustard, *Lunaria,* and stocks are examples.

Indehiscent fruits. These fruits do not split open when mature.

Figure 6–45
Corn, barley, rice, and wheat (clockwise from upper left) are examples of caryopsis (grain) fruits. Source: Margaret McMahon, The Ohio State University.

Achene. Simple, one-seeded, thin-walled fruit attached to an ovary wall. Achenes are very often mistaken for seeds as in the case of strawberry "fruits" (Fig. 6–44), the so-called seeds of the rose hip, and sunflower fruits.

Caryopsis (grain). A one-seeded fruit with a thin pericarp surrounding and adhering tightly to the true seed. Corn, rice, wheat, and barley are examples. (Fig. 6–45)

Nut. A one-seeded fruit with a thick, hard, stony pericarp. Oak (acorn), chestnut, filbert, walnut, and hickory are examples.

Samara. A one-seeded (elm) or a two-seeded (maple) fruit with a winglike structure formed from the ovary wall. Ash is another example. (Fig. 6–46)

Figure 6–46
Maple fruit is an example of a samara, which has a winglike structure formed from the ovary wall. Source: Margaret McMahon, The Ohio State University.

Schizocarp. A fruit formed from two or more carpels that at maturity yield two one-seeded halves. Carrots, dill, caraway, and parsley are examples.

Aggregate and Multiple Fruits Aggregate and **multiple fruits** form from several ovaries. The true fruits are attached to, or contained within, a receptacle or an accessory structure.

Aggregate fruits develop from many ovaries on a single flower. The strawberry, for example, has many achenes (true fruits), each attached to a single fleshy receptacle (Fig. 6–44). The many achenes of the rose hip develop inside the receptacle. The blackberry and raspberry are similar to the strawberry except that individual small drupes, instead of achenes, are attached to the fleshy receptacle.

Multiple fruits develop from many individual ovaries fused into a single structure borne on a common stalk. The fig "fruit" we eat is made up of small drupes (the true fruits) contained inside a fleshy receptacle. The whole structure is termed a syconium. The pineapple is a large accessory structure covered with seedless (parthenocarpic) berries (Fig. 6–47). Mulberries are multiple drupelets borne on a fleshy receptacle.

Seeds

Seeds vary considerably in size, shape, structure, and mode of dissemination. The seeds contained in samaras (Fig. 6–46) take to the air on their small wings. The downy tufts of milkweed and dandelion seeds enable them to be carried great distances on wind currents (Fig. 6–48). The coconut is known to have floated to new land masses on ocean currents. Some seeds are attached to barbs or hooks or are contained within burrs or siliques that can catch in clothing or animal fur and

Figure 6–47
Pineapples are actually many seedless berries covering a large accessory structure. Source: Margaret McMahon, The Ohio State University.

Figure 6–48
The downy tufts on dandelion seeds allow them to be carried long distances by wind currents. Source: Margaret McMahon, The Ohio State University.

be transported great distances. Squirrels and other rodents bury nuts, many of which germinate later. Birds and other animals disseminate seeds by eating the fruit and passing resistant seeds through their digestive tract. Some kinds of seeds are forcefully expelled from dehiscing dried fruits as they mature. Some oxalis species are weeds that are difficult to control because the seeds are widely scattered when the seeds mature and dehisce. These varied means of seed dissemination aid in species survival by promoting plant growth in new locations with possibly different environments.

A seed is a mature ovule. The three basic parts are the embryo, the food storage tissue (endosperm, cotyledons, or perisperm), and the seed coats.

The **embryo** is a miniature plantlet formed within the seed from the union of the male and female gametes during fertilization. Basically, the embryo has two growing points: the **radicle**, which is the embryonic root, and the **plumule**, which is the embryonic shoot. One or two **cotyledons** are located between these two growing points on the root-shoot axis.

Food can be stored in the endosperm, cotyledons, or perisperm in the form of starch, fats, or proteins. Stored food in the cereal grains is largely endosperm. Seeds having a large portion of their food stored as endosperm are called **albuminous seeds**. Those seeds with no endosperm or only a thin layer surrounding the embryo are called **exalbuminous seeds**. Such seeds store food either in fleshy cotyledons, as in beans, or occasionally in the perisperm (developed nucellus), as in beets.

The seed coverings are usually tough, preventing damage to the enclosed embryo. They are also relatively impervious to water to save the embryo from desiccation, but again there are exceptions. There may be one or two **seed coats (testa)**, which form from the **integuments**, the outer layers of the ovule. In dry fruits such as achenes, samaras, and schizocarps, the fruit adheres closely to the seed coat; consequently the true fruit is usually called a seed. The pits or stones of such drupe fruits as peaches and apricots are often called seeds; however, they are actually the hardened inner wall (endocarp) of the pericarp that surrounds the seed.

The scar that remains after breaking the seed from the stalk is called the **hilum**, and the small opening near the hilum is the **micropyle**. The ridge on the seed is the **raphe** (see Fig. 6–4, page 80).

SUMMARY AND REVIEW

Plants are made up of cells, tissues, and organs. Tissues are groups of cells that perform a collective function. Some of the major tissues that make up a plant include meristems, pith, epidermis, cambium, xylem, phloem, palisade layer, and spongy mesophyll. Organs are made up of tissues that are grouped together to form complex systems such as roots, stems, leaves, flowers, and seeds. Each cell, tissue, and organ has a specific function.

Cells are made up of subcellular organelles including, but not limited to, the nucleus, mitochondria, chloroplasts, vacuoles, and endoplasmic reticulum. The organelles of the plant cell and the surrounding fluid matrix called the cytoplasm are contained within the cell membrane or plasmalemma. Surrounding the plasmalemma is the cell wall. Plasmodesmata are strands of cytoplasm that connect cells by extending through the plasmalemma and pits in the cell wall. Cells are the site of DNA and RNA synthesis, respiration, photosynthesis, protein synthesis, and other biochemical processes.

Tissues are collections of cells that have a specific function. Tissues can be simple or complex. Simple tissues are the meristems, epidermis, and parenchyma. Meristematic tissues, including apical, subapical, intercalary, and vascular cambium, are the regions of cell division (mitosis) and contribute to the growth of the plant by adding more cells. The epidermis is the exterior layer of living cells that protects all parts of the plant and includes the guard cells of the stomata. Some epidermal cells produce hairlike projections called trichomes. Parenchyma tissue is made up of thin-walled cells that are found in all parts of the plant, and they perform different functions depending on their location. These cells can become meristematic and regenerate the same or different tissues when a plant is injured.

Complex tissues are the vascular tissues, xylem and phloem, which move liquids and dissolved solids throughout the plant. Xylem tissue is nonliving and can be composed of vessels, tracheids, and fibers. Phloem tissue is living and is composed of sieve tubes, sieve tube members, companion cells, fibers, and parenchyma.

Plant organs are groups of tissues organized to perform very complex functions in the plant. They are the roots, stems, leaves, flowers, fruits, and seeds.

Roots can be tap or fibrous in form. They anchor the plant and provide the mechanism to bring water and nutrients into the plant. They are often the storage site of photosynthesized food. Parts of the root include the root cap, epidermis, root tip, cortex, endodermis, procambium layer, and stele (comprised of the vascular cylinder and pericycle).

The stem makes up the support system of the plant and supports the leaves, flowers, and fruit. Stem tissues vary from one plant to the next, but most, if not all, stems have an epidermis, cortex, vascular system of xylem and phloem and other tissues, and vascular cambium (not found in monocots). Stems can have different forms besides that usually associated with the aerial portion of a plant. These forms include rhizomes, stolons, corms, bulbs, and stem tubers.

Leaves are the site for most of the photosynthetic activity in a plant. Leaves also transpire large quantities of water to help regulate plant temperature and provide

the force for carrying xylem contents through the plant. Leaves are comprised of epidermis, vascular bundles, palisade, and spongy mesophyll cells. Quite often, leaf shape is a distinguishing characteristic of a plant.

Flowers are the sexual reproductive structures of angiosperms. Flower parts include sepals, petals, stamens, and pistils. When a flower has all four parts, it is called complete. Flowers missing one or more parts are called incomplete. The stamens and pistils are the male and female reproductive organs, respectively. Stamens are composed of anthers (site of pollen production) and filaments. Pistils are composed of stigma, style, and ovary (site of ovule production). Perfect flowers have both stamens and pistils. Imperfect flowers lack one or the other. Monoecious plants have both male and female flowers on the same plant. Dioecious plants

have female flowers on one plant and male flowers on another. Flowers can occur singly on a stalk (solitary) or there can be multiple flowers per stem (inflorescence).

Fruits are matured ovaries plus associated parts. Generally fruits are seed-bearing, but some fruits are parthenocarpic and are seedless. The ovary wall (pericarp) can develop in many different ways and often is a significant portion of the fruit. Fruits can be simple (developing from one ovary) and having one, few, or many seeds. Examples of simple fruits are berries (grapes and tomatoes), drupes (peaches and cherries), pomes (apples and pears), pods (beans), achenes (strawberries), caryopsis (grains), nuts, and others. Multiple and aggregate fruits form from several ovaries, for example, blackberries, raspberries, and strawberries (several fruits called achenes on a single swollen receptacle).

KNOWLEDGE CHECK

1. What are the major difference between angiosperms and gymnosperms?
2. What is the difference between prokaryotic and eukaryotic cells? Plant cells are which type?
3. What is the plasmalemma?
4. What process takes place in the chloroplasts? In the mitochondria?
5. How does the inheritance of DNA from the chloroplast and mitochondria differ from inheritance of DNA in the nucleus? What has this difference helped us to understand?
6. What role does the tonoplast play in what is stored in the vacuole?
7. What functions can the cell wall perform?
8. What are the strands of cytoplasmic material that connect individual cells called?
9. What are meristems (meristematic tissue)?
10. What kind of meristems are found in grass leaves?
11. What kind of meristem causes the girth of woody perennial plants to increase?
12. What is the epidermis? What usually covers it?
13. What are parenchyma cells and how are they involved in healing wounds?
14. What does xylem do?
15. What does phloem do?
16. In a root what is the function of the root cap, the Casparian strip, root hairs?
17. What are adventitious roots and where can they originate?
18. Secondary xylem and secondary phloem are produced by what tissue? On what side of that tissue is the xylem formed and on what side is phloem formed?
19. What is the difference between rhizomes and stolons?
20. What two types of leaf cells contain chlorophyll?
21. What are stomates?
22. How do guard cells control such things as the movement of water out of and carbon dioxide into the cell?
23. What is a bud?
24. What are the four parts that can be found in flowers? What is complete flower, a perfect flower?
25. What is the difference between monoecious and dioecius plants?
26. What is the difference between a solitary flower and an inflorescence?
27. What part of the plant is the fruit?
28. What is the pericarp and what are some fruit structures that develop from it?
29. Why is it incorrect to call a peach pit or a rice grain a seed?
30. What are commonly referred to on strawberries as the seeds are really what?
31. Endosperm, cotyledons, and perisperm serve what function in seeds?
32. What are the four parts of a plant embryo?

FOOD FOR THOUGHT

1. If you were talking with someone who seemed very confused because she or he had just heard that tomatoes are berries and not vegetables, what would you tell them to help them understand the difference between fruits and vegetables both botanically and dietarily?

2. Some fertilizers foster the growth of the top of the plant, some promote growth of subterranean plant parts more. Which one would you most likely choose for each of the following crops: carrots, kale, spinach, beets?

3. Some fertilizers foster vegetative growth, some promote reproductive growth. If you were a flower grower which would you choose when the plants were young and you wanted to develop the plants stems and leaves, and which would you choose when the plants got older and you wanted the flowers to develop?

FURTHER EXPLORATION

NABORS, M. W. 2004. *Introduction to botany.* Upper Saddle River, NJ: Pearson/Benjamin Cummings.

RAVEN, P. H., R. F. EVERT, AND S. E. EICHORN. 2005. *Biology of plants.* 7th ed. New York: Freeman.

*7

Plant Growth and Development

James D. Metzger

key learning concepts

After reading this chapter, you should be able to:

- Know the difference between plant growth and plant development and understand ways to measure each.

- Understand the factors that affect plant growth and development and what the effects are.

- Understand how those factors can be manipulated to control plant growth and development.

- Recognize the categories of plant hormones, understand their role in plant growth and development, and how they are used to control plant growth and development.

DEFINITIONS AND MEASUREMENTS

What is **plant growth** and how is it measured? We generally think of growth as an irreversible increase in volume or dry weight (biomass). The swelling of wood after it becomes wet is not growth because the wood will shrink upon drying. Growth can be measured as increases in fresh or dry weigh, or in volume, length, height, or surface area. As its gross size increases, a plant's form and shape change as directed by genetic factors with influence from the environment. Plant growth is a product of living cells, with all their myriad metabolic processes. A definition of plant growth is as follows: size increase by cell division and enlargement, including synthesis of new cellular material and organization of subcellular organelles. **Plant development** is its progress through its lifecyle. Figure 7–1 shows the difference between growth and development. Stages of development include seed germination, growth of vegetative organs and tissues, initiation and maturation of reproductive organs and tissues, fertilization, seed development and maturation, and senescence and death. With perennials, some of these are repeated many times during the complete lifestyle. Asexual propagation would substitute cutting rooting or some other initial stage for seed germination.

HOW THE PLANT GROWS

The importance of meristems in the growth of plants must be clearly understood. In dicotyledonous plants, vegetative buds at the shoot tips (**apical meristems**) and in the axils of leaves contain meristematic (actively dividing) cells that are capable of producing millions of cells along a longitudinal axis. Cell division, together with cell elongation and expansion, causes the shoots to grow. A growing point at a shoot tip, with its meristematic region producing new cells year after year for many hundreds of years, can account for the towering height of the redwood, Douglas fir, and other forest trees. Similar meristematic cells are also located in the root tip, just behind the root cap.

The thickening of the trunks of dicotyledonous trees is due to secondary growth produced by another meristematic region, the vascular cambium. It is a cell

Figure 7–1
The poinsettia on the right has more growth while the one on the left is more developed.
Source: Margaret McMahon, The Ohio State University.

layer that lies between the xylem and the phloem and encircles the tree from the roots to almost the top of the shoot. Cells of this vascular cambium also have the capability of dividing, producing both the permanent woody xylem tissues toward the inside that give the tree its girth and mechanical strength and the more fragile transitory phloem cells to the outside. External to the vascular cambium is another meristematic region, the cork cambium, which produces the cork in the bark layer.

In mature plants, vegetative apical meristems can become reproductive meristems and produce the floral parts needed for seed production.

Factors Affecting Plant Growth and Development

Genetic and environmental factors interact to determine how a plant grows and develops. Some of the most important ways these factors influence plant growth and development follow.

Genetic Factors   One of the marvels of biology is the manner in which the off-spring of plants and animals resemble, yet differ from, their parents. A peach tree, not a cherry tree, grows from a germinated peach seed. In a field planted with wheat seed, wheat plants develop, not oats or barley. The organism developing by cell division and elongation from the fertilized egg—the zygote—in every case is under the genetic control of the genes inherited from the parents at the time of fertilization.

As the plant enlarges from the fertilized egg or zygote to its mature size, many developmental processes take place. Genes direct the synthesis of

proteins that can be enzymes, structural proteins such as the mitotic spindles, and special proteins called **transcription factors** that regulate the activity of structural genes and other regulatory genes. At any given time, some of the organism's genes are transcriptionally active, while others are silent. The control of gene activity depends on the cell type, environmental conditions, or the particular stage of development. As development proceeds, genes are activated and deactivated, depending on signals received in the nucleus. The control of development through selective gene activation and deactivation is mediated mostly (but not entirely) through the action of transcription factors that turn on or turn off genes.

What are the signals that trigger the action of the regulatory genes? Although not clearly understood, biotic signals are believed to include plant hormones, certain inorganic ions, coenzymes, and other metabolites. The classes of hormones that most affect plant growth and development are auxin, gibberellin, cytokinin, and ethylene. A full discussion of the biological function these and other hormones and their uses as chemical plant growth regulators is at the end of this chapter. Environmental factors such as temperature or light can also function as signals during certain developmental stages. Thus, the particular combination of genes directs the form and size that each plant is finally to assume, as altered by environmental influences, either beneficial or deleterious. Many of the measures we use to control plant growth and development ultimately work by activating and deactivating gene transcription.

Environmental Factors   You learned about these factors in other chapters of this book, but for a quick refresher and in the context of affects on plant growth and development, we will review them here.

Light   The sun is the source of energy for photosynthesis (the production of carbohydrates), but a substantial amount of radiation is lost because of absorption and refraction as it passes through the atmosphere. The atmosphere does not absorb very much of the sun's radiation with wavelengths of 400 to 700 nm; this is important because this band of radiation, which is commonly referred to as **visible light** or **photosynthetic active radiation (PAR)**, is the most important to life on earth. Plants have many mechanisms to most efficiently capture light for photosynthesis. These mechanisms often involve changes in the way the plant grows and develops.

Radiation passing through the earth's atmosphere can be refracted, or scattered, resulting in significant reflection of light back into outer space and thus reducing

the amount of light available for photosynthesis. In addition to absorption and scattering, the amount of light reaching the earth's surface depends on the **angle of incidence** of the sunlight, that is, the angle that a beam of sunlight, makes with the earth's surface. An angle of incidence of 90° has the maximum amount of light striking an area. As the angle of incidence decreases from 90°, a greater proportion of the incident light is reflected, which explains why so much more sunlight is reflected around sunrise and sunset than at midday. An angle of incidence of 90° also spreads the light out over a greater region (Fig. 7–2). The angle of incidence of sunlight has two major effects on crop growth. First, light intensity is directly related to plant photosynthetic rates. As the sun rises the angle of incidence increases, light intensity increases, so does photosynthetic activity until the photosynthetic machinery is saturated. Then photosynthetic activity gradually declines as the day progresses toward sunset.

Not only does the angle of incidence affect the intensity of light, it also affects the amount of light entering the leaf and available for absorption by chlorophyll. Light striking a surface has two possible fates: it may penetrate the surface, where it is either absorbed by or transmitted through the material, or it can be reflected. The proportion of incident light that is either absorbed or transmitted by (or reflected from) a surface varies with the angle of incidence.

Light that is reflected from the surface of a leaf is lost for photosynthesis. Some plants such as soybean can compensate for the reduced light intensity and increased reflection at low solar angles through **heliotropic movements**, in which leaf angles are adjusted so that the sun's rays are normal, or perpendicular, *to the leaf* during most of the day. Heliotropic movements are also used to lower the amount of light absorbed by reducing the angle of incidence of sunlight

on the leaf when overheating might occur. Leaves are not the only plant organ that moves in response to solar movement. Perhaps the best known example of heliotropism occurs in *Helianthus annus* (sunflower) which gets both its scientfic and common name from its ability to keep its flowers facing the sun all day.

Another characteristic of light that is important for plant processes is quality. **Light quality** does not refer to how beneficial the light is for the plant, but rather to the relative quantity of individual component wavelengths contained in an incident beam of light. Better terms for light quality are *spectral composition* or *spectral distribution,* but these have not yet gained widespread acceptance by plant scientists. The band of wavelengths that affect plant processes range from about 380 nm to 800 nm, but individual processes such as photosynthesis have much more narrow requirements for the specific wavelengths that are most effective. For example, blue and red light at 440 and 650 nm, respectively, are much more effective in driving photosynthesis than is green light. Other light-controlled processes have different spectral requirements.

Light also affects plant growth and development through photomorphogenesis. **Photomorphogenesis** describes several highly integrated processes that are regulated to produce the shape or form of the plant, and they include seed germination in light-sensitive seeds, de-etiolation (the greening of young seedlings when they emerge from the soil), and stem growth in plants competing for light with other plants. A characteristic that distinguishes most photomorphogenic responses from photosynthesis is the relative insensitivity to light intensity. In fact, most photomorphogenic responses are fully induced by light intensities that are below the compensation point for photosynthesis. In addition, photomorphogenic responses typically have more specific requirements for the spectral composition of the incident light than does photosynthesis.

Most photomorphogenic responses are regulated by the phytochrome pigment system. Phytochrome is a pigment that has two interconvertible forms: a red light absorbing form and a far-red light absorbing form. Far-red light is the region of the spectrum with wavelengths between 700 and 800 nm, but the far-red absorbing form of phytochrome absorbs maximally in the region of 720 to 740 nm. As shown in Figure 7–3, a phytochrome molecule in the red absorbing form is converted to the far-red absorbing form following irradiation with red light. Irradiation with far-red light is required for the phytochrome molecule to be converted back to the red absorbing form. As a consequence, the

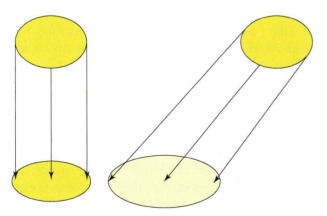

Figure 7–2

The effect of angle of incidence on how much light strikes an area. Source: Margaret McMahon, The Ohio State University.

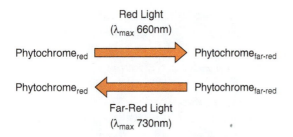

Red Light
(λ_{max} 660nm)

Phytochrome$_{red}$ → Phytochrome$_{far-red}$

Phytochrome$_{red}$ ← Phytochrome$_{far-red}$

Far-Red Light
(λ_{max} 730nm)

Figure 7–3

How red and far-red light affect phytochrome. Phytochrome$_{red}$ can absorb only red light, while phytochrome$_{far-red}$ can absorb only far-red light. The wavelength of light that is most effectively absorbed by the phytochrome molecule is signified by λ_{max}. Slightly longer and shorter wavelengths are also absorbed, but less effectively.

relative amounts of the far-red absorbing form compared to the red absorbing form are proportional to the ratio of red to far-red (R:FR) light in the environment. During most of the day, the R:FR ratio is about 1.2:1, so roughly speaking, about two-thirds of the total pool of phytochrome is in the far-red absorbing form. This is important because the far-red absorbing form resulting from irradiation with red light leads to different biological reactions compared to the red-absorbing form. Thus, plants have a highly sensitive system for sensing changes in the light environment by continuously measuring the R:FR ratio. The R:FR ratio declines dramatically as light penetrates through leaf canopies because red light is efficiently absorbed by chlorophyll, while the far-red light is either transmitted or reflected (Fig. 7–4). In some dense canopies, for example in a cornfield, R:FR ratios can be as low as 0.05. The R:FR ratio of light impinging on an individual plant is highly dependent on how close and, of course,

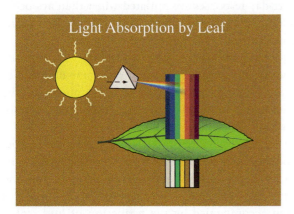

Light Absorption by Leaf

Figure 7–4

Diagram of how a leaf selectively absorbs light for photosynthesis. The dark red line on the far-right represents far-red which is not involved in photosynthesis but has other important effects on plants. Source: Margaret McMahon, The Ohio State University.

how big are the neighboring plants. The information contained in the R:FR ratio provides an accurate means to sense neighboring plants and to respond so that it can compete for resources, especially light (shade avoidance response).

Plants compete for light by redirecting growth and development so that they produce leaves that are above the leaves of their neighbors. It is no surprise then that plant architecture is highly dependent on the R:FR ratio under which the plant grows. Plants grown in light conditions in which the R:FR ratio is high (for example, open sunny areas) are typically compact, with dark green leaves and more branches and tillers. Plants respond to a decline in the R:FR ratio with increased height, reduced branching, and smaller stem diameters (Fig. 7–5). Chlorophyll synthesis declines when the R:FR ratio is low, so plants appear chlorotic even though soil fertility is adequate. These changes in plant architecture result in tall, spindly plants that are more prone to lodge (fall over) under typical field conditions and are more easily damaged during shipment. Such plants are also more susceptible to disease and environmental stress. Reduced branching and tillering under low R:FR ratios also results in lower yields. The practical implication of the neighbor detection system is that a maximum number of plants can be grown in an area; any number above the maximum means that yield or quality is adversely affected.

Another pigment system mediating photomorphogenic responses is the **cryptochrome** system. In contrast to phytochrome, this pigment is a blue light photoreceptor. Cryptochromes are a family of molecules that are involved in photomorphogenic responses as well as circadian (daily) rhythms. A blue light receptor, different than cryptochrome, called **phototropin**, is responsible for **phototropism** (movement in response to light) (heliotropism is a type of phototropism) and it may be involved with stomate opening. For the most part, above-ground organs such as stems, leaves, and flowers bend toward a unidirectional beam of light and are said to exhibit a positive phototropic response. The bending in positive phototropic responses is due to increased cell growth on the side away from the light source. It is believed that the plant hormone auxin accumulates on the shaded side, promoting cell expansion. In some species, the roots exhibit a negative phototropic response; that is, they grow away from the light source. Cryptochrome and phototropin complement the functions of phytochrome in providing the plant with more complete information about the light environment in which it is growing.

Figure 7–5
Right: plant was grown under a filter that blocked far-red but not red light (high R:FR). Center: plant was grown under normal sun light. Left: plant was grown under a filter that blocked red but not far-red (low R:FR). The intensity of photosynthetically active light was the same for all three plants. Source: Margaret McMahon, The Ohio State University.

Another characteristic of light that is important to plant growth and development is duration. **Photoperiodism** is the photomorphogenic response to variations in daylength. Numerous aspects of plant growth and development are controlled by photoperiod including flowering (which will be discussed in more detail later in this chapter); induction of bud dormancy in woody species; and the formation of vegetative propagules such as bulbs, tubers, corms, and runners (stolons).

All photoperiodically controlled processes can be categorized into three basic response types: **long-day plants (LDPs)**; short-day plants (SDPs); and **day-neutral plants (DNPs)**; which are photoperiodically insensitive (there are actually more response types, but the three described here represent the vast majority of species). The designation as a long- or short-day plant is not based on the absolute length of the day, but rather if the photoperiodically controlled process is induced only at daylengths longer or shorter than specific daylength, called the **critical daylength (CDL)**. A plant with photoperiodically controlled process that is induced only when the days are *longer* than the critical daylength is considered an LDP for that process, while an SDP represents the inverse situation: the process is induced only when the daylength is shorter than the CDL. It is important to understand that no direct relationship exists between the response type and the

absolute length of the CDL. For example, consider the flowering response of a typical LDP and an SDP. Red clover is an LDP with a CDL of twelve hours, meaning that it will not flower unless the daylength is longer than twelve hours. In contrast, the CDL for the hardy chrysanthemum, which is an SDP, is fifteen hours, considerably longer than red clover, a so-called long-day plant. In general, whether or not a plant is an SDP or LDP determines when the process is induced relative to the summer solstice. In other words, long-day responses are initiated prior to the longest day of the year, while short-day responses are initiated when the days begin to shorten, so that the shorter the CDL, the later in the summer or early fall the process is induced.

Next to flowering, the collection of integrated processes known as the autumn syndrome observed in many woody plants is the most visible photoperiodic process. The autumn syndrome describes a series of processes that occur in plants in temperate climates at the end of the summer in preparation for winter; these processes include acquisition of freeze tolerance; dormancy of buds; and, in deciduous trees, leaf fall. Dormancy is a temporary cessation of growth and is often accompanied by the formation of bud scales, modified leaves that protect the delicate shoot tips from desiccation during winter.

Dormancy is induced by short days in several familiar trees and shrubs, including red maple (*Acer*

rubrum), redbud (*Cercis canadensis*), American elm (*Ulmus americana*), eastern hemlock (*Tsuga canadensis*), and weigela (*Weigela florida*). However, there are some notable exceptions in which dormancy and other aspects of the autumn syndrome are not controlled by photoperiod. Some examples include ash (*Fraxinus* spp.), mountain ash (*Sorbus* spp.), and common fruit trees such as apple (*Malus*) and pear (*Pyrus*).

The critical daylength for dormancy induction can vary by several hours among members of a species, especially when individuals from different latitudes are compared. This variation maximizes the number of geographical areas a species can colonize. Table 7–1 provides a comparison of the approximate critical daylengths for the induction of bud dormancy in various woody plants at three different latitudes. Typically, the higher the latitude, the longer the CDL for the induction of bud dormancy (and other short-day responses as well). The longer CDL at higher latitudes results in increasingly earlier dates at which buds begin to go dormant, and this process is directly related to earlier onset of winter.

In addition to flowering and bud dormancy, other developmental processes are controlled by photoperiod. Many of these processes are related to strategies for survival in unfavorable environmental conditions such as the cold temperatures of winter or the hot, dry conditions of summer. Many herbaceous perennials survive winter with the formation of tubers that are protected by being buried underground; some examples of plants in which tuber formation is a photoperiodically controlled process are potatoes, dahlias, yams, Jerusalem artichokes, and tuberous begonias. In all these cases, tuber formation is a short-day process that is initiated at the end of the summer. The formation of vegetative rosettes at the base of certain herbaceous perennials and adventitious crown or root buds are often short-day induced processes that are specially adapted to survive winter.

In contrast, processes relating to vegetative reproduction are often induced in late spring and early summer by long days to take full advantage of the growing season. Examples include runner formation and growth in the strawberry plant and bulb formation in onion, garlic, and related species. (Note that bulb and corm formation in spring blooming ornamental species like tulip, hyacinth, and crocus are *not* under photoperiodic control.) In commercial onion production, it is important to match the critical daylength of a cultivar with the latitude in which it will be grown. Table 7–2 provides a comparison of the critical daylength for bulb formation in different onion cultivars. It is obvious that some of these cultivars (e.g., Yellow Zittau) would exhibit poor bulb formation if grown, say, in Orlando, Florida (28°N), in which the longest day of the year is about fourteen hours.

Temperature The seasonal variation of light intensity is responsible for the temperature changes from summer to winter in the various temperature zones. To make sure that the crops they grow have enough time to grow and develop, farmers depend on climatic records in their areas to predict the last day of frost in the spring and the number of available growing days before the first killing frost in the fall. The greater the distance from the equator, the fewer the number of available growing days to mature crops. It is possible to grow temperature-sensitive tomatoes in Alaska or northern Europe, but precautions must be taken to protect the seedlings from frost in early spring. Once summer arrives in these northern latitudes, the days are

Table 7–1
APPROXIMATE CRITICAL DAYLENGTH AND DAY OF THE YEAR WITH THAT DAYLENGTH FOR THE INDUCTION OF BUD DORMANCY IN VARIOUS WOODY PLANTS COLLECTED AT DIFFERENT LATITUDES IN SCANDINAVIA

56°N Latitude		63°N Latitude		70°N Latitude	
CDL	Date	CDL	Date	CDL	Date
14 hours	Sept. 1	16 hours	Aug.16	20 hours	Aug. 5

Source: Adapted from A. Håbjørg, *Horticultural Science* 23, no 3 (1988): 539. American Society of Horticultural Science, Alexandria, VA.

Table 7–2
THE EFFECT OF DAYLENGTH ON BULB FORMATION IN DIFFERENT CULTIVARS OF ONION*

Cultivar	Approximate Critical Daylength for Bulb Formation
Yellow Bermuda	12 hours
California Early Red	13 hours
Ohio Yellow Globe	13.5 hours
Italian Red	14 hours
Sweet Spanish	15 hours
Yellow Zittau	16 hours

*Cultivars bred to be grown at lower latitudes have a shorter critical daylength because maximum daylengths in the summer are shorter than at higher latitudes.

Source: R. Magruder and H. A. Allard, *Journal of Agricultural Research*; (1937): USDA Agricultural Research Service.

so long and the temperatures are warm enough that plants grow, flower, and set fruit rapidly. The growing season may be short, but plants develop quickly.

All plants have optimal temperatures for maximum vegetative growth and flowering, as noted for plants discussed as crops in subsequent chapters. Most temperate-region plants grow between temperatures of 4°C (39°F) and 50°C (122°F), but these are generally the limits of plant growth. The high temperatures destroy the protoplasm of most cells; however, some spores and seeds can withstand the temperature of boiling water for short periods. At the low temperatures, most plants just fail to grow because of a lack of cell activity. However, there are some arctic or mountain plants that function near freezing, but these are rare exceptions.

Plant parts are injured by very high temperatures, even if the exposure is short. Leaves may be **solarized** or sunburned when exposed to high light intensities. In the leaf, light energy converts to heat, which destroys the cells. Young trees in orchards are prone to sunscald, which kills the cambium layer just under the thin bark of the trunk and limbs. Injury can be prevented by whitewashing the bark to reflect the heat. Occasionally, plants suffer heat damage when the relative humidity drops suddenly because of hot, desiccating winds. Heat damage in intensively managed systems such as nurseries, golf course and athletic fields, and in greenhouses can be prevented if the relative humidity is increased by misting in the immediate vicinity of the plant.

The most common low-temperature injury is evident after the night of the first killing frost in the fall. Plant surfaces may be coated with frost depending on the dew point temperature, but soon after the sun shines on the leaves, the damage is evident in blackened leaves. The contents of the cells are damaged by the formation of ice crystals that rupture the cell membranes and walls, allowing water to flow out of the protoplasm and desiccating the cells. Frost damage to plants due to heat loss at night by radiation can be avoided to some extent by placing a covering between the plant and the clear sky on a calm night. This covering prevents radiant heat transfer from the leaves to the cold sky. Citrus and grape growers sometimes use smudge pots burning oil to create heat that radiates to the trees. In locations where there are temperature inversions above the plants (slightly warmer temperatures in a layer some distance above the soil than at ground level), wind machines may be useful. The power-driven propellers mix the warm air above with the cool air below, thus preventing low-temperature injury.

Many woody perennials grown in locations with severe winters enter into a rest period—brought on by shortening days in fall—and become resistant to low winter temperatures. A covering of snow or leaf litter or mulch helps the plants withstand the winter because it acts as a good insulator at extremely low temperatures.

Some plants of tropical origin may be injured at temperatures well above the freezing point—for some plants 13°C (55°F) is low enough to cause injury. This is referred to as a chilling injury. The leaves may wilt and never recover and developing fruits may not mature; some examples are avocado, banana, mango, okra, and tomato. Unripe tomatoes, which show pink color, are injured and do not finish the ripening process if refrigerated at 4°C (39°F) or less. Many ornamental foliage plants are susceptible to chilling injury. These plants need protection when being shipped during the winter from warm production areas such as Florida to colder areas. They should also be kept away from cold areas in homes and office buildings.

While we usually think of low temperatures as being injurious or as negatively affecting plant processes, low, nonfreezing temperatures are sometimes used by plants as cues to coordinate growth and development with the changing seasons. Temperatures for these cold-induced processes are usually in the range of 0°C (32°F) to 10°C (50°F). Examples of cold-induced processes include:

1. *Seed germination.* Some seeds require a period of time (usually several weeks) during which the seeds are imbibed at low temperatures (**stratification**) before germination is possible. The seeds must be kept moist because dry seeds cannot sense the cold temperatures.

2. *Flowering.* The cold induction of flowering is called **vernalization** and will be discussed in more detail later in this chapter.

3. *Dormancy breakage.* As spring approaches, bud dormancy and other processes (such as freeze tolerance induced in the fall in many woody plants) are gradually lost, while the ability to resume growth is regained. The environmental cue initiating these changes is often cold temperature. The duration required for complete loss of dormancy is called the **chilling requirement**. The length of the chilling requirement varies among species and even among cultivars of species.

4. *Acquisition of cold and freeze tolerance.* For many herbaceous perennials, the low, nonfreezing temperatures (0°C to 10°C) that frequently occur during autumn nights induce the physiological

processes responsible for the ability to survive the freezing temperatures of winter. The freeze tolerance acquired in the autumn is quickly lost when temperature rises the following spring. Unlike the first three cold-induced processes above which require several weeks of cold, the duration of the cold period required to induce maximum freeze tolerance is relatively short—about one to two weeks.

Water   Water is very important in the growth and development of plants. As you have or will learn in this book, water is required for: biological processes, as a part of the plant structure, nutrient and metabolite transport, as well as temperature control in the plant. When water is limiting, the above can be affected which impacts how the plant can grow and develop. Water also directly influences how plants grow. When water is plentiful, plants grow more succulently than when water is limited. Much of this difference in growth is due to how the cells that are enlarging are affected. When water is plentiful, the cells are turgid (filled with water). The plasmalemma presses against the developing cell wall. Under these conditions the wall elongates longitudinally and it is it not as thick as when the plasmalemma is not pressing on it. When water is limiting, the cell can undergo plasmolysis because it is not fully turgid. In plasmolysis the plasmalemma pulls away from the cell wall. Under those conditions as long as plasmolysis is not so severe that the cell desiccates and dies, the developing wall becomes thicker and does not elongate as much. These cells usually are shorter and stronger than cells that developed under turgid conditions, resulting in shorter, stronger stems. A strategy for keeping plants short is to carefully withhold water during cell elongation in the stems. This is called drying down. It should be noted that a grower must pay very close attention when using this practice, as it is often a very short time between drying down and drying up (desiccation), which is not good for plant quality.

Gases   The two gases most important to the growth of green plants are carbon dioxide (CO_2) and oxygen (O_2). Carbon dioxide is the third most abundant gas in the atmosphere, behind nitrogen (N_2) and O_2. Although used in great quantities for phototsynthesis, in relative amounts in the atmosphere, CO_2 is very small. Nitrogen is approximately 78 percent ($\approx$ 780,000 ppm), O_2 is approximately 21 percent ($\approx$ 210,000 ppm), and CO_2 is approximately 0.035 percent ($\approx$ 350 ppm). Stomatal opening and closing is regulated in part by the CO_2 level in the leaf, which is influenced strongly by photosynthesis. When the CO_2 concentration in the leaf cells behind the stomata is lower than that found in the atmosphere, stomata open. The absorption of CO_2 for photosynthesis keeps the concentration lower than atmospheric concentration and the stomata remain open if other conditions (especially water availability) are favorable. When sufficient water is not available to support the plant, the stomates close, thus blocking the entry of CO_2 and dramatically reducing the rate of photosynthesis, which can have a negative impact on growth.

Oxygen is important in the respiration of all plant parts. Respiration is the release of energy captured and stored in the carbohydrates (sugars) synthesized during photosynthesis. The released energy is used to drive the complex biochemical reactions needed for the growth and development of not only plants but also all living organisms. While a low concentration of O_2 is rarely a problem for the aerial portion of the plant, roots can experience low O_2 when the soil is flooded or compacted. As a result overall plant growth is negatively affected.

STAGES OF GROWTH AND DEVELOPMENT

Germination and Early Seedling Growth

In general, seed germination occurs in three stages: (1) imbibition (water uptake); (2) increase in biological activity; and (3) radicle (root) and shoot emergence (Fig. 7–6). There are many factors that control the start of germination including water availability, appropriate temperatures, presence or absence of light, as well as others. The specific factors needed vary tremendously among species and even varieties and cultivars within species. However, once a plant's requirements are met, then germination can begin. Germination starts with the seed imibing water. This causes cells to swell and the seed increases in size. Respiration and other biological activity, such as enzyme activity and synthesis, increase, thus decreasing the energy reserves of carbohydrates and lipids. Cells elongate and begin to divide and differentiate. The embryonic root (radicle) and shoot (plumule) emerge. The radicle quickly becomes a functioning root. Leaves emerging from the shoot start to photosynthesize and the new plant becomes independent of the energy reserves in the seed. Many of the techniques used in growing plants are designed to promote seed germination and seedling development.

Figure 7–6
The seed on the left has not begun to germinate, the center seed has imbibed water, the seed on the right has the radicle and plumule emerging.
Source: Margaret McMahon, The Ohio State University.

Vegetative Growth and Development

Shoot and Root Systems   In living plants, we see primarily the shoot system, in all its diverse patterns—from the mosses to the magnificent towering redwoods, oaks, and pines. In many crop plants grown for livestock forage—alfalfa, corn for silage, and pasture grasses—the entire shoot system is harvested. A beautiful lawn results from the shoot and leaf growth of millions of small grass plants. A timber tree cut for lumber is the product of many years of shoot growth. The much-admired potted foliage plants in our homes are the shoot and leaf systems of particular kinds of plants brought into cultivation many years ago after their discovery by plant explorers in tropical lands.

In growing most plants—with the exceptions, perhaps, of bonsai plants and container-grown ornamentals—we are interested in obtaining vigorous **vegetative growth** quickly as we can (Fig. 7–7). Strictly speaking, vegetative growth includes roots, shoots, and leaves but not reproductive structures. However, in most contexts vegetative growth refers to the aerial, nonreproductive parts of the plant. In the latter context, early, vigorous vegetative growth is important to growers of plants because it means a greater amount of shoot growth harvested (hay, silage, and some vegetables such as lettuce and celery) or greater amount of leaf area produced to nourish the developing roots, tubers , flowers, fruits, grains, or seeds, where those are the harvestable or usable parts.

Roots   The roots of plants are largely unseen and tend to be forgotten, but in the higher plants they are essential

Figure 7–7
Growers of bonsai (left) and potted flowering plants (right) usually want to inhibit vegetative growth to create an effect or enhance aesthetic appeal. Whereas, growers of other crops generally prefer at least initial vegetative growth to be vigorous.
Source: Margaret McMahon, The Ohio State University.

for growth. The four principal functions of roots in higher plants are:

1. Anchoring plants in the soil.
2. Absorbing water and mineral nutrients.
3. Conducting water and dissolved minerals, as well as organic materials to other parts of the plant.
4. Storing food materials, for example, in plants such as sweet potatoes, sugar beets, and carrots.

The roots of some plants can also function in vegetative reproduction, as in root cuttings.

Root Growth Patterns Once the radicle emerges and enters the soil, the root grows through the soil bringing it in contact with water and nutrients. Root hairs which are an extension of the epidermal cells of young roots are the site of most water and nutrient absorption. These are fragile and constantly have to be replaced by root hairs from new cells. That is why actively growing roots are required for proper growth and development.

The root system and the shoot system maintain a balance. As the top of the plant grows larger and larger, the leaf area increases and water loss through transpiration increases. This increased water loss is made up by water absorption from an increasing root system. The enlarging shoot and root systems also requires greater amounts of water and mineral absorption for growth.

Various studies of root growth patterns conclude that the roots of mature deciduous woody perennials start to grow in the spring before bud break. In some species, root growth stops at various times in the summer; but in others, the roots continue to grow until after leaf drop in the autumn. Growth peaks in the spring and again in late summer or early autumn. Roots of some species, however, continue to grow even during the winter months whenever soil temperatures and moisture are favorable. Not all roots of a tree may be growing at any one time. While some are actively growing, others may be quiescent. The spring flush of root growth results from the accumulated foods stored in the tree the previous year. When this source is depleted, root growth slows but, following gradual accumulation of carbohydrates from photosynthesis through the summer, root growth again increases in the autumn. It does not appear that the growing points of the roots are under the same control as the shoot buds, which in many woody perennials go into a resting period in late summer and require chilling through the winter to be reactivated.

Shoots Plant shoot growth can be classified as **determinate** or **indeterminate** (Fig. 7–8). In the case of determinate growth, after a certain period of vegetative growth, flower bud clusters form at the shoot terminals so that most shoot elongation stops. Many vegetable species and cultivars grow this way, remaining small bush plants; bush snap beans are an example of a determinate type. Plants with indeterminate growth bear the

Figure 7–8

The tomatoes shown on the left have an indeterminate growth pattern, the poinsettias on the right have a determinate pattern.

Source: Margaret McMahon, The Ohio State University.

flower clusters laterally along the stems in the axils of the leaves so that the shoot terminals remain vegetative and the shoot continues to grow until it is stopped by senescence or some environmental influence. Trailing pole beans, grapevines, and forest trees are examples of plants with an indeterminate growth pattern. The determinate, bush-type plants produce much less vegetative growth than do the indeterminate type.

Shoot Growth Patterns: Annuals, Biennials, and Perennials   **Annuals**, which are herbaceous (non-woody) plants, complete their life cycle (seed to seed) in one growing season. Shoot growth commences after seed germination and continues in a fairly uniform pattern—provided no environmental influences are limiting—until growth is stopped by frost or some senescence-inducing factor. Flowering, followed by fruit and seed production, occurs at intervals through the growing season. General growth curves for the annuals are shown in Figure 7–9, a detailed growth curve for barley, an herbaceous annual, is shown in Figure 7–10.

Figure 7–11 shows various events in the life cycle of a typical **angiosperm** annual plant. All these events occur during a single summer growing season.

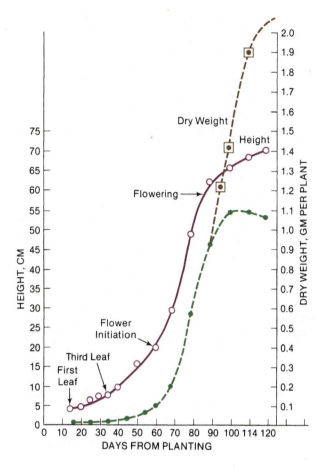

Figure 7–10
Growth curve of a field-grown barley plant from leaf emergence to grain maturity. ○ = plant height. ● = dry weight of plant minus grain weight. ⊡ = dry weight of plant plus grain weight. Source: Adapted from G. R. Noggle and G. J. Fritz, *Introductory Plant Physiology* (Englewood Cliffs, NJ: Prentice Hall, 1976).

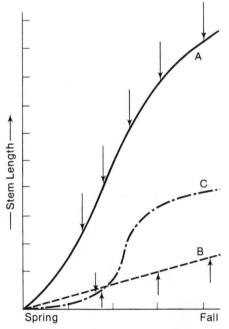

Figure 7–9
Vegetative growth patterns of annual plants.
(A) Indeterminate vine-type plants. (B) Determinate, bush-type plants. (C) Terminal-flowering plants, such as cereals and grasses. Arrows indicate times of flower initiation.
Source: Adapted from L. Rappaport and R. M. Sachs, *Physiology of Cultivated Plants* (Davis, CA: University of California–Davis, 1976).

The **biennials**, which are herbaceous plants, require two growing seasons (not necessarily two years) to complete their life cycle (seed to seed). As shown in Figures 7–12 and 7–13, stem growth is limited during the first growing season. The plants remain alive but dormant through the winter. Exposure to chilling temperatures triggers hormonal changes leading to stem elongation, flowering, fruit formation, and seed set during the second growing season. Senescence and death of the plant follows shortly thereafter. Examples of herbaceous biennials are celery, Swiss chard, beets, and cole crops such as cabbage and Brussels sprouts.

Most annual and biennial plants flower and fruit only once before dying. The production of the flowers and fruits or, perhaps, just the flowering stimulus itself apparently causes the plants to senesce and die. In such plants, continued removal of flowers and fruits often delays senescence.

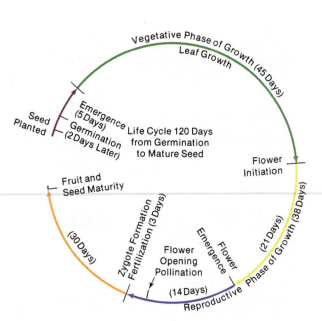

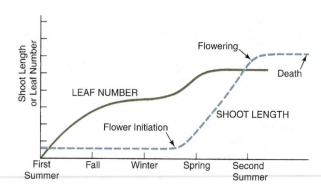

Figure 7–12
Growth curves of biennial plants during the first growing season (vegetative growth only), a required winter chilling-period, and a second growing season (flowering, fruiting, and seed production). Source: Adapted from L. Rappaport and R. M. Sachs, *Physiology of Cultivated Plants* (Davis, CA: University of California–Davis, 1976).

Figure 7–11
Events in the life cycle of a typical annual plant—from seed planting to seed maturity—accomplished in four months.
Source: Adapted from G. R. Noggle and G. J. Fritz, *Introductory Plant Physiology* (Englewood Cliffs, NJ: Prentice Hall, 1976).

Perennials are either herbaceous or woody. In herbaceous perennials, the roots and shoots can remain alive indefinitely but the shoot system may be killed by frosts in cold-winter regions or by senescence-inducing factors. Shoot growth resumes each spring from latent or **adventitious** buds at the crown of the plant. Figure 7–14 shows typical growth patterns for two types of herbaceous perennials. Many tropical, subtropical and warm-temperate herbaceous perennial ornamental plants (e.g., pelargoniums) are grown as annuals in areas with severe winters, such as the Midwest region of the United States. When grown in areas with mild winters such as parts of the South and West (United States), these plants exhibit their perennial characteristics.

In woody perennial plants, both the shoot and root system remain alive indefinitely, each growing to the ultimate size for the particular plant as programmed by its gene complement and the environment in which it is growing. Shoot growth of temperate zone plants takes place annually during the growing season, as indicated by Figure 7–15, adding to the growth accumulated in previous seasons. The magnitude of growth can vary considerably from season to season under the influence of several environmental

Figure 7–13
Hollyhocks are a biennial plant. The plant on the right is two years old and flowering. The one on the left is in its first growing season and will remain vegetative. Source: Margaret McMahon, The Ohio State University.

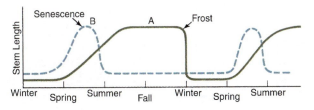

Figure 7–14

Shoot growth patterns of herbaceous perennials over a two-year period. (A) Plants such as garden (outdoor) chrysanthemums, peony, and phlox whose growth is stopped by cold weather in the fall. (B) Bulbous plants such as tulips, narcissus, and hyacinths, whose growth is terminated after spring flowering. Source: Adapted from Rappaport, L., and R. M. Sachs, 1976. *Physiology of Cultivated Plants.* Davis, CA: UCD Bookstore.

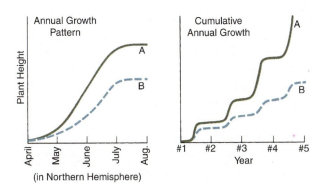

Figure 7–15

(Left) Growth patterns for temperate zone woody perennials in the northern hemisphere during one growing season, and (Right) over a period of several years. (A) Curve for a rapid-growing species such as poplar. (B) Curve for a slow-growing species such as oak. Source: Adapted from L. Rappaport and R. M. Sachs, *Physiology of Cultivated Plants* (Davis, CA: University of California–Davis, 1976).

Table 7–3

COMPARISON OF THE DURATION OF THE JUVENILE PHASE IN VARIOUS SPECIES

Species (Common Name)	Duration of Juvenile Phase
Chenopodium rubrum (coast blite)	0
Pharbitis nil (Japanese morning glory)	0
Perilla crispa (perilla)	1–2 months
Bryophyllum daigremontianum (bryophyllum)	1–2 years
Malus pumila (apple)	6–8 years
Citrus sinensis (orange)	6–7 years
Citrus paradisi (grapefruit)	6–8 years
Pinus sylvestris (Scotch pine)	5–10 years
Betula pubescens (birch)	5–10 years
Pyrus communis (pear)	8–12 years
Larix decidua (European larch)	10–15 years
Pseudotsuga menziesii (Douglas fir)	15–20 years
Fraxinus excelsia (ash)	15–20 years
Acer pseudoplatanus (sycamore maple)	15–20 years
Picea abies (Norway spruce)	20–25 years
Abies alba (white fir)	25–30 years
Quercus robur (English oak)	25–30 years
Fagus sylvatica (European beech)	30–40 years

Source: J. D. Metzger, "Hormones in reproductive development." In P. J. Davies (ed.), *Plant hormones*, 2nd ed. (Berlin: Springer-Verlag, 1995). Used with kind permission of Springer Science and Business Media.

factors. For some tropical trees, and all temperate trees, growth occurs intermittently—that is, in flushes—during the growing season. There can be a single or multiple flush of growth during the growing season depending on the genetic makeup of the species and environmental conditions such as rainfall and drought periods.

Phase Change: Juvenility, Maturation, Senescence

A newly emerged seedling undergoes a phasic development throughout its life that is essentially the same as in animals. It will pass through embryonic growth; juvenility; a transition stage, which in plants is called phase change; maturity or **adult phase**; senescence; and death. The **juvenile phase** is characterized by the inability to reproduce sexually; in angiosperms, plants at this stage cannot flower, even though conditions are permissive. Flowering occurs only in plants that have reached the adult phase following phase change. Adult plants that have not flowered because the conditions are not proper are said to be ripe to flower. The duration of the juvenile phase varies from a week or two up to thirty or forty years in some tree species. It should be noted, however, that plants do not measure time in terms of calendar days, but rather with some factor related to an increase in plant size, probably number of nodes (leaves) produced. From a practical point of view, juvenility is a serious obstacle in breeding programs for economically important fruit and forest trees. A comparison of the duration of the juvenile phase in various plant species is shown in Table 7–3.

The morphology of juvenile and adult plants is often quite different. For example, juvenile acacia leaves are bipinnately compound, while the adult form appears linear (Fig. 7–16). Juvenile eucalyptus trees have opposite leaves that are broad and lack a petiole. Adult eucalyptus leaves assume a different appearance: they become alternate, are narrower, and have a distinct petiole. Juvenile citrus seedlings are thorny, but thorns are not produced after the plant undergoes phase change.

Figure 7–16

Acacia melanoxylon *seedling showing phase change from juvenile to mature form. Lower, juvenile leaves have compound bipinnate structure. Upper, mature leaves are actually expanded petioles (phyllodia). Transition stages are evident in between.*

Source: D. E. Kester and H. T. Hudson, *Plant Propagation: Principles & Practices,* 4th ed. © 1983. Adapted by permission of Pearson Education, Inc., Upper Saddle River, NJ.

The growth form of juvenile and adult plants may also be drastically different. The familiar English ivy (*Hedera helix*) in its juvenile stage has three or five leaves and a hairy vinelike (**plagiotropic** growth) stem. Branches formed after phase change exhibit a much more upright growth habit (**orthotropic** growth) and are hairless; leaves are nonlobed ovate and opposite. In all of these cases, the adult plant is actually a chimera (see Chapter 9 for an explanation of chimeras), with the adult portion located with the youngest portion of the plant, while the juvenile portion is found at the base and oldest part of the plant. A summary comparing morphological features of juvenile and adult plants in various species can be found in Table 7–4. Rooting ability of cuttings is one of these features that is of significant economic importance. Plant propagators would like to maintain the juvenile stage in stock plants as long as possible so that cuttings taken from them will root properly and in high percentages.

Table 7–4

COMPARISON OF MORPHOLOGICAL CHARACTERISTICS ASSOCIATED WITH THE JUVENILE AND ADULT STATES OF VARIOUS SPECIES

Characteristic	Species	Juvenile Form	Adult Form
Growth habit	*Hedera helix*	Plagiotropic	Orthotropic
	Ficus punula	Plagiotropic	Orthotropic
	Euonymus radicans	Plagiotropic	Orthotropic
Leaf shape	*Cupressus* spp.	Acicular	Scalelike
	Acacia spp.	Pinnate	Phyllodes
	Eucalyptus spp.	Oval, sessile	Lanceolate with petioles
	Pinus spp.	Flat, glaucous	Scale- and bractlike
	Hedera helix	Palmate	Ovate, entire
Phyllotaxis	*Eucalyptus* spp.	Opposite	Alternate
	Hedera helix	Alternate	Spiral
Anthocyanin pigmentation in leaves	*Malus pumila*	Present	Absent
	Carya illinoisiensis	Present	Absent
	Acer rubrum	Present	Absent
	Hedera helix	Present	Absent
Thorniness	*Robinia pseudoacacia*	Thorns	No thorns
	Citrus spp.	Thorns	No thorns
Autumn leaf Abscission in deciduous trees	*Fagus sylvatica*	Keep leaves	
	Quercus spp.	Keep leaves	Abscise
	Robinia pseudoacacia	Keep leaves	Abscise
	Carpinus spp.	Keep leaves	Abscise
Rooting ability of cuttings	*Hedera helix*	Will root	Will not root
	Quercus spp.	Will root	Will not root
	Fagus sylvatica	Will root	Will not root
	Pinus spp.	Will root	Will not root
	Pyrus malus	Will root	Will not root

Source: J. D. Metzger, "Hormones in reproductive development." In P. J. Davies (ed.), *Plant hormones,* 2nd ed. (Berlin: Springer-Verlag, 1995). Used with kind permission of Springer Science and Business Media.

REPRODUCTIVE GROWTH AND DEVELOPMENT

Fruit and seed production involves several phases:

1. Flower induction and initiation
2. Flower differentiation and development
3. Pollination
4. Fertilization
5. Fruit set and seed formation
6. Growth and maturation of fruit and seed
7. Fruit senescence

Flower Induction and Initiation

When a plant is mature, a change can occur in an apical meristem that switches it from being vegetative (producing shoots and leaves) to reproductive (producing flowers) (Fig. 7–17). Some annuals mature and can flower in only a few days or weeks after the seeds sprout; some forest and fruit trees require years before flowering. Once mature, the plant can be induced to flower by becoming sensitive to the conditions of its environment. What brings about the formation of flowers? The majority of agricultural plants are self-inductive for flowering; that is, they initiate or form flowers when they reach a certain morphological maturity that is often determined by how much heat the plant has received (accumulated heat). Most fruit trees, shrubs, woody plants, garden perennials, and vegetable crops (beans, peas, tomatoes, peppers, cucumbers) have self-induced flowering. **Growing degree days (GDD)** are a way to measure heat accumulation. For many plants the GDDs needed for flowering or seed maturation is known and can be used to predict or control the time of flowering or harvest. Easter lilies and sweet corn

are two crops where GDDs are important for timing and harvest, respectively. Phenology is the study of life-cycle responses to annual and seasonal variations in temperature and other climate factors. In other plants, flowering is controlled by is daylength (photoperiodic effect) and/or low temperatures (vernalization).

Photoperiodism (Daylength)   Earlier in this chapter, we discussed the general features exhibited by photoperiodically controlled processes. In this section, we will consider in more detail the photoperiodic control of flowering. The influence of the photoperiod on the flowering of several plant species was first studied in detail by the USDA at Beltsville, Maryland, and the results were published by W. W. Garner and H. A. Allard in 1920. They grew plants in containers that could be wheeled into dark sheds at the end of the workday and returned to the sunlight in the morning. They found that Maryland Mammoth tobacco and certain cultivars of soybeans and cosmos required short days for flower induction. A set number of successive short days was required to complete differentiation (change from vegetative to reproductive growing points or shoot terminals). These plants, such as strawberry, poinsettia, and soybean, were called short-day plants. Later studies by many other workers found that long days were necessary to induce flowering of some plants, such as spinach, sugar beets, winter barley, and carnations. These were classified as long-day plants. A third group of plants, including tomato, corn, fruit trees, and cucumber, was those in which flowering was not affected by daylength and were called day-neutral plants. This phenomenon whereby daylength controls certain plant processes, as noted above, was termed photoperiodism. Other workers in later experiments

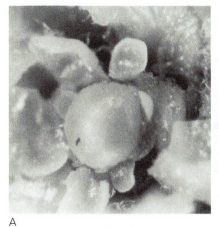
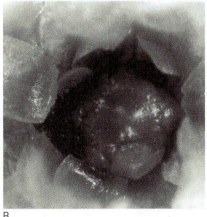
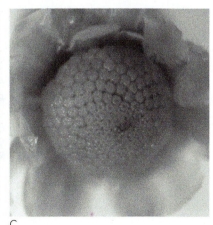

A   B   C

Figure 7–17

Transition of an apical meristem of chrysanthemum from (A) vegetative (note the smooth, rounded surface of the meristem); to (B) floral structures beginning to develop; to (C) fully developed flower bud. Chrysanthemums are in the Asteraceae family, so there are multiple flowers on the flower head or capitulum. Source: Margaret McMahon, The Ohio State University.

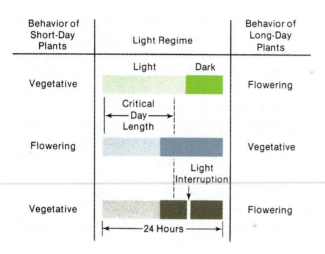

Figure 7–18

The effect of light interruption of the dark period on flowering in short-day and long-day plants. Source: A. W. Galston and P. J. Davies, *Control Mechanisms in Plant Development*, 1st ed. © 1970. Reprinted by permission of Pearson Education, Inc., Upper Saddle River, NJ.

Figure 7–19

These chrysanthemums were grown during the summer and forced into flowering uniformly by drawing blackout cloth over them every night from about 6:00 P.M. until 8:00 A.M. Under normal photoperiods, these plants would not flower until late fall. Source: Margaret McMahon, The Ohio State University.

found this phenomenon to be more complicated and that many plants did not fit nicely into these three categories because of interactions of daylength with temperatures.

Figure 7–18 demonstrates how a flash of light (or night break) of sufficient intensity or duration inhibits flowering of a short-day plant (long-night plant) but may induce flowering of a long-day plant (LDP). This information has been useful to commercial chrysanthemum growers who grow these short-day plants on a year-round schedule. When they want the young plants to reach a size adequate for flowering, the growers use incandescent or fluorescent lamps over the chrysanthemum plants in the middle of the night, each night for one to four hours, depending on time of year and latitude. This night break inhibits flowering until the plants reach the desired height. Conversely, when the natural daylight of summer is too long for chrysanthemum plants to flower, they cover the plants of proper size with black cloth or plastic each evening about 6 P.M. and remove it in the morning at about 8 A.M. This shortens the plant's day (lengthens the night) enough to induce and fully develop the flowers. These manipulations enable growers to have chrysanthemums uniformly in flower for every day of the year (Fig. 7–19). Poinsettias flower the same way and could be produced year-round. However, because of the association with Christmas, the public will purchase poinsettias only for Christmas.

Further studies have shown that there is a daily fluctuation in the sensitivity to light such that, depending on the time of day, light either inhibits or promotes flowering. This cycle of sensitivity to light is one of many **circadian rhythms** exhibited by plants and animals alike. Circadian rhythms are biological rhythms that complete one cycle in approximately 24 hours; they are typically initiated, or entrained, by transitions between darkness and light, as occurs at dawn. (Some circadian rhythms are entrained by the changes in temperature that occur between day and night.) The circadian rhythms of light sensitivity in SDPs and LDPs are qualitatively identical, but they differ in time of day when the promotive and inhibitory stages are maximally expressed.

Flowering Signal

As is the case for other photoperiodically regulated processes, the pigment system responsible for sensing light is phytochrome. The phytochrome system is located in the leaf, yet it is the apical meristem that must go through a transition to produce floral organs instead of leaves. Thus, both positive and negative signals are transported via the phloem from the exposed leaves to the apical meristem, the timing of which is determined by the response type and the amount of time that has elapsed after the start of the day. Most of the attention has been focused on the positive signal, which promotes photoperiodic flowering. This signal is sometimes called **florigen**, or the floral stimulus. The existence of the floral stimulus was discovered by partial leaf removal when plants were placed under the inductive photoperiod for flower formation, then failed to flower. The SDP *Xanthium* (cocklebur) exposed to

long noninductive days initiated flowers if light was blocked from a single leaf. Some experiments in which flowering plants (donor) were grafted to nonflowering plants (receptor) caused the latter group to flower. These and other experiments gave rise to the theory that a flowering hormone (named florigen), which might be similar in all plants, was responsible for flower induction. For around seventy years, researchers looked for florigen with no success. Then, in 2001, researchers discovered a gene called CONSTANS that produces a transcription factor protein that acts as the signal for flowering by turning on the FT (flowering locus T) gene that starts the conversion of apical buds to flower buds. The activity of CONSTANS appears to be mediated by phytochrome and chryptochrome. Current research is investigating more thoroughly how this protein works to influence flowering.

There has been considerable documentation to categorize plants as short-day, long-day, or day-neutral. Although not complete, one such list appears in Table 7–5. The critical daylength of many of the species shown in the table may be changed by a slight shift in temperature above or below the optimum for that species.

Understanding photoperiodic response allows crop producers to select species and cultivars that flower and seed at the right time for their geographic location or market window.

Low Temperature Induction   Some plants, including many of the biennials, require low temperature for flower induction. The term for this is vernalization, which means "making ready for spring." It was first observed in winter wheat over a century ago. Vernalization is any cold temperature treatment that induces or promotes flowering. The temperatures required to vernalize a given plant and the length of the vernalization period vary among species and may even differ among cultivars of the same species. Broadly speaking, however, vernalization temperatures range between 0° and 10°C (32° and 50°F). An example is the olive tree (*Olea europaea*), which needs chilling temperatures for induction of flower parts. In the kiwifruit (*Actinidia deliciosa*), there is no evidence of reproductive structures in the bud until after exposure to chilling temperatures. Some of the biennials that require vernalization are beets, Brussels sprouts, carrots, celery, and some garden flowers such as Canterbury bells and foxglove. Winter annuals—such as the cereal crops, barley, oats, rye, and wheat—also respond to cold by flowering. Some plants, such as lettuce, peas, and spinach, can be induced to flower earlier with vernalization, but

vernalization is not an absolute requirement; they will eventually flower without it. Some species can be vernalized as seeds (beet and kohlrabi), but most plants must reach a minimum size or produce a certain number of leaves to be sensitized by the cold. Bulb plants, such as the hyacinth, narcissus, tulip, and some lilies required low temperatures to induce or promote flower development in the bulb. Growers who produce these as flowering plants for spring sales vernalize the bulbs in coolers then bring them into the greenhouse where the warm temperatures cause the plants to grow and flower. This process is called "**forcing**."

Many herbaceous perennials, as well as plants with corms or tubers, and many flowering shrubs and fruit trees require low temperatures to overcome the rest period, but few require low temperatures for flower induction.

Flower Development

Once flowering is induced for whatever reason, there is a change in the apical meristem from vegetative (producing shoots and leaves) to reproductive (producing floral parts). Once the apex has changed to a flower primordium, the process is not reversible. The floral apex may abort, however, if the subsequent environmental conditions are not favorable for full flower development. For example, short day plants are returned to long day conditions. In such a case, the axillary buds below the aborted floral apex usually grow vegetatively (**bypass growth**) (Fig. 7–20) until daylength or temperature conditions once again are favorable for flower induction.

Figure 7–20
These poinsettias are exhibiting the start of bypass growth that resulted from being placed back in long day conditions after flower induction under short days. The flowers have aborted. Source: Margaret McMahon, The Ohio State University.

Table 7–5
A Partial List of Long-Day, Short-Day, and Day-Neutral Plants

Long-Day Plants	Length of Daily Light Period Necessary for Flowering	Short-Day Plants	Length of Daily Light Period Necessary for Flowering
Althea (*Hibiscus syriacus*)	More than 12 hours	Bryophyllum (*Bryophyllum pinnatum*)	Less than 12 hours
Baby's breath (*Gypsophila paniculata*)	16 hours	Chrysanthemum (*Chrysanthemum × morifolium*)	15 hours
Barley, winter (*Hordeum vulgare*)	12 hours	Cocklebur (*Xanthium strumarium*)	15.6 hours
Bentgrass (*Agrostis palustris*)	16 hours	Cosmos, Klondyke (*Cosmos sulphureus*)	14 hours
Canary-grass (*Phalaris arundinacea*)	12.5 hours	Cotton, Upland (*Gossypium hirsutum*)	14 hours
Chrysanthemum frutescens	12 hours		
Chrysanthemum maximum	12 hours	Kalanchoe (*Kalanchoe blossfeldiana*)	12 hours
Clover, red (*Trifolium pratense*)	12 hours		
Coneflower (*Rudbeckia bicolor*)	10 hours	Orchid (*Cattleya trianae*)	9 hours
Dill (*Anethum graveolens*)	11 hours	Perilla, Common (*Perilla crispa*)	14 hours
Fuchsia hybrida	12 hours	Poinsettia (*Euphorbia pulcherrima*)	12.5 hours
Henbane, annula (*Hyoscyamus niger*)	10 hours	Rice, winter (*Oryza sativa*)	12 hours
Oat (*Avena sativa*)	9 hours	Soybean (*Glycine max*)	12 hours
Orchardgrass (*Dactylis glomerata*)	12 hours	Strawberry (*Fragaria × Ananasia*)	10 hours
Ryegrass, early perennial (*Lolium perenne*)	9 hours	Tobacco, Maryland Mammoth (*Nicotiana tabacum*)	14 hours
Ryegrass, Italian (*Lolium italicum*)	11 hours	Violet (*Viola papilionacea*)	11 hours
Ryegrass, late perennial (*Lolium perenne*)	13 hours		
Sedum (*Sedum spectabile*)	13 hours		
Spinach (*Spinacia oleracea*)	13 hours		
Timothy, hay (*Phleum pratensis*)	12 hours		
Timothy, pasture (*Phleum nodosum*)	14.5 hours		
Wheat, winter (*Triticum aestivum*)	12 hours		
Wheatgrass (*Agropyron smithii*)	10 hours		

Day-Neutral Plants

Balsam (*Impatiens balsamina*)
Bluegrass, annual (*Poa annua*)
Buckwheat (*Fagopyrum tataricum*)
Cape jasmine (*Gardenia jasminoides*)
Corn (maize) (*Zea mays*)
Cucumber (*Cucumis sativus*)
English holly (*Ilex aquifolium*)
Euphorbia (*Euphorbia peplus*)
Fruit and nut tree species
Globe-amaranth (*Gomphrina globosa*)
Grapes (*Vitis* spp.)
Honesty (*Lunaria annua*)
Kidney bean (*Phaseolus vulgaris*)
Pea (*Pisum sativum*)
Scrofularia (*Scrofularia peregrina*)
Senecio (*Senecio vulgaris*)
Strawberry, everbearing (*Fragaria × Ananasia*)
Tomato (*Lycopersicon lycopersicum*)
Viburnum (*Viburnum* spp.)

Source: Modified from *Plant Physiology*, 3rd edition by Salisbury/Ross. © 1985. Reprinted with permission of Brooks/Cole, a division of Thomson Learning: www.thomsonrights.com. fax: 800-730-2215.

The number of days from flower initiation to anthesis (time of flower opening) depends on the species and the cultivar. Generally the time is increased if temperatures are low and decreased with warm temperatures (low and warm being relevant to the temperature requirements of the plant). Extreme changes in temperature however can cause the flowers to abort or be deformed.

Pollination

In the production of most floral crops and flowering shrubs—for example, carnations, petunias, chrysanthemums, roses, and camellias—the flower itself is the desired product. There is little interest in any resulting fruits and seed, except in the case of the plant breeder working with such species. In fact, flower growers often use techniques to prevent pollination in order to slow flower senescence. But in the food crops—the cereals, fruits, and many vegetable species—the postflowering structures are the desired products. It is the fruits, grains, and seeds that are harvested.

Fruit, grain, and seed formation starts with **pollination**, which is the transfer of pollen from an anther to a stigma in angiosperms. The anther and stigma may be in the same flower (self-pollination), in different flowers on the same plant (self-pollination), in different flowers on different plants of the same clone (self-pollination), or in different flowers on plants of different cultivars (cross-pollination).

Pollen grains come in many sizes and shapes and, while essential for sexual plant reproduction, can be devastating to many people as allergy producers. See Figure 7–21.

Figure 7–22 shows the various parts of a simple flower dependent upon pollination for fruit set.

If a plant is **self-fertile**, it produces fruit and seed with its own pollen, without the transfer of pollen from another plant. If it is **self-sterile**, it cannot set fruit and seed with its own pollen, but instead requires pollen from another plant, usually of a different clone. Often this is due to **incompatibility**, where a plant's own pollen will not grow through the style into its embryo sac (Figs. 7–22 and 7–23). Sometimes, too, cross-pollination between two particular cultivars is ineffective because of incompatibility, which is believed to be due to factors that inhibit pollen tube germination or elongation.

Pollen transfer from the anthers to the stigmas is principally by:

1. Insects (see Fig. 7–24). Insect pollination is common among cultivars with white or brightly colored flower parts and attractive nectar. Most

Figure 7–21

Scanning electron micrograph of pollen grains produced in a male cone of red pine (Pinus resinosa). Source: USDA.

fruit crops, many vegetables, and legume forage crops are pollinated by insects.

Adequate pollination is so important in many crops that considerable efforts are made to aid the bees in their pollen distribution. Hives are placed in or near the fields needing pollination. Also pollen inserts are placed at the entrances of hives to coat bees with pollen as they enter and exit.

2. Wind. This is the main pollinating agent for plants with inconspicuous flowers—the grasses, cereal grain crops, and forest tree species, as well as some fruit and nut crops such as the olive, walnut, pistachio, and pecan.

Other pollinating agents are water, snails, slugs, birds, and bats.

Figure 7–23 shows a longitudinal section through the pistil of a flower following pollination. Note the elongated pollen tube. A pollen grain that germinated the sticky surface of the stigma has grown down through the style carrying the male gametes to the embryo sac in the ovary.

Fertilization

In the angiosperms the pollen tube grows through the micropyle opening in the ovule into the embryo sac and discharges two sperm nuclei (1N each). One unites

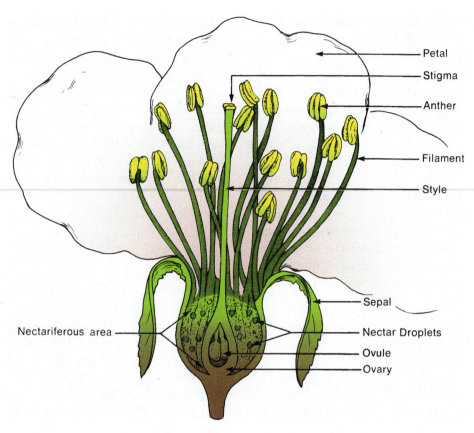

Petal
Stigma
Anther
Filament
Style
Sepal
Nectar Droplets
Ovule
Ovary
Nectariferous area

Figure 7–22
Longitudinal section of a cherry flower showing the structures involved in the transfer of pollen from anthers to stigma (pollination). Source: USDA.

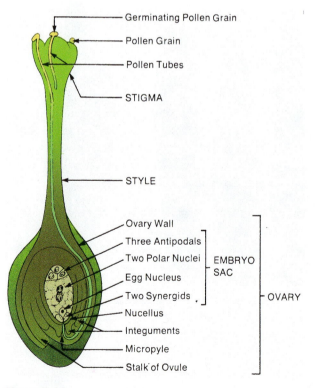

Germinating Pollen Grain
Pollen Grain
Pollen Tubes
STIGMA
STYLE
Ovary Wall
Three Antipodals
Two Polar Nuclei   EMBRYO
Egg Nucleus   SAC
Two Synergids
Nucellus   OVARY
Integuments
Micropyle
Stalk of Ovule

Figure 7–23
A longitudinal section through the carpel of a flower following pollination and just before fertilization.

Figure 7–24
Honeybees, collecting nectar from the flowers, also cause pollination by distributing pollen from the anthers to the stigma. Bees perform a great service in the culture of many crops by their pollination activities.
Source: Margaret McMahon, The Ohio State University.

with the egg (1N) to form the zygote (2N) (**fertilization**), which will become the embryo and eventually the new plant. The other sperm nucleus unites with the two polar nuclei (1N each) in the embryo sac to form the **endosperm** (3N), which develops into food storage tissue. This process is termed double fertilization. The elapsed time between pollination and fertilization in most angiosperms is about twenty-four to forty-eight hours.

In the cone-bearing plants of the gymnosperms, no flowers or fruit is produced. Pollen-producing cones are produced on the tree separately from the ovulate cones. Pollen reaches the egg and fertilization takes place when the sperm travel down the pollen tube through the egg covering (integument) to the egg, but no fruit develops. The seed is formed "bare" on the cone scale.

Parthenocarpy   If pollination and fertilization do not occur, fruit and seed rarely develop. One important exception, however, is fruit that sets parthenocarpically. **Parthenocarpy** is the formation of fruit without the stimulation of pollination and fertilization. Without fertilization, no seeds are produced; therefore, parthenocarpic fruits are seedless. There are many examples of parthenocarpic fruits such as the Washington Navel

orange, the Cavendish banana, the oriental persimmon, and many fig cultivars. Parthenocarpy is induced in some fruits by the application of chemicals such as certain plant hormones, especially gibberillins and auxins. Not all seedless fruits are parthenocarpic. Sometimes, as in certain seedless grapes (Thompson), pollination and fertilization occur and the fruit forms but the embryo aborts, thus no viable seed is produced. This is called stenospermocarpy. Parthenocarpic and stenospermocarpic plants have to be propagated vegetatively.

Fruit Setting

Following formation of the zygote many significant changes occur, leading to the formation of the fruit and (usually) seeds within the fruit. Accessory tissues in the flower are often involved in fruit formation, such as the enlarged fleshy receptacle surrounding the ovary wall in the apple and pear or the receptacle itself in strawberry. Botanically, however, the true fruit is just the enlarged ovary. Figure 7–25 shows various stages in the development of the different tissues in a lettuce fruit. Fruit development is the coordinated maturation of all tissues and organs involved in the formation of the fruit.

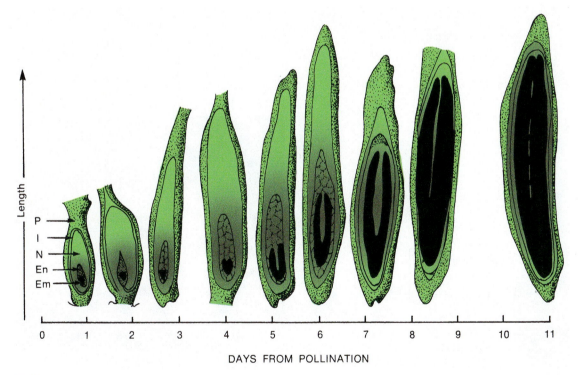

Figure 7–25

Developmental pattern of the tissues in a lettuce fruit, from the fertilized egg to the mature fruit. The ovary wall (the pericarp) is firmly attached to the seed coat (integument), so the structure is correctly considered a fruit and not a seed. P = pericarp; I = integument; N = nucellus; En = endosperm; Em = embryo. Source: Adapted by H. T. Hartmann and D. E. Kester, *Plant Propagation*, 4th ed. (Englewood Cliffs, NJ: Prentice Hall, 1983. From H. A. Jones, "Pollination and Life History Studies of the Lettuce (*Latuca sativa*)," *Hilgardia* 2 (1927): 425–479.

In many plants only a small percentage of the flowers develop into fruits. This is particularly true in fruit crops where a tree could not possibly mature as many fruits as there are flowers. Many of the flowers drop without fertilization of the egg, and many of the flowers with a fertilized egg abort at the zygote stage or later. When the zygote fails to develop and no seed forms, the immature fruit usually drops, except in the case of parthenocarpy and stenospermocarpy.

Plant hormones appear to be involved in fruit setting, but the actual physiological mechanisms are largely unknown. In fruits of some species—tomatoes, peppers, eggplants, and figs—applied auxin can replace the stimulus of pollination and/or fertilization. Fruit set can also be induced in grapes, certain stone fruits—apricots, for example—and apples and pears by gibberellin sprays. Cytokinins also stimulate fruit set in grapes.

One of the chief problems in fruit production is obtaining the optimal level of fruit setting. Too low a fruit set gives a light, unprofitable crop. Too heavy a set leads to undesirable small, poor-quality fruits that mature late, possibly exhausting the tree's food supply and often resulting in little or no crop the following year. To overcome excessive fruit set, half—or more—of the fruits are removed at a very early stage, either by hand thinning, machine shaking, or chemical sprays. Interestingly, fruits of some species (the Washington Navel orange, for example) are self-thinning. Most of the tiny fruits originally forming drop, leaving an optimum number to develop to maturity.

As might be expected, temperature strongly influences fruit set. Temperatures that are too low or too high at this critical period are often responsible for crop failures. Peaches are an example of a crop that is often severely damaged by low temperatures that occur at the time of flowering. Low light intensity and lack of adequate soil moisture can also adversely affect fruit set.

Fruit Growth and Development

Once fruit has set, the true fruit and, sometimes, various associated tissues begin to grow. Food materials move from other parts of the plant into these developing tissues. Hormonal substances, such as the auxins, gibberellins, ethylene, and cytokinins, may be involved in some phases of fruit growth just as they are in fruit set. These materials originate in both the developing seeds and fruit.

An interesting relationship between fruit growth and the presence of auxin has been observed in strawberry fruits. Removing some of the achenes (the fruits usually mistaken for seeds) from the surface of the strawberry (botanically the receptacle) at an early

growth stage causes it to be lopsided; the strawberry fails to develop under the section where the achenes were removed. The stimulatory effect of some mobile material originating in the achenes is lost. Presumably this material is an auxin because application of auxin paste to the area where the achenes were removed allows the strawberry to develop normally.

Evidence of the participation of gibberellins in fruit growth has been shown in the grape. Application of gibberellin to Thompson seedless grape clusters at an early stage of berry development markedly increases the ultimate fruit size. The size increase is so pronounced that almost all table grapes of this cultivar grown in California are now treated with gibberellin. This effect on size also holds true for certain other grape cultivars.

While various plant hormones may be involved in fruit growth, the basic mechanisms are still barely understood. During flower development and in the early stages of fruit growth, there is considerable cell division. Following this period of intense cell division, most fruits increase in size because of cell enlargement.

Fruits have two basic patterns of growth, as shown in Figure 7–26. One is the simple sigmoid growth curve— typical of fruits such as the orange, apple, pear,

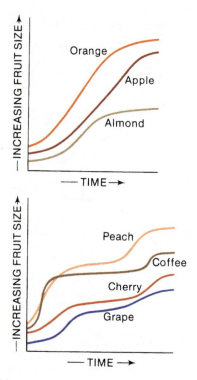

Figure 7–26

Growth curves of representative kinds of fruits showing the two characteristic types. Top: The sigmoid growth curve. Bottom: *The double sigmoid growth curve.* Source: D. E. Kester and H. T. Hudson, *Plant Propagation: Principles & Practices*, 4th ed. © 1983. Adapted by permission of Pearson Education, Inc., Upper Saddle River, NJ.

pineapple, olive, almond, tomato, and strawberry—in which there is a slow start followed by a period of rapid size increase, then a decrease in growth rate near fruit maturity. The second pattern is a double sigmoid growth curve, in which the single sigmoid growth curve is repeated. Near the center of the growth period, the growth curve is flat; the fruit increases little, if at all, in size. The stone fruits—peach, apricot, plum, and cherry—as well as the grape and fig show a double sigmoid growth pattern. In the stone fruits (peach, nectarine, plum), which have a hard endocarp or pit, the pit hardens during the second phase of fruit development. In addition, some important changes take place in seed development within the pit, as illustrated in Figure 7–27.

Some fruits as they ripen experience a burst of respiration and a release of high levels of ethylene. These are called **climacteric** fruits. Apples, apricots, bananas, and tomatoes are example of climacteric fruits. The ripening of many of these fruit can be hastened by treating them with ethylene before or after they are harvested. If these fruit are stored with ethylene sensitive fruit or plants, the ethylene they generate can harm the sensitive plants nearby.

Aging and Senescence

The life spans of the different kinds of flowering plants differ greatly, ranging from a few months to thousands of years. Some of the coniferous evergreen forest trees are the earth's oldest living organisms. Some of the California coast redwoods (*Sequoia sempervirens*) are known to be over 3,000 years old. Olive trees with huge trunks found in the eastern Mediterranean area are believed to be over 2,000 thousand years old (Fig. 7–28).

Senescence is considered to be a terminal, irreversible deteriorative change in living organisms, leading to cellular and tissue breakdown and death. It is a conspicuous period of physical decline, particularly evident toward the end of the life cycle of annual plants (population senescence) and of individual plants (whole plant senescence), but it can also occur in leaves, seeds, flowers, or fruits (organ senescence). Plants exhibit senescence in different ways. In annuals, the entire plant dies at the end of one growing season, after and probably because of fruit and seed production. In herbaceous perennials, the tops of the plants die at the end of the growing season, perhaps killed by frost, but the shoot grows again in the spring and the roots can live for many years. In deciduous woody perennials, the leaves senesce, die, and fall off each year, but the shoot and root systems remain alive, often for a great many years. Vegetatively propagated clones can extend the "life" of a plant for what theoretically could be indefinitely. For example the Winter Pearmain apple that is cultivated in England, was grown there as early as 1200 CE The Black Corinth grape is a clone that has been grown in Greece for thousands of years. The modern plants of both of these clones are in reality branches from their original plant.

Senescence is usually considered to be due to inherent physiological changes in the plant, but it can also be caused by pathogenic attack or environmental

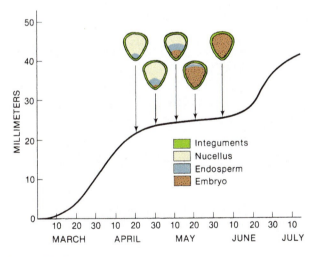

Figure 7–27

Growth curve of an apricot fruit through the growing season. During the second growth period the pit (endocarp) hardens and the seed within the pit develops mostly from nutritive tissue (nucellus and endosperm) to finally consist entirely of the embryo. Source: Adapted from J. C. Crane and P. Punsri, "Comparative Growth of the Endosperm and the Embryo in Unsprayed and 2,4,5-trichlorophenoxyacetic Acid Sprayed Royal and Tilton Apricots," *Proceedings of the American Society of Horticultural Scientists* 68 (1956): 96–104.

Figure 7–28

Ancient olive tree (Olea europaea) growing on the Mount of Olives in Jerusalem. Trees of this species are known to live for several thousand years. Every year to produce olives it has to go through a chilling period in the winter.

stress. As individual trees age, for example, they are more and more vulnerable to lethal attacks by fungi, bacteria, and viruses. The long-lived trees mentioned above characteristically have very durable heartwood, containing high levels of resins and phenolic compounds that resist decay.

Considerable study has been given to senescence in plants, particularly in regard to leaves and their abscission. During leaf senescence, DNA, RNA, proteins, chlorophyll, photosynthesis, starch, auxins, and gibberellins decrease, sometimes drastically. Senescence is not entirely degradation, however; particular mRNAs and proteins are synthesized only in senescing tissues.

A decline in photosynthetic activity of many determinate annuals such as wheat plants following flowering; the decline in photosynthesis, of course, soon leads to senescence and death. Plant senescence is hastened, too, by the transfer of stored nutrients to the reproductive parts—the flowers, fruits, and seeds—as they develop and mature, at the expense of the root and shoot systems. As a result, senescence can be postponed in many plants by picking off the flowers before seeds start to form. Pollination and fertilization can cause senescency of some flowers and the cessation of further flowering. Removing sweet peas flowers as soon as they start to wither and before seeds form prolongs the blooming period. Just as plant hormones are involved in many plant functions, they are also involved in senescence. For example, ethylene plays a major role in fruit ripening and deterioration as well as other senescence processes.

PLANT GROWTH REGULATORS

As you have read in the previous sections, **plant hormones** were mentioned several times as being factors or signals involved in plant growth and development. Because of their importance in plant growth and development, the following section describes both natural and synthetic plant hormones as well some chemical plant growth regulators that act on the hormones.

In plants, as in animals, many growth and behavioral patterns and biological functions are controlled by hormones. Hormones are produced in extremely small amounts at one site in the plant and translocated to other sites where they can alter growth and development in that area. Hormones are essentially chemical messengers, influencing the many aspects of plant development.

A distinction must be made between the terms *plant hormone* and *plant growth regulator*. A plant hormone is a natural substance (produced by the plant itself) that acts to control plant activities. Plant hormones that are chemically synthesized can initiate reactions in the plant similar to those caused by the natural hormones. **Plant growth regulators**, on the other hand, include plant hormones—natural and synthetic—as well as other chemicals not found naturally in plants but that, when applied to plants, influence their growth and development.

There are five traditionally recognized groups of natural plant hormones: **auxins**, **gibberellins**, **cytokinins**, **ethylene**, and **abscisic acid** (Fig. 7–29). Recently additional hormones or hormone classes, including the Brassinolides, salicylic acid, and jasmonates, have been identified. The discovery and subsequent study of plant hormones is one of the most exciting and fascinating chapters in the history of plant physiology, partly because their discovery lead to the creation of many new methods of regulating plant growth and development. Despite considerable study of hormones, however, the mechanism of their actions in the plant is still not completely understood.

In addition to natural hormones, synthetic growth regulators have been developed to allow growers to manipulate plant growth and development. Synthetic growth regulators are used to promote rooting, reduce stem elongation, encourage branching, regulate flowering, and influence other aspects of growth and development.

Auxins

Auxins were the first group of plant hormones to be discovered. The discovery came in the mid-1930s, and for many years thereafter auxins and their activities in plants were studied intensely throughout the world.

The auxins, both natural and synthetic, influence plant growth in many ways, including cell enlargement or elongation, photo- and geotropism, apical dominance, abscission of plant parts, flower initiation and development, root initiation, fruit set and growth, cambial activity, tuber and bulb formation, and seed germination. Auxins operate at the cellular level, affecting activities such as protoplasmic streaming and enzyme activity. Auxins are related to many other chemical control mechanisms and are readily transported throughout the plant, principally in an apex-to-base direction (basipetally).

The natural auxins originate in meristems and enlarging tissues, such as actively growing terminal and lateral buds, lengthening internodes, and developing embryos in the seed. Auxins are produced in relatively high amounts in the shoot tip or terminal growing point of the plant and move down the plant through

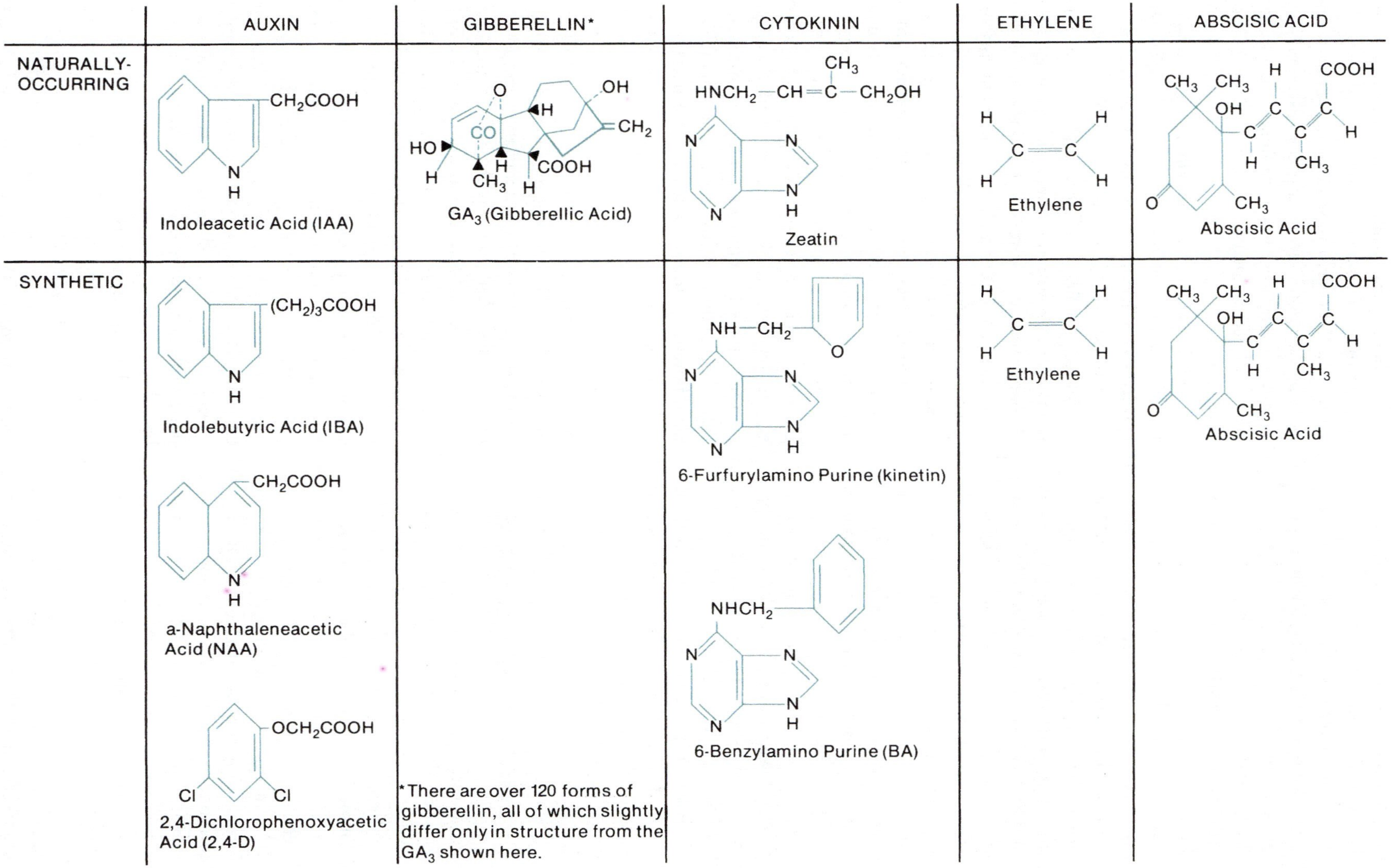

Figure 7–29

Structural formulas of some natural and some synthetic plant growth regulators.

the vascular tissues, causing the phenomenon known as **apical dominance** (blockage of the growth of lateral buds by the presence of terminal buds). High levels of auxin in the stem just above the lateral buds block their growth. If the shoot tip supplying this auxin is broken or cut off, the auxin level behind the lateral buds is reduced and the lateral buds begin to grow. This is part of the reason why, when a shoot tip is removed, many new shoots arise from buds down along the stem.

One of the most widespread auxins that occurs naturally in plants is indoleacetic acid (IAA) (Fig. 7–29). Several other natural auxins have also been identified, and there are others whose chemical structure is yet unknown. Many synthetic auxins induce the same effects as natural auxin. Some of these are indolebutyric acid (IBA), naphthaleneacetic acid (NAA), and 2,4-dichlorophenoxyacetic acid (2,4-D). (See Fig. 7–29.)

Some important commercial uses of these synthetic auxins are:

1. *Adventitious root initiation.* One of the first responses attributed to auxins was the stimulation of root formation in stem cuttings. Two synthetic auxins, indolebutyric acid and naphthaleneacetic acid, are now widely used commercially in treating the bases of stem cutting to stimulate the initiation of adventitious roots.

2. *Weed control.* The synthetic auxin, 2,4-D, is in widespread commercial use as a selective weedkiller that eliminates broad-leaved weeds in grass or cereal fields.

3. *Inhibition of stem sprouting.* Many kinds of woody ornamental trees produce masses of vigorous sprouts from the base of the trunk that, if not removed, would transform the tree into a bush. Continual removal of these sprouts by hand is costly and time consuming. It has been found that treatment of the tree trunks with the auxin naphthaleneacetic acid at about 10,000 ppm (1.0 percent) strongly inhibits the development of such sprouts.

4. *Tissue culture.* The initiation of roots and shoots on small pieces of plant tissue cultured under aseptic conditions has become a standard method of micropropagation of some plant species. Often an auxin, such as IAA or 2,4-D, has to be included in the culture medium for roots to initiate.

Gibberellins (GAs)

The gibberellins are a group of natural plant hormones with many powerful regulatory functions. The most obvious is the stimulation of stem growth dramatically,

far more than auxins can. Gibberellins may stimulate cell division, cell elongation, or both, and they can control enzyme secretion.

In some plants, GA is involved in flower initiation and sex expression (male or female flower parts). Fruit set as well as fruit growth, maturation, and ripening seem to be controlled by gibberellin in some species. Senescence of plant parts, particularly leaves, is also affected by GA. Certain dwarf cultivars of peas and corn, if treated with GA, grow to a normal height, indicating that the dwarfed plants lack a normal level of gibberellin (Fig. 7–30).

Gibberellins are also involved in overcoming dormancy in seeds and in buds. Their role in the germination of barley seed has received much study (Fig. 7–31). After the seed has been moistened and placed at room temperature, a natural gibberellin produced in the embryo translocates to the aleurone layer surrounding the endosperm. Triggered by the GA, cells in the aleurone layer synthesize enzymes such as amylases, proteases, and lipases. These enzymes then diffuse throughout the endosperm, hydrolyzing starches and proteins into sugars and amino acids that then become available to the embryo for its growth and development.

The molecular structure of the gibberellins is well known; a typical one is shown in Figure 7–29. By 2000, more than 100 different gibberellins had been discovered in tissues of various plants. Some common ones are GA_1, GA_3 (gibberellic acid), GA_4, and GA_7.

Figure 7–30
Overcoming dwarfness in corn by spraying with gibberellin. Left: Untreated, genetically dwarf corn plants. Center: Nondwarf corn sprayed with gibberellin. Right: Genetic dwarf corn sprayed with gibberellin. Photographs taken six weeks after spraying. Source: From *Plant Growth Substances in Agriculture* by Robert J. Weaver. W. H. Freeman and Company. Copyright © 1972.

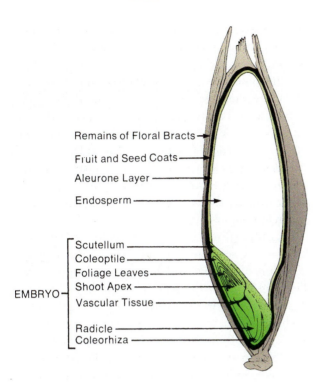

Remains of Floral Bracts
Fruit and Seed Coats
Aleurone Layer
Endosperm

EMBRYO
Scutellum
Coleoptile
Foliage Leaves
Shoot Apex
Vascular Tissue
Radicle
Coleorhiza

Figure 7–31
Longitudinal section of a barley seed. Source: From a drawing by Peter J. Davies. In A. W. Galston, P. J. Davies, and R. L. Satter, *The Life of the Green Plant,* 3rd ed. (Edgewood Cliffs, NJ: Prentice Hall, (1980).

Gibberellins were first discovered in 1926 by Japanese researchers studying a disease of rice plants caused by the fungus *Gibberella fujikuroi.* Plants infected with the fungus grew excessively and abnormally. Extracts from this fungus applied to noninfected plants stimulated the same abnormal growth. By 1939, the active ingredient was extracted from the fungus, crystallized, and named gibberellin. This early work with gibberellin in Japan went unnoticed in the Western world until the early 1950s when a great surge of gibberellin research began, particularly in the United States and England. This research led to the isolation of many different forms of gibberellin extracted from the *Gibberella* fungus and from higher plants.

Gibberellins are synthesized in the shoot apex of the plant, particularly in new leaf primordial. They are also found in embryos and cotyledons of immature seeds and in fruit tissue. In addition, the root system synthesizes large quantities of gibberellin, which moves upward throughout the plant. GA translocates easily in the plant in both directions, unlike auxin, which moves largely in an apex-to-base direction.

Pharmaceutical companies produce crystalline GA3 as the acid or the potassium salt for research studies and certain commercial applications. These preparations are all obtained from growth of the *Gibberella*

fungus in a process similar to that used to produce antibiotics.

Even though it has been demonstrated that gibberellins occur naturally in many of the higher plants, little is known of the physiological mechanisms of gibberellin action or transport.

Although gibberellins are a powerful and important group of plant hormones involved in many of the plant functions, only a few agricultural uses have been found for them. Some are:

1. *Increasing fruit size of seedless grapes.* This is the principal commercial application of gibberellin. Practically all vines of the Thompson Seedless grape grown for table use in California are sprayed each year. Berry size of other grape cultivars, such as Black Corinth, is also increased by gibberellin sprays, as shown in Figure 7–32.

2. *Stimulating seed germination and seedling growth.* Several cases have been reported where soaking seeds in solutions of gibberellic acid before germination greatly stimulates seedling emergence and growth. Such responses have been obtained with barley, rice, peas, beans, avocado, orange, grape, camellia, apple, peach, and cherry. Figure 7–33 shows the stimulation obtained with grape seedlings by gibberellin treatment.

3. *Promoting male flowers in cucumbers.* When pollen is wanted for hybrid seed production, a single application of GA3 to the leaves stimulates maleness of a cucumber. This has proved an important discovery for hybridizers.

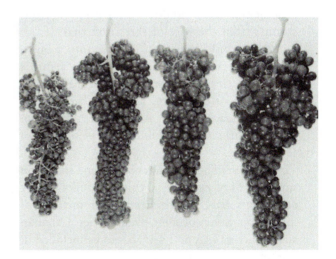

Figure 7–32
Effect of gibberellin sprays on growth of Black Corinth grapes. (A) Untreated control. (B) Stem girdling control. (C) Plants sprayed at an early growth stage with gibberellin at 5 ppm. (D) At 20 ppm. Photos taken 59 days after spraying.
Source: From *Plant Growth Substances in Agriculture* by Robert J. Weaver. W.H. Freeman and Company. Copyright © 1972.

Figure 7–33
Effect of gibberellin on germination and growth of Tokay grape seeds. Seeds soaked (before planting) at 0, 100, 1,000, or 8,000 ppm in potassium gibberellate solution for 20 hr.
Source: From *Plant Growth Substances in Agriculture* by Robert J. Weaver. W.H. Freeman and Company. Copyright © 1972.

4. *Overcoming the cold requirement for some plants.* Azalea plants require six weeks of cool temperatures (8°C or 46°F) to develop flower buds. Several leaf applications of 1,000 ppm gibberellin completely or partially replace this cold requirement for flower bud development. It has been shown experimentally that gibberellins applied to biennial plants that require a cold period before they flower causes early flowering. Most of these treatments, however, have limited commercial value.

5. *Promoting stem elongation.* Some ornamental plants such as poinsettias and geraniums can be grown into a tree form (Fig. 7–34) by applying gibberellins to cause stem elongation. As the stem is growing, side branches are removed for a period of time. Then the stem is allowed to grow until several internodes develop. The apical meristem is pinched to stop apical dominance and allow the lateral branches to form. The plant is allowed to grow and produce flowers. In the case of poinsettia, the plant must be grown in a flower-inducing photoperiod.

Cytokinins

This group of plant hormones primarily promotes cell division but they also participate in a great many aspects of plant growth and development, such as cell enlargement, tissue differentiation, dormancy, different phases of flowering and fruiting, and in retardation of leaf senescence.

Cytokinins interact with auxins to influence differentiation of tissues. As shown in Figure 7–35, externally applied cytokinin alone stimulates bud formation in tobacco stem segments; auxin applied alone causes roots to develop, but when cytokinin plus auxin are applied together, there is a canceling effect—only masses of undifferentiated callus form.

Figure 7–34
Poinsettias being grown as trees. Left: before the start of short days and flowering. Right: after flowering is complete. Gibberellins were used to elongate the stems. Lateral branches were also removed from the lower portion of the stem.
Source: Margaret McMahon, The Ohio State University.

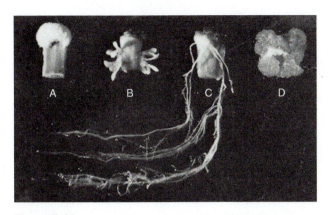

Figure 7–35

*Effects of a cytokinin and an auxin on growth and organ for-
mation in tobacco stem segments. (A) Control, no treatment.
(B) Cytokinin—bud formation but no root formation.
(C) Auxin—root formation with prevention of bud development.
(D) Cytokinin plus auxin—stimulation of callus growth but no
organ formation.* Source: D. E. Kester and H. T. Hudson, *Plant Propagation:
Principles & Practices*, 4th ed. © 1983. Adapted by permission of Pearson
Education, Inc., Upper Saddle River, NJ.

There are both natural cytokinins, such as zeatin,
and synthetic forms, such as kinetin and benzylade-
nine (BA) (see Fig. 7–29 on page 134). There are over
100 known natural and synthetic cytokinins.
Cytokinins occur in many plant tissues as both the free
hormonal material and as a component of transfer
RNA. They are found in abundance in embryos and
germinating seeds and in young developing fruits—all
tissues with considerable cell division. Roots supply
cytokinins upward to the shoots.

The mechanism of cytokinin action in the plant
is not clear. Cytokinins indirectly increase enzyme
activity and increase the DNA produced in some tissues. Their regulatory effects seem to result from interactions with other hormones in the plant.

Cytokinins were discovered when scientists at
the University of Wisconsin in 1955 used a synthetic
material—kinetin, later named a cytokinin—to cause
cell division of tobacco stem pith. After many interesting physiological activities of kinetin became apparent,
plants were examined for possible natural similar
materials. In 1964 such a material was isolated from
young corn seeds by researchers in New Zealand and
was named zeatin. An active promoter of cell division
known to exist in coconut milk was finally determined
to be a zeatin-riboside.

Even though cytokinins are strongly involved in
plant growth regulation, no important agricultural uses
have been developed for them. In media for tissue culture, however, cytokinin usually must be added to
induce shoot development. Applications of cytokinins
to green tissue have been shown to delay senescence.

These materials have also been used experimentally,
and in limited commercial applications, on greenhouse
roses and potted chrysanthemum plants to stimulate
growth of axillary buds by overcoming natural bud
inhibitors.

Ethylene

It has been well established for many years that ethylene
gas (see Fig. 7–29 on page 134) evokes many varied
responses in plants. As long ago as 1924, it was found that
ethylene could induce fruit ripening; in 1925, it was
determined that ethylene could overcome bud dormancy
in potato tubers; in 1931, that it could induce leaf abscission; in 1932, that it could induce flowering in pineapple
plants; in 1933, that it could cause roots to form on stem
cuttings. By the mid-1930s, it was determined that ethylene was itself a plant product, and arguments arose
among plant scientists about whether ethylene, a gas,
should be considered a plant hormone.

Little further attention was paid to possible roles
of ethylene as a natural growth regulator until the
development in the 1960s of gas chromatographic
techniques that permit the detection of ethylene in concentrations as low as one part per billion. Vast amounts
of research of ethylene physiology in the 1960s and
1970s established ethylene as a plant hormone.

Ethylene itself is a tiny molecule ($CH_2 = CH_2$),
compared with the other plant hormones. The pathway
for ethylene biosynthesis in plants has been fairly well
elucidated. Ethylene, as a gas, diffuses readily throughout the plant, moving much like carbon dioxide, and it
can exert its influence in minute quantities. Its solubility in water also enhances its movement through the
plant. The cuticular coatings on external cell surfaces
tend to prevent losses from the plant. Ethylene is
apparently produced in actively growing meristems of
the plant, in ripening and senescing fruits, in senescing
flowers, in germinating seeds, and in certain plant tissues as a response to bending, wounding, or bruising.
Synthetic ethylene, from ethephon, applied to plant
tissues can cause a great burst of natural ethylene
production—an autocatalytic effect.

Just how ethylene exerts its regulatory effects is no
better known than the basic mechanisms involved in the
action of the other plant hormones. One theory is that
ethylene regulates some aspect of DNA transcription or
RNA translation, thus changing RNA-directed protein
synthesis and, consequently, enzyme patterns. But many
other mechanisms are also likely to be in operation.

The possible commercial uses of ethylene were
greatly increased with the development in the 1960s of

ethylene-releasing compounds such as ethephon (2-chloroethyl) phosphonic acid. This compound applied as an agricultural spray gradually releases ethylene into plant tissues. In contrast to some of the other plant hormones, ethylene and ethylene-releasing chemicals have several valuable commercial applications:

1. *Fruit ripening.* Ethylene gas, injected into airtight storage rooms, is used commercially to ripen bananas, honeydew melons, and tomatoes. The ethylene-releasing chemical ethephon also ripens tomatoes that are green but horticulturally mature. To harvest canning tomatoes by machine harvesting equipment, where the entire crop is picked at one time, it is important that most of the fruits be ripe and fully colored at the time of harvest. Spraying the field with an ethylene-releasing material before harvest promotes uniform red color development of the green fruits. Ethephon is used as a preharvest spray to promote uniform ripening of apples, cherries, figs, blueberries, coffee, and pineapple.

2. *Flower initiation.* Ethylene gas released from ethephon has initiated flowers in several ornamental bromeliad species, including *Ananas* spp., *Aechmea fasciata*, *Neoregelia* spp., *Billbergia* spp., and *Vriesia splendens*. Ethephon has been widely used to promote uniform flowering in the cultivated banana.

3. *Changing sex expression.* Ethylene application to certain plants, such as cucumbers and pumpkins, can dramatically increase the production of female flowers. Some cucumber cultivars produce both female and nonfruiting male flowers on the same plant. Spraying the vines with ethephon causes all flowers to be female, which develop into fruits and thus increase yields. This practice gives results similar to previous studies where auxins were used.

4. *Degreening oranges, lemons, and grapefruit.* Sometimes the rind of maturing oranges and grapefruits remains green owing to high chlorophyll levels, even though the eating quality, juice content, and ratios of soluble solids to acid are high enough to meet grade standards for harvest. Citrus packers can treat such fruits with ethylene at about 20 ppm for twelve to seventy-two hours. This breaks down the chlorophyll and allows the orange and yellow carotenoid pigments to show.

5. *Harvest aids.* Certain fruit and nut crops, such as sour cherries and walnuts, are harvested by mechanical tree shakers that shake the trees until the crop falls into catching frames or onto the ground to be picked up later. Often this practice is not completely successful because the fruits or nuts are so tightly attached that the tree shakers do not remove them. However, by spraying the trees about a week before harvest with an ethylene-releasing compound—ethephon, for example—the abscission-inducing effects of the ethylene result in a much higher percentage of crops removed. Mature cotton plants are sprayed with ethephon to cause the leaves to senesce and abscise, making mechanical harvesting much easier.

6. *Growth regulation.* One of the physiological effects of ethylene application to plants is a reduction in the growth of stems and leaves. The greenhouse industry recently made use of this fact in controlling excessive growth in floriculture crops. Florel® is a special formulation of ethephon registered for this purpose. Great care must be exercised when using ethylene as a growth regulator because high concentrations of this hormone can cause leaf abscission and other deleterious effects.

Ethylene can also harm plants. It can cause unwanted leaf abscission and can hasten senescence of most flowers. Some fruits give off ethylene as they ripen. The ethylene can harm any ethylene-sensitive produce, including flowers, stored in the same area. The introduction of a few parts per billion of ethylene into the surrounding air causes carnation flowers to close, rose buds to expand prematurely, and orchid flower petals and sepals to develop a water-soaked appearance. The pollination of an orchid flower can generate sufficient ethylene to cause injury to the other parts of the flower. Ethylene can cause flower bud abortion of bulbs during shipment. A few diseased tulip bulbs give off enough ethylene in a packing crate to stop further development of the flower buds within the bulbs. Gas- or oil-fired heaters situated directly in greenhouses can generate toxic levels of ethylene if the units do not burn properly (incomplete combustion) and are not vented adequately. Incomplete combustion generates other unwanted gases such as carbon monoxide, too. For some commercial flowers such as carnations and geraniums, the deleterious action of ethylene may be blocked with the application of methylcyclopropene (MCP). MCP has also been shown to be effective at slowing the ripening of climacteric fruit.

Abscisic Acid (ABA)

ABA was originally identified as a component of a complex of inhibitory substances associated with the dormant buds of ash trees and with substances that accumulated in abscising leaves. In fact, the original

names for ABA were *dormin* and *abscisin II* because researchers believed there was a primary function for ABA in these two processes. Later, when the chemical structure (see Fig. 7–29 on page 134) was elucidated, dormin and abscisin II were shown to be identical compounds, and the name abscisic acid was adopted. Further research showed, however, that ABA appears to have less influence on dormancy and abscission than the other hormones. Today, ABA is recognized to have two major roles in the life of a plant. First is the regulation of processes in seed development, including the accumulation of seed proteins and the prevention of precocious seed germination, that is, germination while on the mother plant. The second role is one of a mobile stress hormone in which ABA action initiates plants' responses to cold and water stress. One of the most important stress responses mediated by ABA is the closure of stomates when the loss of water from the plant reaches a critical value. Once this critical value is reached, plant cells begin to synthesize large amounts of ABA, which is then transported to the guard cells signaling them to reduce the stomatal aperture, thus minimizing further water loss by transpiration.

ABA is synthesized in both leaves and roots. This characteristic provides the plant with a mechanism to adjust the amount of water loss through transpiration in response to the water status of not only the leaves but also the roots and surrounding soil. For example, during a hot, sunny day, leaves may begin to wilt despite adequate moisture in the soil because more water is lost than can be replaced via the xylem transport system. The reduced leaf water content results in the production of ABA by the mesophyll cells, which signals the guard cells to close the stomates until leaf turgor is restored. On the other hand, as soil moisture reserves are gradually depleted through direct evaporation to the atmosphere and transpiration, the water content of the roots declines commensurately. When a threshold level of water loss is reached, root cells produce ABA, which is transported to the leaves via the xylem, closing the stomates. Thus, plants have a very elegant system for adjusting stomatal apertures to match soil moisture content.

At present, no commercial uses of ABA exist in crop production, although considerable effort was made to see if it could be used as an antitranspirant to minimize transplant shock. The biggest obstacle for such use, besides cost, is the fact that the plant rapidly deactivates ABA, so any effect on depressing transpiration is transient.

Additional Hormones or Hormone Classes

In recent years, four additional hormones or hormone classes have been discovered, indicating that other likely plant hormones are waiting to be discovered. **Brassinolides** are steroids, closely related in structure to animal steroid hormones such as estrogen and testosterone. In plants, brassinolides appear to function in the regulation of cell division and elongation. Plants lacking brassinolides are dwarf and exhibit weak growth.

A second new hormone is **salicylic acid**, which coincidentally is the biologically active component of aspirin. This hormone is an important component of plants' response to pathogen attack; it serves as a signal to activate genes involved in pathogen defense. Through a relatively minor chemical modification, plants use a derivative of salicylic acid as a form of interplant communication in an early warning system that a nearby plant is under attack by a pathogen. This modified version is the methyl ester of salicylic acid, which is wintergreen oil. Methyl salicylate is much more volatile than salicylic acid and readily evaporates from the leaves. The vaporized molecules diffuse through the atmosphere and can be absorbed by the leaves of neighboring plants. Once inside living cells, methyl salicylate is easily hydrolyzed to re-form salicylic acid, which can then induce plant defense systems without the plant actually being infected.

While salicylic acid has a major role in pathogen defense, the **jasmonates** represent a group of compounds involved in systems that defend plants against herbivores. Jasmonates are derived from fatty acids and are similar in structure to the class of animal hormones known as prostaglandins. Jasmonates are also volatile and are the major component of the fragrance associated with jasmine tea and gardenia flowers. The volatile nature of jasmonates provides a mechanism similar to that of methyl salicylate for interplant signaling of attacks by herbivores such as insects.

The last hormone is also involved in herbivore defense. **Systemin** is unique among plant hormones because it is the only one known to be a peptide (there are several animal peptide hormones—insulin is a well-known example) composed of eighteen amino acids. It is produced in tissue wounded by herbivores and is transported to remote tissues and organs, where it induces defense genes.

Synthetic Growth Retardants

A rather diverse group of growth retardants developed since about 1950 has several important commercial uses regarding ornamental plants, principally in obtaining compact, dwarf-type plants. These materials generally act by slowing, but not stopping, cell division and elongation in subapical meristems, usually without causing stem or leaf malformations. The primary effect of these materials is the opposite of gibberellin, often converting a tall-growing plant into a rosette. Most act by blocking gibberellin synthesis. Plants treated with these growth retardants have a compact, scaled-down appearance, which is often more attractive than larger, untreated plants with a loose, open growth. The treated plants also often have darker, more attractive foliage and more flowers than untreated plants (Fig. 7–36). These chemicals only block gibberellin synthesis, they do not stop its activity or the plant's sensitivity to it. Applying gibberellin to growth regulator treated plants will at least partially overcome the effects of the growth regulator.

Some of the better known synthetic growth retardants are described below:

Daminozide (succinic acid-2, 2-dimenthyl hydrazide; Alar, B-Nine). Tests have shown that daminozide effectively retards growth and stimulates flowering of several kinds of herbaceous and woody ornamental plants and enhances the size and color of various fruit species. Those plants that respond well include chrysanthemums (see Fig. 7–36), various bedding plants (2,500 to 5,000 ppm), and azaleas (2,500 ppm). Sometimes two applications two or three weeks apart are required to maintain the desired dwarf form.

Chlormequat [(2-chloroethyl) trimethylammonium chloride; Cycocel, CCC]. Chlormequat is effective in retarding the height of some ornamental plants.

Figure 7–36

Growth retardation in chrysanthemum plants when treated with a growth retardant, Chrysanthemum x morifolium. "Circus" plants sprayed with daminozide (B-Nine) to retard shoot growth; left: control, no daminozide; center: 2,500 ppm; right: 2,500 ppm sprayed Aug. 14 and again on Aug. 21.

The height of poinsettias may be controlled if chlormequat is applied as a drench to the soil or as a spray to the stems and foliage.

Ancymidol (α-cyclopropyl(P-methoxyphenyl)-5-pyrimidine-methanol; A-rest®). This growth retardant is very effective for reducing the height of some bulbous and other potted ornamental crops.

Paclobutrazol [2RS, 3RS]-1-[4-chlorophenyl] 4–4-dimethyl-2–1, 4-triazol-yl-pentan-3-ol; Bonzi® P. and its close chemical relative uniconazole (Sumagic®) are very potent growth-retarding chemicals that effectively control the height of many herbaceous and woody ornamentals. The rates used for these chemicals are much lower than for other growth retardants.

Trinexapac-ethyl [4-(cyclopropyl-α-hydroxy-methylene)-3,5-dioxocycloheanecarboxylic acid ethyl ester Primo®]. Trinexapac-ethyl is a relatively new growth retardant registered for use in turfgrass. It is relatively specific for monocots.

CHEMICAL REGISTRATION AND USE FOR PLANTS

In the United States, the application of chemicals to plants for commercial use is strictly controlled by Environmental Protection Agency (EPA) regulations. Before the chemicals can be used legally, an EPA registration must be obtained for each crop, stating the dosage allowable and the time of year that application is permissible. Application for registration is usually made by the chemical company manufacturing the material after a patent has been obtained. Regulations for obtaining a registration for use of chemicals to be applied to food crop plants are much stricter than are those for ornamentals. Most chemical plant growth regulators are forbidden by law to be used on any plants that are grown for human consumption. Chemical use on plants must be in accordance with the law. Before using any chemical, check the label to be certain that the crop you are treating, the rate you are using, and the intended use, along with any other considerations, are in agreement with the label.

SUMMARY AND REVIEW

Growth is the increase in size of a plant by cell division and enlargement. Plant development is progress through the stages of its lifecycle. Growth in plants results from cell division in the meristems and subsequent cell elongation and expansion. Shoot growth can be determinate (shoot elongation ceases with the formation of reproductive structures) and indeterminate (bearing clusters of fruits and flowers along the stem).

Genetic and environmental factors interact to determine plant growth and development. Genetic factors include the overall genetic (DNA) composition of the plant and the active or inactive state of genes at any particular time.

Environmental factors include light, heat, water, and atmospheric gases. Sun light provides the energy for photosynthesis and the accumulation of carbohydrates needed for growth. Changes in light quality or duration direct the shape of the growing plant (photomorphogenesis), including the flowering of photoperiodic plants. Photoperiodic plants flower in response to changes in daylength. Many photogenic responses are mediated by the phytochrome system which senses changes in the amount of red and far-red light striking the plant.

Heat determines how fast most plants grow and develop. Most plants have an optimum growing temperature. Temperatures much below or above the optimum slow growth rate and may be detrimental or even lethal. However, low temperatures can serve as cues for the plant to coordinate growth with the changing seasons. Low temperatures are involved in seed stratification, vernalization, dormancy breaking, development of cold and freeze tolerance.

Water can indirectly influence plant growth because of the role it has in so many plant processes. The presence or absence of water-induced pressure of the plasmalemma on the walls of expanding cells determines the shape and strength of the cell wall and consequently the size and shape of the plant.

The most important atmospheric gases for plants are CO_2 (needed for photosynthesis) and O_2 (needed for respiration). CO_2 concentrations in the leaf influence stomate opening; when internal CO_2 is low, the stomates are open. The need for a plant to conserve water (stomates close) and the need to bring in CO_2 for photosyntheis (stomates open) can be in conflict. Roots can "drown" in flooded soil because water displaces the O_2 needed for root respiration.

Growth starts with germination (seed imbibition followed by radicle and plumule emergence) and early seedling growth. Development of the root and shoot systems follow. Plant growth patterns include annual (complete life cyle including death in one growing season), biennial (life cycle covers two growing seasons), and perennial (life cycle extends for many years). Plants go through juvenile and adult phases. During the juvenile phase, the plant cannot become reproductive even if condtions are right for the development of reproductive organs. At maturity reproductive growth begins as long as environmental factors are appropriate. When a plant becomes reproductive, some or all stem meristems start producing floral parts. For determinate plants, leaf and shoot production stops when the switch is made.

Once the flower matures (it is producing pollen, ovum, or both), pollination and fertilization occurs. Plants can be self-pollinated or cross-pollinated, depending on whether they are self-fertile or self-infertile. Parthenocarpy is the development of fruit without pollinization and fertilization. Stone fruits and some others such as grape and fig have a double sigmoid pattern where there are periods of rapid growth.

Aging and senescence are the final stages of the life cycle of all plants. Senescence is the terminal and irreversible deterioration in plants leading to death. It is usually caused by natural processes in the plant such as transfer of metabolites from leaves to developing flowers and fruit but can also be caused by pathogen attack or environmental stress. Removing flowers and immature fruit can delay senescence of leaves in some plants and extend the flowering period of others. For temperate biennials and perennials senescence occurs for some but not all tissues and organs at the end of each growing season.

There are five traditionally recognized classes of plant hormones: auxins, gibberellins (GAs), cytokinins, ethylene, and abscisic acid (ABA). Auxins influence cell enlargement, photo- and geotropism, apical dominance, and other growth traits. Gibberellins can promote flower initiation and stem elongation, and can overcome dormancy in seeds and buds. Cytokinins promote cell division and slow leaf senescence and influence other physiological processes. Ethylene promotes senescence and fruit ripening in many species. Unwanted exposure to ethylene can cause severe damage to plants. Abscisic acid (ABA) is involved in seed development and the closing of stomates when plants

are stressed by low water availability. In some plants, ABA also promotes leaf abscission and/or dormancy. In recent years, four more classes of hormones have been discovered and more are likely to be discovered.

Chemical growth regulators generally promote or inhibit the influence of hormones. Most of them act by blocking the synthesis of gibberellins.

KNOWLEDGE CHECK

1. What is the difference between plant growth and plant development?
2. How is plant growth measured?
3. What is the difference between determinate and indeterminate growth?
4. What are meristems?
5. Why do we say that many of the methods we use to manipulate plant growth and development ultimately work by activating and deactivating genes?
6. What is heliotropic movement?
7. What is photomorphogenesis?
8. What plant pigment controls most photomorphogenic responses? What regions of light does this pigment absorb?
9. Describe briefly how phytochrome responds to red and far-red light?
10. How does the shade avoidance response explain why plants grown in full sun look different from the same plants if they were grown close together?
11. What are chryptochrome and phototropin?
12. What are: photoperiodism, short-day plants, long day plants, day-neutral plants?
13. How is it possible for a short day plant to have a critical day length of 14 hours while a long-day plant can have a critical day length of 11 hours?
14. Besides flowering what are 2 other plant processes that can be controlled by photoperiod?
15. What happens when a leaf is solarized?
16. What is chilling injury and at what temperature can it happen to some tropical plants?
17. Describe: stratification, vernalization, chilling requirement
18. How can withholding water help strengthen stems and keep them short?

19. What are the 3 general stages of seed germination?
20. What are the 4 main functions of roots?
21. Why is a balance between root and shoot growth important?
22. What is the difference between determinate and indeterminate stem growth?
23. Describe the lifecycle of an annual, biennial, and perennial plant.
24. Why are many perennials grown only as annuals in areas with cold winters?
25. What determines if a plant is in a juvenile or mature stage of development?
26. Describe how growers manipulate the photoperiod to produce the short-day plant chrysanthemum year-round.
27. What is circadian rhythm?
28. What are: vernalization, forcing?
29. Define the following: pollination, self-fertile, self-sterile.
30. Define parthenocarpy.
31. Botanically, what is a fruit?
32. Why do fruit growers often remove up to half of the developing fruit from their plants?
33. Describe the growth of fruit using the sigmoid curve.
34. What is senescence in plants?
35. How does senescence differ in annual and perennial plants?
36. Give a natural function or production use for: auxin, gibberellins, ABA, cytokinin, and ethylene.
37. What chemical characteristic makes ethylene different from the other traditional hormones?
38. On what hormone do almost all chemical growth regulators act? How do they affect this hormone?

FOOD FOR THOUGHT

1. You are growing zinnias for sale in the spring. Zinnias flower in response to warm temperatures. You want to have them in flower when they are sold but at the same time you want to keep them from getting too big for the containers they are in. Which method would you use to slow growth—cool temperatures or a chemical growth retardant? Explain why you made the choice you made.
2. Soybeans are short day plants and cultivars have been developed for different latitudes. You are a soybean grower in southern Indiana whose crop was flooded out just after planting. The cultivar you usually use is sold out. You found a "bargain-priced" cultivar for sale on the Internet by a farmer in central Minnesota who has some left over from his planting. Explain why you would or would not buy it.
3. You are an amateur plant propagator who for a couple of years has been taking cuttings from a plant that you like that is growing on an empty

lot. This year you noticed the new growth has a different type of leaf shape and growth habit. The plant appears healthy so you take some cuttings from both the old and new type of growth. The old type roots as usual, but the new type fails to root. How can you explain this?

4. You are the produce manager of a grocery store. Why should you be aware of which of the plants you handle are ethylene sensitive and which are climacteric?

FURTHER EXPLORATION

BASRA, A. S. 2004. *Plant growth regulators in agriculture and horticulture.* Florence, KY: Taylor and Francis.

CAPON, B. 2004. *Botany for gardeners*, 2nd ed. Portland, OR: Timber Press.

TAIZ, L., AND E. ZEIGER. 2006. *Plant physiology,* 4th ed. Sunderland, MA: Sinauer.

*8

Plant Chemistry and Metabolism

JOHN STREETER

key learning concepts

After reading this chapter, you should be able to:

- List the major biochemicals found in plants.

- Explain how some of those chemicals are formed and some of their uses.

- Describe how relatively few elements (carbon, hydrogen, oxygen, nitrogen, phosphorus, sulfur) are combined in nearly innumerable ways to create the structures or perform the functions required for plant growth and development.

Plant growth and development occur because of many biochemical and other metabolic processes in the plant. In this chapter a very brief introduction to the biosynthesis, structure, and function of some common plant metabolites including carbohydrates, lipids and fatty acids, nucleic acids, and proteins is provided. Metabolites are the molecules that make up the plant. They can be part of the plant structures or involved in the plant processes you have already learned about or will learn about in later chapters. They are the basis for nearly all of the plant's nutritional needs because they are made from plant nutrients and/or they require plant nutrients to function properly. This chapter provides an overview. Therefore, students should understand that a vast amount of additional information is available from other sources. Advanced students will want to study plant physiology and plant biochemistry to gain a better understanding of these subjects. Some additional information can be found in the references listed at the end of this chapter.

CARBOHYDRATES

The creation (synthesis) and metabolism of carbohydrates or sugars begins with photosynthesis. This process, involving the conversion of light energy into chemical energy, is the beginning point for all the compounds found in plants. Thus, carbohydrates are the compounds from which all other plant biochemistry is derived. Relative to other organisms, plants are carbohydrate-rich organisms, and they depend on carbohydrates for much of their structure and energy storage. The three elements in carbohydrates come from water and carbon dioxide and are not considered nutritional elements. However, if they are not present in the plant in sufficient quantities, plant growth and development can be inhibited. Providing adequate water and sufficient CO_2 is an important part of successfully growing plants.

The general composition of carbohydrates is just what their name implies—hydrate (water) and carbon. Thus, the general formula for carbohydrates is $C(H_2O)$, that is, one carbon atom, two hydrogen atoms, and one oxygen atom. (There can be minor deviations from this general formula.) These three

elements are assembled in various combinations to form many different structures.

Carbohydrates are classified into different structural groups, and these distinctions are important in understanding the function of these compounds:

Monosaccharides Single carbohydrate molecules.

Disaccharides Two carbohydrate molecules linked together.

Oligosaccharides More than two molecules but less than seven molecules linked together.

Polysaccharides Combinations of seven or more molecules linked together.

Monosaccharides fall into two general categories: aldoses (Fig. 8–1A), and ketoses (Fig. 8–1B). The smallest molecules considered to be carbohydrates are three-carbon molecules such as glyceraldehyde. It is important to understand a little about the structural diversity of these compounds, and this information will be presented using the most common and important of the six-carbon sugars (hexoses), namely, **glucose**. The glucose molecule exists in three structural forms (Fig. 8–2). In solution, these structural forms of glucose all exist and are in equilibrium (represented by the double arrows). The carbon atoms in the molecule are numbered to help us keep track of the different structures, and vertical lines in the ring structures represent the hydroxol (–OH) groups. Some important things to notice about carbohydrates are: (1) that the position of the –OH groups is important in determining structure and chemical properties; (2) that the ring structures differ only by the position of the –OH at carbon 1; and (3) that the Greek letters alpha

(α) and beta (β) are used to indicate the position of the OH group at carbon 1. The ring structures are the molecular form involved in metabolism, so it is important to understand their structure and diversity.

If the position of one of the OH groups at carbons 2, 3, or 4 is changed, compounds with different chemical properties and functions are created. Examples of other hexoses important in plant metabolism and similar in structure to glucose are galactose and mannose. Fructose, another very common sugar in nature, is a six-carbon ketose where the carbon at position 1 is –CH$_2$OH and the carbon at position 2 is –C=O; the structure of fructose is otherwise identical to the structure of glucose. Smaller monosaccharides, pentoses (five carbons) and tetroses (four carbons), are also important in plant structure and function. The pentose ribose is particularly important as ribose and its deoxy derivative deoxyribose are important components of the nucleic acids DNA and RNA (Fig. 8–3), which are discussed later in this chapter.

Other carbohydrates that are important in plant structure and metabolism include polyols, in which all carbons have an OH group (e.g., sorbitol, the main form of carbohydrate transported in apple plants), amines (e.g., glucosamine), and uronic acids (e.g., galacturonic acid, the major component of pectin). The ring structures of these compounds are shown in Figure 8–4.

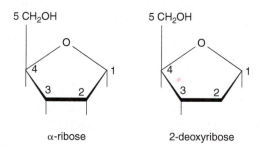

Figure 8–3

Oxygenated and deoxygenated forms of the sugar, ribose.

$$-\overset{\displaystyle H}{\underset{}{C}}=O \qquad -\overset{\displaystyle O}{\underset{}{\overset{\|}{C}}}-$$

A   B

Figure 8–1

Example of an (A) aldehyde group and (B) keto group.

Figure 8–2

Illustration of the ring and planar structures of glucose. Note the difference between the α and β forms.

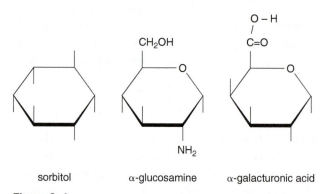

Figure 8–4

Ring structures of sorbitol, alpha-glucosamine, and alpha-galacturonic acid.

The crucial importance of these structural details in defining the function of carbohydrates can be illustrated by the comparison of the structures of **starch** and **cellulose**, two common plant polysaccharides. The general structure of starch is:

—glucose- $\alpha 1–4$—glucose- $\alpha 1–4$—
glucose- $\alpha 1–4$—glucose- $\alpha 1–4$— . . .

The 1–4 notation means that the linkage between the two glucose molecules is from carbon 1 of one molecule to carbon 4 of the next molecule. The α notation means that the OH group at position 1 of the ring structure is down (Fig. 8–2).

Only one simple change in the structure of starch is needed to make the structure of cellulose:

—glucose- $\beta 1–4$—glucose- $\beta 1–4$—
glucose- $\beta 1–4$—glucose- $\beta 1–4$— . . .

Thus, the two structures differ only in the position of the OH group at carbon 1. But the chemical nature of the two molecules is very different. Starch is slightly soluble in water and can be broken down easily by various organisms (including humans and plants) to the simple glucose molecule. Thus, starch is the main form of carbohydrate stored in plants and is very important in animal nutrition. In contrast, cellulose is essentially insoluble in water and, when assembled into fibrils, provides the rigidity required for cell walls. Cellulose is composed of long linear chains of up to 2,000,000 glucose units. Cellulose is the main structural component in plants and is essentially inert to digestion by most higher organisms. It can be utilized for energy only by microorganisms—bacteria and fungi and a few other higher organisms such as ruminant animals and termites. Examples of substances that are important in our everyday lives and are essentially pure cellulose are paper products and cotton cloth.

Another common carbohydrate in plants is **sucrose**, the sugar that is most common form of carbohydrate used in carbon transport in plants. Sucrose is a dissaccharide comprised of a glucose and a fructose molecule (Fig. 8–5). Sucrose, also known as table sugar, is obtained in large quantities from sugar cane and sugar beets and is another carbohydrate that plays an important role in our everyday lives.

LIPIDS

Fatty acids and **lipids** are compounds that are derivatives of glycerol, a simple 3-carbon molecule. They are hydrophobic, meaning they are not soluble in water. However, they are soluble in fats. They are important in cell membrane structure and for energy storage, especially in seeds.

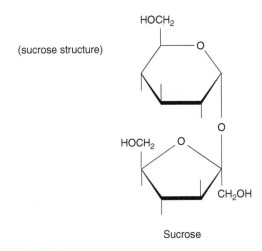

(sucrose structure)

Sucrose

Figure 8–5
Ring structure of sucrose. Sucrose is made up of a glucose sugar attached to a fructose sugar.

Plants and animals have a wide variety of fatty acids, which have the following general structure.

$$O = C – (CH_2)_n — CH_3$$
with OH above the C.

The subscript n after the parenthesis means that the CH_2 group is present a certain number of times, depending on the fatty acid; this number is generally between 10 and 22. Some examples are:

Fatty Acid	n
Lauric acid	10
Palmitic acid	14

In addition to having different chain lengths, fatty acids may be either saturated or unsaturated. The term *unsaturated* refers to the presence of C=C double bonds instead of C–C single bonds. The above listed acids are all saturated, that is, linear chains of nothing but CH_2 groups. Oleic acid is an example of an unsaturated fatty acid in plants.

Oleic acid $CH_3(CH_2)_7CH=CH(CH_2)_7COOH$

Saturated or unsaturated has serious dietary implications. Unsaturated fatty acids are generally considered to be more healthy than saturated in the human diet.

To complete the structure of the fundamental lipid molecule, linkage of fatty acids to the glycerol molecule is required to give a triglyceride molecule (Fig. 8–6). The fatty acids have been written out to convey their true molecular size and complexity. Depending on the molecular weight of the fatty acid chains and their degree of unsaturation, a wide variety

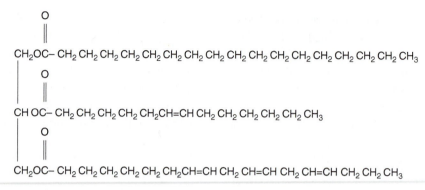

Figure 8–6

A triglyceride molecule is comprised of three fatty acids attached to a glycerol molecule.

of molecules with different chemical properties can be constructed. Thus, materials such as fats, waxes, and oils—materials with very different physical properties—all belong to this class of compounds.

Another type of lipid is an important part of membrane structure (see Chapter 13). The membrane lipids are described as **phospholipids**, because one of the OH groups on the glycerol molecule is linked to a phosphate group, not a fatty acid (Fig. 8–7).

In addition, another chemical group is almost always attached to the phosphate group. Here, a compound named phosphatidyl choline is used to illustrate the point (Fig. 8–8). Because a wide variety of other chemical groups can be attached to the phosphate, this diversity, coupled with the presence of different fatty acids, means that the variety of structural and chemical properties of these phospholipid molecules is enormous.

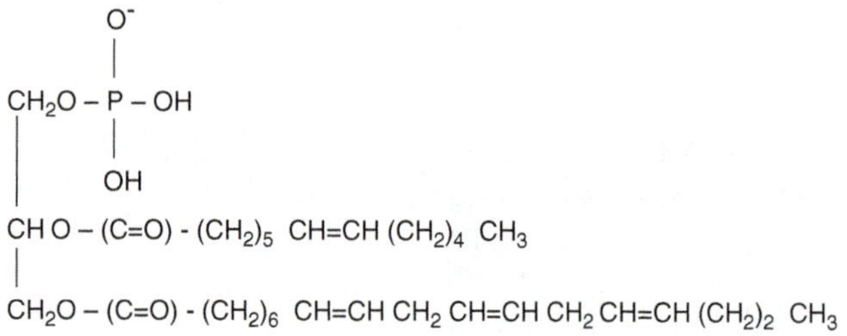

$$CH_2O - P - OH$$

with O^- above and OH below

$$CHO - (C=O) - (CH_2)_5 \ CH=CH \ (CH_2)_4 \ CH_3$$

$$CH_2O - (C=O) - (CH_2)_6 \ CH=CH \ CH_2 \ CH=CH \ CH_2 \ CH=CH \ (CH_2)_2 \ CH_3$$

Figure 8–7

A phospholipid molecule is comprised of two fatty acids and a phosphate group attached to a glycerol molecule.

$$CH_2O - P - OCH_2CH_2N^+(CH_3)_3$$

with O^- above and OH below

$$CHO - (C=O) - (CH_2)_5 \ CH=CH \ (CH_2)_4 \ CH_3$$

$$CH_2O - (C=O) - (CH_2)_6 \ CH=CH \ CH_2 \ CH=CH \ CH_2 \ CH=CH \ (CH_2)_2 \ CH_3$$

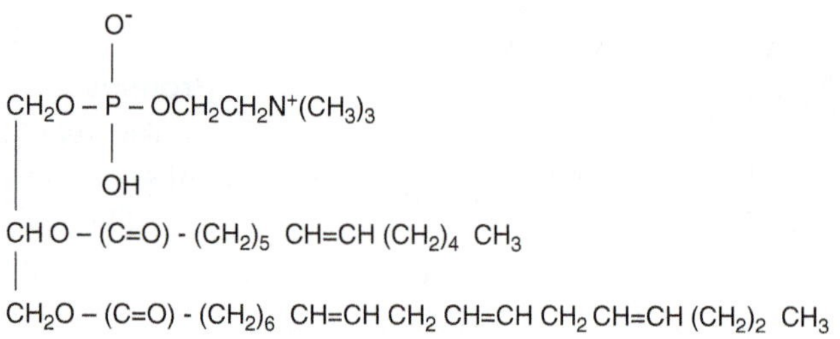

Figure 8–8

Structure of phosphatidyl choline.

In the actual structure of the membrane, the phosphate and attached group are oriented to give two very different chemical surfaces; the phosphate and other attached group are charged and hydrophilic (water-loving), while the long hydrocarbon chains of the fatty acids are hydrophobic. When molecules with these two different chemical properties are combined, they position themselves so that the hydrophobic groups are in close contact with other hydrophobic groups, and the hydrophilic groups are in contact with other hydrophilic groups. In addition, this single layer of lipid molecules is arranged into a double layer because of the mutual attraction among the hydrophobic hydrocarbon chains. This **bilayer** arrangement gives the basic structure of a membrane, as illustrated in Figure 13–3 on page 253.

The type of fatty acids found in the membrane affect the ability of the plant to withstand cool temperatures. Plants that have evolved in cold or temperate climates have more unsaturated fatty acids in their membranes than tropical and subtropical plants. The unsaturated fatty acids are less sensitive to cold and remain "fluid" when temperatures drop, thereby allowing the membrane to function normally. Saturated fatty acids start to solidify in cool temperatures, often at or below 55°F. This causes dysfunction of the membrane and damage or death to the cells and possibly the whole plant. That is why tropical and subtropical plants suffer from "chilling injury" at temperatures well above freezing. The difference in saturated versus unsaturated fatty acids in temperate and tropical plants is also important in the types of oils they provide for human nutrition.

The two different regions of the membrane bilayer also attract proteins that sit on the surface of the membrane, or are imbedded in the membrane, or transverse the entire membrane structure. These proteins are also very important in membrane function. A sketch of a bilayered phospholipid membrane with imbedded proteins is shown in Figure 13–3 on page 253.

One other feature of membrane structure should be mentioned—a group of molecules called sterols. When sterols are inserted into the membrane, they increase the stability of the membrane structure. The sterol content of some plant membranes reaches as high as 50 percent. Sitosterol is the most common sterol in plant membranes. It is very closely related to cholesterol, which is the main sterol in animal membranes (Fig. 8–9). Cholesterol causes much concern because of its relation to cardiovascular diseases in humans. Plants, however, contain no cholesterol.

PROTEINS

Proteins are long chains of amino acids linked together that create complex three-dimensional molecules. Proteins can be structural, serve as storage units for N,

or act as catalysts (enzymes) that direct the biochemical reactions in all living organisms. Amino acids all have a common structure consisting of N, H, C, and O, but they differ from each other by a wide variety of chemical groups attached to the common structure.

Amino acids are synthesized when ammonium is combined with a carbon compound to create glutamate. The process is called transamination (Fig. 8–10). Through transamination, the amino group of glutamate can be passed to various organic acids to form the complete array of amino acids required for protein synthesis.

There are twenty different amino acids that can be found in proteins. These are: glycine, alanine, serine, cysteine, methionine*, valine*, leucine*, isoleucine*, threonine*, aspartic acid, asparagine, glutamic acid, glutamate, phenylalanine*, tyrosine, tryptophan*, histidine, proline, arginine, lysine*. They are arranged in precise order to

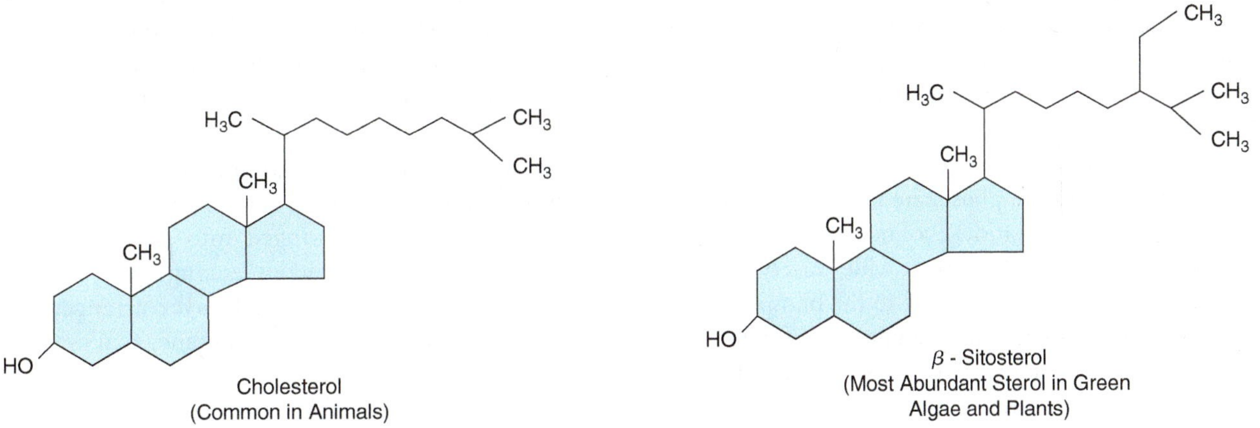

Figure 8–9
Structures of cholesterol and beta-sitosterol.

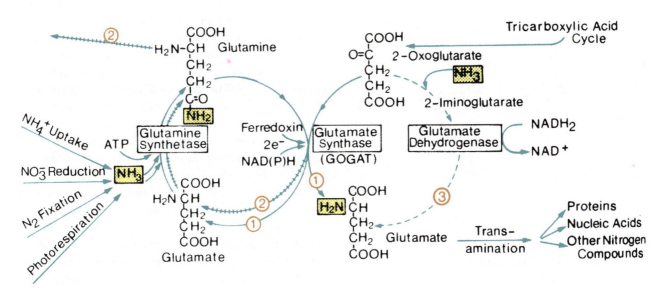

Figure 8–10
Incorporation of ammonium into the amino acid glutamate via two reactions. Ammonium may be the product of more than one process, but incorporation into biologically important compounds is almost entirely via this route.

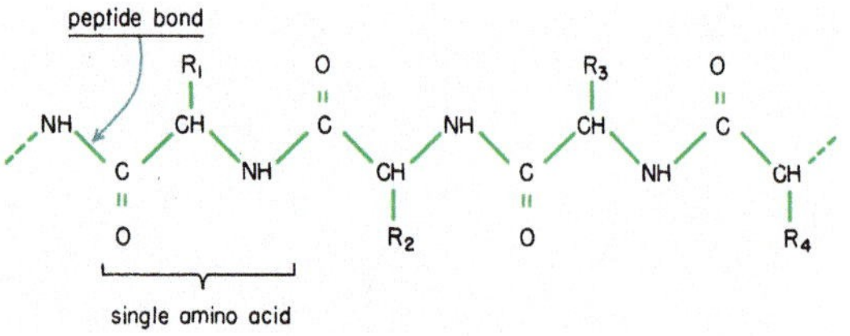

Figure 8–11

Linkage of one amino acid to another to provide the primary structure of proteins. R1, R2, R3, and so on, indicate side groups on different amino acids.

give each protein its special characteristics and function. Considering that proteins can contain hundreds if not thousands of amino acids and the array of 20 amino acids available, the complexity of protein structure can be seen. The precise sequence of the amino acids in proteins is determined by the makeup of the DNA of genes.

The eight amino acids indicated by asterisks (*) are called the essential amino acids. They cannot be synthesized by animals; thus, it is essential that animals (including humans) obtain these amino acids from plant protein, either directly or through the consumption of meat or fish. In the primary structure of proteins, amino acids are linked together between the carbon of one amino acid and the amino (NH) group of another amino acid to form what is known as the peptide bond (Fig. 8–11). After the series of peptide bonds is completed, the chain can be folded to give two general types of secondary structure—an α helix or a β pleated sheet (Figs. 8–12B and 8–12C). The α helix and β pleated sheet structures fold again to form what is called the tertiary structure of the protein (Fig. 8–12D). The tertiary structure is very important because it can change conformation depending on conditions in the cell. The different conformations of the same protein give it very different biochemical properties. This is especially important in enzymes where one conformation makes it active and another makes it inactive. Many enzymes have a fourth level of structure wherein two or more of the tertiary structures (or subunits) combine in a specific way to form a quaternary structure (Fig. 8–12E).

Enzymes do not create chemical reactions; they simply facilitate or accelerate chemical reactions between other molecules or substrates. All enzymes have an active site where the substrate molecule is bound and altered to some other chemical structure (Fig. 8–13). The altered structure is what determines how a biochemical reaction will proceed. Living cells have hundreds of different enzymes, one for each of the many metabolic reactions that are needed to sustain life and growth.

Storage of N represents another critical role of proteins. A young seedling requires inputs of nitrogen (N) and other nutrients before it can begin active assimilation of N by itself. Seeds contain various storage protein structures. Protein concentration in seeds varies widely but is usually between 5 and 25 percent of dry weight. Seeds containing high levels of protein are important in agriculture because they serve as sources of protein for animal feeds. They also can be a source of protein in human diets. Peanuts and soybeans are examples of high protein seeds. Depending on the balance of amino acids in a seed storage protein, it may be a more or less desirable source of protein for animal feeds. In particular, the balance of essential amino acids, which animals cannot synthesize by themselves, is critical for nutritional value, and it is not uncommon for supplemental methionine or lysine to be required in the diet because the content of these amino acids in the seed storage protein is too low.

In plant nutrition, the N needed for proteins is so great that of all the nutrients taken up by the roots, it is the one taken up in the largest quantities. It is also the nutrient that is often deficient and as such has to be supplied in the greatest quantity by fertilizer.

NUCLEOSIDES, NUCLEOTIDES, AND NUCLEIC ACIDS

Nucleosides are comprised of the sugar ribose coupled to an N-containing organic base; the organic bases that are important in metabolism are derivatives of compounds known as purines or pyrimidines. The structure of purine and pyrimidine, along with one example of each type of base, is shown in Fig. 8–14. The nucleoside adenosine (ribose + adenine) is not important in plant metabolism until the molecule is coupled to a phosphate group to form a nucleotide.

The structure of adenosine-3-monophosphate (AMP) is shown in Fig. 8–15, alongside the structure of adenosine triphosphate (ATP), one of the most important molecules in all of metabolism: The importance of ATP lies in the energy stored in the $-P-O-P-O-PO(OH)_2$ bonds. In particular, the terminal P–O–P bond, when broken, yields about 8,000 calories/mole—a very large energy yield. The energy that is released is used to provide the energy for other biochemical reactions. The other important feature of ATP is that it is very mobile within a cell. Thus, ATP moves readily from place to place to participate in a wide variety of enzymatic reactions.

An equally important role of these organic bases is their function in the nucleic acids—deoxyribonucleic

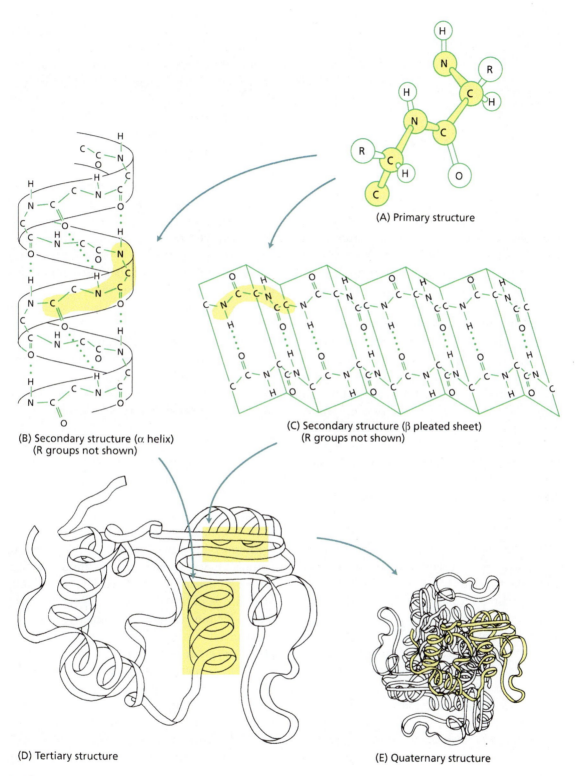

(A) Primary structure

(B) Secondary structure (α helix)
(R groups not shown)

(C) Secondary structure (β pleated sheet)
(R groups not shown)

(D) Tertiary structure

(E) Quaternary structure

Figure 8–12

Schematic picture of the secondary, tertiary, and quaternary structure of proteins. Follow the shaded portions in the diagrams to see the progression of structural complexity. (A) The peptide bond linkage shown in Figure 8–11 but with correct atomic configuration. (B) and (C) Helix and sheet structures formed by the attraction of one peptide chain to another. (D) Assembly of helices and sheets into a globular protein molecule (tertiary structure). Most of this assembly occurs because of the attraction of various side groups to each other. (E) Combination of several identical proteins (termed subunits) to form the functional protein.

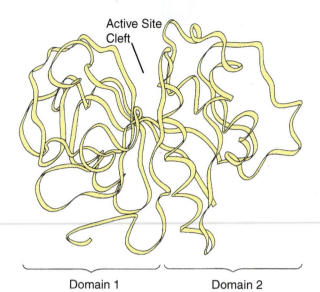

Active Site
Cleft

Domain 1   Domain 2

Figure 8–13
Schematic representation of an enzymatic protein and an active site on the enzyme. The active site is the point at which the substrate is bound and the chemical reaction takes place to form the product.

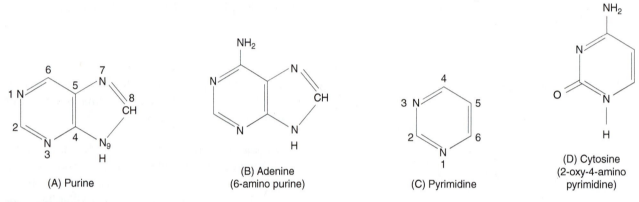

(A) Purine

(B) Adenine
(6-amino purine)

(C) Pyrimidine

(D) Cytosine
(2-oxy-4-amino
pyrimidine)

Figure 8–14
Structures of the organic base (A) purine, (B) adenine, a type of purine, the organic base (C) pyrimidine and (D) cytosine, a type of pyrimidine.

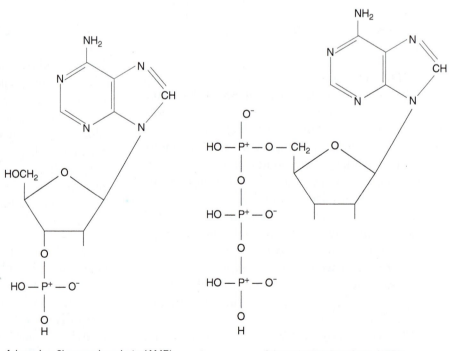

Adenosine-3'-monophosphate (AMP)
(A)

Adenosine triphosphate (ATP)
(B)

Figure 8–15
Structures of (A) adenosine-3'-monophosphate (AMP) and (B) adenosine triphosphate (ATP).

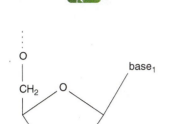

Figure 8–16
Structure of ribonucleic acid (RNA) showing its chain-like configuration.

acid (DNA) and ribonucleic acid (RNA). RNA is a chainlike molecule with a ribose + phosphate backbone (Fig. 8–16). In DNA, one of four **organic bases** (adenine, guanine, thymidine, or cytosine) is attached to each ribose at carbon 1. Long chains of DNA molecules make up the chromosomes in a cell nucleus. Chromosomes carry the genetic information to create all the proteins needed to build and maintain the organism. To provide enough information to build and maintain a plant between 10^7 and 10^{12} bases, chromosomes are arranged in a very specific way.

Most plants have their own unique arrangement of these bases giving them their individual characteristics, although there are some exceptions that are discussed in Chapter 9.

The plant's need for P for the phospholipids and nucleotides explains why P is a common ingredient in fertilizers. As with proteins, the N in the organic bases contributes to a plant's need for N as a nutritional element.

SECONDARY PRODUCTS

The term *secondary* is used because these compounds are not part of the main processes of metabolism in plants; that is, they are not part of primary metabolism, including carbohydrate metabolism, energy production, protein production, and so on. The story of these compounds is difficult to present because of their chemical complexity and diversity. However, these metabolites are an extremely important feature of plant chemistry. Animals do not synthesize these compounds, but humans have exploited some of these chemicals for a wide variety of uses.

Alkaloids

This class of compounds includes secondary metabolites that should be familiar to students as a result of daily life and general knowledge:

- *Morphine.* This first alkaloid to be identified (in the early nineteenth century) is synthesized and extracted from the opium poppy (*Papaver somniferum*). It has long been used in medicine as a pain killer, but continued use of the compound can lead to addiction.
- *Cocaine.* An alkaloid produced by the coca plant (*Erythroxylum coca*), which grows in the Andes mountains of South America. It is relatively harmless in small doses (used by South American natives as a stimulant), but purified, concentrated cocaine is strongly addictive and dangerous.
- *Nicotine.* This alkaloid is obtained from the tobacco plant (*Nicotiana tabacum*) and is very toxic when consumed in large quantities. It acts to constrict the blood vessels when consumed in small doses via tobacco smoke.
- *Caffeine.* This compound is obtained from the coffee plant (*Coffea arabica*) and is common in many popular beverages, carbonated soft drinks in particular. It acts as a stimulant in warm-blooded animals.

So why should plants synthesize these noxious compounds? Most of the alkaloids are formed rapidly from precursors upon wounding of the plant. This formation occurs when the plant is attacked by a chewing insect or is grazed by an animal. Most of these compounds have a bitter taste and act as deterrents or protective chemicals. Because the plant is a carbohydrate-rich

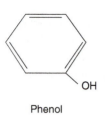

Figure 8–17
Structure of phenol.

organism and cannot move to avoid attackers, it is only natural that compounds of this type are produced by plants.

Phenolics

Another very important group of secondary metabolites is termed phenolics. This term results from the presence of the phenol molecule in some form in all of these compounds (Fig. 8–17). By far the most important phenolic in plants is lignin, the most abundant compound in plants after cellulose. The structure of lignin molecules is difficult to determine because of their high molecular weight and complex structure. Most lignin molecules probably contain twenty to twenty- four phenol units linked together in many different ways. Lignin is deposited in many cell walls to provide additional rigidity and strength. A good example is the lignification of secondary xylem tissues in woody plants; that is, the polymer cellulose is rigid, but it is insufficiently strong by itself to permit long-distance upright growth. Woody plants have highly lignified secondary xylem tissues to stand tall for the acquisition of light energy.

Another important group of phenolics is the flavonoids. Probably the most important role of the flavonoids is acting as pigments produced in flowers to attract pollinating insects and birds. Most of these pigments are shades of red, but blue and purple pigments are also in this class. One of the most intriguing stories regarding flower pigmentation has emerged only recently: flowers produce flavonoid pigments that absorb ultraviolet (UV) light. This phenomenon went unnoticed for many years because humans cannot see in the ultraviolet range; however, honeybees can. Thus, many flowers appear different to a bee than they do to a human (Fig. 8–18). The presence of UV-absorbing pigments is assumed to present a more easily recognized target for the beneficial insects.

Terpenoids

This largest of all the classes of secondary metabolites are all polymers of the hydrocarbon isoprene (Fig. 8–19). One of the most important group of terpenoids is the carotenoids—the yellow pigments in leaves. These pigments serve a protective role in chloroplasts but have no

Figure 8–18
Two photographs of the golden eye flower (Viguiera). The photo on the left was taken with regular film, which records an image that can be seen by the human eye. The photo on the right was taken using a different light source and film that is sensitive to ultraviolet (UV) light. Because honey bees can see light in the UV range, the picture seen by the bees (right) is different from that seen by humans (left). The pattern on the right is due to the presence of UV-absorbing flavonoids in the flower petals.

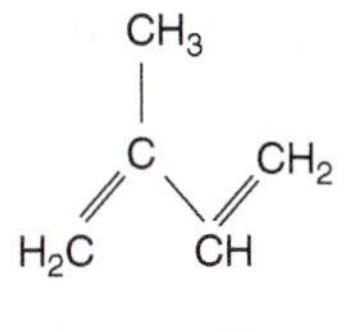

CH₃
|
C —— CH₂
H₂C   CH

Isoprene (C_5H_8)

Figure 8–19
Structure of isoprene.

further use in senescing leaves, leading to the yellow leaf colors we see in the autumn. Other terpenoids exude from insect-damaged trees and appear to play a protective role against insect invasion. Some terpenoids are used in flavoring food; this group includes menthol, a compound that provides a strong aroma in many products. Perhaps the

most exploited terpenoid is rubber, a very high molecular weight compound obtained from latex. The latex is obtained from the tropical tree *Hevea brasiliensis*.

Finally, some terpenoids are poisonous to many insects and are clearly deterrents to feeding. However, one insect, the Monarch butterfly in the caterpillar and adult stages, has taken advantage of the toxic flavonoids in the milkweed plant (*Asclepia* species) by ingesting sufficient quantities and storing them inside its body. When ingested by birds, these insects provide a bitter taste and will upset the bird's stomach. Birds soon learn to leave these particular insects alone.

Other secondary products have been described, but the short discussion above serves as an introduction to this plant-specific group of compounds. Much more detail is available in the references cited below.

SUMMARY AND REVIEW

Carbohydrates are the product of photosynthesis, they contain only three elements—C, H, and O—and are the beginning point for all other plant metabolism. Most carbohydrates are five- or six-carbon compounds and are known as monosaccharides. The hundreds of carbohydrates found in plants consist of various combinations of monosaccharides. Depending on the structure of the monosaccharides, the number and types of monosaccharides linked together, and the type of bonding between monosaccharides, carbohydrates have a wide variety of chemical properties. Some plant carbohydrates like sucrose and starch are very important in animal nutrition. Another carbohydrate, cellulose, provides the fiber for clothing and paper.

Lipids are compounds that are present in fats, waxes, and oils. They are constructed by attaching fatty acids to glycerol. Lipids have a wide variety of different chemical properties. Some lipids are used as energy storage compounds (e.g., corn oil, soybean oil) but perhaps the most important role of lipids is in the formation and function of cell membranes. The lipids in membranes also usually contain a phosphate molecule (phospholipids) and, in addition, a variety of other compounds may be appended to the phosphate. The arrangement of phospholipids in a membrane creates a bilayer with hydrophilic outer edges and a hydrophobic center. The type of fatty acids in the hydrophobic center affects the functioning of the membrane.

Proteins are long chains of amino acids linked together by a unique chemical bond—the peptide bond. All amino acids have a common sequence of N, H, C, and O atoms but amino acids differ in the wide variety of chemical groups attached to the common structure. Protein molecules are very complex structures, and their shape plays an important role in their function, especially as enzymes. Enzymes are proteins that are biochemical catalysts and facilitate most, if not all, biochemical reactions. Other proteins serve as food reserves for seeds. Examples of protein-rich seeds are peanuts and soybeans. The large amount of N needed for proteins is one of the reasons plants often require additional N in the form of fertilizer.

Nucleosides, nucleotides, and nucleic acids form the basis for many important molecules. The nucleoside adenosine triphosphate (ATP) is one of the most important metabolic molecules because of the high energy levels it carries in its phosphate bonds. ATP movement enables energy to be transported from one area in the plant to another. Chromosomes consist of long chains of nucleic acids known as DNA. DNA leads to formation of all of the proteins that build and maintain an organism.

So-called secondary plant chemicals include the alkaloids, phenolics, and terpenoids, whose specific functions and characteristics provide protection, structural strength, and coloration along with other features for the plant. Humans have long used many of these products for industrial, pharmaceutical, and recreational uses.

KNOWLEDGE CHECK

1. What elements make up carbohydrates? What has to be supplied in sufficient amounts to plants to make sure carbohydrates can be synthesized by the plant?
2. What is the general formula for carbohydrates?
3. Why is the ring formation of monosaccharides considered to be so important?
4. What is the structural difference between starch and sugar? How does this affect their ability to be digested by humans and most other animals?
5. Sucrose or table sugar is a disaccharide composed of what two monosaccharides?
6. From the previous question, how do the two monosaccharides differ structurally from each other?
7. What are the two main functions of lipids in plants?
8. What is the difference between a saturated and unsaturated fatty acid?
9. From the previous question, which one is more likely to remain fluid at temperatures between 32°F and 55°F? Which one is more likely to be found in temperate plants than in tropical plants?
10. Where are phospholipids most commonly found?
11. Describe how phospholipids are arranged in a bilayer.
12. What is the main sterol found in plant membranes? In animal membranes?
13. What are the three main functions of proteins?
14. How are amino acids related to proteins and what do all amino acids have in common?
15. Why are some amino acids considered to be essential?
16. Why is the shape that an enzyme assumes after it is synthesized so important?
17. What are two protein-rich seeds?
18. What is the function of adenosine triphosphate (ATP) in organisms?
19. What are the four organic bases found in DNA? How does the arrangement of the base pairs relate to the characteristics of an organism?
20. Explain how proteins and organic bases are related to a plant's need for nitrogen.
21. Explain how RNA and some lipids help explain a plant's need for nitrogen.
22. What are the three main types of plant secondary products?
23. What are some functions of the secondary products in plants?
24. What are some ways humans have used plant secondary products?

FOOD FOR THOUGHT

1. If a plant has a severe nitrogen deficiency, how do you think biochemical reactions in the plant could be affected? What would be the effect on the plants?
2. Straw is mostly the structural material of grain plant stems that is left behind after the seed heads are harvested. In some areas where starvation is prevalent, people eat straw to try to survive. Explain why or why not straw would be effective at providing these people with nutrition.

FURTHER EXPLORATION

BOWSHER, C., M. STEER, AND A. TOBIN. 2008. *Plant Biochemistry*. London: Garland Science.

HOPKINS, W. G., AND N. A. P. HÜNER. 2004. *Introduction to Plant Physiology*, 3rd ed. Hoboken, NJ: John Wiley & Sons.

*9

Genetics and Propagation

BASIC GENETIC CONCEPTS IN PLANT SCIENCE

The plants we cultivate for our survival and pleasure all originated from wild plants. However, most of our domesticated plants appear very different from their wild relatives. These differences have come about mainly through the recombination of genes and the redistribution of heritable traits through the generations separating the wild types from the domestic plants. Manipulating natural genetic processes has been the basis for crop improvement since the time of the first plant scientists. Understanding basic genetic principles is crucial to understanding the principles of the traditional methods of plant breeding and plant propagation and the most recent tool for crop improvement: bioengineering.

Chromosomes

Some plants consist entirely of living cells, whereas others, such as woody perennials, are made up of both living and dead cells. The living cell consists of a **cell wall** containing a fluid, the **cytoplasm**, in which the **nucleus** is suspended, along with many other structures. The nucleus contains the **chromosomes**, which carry most of the genetic information and transmit that information from one generation of cells to the next. The number of chromosomes is the same in all vegetative cells of an entire plant species. This number is usually the **2n**, or **diploid**, chromosome number (although higher numbers—tetraploid or octaploid, for example—also occur). In the sex cells—the egg and sperm—the number is reduced by half and is termed the **haploid**, or **1n**, chromosome number.

Although usually constant for a given species, the number, size, and appearance of chromosomes vary considerably between different plant species. The chromosomes can be counted by microscopic examination of the nucleus at a stage just before cell division. The chromosome numbers are known for most plant species. For example, the diploid chromosome number for alfalfa (*Medicago sativa*) is 32; for barley (*Hordeum vulgare*), 14; for corn (*Zea mays*), 20; and for sugar beet (*Beta vulgaris*), 18.

Chromosomes change in appearance during cellular development and division, but basically each is a long, threadlike structure consisting of

deoxyribonucleic acid (DNA) plus associated **ribonucleic acids (RNA)** and various proteins. DNA can replicate itself and it can transmit information to other parts of the cell. DNA is a polymer—a very large molecule made up of many repeating units, but the repeating units called **nucleotides** can vary. DNA, the active genetic material, is a very large molecule composed of two spiral strands (Fig. 9–1). The "backbone" of these strands is composed of sugar residues (S) linked by phosphates (P) on each side. A sugar residue on one strand is connected with a sugar residue on the other by two bases that are linked to each other by hydrogen bonds (see Fig. 9–1). These bases are cytosine (C), guanine (G), adenine (A), and thymine (T). The sugar residues in each strand are held tightly together by the phosphate radicals, but the two spiral strands are bound together more loosely by the hydrogen bonds.

One of the characteristics of DNA that makes it possible for the chromosomes to transmit genetic information from one cell generation to the next is its ability to replicate itself. This ability arises from the double-strand structure of the molecule and the properties of the four bases. Adenine and thymine are held together by two hydrogen bonds (A = T). Their molecular structure is such that only they can join together. In the same manner, only cytosine and guanine can join together, and they are held by three hydrogen bonds (C = G). When the chromosome divides during cell division, the two spiral strands of the DNA molecule unravel and separate at the position of these hydrogen bonds. Every base (cytosine, guanine, adenine, or

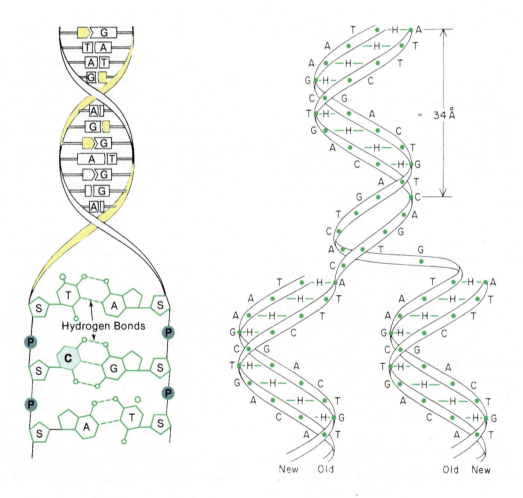

Figure 9–1

Left: *Schematic diagram of the DNA molecule showing the helical structure and the base pairing. Lower detailed view shows the alternate attachment of sugars (S) and phosphates (P) on each strand. Connecting the two strands are the base pairs adenine (A) —thymine (T) guanine (G) —cytosine (C), which are held together by hydrogen bonds and attached to the sugars on each strand. Right: During cell division and chromosome splitting, the DNA molecule replicates itself by unraveling, separating at the hydrogen bonds. Then each strand quickly becomes a new double strand just like the original, with each base attracting to itself its complementary base (A = T and G = C).* Source: *Left:* William D. McElroy and Carl P. Swanson, *Modern Cell Biology,* 2nd ed. © 1976. Reprinted by permission of Pearson Education, Inc., Upper Saddle River, NJ. *Right:* J. W. Wright, *Introduction to Forest Genetics* (New York: Academic Press, 1976).

thymine) attached to each strand attracts its complementary base, so that each single strand immediately becomes a new double strand exactly the same as the original double strand (Fig. 9–1).

Although DNA includes the self-perpetuating **genetic code**, a related substance—ribonucleic acid (RNA)—actually controls the growth processes in the cell. DNA and RNA have certain differences. DNA is double-stranded, whereas RNA is a single strand. RNA sugars have one more oxygen atom than DNA sugars. RNA has uracil (U) as a base in place of thymine (T). The DNA molecule acts as a template from which a complementary strand of RNA is formed (in the same manner by which a complementary strand of DNA forms from a single strand).

Using the RNA as a template, proteins are constructed from amino acids aligned in a specific arrangement determined by the bases of the original DNA template.

Genes

Structurally, a gene[1] is a sequence of triplet organic bases (cytosine, guanine, adenine, and thymine) along a DNA molecule. The gene is the ultimate hereditary unit that functions as a certain part of a chromosome determining the development of a particular characteristic in an organism. Thus, one gene (or several interacting genes) may determine plant height, leaf shape, flower color, or fruit size. Genes are much too small to be seen, even with an electron microscope. There are, of course, a great many genes in each cell of the higher plants; they number in the thousands. Any individual gene may have a large effect or a small effect. Some genes act independently, whereas others act only in conjunction with other genes. Because genes are arranged along the chromosome, genes on the same chromosome are linked—that is, genes on the same chromosome move from one cell generation to the next as a unit. Linkage is not perfect, however; sometimes during meiosis, chromosomes break and exchange parts, as we will soon study in more detail.

Homologous Chromosomes

In each vegetative cell are pairs of each individual chromosome. These pairs are called **homologous chromosomes** and are properly defined as such if they each have the same gene or genes affecting the same traits at

corresponding positions. Genes are termed **alleles** to each other if they occupy the same position on homologous chromosomes and affect the same trait. If a plant had the genes ABCDEFG paired with abcdefg on a certain pair of homologous chromosomes, the genes A and a would be alleles for the same trait, B and b would be alleles, and so forth. Alleles can be the same or different. For example, A could be for yellow flower color and b for white.

Allelic genes can be dominant or recessive to each other. A **dominant gene**, A, is one that causes a certain characteristic to be expressed whether the plant is **homozygous**, AA—both alleles the same—or **heterozygous**, Aa—the two alleles different. A **recessive gene** causes the character it controls to be expressed only if both alleles are recessive, aa. (By convention, capital letters express dominant genes; lowercase letters express recessive genes.)

Mitosis

Cell division in the shoot and root tips, axillary buds, leaf primordia, and the vascular cambium—all of which increases plant size—is called **mitosis**. These vegetative cells usually contain two sets of homologous chromosomes—the 2n or diploid number. During cell division, the chromosomes split longitudinally, replicating to produce two chromosomes that are identical to each other. One of each pair goes to one daughter cell, and one to the other. An equatorial plate, as it is called, develops between them to form a new cell wall and thus two new cells, each with its full complement of chromosomes—and genes. So each daughter cell has a genotype identical to that of the mother cell.

Meiosis and Fertilization

Meiosis refers to the type of cell division, sometimes called **reduction division**, that occurs in the flower—in the angiosperms—to form the cells from which the pollen grains and the embryo sac (which contains the egg) develop (Figs. 9–2 and 9–3). In this type of cell division, the homologous chromosomes separate from each other without replicating, one going to one daughter cell and one to the other, thus reducing the number of chromosomes—the 1n or haploid number. Half of the original chromosomes come from the male parent of the plant and half from the female parent. During meiosis, when the chromosomes line up to separate, although the maternal and paternal chromosomes line up across from each other, they do not line up on the same side of the dividing line. They line up randomly, meaning that the haploid cells can have all maternal chromosomes or all paternal chromosomes or

---

[1]Virus particles are similar to genes. Because viruses can be isolated and easily studied, much knowledge of gene structure is obtained from the study of viruses.

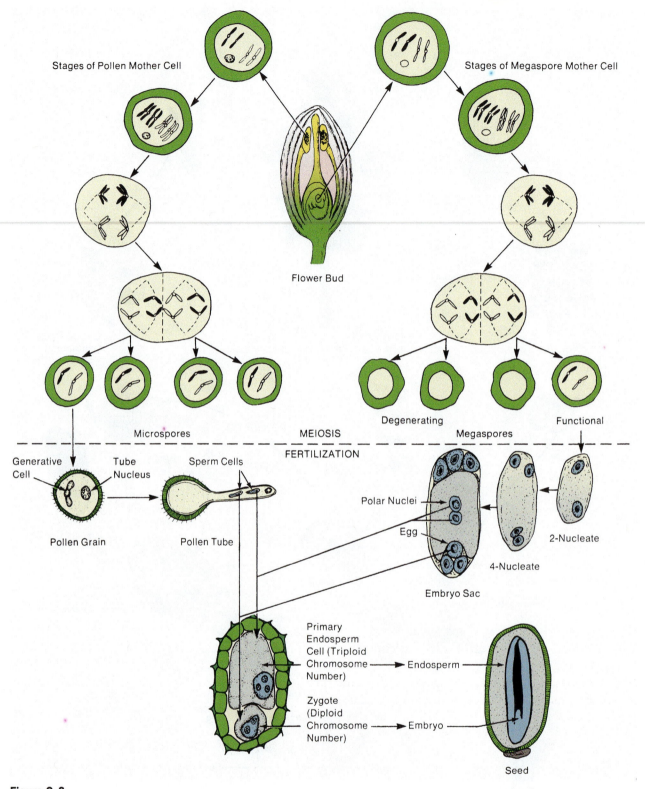

Figure 9–2

Diagrammatical representation of the sexual cycle in angiosperms. Meiosis occurs in the flower bud in the anther (male) and the pistil (female) during the bud stage. During this process, the pollen mother cells and the megaspore mother cells, both diploid, undergo a reduction division in which homologous chromosomes segregate to different cells. This is immediately followed by a mitotic division, which produces four daughter cells, each with half the chromosomes of the mother cells. In fertilization, a male gamete unites with the egg to produce a zygote, in which the diploid chromosome number is restored. A second male gamete unites with the polar nuclei to produce the endosperm. Source: Dale E. Kester and Hudson T. Hartmann, *Plant Propagation: Principles & Practices,* 4th ed. © 1983. Adapted by permission of Pearson Education, Inc., Upper Saddle River, NJ.

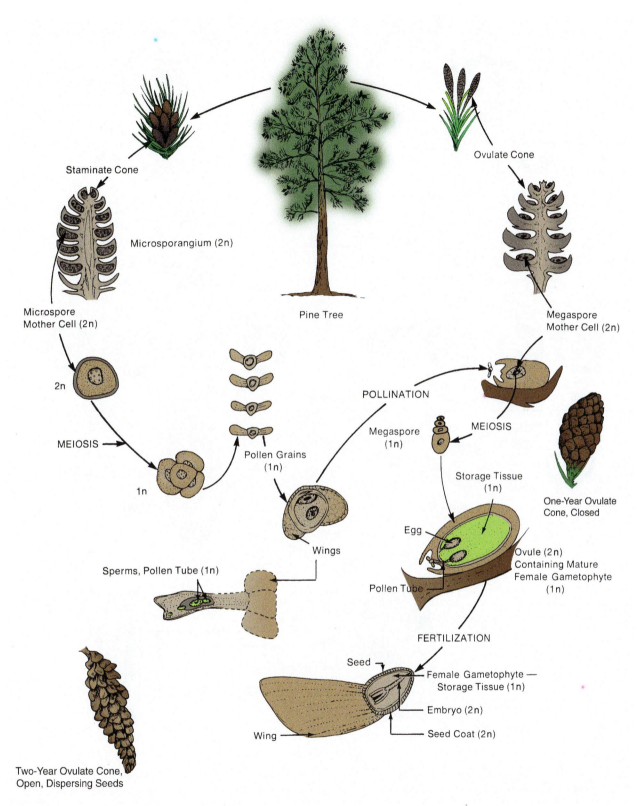

Staminate Cone

Microsporangium (2n)

Microspore
Mother Cell (2n)

2n

MEIOSIS

1n

Pine Tree

Ovulate Cone

Megaspore
Mother Cell (2n)

Pollen Grains
(1n)

POLLINATION

Megaspore
(1n)

MEIOSIS

Storage Tissue
(1n)

One-Year Ovulate
Cone, Closed

Egg

Wings

Ovule (2n)
Containing Mature
Female Gametophyte
(1n)

Sperms, Pollen Tube (1n)

Pollen Tube

FERTILIZATION

Seed

Female Gametophyte —
Storage Tissue (1n)

Embryo (2n)

Wing

Seed Coat (2n)

Two-Year Ovulate Cone,
Open, Dispersing Seeds

Figure 9–3

Diagrammatical representation of the sexual cycle in a gymnosperm (pine), showing meiosis and fertilization. Source: Dale E. Kester and Hudson T. Hartmann, *Plant Propagation: Principles & Practices*, 4th ed. © 1983. Adapted by permission of Pearson Education, Inc., Upper Saddle River, NJ.

a mixture of maternal and paternal chromosomes. This is called genetic segregation.

In **fertilization** in angiosperms, one male gamete (1n) from the pollen grain unites with a female gamete—the egg (1n)—to form the zygote (2n), which develops into the embryo and finally the new plant (see Figs. 9–2 and 9–3). The manner in which these gametes can segregate and reunite to form new combinations is illustrated in Figure 9–4 for both a monohybrid and a dihybrid cross.

Also during fertilization in the angiosperms, one male gamete (1n) unites with the two polar nuclei (1n each) in the embryo sac to form a food storage tissue, the endosperm (3n), which can serve as a nutritive material for the developing embryo. The seed coats, developing by mitosis from the female parent, cover

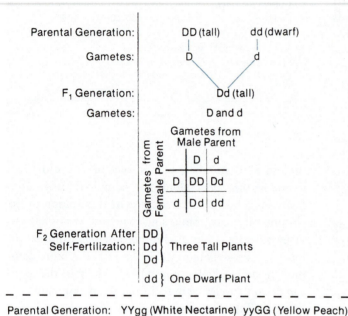

Figure 9–4

Top: *Inheritance involving a single gene pair in a monohybrid cross (one characteristic involved). In garden peas, tallness (D) is dominant over dwarfness (d). A tall pea plant is either homozygous (DD) or heterozygous (Dd). Segregation occurs in the F_2 generation to produce three genotypes (DD, Dd, dd) and two phenotypes (tall and dwarf).*

Bottom: *Inheritance in a dihybrid cross (two characteristics involved) of the peach (Prunus persica). Fuzzy skin (G) of a peach is dominant over the glabrous (i.e., smooth) skin of the nectarine (g). White flesh color (Y) is dominant over yellow flesh color (y). In the example shown, the phenotype of the F_1 generation is different from either parent. Segregation in the F_2 generation produces nine genotypes and four phenotypes.* Source: Dale E. Kester and Hudson T. Hartmann, *Plant Propagation: Principles & Practices*, 4th ed. © 1983. Adapted by permission of Pearson Education, Inc., Upper Saddle River, NJ.

both the endosperm and the embryo. In gymnosperms, the endosperm tissue—more correctly termed the female gametophyte—is 1n tissue, rather than 3n, as in the angiosperms (Fig. 9–3).

Mutations

As plants grow, hundreds of thousands of cells may be dividing constantly. At each cell division, DNA must replicate itself—the double strand must untwist, the millions of nucleotide triplets must be reproduced exactly on new single strands, and the single strands must then twist around each other to form new double strands. It is a marvel that this complicated process takes place on such a grand scale. With so many cells involved, however, errors can and do occur during replication. When they do, they are called **mutations**, and the altered genes may possibly result in changes in the characteristics of the plant—although most mutations have such slight effects that they go unnoticed. The great majority of mutations are deleterious, but some are not and they provide a source of variability that aids the plant breeder in developing new cultivars. Mutation rates can be vastly increased by treatment with ionizing radiation and by certain chemicals such as colchicines, which comes from a type of crocus. Mutations that occur during the formation of pollen grains and egg cells and appear in the plant's seedling offspring are particularly important to the plant breeder.

In addition to gene mutations, hereditary modifications can also be caused by gross changes in chromosome number or structure. This can involve the doubling of chromosome numbers, the addition or subtraction of an entire chromosome, or some other structural change in a chromosome such as crossing over. Crossing over occurs during the pairing of the two sets of homologous chromosomes. The chromosomes may break at the same location on each and exchange segments. Thus, if the original gene sequence on the homologous chromosomes was ABCDEFGHI/abcdefghi and crossing over occurred between E and F, then the gene sequence for this particular chromosome appearing in the pollen grain or egg cell would be ABCDEfghi or abcdeFGHI. This gene alteration would, of course, be expressed in altered characteristics of the new plants. Gross chromosome changes may cause pronounced changes in the plant's characteristics.

Polyploidy

Polyploidy is a condition in which individual plants have more than two sets of homologous chromosomes in their somatic (vegetative) cells. Beyond the normal

Table 9–1
POLYPLOIDISM IN OATS, WHEAT, AND TOBACCO

Common Name	Species	Somatic Chromosome Number
Sand oats	Avena strigosa	14
Slender wild oats	Avena barbata	28
Cultivated oats	Avena sativa	42
Einkorn	Triticum monococcum	14
Emmer	Triticum dicoccum	28
Common wheat	Triticum aestivum	42
Wild tobacco	Nicotiana sylvestris	24
Cultivated tobacco	Nicotiana tabacum	48

diploid (2n) number, plants may be triploid (3n), tetraploid (4n), pentaploid (5n), hexaploid (6n), and so forth. Polyploid plants may arise by duplication of the chromosome sets from a single species—autoploidy—or by a combination of chromosome sets from two or more species—alloploidy. The latter is the more common type of polyploidy in nature. Many of the cultivated crop species evolved in nature as polyploids, as shown in Table 9–1 for oats, wheat, and tobacco.

Cytoplasmic Inheritance

While most inherited characteristics are transmitted by the genes in the nucleus, certain characteristics in some herbaceous plants can be controlled by genes in the mitochondria and chloroplast (mDNA and cDNA, respectively), which are contributed only by the female parent. Mitochondrial and chloroplast DNA are often very useful in determining hereditary relationships among organisms.

Genotype and Phenotype

The term **genotype** refers to the genetic makeup of the plant—its genetic constitution—the kinds of genes it has on the chromosomes and the order in which they are situated. **Phenotype** refers to the plant's appearance, behavior, and chemical and physical properties. The phenotype expressed is the result of gene activity and interaction with the environment. A plant may have genes for very vigorous growth but when grown under a deficiency of soil nitrogen, for example, its inherent vigor is not expressed. However, the genes still control the plant's characteristics even if an environmental factor

drastically alters the phenotype. Therefore, if seeds are taken from stunted plants grown under deficient soil nitrogen conditions and planted in rich fertile soil, the plants again show the strong, vigorous growth called for by their genotype.

PLANT SELECTION

One of the first steps in plant improvement is selecting plants with desirable characteristics (phenotypes) and increasing their numbers by seed or asexual propagation. Many plants grown today started from a single plant selected for its attributes. However, selection has limited capacity to generate plants with new characteristics. To get those characteristics, selective breeding or genetic engineering is often required.

Breeding

Breeding means selecting a female and male parent with unique desirable traits and crossing them to produce offspring with both sets of desirable traits. Although this sounds easy, it can be very complicated and take many years.

Breeding starts with finding a parent that has one or more desired traits and another parent that has another desired trait or traits. For example, a commercial variety that produces a high number of quality fruit is susceptible to a disease. We select the commercial variety as one parent for its production characteristics and a disease-resistant plant as the other parent. The resulting generation is called **F1**.

The F1 generation from a cross like the one described above is highly variable and none may be commercially acceptable. But some may exhibit disease resistance. The F1s can be crossed and/or self-pollinated to create the **F2** generation. A series of back-crosses (crossing an offspring with a parent or other direct ancestor) and selections is performed to try to get the desired product. For example, offspring from the F1 or F2 generations with good disease resistance and some commercially acceptable traits are selected and then backcrossed with the commercial parent to enhance the commercially desirable traits while retaining the disease resistance. This may have to be done several times to get a disease resistant variety with good commercial characteristics.

In the past, breeding programs like the one just described could take years and many, many crosses to develop a cultivar with the desired traits because plants had to be grown to maturity to evaluate their characteristics. Now DNA analysis for many traits can be done in very young plants and embryos to determine if desired traits are present and only those with the traits are grown to maturity, thereby making the process much more efficient.

In the breeding process, pollen is collected from the male plant and then used to fertilize the female plant (Fig. 9–5). Breeding requires meticulous attention to small details. Usually the pollen is removed from the male plant and carefully brushed on to the stigma of the female parent. Care has to be taken to keep the female from receiving unwanted pollen. Using male-fertile plants to produce hybrid seed on self-pollinating plants means the male parts have to be emasculated before they mature. Using male-sterile plants that produce seeds but do not produce viable pollen eliminates this problem. An alternative is protecting the female from unwanted pollination.

Single gene traits are the easiest to breed for. The more genes involved, the more difficult it is to get all desired traits into a single offspring, especially if those genes are on separate chromosomes. In an organism with 3 chromosomes, the possible different gametes in the F1 generation is 8. For corn, which has 20 chromosomes, the F1 generation has 1,024 possible gene combinations

Figure 9–5

Left: *Removing pollen from the male parent.* Right: *Applying pollen to the female plant.* Source: Margaret McMahon, The Ohio State University.

in its gametes. The gamete possibilities for the F2 generation is over a million different combinations.

Once a breeding program has been started, after careful breeding and several generations, it is possible to develop a homozygous line from a single, self-pollinating plant. However, when a population becomes completely or nearly homozygous, reduced vigor can occur.

The production of plants from **hybrid seeds** has been one of the outstanding scientific breakthroughs in agricultural history. Hybrid seed produces the F1 generation from a cross of two different yet homozygous parents (see Fig. 9–4, top). The F1 plants are very uniform because although they are heterozygous they have the same genotype. The use of hybrid seed has more than doubled the yield of both sweet corn and field corn. In the United States, almost all corn is now produced from hybrid seed. This method of producing hybrid seed was developed in 1918, but it was not used commercially until about 1935. Now, many other crops for food, forage, and ornamental use are produced from hybrid seeds to insure high yield, better vigor, color, or other traits. However, unlike open-pollinated crops, the seed from the hybrid crop cannot be saved and used to produce future crops because the F2 generation segregates into several different genotypes and phenotypes (see Fig. 9–4, bottom).

BIOTECHNOLOGY

Biotechnology, in a broad sense, can be defined as the management of biological systems for the benefit of humanity. By this definition, it is obvious that biotechnology has been practiced since the days when humans first began cultivating and selecting plants for desirable traits. But the word *biotechnology*, as generally used now means organisms developed by molecular biology and molecular genetics using advanced genetic engineering techniques. Because the genetic code is the same for all DNA-carrying organisms, DNA from one organism will be coded the same by another, even though the two organisms may be biologically very different. Scientists from many different disciplines have worked since the 1970s to perfect methods of identifying genes, separating them from the host DNA, and transferring them to new hosts. It is a complicated process and it takes many years to introduce a plant or animal that has the foreign DNA in it. However, improvements in the process have enabled plant scientists to introduce many plants with foreign DNA that gives a desirable trait such as resistance to insects, herbicides, certain environmental stresses, enhanced dietary attributes, and other traits.

When a gene is introduced into a species that otherwise does not have that gene, we call it **genetic engineering**, and the plants are called **transgenic plants**. Often the term **genetically modified organism (GMO)** is used to describe genetically engineered plants, but *genetically engineered* is a more accurate term. A GMO is created each time sexual reproduction occurs and there is a different combination of traits in the offspring. Traditional breeding creates GMOs but does not create transgenic plants because breeding can only impart the traits (genes) that exist in the gene pool of those crops or species with which they are breeding compatible. However, not all genetically engineered plants are the result of gene transfer. Sometimes an **endogenous** gene is altered. An example is the Flavr Savr® tomato. In these tomatoes the gene that produces the enzyme that softens the ripening fruit is "scrambled" so that the enzyme produced is dysfunctional. The fruit ripens but softens much more slowly than unaltered fruit. The shelf life of the harvested fruit is increased while the traditional tomato flavor is maintained.

Genetic engineering complements but does not replace traditional breeding techniques. Most genetic engineering starts with the identification of a gene that imparts a desired trait such as insect resistance or more nutritive content. The gene is removed from the chromosomes of cells taken from the host organism. Removal is done by the use of enzymes that can clip the chromosomes into sections at specific locations. If insufficient numbers of the gene cannot be extracted, the gene is amplified, meaning it is increased in great numbers, usually by a process called polymerase chain reaction (PCR). The genes are then inserted either by a bacterium or gene gun into the nuclei of cells of the plant being transformed. These "target" cells of the have been separated from the host plant as single cells. The plants that then develop from these cells are tested to be sure they carry the gene. The testing is done by inserting a **marker gene** along with the desired gene. The marker gene gives a signal that indicates it is present in the new organisms and it is assumed that the desired gene is also present. A common marker gene is for antibiotic resistance. The transformed cells are placed in petri dishes containing an antibiotic in the media. Only those cells with the antibiotic resistant marker (and presumably the desired gene) will survive to produce new plants. Then through traditional breeding or asexual propagation, large numbers of the transformed plant are produced (Fig. 9–6). To date, only existing genes have been modified or transferred, no entirely new genes have been created. To create an entirely new gene would be an extremely difficult task.

Steps of Genetic Engineering

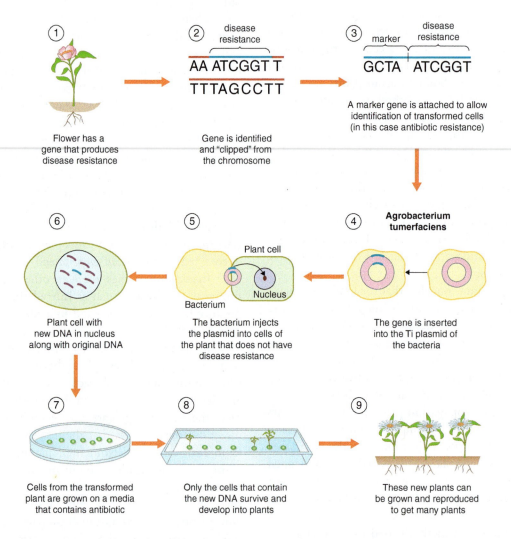

Figure 9–6

The steps of genetic engineering: 1) and 2) Identification and removal of gene, this case a disease resistance gene, from a chromosome, 3) attachment of a marker gene, in this case antibiotic resistance, 4) insertion of the gene complex into the Ti plasmid in the Agrobacterium 5 and 6) the bacteria inserts the gene into cells of the plant without disease resistance, 7) the cells are grown on a media containing antibiotic, 8) only the cells that have the marker gene (and presumably the disease resistant gene) survive and, 9) develop into whole plants with disease resistance. Source: Margaret McMahon, The Ohio State University.

When a foreign gene is inserted into a cell, it then can be induced to produce the protein or proteins that confer the desired trait. Sometimes the gene is attached to a **promoter** that has the gene turned on constantly. In other situations, the promoter turns the gene on only under specific conditions or in certain tissues or organs. Keep in mind, however, that even though a protein is generated, its function may or may not be the same as in the original organisms, depending on what other factors are influencing the protein. The inability to have a protein respond to the same in the modified organism as it did in the original organism is often a challenge to genetic engineers. For example, the pH of a rose flower is such that it makes most blue pigments found in plants turn a gray/brown color. Although genes for a blue pigment have been successfully placed in roses, expression of blue color in roses has not been completely successful to date because of flower pH.

In spite of the challenges, genetically engineered plants now play an enormous role in the production of food and other plants. In 2008, 86 percent of the cotton, 80 percent of corn, and 92 percent of the soybeans grown in the United States were genetically engineered to be resistant to herbicides or attacks from certain insects (USDA Economic Research Service). The use of genetically engineered plants is expected to increase as their acceptance improves and more traits are capable of being transferred.

PROPAGATION OF PLANTS

Once we have a plant with desired characteristics, we have to increase the numbers of that plant. The process of increasing plant numbers is called **propagation**. Plants are propagated by either sexual (seed) or asexual (vegetative) methods. Some kinds of plants are almost always propagated by one method or the other; other kinds can be propagated successfully either way. The various ways of propagating plants are outlined in Table 9–2. A successful propagation method is one that transmits all the desirable characteristics of the original plant or plants to all the progeny. If the characteristics of the original plant/s are lost or changed during the propagation procedures, that particular method is unsatisfactory for that type of plant. It is then necessary to use another propagation procedure that will preserve these characteristics.

It is important to realize that the kinds of agricultural and ornamental plants being grown today are plants with particularly desirable characteristics. Such plants originated from a plant or plants found growing in the wild, in cultivated plant populations, from mutations, or from breeding programs conducted by government agencies or private plant breeders. With the appropriate propagation procedures, such plants can become the starting point for populations of many millions of individuals, all just like the original mother plant. The groups of plants that people have developed and cultivated have been given cultivar names; for example, Redhaven peach, Bing cherry, Pawnee wheat, Imperial Blue pansy, Peace rose, Ranger alfalfa, and so on. The propagation methods used for increasing the populations of these groups of plants must do so without changing their characteristics.

There are several kinds of cultivars (*culti*vated *var*iety). If the plant group reproduces "true" by seeds—with no characteristics changed—the cultivar is termed a **line**. A line is homozygous[2] and, if self-pollinated (or if cross-pollination is prevented), seed propagation yields progeny like the original plant. Many of the world's leading economic plants including some cereals, vegetables, and garden flowers—are made up of groups of these lines. They are seed propagated and faithfully maintain their characteristics when propagated in this manner. Many forest species, although seed propagated, are not considered lines because of their higher level of variability. They are known by their species names, such as Douglas fir (*Pseudotsuga menziesii*), ponderosa pine (*Pinus ponderosa*), or coast redwood (*Sequoia sempervirens*).

In addition to lines, there are other types of seed-propagated cultivars, such as **inbred lines**—used to produce hybrid cultivars—and **hybrids**—as in hybrid corn (Fig. 9–7).

Many groups of plants are heterozygous[3] rather than homozygous. These plants have many dissimilar genes controlling their characteristics. During meiosis and fertilization leading to embryo formation in the seed, these genes segregate and recombine in a great many different ways so that the resulting plants differ from each other and from their parent. With these plants, seed propagation cannot maintain the characteristics of the female parent and cannot be used as a successful propagation procedure. With such heterozygous plants, vegetative (asexual) propagation is usually used. A piece of vegetative tissue—a section of a stem, root, or leaf—is placed in a suitable environment, such as a warm, humid rooting frame in the greenhouse. In time, the piece of tissue may regenerate the missing part: a stem piece forms roots, a root piece forms shoots, or a leaf forms both shoots and roots. By these means, new plants form that are exactly the same genetically as the plant from which the piece of vegetative tissue was taken; therefore, the new plant has all the same characteristics as the parent. The flower is not involved in vegetative propagation, allowing no opportunity for genetic change (unless, perhaps, a mutation has occurred, which does happen, but rarely). When such vegetative propagation is used, cultivars, even though heterozygous, can be perpetuated generation

---

Table 9–2
METHODS OF PROPAGATING PLANTS, WITH TYPICAL EXAMPLES

I. Sexual—Propagation by seed
II. Asexual (vegetative)
   A. Apomictic embryos—citrus mango
   B. Cuttings—geraniums
   C. Grafting—dwarf fruit trees
   D. Budding
   E. Layering
   F. Runners—strawberry
   G. Suckers—red raspberry, blackberry
   H. Separation
   I. Division
      1. Stem tubers—white potato
      2. Tuberous roots—sweet potato, dahlia
      3. Rhizomes—iris, canna
   J. Micropropagation—orchid, fern

---

[2]Having similar genes of a Mendelian pair present in the same cell as, for example, a dwarf pea plant with genes (tt) for dwarfness only.
[3]Having different genes of a Mendelian pair in the same cell as, for example, a tall pea plant with genes for tallness (T) and genes for dwarfness (t).

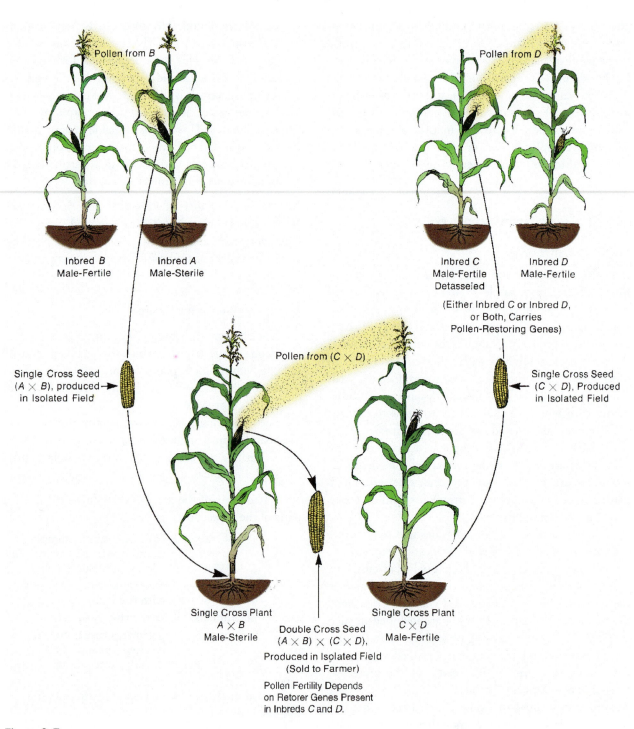

Figure 9–7

Hybrid seed corn production by the utilization of cytoplasmic male sterility in the production of single cross and double cross. In this example, only one inbred, A, is male sterile. The cytoplasmic male sterility is transmitted to the single cross A × B. Pollen-restoring genes carried by the inbreds C or D give pollen fertility to the double cross (A × B) × (C × D) also.
Source: USDA.

after generation, involving hundreds of thousands or more individual plants. Cultivars originating from a single plant or plant part and maintained in this manner by vegetative propagation are called **clones**.

The vegetative procedures just described—where a piece of tissue regenerates the missing part—is termed cutting propagation. There are a number of other, somewhat more complicated vegetative propagation methods, such as grafting, budding, layering, and runners.

SEXUAL (SEED) PROPAGATION

Seed Production

If a cultivar—Great Lakes head lettuce, Calrose rice, or Kombar barley, for example—is to be maintained by seed propagation, careful control of the seed source is essential. If the cultivar is homozygous and self-pollinated, seed purity is generally assured. If the cultivar is homozygous but cross-pollinates with other cultivars or with other species, then the plants used to produce the seeds must be separated by considerable distances from other plants with which they are likely to cross to prevent pollen contamination and loss of genetic purity.

With certain plants propagated by seed, some variability may be tolerated, or it may even be desirable. Examples are reforestation projects, tree and shrub plantings in shelter belts, plantings for wildlife cover, and plant breeding projects.

Most states in the United States and many other countries have established seed certification programs to protect and maintain the genetic quality of cultivars. Government agencies are often at the state level and set the standards for seed quality and characteristics. Field inspections and final seed testing are included in such programs. Thorough cleaning of seed harvesting equipment between seed lots is required. Certified seed usually costs more than uncertified seed, but the grower is assured of better quality. Farmers in North America have become accustomed to using certified seed obtained from recognized seed producers rather than relying on their own or a neighbor's "bin-run" seed of dubious quality. Seed certification programs in the United States recognize four classes of seeds in agronomic crops, such as cotton, alfalfa, soybeans, and cereal grains:

1. *Breeder seed* This is produced only in small amounts and is under the control of the plant breeder. It is planted to produce foundation seed. Breeder seed is labeled with a white tag.
2. *Foundation seed* This is multiplied from breeder seed; it is available only in limited amounts and is planted to produce registered seed. It is controlled by public

or private foundation seed stock organizations. Foundation seed is also labeled with a white tag.
3. *Registered seed* This is the seed source for growers of certified seed and is under the control of the registered seed producers. It is the progeny of either breeder or foundation seed. It is labeled with a purple tag.
4. *Certified seed* This seed is available in large quantities and is sold to farmers for general crop production. It is labeled with a blue tag. Certified seed is of known genetic identity and purity.

Genetic quality of vegetable and flower seeds, however, is largely regulated by the seed companies, which maintain careful control over their seed plantings and continually test seed purity in special test gardens.

Seed Formation

A seed has three essential parts:

1. The embryo, which develops into the new plant.
2. Food storage material, which is available to nourish the embryonic plant. This may be either endosperm tissue or the fleshly cotyledon(s), part of the embryo itself.
3. Seed coverings, which are usually the two seed coats but may include other parts of the ovary wall. Coverings protect the seed and may help control seed dormancy.

The parts of the seed as they develop from the parts of the flower are:

Ovary → fruit (sometimes composed of more than one ovary, plus additional tissues)

Ovule → seed (sometimes coalesces with the fruit)

Integuments → seed coats (two)

Nucellus → perisperm (usually absent or reduced but sometimes develops into storage tissue)

2 polar nuclei + 1 sperm nucleus → endosperm (triploid—3n)

Egg nucleus + 1 sperm nucleus → zygote → embryo (diploid—2n)

After pollination and fertilization are completed, many changes occur in the flower to produce the fruit and the seed. Fruits and seeds appear in innumerable forms, depending on the species, but they all contain the same essential parts listed above.

Seed Storage and Viability Testing

For a seed to germinate, the embryo must be alive. In some plants, such as the willow, maple, and elm, the embryos are very short-lived, remaining viable for only

a few days or months. In others, such as the hard-seeded legumes, the embryo generally remains alive for a great many years. Seeds of other kinds of plants range between these extremes, the length of embryo viability often depending on seed storage conditions. For example, seeds stored in a sealed container under refrigeration at 0°C to 4°C (32°F to 40°F) and at low relative humidity (e.g., 15 percent) generally retain viability considerably longer than seeds sorted at room temperature and high relative humidity. Seeds of certain plants, however, soon become desiccated if stored under dry conditions, and the embryos die. Examples are citrus, maple, oak, hickory, walnut, and most tropical species.

It is often advisable before planting seeds to test the viability of a representative sample of the seed lot to be planted. There are several seed viability tests.

One easy means of determining the possible germinability of a seed lot is a cut test. Seeds of a representative sample are simply cut in half to see whether there is an embryo inside. Often the embryo has aborted or has been eaten by insects and, of course, the seed would not germinate. The mere presence of an embryo, however, does not mean it is alive.

Another simple test is to float the seeds in water. Quite often the "floaters" are empty seeds and can be skimmed off. The sound, full seeds sink and are the ones to be planted.

X-ray photographs of seeds do essentially the same thing as a cut test and are used in some seed laboratories to determine if the seed is empty or the embryo is shrunken. X-ray tests can also be used to determine the optimum time to harvest seeds by observing when the embryos have completely filled the seed.

These tests are not, strictly speaking, viability tests but are useful to rule out seeds that have no possibility of germinating. They still do not give the viability status of seeds with full-sized, apparently sound embryos.

Germination Test   The germination test is useful for seeds that have no dormancy problems, such as flower, vegetable, and grain seeds. They can be tested for germinability by several methods, such as rolling them in several layers of moist paper toweling, or actually planting the seeds in flats of a suitable growing medium in a greenhouse. Seed-testing laboratories use elaborate seed germination boxes with controlled lighting and temperature. After several days or weeks, viability is calculated as the percentage of seedlings developing from the number of seeds planted.

Tetrazolium Test   The chemical 2,3,5-triphenyl tetrazolium chloride, which is colorless when dissolved in water, changes to the red-colored chemical, triphenyl formazan, whenever it contacts living, respiring tissue. In this test, the seeds are usually soaked in water to allow them to become completely hydrated, then cut in half lengthwise to expose the embryo. They are then placed in a tetrazolium solution and held at room temperature for a period of time depending on the species of seed. The parts of the seed that are living (and respiring) become red; the nonliving parts remain white. If the embryo turns red, the seeds are viable; if the embryo remains white, the seeds are nonviable.

Excised Embryo Test   The embryos in the seeds of many woody plant species have dormancy conditions and do not respond in a direct germination test. However, if the embryos are carefully excised from the seed and placed on moist paper in a covered dish, viable embryos will show some activity—possibly a greening and separation of the cotyledons with definite indications of life.

Seed Dormancy

Seeds of many plant species do not germinate when they are extracted from the mature fruit and planted, even though all temperature, light, and moisture conditions favor germination. This is an important survival mechanism for the species and a result of evolutionary development. These species have survived because their seeds have not germinated just before adverse weather conditions that would kill the young, tender seedlings. Thus, in nature, these dormancy factors prevent seed germination of woody perennials in the autumn, allowing the embryonic plant within the seed to overwinter in a very cold-resistant form. Any plant species whose seeds did germinate in the fall in an area with severe winters would likely not survive in that region. Often, the causes for dormancy can persist indefinitely in the seed and require specific treatments to overcome them before germination can take place. This poses problems for the propagator and requires a knowledge of seed dormancy and how to overcome it. Seed dormancy can result from structural or physiological conditions in the seed coverings, particularly the seed coats, or in the embryo itself, or both.

Seed Coat Dormancy   Seed coats or other tissues covering the embryo may be impermeable to water and gases, particularly oxygen, which therefore cannot penetrate to the embryo and initiate the physiological processes of germination. This situation usually occurs in species whose seeds have hard seed coats, such as alfalfa, clover, and other legumes as well as in some pine, birch, and ash species. In nature, continued weathering, the action of microorganisms, or passage

through the digestive tract of animals can soften the seed coverings sufficiently so that they do become permeable and germination can proceed.

Various artificial methods of softening seed coats are widely used to enhance germination. Three principal procedures are:

Scarification The surface of the seed is mechanically scratched or ruptured. This is often done by rubbing the seed between sheets of sandpaper.

Heat treatment In many kinds of seed, exposure to heat, usually boiling water, for a short time sufficiently disrupts the seed coat to permit passage of water and gases.

Acid scarification Soaking seeds with impervious coverings in concentrated sulfuric acid for the proper length of time etches their coats enough for germination.

Embryo Dormancy   Embryo dormancy is very common in seeds of woody perennial plants. It is due to physiological conditions or germination blocks in the embryo itself that prevent it from resuming active growth even though all environmental conditions (temperature, water, oxygen, light) are favorable.

It has been known for hundreds of years that dormant seeds will germinate readily in the spring if they are allowed to winter outdoors in regions with cold climates so that they received some chilling while being kept moist. From this arose the practice known as **stratification**, in which boxes are filled with alternate layers of moist sand and seed and set outdoors in a protected shady place to overwinter. The following spring, the seeds are removed from the box and planted. The critical conditions in seed stratification are:

1. *Chilling temperatures* From about 1°C to 7°C (34°F to 45°F).
2. *Moisture* The seeds should be soaked in water to start and then kept moist.
3. *Adequate oxygen* The seeds should have adequate air and not be kept in an airtight container.
4. *Period of time* The optimum stratification time varies considerably among species. Seeds of the American plum (*Prunus americana*), for example, require at least ninety days chilling while, at the other extreme, apricot seeds (*Prunus armeniaca*) require only twenty to thirty days.

In nursery practice, many propagators sow seeds having embryo dormancy in outdoor nursery beds in the fall, allowing the natural winter chilling to satisfy the embryo's chilling requirement.

There is evidence that during the stratification treatment, growth-promoting hormones (e.g., gibberellins and cytokinins) in the seeds increase while the level of growth-inhibiting hormones (e.g., abscisic acid) decreases, thus permitting germination.

The term *after-ripening* is often used to describe the physiological changes in the mature seed that allow germination to take place.

Double Dormancy   Seeds of some species have both seed coat and embryo dormancy. An example is redbud (*Cercis occidentalis*). To obtain good germination of such seeds, they should be first treated in some manner as described above to soften the seed coats and then given a cold stratification treatment to overcome the embryo dormancy.

Seed Germination

If the seeds have viable embryos, have all germination blocks removed, and are placed under proper environmental conditions of moisture, temperature, and (sometimes) light, the quiescent embryos in the seeds can resume their growth. The nutrients stored in the endosperm or cotyledons of the seed nourish the developing embryo until the new shoot rises above ground, develops leaves, and produces its own food by photosynthesis.

Germination can proceed in several ways, depending on the species (Figs. 9–8 and 9–9). Sometimes the cotyledons are pushed above ground (**epigeal germination**) and sometimes they remain below ground (**hypogeal germination**). The sequence of events during seed germination is as follows:

1. *Imbibition of water by the seeds* The colloidal properties of seed tissues give them great water-absorbing properties. Moist seeds may swell to a size much larger than the dry seeds. The cells become turgid and the seed coverings soften and rupture, permitting easy passage of oxygen and carbon dioxide.
2. *Activation of hormones and enzymes* After water is absorbed, various enzyme systems are activated or synthesized, often as a result of stimulation by hormones. The enzymes convert complex food storage molecules into simpler food materials that can be readily translocated and used for growth. Other enzymes are involved in the respiratory processes, which release energy for cell division and growth. Food materials are translocated to root and shoot growing points.
3. *Embryo growth and development* The root-shoot axis (plumule, epicotyl, hypocotyl, and radicle) grows

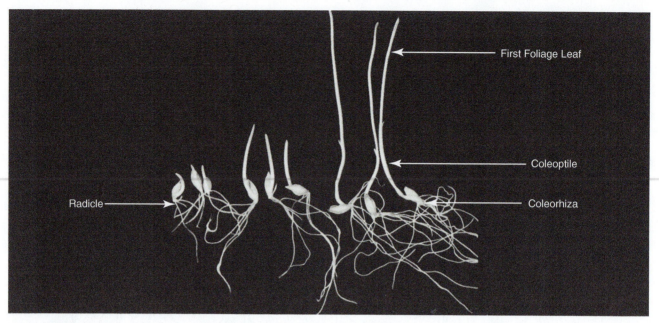

Figure 9–8
Seed germination in a monocotyledonous plant, barley.

by cell division and enlargement. At the same time, food materials translocate to the growing points from the storage tissues, which gradually become depleted. The seed coats rupture and photosynthetic tissue (green leaves and shoots) emerge into the light to carry on photosynthesis. In addition, the embryonic root (radicle) emerges and grows into moist soil to supply the newly developed leafy tissues with water for growth and transpiration. By this time, if no unfavorable environmental influences interfere, the seedling has become established and can exist as an independent plant.

Environmental Factors Influencing Seed Germination
For successful seed germination and seedling growth, certain environmental conditions are required:

1. Adequate moisture
2. Proper temperature
3. Good aeration
4. Light (in some cases)
5. Freedom from pathogenic organisms
6. Freedom from toxic amounts of salts

Moisture   It is essential that water be available in adequate amounts to initiate the physiological and biochemical processes in the seed that result in reactivation of embryo growth.

Temperature   The temperature can strongly influence the percentage and rate of seed germination, varying with the kind of seed. Seeds of the cool-season crops germinate best at relatively low temperatures of 0°C to 10°C (32°F to 50°F). Examples are peas, lettuce, and celery. Seeds of warm-season crops germinate best at temperatures ranging from 15°C to 26°C (59°F to 79°F); examples are soybeans, beans, squash, and cotton.

Aeration   Respiration rates are high in germinating seeds, which thus require adequate oxygen. The usual amount in the air is 20 percent. If this concentration is decreased, germination rate and percentage germination of most kinds of seeds is retarded. In seedbeds that are overwatered and poorly drained, the soil pore spaces may be so filled with water that the amount of oxygen available to the seeds becomes limiting and germination of most kinds of seeds is retarded or prevented.

Light   Light is essential to the germination of some kinds of seeds, such as lettuce, celery, most grasses, and many herbaceous garden flower plants. Such seeds should be planted very shallow for good germination. However, seeds of other plants—onion, amaranth, nigella, and phlox—are inhibited by light and do not germinate unless planted deep enough to avoid light.

Pathogenic Organisms   **Damping off** describes the situation in which the seedlings die during or shortly after germination. Damping off is caused primarily by attacks of certain universally present and very destructive fungi—*Pythium ultimum* and *Rhizoctonia solani*, and to a lesser extent, *Botrytis* spp. and *Phytophthora* spp. The best control methods are not overwatering the seeds, fumigation or heat pasteurization of the

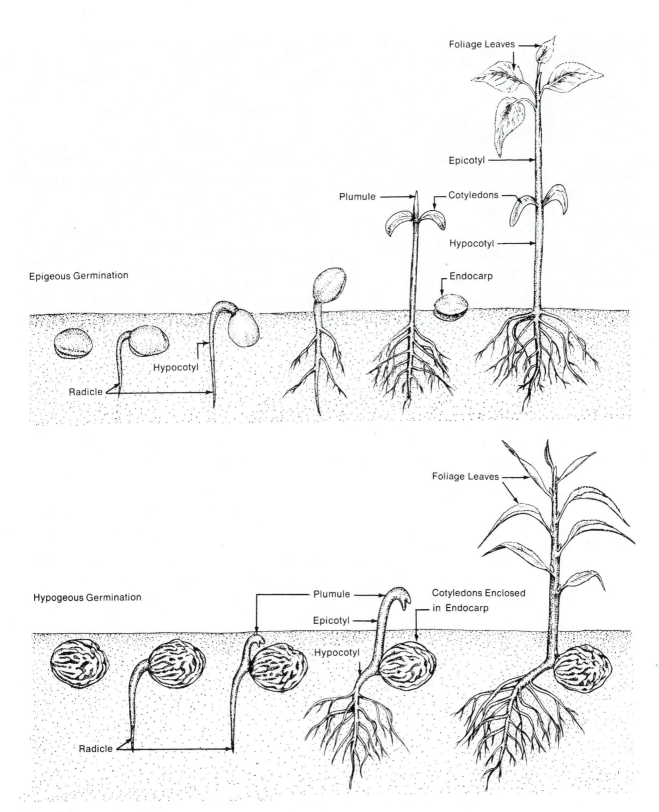

Figure 9–9

Seed germination in dicotyledonous plants. Above: *Epigeous germination as shown in the cherry. The cotyledons are above ground.* Below: *Hypogeous germination as shown in the peach. The cotyledons remain below ground.* Source: Dale E. Kester and Hudson T. Hartmann, *Plant Propagation: Principles & Practices*, 4th ed. © 1983. Adapted by permission of Pearson Education, Inc., Upper Saddle River, NJ.

germination medium, surface treatment of the seeds with fungicides before planting, and good sanitation procedures.

VEGETATIVE PROPAGATION

Vegetative or asexual propagation is accomplished entirely through **mitosis**, the same nonreductive cell division process by which the plant grows. Each daughter cell is an exact replica of its mother cell. Chromosome numbers and composition do not change during cell division. Mitotic cell division, as illustrated in Fig. 9–10, produces the adventitious roots and shoots as well as the callus (parenchyma cells required for healing of a wound or graft union) necessary for successful vegetative propagation (Fig. 9–11). **Adventitious shoots** are those appearing any place on the plant other than from the shoot terminals or in the axils of leaves. **Adventitious roots** appear any place on the plant other than from the radicle (root tip) of the seed or its branches. **Callus** is a mass of undifferentiated and proliferating parenchyma cells.

Vegetative propagation is used primarily for plants that are highly heterozygous; that is, those that do not "breed true" from seed. In these plants, the mother plant's desirable characteristics are lost if seed propagation is used. To maintain the genetic identity of the stock or "mother" plant in such cases, it is necessary to avoid use of the seed altogether in reproducing the plant. Vegetative tissues (stem, root, or leaf) are used to develop new plants by inducing the formation of adventitious shoots, roots, or both. If pieces of stem tissue will not produce adventitious organs (roots, shoots, or both), or if a particular kind of rootstock is required, it is necessary to graft or bud two pieces of tissue together, one to become the top part of the plant (the scion) and one to become the root system (rootstock). It should be noted that a stock plant can be a male plant.

As noted previously, a cultivar that must be reproduced by asexual methods to maintain its characteristics is termed a *clone,* as distinguished from a *line,* which maintains its characteristics without change by seed propagation. Almost all fruit and nut cultivars and many ornamental cultivars are clones. All plants that are members of the same clone have the same genetic makeup and are, in reality, extensions of the stock plant from which the clone originated, although the original plant may have died many years earlier. Some clones, such as the Thompson Seedless grape, have been in existence since ancient times and consist of many millions of individual plants scattered all over the world.

Cultivated clones originate in two ways. The first, and usual, way is as seedling plants that some person recognizes as having some superior qualities and proceeds to propagate vegetatively. For example, the world-famed Golden Delicious apple originated as a seedling tree that A. H. Mullins found growing on his farm in West Virginia. Mullins recognized this apple's value. Later it came to the attention of Stark Brothers Nurseries of Louisiana, Missouri, who in 1912 paid Mullins for the propagation rights to the tree. They named it Golden Delicious and introduced nursery trees for orchard planting in 1916. Although the original tree died long ago, millions of Golden Delicious apple trees are now growing throughout the world, all with the same genetic makeup as the one original tree and all producing the same kind of apples. Although many clones have originated as chance seedlings found growing in the wild, they can also originate from controlled crosses made by plant breeders.

A second way clones originate is from mutations (bud sports). A single cell on a plant may have its genetic makeup altered during cell division so that, as the cell grows and divides, the characteristics of the tissues from that cell differs from those of the rest of the plant. Many mutations are minor or inferior or go unnoticed. But occasionally some strikingly superior characteristic appears; someone notices it and propagates new plants taken from the shoot on which the mututation is seen. This, then, becomes the start of a new clone. The desirable pink-fleshed Ruby grapefruit originated in 1929 as a mutated branch on the Thompson Pink grapefruit, which in turn had originated in 1913 as a mutated branch on a white Marsh grapefruit tree.

The primary advantage of clones is the uniformity of the member plants. All members have the same genetic makeup (**genotype**) and potentially they can all be exactly alike. However, environmental factors can so modify the appearance and behavior of the plant (**phenotype**) that individual plants can differ strikingly. For example, a vineyard of Concord grapevines, pruned, irrigated, sprayed, and fertilized properly, would appear totally different from an adjacent abandoned vineyard of the same cultivar, yet the vines in both would be genetically identical.

The mutation process can also cause undesirable genetic changes in clones. This happens frequently in many species and must be avoided to prevent deterioration of the clone. Sometimes the mutations are unstable and revert back to the original type. This is often seen as a green branch on a variegated plant. The green branch is a reversion to the original form.

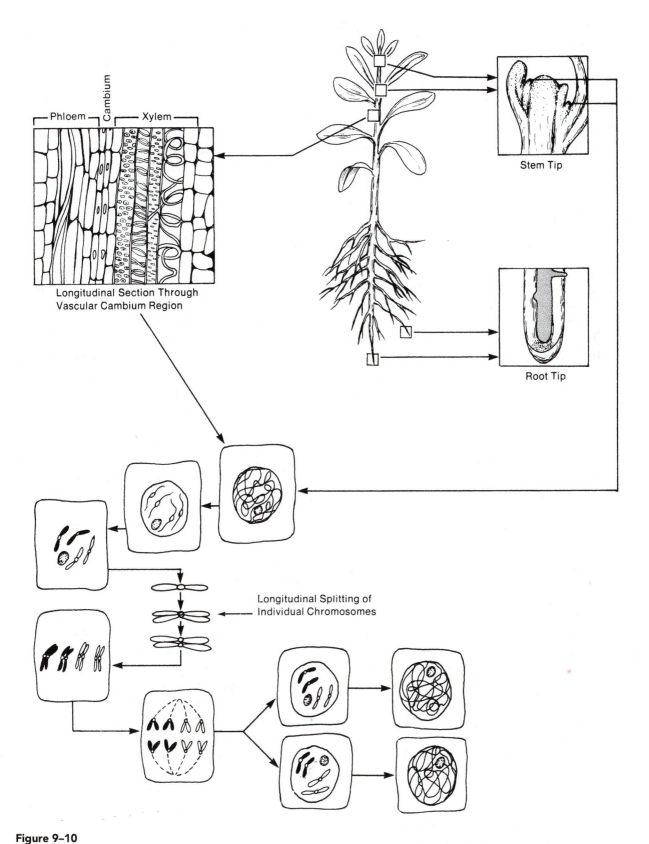

Figure 9–10

The process of growth and asexual reproduction in a dicotyledonous plant. Mitosis occurs in three principal growing regions of the plant: the stem tip, the root tip of primary and secondary roots, and the cambium. A meristematic cell is shown dividing into two daughter cells whose chromosomes are (usually) identical with those of the original cell. Source: Dale E. Kester and Hudson T. Hartmann, *Plant Propagation: Principles & Practices*, 4th ed. © 1983. Adapted by permission of Pearson Education, Inc., Upper Saddle River, NJ.

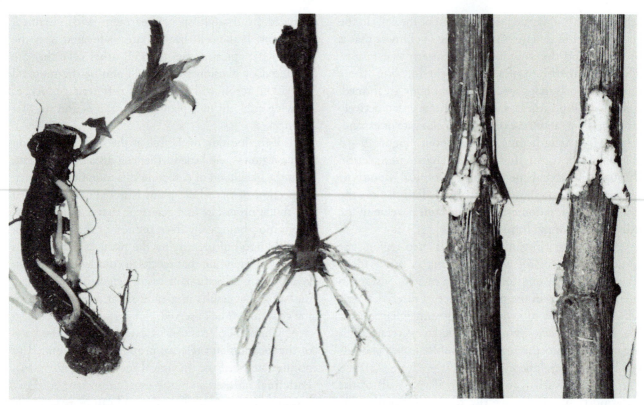

Figure 9–11
Regeneration in asexual propagation. Left: Adventitious shoots growing from a root cutting. Center: Adventitious roots developing from the base of a stem cutting. Right: Callus tissue produced to give healing of a graft union. Source: Dale E. Kester and Hudson T. Hartmann, *Plant Propagation: Principles & Practices,* 4th ed. © 1983. Adapted by permission of Pearson Education, Inc., Upper Saddle River, NJ.

Some mutations, called **chimeras**, genetically change only a portion rather than the entire shoot. Some of the variegated leaf patterns found in the foliage of certain plants such as *Pelargonium* or citrus and the bracts of poinsettias are due to chimeras (Fig. 9–12). The word *chimera* comes from Greek mythology. The chimera was a fire-breathing beast with the head of a lion, the body of a goat, and the tail of a dragon. Chimeras can often be the source of desirable traits such as thornlessness, reduced fuzziness in peaches, and variegated colors in leaves and flowers. However, chimeras can also be very difficult to propagate true to form. Because many of our fruit and ornamental crops are chimeras, a more detailed explanation of them is presented here.[4]

No discussion of the origin of chimeras would be complete without a review of the organization of the shoot apex. The pattern of cell division, frequency of

[4]The following discussion is adapted with permission from Dr. Dan Lineberger, Texas A&M University.

Figure 9–12
A variegated pink Eureka lemon, a type of chimera.

cell division, and layered organization of the cells in the apex interact in determining the type of chimera that is produced and the stability of the pattern that results. The apical meristem of a shoot is the location where most of the cells that produce the plant body are formed. Cell division occurs at a very rapid rate in an actively growing shoot, and these cells in turn elongate or expand, resulting in lengthening of the shoot. Figure 9–13 is a sketch of a longitudinal (or lengthwise) section through the apical meristem of a typical woody or herbaceous dicot shoot. Leaf primordia arise on the sides of the apical dome, and lateral buds develop in the axils of these young leaves.

The apex is organized into a layered region (the tunica) and a region where layering is not evident (the corpus). The controlled pattern of cell divisions in the tunica results in the maintenance of discrete layers, with the number of layers varying somewhat among the different species. Note that the layers retain their organization into the region where the leaves and lateral buds are developed.

The derivatives (or progeny, if you will) of the outermost layer (L.I) give rise to the epidermis. The epidermal layer is continuous as an outer covering over all tissues of the leaf, stem, flower petals, etc. Derivatives of layer II (L.II) give rise to several layers within the stem and a large proportion of the cells in the leaf blade. Derivatives of layer III (L.III) give rise to most of the internal tissue of the stem and a number of cells around the veins within the leaf. The significance of the cell layers and their resulting progeny will be discussed in more detail.

Chimeras arise when a cell undergoes mutation. This mutation may be spontaneous or it may be induced by irradiation or treatment with chemical mutagens. If the cell that mutates is located near the crest of the apical dome, then all other cells that are produced by division from it will also be the mutated type. The result will be cells of different genotypes growing adjacent in a plant tissue, the definition of a chimera.

If the location of the cell at the time of mutation is in a region where little further cell division will occur, then the likelihood of detecting this mutation by visual inspection of the whole plant is low. Furthermore, if the mutation results in a genotype that is not very different morphologically from the rest of the plant, then the likelihood of identifying the plant as a chimera is also low. A mutation that results in colorless rather than green cells (variegation) is easily detectable, whereas a mutation that results in greater sugar accumulation in the cells cannot be observed.

Chimeral plants can be categorized on the basis of the location and relative proportion of mutated to nonmutated cells in the apical meristem (Fig. 9–14). **Periclinal chimeras** are the most important category because they are relatively stable and can be vegetatively propagated. A mutation produces a periclinal chimera if the affected cell is positioned near the apical dome so that the cells produced by subsequent divisions form an entire layer of the mutated type. The resulting meristem contains one layer that is genetically different from the remainder of the meristem. If, for example, the mutation occurs in L.I, then the epidermal layer of the shoot that is produced after the mutation is the new genetic type.

A classical example of an L.I periclinal chimera is the thornless blackberry. The epidermal layer of this type produces no thorns (the modified epidermal cells are correctly called prickles). The thornless epidermis covers a stem whose cells contain the information for the thorny genotype. This can be demonstrated by taking root cuttings. The adventitious shoots that differentiate on the root cuttings are not chimeral and therefore revert to the thorny genotype. Other common periclinal chimeras we see in cultivated crops involve loss of epidermal appendages (thornlessness in blackberries, "fuzzless" peaches), alteration in bract color in poinsettia, and various petal or flower color patterns in carnation and chrysanthemum.

Mericlinal chimeras and **sectorial chimeras** result in the development of an epidermis that is not entirely chimeric. Further cell division and development can give rise to shoots and leaves that are not chimeras. Reliable propagation of these types of chimeras is therefore nearly impossible. For that reason, they will not be discussed here.

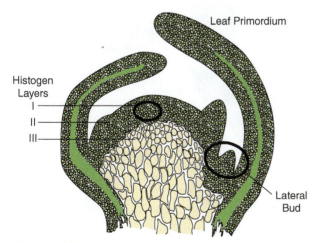

Figure 9–13

Typical angiosperm meristem showing the histogen layers: I, II, and III. Note the lateral bud in the axil of the immature leaf.

Source: Courtesy of Dr. Dan Lineberger, Texas A&M University.

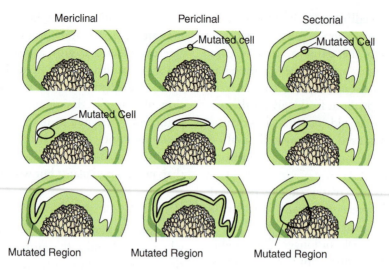

Mericlinal   Periclinal   Sectorial

Mutated Cell   Mutated cell   Mutated Cell

Mutated Cell

Mutated Region   Mutated Region   Mutated Region

Figure 9–14

Development of mericlinal, periclinal, and sectorial chimeras. The location of the mutated cell determines the type of chimera that will develop. Source: Courtesy of Dr. Dan Lineberger, Texas A&M University.

Although periclinal chimeras cannot be reproduced by sexual propagation, the most commonly used techniques for asexual plant propagation (described later in this chapter) result in the formation of true-to-type periclinal chimeras. However, important departures from the rule are found in cases where the propagules differentiate from adventitious shoots. A case discussed previously that illustrates this point is the thornless blackberry. Adventitious roots that form on a blackberry stem cutting or tip layer originate in the subepidermal tissues of the stem (L.II and L.III). If root cuttings are then taken from these plants, the adventitious shoots do not contain the layer with the thornless genotype (L.I) and the propagules are of the thorny type. Likewise, the pinwheel-flowering African violets similarly cannot be propagated in a true-to-type fashion from leaf cuttings. Shoots that originate on the leaf cuttings are either a single color or irregularly mottled bicolors without the characteristic flowering pattern. Propagation of the pinwheel-flowering African violets is currently achieved by separating suckers, which originate from lateral buds. Because micropropagation generates plants from a small number or even single cells, segregation of plantlets with and without the chimeric trait is likely to occur, making micropropagation unsuitable for use with chimeras.

Apomixis

Apomixis is an interesting phenomenon in which the genetic identity of the mother plant is transmitted to daughter plants that develop by seed formation and germination. Apomixis is a form of asexual propagation because there is no union of male and female gametes before seedling production. There are several types, but a common one is found in citrus seeds where, in addition to the sexual embryo formed through the usual pollination and fertilization processes, embryos also arise in the nucellar tissue (nucellar budding). The nucellar tissue enclosing the embryo sac has not undergone reduction division and has the same genetic makeup as the female parent. So the nucellar embryos, although developing in a seed, are exactly the same genetically as the mother plant and thus maintain the clone.

Such seeds can contain several nucellar embryos in addition to the sexual embryo. Thus, several seedlings are obtained from one seed, a situation known as polyembryony.

Even though plants arising by apomixis from nucellar embryos maintain the clone, they go through the juvenile to mature transition stages just as any plant seedling would.

Disease Problems in Clones

Propagating plants vegetatively with clonal material has one great disadvantage: clones can become infected with systemic viruses and mycoplasmalike organisms that are passed along to the daughter plants during asexual propagation procedures. In time, all clonal members may become infected with viruses. Some viruses are latent in particular nonsusceptible clonal material, but if this material is used in a graft combination where the virus can move through the graft union to the graft partner—which is susceptible—then the entire grafted plant will be killed by the virus. On the

other hand, virus-free seedlings can be obtained in many species by seed propagation because the virus is not transmitted through the embryo. In one case though, an infection by a mycoplasmalike organism has great commercial value. Most of the poinsettia cultivars grown today branch much more freely, giving them a much nicer appearance than old, nonbranching cultivars. The branching is caused by a mycoplasmalike organism that has been deliberately transferred from one cultivar to another.

Some disease-causing viruses can be removed from clonal material by heat treatment. The virus-infected plant material, is held at 37°C to 38°C (98°F to 100°F) for two to four weeks or longer. This combination of time and temperature eliminates the virus. After treatment, cuttings can be taken for rooting, or buds may be taken for budding into virus-clean seedling root stocks.

Another procedure to eliminate viruses from clones is shoot-tip culture. In virus-infected plants, the terminal growing point is often free of the virus. By excising this shoot apex aseptically and growing it on a sterile medium, roots will often develop, producing a new plant free of the virus. Here again, a starting point becomes available for continued propagation of the clone but without the virus. Often, both procedures are required—heat treatment of the plant, followed by excision and culture of the shoot tip—to free the plant of known viruses.

In recent years certification programs have been established by government agencies in many states in the United States and in other countries to provide nurseries with sources of true-to-name, pathogen-free propagation material. Elaborate programs, for example, have been established for citrus in Florida and California and for deciduous tree fruits, grapes, strawberries, potatoes, and certain ornamentals in many states.

In such programs, mature flowering or fruiting plants known to be true-to-name and true-to-type are selected as mother plants. These are "culture-indexed" by certain grafting procedures or other tests such as enzyme-linked immunoassays (ELISAs) to be sure that no known viruses or other diseases are present. If no pathogen-free source plants can be located, then procedures like those described above for eliminating viruses must be used to obtain pathogen-free plants. Once a "clean" source is obtained, it must then be maintained under isolated and sanitary conditions, with periodic inspection and testing to ensure that it does not again become infected. Sometimes it is necessary to grow the plants in insect-proof screenhouses or greenhouses or in isolated areas far from commercial production fields.

Distribution systems from these foundation plantings are necessary, sometimes requiring plots of land, "increase blocks," to grow a greater amount of propagating material that may be needed. This material can be termed "certified stock" if it is grown under the supervision of a legally designated agency with prescribed regulations designed to maintain certain standards of cleanliness and clonal identity. As with certified seed, culture-indexed or certified cuttings are more expensive than nonindexed plants but worth the extra cost.

Propagation by Cuttings

A cutting is essentially a piece of vegetative tissue that, when placed under the proper environmental conditions, regenerates the missing parts—roots, shoots, or both—and develops into a self-sustaining plant.

Cuttings can be classified according to the part of the plant from which they are obtained:

Stem cuttings

Leaf cuttings

Leaf-bud cuttings

Root cuttings

Stem cuttings already have terminal or axillary buds (potentially a new shoot system), but new roots must develop at the base of the cutting before a new plant will be formed. Stem cuttings can be prepared to include the shoot tip, or cuttings can be made from the more basal parts of the shoot that have only axillary buds. Leaf cuttings have neither buds nor roots, so both must form. Leaf-bud cuttings have a bud at the base of the petiole—for the new shoot system—so only new roots must form. Root cuttings must produce a new adventitious shoot and continue growth of the existing root piece, or develop roots from the base of the new shoot. Figure 9–15 illustrates types of cuttings.

Plant species and cultivars vary markedly in their ability to develop adventitious roots. Cuttings from some kinds of plants root easily even when the simplest procedures are used. Cuttings of others root only if the influencing rooting factors are carefully observed. Cuttings of still other kinds of plants have never been rooted, or rooted only rarely and in meager numbers, despite great efforts and much research. The basic reasons for such differences in rooting ability among different kinds of plants are not well understood.

Stem Cuttings

Cuttings are harvested by removing the tops of branches from the plant. How long the cutting is depends on the species but in general they are from 1 to 4 in. long.

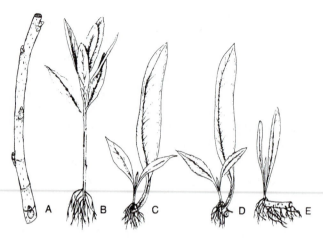

Figure 9–15

Types of cuttings: (A) hardwood stem cutting, (B) leafy stem cutting, (C) leaf cutting, (D) leaf-bud cutting, and (E) root cutting.

Figure 9–16

Mist propagation bed operating in a greenhouse. The mist is on only a few seconds—just long enough to wet the leaves— then turns off. When the leaves start to dry, the mist is turned on again. The mist is generally applied only once or twice during the night or left off altogether.

Source: Margaret McMahon, The Ohio State University.

For many species, cutting through a node helps to promote more rapid root formation. A few leaves should be on the cutting stem. If there are many leaves, it is usually best to remove the lower leaves, especially those that might be under or at the ground when stuck in the propagation media. For long-stemmed or vining herbaceous plants, the stem can be cut into several cuttings, each containing an axillary bud. The axillary bud will form the new shoot for the lower sections. If the mother plant is to be discarded after harvest, what is left behind on the plant is not important. However, if the mother plant is needed to produce more cuttings, then be sure to leave enough branches and leaves on the plant to produce future cuttings.

Some woody species can be propagated outdoors in nursery beds. However, most cuttings are propagated in greenhouses where they are inserted into rooting containers called flats or directly into benches. A porous rooting medium, such as equal parts of perlite and peatmoss or of perlite and vermiculite, is used. The cuttings are placed in a humid location, with reduced light to minimize water loss from the leaves. To maintain humidity, a fine mist is sprayed over the cuttings intermittently during the day (Fig. 9–16). Light is reduced by using a shade material that blocks 35 to 75 percent of the normal greenhouse light. The amount of shade varies by species. Heating the root zone area to 75° to 80°F decreases the time of root formation and makes rooting more uniform. Misting frequency should decrease and light levels increase as callus and roots form to reduce the risk of disease and increase hardiness of the cutting.

Rooting time depends on species and cultivar, taking several days to several weeks.

Leaf Cuttings   There are various types of leaf cuttings, as shown in Figure 9–17. A common one consists of a single leaf blade and petiole, as might be taken from an African violet. As with other types of leafy cuttings, the humidity must be kept very high. Roots and shoots generally develop from the same point at the base of the petiole and grow into a plant independent of the leaf blade, which functions to nourish the new plant. African violet, peperomia, begonia, and sansevieria are examples of plants commonly started by leaf cuttings.

Leaf-Bud Cuttings   Leaf cuttings of some species form roots at the base of the petiole but do not develop a shoot, resulting only in a rooted leaf that may stay alive for months (or years). To avoid this, a leaf-bud cutting can be prepared. This cutting consists of a short piece of stem with an attached leaf and a bud in the axil of the leaf, as shown in Figure 9–18. Such cuttings are rooted under the same conditions as described above. The axillary bud develops into the new shoot system. Leaf-bud cuttings are useful as substitutes for stem cuttings in obtaining as many plants as possible from scarce propagating material. Leaf-bud cuttings give one and perhaps two (if the plant has opposite leaves) new plants from each node, whereas each stem cutting generally requires a minimum of two nodes.

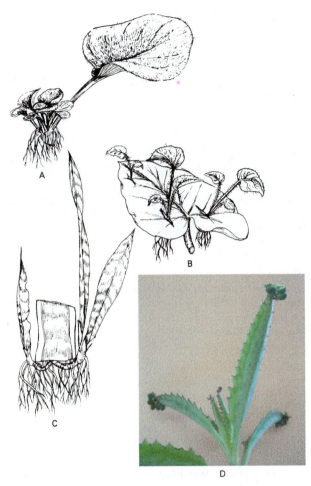

Figure 9–17

Types of leaf cuttings: (A) new plants arising from base of petiole in African violet (Saintpaulia); (B) new plants arising from cuts in veins of a begonia leaf; (C) new plant arising from base of leaf blade in sansevieria; and (D) new plants growing from notches of leaf in Kalanchoe (Bryophyllum).

Source: (D) Margaret McMahon, The Ohio State University.

Figure 9–18

Leaf-bud cuttings of peperomia. Each cutting consists of a leaf blade, petiole, axillary bud, and a piece of stem. Arrows show axillary buds starting to grow.

Root Cuttings   Many plant species can be propagated by cutting the small, young roots into pieces and planting them horizontally in soil about 1.3 cm (0.5 in.) deep or vertically with the upper end (nearest the crown of the plant) just below the soil level. The soil is kept moist, but frequent misting is usually not necessary. One or more new adventitious shoots form along the root piece, and either this shoot forms roots or the root piece itself develops new branch roots, thus producing a new plant.

Origin of Adventitious Roots in Stem Cuttings   In stem cuttings of herbaceous plants, adventitious roots generally originate laterally and adjacent to the vascular bundles, whereas in cuttings of woody perennials, the roots originate in the region of the vascular cambium, often in young phloem parenchyma (Fig. 9–19). In each case, the new roots are in a position to establish a vascular connection with the conducting tissues of the xylem and phloem in the cutting.

Factors Influencing the Rooting of Cuttings   There is a large group of plants whose cuttings only root with considerable difficulty. It is necessary to consider carefully the factors described below to satisfactorily root cuttings in this group.

Source of Cutting Material   Generally, the cuttings most likely to root come from stock plants that are growing in full sun at only a moderate rate and that have thus accumulated carbohydrates in their tissues. If the cutting material for woody plant species can be taken from young (juvenile), nonflowering plants only a few years away from a germinated seed, rooting will be much better than when the cuttings are taken from old, mature flowering and fruiting plants. The juvenility influence in the young material can sometimes be retained by keeping the stock plants cut back heavily each year to force new shoot growth out from the lower part of the plants near ground level.

Time of Year the Cutting Material Is Taken   In woody perennial plants, cutting material can be taken at any time of the year. In some species, the time the material is taken can dramatically influence rooting. Hardwood cuttings often root best if the material is gathered in late winter, while softwood cuttings usually root best if taken in the spring shortly after new shoot growth has attained a length of 10 to 15 cm (4 to 6 in.). Semihardwood cuttings are best taken in midsummer after the spring flush of growth has matured somewhat. Herbaceous cuttings can be easily rooted any time of the year, especially when succulent.

Figure 9–19

Cross section of euonymus shoot (cutting) after an adventitious root has formed: (1) primary bark, (2) phloem, (3) newly formed adventitious root, (4) cambium, (5) xylem formed after rooting, (6) primary xylem. Source: Prof. Bojinov Bogdanov and *Proceedings International Plant Propagators' Society* 35 (1985): 449–53.

Etiolation   It has long been known that stem tissue developing in complete darkness is more likely to initiate adventitious roots than tissue exposed to light. Thus, if the basal parts of shoots that are later to be made into cuttings can be kept in darkness, they are likely to form roots. Such techniques are used successfully in rooting cuttings of difficult species, such as the avocado. (**Etiolation** is the growth of shoots in the absence of light or in low light, causing them to be abnormally elongated and colored yellow or white due to the absence of chlorophyll.)

Treatment of Cuttings with Auxins   In the mid-1930s, it was discovered that one of the natural plant hormones, auxin (indoleacetic acid [IAA]), stimulated the initiation

of adventitious roots on stem cuttings. It was soon discovered that other closely related synthetic auxins—indolebutyric acid (IBA) and naphthaleneacetic acid (NAA)—were even more effective. This knowledge was quickly picked up by plant propagators, who now routinely treat the base of cuttings with one of these materials, particularly IBA, just before the cuttings are stuck in the rooting medium. The hormone can be applied as a light dusting of powder or a liquid on the base of the cuttings. If applied as a liquid, caution must be taken to not allow the spread of pathogens. The cut end of the stem is an open wound through which pathogens can enter and leave. The optimum concentration for promoting rooting varies with the species but ranges from about 1,000 to 10,000 parts per million (ppm).

Misting   Misting is an essential factor in the propagation of most cuttings, especially herbaceous cuttings. The frequency of misting is most critical during the early stages of rooting, before callus formation. As rooting takes place, frequency tapers off. The primary purpose of misting is to prevent cuttings from dehydrating by keeping relative humidity near 100 percent around the cutting. Misting also helps to keep the cutting cool, which reduces transpirational loss of water. Frequency is determined not only by root development, but by environmental factors such as light intensity, relative humidity, and air temperature. Low light, high relative humidity, and cool temperatures reduce the need for mist. Timers can be used to control the frequency of misting but are not the best mechanism. Timers do not sense changes in environment and do not make adjustments as conditions change. Using light or relative humidity sensors to set misting frequency allows misting frequency to change as environmental conditions that affect the need for mist change.

Bottom Heat in the Cutting Beds   To force rooting at the base of cuttings before shoot growth starts, it is advisable to maintain temperatures at the base of the cuttings at about 24°C (75°F)—or about 6°C (10°F) higher than that at the tops of the cuttings, 18.5°C (65°F). This is best done by providing bottom heat from thermostatically controlled electric heating cables or hot water pipes under the rooting frames. Bottom heat under the cuttings often greatly stimulates rooting.

All of the environmental factors that influence rooting can be efficiently controlled with the appropriate sensors and computer system. These systems are now used in almost every propagation facility.

Propagation by Grafting and Budding

Grafting and budding are vegetative methods used to propagate plants of a clone whose cuttings are difficult to root. These methods are also used to make use of a particular rootstock rather than having the plant on its own roots. Certain rootstocks are often utilized to obtain a dwarfed or invigorated plant or to give resistance to soil-borne pests.

Grafting   Grafting can be defined as the art of joining parts of plants together so that they will unite and continue their growth as one plant. The **scion** is that part of the graft combination that is to become the upper or top portion of the plant. Usually the scion is a piece of stem tissue several inches long with two to four buds. If this piece is reduced in size so there is just one bud, with a thin layer of bark and wood under it, then the operation is termed **budding**. The **rootstock**

(or **understock** or **stock**) is the lower part of the graft combination, the part that is to become the root system. Root grafting is a common method of propagation in which a scion is grafted directly onto a short piece of root and the combination is then planted. Sometimes grafting is used to change the fruiting cultivar in a fruit tree or grapevine to a different one (**top-grafting** or **top-working**). Grafting may be used to repair the damaged trunk of a tree (bridge-grafting) or to replace an injured root system (inarching). Grafting, or budding, is sometimes used to study the transmission of viral diseases (indexing). The indexing test involves inserting a bud from a plant suspected of carrying a virus into another indicator plant. A definite visual response results from the presence of a virus, such as gumming around the inserted bud or expression of disease in the scion.

Several standard methods of grafting and budding have been widely used over the world for a great many years.

Whip Grafting   The whip graft (Fig. 9–20) is useful in grafting together material about 0.6 to 1.3 cm (0.25 to 0.5 in.) in diameter. A small stem piece of the scion cultivar is grafted onto a root piece. The completed grafts are buried for two or three weeks in a moist material, such as wood shavings, at about 21°C (70°F) to encourage callusing or healing of the union. After the union has healed, the graft can be planted in the nursery.

Various grafting machines can be used as a substitute for the whip graft and are considerably faster than hand grafting. Such machines have been used mostly in grape grafting but are also satisfactory for making fruit tree root grafts.

Cleft Grafting   The cleft graft (Fig. 9–21) is mostly used in top-grafting, where scions 0.6 to 1.3 cm (0.25 to 0.5 in.) in diameter are inserted into stubs of older limbs that are 8 to 10 cm (3 to 4 in.) in diameter after they have been cut off a foot or so out from the trunk of the tree. It is very important in cleft grafting to match the cambium layer of the scion as closely as possible with the cambium layer of the stock so that the two pieces heal together. Cleft grafting is usually done in late winter or early spring.

In all types of grafting and budding, the two parts must be held together very tightly and securely by wedging, tying, nailing, or wrapping with string, rubber bands, or with plastic or cloth tape. The graft union must also be completely covered by grafting wax to prevent the cut surfaces from drying out. Polarity must be observed: the scions must be inserted so that the buds point upward.

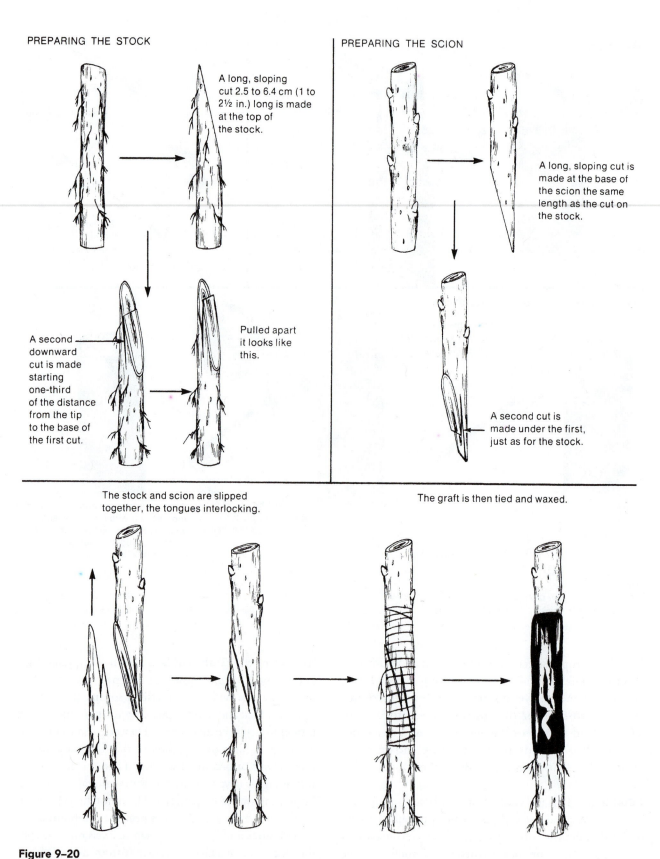

PREPARING THE STOCK

A long, sloping cut 2.5 to 6.4 cm (1 to 2½ in.) long is made at the top of the stock.

A second downward cut is made starting one-third of the distance from the tip to the base of the first cut.

Pulled apart it looks like this.

PREPARING THE SCION

A long, sloping cut is made at the base of the scion the same length as the cut on the stock.

A second cut is made under the first, just as for the stock.

The stock and scion are slipped together, the tongues interlocking.

The graft is then tied and waxed.

Figure 9–20

The whip, or tongue, graft. This method is widely used in grafting small plant material and is especially valuable in making root grafts, as illustrated here. Source: Dale E. Kester and Hudson T. Hartmann, *Plant Propagation: Principles & Practices*, 4th ed. © 1983. Adapted by permission of Pearson Education, Inc., Upper Saddle River, NJ.

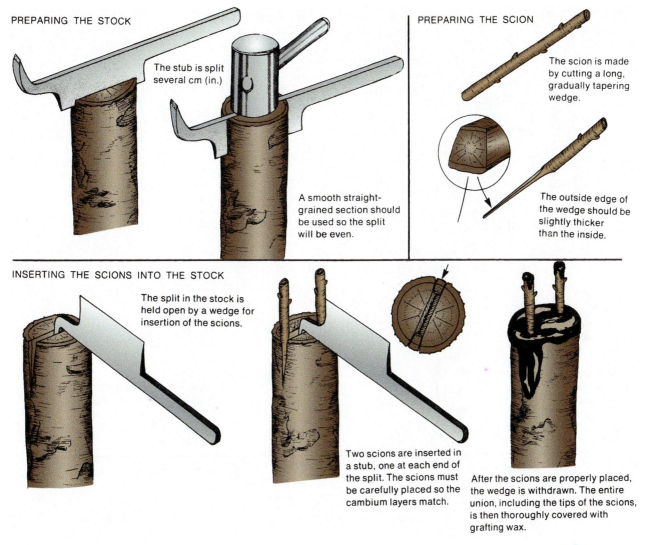

PREPARING THE STOCK

The stub is split several cm (in.)

A smooth straight-grained section should be used so the split will be even.

PREPARING THE SCION

The scion is made by cutting a long, gradually tapering wedge.

The outside edge of the wedge should be slightly thicker than the inside.

INSERTING THE SCIONS INTO THE STOCK

The split in the stock is held open by a wedge for insertion of the scions.

Two scions are inserted in a stub, one at each end of the split. The scions must be carefully placed so the cambium layers match.

After the scions are properly placed, the wedge is withdrawn. The entire union, including the tips of the scions, is then thoroughly covered with grafting wax.

Figure 9–21

Steps in making the cleft graft. This method is very widely used and is quite successful if the scions are inserted so that the cambium layers of stock and scion match properly. Source: Dale E. Kester and Hudson T. Hartmann, *Plant Propagation: Principles & Practices,* 4th ed. © 1983. Adapted by permission of Pearson Education, Inc., Upper Saddle River, NJ.

The proper selection of scionwood and budwood is very important in all types of grafting and budding. Scionwood and budwood should be taken from source trees true-to-type for the cultivar to be propagated. They should be free of known viruses and any other diseases. With woody plants, grafting and budding must be done with dormant buds on the scion.

T-Budding   This technique is widely used in propagating fruit trees and roses. Buds taken from budsticks are inserted under the bark through "T" cuts in the bark of small seedling rootstock plants a few inches above ground level. Buds are then tied into place with budding rubbers, but the tops of the seedling rootstocks are not cut off above the inserted bud until just before growth starts the next spring.

Healing of the Graft and Bud Union   In preparing a graft or bud combination, the two parts are joined by one of the methods just described so that the cambial layers of stock and scion exposed by the grafting cuts are brought into intimate contact. They are held in place by wedging, nailing, or wrapping so that the parts cannot move about or become dislodged. Then the graft union is thoroughly covered with plastic or cloth tape or, better, by grafting wax to keep out air. The union heals by callus production from young tissues near the cambium layers of both stock and scion. Temperature levels must be conducive to cellular activity—generally from about 10°C to 30°C (50°F to 86°F)—and no dry air must contact the cut surfaces because it would desiccate the tissues.

The steps in healing of a graft union are illustrated in Figure 9–22. Healing usually takes about two

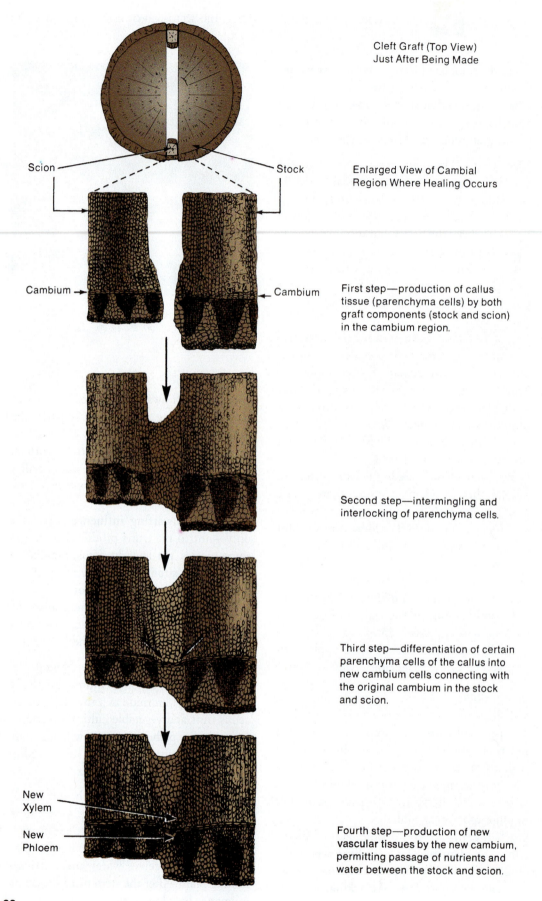

Cleft Graft (Top View)
Just After Being Made

Scion ——— Stock

Enlarged View of Cambial
Region Where Healing Occurs

Cambium ——→ ←—— Cambium

First step—production of callus
tissue (parenchyma cells) by both
graft components (stock and scion)
in the cambium region.

Second step—intermingling and
interlocking of parenchyma cells.

Third step—differentiation of certain
parenchyma cells of the callus into
new cambium cells connecting with
the original cambium in the stock
and scion.

New
Xylem

New
Phloem

Fourth step—production of new
vascular tissues by the new cambium,
permitting passage of nutrients and
water between the stock and scion.

Figure 9–22

Developmental sequence during the healing of a graft union as illustrated by the cleft graft. Source: Kester Dale E. Kester and Hudson T.
Hartmann, *Plant Propagation: Principles & Practices*, 4th ed. © 1983. Adapted by permission of Pearson Education, Inc., Upper Saddle River, NJ.

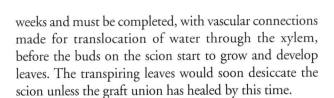
weeks and must be completed, with vascular connections made for translocation of water through the xylem, before the buds on the scion start to grow and develop leaves. The transpiring leaves would soon desiccate the scion unless the graft union has healed by this time.

Limits of Grafting   There are certain limits to the combinations that can be successfully established by grafting or budding. The partners (stock and scion) in the combination must have some degree of botanical relationship—the closer the better. For instance, woody perennial plants in different botanical families have never, as far as is known, been successfully grafted together. It is useless to attempt to graft a scion taken from a grapevine (VITACEAE) onto an apple tree (ROSACEAE).

In only a few cases have completely successful graft combinations been made between plants in the same family but different genera. For example, the thorny, deciduous large shrub trifoliate orange (*Poncirus trifoliata*) is widely used commercially as a rootstock for the common sweet orange (*Citrus sinesis*), a large evergreen tree. Both are in the family RUTACEAE, but different genera—*Poncirus* and *Citrus.*

If the two graft partners are in the same genus but different species, then the chances of success are greatly improved. Still, many such graft combinations will not unite. Plant propagation books should be consulted for information that has been accumulated by trial and error over the years.

If the two partners are different cultivars (clones) within a species, the chances are almost 100 percent that the graft combination will succeed. For instance, if you have a Jonathan apple tree (*Malus pumila*), you could successfully top-graft on it any other apple cultivar, for example, the Golden Delicious (*Malus pumila*).

Graft Incompatibility   There are many puzzling, unexplained situations among various graft combinations. Some pear cultivars are commercially grafted onto quince roots (an intergeneric combination), but scions of other pear cultivars grafted on quince roots soon die. The reverse, quince on pear roots, always fails. Plums can be successfully grafted on peach roots but peaches on plum roots are a failure.

Even when grafts are compatible, grafted plants can sometimes be identified by a swelling where the graft was made (Fig. 9–23). This can become a weak point and susceptible to breaking or cracking.

Effect of Rootstock on Growth and Development of the Scion Cultivar   Tree fruit growers often select a certain rootstock for a particular fruiting cultivar

Figure 9–23
Swelling of graft union. Source: Margaret McMahon, The Ohio State University

because it will dwarf the tree to some extent and thus make harvesting easier. This is particularly true in apples, where an entire series of clonal rootstocks is available to produce apple trees with any desired degree of dwarfness. Quince roots will dwarf pear trees. Trifoliate orange roots will dwarf sweet orange trees.

These dwarfing influences extend to the tree only—not to the fruits produced by the trees. But in some species, particularly citrus, the kind of rootstock used can also strongly influence the quality of fruit produced by the scion cultivar. For instance, when sweet orange seedlings are used as the rootstock for orange trees, the fruits will be of much higher quality than when rough lemon is used as the rootstock.

Herbaceous Grafting   Recently grafting of herbaceous plants has become common as the result of the ban of methyl bromide as a soil sterilant. Some herbaceous fruit and vegetable cultivars that are susceptible to soil-borne pathogens are now being grafted to rootstock from resistant plants. The procedures are very similar to grafting woody plants.

Layering   A simple and highly successful method of propagating plants is layering. Layering is similar to propagation by cuttings except that, instead of severing the part to be rooted from the mother plant, it is left attached and receives water and nutrients from the mother plant. After the stem piece (layer) has rooted— no matter how long it takes—it is cut from the mother plant and transplanted to grow independently. Various layering procedures can be used for many different kinds of plants (Fig. 9–24).

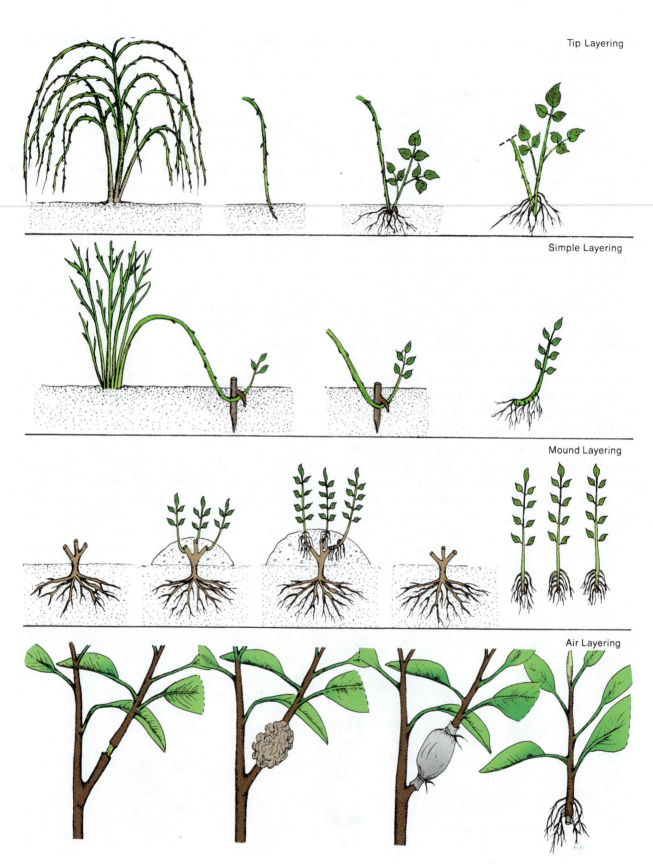

Tip Layering

Simple Layering

Mound Layering

Air Layering

Figure 9–24
Steps in preparing four kinds of layers. See text for details.

Simple Layering   This method can be used with plants that produce long shoots arising from the plant at ground level. In early spring the ends of these shoots can be bent over, placed in a hole in the soil several inches deep, and recurved so that the shoot tip is exposed above ground. The curved section to be buried should be nicked or twisted slightly to retard translocation of food materials through the stem; this promotes rooting. The hole is then filled in with loose, moist soil tamped firmly in place. After layer has rooted, it can be cut from the parent plant, dug, and transplanted.

Air Layering   In this method, the rooting medium is brought up to the stem to be rooted, rather than bringing the stem down to the soil. The leaves of the branch to be air-layered are removed from the area to be layered. The stem is girdled by cutting the bark away down to the wood for a width of 2.5 cm (1 in.) and a root-promoting auxin powder is rubbed into this cut. In some plants, cuts that go through the bark can be used instead of girdling. A ball of moist (not wet) sphagnum moss is wrapped around the cut, then a sheet of plastic or aluminum foil is wrapped snugly around the sphagnum moss and tied tightly above and below the ball. No further watering is needed because the moss absorbs moisture from the plant itself and the covering retards water loss. After several weeks, roots start developing. When a good root system has formed, the layer can be cut off just below the root ball, removed from the plastic, and, without disturbing the roots, planted in a large pot of soil. A few leaves should be removed to reduce water loss and the plant should be put in a cool, humid, and shady place until it becomes well established on its own new roots system.

Other Plant Structures Providing Natural Propagation Methods

Runners   Some plants, such as the strawberry (Fig. 9–25), grow as a rosette crown, with runners (stolons) arising from the crown. New plants arise from nodes at intervals along these runners. From these runner plants additional new runners arise, thus developing a natural clonal multiplication system. The runner plants must have favorable moist soil conditions to root. The strawberry produces runners in the summer in response to long days, and stops producing runners as the days shorten in the fall. Then the strawberry runner plants can be dug, packed in polyethylene-lined boxes, and placed in cold storage (–2°C; 28°F) for planting later, usually the following spring. Other plants with this natural asexual reproduction system are the ground cover (*Duchesnea indica*) and the strawberry geranium (*Saxifraga stolonifera*).

Suckers   Some plants, such as the blackberry and red raspberry, produce **adventitious** shoots—or suckers—from their horizontal root system, which eventually spread to form a dense thicket of new plants. These individual shoots with a piece of the old root attached can be dug and replanted. This is a simple, highly successful asexual propagation method.

Figure 9–25
Strawberry plant with new plants developing from nodes on the runners (stolons).

Figure 9–26
Crown of a herbaceous perennial, the Shasta daisy. Lateral shoots develop from the underground portion of older stems and root. These rooted shoots can be cut from the mother plant and replanted. Underneath is a rooted shoot cut from the mother plant.

Crowns   Many perennial plants exist as a single unit, becoming larger each year as new shoots arise from the crown (root-shoot junction) of the plant. Vegetative propagation of such plants consists of crown division—cutting the crown into pieces, each having roots and shoots, and transplanting to a new location (Fig. 9–26).

Propagation Using Specialized Stems and Roots

A number of herbaceous perennial plants have structures such as **bulbs**, **corms**, **tubers**, **tuberous roots**, or **rhizomes**. These structures function as food storage organs during the plant's annual dormant period and also as vegetative propagation structures. In bulbs and corms, the newly formed plants break away from the mother plant naturally. This type of propagation is termed **separation**. The remaining structures—tubers, tuberous roots, and rhizomes—must be cut apart; this is termed propagation by **division**.

Bulbs   These are short, underground organs having a basal plate of stem tissue with fleshy leaf scales surrounding a growing point or flower primordium. In the axils of some of these leaf scales, new miniature bulbs develop that eventually grow and split off from the parent bulb to form a new plant. There are two types of bulbs: (1) the tunicate bulb has a solid tight structure with fleshy scales arranged in concentric layers and covered by a dry membranous protective layer; examples are onion, daffodil, tulip, and hyacinth; and (2) the scaly bulb, such as the lily, has loose, separate scales with no protective layer to prevent desiccation. Tunicate bulbs can be propagated by harvesting and planting the offsets (small bulbs that develop at the base of the parent bulb). In some species, offsets are not produced or only a few are produced. Offsets can be encouraged to form by scooping out or scoring the base of the bulb. Offsets will form at the base of the fleshy leaves. For scaly bulbs, the scales can be removed from the bulb and placed under moist, humid conditions to promote the development of one or more new miniature bulbs at the base of each scale.

Corms   A **corm** is the swollen underground base of a stem axis; it has nodes and internodes and is enclosed by dry scalelike leaves. The gladiolus, freesia, and crocus are examples of plants having a corm structure. Following bloom in gladiolus, one or more new corms develop just above the old corm, which disintegrates. In addition, several new small corms called cormels are produced just below each new corm. These may be detached and grown separately for one or two years to reach flowering stage.

Stem Tubers   The edible Irish (white) potato is a good example of a plant having a tuber structure. The underground tuber consists of swollen stem tissue with nodes and internodes. When a potato tuber is cut into sections and planted, shoots arise from the buds (eyes) on the potato piece, which itself serves as a food supply for the developing shoot. From the lower portion of such shoots, adventitious roots develop, along with several underground horizontal shoots that are stem tissue. The terminal portions of these horizontal shoots enlarge greatly to form the fleshy potato tuber. Potato cultivars are clones and are propagated by dividing each tuber into several sections. The crop eventually developing from each section usually consists of three to six new tubers.

Tuberous Roots   Tuberous roots are root tissue. The dahlia, tuberous-rooted begonia, and the sweet potato (Fig. 9–27) are examples. Sweet potatoes are propagated vegetatively by placing the tuberous roots in beds so they are covered with about 5 cm (2 in.) of soil. Adventitious shoots (slips) develop from these root tubers; adventitious roots form from the base of the shoots. These rooted slips are then pulled from the tuberous root and planted. As the sweet potato vines grow, some of their roots swell to form the familiar edible sweet potato.

Rhizomes   Certain plants, such as the German iris and bamboo (Fig. 9–28), have the main stem axis growing horizontally just at slightly below the soil surface. This stem is a rhizome. Some plants have thick and fleshy rhizomes, while others have thin, slender rhizomes. Like any stem, rhizomes have nodes and internodes. Leaves and flower stalks and adventitious roots develop from the nodes. Propagating plants with rhizomes is easy. At a time of year when the plant is not actively growing, the rhizome can be cut into pieces several inches in length and transplanted. Noxious weeds, such as Johnson grass, that have a rhizome structure cannot be controlled by cultivation because it merely breaks up the rhizomes and spreads the pieces about, each piece developing a new plant. Many important economic plants are propagated from rhizomes; examples are banana, ferns, ginger, and many grasses. Aboveground horizontal stems are termed stolons and can be propagated the same as rhizomes.

MICROPROPAGATION (TISSUE CULTURE)

A major advance in plant propagation involves the use of very small pieces of plant tissue grown on sterile nutrient media under aseptic conditions in small glass or clear plastic containers. These small pieces of tissue, called **explants**, are used to regenerate new shoot systems, which can be separated for rooting and growing into full-size plants. Some of the pieces of tissue are retained for further regeneration. The increase is geometrical, giving rise to fantastically large numbers of new individual plants in a short time. Some nurseries produce millions of plantlets a year by tissue culture methods, which require the use of highly trained technicians working under sterile conditions.

Micropropagation or tissue culture was used at first with herbaceous plants, such as ferns, orchids, gerberas, carnations, tobacco, chrysanthemum, asparagus,

Figure 9–27
Left: *Sweet potato tuberous root, producing rooted adventitious shoots.* Right: *Two detached shoots (slips) ready for planting.*

Figure 9–28
Rhizome of bamboo with lateral buds and adventitious roots arising at nodes. Plants with rhizomes are propagated simply by cutting the rhizome into pieces and planting them.

gladiolus, gloxinia, strawberry, and many others. Later, with modifications of the media, it was found that many woody plants, such as rhododendrons, kalmias, deciduous azaleas, roses, plums, apples, and many others could also be successfully propagated by tissue culture methods (Fig. 9–29).

Different parts of the plant can be taken as the explant. Entire seeds themselves can be used; for example, very tiny orchid seeds have been commercially germinated in sterile culture for many years. Embryos can be extracted from seeds and grown on a sterile nutrient medium. Shoot-tip culture involves excision of the growing point, which increases in size when it is kept in a nutrient medium and is divided over and over, greatly increasing plant numbers. This method has been successful with orchids, ferns, apples, and carnations. In other cases, tissue culture involves, for example, the excision of a piece of stem tissue and placing it on a nutrient medium. The explant then develops masses of **callus** by continuous cell division. From these callus clumps, roots, and shoots may differentiate to form new plants. This method has been used with carrot, tobacco, asparagus, endive, aspen, Dutch iris, and citrus.

Single pollen grains of some plants, such as tobacco, have been germinated in sterile culture, when taken at just the right stage, to form **haploid** (1n) plants. Doubling the chromosome number with colchicine treatments yields diploid homozygous plants.

In addition to propagation, the following aseptic tissue cultures procedures are used for pathogen elimination and germplasm preservation. The nutrient media used for micropropagation are also favorable substrates for the growth of bacteria, fungi, and yeasts,

Table 9–3
COMPONENTS OF THE MURASHIGE AND SKOOG MEDIUM FOR GROWING TISSUE EXPLANTS UNDER STERILE CULTURE

NH_4NO_3	400 mg/l	Indoleacetic acid	2.0 mg/l
$Ca(NO_3)_2 \cdot 4H_2O$	144	Kinetin	0.04–0.2
KNO_3	80	Thiamin	0.1
KH_2PO_4	12.5	Nicotinic acid	0.5
$MgSO_4 \cdot 7H_2O$	72	Pyridoxine	0.5
KCl	65	Glycine	2.0
NaFe—EDTA	25	Myo-inositol	100
H_3BO_3	1.6	Casein hydrolysate	1,000
$MnSO_4 \cdot 4H_2O$	6.5		
$ZnSO_4 \cdot 7H_2O$	2.7	Sucrose	2%
KI	0.75	Powdered purified agar	1%

Micronutrient elements may or may not be required but are usually added routinely. The following stock solution provides the required materials. One milliliter of this solution is added per liter of culture medium.

$MnSO_4 \cdot 4H_2O$	1.81 g	$CuSO_4 \cdot 5H_2O$	0.08 g
H_3BO_3	2.86 g	$(NH_4)_2MoO_4$	0.09 g
$ZnSO_4 \cdot 7H_2O$	0.22 g	Distilled water	995 ml

Iron is usually essential and can be supplied in several ways: iron tartrate (1 ml of a 1 percent stock solution), inorganic iron ($FeCl_3 \cdot 6H_2O$, 1 mg/l) or $FeSO_4$, 2.5 mg/l; or chelated iron. Chelated iron can be supplied as NaFeEDTA, 25 mg/l, or by mixing Na_2EDTA and $FeSO_4 \cdot 7H_2O$ in equimolar concentrations to give 0.1 mM Fe.

so the prepared nutrient medium and its containers must be sterilized. The surface sterilization of the explants with a material such as a 10 percent Clorox® solution is necessary, followed by rinsing with sterile water. Excision and insertion of the plant tissues onto the nutrient medium in the sterilized containers must be done with laboratory skill and procedures to prevent recontamination.

The nutrient media on which the excised plant parts are grown include mineral salts, sugar, vitamins, growth regulators (auxins and cytokinins), and sometimes certain organic complexes such as coconut milk, yeast extract, or banana puree. Many different media have been developed for obtaining shoot and root proliferation on explants of various species. Varying the ingredients can influence the development of the new plants. For example, one media promotes callous, another might promote roots, and yet another promotes shoots. The constituents of one widely used medium are given in Table 9–3. A mixture of micronutrient elements, including iron, is often added.

Figure 9–29
Small plantlets being propagated in a sterile nutrient medium in a micropropagation facility. Source: Margaret McMahon, The Ohio State University.

SUMMARY AND REVIEW

Plants are domesticated and improved by selecting, breeding, genetically engineering, and propagating plants with superior characteristics.

Selecting plants with desirable characteristics and increasing their numbers is the first step in improvement. When selection is not enough, selective breeding is used to produce plants with desirable traits. Breeding starts with selecting parents that each have unique, desirable traits and crossing them to produce offspring with all the desirable traits. This can take several generations of crosses and backcrosses, the evaluation of thousands of plants, and require many years. The use of DNA markers has dramatically reduced the number of plants that have to be evaluated, but the process is still time-consuming. Selective breeding can only be done with sexually compatible plants, which limits the traits that can be passed on to offspring.

Genetic engineering allows genes to be transferred directly from one organism to another related or unrelated organism without sexual recombination of chromosomes. A gene is a section of a chromosome that determines a trait. Traits can be controlled by one or more genes. Single gene traits are the easiest to transfer through genetic engineering. Usually a promoter and marker gene are transferred along with the trait gene to control expression of the trait and to indicate that the gene has been transferred.

Propagation can be sexual (seed) or asexual (vegetative). With many plants, using seeds is the most practical method of propagation. As long as the seeds produce plants that maintain the desired characteristics of the parent(s) (true to type), the seeds can be saved from one generation to produce the next. When seeds do not produce true to type, then hybrid seeds that are obtained from specific parental crosses must be obtained for each crop. The development of hybrid seed has been one of the more important breakthroughs in crop improvement. Reliable production of crops from seeds requires careful control of the seed source. Seed certification programs are designed to maintain the genetic quality of each generation of seed.

In many cases, asexual propagation allows desirable traits to be passed easily from one generation to the next. Asexual propagation involves the production of an entire plant from a cell, tissue, organ, or other part of another plant. Except in the case of mutation or chimera, the new plant has the same genetic makeup as the original plant. The technique is very useful when genetic integrity cannot be maintained by seed production or when seed production is inefficient.

Cutting production involves removing a part of a stem or root of a plant and regenerating the missing organs and tissues by providing an environment that favors the regeneration process. Generally, cutting propagation requires warm temperatures, reduced light intensity, and high humidity. As rooting occurs, temperatures are reduced, light levels are increased, and humidity is reduced to normal growing conditions.

Grafting is the attaching of a twig or bud called the scion from a plant with one genetic complement to the rootstock from a plant with a different genetic complement. The scion and rootstock must be related closely enough for grafting compatibility. There are several different kinds of grafts, including whip and cleft.

Layering is similar to cutting propagation except the part to be propagated is not removed from the parent plant until the missing stem or roots have regenerated. Layering can be done by covering part of the stem with soil or wrapping a section of stem with damp sphagnum moss and plastic film until rooting occurs.

Other plant structures that can be used for propagation are plantlets that can be separated from the parent plant, or crowns, bulbs, corms, and tubers that can be divided. Plantlets can come from runners, which extend from the main portion of the plant or form naturally in some species along leaf margins and at leaf bases. Care must be taken to ensure that, over time, the new generations have not acquired undesirable mutations or become infected with pathogens.

Micropropagation is the rapid multiplication of plants by using small pieces of tissue (explants) and placing them on a specific type of medium in petri dishes or clear jars. Callus forms on the explants and from the callus many new plants develop that are genetically identical to the original plant.

Crop improvement has progressed from its early stages, when superior plants were selected by a farmer and the seeds of those plants were used for the next crop, to modern methods that involve very rigid and competitive breeding and genetic engineering programs throughout the world. The potential for crop improvement through traditional breeding and genetic engineering is inestimable. No longer is the genetic information in one plant limited to what can be obtained through compatible sexual crossing. The ability to impart resistance, productivity, or nutrition in plants appears to be limitless from a scientific basis.

KNOWLEDGE CHECK

1. Indicate which of the following are homozygous and which are heterozygous: AABB, Aabb, aaBB, AaBb.
2. Which of the above can produce all true-to-type seed if they are self-pollinated?
3. What is the role of each of the following in determining an organism's traits: DNA, RNA, proteins?
4. In question 1, which are the dominate alleles and which are recessive? Which group or groups would express both dominate traits? Which would express both recessive traits?
5. Although many genetic mutations can be deleterious, mutations can be useful to plant breeders for what reason?
6. What are genotype and phenotype and how are they related to each other?
7. Explain why it can take several generations of selective breeding to get the desirable traits from two genetically different parents into an offspring.
8. When can sexual propagation be used to produce the next generation of plants? Under what circumstances must asexual propagation be used?
9. What is seed certification and why is it worth the extra cost to growers to buy certified seed?
10. List the tests that can be used to determine if a seed is likely to germinate or is viable.
11. Describe how dormancy can be a survival mechanism for seeds.
12. What are some methods used to break seed dormancy?
13. List the three steps of seed germination.
14. How does light affect seed germination?
15. What is damping off and how can it be prevented?
16. What are adventitious roots, adventitious shoots, and callus?
17. What are clones?
18. What are chimeras? What type of chimera is the most stable and can be propagated vegetatively?
19. What is apomixis?
20. Why is disease transmission of particular concern in asexual propagation?
21. What is culture-indexing of stock plants?
22. What is a cutting and what are the parts of a plant from which they can be obtained?
23. Briefly describe the environmental conditions for stem cutting propagation.
24. List the factors that influence how easy or difficult it is to root cuttings.
25. Why are auxins used in cutting propagation?
26. In grafting, what is the scion and what is the rootstock? How does the genetic relationship between two plants affect the ability of a graft to be successful?
27. What is graft incompatibility?
28. How is grafting used to create some dwarf trees?
29. What is layering and what are the different types of layering?
30. How can corms and bulbs be used for asexual propagation?
31. What is micropropagation and why has it been a very effective propagation technique for several plants?
32. In the broad sense, what is biotechnology?
33. Why are "genetically engineered" or "transgenic" better terms than "GMO" for plants that have had genes inserted by non-sexual methods?
34. Although all organisms translate DNA into the same proteins, why may the traits be expressed differently in a transgenic plant than the donor plant?
35. Why are promoters and marker genes transferred along with the trait gene?

FOOD FOR THOUGHT

1. When seed companies first introduced hybrid seed, farmers were very skeptical and thought that companies were just trying to get them to buy new seed every year instead of using seed they saved from the previous harvest. What do you think changed their minds after a few years and made them overwhelmingly favor buying the hybrid seed every year?
2. How might herbaceous grafting be used to produce full-sized tomatoes on dwarf plants?
3. If you were a plant breeder and needed to introduce greater heat tolerance into a potato plant from tobacco (both are in the Solanaceae family) would you likely be able to do it with traditional breeding methods? If not, what method could you use?

FURTHER EXPLORATION

OMOTO, C. K., AND P. F. LURQUIN. 2004. *Genes and DNA: A beginners guide to genetics and its applications.* Irvington, NY: Columbia University Press.

SCHLEGEL, R. H. J. 2003. *Encyclopedic dictionary of plant breeding and related subjects.* Florence, KY: Haworth Press/Taylor and Francis.

TOOGOOD, A. 1999. Plant *Propagation: The Fully Illustrated Plant-by-Plant Manual of Practical Techniques.* New York, NY: Dorling Kindersley Publishing.

10

Cultivated Plants: Naming, Classifying, Origin, Improvement, and Germplasm Diversity and Preservation

DAVID TAY

There are over 500,000 different kinds of plants, and for humans to be able to communicate about them, some method of classifying and naming them had to be developed. The process of developing plant names started over 4,000 years ago, but the first recorded names were attributed to Theophrastus (370–285 BCE). Naming plants, no doubt, began with simple names that referred to the plant's use, growing habit, or other visible attribute. One example is the milk weed—so named because its sap (latex) is milky in appearance. One difficulty with such names is that often they are only used locally. People in one place know the plant by one name, while elsewhere the same plant is known by a different name. The weed *Tribulus terrestris* is known as the puncture vine in some areas because the seeds have sharp spines that puncture tires or bare feet; another common name for the same plant in other areas is goat-head because the shape of the seed resembles a goat's head. As plant knowledge expanded and the exchange of this knowledge became desirable, it was obvious that a uniform and internationally acceptable system was needed to name and classify plants.

There are many ways to classify plants, and any system depends on how the classification is to be used. Some classifications relate directly to specific environmental requirements of the plant for satisfactory growth. For example, such a classification could categorize plants according to their climatic requirements.

CLIMATIC AND RELATED CLASSIFICATIONS

Farmers and others who grow plants commercially have to be able to identify and name crops. But this alone is not enough. They also have to distinguish which of many crops and plants suit their climate. Most commercial growers in the United States are working in the temperate zone. For example, some

key learning concepts

After reading this chapter, you should be able to:

- Explain how plants are named and classified.
- Use the nomenclature and system of taxonomic classification to identify plants and their relationship to each other.
- Explain how several crops originated and where they were domesticated.
- Discuss the importance of saving germplasm from extinction and the global system created to preserve germplasm.

fruit and nut crops grown in the temperate zone are almond, apple, apricot, cherry, peach, pear, pecan, and plum. Fruit growers in a tropical region would have an entirely different choice of crops, such as cacao, cashew and macadamia nuts, banana, mango, papaya, and pineapple. The fruits of the subtropical region, between the temperate zone and the tropics, cannot withstand the severe winters of the temperate zone but may need some winter chilling. Some of these subtropical plants are citrus, date, fig, olive, and pomegranate. Thus, by using climate as a criterion, plants can be classified into distinct groups. The USDA has divided the United States into several different cold hardiness zones, and many plants are classified by their ability to grow in the different zones.

Agronomic crops like grains, forage, fiber, and oil crops and the vegetable and ornamental plants can also be classified by their temperature requirements. For example, some annuals have specific climatic requirements for growth and flowering and are distinguished as winter or summer annuals. Winter annuals are planted in the fall and bloom early the following spring. Summer annuals are planted in the spring and bloom through the summer and fall.

Some crops grow best in certain seasons and are thus classified. For example, warm-season plants such as corn, beans, tomatoes, peppers, watermelons, petunias, marigolds, zinnias, and Bermuda grass grow best where monthly temperatures average 18°C to 27°C (65°F to 80°F), while broccoli, cabbage, lettuce, peas, flowering bulbous plants, snapdragons, cyclamen, and bluegrass are cool-season crops growing best at average monthly temperatures of 15°C to 18°C (60°F to 65°F). Plants can be classified by the seasons in which they are most likely to flower and fruit or when the quality of the product can be expected to be at its maximum. Numerous flower and vegetable cultivars can be classified as early, midseason, or late maturing.

Vegetables are classified into groups according to their edible parts. Some are grown for fruits and seeds, such as the tomato, bell pepper, string bean, pea, and corn. Many are grown for their shoots or leafy parts, such as asparagus, celery, spinach, lettuce, and cabbage. Others are grown for their underground parts (either roots or tubers), such as the carrot, beet, turnip, and potato.

Ornamentals are sometimes classified by use—that is, as houseplants, greenhouse plants, cut flowers, garden plants, street trees, and various classes of landscape plants. Houseplants, which have become very popular, are often classified according to their foliage, flowers, or growth habits. Common foliage plants,

Figure 10–1
Jeffrey pine (Pinus jeffreyi Grev. and Balf.) tree growing in the Sierra Nevada mountains of northern California near Lake Tahoe. This tree is about 2 m (6.5 ft) in diameter at breast height (DBH), the point where lumber trees are measured.
Source: Robert A. H. Legro.

which stay green all year, are the philodendrons, dieffenbachias, and ferns, to name a few. Blooming of some flowering plants may be only seasonal, as in the Easter lily and poinsettia, but the African violet and chrysanthemum are available in flower year-round. Plants outside the home are typically classified by use, for example, bedding plants such as petunias, marigolds, and zinnias, and landscape plants like trees and shrubs.

The forester classifies trees into two broad groups: the hardwoods and softwoods. Some hardwood types are oaks (*Quercus*), maples (*Acer*), birch (*Betula*), and beech (*Fagus*). Some softwood trees are pines (*Pinus*) (Fig. 10–1), firs (*Abies*), redwood (*Sequoia*) (Fig. 10–2), cedars (*Cedrus*), and spruce (*Picea*). Trees are also classified according to the hardiness zones in which they can survive. Some can withstand very low temperatures during the winter, whereas others are subject to frost damage and therefore must be grown in a subtropical climate.

COMMON AND BOTANICAL NAMES

Most plants are generally known by their common names because common names are often easier to remember, pronounce, and use. Maple or elm trees

Figure 10–2
A grove of coast redwoods (Sequoia sempervirens D. Don, Endl.) along Highway 101 on "The Avenue of the Giants" in northern California. The camper truck is dwarfed by these majestic trees.

growing along the streets are referred to by common names in everyday conversation. Common names often evolve because of certain plant characteristics.

A common name has value in conversation only if both persons know exactly what plant is being discussed. This is most likely when persons are from the same community and the common name of the plant cannot be mistaken for another. But take the case of the jasmine, a plant known all over the world and prized for its fragrance, flavoring (tea), and landscaping. Many common plant names contain the word *jasmine,* but such plants do not resemble one another and may not even be closely related botanically. Some of the so-called jasmines are listed below, with the botanical (italicized) name following the common name:

Star jasmine	(*Jasminum gracillimum*)
Star jasmine	(*Trachelospermum jasminoides*)
Blue jasmine	(*Clematis crispa*)
Cape jasmine	(*Gardenia jasminoides*)
Crape jasmine	(*Tabernaemontana divaricata*)
Night jasmine	(*Cestrum nocturnum*)
Night jasmine	(*Nyctanthes arbor-tristis*)

It is obvious from these examples that common names have their limitations for universal written or verbal communication. There are too many and they are too variable to serve most scientific purposes.

DEVELOPMENT OF BOTANICAL CLASSIFICATIONS

Theophrastus (370–285 BCE), a student of Aristotle, classified plants by their texture or form. He also classified many as herbs, shrubs, and trees. He noted the annual, biennial, and perennial growth habits of certain plants, and described differences in flower parts that enabled him to group plants for purposes of discussion. He is known as the father of botany for these significant contributions.

Carl von Linné (1707–1778), better known as Carolus Linnaeus, devised a system of categorizing plants that led to the modern taxonomy or nomenclature of plants.

Scientific Classification

The scientific system of classification has all living things divided into groups called **taxa** (sing. taxon) based on physical characteristics. The first taxon, called **Domain**, divides all living things into two Domains: Prokaryotes (cells having no separate subcellular units) and Eukaryotes (cells having subcellular units). The Eukaryote Domain is divided into the four Kingdoms of Fungi, Protista, **Plantae**, and Animalia. The Plantae Kingdom is divided into two groups: **bryophytes** (includes mosses and liverworts) and **vascular plants**. The vascular plants are divided into two subgroups: seedless and seeded. Seedless and seeded plants are further classified by **Phyla**. Seedless phyla include the Pterophyta (ferns). Seeded phyla include Cycadophyta (cycads), Ginkgophyta (ginkgo), Coniferophyta (conifers), and Anthophyta (angiosperms, which are subdivided into monocotyledon and dicotyledon). Almost all commercially important crop plants are in the seeded group. After phylum, plants are classified in descending rank by **class**, **order**, **family**, **genus**, and **species**. Each descending rank more closely defines the physical characteristics common to members of that rank.

PLANT IDENTIFICATION AND NOMENCLATURE

The family is usually the highest taxon commonly included in plant identification or study. Students of plant science are usually required to learn the family, genus, and species of some plants as well as their common names.

Since the early Christian era, naturalists wrote their books in Latin, which was the language of all educated people in Europe. Thus, Linnaeus used names of Latin form. Most of the names he gave, which describe morphological characteristics of the plants, came from Latin words, although some were derived from Greek and Arabic.

The names are usually phonetic and often give a clue to the plant's characteristics, its native habitat, or for whom it is named. Such derivatives are numerous. Names that refer to leaves include *folius, phyllon,* or *phylla,* usually as suffixes. The names can also have prefixes, such as *macro* or *micro.* Thus, words are created, such as *macrophylla* (large leaf), *microfolius* or *microphylla* (small leaf), *illicifolius* (holly leaf), and *salicifolius* (willow leaf). The Latin for flower is *flora;* add the prefix *grand* and it becomes *grandiflora* (large flower), as in *Magnolia grandiflora* L., the southern magnolia (Fig. 10–3). Shapes or growing habits of plants can be described with *altus* or *alta* (tall), *arboreus* (treelike), *compactus* (dense), *nanus* or *pumilus* (dwarf), *repens* or *reptans* (creeping), and *scandens* (climbing). Names based on flower or foliage color include *albus* or *leuco* (white), *argentus* (silver), *aureus* or *chryso* (gold), *rubra, rubens,* or *coccineus* (red), and *croceus, flavus,* or *luteus* (yellow). Species names sometimes reflect the plant's place of origin. Examples are *australis* (southern), *borealis* (northern), *canadensis* (from Canada), *chinensis* or *sinensis* (from China), *chilensis* or *chileonsis* (from Chile), *japonica, nipponica,* or *nipponicus* (from Japan), *campestris* (field), *insularis* (island), and *montanus* (mountain). Sometimes Linnaeus took names, like *Narcissus,* from classical mythology, or devised names to honor other scientists, like *Rudbeckia,* honoring Linnaeus' botany professor, Rudbeck.

Each plant has a two-word, or **binomial**, name[1] given in Latin. The first name refers to the plant's genus, the second to its species. The Latin binomial name is international and understood universally.

Figure 10–3

*The southern magnolia (*Magnolia grandiflora *L.). The name* grandiflora *is more than justified; the flower measures 15 cm (6 in.) across.*

Complete Linnaean names have a third element—the **authority**, or the abbreviated name of the scientist who named the species. Consider the name for the common white (Irish) potato, *Solanum tuberosum* L. The "L." means Linnaeus, and his initial appears commonly, because Linnaeus named so many species. Books and journals often omit the authority for brevity and simplicity, but it is important in determining which taxon is being referred to in situations where different botanists have used different binomials for the same plant. In such cases, the botanical or scientific name that was published first takes precedence.

Wild or naturally occurring plants are named under the rules of the *International Code of Botanical Nomenclature.* Cultivated plants are named according to the same principles but are covered by the *International Code of Nomenclature for Cultivated Plants.* Some of the basic rules of nomenclature follow.

The generic name always begins with a capital letter; it is underlined when written by hand or typewriter and italicized in print. Thus, the genus name for potato is *Solanum.* The specific epithet *tuberosum* is likewise underlined or italicized. The specific epithet usually begins with a lowercase letter, but it *may be* capitalized if it is a person's name; it is always correct if written entirely in lowercase. Many of the original species names are frequently capitalized (e.g., *Pinus Jeffreyi* Grev. and Balf. or *Pinus jeffreyi* Grev. and Balf.).

To complete the binomial name, the authority for describing and naming the plant is given after the genus and species; thus, *Solanum tuberosum* L. In this text, unless otherwise specified, both the binomials and the authorships agree with those given in *Hortus Third,* which is a widely recognized compilation of the cultivated plants

---

[1] The binomial is a binary combination of the name of the genus followed by a single specific epithet. If an epithet consists of two or more words, these are to be united or hyphenated (Article 23 *International Code of Botanical Nomenclature* [2]). Although the second word of the binomial is technically the "specific epithet," many persons refer to that second name as the "species." This shorter and therefore more convenient form is the one used in this text.

for the United States and Canada. These authority names are often abbreviated. Each taxonomy book has a list of the full names of these authorities.

When several plants in the same genus are listed, the genus name is given in full for the first plant, then shortened to the first initial (which is always capitalized) for the other plants in the list. As an example, the apricot, the European plum, and the peach are all members of the genus *Prunus.* Listed as scientific binomials, they would be *Prunus armeniaca, P. domestica,* and *P. persica,* respectively. This procedure should not be used if there is any chance of confusion with another genus with the same first initial.

The singular and plural spelling of *species* is the same. Occasionally the plant genus is known but the exact species is not known because it is difficult or impossible to identify. In such a case, the genus name is given and followed by the lowercase letters "sp." for species (singular) and "spp." for species (plural). An example is *Prunus* spp. The "spp." usually refers to all of the many species in the genus. However, the "sp." refers to a definite plant whose specific epithet is not known. The "sp." or "spp." is never underlined or italicized.

Subspecific Categories

Sometimes a botanical binomial is not sufficient to identify a species—wild or cultivated. Botanists and horticulturists may form subspecific categories, such as botanical variety, cultivar, and group.

Botanical Variety

A plant group can be so different in the wild from the general species described originally that it warrants a botanical variety classification below that of species. An example of this is *Buxus microphylla* Sieb. and Zucc. var. *japonica* Rehd. and Wils., which is native to Japan. The "var." stands for *varietas,* Latin for "variety." Another botanical variety originated in Korea; it is *Buxus microphylla* var. *koreana* Nakai. These botanical varieties are sufficiently different to warrant unique names and authorities to distinguish them from one another. In this case, the name of the variety *japonica* or *koreana* is underlined or italicized. When a varietas epithet is formed from a surname, it may or may not be capitalized depending on the personal preference of the author. However, the trend is to not capitalize them, as recommended by the International Codes.

Cultivar

Many kinds of plants that are valuable in agriculture must be propagated with little or no genetic change in the offspring. In agronomy and horticulture, there are cultivated varieties that remain genetically true. These cultivated varieties may be different from botanical varieties and are called **cultivars**, a contraction of *culti*vated *vari*ety. There are two main categories of cultivars—the clones and the lines. If propagated by vegetative methods, they are called *clones*; if by seeds (under certain specified conditions), they are called *lines.* The word *cultivar* is abbreviated "cv." and the plural is "cvs." A cultivar is often a distinct variant selected by someone who believed it was uniquely different from any plant already in cultivation. The flower color may have changed from red to white because of a mutation, as in some carnations. Perhaps a plant has fewer spines or thorns than does the ordinary species; an example is a Chinese holly (*Ilex cornuta* Lindl. and Paxt.) found to have few or no spines. It was named *Ilex cornuta* cv. Burfordii. The cultivar name is always capitalized but never underlined or italicized. The term "cv." after *cornuta* may be dropped in favor of single quotes around the cultivar name. Either way of expressing the cultivar name is acceptable. Either single quotes or the term "cv." is used, but *never both.* Tables or lists usually use "cultivar" or "cv." in the heading to avoid single quotes around each cultivar name.

Many annual flowers, vegetables, grains, and forage crops are cultivars that are propagated by seed from open pollination. Others are F_1 hybrids. An example is *Petunia* × *hybrida* Hort. Vilm.-Andr.—the hybrid garden petunia. A breeder may develop a new strain that is believed to warrant a cultivar name such as 'Fire Chief' or 'Pink Cascade.' The parent plants can be maintained and crossed to produce the same F_1 hybrid cultivar year after year. Many vegetables and flowering annuals are maintained as cultivars in this manner, with the parents maintained to produce new crops of seed each year for planting. Cultivars of fruit trees, grapes, and woody ornamentals are usually maintained as true-to-type **clones** by vegetative propagation methods.

Group

The **group** category is used for some vegetables and some ornamentals such as lilies, orchids, roses, and tulips. It is a category below the species and not used as frequently as the cultivar category. A group includes more than one cultivar of a particular kind of plant. For example, when there are evident differences among plants of the same species, they can be further categorized by a group name. When a species has many cultivars, cultivars that are similar are categorized into groups. For example, cultivars of *Brassica oleraceae* can be grouped into the Acephala Group (kale and collards), the Alboglabra Group (Chinese broccoli), the Botrytis Group (cauliflower), the Italica Group (broccoli), the

Capitata Group (cabbage), and other groups depending on their morphological characteristics. These groups have the same botanical name—*Brassica oleracea*. The name of the group is written within parentheses between the species name and the cultivar name, as *Brassica oleraceae* (Capitata) 'King Cole,' and the group name is always capitalized but not enclosed in single quotes.

Family

The family is a group of closely related genera. The relationship can be based on certain plant structures or on chemical characteristics, such as the presence of latex in the milk weed family ASCLEPIADACEAE, but flower structure is the usual basis for association. The nightshade family SOLANACEAE contains not only *Solanum* (potato) but also *Lycopersicon* (tomato), *Capsicum* (pepper), *Nicotiana* (tobacco), *Datura* (deadly nightshade), *Petunia*, and many others. This is a large family (about ninety genera and more than 2,000 species), most of which are native to the tropics. All species in this family have similar flower structures; the similarities between a tobacco, a tomato, and a potato flower, for instance (Fig. 10–4), are readily seen.

The first letter of family names is always capitalized and the names are sometimes underlined or italicized.

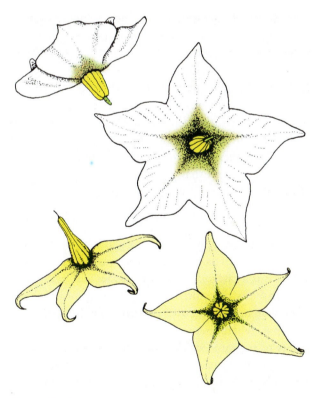

Figure 10–4
Side and front views of potato (upper) and tomato (lower) flowers. These views show the similarities of the flowers of plants in the same family, solanaceae. Source: Moira Tanaka.

The family names may be written entirely in capital (uppercase) letters. Most families' names end with *-aceae* (pronounced ace-ay-ee) attached to a genus name; for example, SOLANACEAE, ROSACEAE, AMARYLLIDACEAE, LILACEAE, and MAGNOLIACEAE. Eight families, however, did not follow this standard rule. For the sake of uniformity, new names have been adopted for these families. The old names appear in parentheses following the new names:

> ASTERACEAE (COMPOSITAE)
>
> BRASSICACEAE (CRUCIFERAE)
>
> POACEAE (GRAMINAE)
>
> CLUSIACEAE (GUTTIFEREAE)
>
> LAMIACIACE (LABIATE)
>
> FABACEAE (LEGUMINOSAE)
>
> ARECACEAE (PALMAE)
>
> APIACEAE (UMBELLIFERAE)

Plant classifications can be studied in detail in various plant biology or taxonomy books and references.

PLANT IDENTIFICATION KEY

See Table 10–1 for a simplified dichotomous key used to identify some commonly known seed-bearing plants (Spermatophyta). To use a key you need to know the vocabulary of plant structure. Examine the plant in question to decide if its characteristics fit in one category or the other offered by the key. Keying plants is a process of elimination by making yes or no decisions to characteristics offered in the key—rejecting those that do not apply (dichotomy).

To use a simplified key (Table 10–1), eliminate the alternative that does not pertain to the plant in question, and then proceed to the next pair of numbers (choices) directly under the proper choice. As an example, the first choice in this key is to determine whether the seeds of the plant being identified are borne naked, as in the cone-bearing plants of the gymnosperms (Fig. 10–5), or enclosed in an ovary, as in the angiosperms. Once this yes or no decision has been made, the next step is to compare the next pair of descriptions directly under the previously chosen characteristic. If the seeds are enclosed within an ovary, the plant in question is an angiosperm and the next pair of numbers to compare is 8. Note that the two 8s are separated by several pairs of subsidiary descriptions. A choice between the two 8s must be made before proceeding further. If the plant in question has parallel-veined leaves and the seeds have one cotyledon, the

Table 10–1
A SIMPLIFIED DICHOTOMOUS KEY FOR IDENTIFYING SOME SEED-BEARING PLANTS (SPERMATOPHYTA)

1. Ovules and seeds borne naked on scales in cones without typical flowers (see Fig. 10–5); trees or shrubs, often evergreen. Gymnospermae
   2. Plant foliage palmlike. CYCADACEAE (see Fig. 10–7)
   2. Plant foliate not palmlike
      3. One seed in a cup-shaped, drupelike fruit. TAXACEAE (see Fig. 10–8)
      3. Many seeds in a dry woody cone
         4. Leaves alternate and single
         4. Leaves alternate and in clusters; needle-shaped
            5. Cone-scale without bracts with two to nine seeds
            5. Cone-scales in axils of bracts, flattened, with two seeds. PINACEAE
               6. Cones upright on top of branchlets. *Abies*
               6. Cones not upright on branchlets. *Pinus*
                  7. Twigs not grooved. White pines or soft pines
                  7. Twigs grooved. Pitch or hard pines
1. Plants with seeds borne in an ovary (base of pistil) with typical flowers. Herbs, trees, and shrubs. Angiospermae
   8. Leaves usually parallel veined, flower parts usually in multiples of three. One seed-leaf or cotyledon. Do not form annual rings when increasing in stem girth. Monocotyledonae
      9. Plant with palmlike leaves. ARECACEAE
         10. Leaves fanlike
         10. Leaves featherlike. Feather and fishtail palms
            11. Lower feathery leaves not spinelike
            11. Lower feathery leaves spinelike, fruit fleshy with long grooved seed. *Phoenix* spp.
               12. Plant is a tree with shoots at base, trunk about 50 cm (20 in.) in diameter. Fruit edible. *Phoenix dactylifera*, date palm
               12. Plants not as above. Other *Phoenix* spp.
      9. Plants without palmlike leaves
         13. Perianth none or rudimentary
            14. Stems solid. CYPERACEAE
            14. Stems mostly hollow. POACEAE
               15. Plants woody, bamboolike. Bamboos
               15. Plants herbaceous, not bamboolike
                  16. Grasses that produce sugar. *Saccharum* spp.
                  16. Grasses that produce little sugar
                     17. Small grains and their kin (rice, wheat, etc.)
                     17. Cornlike plants and their kin
                        18. Plants monoecious. *Zea mays*, corn
                        18. Plants not monoecious. *Sorghum* spp.
          13. Perianth present
            19. Pistils several, not united. APONOGETONACEAE
            19. Pistils one, carpets united, ovary and fruit superior. AMARYLLIDACEAE
               20. Anthers six, stem a fibrous rhizome. *Agapanthus* spp.
               20. Anthers six, stem a corm or bulb *Allium* spp.
                 21. Leaves large, usually hollow and cylindrical. Bulb rounded and large. *Allium cepa*, onion
                 21. Leaves large, usually hollow and cylindrical. Bulb slightly thicker than neck. *Allium fistulosum*
   8. Leaves usually without parallel venation, two cotyledons. Herbs, trees, and shrubs with stems increasing in thickness with cambium cells, which form annual rings in woody plants. Dicotyledonae
      22. Corolla absent or not apparent, calyx present or lacking
      22. Corolla present, calyx usually forming two series of calyxlike bracts
         23. Petals united
         23. Petals separate
            24. Ovary inferior or partly so
            24. Ovary superior
               25. Stamens few, not more than twice as many as petals
               25. Stamens numerous, more than twice as many as petals
                 26. Habit aquatic. NYMPHAEACEAE, water lilies
                 26. Habit terrestrial

(Continued)

Table 10–1 (Continued)

    27. Pistils more than one, filaments of stamens united into a tube. MALVACEAE
        28. Style—branches slender, spreading at maturity, seeds kidney-shaped. *Hibiscus* spp.
        28. Styles united, ovary several carpels, calyx deciduous, seed angular. *Gossypium* spp.
            29. Staminal column long, anthers compactly arranged on short filaments.
                G. barbadense, sea-island cotton
            29. Staminal column short, anthers loosely arranged and of varying lengths.
                G. hirsutum, upland cotton Fig. 10–9)
    27. Pistil more than one, filaments not united into a tube. ROSACEAE
        30. Ovaries superior, fruit not a pome
            31. Pistils one, leaves simple and entire. Prunus spp.
                32. Fruit soft and pulpy. *P. armeniaca*, apricot
                32. Fruit dry and hard. *P. mume*, Japanese apricot
            31. Pistils two to many, leaves compound (at least basal leaves)
                33. Plants woody shrubs. *Rosa* spp.
                    34. Styles not extended beyond mouth of hip. Stamens about one-half as long
                        as styles. *R. odorata*
                    34. Styles extend beyond mouth of hip, stamens about as long as styles.
                        R. multiflora
                33. Plants herbaceous. *Fragaria* spp.
                    35. Underside of leaves are bluish white. *F. chiloensis*, wild strawberry (Fig. 10–10)
                    35. Underside of leaves are green. Other Fragaria spp.
        30. Ovaries inferior, fruit a pome
            36. Fruit with stone cells. *Pyrus* spp., pears
            36. Fruit without stone cells. *Malus* spp., apples

Figure 10–5
Cones of the sugar pine (Pinus lambertiana Dougl.) in closed and open conditions. When the cones dry, the winged seeds, shown on the 17.8-cm (7-in.) ruler, are released and can be disseminated by the wind.

plant belongs to the monocotyledon subclass. This procedure is followed until the plant to be identified "fits" a given set of plant characteristics.

A slight variation of a dichotomous key appears in Chapter 25 for the identification of turfgrasses; the number at the end of the phrase in that case indicates the next number to pursue.

Certain plant parts might not be available because they may not be in season—when needed to identify the plant. An example of this occurs in the simplified key in Table 10–1. In the family ROSACEAE (the second item), both flowers and fruit are needed to be certain of making the correct choice. In the case of strawberries, both fresh flowers and fruits can be obtained at the same time; however, it is not possible to have flowers and fruits at the same time for the apricot or the pear (Fig. 10–6). In such cases, it might be necessary to preserve flowers for future examination or simply describe them thoroughly in the spring when they are abundant, and then wait for the fruit to develop to accurately determine its characteristics. See Figures 10–7 through 10–10 for additional examples of hard-to-classify plants.

This simple key illustrates how choices are made between two characteristics—eliminating alternatives that are not pertinent to finally identify the plant in question. Much of the upper portion of the key can be ignored by a trained taxonomist or a good botany student because, to them, it is easy to identify a gymnosperm or an angiosperm. The family characteristics are determined by observing and studying certain flower characteristics. Sometimes it is difficult but necessary to distinguish between an inferior or superior ovary (the characteristic that divides the

Figure 10–6
An inflorescence of the pear (Pyrus communis L.) showing the five petals and the many stamens typical of the rosaceae. Mature fruits do not develop for another four months.

Figure 10–7
The inflorescence of a cycad (Dioon edule Lindl.) native to Mexico. The inflorescence is about 23 cm (9 in.) tall. The pinnate leaves might erroneously lead one to call this plant a palm.

Figure 10–8
The fruits of Taxus baccata L. 'Lutea'. The single seed (arrow) is within the cuplike fruit, which is poisonous.
Source: Margaret McMahon, The Ohio State University.

ORIGIN OF CULTIVATED PLANTS

Most of the crop plants important today were cultivated in a primitive way long before recorded history. Agriculture began about 10,000 years ago when ancient peoples selected certain plant types they found growing about them and thus enlarged their food sources beyond hunting and fishing. Most of these early food plants are still cultivated but in much improved forms. Others that were unknown to early humans have been added over the centuries to make up our present-day selection of food plants. Then, much later, as people became more concerned about the aesthetics of their environment, they also added shade trees, shrubs, flowering plants, and lawn grasses to the list of cultivated plants.

Several botanists have brought together fascinating information on the subject of where and when the crops that now feed the world's peoples, their livestock, and other domestic animals originated. The Swiss botanist Alphonse DeCandolle wrote *Origin of Cultivated Plants,* first published in 1833. Later, the famous Russian plant geneticist Nikolai Vavilov also studied the origin of cultivated plants. The results of his studies were published in *The Origin, Variation, Immunity, and Breeding of Cultivated Plants,* translated

LILIACEAE from the AMARYLLIDACEAE families; see Fig. 6–36), but one finds these determinations easier after some experience is gained. Once the family is known, the correct genus and species are identified by referring to taxonomic books with detailed keys.

Obviously one must be familiar with plant parts and structures (Chapter 6) to use a botanical key and determine the identity of the plant. Various plant parts and shapes are useful in illustrating plant identification (see Figs. 6–35, 6–36, and 6–37).

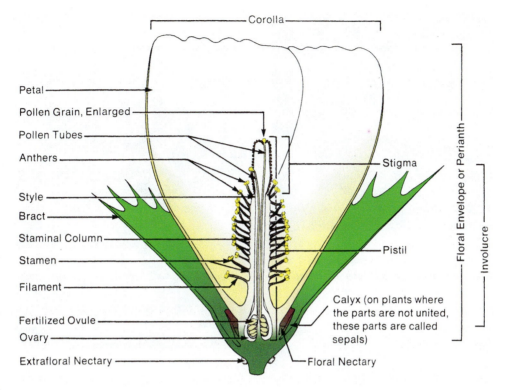

Figure 10–9
Cotton (Gossypium hirsutum L.) flower in a longitudinal section. Source: USDA.

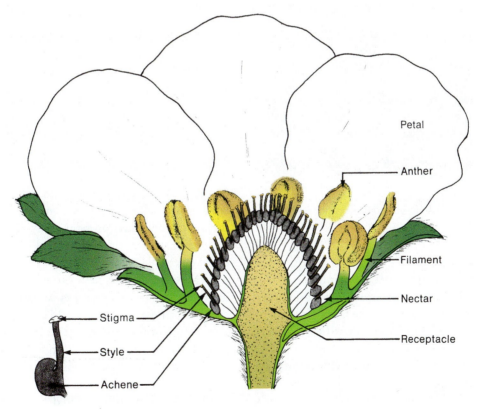

Figure 10–10
A strawberry flower in a longitudinal section. Also shown is an individual achene. Source: USDA.

from the Russian and published in English in 1951. Vavilov concluded from his studies that the various cultivated plants originated in eight independent centers: (1) central China, (2) India, (3) Indochina and the Malay Archipelago, (4) the Turkey–Iran region, (5) the Mediterranean area, (6) the Ethiopia–Somaliland area of east Africa, (7) Mexico and Central America, and (8) the Peru–Ecuador–Bolivia and the Brazil–Paraguay area of South America. Many archeological studies carried out in later years generally confirm Vavilov's locations for these centers.

More recent studies and theories on the origins and movements of the world's agricultural crops have been summarized by Carl Sauer in *Agricultural Origins and Dispersals* (1952), by Jack Harlan in *Crops and Man* (1975), by Jack Hawkes in *The Diversity of Crop Plants* (1983), by Barbara Bender in *Farming in Prehistory (1975),* and by Joseph Smartt and Norman Simmonds in *Evolution of Crop Plants* (1976).

Because breeding is most often accomplished between closely related species, many times plant breeders use close **wild-type** relatives of commercial plants to introduce or re-introduce desirable traits. One of the best regions to look for wild-types is in the region where a domesticated plant originated.

DOMESTICATION OF PLANTS

Toward the end of the Ice Age, about 11,000 to 15,000 years ago, when the glaciers were in full retreat and early humans were wandering about the earth, the stage was being set for the initiation of food production (Table 10–2). Before this time, tool-using hunter-gatherers had been on earth for about 4 million years. Then—only about 400 generations ago—there was a gradual transition to food-producing activities. This transition phase suggests several interesting questions, such as why it took so long for humans to become food producers and why such activities began in various parts of the world—southwest and southeast Asia, middle America, and western South America—at about the same time.

There is definite evidence from archeological sites that agricultural villages existed about 8000 to 9000 BCE in the area of southwest Asia known as the Fertile Crescent. It extends from the alluvial plains of Mesopotamia (now Iraq) across Syria and down the eastern coast of the Mediterranean sea to the Nile Valley of Egypt. Radiocarbon dating suggests that plants and animals were being domesticated at several places in this area by at least 5000 to 6000 BCE. There is evidence from site excavations that einkorn wheat, emmer wheat, barley, lentil, chickpeas, oats, and vetch were being cultivated, as well as dates, grapes, olives, almonds, figs, and pomegranates. Excavations at Jarmo in Iraqi Kurdistan, at ancient Jericho in the Jordan Valley, and many other sites in the Fertile Crescent have supplied much of the evidence to support these conclusions.

An indigenous savanna type of agriculture apparently was developing from domesticated native plants about 4000 BCE in a belt across central Africa (Fig. 10–11). This area also was the first home of the human race, as we understand it now. The genus *Homo* originated here, where most of human evolution subsequently occurred. Some important world crops brought under cultivation in this area include coffee, sorghum, millet, cowpeas, yams, and oil palm.

The Chinese center of agricultural origins became important about 4000 BCE according to radiocarbon dating. Crops domesticated include millet, chestnuts, hazelnuts, peaches, apricots, mulberries, soybeans, and rice. The West did not learn of the all-important rice plant until about 350 BCE, and the peach was unknown outside China until about 200 CE.

A farming culture in southeast Asia and what is now Indonesia apparently had domesticated rice around 6000 BCE. Other important crops appearing later under cultivation in this area were sugar cane, coconut, banana, mango, citrus, breadfruit, yams, and taro.

In the New World, evidence from archeological sites shows agricultural beginnings in two areas. One is present-day southern Mexico and Central America, where plant cultivation began about 5000 to 7000 BCE. The plants grown were early forms of maize (corn), sweet potato, tomato, cotton, pumpkin, peppers, squash, runner beans, papaya, avocado, and pineapple.

The second American region is a broad "non-center" of agricultural origins stretching from Chile northward to the Atlantic Ocean and eastward into Brazil. There is evidence that both the snap and lima beans were cultivated here by about 6000 BCE or earlier. Other important cultivated crops from this region are the potato, peanut, cacao, pineapple, cashew, papaya, avocado, Brazil nut, peppers, tobacco, guava, tomato, yam, cassava (manioc), and squash.

No major cultivated crop originated in the area of the present-day United States. Agriculture here relies in a large measure on introduced crops. There are, however, many minor native American fruit and nut crops, such as the American grapes and plums, the pecan,

Table 10–2
BEGINNINGS OF AGRICULTURAL DEVELOPMENT IN THE OLD AND NEW WORLD

The Postglacial Time Scale	Climatic Changes Date	Old World in Northern Europe	New World Cultural Stages	Cultural Stages
	10,000 BCE	Last glacial stage (Würm-Wisconsin ice)	Late Paleolithic hunting cultures (Cro-Magnon, etc.)	
	9000 BCE	Retreat of the glaciers (Pre-boreal period, cold dry)	Mesolithic fishing, hunting, collecting cultures	
	8000 BCE			Hunting cultures established
	7000 BCE		Agricultural beginnings	
	6000 BCE	Boreal period (Warm dry)		
	5000 BCE			
		Atlantic period (warm moist)	Neolithic agriculture established and spreading	
	4000 BCE		Beginnings of civilization (Egyptian–Sumerian)	
	3000 BCE		Neolithic agriculture in northern Europe	
	2000 BCE			American agricultural beginnings
		Subboreal period (colder dry)	Babylonian Empire Invasions: Aryans to India; Medes and Persians to southwest Asia and Mesopotamia	
	1000 BCE		Rise and flowering of Greek civilization	Early Mexican and Mayan civilizations
	1000 BCE–1000 CE	Sub-Atlantic period (cool moist)	Roman empire	
			Invasions: Goths, Huns	Decline of Mayans
			Rise of Islam	
	1000 CE		Norsemen to America	
			Mongol and Tartar invasions	Aztecs and Incas
			Voyages of discovery and colonization by Europe	
			Industrial revolution and modern period	
	2000 CE			

Source: R. F. Dasmann, *Environmental Conservation*, 5th ed. (New York: John Wiley & Sons).

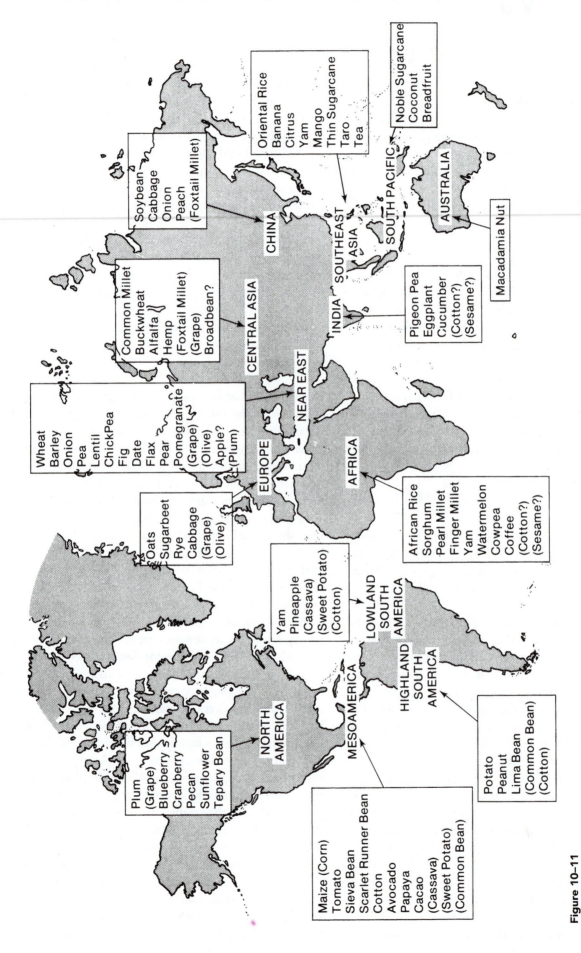

Figure 10–11

Regions of the world where major food crops were domesticated. Crops that apparently originated in several different areas are shown in parentheses. Question marks after the name indicate doubt about the location of origin.

Source: J. R. Harlan, "The Plants and Animals that Nourish Man," *Scientific American* 235, no. 3 (1976):88–97. © September 1976 by Scientific American, Inc. All rights reserved.

chestnut, hickory nut, hazelnut, black walnut, persimmon, blueberry, raspberry, blackberry, and cranberry. The sunflower (which has long been an important oil crop in the former Soviet Union and Eastern Europe) originated in the United States, as did hops, the tepary bean, Jerusalem artichoke, and some grasses. Many plants now used as ornamentals, mostly in improved forms, and many of the world's great timber tree species also originated in the United States.

The only food crop to originate in the entire continent of Australia was the macadamia, or Queensland nut.

Methods of Plant Domestication

When ancient peoples started to domesticate plants—that is, propagating and growing plants under their own control—they would have had the choice of vegetative propagation or seed propagation. Both methods have advantages and disadvantages in crop improvement, but because of genetic segregation between generations, the offspring from sexual propagation are often different (not true to type) from parents, and desired traits may not be present in the offspring. For this reason, vegetative propagation is the more immediate and direct approach to selecting and maintaining superior plants.

Vegetative or Asexual Propagation Methods

It is probably no coincidence that some of the oldest cultivated woody plants are the easiest to propagate by vegetative methods such as hardwood cuttings. These plants include the grape, fig, olive, mulberry, pomegranate, and quince.

An observant person in prehistoric days might have noticed among a wild stand of seedling grapevines, for example, one single plant standing out from the others in its heavy production of large, high-quality fruits. From past experience, too, the person might have known that a piece of the grape cane stuck into the ground in early spring would take root and produce a new plant just like the original. The next logical step, of course, would be to break off and plant many cane pieces (cuttings) from this one very desirable grape plant, then to protect them against animals and the competition of weeds. This action would establish a vineyard based on cultivating a superior seedling and disregarding all seedlings with inferior characteristics.

Many of the tree fruit species—fig, almond, quince, apple, pear, cherry, pomegranate, and walnut—are native to areas of the Near East on the southern slopes of the Caucasus Mountains. Today, there are still forests in the area consisting entirely of seedling wild trees of these species, showing great variability in many characteristics. One may also see trees in which the local farmers have grafted the superior wild types onto inferior seedlings, making use of them only as rootstocks. Also, while clearing the brush and forests to make space for grain fields, the farmers have left superior individual wild apple, pear, and cherry seedlings in the fields for harvesting.

Many types of ancient plants brought under domestication by early people could have been—and probably were—easily maintained and increased by unsophisticated vegetative methods. Such plants would have included the potato and sweet potato, propagated by dividing the tubers; banana, bamboo, and ginger, propagated by cutting up rhizomes; pomegranate, quince, fig, olive, and grape, propagated by stem cuttings; filbert, propagated by layering; and pineapple and date, propagated by a form of suckering.

Seed or Sexual Propagation Methods

Domestication of seed-propagated plants, such as the cereal crops, by the ancient agriculturists probably began as the purposeful harvesting of wild grass seeds, some of which were sown to produce the next year's crop. This established two population types for the next harvest—one that was not harvested and reseeded itself, and one that was sown from the harvested seed. In cereals, this procedure immediately starts to separate the "shattering" types (the seed separates from the head) from the "nonshattering" types. Most of the seeds that shatter fall to the ground, while nonshattering seeds are harvested and can be resown. Thus, seed with the desirable nonshattering trait is easily obtained without intentional selection. The mere practice of harvesting sets up a selection pressure for improved forms.

Planting the harvested seeds close together in a cultivated plot kept free of competing weeds automatically selects for the stronger, more vigorous plants. In some species, seedlings derived from the larger seeds—those having a considerable amount of stored foods—are likely to germinate and grow rapidly, thus suppressing seedlings arising from the smaller seeds with low vitality. The harvested crop would, therefore, come from the more vigorous plants producing the desirable large grains, and the plants producing the smaller, inferior grains would be eliminated. Thus, seedling competition sets up an automatic selection pressure for an improved form.

By these procedures, early peoples unconsciously developed superior forms of their cereal crops. Other desirable characteristics arising from selection pressures were loss of seed dormancy, increased flower numbers and larger inflorescences, and a trend toward determinate

rather than indeterminate growth.[2] These superior characteristics in the cultivated races over the wild types can usually be obtained with only small genetic differences between the two.

Plant size, productivity, seed output, and ecological adaptations are often complex quantitative traits and involve many small genetic changes. Their evolution is a slow, continuous process. Often some other characteristics, such as nonshattering seeds, resistance to natural enemies, or novel fruit or seed color and shape might be controlled by one or very few genes. Such traits can evolve rapidly by early humans' observation and by harvesting separately.

EXAMPLES OF IMPROVEMENT IN SOME IMPORTANT CROP PLANTS[3]

The following examples illustrate patterns in plant improvement, beginning with: (1) the harvest of crops from wild plants by primitive humans, followed by (2) selection of superior types from prehistoric eras to the present time, and going on to (3) modern methods of plant breeding that can dramatically increase crop yield and quality by applying genetic principles and gene transfer to developing improved cultivars. It should be realized that the present, improved cultivars of important crops such as wheat, corn, rice, potato, sweet potato, and all fruit crops have been so adapted to conform to human cultural practices that they all now completely depend on our care for their continued existence.

Grains and Vegetable Crops

Wheat (*Triticum aestivum* and *T. turgidum* Durum group)

Wheat is the most widely cultivated plant in the world today, the chief cereal, and is used worldwide for making bread. Present wheats evolved from wild wheatlike grass plants found and cultivated by ancient humans in the Near East region about 7000 BCE. Native wheatlike plants can still be found in this area. Even in early prehistoric times, wheat probably improved naturally by spontaneous hybridizations, by chromosome doubling, and by mutations to increase fertility. Wheat species occur in a series with increasing chromosome numbers

(polyploidy): First is the small primitive diploid einkorn wheat (7 pairs of chromosomes); second is the much larger tetraploid emmer wheat (14 pairs of chromosomes); and third is the hexaploid bread wheats (21 pairs of chromosomes), the ones grown today. Humans probably had no role in this early advancement of wheat by polyploidy. From the Near East, these early wheat forms were taken into ancient Egypt, the Balkans, and central Europe. The Spanish brought wheat to the Americas; eventually, the United States, Canada, and Argentina became the world's largest wheat producers. A chance introduction of "Turkey Red" wheat into central Kansas, by a small group of Mennonite immigrants from Russia in 1873, established the basis for the tremendous hard red winter wheat industry of the central Great Plains area of the United States.

The two major wheat species today are *Triticum aestivum,* used for flour in making breads and pastries, and *T. turgidum* (Durum group), used for products such as macaroni, spaghetti, and noodles. In the twentieth century, plant breeders, through hybridization and selection, have produced perhaps a thousand cultivars of bread wheat alone designed for certain climates, high productivity, special milling and cooking properties, and particularly disease resistance. Wheat cultivars resistant to the devastating stem rust disease must continually be developed to cope with the mutating stem rust pathogen.

Today's plant breeders, too, have developed dwarf forms of wheat, which can produce high seed yields without falling over (lodging) when heavily fertilized. Some of these new dwarf cultivars were developed by Norman E. Borlaug, working at the Rockefeller Foundation's International Maize and Wheat Improvement Center in Mexico. Borlaug was awarded the Nobel Peace Prize in 1970 for his work. Certain of these cultivars have been so successful with fertilizers and irrigation in Pakistan and India that both countries have become exporters rather than importers of wheat.

Wheat plants are self-pollinated, allowing farmers to save their seeds for future planting. F_1 hybrid wheat for increased plant vigor and yields has not been developed to the extent it has with corn, owing to difficulties arising from the wheat's flower structure. Wheat has perfect (bisexual) flowers, making cross pollination difficult, whereas corn has the male and female flowers separate on the same plant and is cross-pollinated easily.

Corn (*Zea mays*)

Corn (or maize) originated in the New World about 5000 to 6000 BCE, but its earliest history is still a mystery. Maize is known only as a domesticated plant.

---

[2]In determinate growth, shoot elongation stops when flowers form on the shoot terminals. In indeterminate growth, elongation continues after flowering.

[3]For a discussion of the genetic terminology used in this section, see Chapter 9.

There is no wild form except, apparently, teosinte, a close relative. The economic life of the ancient American civilizations—the Aztecs, Mayas, and Incas—depended on corn. At the time of Columbus's expeditions to the New World, corn was being grown by Indian tribes from Canada to Chile. Corn probably originated in several places in both Mexico and South America. An early form of corn is still growing in South America (Fig. 10–12).

Several hypotheses have been advanced to explain the origin of corn: (1) it developed from "pod corn," a type in which each individual kernel is enclosed in floral bracts—as in the other cereals; (2) it originated from teosinte, corn's closest relative, by gradual selection under the influence of harvesting by humans; (3) corn, teosinte, and *Tripsacum* (a perennial grass) descended along independent lines directly from a common ancestor; or (4) there is the tripartite theory that (a) cultivated corn originated from pod corn, (b) teosinte is a derivative of a hybrid of corn and *Tripsacum*, (c) the majority of modern corn cultivars are the product of an admixture with teosinte or *Tripsacum*, or both.

Corn, even today, is an extremely variable species, from the color of the grains to the size and shape of the grains and ears (Fig. 10–12). Corn mutates easily, forming new types. The Native Americans in Mexico must have made considerable conscious selections of corn for so many types to be introduced to European agriculture following the New World explorations.

The development of hybrid corn in the 1930s is one of the outstanding achievements of modern agriculture (see Fig. 9–7). High-yielding F_1 corn hybrids were developed for different climatic zones. In 1935, 1 percent of the corn planted in the United States was of the hybrid type. By 1970, almost all corn produced in the United States was of the hybrid type. The planting of hybrid cultivars, along with fertilizers, irrigation, and mechanization, dramatically improved production. Corn is now one of the world's chief food crops for both humans and domesticated animals.

The latest development in corn improvement is bioengineered germplasm, including the Bt® corn for corn borer resistance, and the Roundup Ready® and Liberty Link® corn for herbicide tolerance. Specialty corns—such as white, waxy, hard endosperm food grade, high oil, nutritionally enhanced, high amylase and high extractable starch corn—are being developed. Ethanol production from high-starch corn as a replacement for some petroleum fuels has become a major use of corn.

Figure 10–12

Improvement in corn (maize) from a primitive type (left) still growing on the eastern slope of the Andes mountains in South America, to modern hybrid corn (right). The primitive type corn contains a valuable trait—multiple-aleurone layers in the grain—that is being transferred by plant breeders to modern dent corn. This example illustrates the need for maintaining the germplasm of seemingly worthless plant types. Source: USDA.

Rice (*Oryza sativa*)

Rice is the basic food for more than half the world's population and one of the oldest cultivated crops. It is believed to have originated in southeast Asia about 5,000 years ago, or even earlier, and it spread to Europe and Japan by the second century BCE.

There are about twenty-five species of *Oryza*, but *O. perennis*, widely distributed throughout the tropics, is probably the one from which cultivated rice was developed. Primitive humans most likely collected seeds and cultivated the wild types. During cultivation of rice, mutations, plus hybridization with other *Oryza* species, probably occurred, leading to improved forms with larger grains and nonshattering fruit stalks. Rice was first cultivated in America along the coast of South Carolina about 1685. Four rice experiment stations were established by the US government in the early 1900s. Breeding work at US government rice experiment stations has resulted in many superior cultivars

being introduced to the rice-growing areas of the United States.

Much of the rice harvested in Asia was from old, unproductive types grown under primitive conditions. The Ford and Rockefeller Foundations established the International Rice Research Institution (IRRI) in the Philippines in 1962, bringing together scientists from eastern and western nations for the purpose of improving rice culture. This group of researchers developed new early, high-yielding dwarf cultivars by hybridization that did not lodge when heavily fertilized and allowed up to three crops to be grown in the same field each year. These cultivars dramatically increased yields, and some are highly resistant to insects and disease pathogens native to the Far East, South America, and Africa. Hybrid rice is now commonly grown in China and other parts of Southeast Asia.

A genetically engineered rice has been developed that produces beta-carotene in its kernels, giving the rice a yellow color. Adding beta-carotene, a precursor for vitamin A (which is essential for human health), to rice increases the nutritional value of the rice significantly. It is predicted that the modified rice would help prevent blindness in millions of children in developing countries where rice is a dietary staple.

Soybean (*Glycine max*)

The soybean has risen spectacularly to prominence in the United States with an increase in production from 135,000 metric tons (5 million bushels) in 1925 to about 75 million tons (3 billion bushels) in 2008. This great increase is due, in part, to the availability of more productive, disease-resistant cultivars. From about 1910 to 1950, large numbers of new strains and seed lots of soybeans were introduced into the United States from the Orient, its native home, largely by United States Department of Agriculture (USDA) plant explorers. From these diverse sources of germplasm, hybridization programs developed many superior cultivars. Much of this work was done at the USDA Regional Soybean Laboratory at Urbana, Illinois, in cooperation with various midwestern agricultural experiment stations. These cultivars give high yields, proper bean maturity for the particular area, strong erect plants that hold their seeds until harvest, and high disease resistance and bean quality. The soybean is particularly well adapted to the United States' Midwest corn belt and the southeastern states, which together account for about 40 percent of the world's soybean production (2008 data).

Sugar Beet (*Beta vulgaris*)

The modern sugar beet is a plant developed entirely by human efforts in plant breeding. It is our only major food crop that was not grown in some primitive form in ancient times. It was developed only about 250 years ago in Europe as a source of sugar to compete with the then very expensive cane sugar. Sugar beets and sugar cane contain the same kind of sugar. Breeding and selection increased the sugar content in the root from about 2 percent to about 16 to 20 percent.

In the early part of the nineteenth century, Napoleon encouraged the development and production of the sugar beet industry in France to free that country from the British monopoly on cane sugar. By the end of the nineteenth century, sugar beets were being grown in North America, and they have now become an important temperate zone crop in many areas of the United States and southern Canada, and even more so in other parts of the world. All production phases are now completely mechanized, permitting the sugar beet to compete favorably with the tropical sugar cane plant. To maintain profitable production, however, plant breeders have had to continue to develop sugar beet cultivars resistant to virus and fungus diseases.

Potato (*Solanum tuberosum*)

Potatoes are one of the big four crops that feed the world's population. Wild potato species are widespread in South America, particularly in the Andes Mountain region. Potatoes were probably cultivated by primitive peoples in this area more than 4,000 years ago, but they became a more productive crop over the years with selection of superior types. The potato was cultivated over the length of the Andes Mountain area at the time the Spanish explorers arrived. About 1575, they carried it back to the European continent, where today 90 percent of the world's potato crop is grown.

The potato plant produces pink, white, or blue flowers that develop into small green berries containing seeds. The seeds, when planted, produce new types of potato plants much different from the parent plant and from each other in many respects. In the early days of potato culture, the South American Indians undoubtedly selected superior plant types resulting from natural crosses. Once one single such superior plant was obtained, it could be perpetuated and increased in great numbers as a clone by tuber division. Some of these superior selections propagated by vegetative methods over the years became commercially successful cultivars. It was noted, however, that certain of these cultivars

would degenerate after many generations of such asexual propagation and yield only weak unproductive plants. It was observed, too, that sowing seeds from such plants gave progeny plants with changed characteristics, including renewed vigor and productivity. It is now known that these clonal cultivars had become infected with viruses that passed along through the tubers to the new plants generation after generation. Such viruses did not pass through the seed to the new seedling plant, so that its growth was no longer inhibited by the virus.

In modern commercial potato growing however, pieces of tubers called seed tubers or seed potatoes are planted to maintain clonal uniformity. Seed potatoes are produced under carefully observed conditions in regions where any viruses can readily be detected and the infected stock discarded. Growers who plant only "certified" stock can be reasonably sure that their fields will not be infected with viruses or other pathogens and will be true to type. In the United States, certified "seed" potatoes are produced by commercial growers in the northern states and are strictly inspected by state government agencies. This is also true for many other, but not all, countries.

Many potato cultivars now being grown have been developed by plant breeders who have introduced superior characteristics into an existing cultivar. An example is resistance to *Phytophthora infestans* (late blight)—from *Solanum demissum,* a wild potato species, obtained from collections in Mexico. Many cultivars are being grown today for different purposes and for different climatic zones. New cultivars are constantly being introduced by plant breeders and older ones are being discarded.

Tomato (*Lycopersicon esculentum*)

The cultivated tomato originated from wild forms in the Peru–Ecuador–Bolivia area of the Andes Mountains in South America. Prehistoric Indians carried it to Central America and Mexico. Early explorers introduced the tomato to Europe about 1550, and it was brought back west, to the Carolinas in North America, about 1710. Thomas Jefferson grew tomatoes on his plantation around 1780. In those days, most people considered the tomato poisonous, but in the United States it started gaining acceptance as a food plant about 1825. The early Indians undoubtedly improved the tomato by planting seeds taken only from the best fruits on the most productive plants. This selection process continued in the early days of tomato culture in Europe and the United States. Early cultivars introduced by US seed companies were 'Stone' and 'Globe' in 1870. In the early 1900s, the USDA and the state experiment

stations began breeding tomato cultivars to include specific characteristics. 'Marglobe' was introduced by the USDA in 1925 and 'Rutgers' by New Jersey in 1934. Some tomato cultivars, more recently introduced, carry resistance to fusarium wilt, verticillium wilt, and nematodes. Cultivars developed especially for machine harvesting have firm-fleshed fruits that all ripen at one time. Vigorous, highly productive F_1 hybrids, marketed both as seeds and as bedding plants, are recent developments by seed companies.

Fruit Crops

The major fruit crops are all heterozygous. They do not "come true" when propagated by seed, so vegetative propagation must be used to maintain an improved seedling selection. In ancient times, most kinds of fruit—other than those very easily propagated vegetatively, such as bananas, grapes, figs, pomegranates, and olives—were probably seed propagated and the variability in offspring accepted by farmers. Later, however, as more sophisticated methods of vegetative propagation were developed, such as budding and grafting, it would have been found that certain superior individual fruit plants could be maintained and increased by these methods. Nevertheless, considerable seed propagation of fruit species was undoubtedly practiced in the early days of fruit growing.

Apple (*Malus pumila*)

The early US colonists planted many seedling apple trees probably because it was easier to bring seeds taken from their favorite apple trees in their native homes rather than material for grafting. As a result, they continued to increase their apple orchards by planting seedlings. In these early days, much of the apple crop was preserved as cider, for which fruit from seedling trees was quite satisfactory. Certain individual seedling trees, no doubt, were much superior to the others and formed the starting point for vegetative propagation and the origin of the many hundreds of apple cultivars grown in the United States up to the early part of the twentieth century. These numbers dwindled, however, until the 1980s, when only fifteen cultivars accounted for over 90 percent of the apples produced in the United States. Important cultivars are 'Delicious,' 'Golden Delicious,' 'McIntosh,' 'Rome Beauty,' 'Jonathan,' 'Winesap,' 'York,' 'Stayman,' 'Cortland,' and 'Granny Smith.' Although apple breeding programs have been conducted by the USDA and by some state agricultural experiment stations, all the major apple cultivars now being grown originated as chance seedlings many

years ago. For example, the 'McIntosh' apple was found growing as a seedling tree near Dundela, Ontario, Canada by John McIntosh in 1796. The 'Delicious' apple started as a single chance seedling near Peru, Iowa, about 1870. The 'Golden Delicious' also originated as a chance seedling in Cass County, West Virginia, about 1910. Most of the preceding cultivars remain popular, but more have been added, among them 'Jonagold,' 'Fuji,' 'Gala,' and 'Braeburn.'

Pear (*Pyrus communis*)

In pears, too, cultivars originating as chance seedlings have dominated the markets. The 'Bartlett' (Williams Bon Chretien) originated in England as a chance seedling in 1796 and has been the world's leading pear cultivar ever since. Other leading pear cultivars in the United States, such as 'Beurre d'Anjou' and 'Beurre Bosc,' originated in Belgium as open-pollinated seedlings of cultivars then grown there. No pear cultivar from a controlled breeding program in the United States has become commercially important, although European pear breeders have developed several superior cultivars that produce well in France, Italy, and Belgium.

Peach and Nectarine (*Prunis persica*)

In contrast to the apples and pears, the important peach and nectarine cultivars grown today are the products of public and private plant breeding programs. Plant breeders have provided the consuming public with truly outstanding peaches and nectarines, in contrast to the small-fruited, nonproductive cultivars of earlier days that got their start as chance seedlings.

Strawberry (*Fragaria* × *Ananassa*)

Today's garden strawberry first originated in France about 1720 as a natural hybrid between two native American *Fragaria* species. From this and subsequent hybridizations, a number of cultivar selections were made and maintained vegetatively by runners in the early days of strawberry culture in Europe. However, many of these early cultivars were susceptible to viruses, verticillium wilt, and other diseases and were low in productivity, so that around 1945 the entire US strawberry industry was falling into a precarious position.

Since World War II, a parade of new strawberry cultivars has been replacing older ones, coming from USDA and several state strawberry breeding programs and from similar programs in other countries. New cultivars have been developed for specific climatic regions and for characteristics such as adaptability of fruits to be used for freezing or for fresh shipping, resistance of plants to viruses and fungi and to winter cold, fruit appearance and flavor, and extended fruit-producing period.

PLANT IMPROVEMENT PROGRAMS

From the time of Neolithic humans up to the early 1900s, sexual plant improvement mostly consisted of selecting seeds from those individual plants in a mixed population that had the desired characteristics. The seed was planted and from that population seeds were again taken from the most desirable plants, and so on. While improvements resulted, this method offered no way to transfer desirable characteristics from one line to another.

Evolution and Darwinism

Charles Darwin, the great English naturalist working in the middle of the nineteenth century, provided scientific explanations of how evolution occurred, published in his monumental work (1859), *On the Origin of Species by Means of Natural Selection, or the Preservation of Favoured Races in the Struggle for Life.* Darwin's concept of evolution, generally accepted today, is:

1. Variation exists in an initial population of plants or animals.
2. Environmental stresses place certain individuals at an advantage.
3. Because certain individuals survive and reproduce more successfully, they leave more offspring, which then carry the same genetic traits.
4. The abundance of the advantageous traits increases in this way in every generation, but variation still persists.

In Darwin's time, nothing was scientifically known about heredity. Darwin, in his proposals, had great difficulty in accounting for a sufficient supply of variation. But the discoveries by the Austrian monk Gregor Mendel, in the 1860s, demonstrated the genetic mode of plant inheritance and developed the foundation for the science of genetics. His published works lay unappreciated in an obscure journal for thirty-four years, but in 1900 his papers were discovered independently by three European botanists. The discoveries of Mendel provided exactly the mechanisms needed by Darwin, and removed his difficulties in explaining variation. The integration of Darwinian selection and Mendelian genetics are now generally accepted as the proper explanation of evolution.

Many crop scientists have discovered patterns of evolution in various crop genera, primarily from two points of view: (1) relationships between crops and their related wild and weedy species, and (2) adaptive variation in geographical races. Plant breeders have made fascinating and useful applications of this knowledge in terms of collecting and utilizing many germplasm resources. In fact, all of the processes of evolution in native plants have direct analogues in breeding methods. For example, induced mutations and hybridization follow natural sources of novel variation; selection and cycles of recombination in hybrid materials provide the genetic changes in both natural and breeders' populations.

Based on Mendel's work and utilizing the expertise of trained geneticists, modern breeding programs have produced an array of new cultivars for many crops. These have been bred for characteristics such as resistance to disease, insects, and cold, and for productivity, flavor, and nutritive value. Today's breeding programs have been called directed and accelerated evolution.

Since the start of the twentieth century, public and private plant breeding programs have had a tremendous beneficial impact on our food supply and range of ornamental plants. These programs have produced a great many new superior cultivars for almost all cereal, vegetable, forage, fruit, and ornamental crops. The USDA and most state agricultural experiment stations in the United States maintain such programs. Most other countries also have plant breeding programs, often specializing in certain crops. Private plant breeding has mostly been done by seed companies producing new agronomic, vegetable, and flower cultivars.

Innovations by plant breeders include the development of F_1 hybrid corn and new vegetable and flower cultivars. Most of the new vegetable and flower lines are far superior to previous cultivars in vigor and in insect and disease resistance; the new vegetable cultivars are also superior in flavor, appearance, and productivity.

By developing plants that show strong resistance to insects and disease, plant breeders are reducing the need for insecticides and fungicides. This, in the long run, would be the best method of pest control. For example, potato cultivars have been developed that are resistant to the late blight disease (*Phytophthora infestans*), which was responsible for the nineteenth-century Irish potato famine. Other potato cultivars have been developed that are resistant to the golden nematode, permitting potatoes to be produced in soils infested by these worms without expensive soil fumigation. Wheat breeders must continually develop new wheat cultivars resistant to stem rust (*Puccinia graminis tritici*) because this fungus continually changes to attack formerly resistant cultivars.

Plant breeders have a useful procedure for obtaining improved plant forms by spontaneous or induced mutations resulting from chromosome or gene changes. These changes can be induced by chemical treatment with colchicine or by irradiation with gamma or X-rays.

In one instance, plant breeders have gone beyond just improving native plants. They have created a new humanmade cereal, triticale, by hybridizing the ancient grains, wheat and rye. The name *triticale* derives from the generic names of these two grains: *Tritium* and *Secale*. The hybrid combines the high yield and protein content of wheat with the winter hardiness of rye. The triticale plant is disease-resistant and thrives in some unfavorable soils and climates.

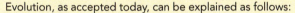

THE MECHANISM OF EVOLUTION

Evolution, as accepted today, can be explained as follows:

1. Genes, in the chromosomes, are largely responsible for the structure, metabolism, and development of plants and animals.

2. The complement of genes does not remain constant because mutations occur that modify the metabolism and structure of the individuals that contain them.

3. Mutations, which may cause considerable change, may kill or greatly weaken the plant because they can upset the equilibriums that exist between the plant and its environment.

4. Hybrids differ from their parents because of the resulting new combinations of genes.

5. If the variants, as developing from mutations or by hybridization, are better adapted to the existing environment than the parent plants, then the parent plants may be replaced by the new forms.

6. As the earth's surface and climate change, or as plants' habitat may be changed in other ways, those plants best adapted to the new environment replace those poorly adapted.

7. Evolution thus results from slow changes in the environment, variations occurring in plants and animals, and adjustments taking place between the changes in the environment and changes in the living organisms.

SEARCHING FOR AND MAINTAINING NEW GERMPLASM

There is no assurance that we are at present cultivating all the useful food and ornamental plants in existence on earth, or that all the germplasm containing useful genes has been found. **Germplasm** is the protoplasm of the sexual reproductive cells containing the units of heredity (chromosomes and genes). Plant explorers have long roamed the world and they continue their searches. They have found many plants that have subsequently made a major impact on the world's agriculture, often in different parts of the world from the plants' native regions. Accessible plants have always been moved about over the world by explorers on land and sea, armies, immigrants, and travelers. Plants moved into a new region often perform much better than they did in their original home. For example, the coffee plant (*Coffea arabica*) is native to the Ethiopian area of eastern Africa. It never developed into much of a crop there, but when moved to Brazil and Colombia in South America, the coffee tree prospered so well that these countries now produce the bulk of the world's coffee supply.

Several prominent plant explorers have contributed much to the wealth of available plant materials. Many of the early plant collecting trips were less than successful because the plants did not survive the long trip back. A London physician who was an amateur horticulturist, Nathanial Ward, invented the Wardian case early in the 1800s. The case was a small glass-enclosed box containing soil in the bottom. Plants kept in the Wardian case could survive long sea voyages, permitting the importation of species never before received alive. Large, magnificent, ornate glasshouses were built in England about this time to house the many tropical and subtropical plants brought back by the plant hunters.

Many plant explorers from the United States brought back plants that would later be the foundation for several crops grown in the United States. Colonel Agoston Haraszthy (wine grapes), N. E. Hansen (cold-resistant fruit and cereal plants), Mark Carleton (hard red winter wheat), David Fairchild, Frank Meyer, and others brought back many different species that have become the genetic basis for many of our edible and ornamental crops.

Plant-collecting trips to the native homes of certain desirable plant types are still being made by plant explorers. Plant explorers may be looking for an entirely new plant species to serve as a crop in a particular climatic region of their own country. Safflower came to the northern Great Plains of the United States this way, and it has proven most profitable. Plant explorers may also be searching for new germplasm for existing crops. Closely related types can be used by plant breeders to introduce, for example, genes for insect or disease resistance or improved vigor or quality into cultivars already being grown.

International Conventions Relating to Plant Genetic Resources, Conservation, and Utilization

Currently, two international conventions relating to the promotion of conservation and protection of biological diversity are in effect. The Convention on International Trade in Endangered Species of Wild Fauna and Flora (CITES), which came into operation in 1975, helps to ensure that international trade in specimens of wild animals and plants does not threaten their survival. CITES has about 28,500 species of plants listed under protection, and this list is growing as more and more plant habitats are damaged. For example, many of the orchids are on this list. Further information on CITES can be found at http://www.cites.org/. The second is the Convention on Biological Diversity (CBD) which is an outcome of the 1992 United Nations Environment Programme's Earth Summit in Rio de Janeiro, Brazil, and became effective in 1993. Its objectives are

> the conservation of biological diversity, the sustainable use of its components and the fair and equitable sharing of the benefits arising out of the utilization of genetic resources, including appropriate access to genetic resources and appropriate transfer of relevant technologies, taking into account all rights over those resources and to technologies, and appropriate funding.

This also means that plant germplasm becomes a country's natural resource, and the use of a germplasm by another country now requires prior official consents, collection permits, and specific material transfer agreements, including benefit sharing when a germplasm is commercialized. Annually, an estimated $500 to $800 billion of genetic resources derived products is traded globally. In the past, plant germplasm was shared and exploited globally as common properties of humankind. Thus, CBD can be seen as a response of germplasm-rich countries to the intellectual property rights protection systems in place in the world to protect developed germplasm. The International Treaty on Plant Genetic Resources for Food and Agriculture (ITPGRFA), a Food and Agriculture Organization (FAO) of the United Nations initiative, is the first endeavor to create a common multilateral agreement to ensure the conservation of sixty-four food and feed crops in harmony with CBD for sustainable agriculture and food security. It covers

approaches in germplasm exploration and conservation, measures to induce sustainable germplasm uses, recommendations on international cooperation, measures to uphold farmers' rights, sovereign rights of contracting states, a standard material transfer agreement, benefit sharing, and so on. To implement ITPGRFA, a $260 million endowment—the Global Crop Diversity Trust—was instituted to provide the necessary funds for protecting and conserving the most threatened and valuable collections of crop diversity. Detailed information on CBD is available at http://www.cbd.int/

Moving plant materials about the world can also introduce devastating insect or disease pests into a country where they have never before appeared. For example, the chestnut blight fungus (*Endothia parasitica*) was inadvertently introduced on imported plant material to the New York area from the Orient in the late 1800s. By about 1935, this fungus had practically eliminated all varieties of the beautiful native American chestnut trees (*Castanea dentata*) from the eastern United States. To guard against the introduction of such pests, most countries have set up elaborate inspection, fumigation, and quarantine procedures. The 1997 International Plant Protection Convention (https://www.ippc.int/), an international treaty under the auspice of FAO and signed by 127 governments in 2004, helps to harmonize and standardize the rules and procedures, and agreements in plant quarantine to prevent the spread and introduction of pests of plants and plant products, and to promote appropriate measures for their control in the world. Where appropriate, storage places, packaging, conveyances, containers, soil, and any other organism, object, or material capable of harboring or spreading plant pests are included, particularly where international transportation is involved. As a result, regional plant protection organizations, for example, the North American Plant Protection Organizations (NAPPO), were established to promote the development and use of relevant international standards for phytosanitary measures, and to encourage interregional cooperation in promoting harmonized phytosanitary measures for controlling pests and in preventing their spread and/or introduction.

Plant material can be introduced as seeds, bulbs or corms, rooted cuttings, or as scions or budwood for grafting or budding onto related growing plants. Seeds are the easiest to ship and pose the least danger of carrying pathogens. Rooted plant parts with soil particles around the roots are particularly suspect because the soil may contain nematodes or soil-borne diseases.

Vegetative plant material coming into the United States is usually fumigated, then held under post-entry quarantine for as long as two years before distribution to nurseries is permitted.

Shipment of plant material is now easy and highly successful because of specialized packing materials, refrigeration, and frequent worldwide air flights. This contrasts to earlier days when only slow ship transport, often through hot tropical seas, was possible.

Preservation of Desirable Germplasm

A concerted worldwide effort is needed to ensure the survival of the earth's endangered plant species and thus preserve genetic diversity. International germplasm networks operate under the auspices of the Consultative Group on International Agricultural Research (CGIAR) and the World Conservation Union (IUCN). These networks interact with, and are supplemented by, many national seed banks and agricultural centers. In 1998, FAO estimated that 6.1 million accessions of primary food, forage, and industrial crops had been collected and conserved globally. Such gene pools are needed for developing improved crops in the future, for introducing beneficial genes into existing crops from close relatives, for maintaining attractive plants and trees valued for their aesthetic purposes, and for keeping plants intact as part of ecosystems where their presence is necessary to the survival of other plant and animal species. The genetic diversity of plants, as well as animals and microbes, is of fundamental importance to our survival on earth. Food and other agricultural crops are derived from the genetic diversity of natural plant populations.

To help prevent eradication of many plant species, the US Congress in 1973 passed the Endangered Species Act, which directed the Smithsonian Institution to prepare a list of endangered plant species and to recommend measures for saving them. As many as one in ten plant species is now extinct or endangered because of the encroachment of agricultural operations, the removal of rare plants by plant collectors, and the general destruction of vegetation from various causes. Such endangered germplasm can be saved and stored for future use as seeds or as living plants in special protected locations.

The CGIAR international research centers, such as the International Rice Research Institute (IRRI) in the Philippines and the International Maize and Wheat Improvement Center (CIMMYT) in Mexico, have assembled one of the most comprehensive collections of our major food and forage crops (Table 10–3). This endeavor

Table 10–3
COLLECTIONS OF FOOD, FORAGE, AND FORESTRY CROPS HELD IN TRUST FOR THE WORLD COMMUNITY AT THE ELEVEN CGIAR CENTERS IN 2002

Center	Crop(s)	Number of Accessions
International Center for Tropical Agriculture (CIAT) Cali, Colombia	Cassava	15,728
	Forages	18,138
	Bean	31,718
International Maize and Wheat Improvement Center (CIMMYT) Mexico	Maize	20,411
	Wheat	95,113
International Potato Center (CIP) Lima, Peru	Andean roots and tubers	1,112
	Sweet potato	6,413
	Potato	5,057
International Center for Agriculture in the Dry Areas (ICARDA) Aleppo, Syria	Barley	24,218
	Chickpea	9,116
	Faba bean	9,074
	Wheat	30,270
	Forages	24,581
	Lentil	7,827
International Crops Research Institute for the Semi-Arid Tropics (ICRISAT) Patancheru, India	Chickpea	16,961
	Groundnut	14,357
	Pearl millet	21,250
	Pigeonpea	12,698
	Sorghum	35,780
	Minor millets	9,050
International Institute for Tropical Agriculture (IITA) Ibadan, Nigeria	Bambara groundnut	2,029
	Cassava	2,158
	Cowpea	15,001
	Soybean	1,909
	Wild *Vigna*	1,634
	Yam	2,878
International Livestock Research Institute (ILRI) Nairobi, Kenya	Forages	11,537
International Plant Genetic Resources Institute (IPGRI) Maccarese, Italy	*Musa*	931
International Rice Research Institute (IRRI) Los Banos, Philippines	Rice	80,617
West Africa Rice Development Association (WARDA) Bouaké, Cote d'Ivoire	Rice	14,917
World Agroforesty Center Nairobi, Kenya	Sesbania	25
TOTAL		532,508

Source: CGIAR, "Research & Impact: Genebanks & Databases—Accessions," http://www.cgiar.org/impact/accessions.html

is coordinated by its specialized plant germplasm conservation center, the International Plant Genetic Resources Institute (IPGRI). IPGRI also provides technical leadership to national gene bank programs. The IUCN has the mission "to influence, encourage and assist societies throughout the world to conserve the integrity and diversity of nature and to ensure that any use of natural resources is equitable and ecologically sustainable." The focus of IUCN is the conservation of the whole ecosystem and the total biodiversity. It monitors the state of the world's species in the IUCN Red List of threatened and endangered species (http://www.redlist.org/). This list is adopted globally by national governments, nongovernmental organizations, and scientific institutions in their biodiversity conservation efforts. An estimated 60,000 to 100,000 plant species, representing about one-third of the world's plants, are currently threatened or facing extinction in their native habitats. In 2002, the Global Strategy for Plant Conservation (http://www.cbd.int/gspc/) was adopted by CBD and in 2004, the Global

Partnership for Plant Conservation (http://www.bgci.org.uk/files/2/786/GlobalPartnershipstatement.doc), consisting of international and national agencies and organizations active in plant conservation, was established to support the worldwide implementation of the strategy. The Botanic Gardens Conservation International (http://www.bgci.org/), founded in 1987, is one of such organizations that is taking the forefront in this attempt. Over 500 member institutions in 112 countries are working together to implement a worldwide botanic gardens conservation strategy for plant conservation. For example, the Royal Botanic Gardens, Kew, England, has established the collaborative Millennium Seed Bank Project to collect and conserve 24,000 plant species from around the world and protect them against extinction.

In the United States, the National Plant Germplasm System (NPGS) is a coordinated network of four main plant introduction stations in four geographic regions of the country, a long-term storage gene bank in Fort Collins, Colorado, and about twenty-three repositories and active collection sites. The functions of the repositories and active sites are to collect, introduce, maintain, characterize, evaluate, catalog, and distribute plant germplasm. Financial support comes from the USDA and from state agricultural experiment stations as well as from commercial plant breeding and seed trade organizations. The general mission of the system is to provide plant scientists with the germplasm needed to carry out their work, for example, in breeding new cultivars resistant to certain insects, diseases, smog, or high soil salinity. In 2002, more than 450,000 accessions are in conservation (Table 10–4).

The activities of the National Plant Germplasm System are:

1. Introduction of plant materials into useful scientific channels is done by planned foreign and domestic exploration trips, by exchanges with foreign agencies, or by traveling scientists. There may also be useful domestic germplasm that should be maintained—for example, mutations, species hybrids, or germplasm resistant to a certain insect or disease—and may be valuable for future crop development. These, along with introduced foreign material, are eligible to enter the National Germplasm System.

2. Maintenance of this potentially valuable germplasm for future research programs is the responsibility of the regional and interregional plant introduction stations, the National Center for Genetic Resources Preservation, and the curators of collections of specific crops.

3. Characterization and evaluation of the plant genetic resources is done by initial screening and subsequent tests in the field, greenhouse, and laboratory by cooperating state, federal, and private scientists.

4. Distribution of plant germplasm is made free of charge to all qualified scientists and institutions requesting it, in sufficient amounts to enable them to initiate their research program.

The decentralized system is held together by a common web-based database management system, the Germplasm Resources Information Network (GRIN), which helps NPGS curators to handle both the plant and store inventory data. Another notable component of NPGS is about forty crop germplasm committees, which are made up of crop specific experts for each crop from USDA, universities, public institutions, and industry to advise on policy and coordinating activities to meet the immediate and long-term national goals of agriculture in the United States.

BROADENING THE BASE OF AGRICULTURAL PRODUCTION

The world's peoples are largely fed today by only about twenty crops. Reliance on so few crops could lead to a catastrophic famine if but a few of them were obliterated by insect or disease attacks or by climatic changes. The ravage of U.S. corn plantings by the corn blight disease in 1970 is an example of such a possibility.

In an attempt to broaden the base of agricultural plants in the tropics and to promote interest in neglected but seemingly useful tropical plants with economic potential, the U.S. National Academy of Science promoted a compilation of plants nominated by plant scientists around the world and published an account of thirty-six plants selected from the 400 proposed. Each plant was described along with its special requirements, research needs, selected readings, research contacts, and germplasm sources.

All thirty-six plants were thought to have considerable potential, but most have not been cultivated out of their own limited region of origin. Among the cereals, for example, the report cited an almost completely neglected grain species in the genus *Amaranthus*, native to Central America, that has very high levels of protein and the essential amino acid lysine, which is usually deficient in plant proteins. Among the vegetables studied, the wax gourd (*Benicasa hispida*) gives three crops a year of a large melonlike fruit that can be stored for twelve months without refrigeration. The mangosteen

Table 10–4

THE NUMBER OF ACCESSIONS MAINTAINED AT THE NPGS LOCATIONS IN 2002

Site	Accession	Country	Genus	Species
Barley Genetic Stocks Center	3,044	3	1	1
Clover Collection	246	30	1	118
Cotton Collection	9,308	124	1	41
Desert Legume Program	2,585	56	198	1279
Maize Genetic Stock Center	4,710	2	1	1
National Arboretum	1,909	51	257	861
National Arctic Plant Genetic Resources Unit	493	31	50	145
National Arid Land Plant Genetic Resources Unit	961	32	12	124
National Center for Genetic Resources Preservation	23,299	106	198	493
National Small Grains Collection	126,563	170	15	148
National Germplasm Repository—Brownwood	881	3	2	23
National Germplasm Repository—Corvallis	11,687	92	56	757
National Germplasm Repository—Davis	5,105	79	19	202
National Germplasm Repository—Geneva	5,136	58	6	91
National Germplasm Repository—Hilo	675	41	22	76
National Germplasm Repository—Mayaguez	560	40	137	229
National Germplasm Repository—Miami	4,606	90	213	527
National Germplasm Repository—Riverside	1,167	30	38	152
North Central Regional PI Station	47,032	177	319	1767
Northeast Regional PI Station	11,730	126	32	196
Ornamental Plant Germplasm Center	967	58	62	287
Pea Genetic Stock Collection	501	3	1	2
Plant Germplasm Quarantine Office	4,827	59	19	76
Potato Germplasm Introduction Station	5,503	38	1	168
Southern Regional PI Station	82,579	184	246	1433
Soybean Collection	20,415	91	1	16
Tobacco Collection	2,106	67	1	65
Tomato Genetic Stock Center	3,287	18	2	22
Western Regional PI Station	69,946	162	368	2423
Wheat Genetic Stocks Center	334	1	1	1
Total	452,162	N.A.		

Note: The twenty-six NPGS germplasm stations, centers, and repositories listed on the NPGS website (http://www.ars-grin.gov/npgs/) are as follows:

1. Barley Genetic Stock Center (GSHO), Aberdeen, Idaho—http://www.ars-grin.gov/ars/PacWest/Aberdeen/hang.html
2. C. M. Rick Tomato Genetics Resource Center. Davis, California—http://tgrc.ucdavis.edu/
3. Desert Legume Program. Tucson, Arizona—http://ag.arizona.edu/bta/bta20.html
4. Maize Genetics Cooperation—Stock Center (GSZE), Urbana, Illinois—http://w3.aces.uiuc.edu/maize-coop/
5. G. A. Marx Pea Genetic Stock Center (GSPI), Pullman, Washington—http://www.ars-grin.gov/ars/PacWest/Pullman/GenStock/pea/MyHome.html
6. National Clonal Germplasm Repository (COR), Corvallis, Oregon—http://www.ars-grin.gov/ars/PacWest/Corvallis/ncgr/ncgr.html
7. National Clonal Germplasm Repository for Citrus and Dates, Riverside, California—http://www.ars.usda.gov/main/site_main.htm?modecode=53-10-30-00
8. National Clonal Germplasm Respository for Tree Fruit/Nut Crops and Grapes (DAV), Davis California—http://www.ars-grin.gov/ars/PacWest/Davis/
9. National Germplasm Resources Laboratory (NGRL), Beltsville, Maryland—http://www.barc.usda.gov/psi/ngrl/ngrl.html
10. National Center for Genetic Resources Preservation (NSSL), Fort Collins, Colorado—http://www.ars-grin.gov/ars/NoPlains/FtCollins/nsslmain.html
11. National Small Grains Collection (NSGC), Aberdeen, Idaho—http://www.ars-grin.gov/ars/PacWest/Aberdeen/nsgc.html
12. National Temperate Forage Legume Genetic Resources Unit, Prosser, Washington—http://www.forage.prosser.wsu.edu/
13. North Central Regional Plant Introduction Station (NC7), Ames, Iowa—http://www.ars-grin.gov/ars/MidWest/Ames/
14. Ornamental Plant Germplasm Center (OPGC), Columbus, Ohio—http://opgc.osu.edu
15. Pecan Breeding & Genetics, Brownwood and Somerville, Texas—http://extension-horticulture.tamu.edu/carya
16. Plant Genetic Resources Conservation Unit (S9), Griffin, Georgia—http://www.ars-grin.gov/ars/SoAtlantic/Griffin/pgrcu/
17. Plant Genetic Resources Unit (NE9), Geneva, New York—http://www.ars-grin.gov/ars/NoAtlantic/Geneva/
18. Plant Germplasm Quarantine Office (PGQO), Beltsville, Maryland—http://www.barc.usda.gov/psi/fl/pgqo.html
19. Soybean/Maize Germplasm, Pathology, and Genetics Research Unit, Urbana, Illinois—http://www.life.uiuc.edu/plantbio/ars/ppgru.html
20. Subtropical Horticulture Research Station (MIA), Miami, Florida—http://ars-grin.gov/ars/SoAtlantic/Miami/homeshrs.html
21. Tropical Agriculture Research Station, Mayagüez, Puerto Rico—http://www.ars-grin.gov/ars/SoAtlantic/Mayaguez/mayaguez.html
22. Tropical Plant Genetic Resource Management Unit (HILO), Hilo, Hawaii—http://pbarc.ars.usda.gov/pages/research/tpgrmu/germplasm.shtml
23. United States Potato Genebank—NRSP-6, Sturgeon Bay, Wisconsin—http://www.ars-grin.gov/ars/MidWest/NR6/
24. Western Regional Plant Introduction Station (W6), Pullman, Washington—http://www.ars-grin.gov/ars/PacWest/Pullman/
25. Wheat Genetic Stock Center (GSTR), Aberdeen, Idaho—http://www.ars-grin.gov/ars/PacWest/Aberdeen/hang.html
26. Woody Landscape Plant Germplasm Repository, Washington, D.C.—http://www.usna.usda.gov/Research/wlpgr.html

(*Garcinia mangostana*) is perhaps the world's best-tasting fruit, but it is little known outside the very humid tropics of southeast Asia.

Since 1990, crop scientists have met regularly in the United States to share information on new crops and the potential for new crop development. These scientists are looking for solutions to global problems such as hunger, malnutrition, deforestation, desertization, and agricultural sustainability. A series of symposia through these years has produced several volumes of accumulated information on new crops, with the aim of broadening crop diversity in agriculture. These volumes include *Advances in New Crops* (1990), published by Timber Press; *New Crops* (1993), published by John Wiley & Sons; and *Progress in New Crops* (1996), *Perspectives in New Crops and New Uses* (1999), and *Trends in New Crops and New Uses* (2002), published by ASHS Press.

Other crops, native to the Americas, which have undeveloped but great potential are jojoba (*Simmondsia chinensis*), guayule (*Parthenium argentatum*), groundnut (*Apios americana*), and the tepary bean (*Phaseolus acutifolius*), as well as *Cuphea* species, whose seeds contain valuable oils. In recent years, an ANNONACEAE, pawpaw (*Asimina triloba*), the largest tree fruit native to the United States, is gaining interest among researchers and growers. It is the only temperate annona in the world and has tremendous potential to become an important nutritious fruit in North America and other temperate regions of the world. The leguminous tree *Leucaena leucocephala,* native to Central America, already has its own success story. It is fast growing, with high-protein leaves, and is used in India for feeding dairy cattle. In Africa, the leaves are used as a much-needed fertilizer, while in Sumatra and the Philippines, it is used as a commercial regenerative timber source.

Numerous exotic tropical fruits are found in America's supermarkets, many being grown in southern Florida. Others arrive by plane from various tropical regions.

SUMMARY AND REVIEW

Plants are classified in many different ways. They may be classified by taxonomy (common physical characteristics), climate (temperate, tropical, desert, etc.), season (warm, cool), temperature or light requirements, edible or usable parts, or other criteria.

Plants usually have two different names: common and scientific. The common name is usually in the language of the region where the plant is growing. The same plant can have many different common names. Likewise, different plants can have the same common name. To avoid confusion, every species of plant is now given a scientific name based on its taxonomic ranking. Taxonomic ranking goes from general to very specific common characteristics. The most general category is Domain. The most specific is species. The scientific name of a species is composed of the genus and specific epithet. Assigning a plant a scientific name is a formal process that ensures each species has a name that is distinct from any other species. Most scientific names are in Latin, although Greek and Arabic and other words are sometimes used.

The use of some species of plants for agricultural purposes (domestication) goes back many thousands of years. Although theories on how species came to be domesticated vary, in general, everyone agrees that certain regions gave rise to crops that developed from plants native to those regions. Some crops originated in more than one region. Few major crops originated in the United States, but several minor crops can be traced back to that country.

Crops are domesticated and improved by continually selecting and propagating plants with superior characteristics. Crop improvement has evolved from simple selection by a farmer to a complex system using selective breeding and bioengineering techniques. When plants are domesticated, the result is often a loss of genetic traits that at the time of selection were not considered important. Many times, however, these traits were what allowed the plant to survive in the wild. With time and changes in where and how the crop is grown, these traits regain their importance, but the genetic information is no longer available in the domesticated lines used for crop improvement. We have come to understand the importance of preserving the genetic information that was found in the ancestors of our crops. To facilitate the collection and preservation of germplasm, centers have been established around the world.

KNOWLEDGE CHECK

1. List three different categories that are used to classify plants.
2. What is a common name and what is a scientific name?
3. Why are plants usually referred to by their common name locally but by their scientific name globally?
4. What was Carl von Linneé's (Linneaus) role in classifying plants?
5. The scientific system of classification of all living things divides them into taxa (groups) based on what?
6. What are the three levels of taxa that most growers of plants need to know for the plants that they grow?
7. What are the two main parts of the binomial scientific name?
8. Which two of the following are in the correct form?
   a. Rudbeckia Hirta
   b. *Rudbeckia hirta*
   c. Rudbeckia "hirta"
   d. Rudbeckia hirta
9. *Brassica oleraceae* contains many groups that contain common vegetables. List the group for each one of the following: cabbage, broccoli, cauliflower.
10. What is the usual characteristic that defines botanical families?
11. Briefly describe how to use a dichotomous key to identify plants.
12. Why is knowing where a plant originated useful to plant breeders?
13. Describe one of the ways that crop improvement could have happened "accidentally" by early humans when they first started growing plants for harvest.
14. Although modern, improved plant cultivars provide a major source of food for humans and domesticated animals as well as for ornamental use, what would likely happen to the cultivars without human attention and cultivation?
15. Why was Norman Borlaug awarded the Nobel Peace Prize in 1970?
16. What species of wheat is used for breads and pastries? For pasta?
17. What is teosinte?
18. What is the rice species from which most cultivated rice probably originated?
19. How much of the world's soybean crop is produced in the US (2008 data)?
20. What influence did Napoleon have on sugar beet production?
21. What are certified seed potatoes and why are they source of most commercially grown potatoes?
22. Thomas Jefferson was ridiculed for growing tomatoes in his garden because most people in his day thought tomatoes were what?
23. Most crop cultivars grown today are relatively new. What makes many currently grown apple cultivars including McIntosh, Delicious, and Golden Delicious different from those modern cultivars?
24. How did the integration of Darwin's and Mendel's discoveries lead to the explanation of evolution?
25. Why is breeding called directed and accelerated evolution?
26. What is tricale?
27. How have plant explorers impacted agriculture?
28. It is much more difficult now to take germplasm from native plants out of the country where they are found. List the two conventions that establish the criteria that restricts the removal of germplasm.
29. What is the danger of moving plant materials into new areas?
30. What is the easiest and safest way to introduce plant material to a new area and why is it the safest?
31. What are the purpose of germplasm networks?
32. What is the US germplasm network called?
33. Why has there been a recent interest in broadening the base of agriculturally important crops?
34. List 10 crops that are considered to have the potential to become agriculturally significant.

FOOD FOR THOUGHT

1. Plant classifications and naming have traditionally relied on physical characteristics. With the recent advances in DNA analysis, there is expected to be a reclassification and/or renaming of many plants. Why do you think this is expected?

2. Look at the label of ingredients on a box of your favorite cereal and decide which are plant derived (including sugar, flavors, and other additives) then using the tables and figures from this chapter:
   a. Determine the likely region of origin of each component.
   b. Determine where the germplasm for those ingredients is most likely to be kept in preservation.

3. The germplasm of native plants is considered to be a natural resource of the country where the germplasm is found. Do you think a country has the right to sell that germplasm to other countries without regard to the Convention on International Trade in Endangered Species of Wild Fauna and Flora?

FURTHER EXPLORATION

BAILEY, L. H., E. Z. BAILEY, and staff of the L. H. Bailey Hortorium. 1976. *Hortus third.* New York: MacMillan.

BRICKELL, C. D., B. R. BAUM, W. L. A. HETTERSCHEID, A. C. LESLIE, J. MCNEILL, P. TREHANE, F. VRUGTMAN, J. H. WIERSEMA (eds.). 2004. *International code of nomenclature for cultivated plants,* 7th ed.

United Nations Environmental Programme and the Convention on Biological Diversity (CBD). 1992. Text of the convention on biological diversity http://www.biodiv.org/convention/articles.asp. Note: Treaty no. 30619, the Multilateral Convention on Biological Diversity (with annexes), concluded at Rio de Janiero on June 5, 1992.

Convention on International Trade in Endangered Species of Wild Fauna and Flora (CITES). 1979. Text of the convention, http://www.cites.org/eng/disc/text.shtml

Food and Agriculture Organization of the United Nations. 2003. *International treaty on plant genetic resources for food and agriculture.* Rome.

GREUTER, W., J. MCNEILL, F. R. BARRIE, H.-M. BURDET, V. DEMOULIN, T. S. FILGUEIRAS, D. H. NICOLSON, P. C. SILVA, J. E. SKOG, P. TREHANE, N. J. TULAND, D. L. HAWKSWORTH (eds.). 2000. *International code of botanical nomenclature.* Königstein: Koeltz Scientific Books.

GRIFFITHS, M. 1994. *Index of garden plants.* Portland, OR: Timber Press.

HARLAN, J. R. 1975. *Plants and man.* St. Paul, MN: American Society of Agronomy.

Iucn (The World Conservation Union). 2004. *IUCN Red List of threatened species,* http://www.iucnredlist.org/

KANG, M. S. 2002. *Crop improvement: Challenges in the twenty-first century.* Binghamton, NY: Food Products Press.

RAY, D., J. JANICK, D. DIERIG, R. MYERS, and C. BAILEY. 2002. *Trends in new crops and new uses.* Alexandria, VA: ASHS Press.

STEARN, W. T., and E. R. STEARN. 1992. *Botanical Latin,* 4th ed. Portland, OR: Timber Press.

VAVILOV, N. I. 1951. *The origin, variation, immunity and breeding of cultivated plants* (trans by K. S. Chester). New York: Ronald Press.

WOODLAND, D. W. 2000. *Contemporary plant systemics,* 3rd ed. Berrien Springs, MD: Andrews University Press.

*11
Photosynthesis and Respiration

The Carbon Cycle and Energy Transfer

The **carbon cycle** is essential for life on earth (Fig. 11–1). It is the means by which radiant energy from the sun is captured, converted to the chemical energy, and then released to support the metabolism of all living organisms. Carbon is the molecule that is cycled and recycled through these processes. The cycle starts with **photosynthesis**, which captures the solar energy. The radiant energy is used to synthesize carbon-based molecules called **carbohydrates** from carbon dioxide (CO_2) and water, releasing oxygen in the process. Some of the energy that was used to synthesize the carbohydrates is now "stored" in the chemical bonds of the carbohydrate molecules. The process of respiration breaks down the carbohydrate bonds, thereby releasing energy along with CO_2 and water vapor into the atmosphere. The energy that is released fuels all the biological processes necessary for life.

Organisms such as plants and the bacteria that can photosynthesize are called **autotrophs** because they can synthesize the compounds that release energy during respiration. Organisms, including humans, that cannot synthesize those compounds are called **heterotrophs**. Heterotrophs must consume autotrophs or other heterotrophs that have fed on autotrophs in order to get the energy-releasing carbohydrates.

Carbon (along with water and oxygen) is constantly being recycled by photosynthesis and respiration working together. As long as there is an equal or nearly equal amount of carbon being taken up by photosynthesis as is being respired, atmospheric CO_2 levels do not vary. Globally, even today with rising levels of atmospheric CO_2, the CO_2 released by respiration of all living creature is roughly equal to the amount taken up by the world's plants for photosynthesis. The atmospheric CO_2 levels are rising because much more CO_2 is being released into the atmosphere than plants are using for photosynthesis. The excess CO_2 is coming mostly from the burning of fossil fuels, which releases the CO_2 stored millions of years ago as carbohydrates in the plants that are the basis of those fuels. The problem is made worse by the loss of forests and other areas where large masses of plants once thrived. These plants are no longer available to absorb carbon. In addition, as they are burned or decompose they release the carbon that was stored in them.

key learning concepts

After reading this chapter, you should be able to:

- Explain the importance of the carbon cycle to life on Earth.
- Describe the process of photosynthesis and how radiant energy is converted to chemical energy.
- Explain the process of respiration and how it releases the chemical energy from photosynthesis.
- Discuss how the carbon cycle relates to practices used in growing plants.

Carbon Cycle

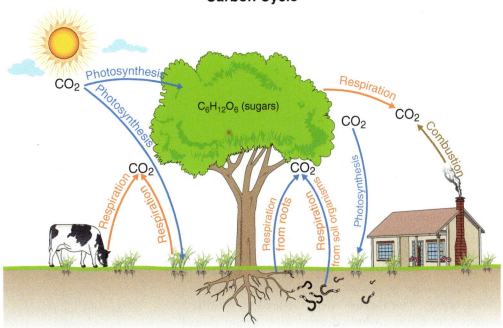

Figure 11–1
The carbon cycle. Source: Margaret McMahon, The Ohio State University.

As you can see, plants play the leading role in the carbon cycle. Without them, no other life can survive. We grow grains, fruits, and vegetables to provide us directly with the carbohydrates we need for energy. When we consume animal products, we are extracting the energy that those animals got from the plants they ate. We utilize or enjoy the "fruits" of photosynthesis in other ways too, such as the vitamins, minerals, and other compounds we need for nutrition; fibers for cloth; flowers and other ornamental plants for decoration; turfgrass for recreation; and wood for construction.

When we grow the plants we need for food, construction, pleasure, or other purposes, much of what we do intentionally or unintentionally affects the rate of photosynthesis and respiration in those plants. What we do to increase or decrease photosynthesis and respiration can significantly change how the plant looks and how it grows. To know how growing practices affect plant photosynthesis and respiration, we first have to understand both processes.

Two fundamental concepts of chemistry underlie an understanding of photosynthesis and respiration (and metabolism in general). The first is the uniqueness of the carbon atom. Carbon can form up to four bonds, each bond representing two electrons. Shared pairs of electrons are indicated as short lines in diagrams of molecules, as in ethane (a simple carbon-based molecule):

$$H-\overset{\overset{\displaystyle H}{|}}{C}-\overset{\overset{\displaystyle H}{|}}{\underset{\underset{\displaystyle H}{|}}{C}}-H$$

Bonds formed by carbon with another carbon or other elements have sufficient strength to make them relatively stable. The strength of the C–C bond is 82.6 kcal/mole (an estimate of how much energy is needed to break the bond). Compared to the bond strength of N–N (39 kcal/mole), S–S (54 kcal/mole), and O–O (35 kcal/mole), the C–C bond is relatively strong. Other examples of carbon bonds are C–O at 85.5 kcal/mole and C–H at 98.7 kcal/mole which are also in the right range for a combination of shareability and stability. *No other element has this ideal combination of shareability of electrons plus stability of the bonds formed.*

The second fundamental concept is the transfer of electrons (energy) between molecules (oxidation/reduction), the basis of many biochemical reactions. Reduced compounds easily donate electrons to other compounds, whereas oxidized molecules easily accept electrons. Carbon is at the center of this process in

living organisms because it is so versatile in donating and accepting electrons. *This transfer of electrons from relatively reduced to relatively oxidized carbon compounds drives all of metabolism and underlies life itself.* Thus, the one way to look at photosynthesis is to say that it is the process of converting oxidized carbon (CO_2) to reduced carbon, that is, carbohydrate or $C(H_2O)$. Respiration is the conversion of reduced carbon to oxidized carbon to generate other forms of energy that are central to the function of living organisms.

A familiar, nonbiological example of oxidation/reduction involving carbon is when we mix methane (natural gas, fully reduced carbon) with oxygen and light a match to start the reaction. As the methane is oxidized to carbon dioxide, the process is accompanied by the generation of heat (fire) energy:

$$4\,CH_3 + 7\,O_2 \rightarrow 4\,CO_2 + 6\,H_2O + \text{HEAT ENERGY}$$

It is important to qualify this example by noting that in the conversion of reduced carbon to oxidized carbon (respiration) in organism metabolism, heat is rarely released which is a good thing or we would all go up in flames with our first breath. Instead, the released energy is used to create other high energy compounds. An example of a high-energy compound produced in metabolism is adenosine triphosphate (**ATP**), which has a phosphate bond that has a lot of stored chemical energy.

PHOTOSYNTHESIS: LIGHT ENERGY → CHEMICAL ENERGY

Since the carbon cycle starts with photosynthesis, we will start with that process. There are several ways to define photosynthesis. Probably the best way is to say that photosynthesis is the process of converting of light (radiant) energy to chemical energy in the form of reduced carbon compounds. The simplest chemical reaction that can be written to represent photosynthesis is:

$$CO_2 + H_2O \xrightarrow{\text{light}} C(H_2O) + O_2$$

Note that this process involves the conversion of carbon dioxide (oxidized carbon) and water to carbohydrate (CH_2O, relatively reduced carbon) and molecular oxygen. However, this equation is a gross oversimplification of what really happens in photosynthesis. There are many steps along the arrow to get to the end products. Plants and photosynthesizing bacteria have very complex systems to accomplish this complicated chemical process.

In plants, photosynthesis occurs in chloroplasts. An electron microscope picture conveys the considerably specialized structure of the chloroplast (Fig. 11–2). The chloroplast contains a complex array of membranes called thylakoids surrounded by a fluid material called the stroma (Fig. 11–3). Embedded in the thylakoid are many proteins that also have an important role to play. The stroma and thylakoids are the site of Photosystems I and II, indicated in the figure. The essential function of these two photosystems is to utilize light energy to obtain electrons from water:

$$2H_2O \rightarrow O_2 + 4H^+ + 4e^-$$

Note that the oxygen produced in photosynthesis comes from water, not carbon dioxide. The photosystems use light energy to raise the electrons to a very high energetic state so that they can be used to start the cascade of reduction reactions. The protons (H^+) generated in the water-splitting reaction also set up a proton (H^+) gradient across the thylakoid membranes. The gradient means that there are more protons on one side of the membrane than the other. This gradient along with some of the proteins embedded in the thylakoid form a mechanism for ATP synthesis which is the first high energy compound generated by photosynthesis.

Although ATP is a high-energy compound, no reduction has yet occurred. There are many biochemical steps (which we will not discuss in this text) between the release of electrons from water and the formation of the first stable, reduced, high-energy chemical. It is enough to know that the reduced chemical that is formed is called nicotinamide dinucleotide phosphate (NADPH). The oxidized and reduced forms of NADP are shown (Fig. 11–4). The energy in NADPH and ATP is used in subsequent biochemical steps to synthesize carbohydrates. The process of generating NADPH is called the light reaction or light dependent step of photosynthesis because light is required for it to occur.

PHOTOSYNTHESIS: CARBON DIOXIDE → CARBOHYDRATE

So far, the discussion has shown that the light-requiring reaction of photosynthesis results in the conversion of water to O_2, and the generation of the energy-rich NADPH and ATP. As noted above, carbon dioxide has not yet been reduced to a carbohydrate. That process comes next in the **Calvin cycle**.

The following reactions, known as the photosynthetic carbon reduction cycle, or Calvin cycle, generate carbohydrates. They do not require light but they do depend on the NADPH and ATP produced by the light reaction. The Calvin cycle is sometimes referred to as the dark reaction which can be misleading as it can

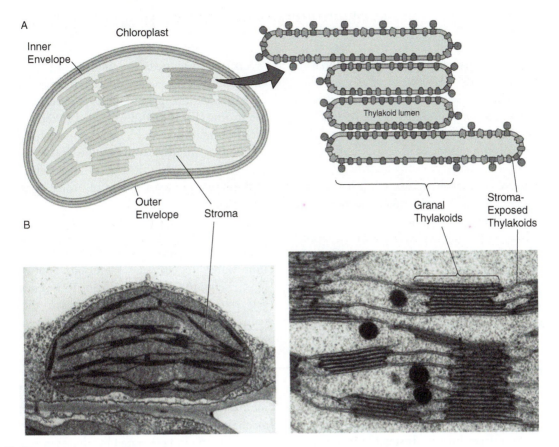

Figure 11–2

(A) Sketch of a chloroplast (left) with emphasis on membrane structures (thylakoids) involved in photosynthesis. More detailed sketch (right) of "grana," or stacks of membranous structure, in which are imbedded many catalytic and transport proteins.
(B) Electron micrographs showing a whole chloroplast (left) and the membrane networks at higher magnification (right). The circles in the right-hand micrograph are lipid bodies. Source: (A) W. B. Buchanan, W. Gruissem, and R. L. Jones, *Biochemistry and Molecular Biology of Plants* (Rockville, MD: American Society of Plant Biologists, 2000). (B) L. A. Staehelin and G. W. M. van der Staay. "Structure, Composition, Functional Organization and Dynamic Properties of Thylakoid Membranes," Chapter 2 in, *Oxygenic Photosynthesis: The Light Reactions,* edited by D. R. Ort and C. F. Yocum. (Dordrecht, The Netherlands: Kluwer Academic Press, 1996). Used with kind permission of Springer Science and Business Media.

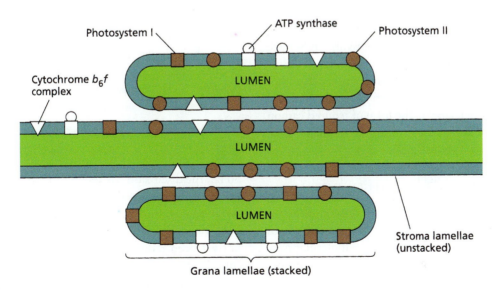

Figure 11–3

Detailed sketch of the structure of grana. Photosystem II (large circles) is located in the inside portion of the grana, whereas photosystem I (squares) and the enzymes for ATP synthesis are located mainly in the thylakoid membranes between grana (called stroma). Other components of the electron-energizing system (cytochromes, triangles) are found in both locations.
Source: L. Taiz and E. Zeiger, *Plant Physiology* (Redwood City, CA: Benjamin Cummings, 1991). Used with permission.

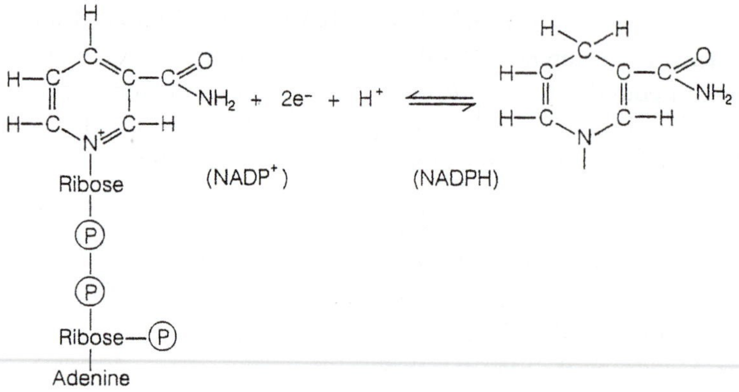

Figure 11–4

Electrons from water are finally utilized in the reduction of the nicotinamide ring of NADP to form the first chemically stable reductant. In a sense, NADPH is at the interface between the photochemical reactions of photosynthesis and the use of the electrons to reduce to carbohydrate. Source: L. Taiz, L. and E. Zeiger, 1991. *Plant Physiology.* (Redwood City, CA: Benjamin Cummings, 1991). Used with permission.

occur in dark or light. A better concept is to think of it as light independent.

The first reaction in the Calvin cycle is the attachment (fixing) of CO_2 to the 5 carbon compound ribulose diphosphate. This reaction, often called **carbon fixation**, is one of the most important in all of biochemistry. It is catalyzed by the enzyme ribulose diphosphate carboxylase, usually referred to as **RUBISCO**. Because of the importance of this reaction, it is shown in detail here:

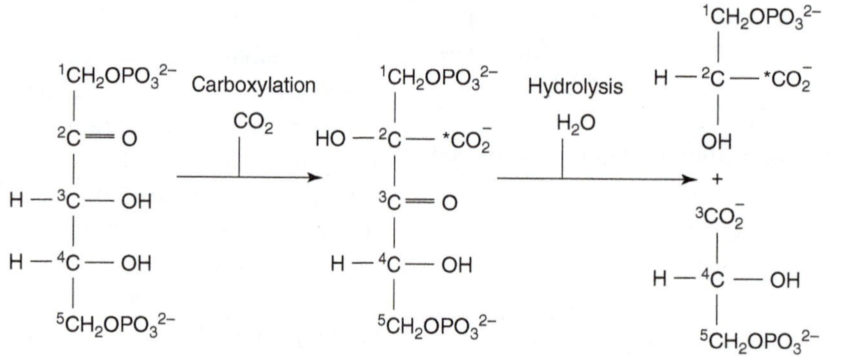

Note that an unstable 6-carbon compound is formed first and that this compound is immediately split into two molecules of 3-phosphoglycerate.

At this point, the NADPH and ATP generated in the light reactions have still not been used. The next two reactions do require ATP and NADPH. They are presented here in word form only:

$$\text{3-phosphoglycerate} + \text{ATP} \rightarrow$$
$$\text{1,3-diphosphoglycerate} + \text{ADP}$$

$$\text{1,3-diphosphoglycerate} + \text{NADPH} \rightarrow$$
$$\text{glyceraldehyde 3-phosphate} + \text{NADP} + \text{P}$$

Glyceraldehyde is a reduced form of glycerate and is the end product of the Calvin cycle.

Because a new carbon atom is obtained from CO_2 with each RUBISCO reaction, a totally new molecule of glyceraldehyde 3-phosphate can be made with every three revolutions of the cycle. Some of the glyceraldehyde 3-phosphate is used to regenerate ribulose diphosphate so that more CO_2 can be fixed. The rest of the glyceraldehyde 3-phosphate is used for the next step in photosynthesis which is the formation of glucose, a 6-carbon sugar. Glucose is a stable, high energy molecule and is the source of energy for metabolism and

provides some of the "raw material" for other metabolites throughout the plant. Glucose can combine with other sugar molecules to form compounds such as sucrose, starch, and cellulose. Sucrose is a disaccharide (a molecule made up of two sugars) comprised of a fructose and a glucose molecule. Starch and cellulose are chains of glucose molecules linked together. Sucrose is the most common form of sugar transported in the plant and is the sweetener that we commonly call table sugar. Starch is important in plants for energy storage and for humans for food and in industrial uses. Cellulose is the basis of many of the fibers that we use for clothing, rope, and the like. The difference between starch and cellulose is how the glucose molecules are bonded together.

As with nearly all biochemical reactions, glucose synthesis takes a series of steps to get to the end product. Each step is usually one simple reaction mediated by an enzyme. By having a stepwise, enzyme-mediated process, the process can be regulated and the energy needed for each reaction is reduced. The steps in glucose synthesis are described here so you can see the complexity of a biochemical process. First, a glyceraldehyde 3-phosphate molecule is combined with another 3-carbon phosphorylated sugar to give fructose-1,6-diphosphate (reaction 1 below). A phosphate is removed from fructose 1,6-diphosphate and the product is fructose-6-phosphate (reaction 2). The fructose 6-phosphate is then converted to glucose 6-phosphate (reaction 3) and, in turn, the phosphate on carbon 6 of glucose is moved to carbon 1 (reaction 4). Through another series of steps not shown the phosphorus is removed.

The reactions of glucose synthesis:

1. glyceraldehyde 3-phosphate + dihydroxyacetone 3-phosphate $\rightarrow$ fructose-1,6-diphosphate
2. fructose-1,6-diphosphate $\rightarrow$ fructose 6-phosphate + inorganic phosphate
3. fructose 6-phosphate $\rightarrow$ glucose 6-phosphate
4. glucose 6-phosphate $\rightarrow$ glucose 1-phosphate
5. Net: glyceraldehyde 3-phosphate + dihydroxyacetone 3-phosphate $\rightarrow$ glucose 1-phosphate + PO_4^{3-}

It is not important to memorize these reactions. Instead, remember two important concepts. The first is to appreciate how reduced carbon compounds are shuffled around to generate the compound required for a specific purpose. Second, phosphorylated compounds are often involved in metabolism because, phosphorylated versions of the sugars are more reactive than the nonphosphorylated forms. This is one reason why phosphorus is needed in relatively large quantities by plants and is a component in many fertilizers. Another need for phosphorus in a plant is for all the ATP and NADPH that is needed for photosynthesis and many other biochemical reactions.

The compounds shown in reactions 3 and 4 participate in many other reactions involved in carbohydrate metabolism.

FACTORS AFFECTING THE RATE OF PHOTOSYNTHESIS IN HIGHER PLANTS

Environmental and plant growth factors that affect the rate of photosynthesis are:

1. Light quality (wavelength)
2. Light intensity (the amount of incident light energy absorbed by the leaf)
3. Carbon dioxide concentration
4. Heat
5. Water availability
6. Plant development and source-sink relationships

Light Quality

The plant must absorb light to keep the photosynthetic mechanism running. As noted above, photosynthesis occurs in chloroplasts. The chloroplasts contain pigments capable of intercepting light and converting electromagnetic energy into the chemical energy necessary to drive the photosynthetic processes. When these pigments (chlorophyll a, chlorophyll b, and some carotenoids) are irradiated with light containing all visible wavelengths, they absorb mostly from the red and blue portions of the spectrum and reflect or transmit much of the green portion which is why leaves are seen as being green (Fig. 11–5). Light quality is particularly

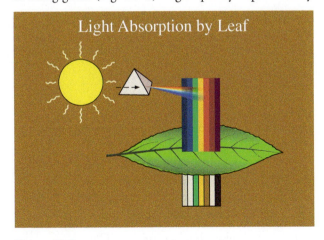

Figure 11–5
Leaves selectively absorb light for photosynthesis. Red and blue wavelengths are the most efficient for photosynthesis and are absorbed the most by the leaf. Source: Margaret McMahon, The Ohio State University.

important when plants are grown under artificial light. Sufficient radiation from the red and blue wavelengths must be provided by the lamps for photosynthesis.

Light Intensity

In addition to the quality of light that is active in photosynthesis, the amount of photosynthetically active light that a plant receives is also very important. The intensity of photosynthetically active light is referred to as **photosynthetic photon flux (PPF)**. PPF is usually expressed as micromoles of photons per m^2 per second $(\mu mol/m^2/s)$.[1] Sensors that accurately measure PPF over the 400 to 700 nm range are used in plant science research. Until recently their cost prohibited their use by growers; however, models are now available that growers can afford. For reference, the PPF of full sunlight is approximately 2,000 $\mu mol/m^2/s$.

Light intensity affects plant growth, mainly by influencing the rate of photosynthetic activity. The effect however varies with different plants. Some species evolved in regions of high solar intensity and require high light intensities to grow well. They are often termed sun-loving plants; examples are corn, potatoes, sugarcane, and most turfgrasses. On the other hand, plant species that do not grow well in high light intensities are sometimes referred to as shade-loving plants. Many such plants grow in the shade of the forest floor. Their tolerance of low light makes several of them useful as indoor ornamental plants. Examples are dieffenbachia, aglaonema, and several ferns. Other plant species are intermediate between the sun-loving and shade-loving types and grow well in moderately intense light or partially shaded areas.

Light intensity noticeably affects the size and shape of leaves. Generally, the leaves of a given plant species grow thinner and larger in area under low light intensity than under high light intensity. The larger leaves allow for a greater area to absorb light. Also, leaves of plants grown in high light intensities tend to be darker green than those grown in low light intensities because the chlorophyll cells are more closely packed together to allow better absorption of the high amount of photons available per area. These responses highlight the role of leaf blades in collecting solar energy.

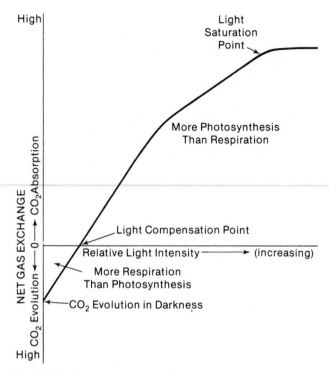

Figure 11–6
The effect of light intensity on net photosynthetic activity.

In a given plant, there is a light intensity at which photosynthesis and respiration rates are equal. This intensity is the light compensation point (Fig. 11–6). At the light compensation point CO_2 is still exchanged but the CO_2 evolved in respiration is the same amount that is used in photosynthesis. The plant is said to be light saturated when further increases in light intensity increase photosynthesis little or not at all. When light intensity is below the compensation point, there is greater respiration than photosynthesis. The plant has to use stored carbohydrates to maintain respiration. What this also means is that during the day, the plant must photosynthesize enough to not only compensate for its respiration at that time but during the night too when respiration but no photosynthesis takes place.

Although photosynthesis is the means by which radiant energy is converted to chemical energy, only about 5 percent of the solar energy that reaches the earth's surface is used directly in photosynthesis (Fig. 11–7) mainly because the photochemistry in chloroplasts uses only a small portion of the sun's energy spectrum. The portion of the light energy that can be used for photosynthesis is absorbed by the chlorophyll molecules in the thylakoid membranes. The reception of a single photon of light by a chlorophyll molecule is not sufficient to energize an electron. Instead, multiple photons of light of the desired

---

[1]Units of footcandles and lux are sometimes used to express light "intensity," but these measurements and units refer to the entire visible spectrum. Although these light measurements are useful for some activities (e.g., photography), they are not very relevant to plant growth. These units of light are no longer accepted in scientific and professional literature relating to photosynthesis.

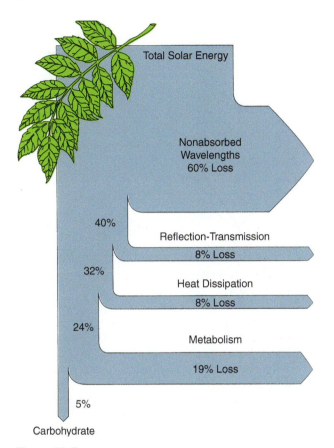

Figure 11–7

Fate of light energy reaching the earth from the sun. Most of the energy is reflected by the earth's atmosphere and, of the energy that actually reaches the earth's surface, much is lost in reflection or in the generation of heat. Because of wavelength restriction of the light reception pigments, only a small portion of the light energy reaching the earth is used in photosynthesis. Source: L. Taiz and E. Zeiger, *Plant Physiology*. (Redwood City, CA: Benjamin Cummings, 1991). Used with permission.

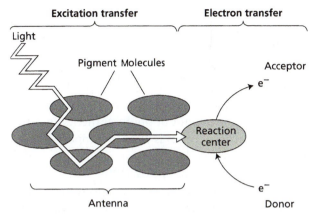

Figure 11–8

Sketch of the photosynthetic apparatus emphasizing the antennalike properties of the chlorophyll molecules. Multiple photons of light are transferred to a reaction center, where an electron is obtained from water or converted to a high-energy electron. Source: L. Taiz and E. Zeiger, *Plant Physiology*. (Redwood City, CA: Benjamin Cummings, 1991). Used with permission.

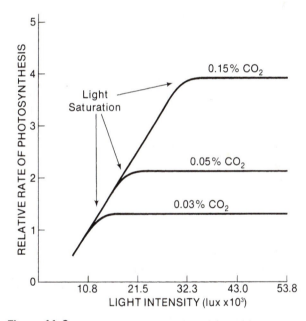

Figure 11–9

Light saturation at different CO_2 concentrations, and at a constant temperature of 25°C (77°F). Increasing CO_2 concentration increases photosynthetic activity. Note that the point of light saturation is also increased at higher CO_2 concentration.

wavelength range must be received by multiple chlorophyll molecules. The analogy of chlorophyll as an antenna molecule is appropriate. So, multiple photons are received and shunted to a reaction center where they energize the electrons from water (Fig. 11–8).

Carbon Dioxide

When a plant is placed in a closed system under lights, CO_2 is assimilated until the CO_2 concentration of the atmosphere reaches equilibrium at some low value. The carbon dioxide concentration at equilibrium is such that the amount of carbon dioxide evolved in respiration exactly equals the amount consumed in photosynthesis. This CO_2 concentration is called the **carbon dioxide compensation point**. At the CO_2 compensation point, no further assimilation (photosynthesis) occurs until more CO_2 becomes available.

The concentration of CO_2 in the air surrounding the leaves markedly affects photosynthesis (Fig. 11–9). Normally the atmosphere contains an average of about 0.03 percent CO_2 and 21 percent O_2. Plant physiologists have found that increasing the CO_2 concentration in a closed system, such as a sealed greenhouse or an acrylic plastic chamber, to about 0.10 percent approximately doubles the photosynthetic rate of certain crops such as wheat, rice, soybeans, some vegetables, and

THE NATURE OF LIGHT

The sun's energy is derived from thermonuclear reactions that convert hydrogen atoms to helium atoms with the release of tremendous amounts of energy. This energy travels through space to the earth as electromagnetic radiation waves at the speed of light, about 300,000 Km/s (186,000 mi/s). The wavelengths of electromagnetic radiation vary from very long radio waves of more than 1 kilometer (0.62 mi) to very short cosmic rays of less than 10^{-4} hm (3.9×10^{-10} in). Visible light is that portion of the electromagnetic spectrum between 400 and 700 nm; however, plants respond to the somewhat wider spectrum of about 300 to 800 nm (Fig. 1). The distance between the peaks of the waves is the wavelength. Light traveling in shorter wavelengths is more energetic than light at longer wavelengths and lower frequencies. Thus, blue light is more energetic than red light because blue has a shorter wavelength. Figure 2 shows different wavelengths. Light has a dual nature, meaning that it acts as both a wave and a particle. Both aspects are important in photosynthesis.

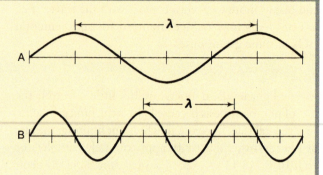

2.

The wave nature of light. The wavelength (A) is twice that of (B). The short wavelength make (B) more energetic than the long wavelength of (A).

The light has to be of the right wavelength to be absorbed by chlorophyll and there have to be enough particles of that absorbed light to drive photosynthesis.

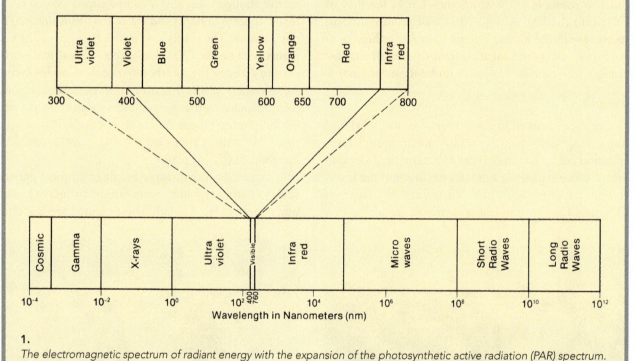

1.

The electromagnetic spectrum of radiant energy with the expansion of the photosynthetic active radiation (PAR) spectrum.

fruits. Many greenhouse crops, such as carnations, orchids, and roses, are grown commercially in a CO_2-rich atmosphere. While increasing the CO_2 concentration is done in the greenhouse or laboratory, it is not possible to markedly increase the CO_2 concentration in the air above a corn or wheat field.

A few practices, however, can improve CO_2 availability in some field situations. For example, the density of the crop and the height of the crop canopy can sometimes be altered to increase the diffusion rate of CO_2 into the canopy, which in turn increases its concentration in the vicinity of the leaves. Applications of organic matter in the form of crop residues or green manure crops to the soil tends to increase CO_2 levels in the atmosphere above the soil. On a warm, sunny day after a rain and with no air movement, some plants grow so rapidly that CO_2 availability at the leaf surfaces become limiting. Under such conditions, wind

machines can increase the available CO_2 by circulating and mixing the air. In greenhouses, the use of horizontal airflow fans (HAFs) increases airflow around plants and brings more CO_2 into contact with the leaves. The use of HAFs may reduce the need for supplemental CO_2 in some greenhouses.

Heat

At low light levels, temperature has little effect on the rate of photosynthesis because light is the limiting factor. As a general rule, however, if light is not limiting, the rate of photosynthetic activity approximately doubles for each 10°C (18°F) increase in temperature for many plant species, up to a point. The effect of temperature varies with species; plants adapted to tropical conditions require a higher temperature for maximum photosynthesis than those adapted to colder regions. However, excessively high temperatures reduce the photosynthetic rate of some plants not accustomed to such high temperatures by causing the stomata of the leaves to close in order to conserve water which then blocks the flow of CO_2 into the leaf. A reduced rate of photosynthesis, together with the increased respiration rate at high temperatures, lowers the sugar content of some fruits (for example, the cantaloupe) grown under these conditions.

Water

Under conditions of low soil moisture (drought) and hot, drying winds, plants often lose water through transpiration faster than their roots can absorb it. The rapid loss of water causes the stomates to close and the leaves to wilt temporarily. When this occurs, the entry of CO_2 and O_2 into the leaf is restricted, resulting in a dramatic drop in photosynthesis. Thus, water deficit reduces photosynthesis, in part, by markedly reducing CO_2 entry into the leaf.

Excessive soil moisture sometimes creates an anaerobic condition (lack of oxygen) around the roots, reducing root respiration and mineral uptake and transport of water and minerals to the leaves—thus indirectly depressing photosynthesis in the leaves.

Different Photosynthetic Mechanisms

Photosynthesis requires the leaf to take in CO_2. Usually this is accompanied by the loss of water vapor (transpiration). When sufficient water is available, transpiration and photosynthesis work together. During times of drought and high light conditions, however, the two may be in conflict. The stomates need to open to allow CO_2 in for maximum photosynthesis. But excessive water loss through the stomates may threaten plant survival.

In many areas where plants grow, there is sufficient rainfall during the growing season that plants receive enough water that a compromise between CO_2 intake (photosynthesis) and water loss (transpiration) is not an issue. The plants that evolve here are called C3 plants because the CO_2 that enters the leaf is directly used to generate the 3-carbon molecule phosphoglycerate. However, plants that evolved in more arid regions have adaptations that help conserve water but do not inhibit photosynthesis during dry periods. Plants with these adaptations are called C4 and crassulacean acid metabolism (CAM) plants.

The C4 mechanism is found in tropical grasses such as corn, sorghum, warm-season turfgrasses, and other species (Fig. 11–10). These plants have a special

A   B

Figure 11–10

C-4 (warm season) grasses: (A) corn and (B) Bermuda grass.

Source: (A) USDA-NRCS, http://photogallery.nrcs.usda.gov/ (B) Dr. David Gardner, The Ohio State University.

anatomy that concentrates CO_2 in special areas of the leaf called bundle sheaths that allow CO_2 to be fixed more efficiently into glucose. They are called C4 plants because CO_2 entering the leaf is temporarily attached to a 3-carbon organic acid making a 4-carbon organic acid. The acid is shuttled to the bundle sheath cells and the CO_2 is released at the areas where it can be used efficiently by RUBISCO to generate glyceraldehyde. The 3-C acid goes back to the stomatal opening to get another CO_2 molecule as it enters. Because of the increased efficiency of capturing CO_2, not as much CO_2 has to enter the leaf, so the stomates can be more tightly closed to reduce water loss.

The CAM plants (named for the species, *Crassula*, where the mechanism was first discovered) are water-storing desert plants such as succulents and cacti (Fig. 11–11). The CAM mechanism is also found in some genera of other families such as ORCHIDACEAE and BROMELIACEAE. As with C4, the CAM process also attaches CO_2 to a three-carbon organic acid, but this time it happens at night when the stomates can open without extreme water loss. The CO_2 is held on the acid until light is available for photosynthesis, then it is released from the acid and follows the normal photosynthetic pathway. The water stored in the plant absorbs heat and keeps the plants cool during the day when the temperatures are hot but the stomates are closed.

Adaptations for both C4 and CAM plants are expensive in terms of the energy required to construct and maintain them. But because they grow in areas where light is plentiful and photosynthesis rates can be high, the plants can "afford" the adaptations. The prob-lem arises when we try to grow these plants in areas where light is not sufficient to support their growth. This is part of the reason why corn cannot be grown as far north as other grains and why warm-season turf-grasses do not do well in the middle and northern latitudes of the United States.

Plant Development and the Source–Sink Relationship

Provided that all other environmental factors are con-stant, the growth of the plant also significantly influ-ences the rate of net and total photosynthesis, both in single leaves and in the total canopy of leaves on a plant. As the leaves expand and grow larger, chloro-plast development and replication proceed in the new cells until the leaf has fully expanded and has reached maturity.

As single leaves on a plant develop, their photo-synthetic rate rises in step with the expansion of the leaf until the leaf has matured. For example, when a soybean or spinach plant leaf begins expansion, its chlorophyll content per square centimeter of leaf area and its rate of net photosynthesis per square centime-ter of leaf area both rise to maximum values just after the leaf reaches full expansion. When the leaf reaches full expansion, it is called a source leaf because the carbohydrate synthesized in that leaf is in excess of what the leaf itself needs. The excess is exported to other parts of the plant that are actively growing. The sites to where the carbohydrates are transported are often called sink tissues. Examples of sinks are roots and reproductive organs (i.e., fruits and seeds). Note that juvenile leaves are also sinks until their photo-synthetic capacity provides carbohydrate above amounts required for local growth. Carbon partition-ing describes the net movement of carbohydrates into and out of tissues and organs. Some growing practices are designed to control carbon partitioning in order to increase the flow of carbohydrates into the harvestable part of the plant.

RESPIRATION

The next step in the carbon cycle is the oxidation of the reduced carbon (carbohydrates) created in photosyn-thesis back to the oxidized form (CO_2) and the release and use of the chemical energy stored in the carbohy-drates. All organisms require the carbohydrates formed in photosynthesis to survive. The building and mainte-nance of cells and tissues require not only carbohy-drates, but also proteins, lipids, nucleic acids, and

Figure 11–11

The Jade plant (Crassula ovata) *is a CAM plant.*
Source: Margaret McMahon, The Ohio State University.

so forth. Therefore, it is necessary to extract the energy stored in glucose and other carbon compounds and convert it into other chemical forms that can be used to synthesize the other biochemicals needed to accomplish the thousands of metabolic tasks required for survival. Respiration is the process that extracts the energy needed for the conversions.

A simple way to think about respiration is to consider the process as the transfer of carbohydrate energy to energy-rich ATP which then provides the energy for the biochemical processes. The transfer of energy to ATP requires carbohydrates, oxygen, adenosine diphosphate (ADP, a low-energy molecule), and can be written as:

$$C_6H_{12}O_6 + 6O_2 + 36\,ADP + 36\,PO_4^{3-}$$
$$\rightarrow 6CO_2 + 6H_2O + 36\,ATP$$

So, in a sense, respiration is the reverse of photosynthesis with respect to carbon dioxide and carbohydrate. The above reaction pertains to all organisms, including plants. *The CO$_2$ that is released by respiration can once again be used in photosynthesis, thus completing the carbon cycle.*

Glucose is the starting molecule for respiration so prior to the start of respiration, carbon based compounds such as carbohydrates and lipids must be converted to glucose. The *first step of respiration* is a process called glycolysis—the conversion of glucose to a three-carbon organic acid named pyruvic acid (or pyruvate), which takes place in the cytoplasm of cells. Starting with glucose, the carbohydrate is phosphorylated to form fructose diphosphate, using phosphorus from two ATP molecules. Fructose diphosphate is then split into two three-carbon pieces—namely, glyceraldehyde 3-phosphate, the same compound involved in the photosynthetic cycle. Further transitions lead to the generation of four high-energy ATPs and two NADH molecules which are very similar to the NADPH generated in photosynthesis. However, the two ATPs were used to create fructose diphosphate so the overall gain is two ATPs. The overall process of glycolysis results in the following:

$$2\,\text{Pyruvic acid} + 2\,ATP + 2\,NADH$$

The *second step of respiration* takes place in the mitochondria of a cell and is called the tricarboxylic acid (TCA) cycle because several of the reactions involve acids with three carboxyl groups. Before entry into the mitochondria and the TCA cycle, the pyruvate is converted to acetyl coenzyme A (acetyl CoA), a two-carbon molecule, and a completely oxidized CO$_2$ molecule is given off. Another NADH is also produced (Fig. 11–12). As the acetyl CoA enters the mitochondria and moves through the cycle it is completely oxidized to two more CO$_2$ molecules coupled to the generation of three more NADH molecules as well as a molecule of FADH, another type of electron accepter.

In summary then, for every glucose molecule that is processed through glycolysis, two pyruvate molecules are formed along with two ATPs. Another ATP is formed from each acetyl CoA in the TCA cycle making a total of four ATPs formed, still a long way to go to reach thirty-six. But much of the reducing power in the original glucose molecule has been retained in the NADH and FADH$_2$ molecules formed: two NADH in glycolysis + eight in the formation of acetyl CoA and in the TCA cycle + two FADH$_2$.

These reduced nucleotide molecules are oxidized to generate ATP in the *third step of respiration*—called the electron transport system. The system consists of a series of cyclic reactions (Fig. 11–13). One way to describe this process is to imagine that the electrons in NADH are passed down a cascade of stairs, with energy being released at each step. During this process, those valuable electrons in NADH are cashed in for ATP production, one NADH for three ATPs. FADH$_2$ enters the cascade after the first ATP synthesis step, so it yields only two ATPs. Note that molecular oxygen is the final electron acceptor (O$_2$ is relatively easily reduced), and that H$_2$O is the final product. Carbon dioxide—the substrate for photosynthesis—is also released in respiration, but this occurs in the operation of the TCA cycle (Fig. 11–12).

Before making the final tally for recovery of ATPs from the reducing power in glucose, localization of the three steps of respiration needs to be considered. Enzymes for glycolysis, the first step, are located in the cell cytoplasm. The remaining two steps are localized in mitochondria. Thus, pyruvate—the product of glycolysis—must be transported through the mitochondria membrane into the mitochondrion (Fig. 11–12).

The enzymes for the electron transport system are localized in the mitochondrial inner membranes—the cristae—the site of NADH consumption and ATP generation. To calculate the theoretical final number of ATP molecules, it is important to note that, while ATP can pass through membranes without cost, there is a cost for the transport of NADH, namely, one ATP per NADH transported into the mitochondrion. So each NADH produced in the cytoplasm yields only two ATPs.

Glycolysis and TCA Cycle

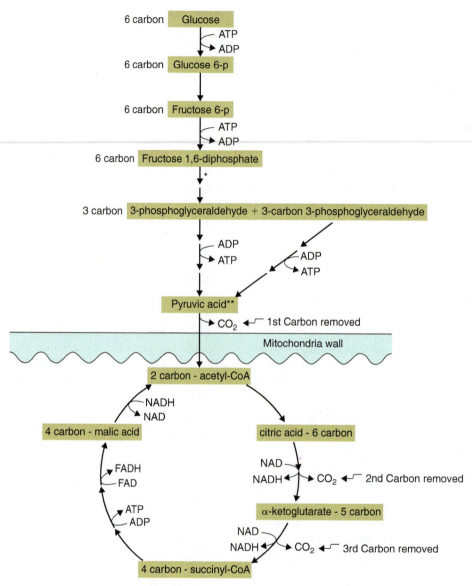

*Multiple arrows indicate several steps occur.

**Because there are 2 pyruvic acid molecules entering the cycle, the numbers of CO_2, ATP, NADH, and FADH generated are doubled to get the yield from a glucose molecule.

Figure 11–12

A diagram of glycolysis and the TCA cycle. Glycolysis takes place in the cytoplasm of the cell, the TCA cycle takes place in the mitochondria. Each glucose molecule generates six CO_2 molecules, six H_2O molecules, and ultimately, thirty-six ATP molecules.

With all of the above information, it is finally possible to calculate the theoretical yield of energy-rich ATP molecules from one glucose:

It is estimated that nearly 60 percent of the energy in a glucose molecule is conserved in ATP production.

| Step | NADH Produced | ATP Synthesis | | | Total ATP Production |
		Direct	From NADH	From FADH$_2$	
Glycolysis	2	2	4*	—	6
TCA cycle + electron transport	8	2	24	4	30
				Grand total	36

*Six are produced but two must be subtracted to support the transport of NADH into the mitochondrion.

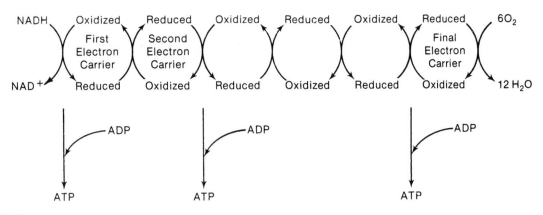

Figure 11–13

A sketch of the electron transport system for NADH. Electrons from NADH pass through a cascade of enzymatic electron transfer reactions, and the high energy compound ATP is synthesized at three points in the cascade. The final step involves the reduction of molecular oxygen to form water. (FADH enters further down the transport system than NADH so only two ATPs are produced for each FADH.)

One might ask: Why not oxidize glucose more directly rather than going through all these metabolic gyrations? The answer is in the above table; that is, metabolism has evolved to get the absolute most out of the reducing power of a molecule of carbohydrate. All these many reactions represent the most efficient possible conversion of one form of chemical energy to another, without the production of other energy forms (e.g., heat, light) that cells cannot use for growth and reproduction. ATP is the energy currency of all living organisms. It is transported to other places in the cell to be used in a wide variety of other chemical reactions. The respiratory steps are well tuned to achieve maximum ATP production.

It is very important to remember that although plants can use light energy and animals cannot, a plant must also respire to support the synthesis of all of the compounds needed for growth and maintenance of its own structure. Any accumulation of photosynthesis products in plants can only occur after the plants own needs for those products have been met. Net photosynthesis is the term for the amount of photosynthates produced above the amount required by the plant itself.

Another consideration that is important to anyone storing live plants or live plant parts, such as lettuce or fruit, is the effect that respiration has on stored carbohydrates. If living plant material is allowed to respire at normal levels during storage when photosynthesis cannot take place, reserve carbohydrates are used for respiration. The loss of these carbohydrates will shorten the storage life and lower the quality of the product. Reducing respiration either by cooling or modifying the atmosphere is an important part of postharvest handling of plants (Chapter 16).

SUMMARY AND REVIEW

The carbon cycle is the reduction of carbon (as CO_2) to carbohydrates through photosynthesis and the oxidation of carbohydrates back to CO_2 through respiration. Autotrophs are those organisms that can photosynthesize carbohydrates. Heterotrophs cannot synthesize carbohydrates, they survive by utilizing the carbohydrates produced by autotrophs. Photosynthesis and respiration are generally in balance with the amount of carbon being reduced and oxidized being nearly equal resulting in little change in atmospheric CO_2 levels. The current rise in atmospheric CO_2 is the result of the oxidation of carbohydrates produced long ago by the plants that make up fossil fuels when those fuels are burned.

The carbon molecule has some very unique bonding properties that make it the molecule best suited for energy transfer. It can share electrons easily with other carbon atoms as well as the atoms of many other elements forming bonds that are relatively strong and stable. It is the unique combination of shareability and stability that makes carbon ideal for energy transfer.

Photosynthesis is the conversion of light (radiant) energy to chemical energy in the form of reduced carbon compounds that takes place in the intricate structure of the chloroplast. It is a complex process that starts with the synthesis of ATP and NADPH in the light reaction. These energy-rich compounds provide the

energy for the Calvin cycle. The Calvin cycle is the formation of glyceraldehyde 3-phosphate and the regeneration of ribulose 1,5-diphosphate, the molecule to which the enzyme RUBISCO attaches CO_2 at the start of the cycle. Glucose is then synthesized from glyceraldehyde 3-phosphate. Glucose is a stable, high energy molecule and is the source of energy for metabolism and provides some of the "raw material" for other metabolites throughout the plant. Glucose can combine with other sugar molecules including other glucose to form compounds such as sucrose, starch, and cellulose.

Several factors can affect photosynthesis. The different wavelengths (light quality) of photosynthetically active light affect photosynthetic efficiency unequally. Blue and red are very efficient; yellow and green are less efficient. Light intensity affects photosynthetic productivity, with higher intensities generally producing more photosynthesis. The light environment in which plants evolved can determine whether plants grow well in low-, moderate-, or high-light conditions. The light compensation point is the intensity of light is such that the rate of photosynthesis equals the rate of the plant's respiration. Light saturation occurs when an increase in light no longer results in an increase of photosynthesis. Carbon dioxide availability is necessary for maximum photosynthesis. If there is not enough CO_2 available for photosynthesis, the rate of photosynthesis is inhibited even if enough light is available for greater photosynthesis. Heat influences photosynthetic rate. Generally if light is not limiting, photosynthesis doubles for each 10°C increase in temperature. However, temperatures that are too low or high can slow photosynthesis. Sufficient water must be available for the plant to keep its stomates open to allow CO_2 to enter the leaf. C4 and CAM plants have special adaptations that let them conduct photosynthesis relatively efficiently when the water supply is limited and stomates have to be completely or partially closed.

The source-sink relationship describes the relationship between photosynthesizing leaves (source) and the developing tissues and organs that use photosynthetic products tissues (sink). In other words, source tissues export carbohydrates, sink tissues import them. Carbon partitioning is the net movement of carbohydrates into and out of tissues and organs.

Respiration, or the release of energy stored in the reduced carbon compounds, takes place in three complex steps. The first step is glycolysis, and it occurs in the cell cytoplasm. In glycolysis, glucose is oxidized into two molecules of pyruvate. Two NADHs and two ATPs are also produced. The pyruvate is converted to acetyl coenzyme A and a CO_2 molecule is given off. The two carbon acetyl groups enter the TCA cycle, which is the second step of respiration and occurs in the mitochondria. In the TCA cycle, NADH, FADH, and ATP are produced and more CO_2 is given off. The third respiration step is the electron transport system, which occurs on the inner membrane of the mitochondria. Here the energy from NADH and FADH is transferred to ATP, and H_2O is generated. In all, thirty-six ATPs are the net energy product of respiration. ATP is the energy currency of all organisms.

KNOWLEDGE CHECK

1. What process is the means by which radiant energy from the sun is captured, converted to the chemical energy, then released to support the metabolism of all living organisms?

2. What is the difference between an autotroph and a heterotroph? Indicate what each of the following would be (autotroph or heterotroph):

   orchid   polar bear   fungus   maple tree   rabbit

3. Why is it not likely that the respiration of all living creatures on earth is a cause of increasing atmospheric CO_2?

4. Why is it important for anyone who grows plants to understand the processes of photosynthesis and respiration?

5. What is the difference between reduced and oxidized compounds?

6. What is it about carbon that makes it suited to be the basis of metabolic energy?

7. The oxygen that is released in photosynthesis comes from what compound?

8. What is the first high energy chemical compound produced in photosynthesis? What is the first reduced high energy produced?

9. During what process does the reduction of CO_2 to a carbohydrate occur?

10. What is carbon fixation and what enzyme is responsible for it?

11. What are the two things that can happen to a glyceraldehyde molecule produced in the Calvin cycle?

12. Describe how glucose is a part of the structures of sucrose, starch, and cellulose? What is the structural difference between starch and cellulose?

13. What are the advantages of having a series of enzyme-mediated steps to get to an end product in a biochemical reaction over having a single step such as the nearly instanteous combustion of methane to get CO_2 and water?

14. Why is phosphorylation such a common factor in metabolic reactions?

15. What regions of light are most effective for photosynthesis?

16. What is photosynthetic photon flux (PPF) and in what units is it expressed?

17. How would the leaves of a plant grown under low light usually differ from leaves on the same plant if it was grown under a higher light intensity? Why would the leaves in high light appear darker green?

18. Why are both wavelength and number of photons important in photosynthesis?

19. What happens to a plant's carbohydrate reserves when light intensity is below the light compensation point at night?

20. Why can increasing CO_2 concentrations around leaves under high light conditions increase photosynthesis?

21. What role do stomates play in photosynthesis?

22. Why is keeping C3 plants well watered during hot, sunny conditions important to keeping photosynthesis levels high?

23. By what mechanism do C4 plants keep photosynthesis going under hot sunny conditions? CAM plants?

24. What are sink and source tissues? What is carbon partitioning?

25. Why can respiration be considered the opposite of photosynthesis?

26. What is the last molecule formed from glucose in glycolysis?

27. What is the TCA cycle? Where does it occur?

28. Where in the respiration process is CO_2 given off?

29. What are the 3 kinds of energy carrying molecules generated in glycolysis and the TCA cycle?

30. Of the three molecules mentioned above, which 2 are oxidized in the electron transport system?

31. What is the high energy product is generated by electron transport in respiration?

32. What is the total number of this product generated by the entire respiration process?

33. What happens to oxygen electron transport at the end of electron transport?

34. In terms of respiration and carbohydrate reserves, what is the reason live plants are shipped under low temperatures?

FOOD FOR THOUGHT

1. Using the concept of the carbon cycle, how would you convince someone who said that plants and photosynthesis are not necessary—because we can get all the nutrition we need by eating animal products and taking vitamin and mineral supplements—that they are wrong?

2. You are a greenhouse grower and you want to buy some shade screen to reduce the amount of light reaching the plant to help keep them cool during the summer. You also want to keep photosynthesis as high as possible. You have a choice of two colors, *green* (which absorbs blue and red) or *purple* (which absorbs yellow and green). Which one would you choose and why?

3. Your plants have been diagnosed with a phosphorus deficiency. In terms of *energy transfer* in both photosynthesis and respiration, what could be happening in the plant if there is not enough phosphorus? How do you think the availability of compounds such as phosphoglycerate for the Calvin cycle would be affected?

FURTHER EXPLORATION

Bowsher, C., M. Steer, and A. Tobin. 2008. *Plant biochemistry.* London: Garland Science Textbook.

Hopkins, W. G., and N. A. P. Hüner. 2004. *Introduction to plant physiology,* 3rd ed. Hoboken, NJ: John Wiley & Sons.

*12
Water Relations
Soil, Plant, Air

JAMES D. METZGER AND MARGARET MCMAHON

WATER

Uses in Plants

Water makes up approximately 90 percent of a plant's mass and performs many functions in plants. It is required for seed germination, serves as part of the plant's structure, carries minerals into and through the plant, transports photosynthates and other biochemicals through the plant, and cools the plant by evaporation. Without sufficient water, plant growth and development is inhibited and if the insufficiency is great enough, the plant dies. Almost all the water a plant takes in comes from the soil through the roots. A small amount may be absorbed by other organs and tissues, but usually only under special circumstances, such as during the propagation of stem and leaf cuttings.

Characteristics

Water is the universal solvent; it dissolves more substances than any other liquid. The characteristics of water arise from its unique structure, which also accounts for its remarkable stability. Water is one of nature's most stable compounds, so much so that for centuries it was considered a single element, not the compound it is. The water molecule in fact comprises two hydrogen atoms attached to one oxygen atom. The water molecule is not symmetrical. This lack of symmetry creates a **dipole** with the molecule having more positive electrical charge and the opposite end a more negative charge (Fig. 12–1). This phenomenon of polarity creates an attraction (cohesion) between water molecules: the positive end of one molecule attracts the negative end of an adjacent molecule. Water molecules can also attract or be attracted by cations, such as Na^+, K^+, and Ca^{++}, anions or clay colloids in the soil, or other molecules.

    Surface tension is that physical property of water in the liquid state that is due to the intermolecular attraction (hydrogen bonding) between the water molecules. The molecules in the surface film in water are inwardly attracted, resulting in a strong surface tension. Were the water's surface tension not so strong, soil would hold little water, water could not

key learning concepts

After reading this chapter, you should be able to:

- Describe the forces that move and hold water in the soil.
- Describe the forces that move water from the soil into and through the plant and into the air.
- Explain the function of water in plants.

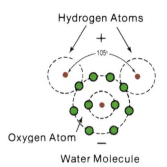

Figure 12–1

Two hydrogen atoms join one oxygen atom at an angle of 105°, giving the water molecule an asymmetrical configuration that accounts for many of its unique properties. Note the electrical polarity of the molecule.

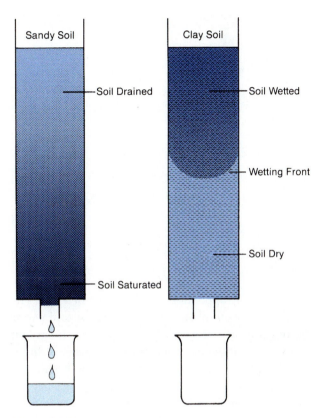

Figure 12–2

Fifty ml of water were added to each column. After one hour, water has drained to the bottom and some dripped out of the sandy soil (left), while the water remained in the top half of the clay soil (right). Obviously, clay soil can hold more water.

reach the top of a tall tree, and blood would not flow through our bodies. This strength of attraction between water molecules is illustrated by the fact that a steel needle can float on the surface of water.

In the liquid state, the attraction among water molecules is chaotic and random, but as the liquid freezes, a rigid symmetrical lattice with an open porous structure forms. This arrangement accounts for the reduction in density and increase in volume as the water solidifies. Water also has an unusually high **specific heat**, which allows it to absorb a large amount of heat without a large increase in temperature. As a result, water can moderate the temperatures of nearby areas and objects that contain water.

The movement of water from the soil to the plant to the air (soil, plant, air continuum) is an important factor in plant biology, and the relationship has to be understood by anyone who grows or works with plants.

SOIL WATER

Availability

Even though water is present in the soil, it sometimes is not available to the plant for reasons that will be discussed later in this chapter. Water held by and moving within the soil supplies the plant with mineral nutrients and oxygen, as well as water. It moves through the soil pore spaces in the liquid or gaseous state. The pore spaces are always filled with water, air, or a mixture of both. When the pore spaces are filled with water, the soil is said to be saturated. Saturation is an unhealthy condition for plants if it lasts too long because the oxygen needed for respiration in the rots is missing. On the other hand, when the pore spaces are filled mostly with air, the soil is too dry for good plant growth. The number and size of the soil pores vary with the soil's texture

and structure. Clay soils have smaller but more numerous pores than sandy soils. Thus, an equal volume of clay soil holds more water than a sandy soil when the pores are filled (Fig. 12–2). The ability of the soil to retain water is called its water-holding capacity. The capacities of different textured soils for holding water are plotted in Figure 12–3.

Water Movement and Retention in Soil

Three forces—gravity, adhesion, and cohesion—are responsible for water movement within the soil. Gravity causes water to move downward and is the principal force when a soil is saturated. **Adhesion** is the force of attraction between unlike molecules (soil particles and water). **Cohesion** is the force of attraction between like molecules (water and water). The latter two forces can cause water to move by capillarity in any direction—upward, downward, or laterally—and are the principal forces that move water in an unsaturated soil.

The upward movement of water, called capillary rise, is responsible for the loss of water from the soil surface by evaporation. Capillary rise can be demonstrated

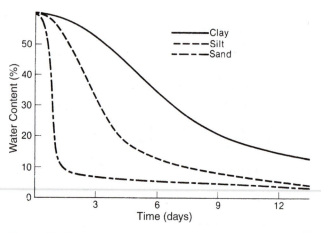

Figure 12–3

The speed with which soil water moves into or out of a soil is determined by the soil's texture. For example, assume that a sandy soil (coarse texture), a silt soil (medium texture), and a clay soil (fine texture) are saturated at the same time (t = 0) and that the moisture contents of the three soils are determined and plotted with time. The graph shows that the sandy soil loses moisture very quickly, the silt loses water moderately, and the clay soil loses it slowly. Thus, sandy soils have to be irrigated more frequently than do silt or clay soils.

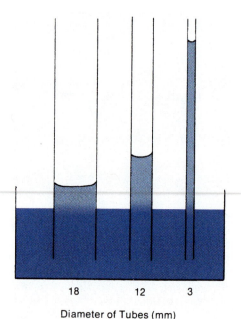

Figure 12–4

Capillary rise: water ascends in capillaries, reaching higher levels in smaller tubes. Doubling the diameter of the tube doubles the area for the water molecules to adhere to—but it also quadruples the weight of the water to be pulled up. Therefore, water does not rise as high in the columns with larger diameters.

with one end of a strip of blotting paper inserted partway into water or with capillary tubes (Fig. 12–4).

As soil dries, the water film surrounding each soil particle thins. Consequently, the adhesive and cohesive forces of attraction increase rapidly, making it more difficult for the plant to extract water that is held tightly in soil particles. The forces that act on soil water create changes in energy level, called water potential, symbolized by the Greek letter psi (Ψ). Water potential determines the direction in which water flows. Water flows from high to low water potential. Think of a waterfall, where water flows from a high to a low level (potential), or a water balloon, where pressure inside the balloon creates a very high water potential, forcing water out of the opening of the balloon if you let go of it. Conversely, negative pressure (tension or suction) lowers water potential, which draws water into an area. An example is sucking on a straw in a glass of iced tea. The greater the suction (tension), the more negative the pressure, and the lower the water potential. Water potential is measured in units of pressure, most commonly atmosphere (atm) and MPa (megapascal). One atm = 0.101 MPa, which also equals 0.101 MPa Ψ. Saturated soil with the atmosphere at 100 percent relative humidity has a water potential of zero. As water drains away, soil water potential decreases (becomes more negative). When the negative potential is stronger than gravitational pull, water no longer drains from the soil. Negative potential, or pressure, is called tension.

Soil water potential is determined by several factors: (1) gravity; (2) matric, from the adhesion and cohesion between soil particles and water; and (3) osmotic, caused by dissolved salts in the soil water.

Imagine an irrigation canal in the soil surface. Water is applied at a constant rate to the soil. The water immediately begins to wet the soil by moving downward through the profile as a result of the force of gravity. As water is added to the channel, it also moves slowly both laterally and downward. Gravity does not move water upwards or laterally, so a different energy source must be responsible. The upward and lateral movement of water (capillarity) is under the influence of matric potential resulting from the attraction between soil particles and water molecules and water molecules for each other (Fig. 12–5A and B).

When irrigating, a grower has to remember that water will move laterally only if there is enough matric potential to pull the water in that direction. This is usually more of a problem with drip irrigation systems, where the water comes in at a single point at the same time the media in the pot that contains a high percentage of peat moss. When the peat moss gets very dry, there may not be enough matric force necessary to move water laterally. In that case, the water will run

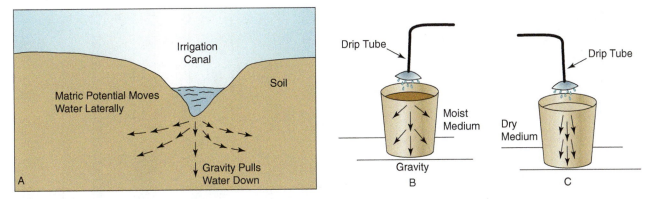

Figure 12–5

Water from an irrigation source has to move from that source into the soil. In the case of an irrigation canal system (A), water moves down because of gravity and laterally because of the matric potential of the soil. The same forces work on a drip irrigation system (B). A wetting problem can occur when using drip irrigation in pots containing a high percentage of peat moss. If the medium is very dry, the water does not move laterally (C), and the irrigation water is not distributed evenly. Only the area under the tube becomes wet. Source: Margaret McMahon, The Ohio State University.

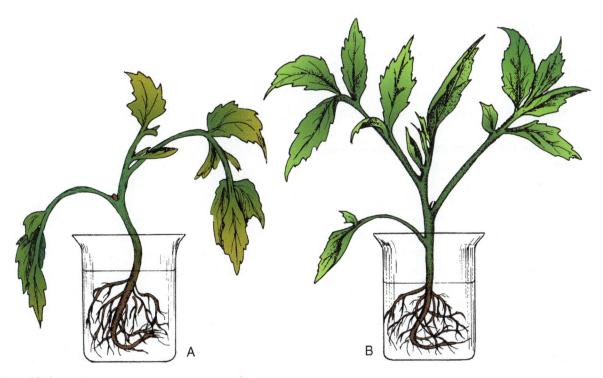

Figure 12–6

Effect of osmotic energy. Even though the plants in both beakers started with the same quantity of water, it was not equally available. In beaker (A), salt was added, creating an osmotic pressure too great for the plant to overcome. Since water was limited in its availability, the plant wilted. Beaker (B) contained pure water, which was readily available to the plant. Remember that fertilizer is a salt. High fertilizer concentrations can create an osmotic potential in soil, which in turn can inhibit water uptake by plants.

straight down through the medium, and the medium to the sides will not be wet (see Fig. 12–5B and C). A thorough surface watering using a hose or other overhead irrigation system or immersing the pot in a bucket of water is needed to rewet the medium and establish the cohesive forces.

Salts are present in soil water. The salts create osmotic energy, and if the salts are present in a sufficiently high concentration, the osmotic energy prevents water movement into the plant (Fig. 12–6). Fertilizer is a salt and, in high concentrations, can create an osmotic potential that inhibits water uptake by the plant. It can even lower the water potential enough to pull water out of the roots, causing root damage often referred to as fertilizer "burn." Container-grown plants that receive fertilizer in the irrigation water

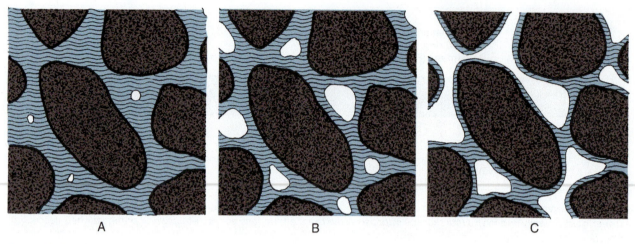

Figure 12–7

(A) In a saturated soil, all the void spaces are filled with water, indicated by the horizontal lines, and the tension (negative pressure) needed to remove the water from the soil particles is zero. (B) As the soil drains, the air:water ratio becomes larger and air occupies some of the void spaces. The tension at field capacity (i.e., when drainage ceases) is about 0.33 atm. (C) If the soil continues to dry, then the water film becomes so thin that extracting the water requires a tension of 15 atm (15.2 bars). This is more than most plants can exert. Therefore, the plants wilt permanently.

(fertigation) can be especially susceptible to this problem when the containers become very dry between fertigations. It is good practice to apply enough clear water shortly ahead of fertigating to make sure the soil or medium is moistened and has a sufficiently high water potential before applying the fertilizer solution.

After a prolonged rain or irrigation, the air in the soil pores is displaced with water. As mentioned earlier, in this condition, the soil is saturated, the soil moisture potential (tension or suction) is zero, and no energy is required to remove water from the soil particles (Fig. 12–7). This state will prevail as long as water is applied at a rate equal to the rate of water loss by drainage, by plant use, and evaporation. When no more water is added, losses continue, first from the larger macropores and then from the smaller micropores. Loss of water continues until the adhesive and cohesive forces equal gravity. This usually takes two to three days in a loam soil. At this moisture content, the soil is said to be at **field capacity**. The water has drained from the macropores but most of the micropores still contain water. If plants are growing in the soil, water loss continues as the plants take up water. If no water is added, eventually the soil reaches a moisture content that has a water potential less than plant roots so the plants cannot take up any more water. The water content of the soil cannot sustain plant life and the plants permanently wilt. The soil moisture content at which a plant wilts (the sunflower is often used as a reference) and cannot recover when placed in an environment of 100 percent relative humidity is termed

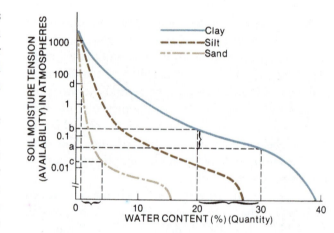

Figure 12–8

Soil moisture release curves showing the relationship between water content and soil moisture tension with three different soil textures. For example, compare the 10 percent loss of water from the clay soil (30 percent – 20 percent) and the accompanying slight increase in soil moisture tension (from a to b), with the 3 percent loss of water from the sandy soil (4 percent – 1 percent) and the accompanying large increase in soil moisture tension (from c to d). A similar analysis can be made on this graph with the same textured soil in different soil moisture ranges. In the wet range (high soil moisture percentage), a large loss of water produces a small increase in soil moisture tension, but in the dry soil moisture range, a small loss of water produces an increase in soil moisture tension that could easily go from wet enough to sustain plant growth to the permanent wilting point.

the **permanent wilting point (PWP)**. Soil texture determines the relationship between soil water content and soil moisture tension (Fig. 12–8).

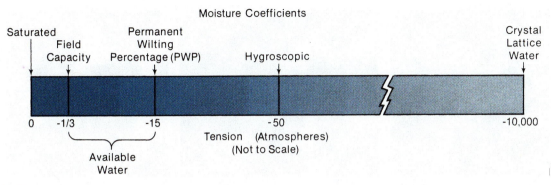

Figure 12–9

Relationship between soil moisture tension and the moisture coefficients. Water available to plants is defined as the water between field capacity and PWP. Note that soil scientists often use the term tension rather than water potential when describing soil water.

Note in Fig. 12–8 that the permanent wilting point of approximately 15 atmospheres of tension occurs at different soil moisture contents, depending on the soil texture. For an average loam soil at the PWP, all macropores and all but the smallest micropores are emptied of water. If the soil moisture depletes further, the soil becomes air dry. In this condition, all liquid water is gone except that held as a thin, tightly bound layer around the soil particles. This adsorbed water is called **hygroscopic water**, and the soil moisture tension is far beyond the availability range for plant use (Fig. 12–9). Another form of water, even less available to plants and called the **crystal lattice water**, is held in the soil's crystalline structure. This water can be removed only by applying sufficient heat to destroy the crystalline structure.

Anyone who grows plants for a living is interested mainly in the water available for plant growth. **Available water (AW)** is defined as the soil moisture between field capacity (FC) and the permanent wilting point (PWP):

$$AW = FC - PWP$$

Water between FC and saturation is not considered available because it is lost through drainage. Water at tensions greater than PWP is held too tightly by the soil for plants to remove. The definition of AW could imply that the water between FC and PWP is equally available to all plants, but this is not necessarily true. Water near PWP is not as available to most plants as that near field capacity. Also, the PWP is not the same for all plant species. Deep-rooted crops may have an advantage when soil closer to the surface is dry because soil moisture in the area of the deeper roots may be adequate for growth. Water in pots and planters is subject to the same influences as that found in field soils, although on a much smaller scale.

PLANT WATER

The role of water in the absorption and transport of raw materials used for photosynthesis in the green parts of the plant and the translocation of the products of photosynthesis to areas of storage or consumption is important for an understanding of plant growth and development.

Absorption and Conduction of Water

Water is absorbed by the roots from the soil and distributed throughout the plant, even to its highest leaves—perhaps 60 m (200 ft) or more above the source. This formidable task requires considerable energy, estimated to equal about 16 kg/cm^2 (225 lb/in^2) of pressure. The energy to move the water comes from the water potential gradient that develops in the soil, plant, and air continuum. In most cases, the water potential of moist soils is greater than that of the roots, so water moves into the roots (some anatomical features of the root aid or deter this process), which have a lower water potential than soil. The water then enters the root xylem by active and passive forces. The upward movement of water occurs mainly in the xylem. Xylem is composed of a thin, hollow tube that extends from the root through the stem and into the leaves. Once inside the xylem of the root, the potential or tension continues to decrease from the root to the stem, to the leaves, to the air (Fig. 12–10). The tension, along with the cohesive (hydrogen bonding) properties of water, creates a chain of water molecules that is pulled through the plant and out the stomates to the air surrounding the leaf. This combination of forces is called **cohesion/tension** and explains how 300 foot tall Coast Redwoods (*Sequoia sempervirens*) can get water to their uppermost needles.

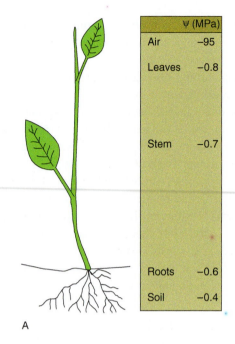

	ψ (MPa)
Air	−95
Leaves	−0.8
Stem	−0.7
Roots	−0.6
Soil	−0.4

A

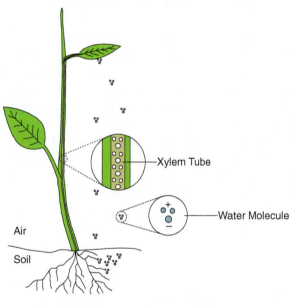

B

Figure 12–10

(A) Decreasing water potential (ψ) from the soil through the plant to the atmosphere creates tension that pulls the cohesive chain of water molecules (B) from the soil into the roots, through the xylem of the stem and leaves, and out into the atmosphere. Source: Margaret McMahon, The Ohio Sate University.

Minerals dissolved in the water are also carried to leaves, stems, and fruits via the xylem stream. Examples of important minerals are nitrogen used for protein synthesis, phosphates used to make ATP during photosynthesis, and magnesium that becomes part of the chlorophyll molecule.

Transpiration

The flow of water from the roots through the xylem in the stems to the uppermost leaves is called the **transpiration stream**. The transpiration pull in the leaves is created by the evaporation of water molecules from the outer surfaces of the mesophyll cells. As this water is lost through the stomates to the atmosphere, the mesophyll cells become water deficient, and a water potential difference is created between the dry mesophyll cells and the walls of adjacent moist cells. Because of water's cohesive properties, water from the wetter adjacent cell walls begins to diffuse into the less hydrated cells. This process moves from cell to cell deeper into the leaf where eventually water from the xylem moves into the most interior cells as they dry out. Continued loss of water molecules at the leaf surface establishes a flow of water throughout the plant from the roots to the leaves to the air.

The evaporation of water from plant leaves is one of the most important ways a plant regulates its temperature. As the water evaporates from the plant, evaporative cooling takes place. Evaporative cooling occurs because for water to change from a liquid to gas requires an input of energy in the form of heat and the resultant cooling of the surrounding area. That is why on a hot, sunny day the leaves of plants that are well-watered can be several degrees cooler than the surrounding air. It also reduces considerably the heating of the leaf by radiant (solar) energy.

When there is not enough water available to allow the plants to take in water, the plant's stomates close to conserve water. This reduces the rate of transpiration and can cause an increase in plant temperature. However, because plants contain a large amount of water, that water can still help to cool the plant because, as you learned earlier in this chapter, water can absorb a lot of heat before it increases in temperature. That is why desert plants such as cacti and succulents can survive under the very hot and dry conditions of the desert. They contain a huge amount of water that acts as a heat absorber. That water also cools down very slowly so during the cold nights that occur in deserts, the plants are kept relatively warm.

In addition to the transpiration stream, a small amount of water is absorbed into the roots by osmosis. **Osmosis** is the flow or diffusion that takes place through a semipermeable membrane (as in a living cell) that typically separates either a solvent (water) and a solution, or a dilute solution and concentrated solution. These concentration gradient differences create flow conditions through the semipermeable membrane

until equilibrium is established. At this equilibrium concentration, osmosis ceases. The proportion taken in by osmosis compared with the amount taken in by the transpiration stream depends on the rate of transpiration. Normally, however, the amount entering by osmosis is relatively small.

Not all water evaporates from the leaf. Some water is part of the plant's structures or held in the cytoplasm, some is used for biochemical processes, and some is stored in the tonoplast. The high volume water inside a cell creates turgor pressure, which gives plants rigidity. When there is insufficient water to create turgor, the plant wilts.

Absorption and Transport of Mineral Nutrients

In higher plants, the minerals initially needed to start growth of a new plant are normally provided by the seed or stored in the propagating tissue. But as these are used, additional nutrients for the plant's continued growth must be absorbed from the soil.

Chemical analyses of the sap in root cells for various mineral nutrients reveal that these cells have concentrations 500 to 10,000 times higher than those of the same element in the soil solution. If simple diffusion were the only mechanism involved in taking up soil nutrients, the mineral nutrients could not move into the roots against such a high concentration gradient. In addition, the plasma membranes of cells are largely impermeable to the movement of ions. Thus, energy is required to move ions against the concentration gradient and through the impermeable membranes. In the

root cells, this energy is obtained from the respiration of starches and sugars that originated in the photosynthetic processes.

A plant absorbs mineral nutrients against a high concentration gradient by a process called **active transport**. Active transport is facilitated by special proteins present in the plasma membranes of cells. Various models for these "carrier proteins" have been proposed, and one model is shown in Figure 12–11. In this model, a potassium ion, which has moved through the cell wall, is bound by the carrier at the outer surface of the plasma membrane. In the energy-requiring step, the protein moves or changes shape so that the K^+ ion is moved to the inside of the membrane. Following release of the ion, the protein resumes its original shape or position and is able to carry another potassium ion.

Translocation of Sugars

Sugars that are synthesized during photosynthesis and other metabolites synthesized by the plant using photosynthetic energy move throughout the plant, primarily in the phloem tissues. The movement is mostly downward from leaves to roots, but lateral or even upward movement from leaves to fruits or buds or other storage organs also occurs. In woody perennials, the phloem tissue is just underneath the bark following development of secondary vascular tissue. Thus, when this phloem tissue is severed by girdling, the tissue above the cut proliferates whereas the tissue below the cut is starved for photosynthates (Fig. 12–12). Accidental or unintentional girdling, which can be caused by deer feeding on bark during the winter,

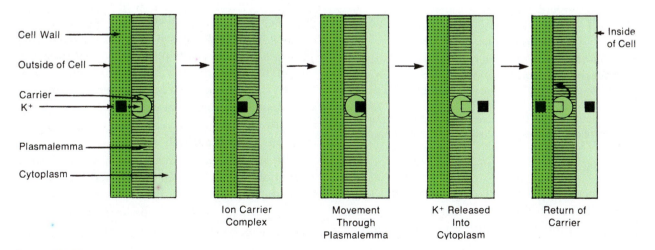

Figure 12–11

A schematic diagram of carrier ion transport. Source: Figure drawn using data from T. E., Weier, C. R. Stocking, M. G. Barbour, and T. Rost, *Botany: An Introduction to Plant Biology,* 6th ed. (New York: John Wiley & Sons, 1982).

Figure 12–12
Grape shoot girdled by removing the bark and the phloem and leaving the xylem intact causes sugar and other carbohydrates to concentrate above the girdle. Grape shoots are often girdled commercially to increase fruit size.

constricting wires or ties left around a tree trunk, or roots circling around a tree trunk below the ground, can severely stunt and even kill a tree by starving the roots for photosynthates. In some cases, trees are intentionally girdled to kill them.

The rate of translocation of sugars in the phloem is, in some instances, more than a thousand times faster than simple diffusion of sugar through water. The rate of translocation has been measured in many plants, and average values of 1 to 6 $g/cm^2/hr$ have been found in developing fruits and tubers. The forces that create these high rates of movement are called active transport, as already mentioned.

Metabolites, ions, and water move through the plant mainly by the processes described above but there are some other types of movement too. Some of the processes are relatively slow and some relatively fast. The processes are:

1. Diffusion, which transports ions and molecules slowly.
2. Cytoplasmic streaming, which transports molecules and ions within the cytoplasm at a considerably faster rate than diffusion.
3. Multi-directional mass flow translocation of material in the phloem.
4. Very rapid upward movement of water and mineral nutrients through the xylem.

SUMMARY AND REVIEW

Water makes up much of a plant's mass and is required for the plant's structure and nearly all plant processes. The dipole characteristic of water is what gives it its unique properties and stability. Water can dissolve many materials and its molecules are attracted to each other (adhesion) and to other molecules (cohesion). It has the ability to absorb a high amount of heat without a large increase in temperature.

Water is held, removed from, and moved around in the soil by several forces including gravity, matric (cohesive and adhesive), and osmotic. The movement of water in soil depends on water potential (symbolized by the Greek letter Ψ) which is the sum of the forces acting on the water. Pressure raises water potential and tension (suction) lowers it. Water moves from high to lower water potential so changes in water potential direct the flow of water. Water movement in the soil also moves oxygen and minerals through the soil. Good irrigation management is actually good management of water potential. Fertilizer can increase osmotic potential so applying liquid fertilizer to a very dry soil can inhibit the water from entering plant roots.

Soil that is filled with all the water it can hold is saturated and has zero water potential. The first water to leave a saturated soil is removed by the force of gravity. When gravity has removed all the water it can, the soil is at field capacity. Further removal happens by plants taking up the water or evaporation from the soil surface. When enough water has been removed from the soil that plants can no longer take up water, the soil is at its permanent wilting point (PWP). The water held in the soil between saturation and PWP is called available water.

Cohesion between water molecules and the tension created from decreasing water potential create the force that moves water from the soil through the plant to the air. This cohesion/tension force is strong enough to move water to the tops of the tallest trees. The flow of water into, through, and out of plants is called the transpiration stream. Transpiration is very important for many reasons. It is a major factor in keeping plants cool under hot conditions and it moves mineral nutrients taken in by the roots to other parts of the plant.

Water that does not leave the plant through transpiration is used as part of the plant's structure, held in

the cytoplasm and tonoplast, and used for biochemical processes. The water stored in the cell creates turgor pressure, which helps to give plants rigidity.

Water is also the solute that carries the sugars and other products of photosynthesis throughout the plant through the phloem tissue. Blocking the flow of phloem sap can severely injure and even kill a plant. Other methods of moving water and solutes within a plant are diffusion and cytoplasmic streaming.

KNOWLEDGE CHECK

1. List four functions of water in plants.
2. Why is the water molecule called a dipole? How does being a dipole explain why water molecules are attracted to each other?
3. What force causes water to move just downward in soil? What force causes it to move in any direction, not just downward?
4. What is the term for the force that determines the direction in which water flows?
5. In each of the following, place an arrowhead on the line to indicate the direction in which water will flow.
   a. 10 Mpa —— 2 MPa
   b. –10 Mpa —— –2 MPa
   c. –0.1 Mpa —— –2 MPa
6. Why may it be necessary to use a hose to thoroughly wet a very dry pot that is watered by a drip tube?
7. Why is it recommended to apply some clear water to pots before applying a liquid fertilizer?
8. Describe the following as they relate to the amount of water in the soil:
   a. saturated soil
   b. field capacity
   c. permanent wilting point
   d. hygroscopic water
   e. crystal lattice water
   f. available water
9. What are the two forces that move water to the top of plants?
10. What tissue carries the water from the root to the leaf?
11. Describe how the transpiration stream is created.
12. How does transpiration cool plant leaves?
13. Why is transpiration important in the uptake of mineral nutrients from the soil?
14. What tissue carries the water in which sugars and metabolites are dissolved?
15. In what direction can the tissue described in question 14 carry the water and solutes?

FOOD FOR THOUGHT

1. To provide better water holding capacity to soil and growing media, gels that hold water have been added to them. Some of these appear to work better than others. If you were to investigate the reason for the differences, how would information about the water potential of each gel be useful to you?
2. A farmer has noticed that the roots of his young plants appear to be burnt. He had recently irrigated the plants because the soil was getting very dry and he didn't want them to wilt. He also applied fertilizer granules around them just before he irrigated. How would you explain that the fertilizer might have caused the problem?
3. Growing tomato plants upside down in containers has become popular for homeowners. Why would an advertisement for the containers that stated that gravity would help the plants take in water probably be inaccurate?

FURTHER EXPLORATION

BROOKS, K. N., P. F. FOLLIOTT, H. M. GREGORSON, AND L. E. DEBANO. 2003. *Hydrology and watershed management,* 3rd ed. Ames, IA: Iowa State University Press.

REED., D. W. (ed.). 1996. *Water, media, and nutrition for greenhouse crops.* Batavia, IL.: Ball Publishing.
TAIZ, L., AND E. ZEIGER. 2006. *Plant physiology,* 3rd ed. Sunderland, MA: Sinauer Associates, Inc.

*13
Mineral Nutrition

HOW PLANTS ABSORB NUTRIENTS FROM THE SOIL

Before discussing plant nutrients and their function, some understanding of how these nutrients get into the plant is needed. Nutrients do not simply flow into the cytoplasm of a plant cell in the water absorbed. Instead, there are significant barriers to the uptake of nutrients into the plant. These barriers require systems to move nutrients into the roots. These systems for nutrient movement have a strong attraction for a single type of nutrient (e.g., nitrogen, potassium), so the uptake and movement of individual nutrients is regulated by the plant.

It is important to emphasize that the plant nutrients are the charged forms (ions) of the elements. Some ions may have one or more different charges and chemical forms (Table 13–1). In addition, most of the ions in the soil solution are hydrated; that is, they are surrounded by water molecules.

Long-distance mineral transport in plants is accomplished by two tissues: xylem and phloem (Figs. 13–1 and 13–2). In general, the xylem is the vascular tissue that provides upward movement of water and ions to the shoots, whereas the phloem is the vascular tissue that provides downward and lateral movement of nutrients from leaves to other parts of the plant. The process of mineral transport starts with getting the nutrient elements into the xylem of the plant.

Plant cell walls are composed of cellulose, hemicellulose, and pectin. You can think of the cell wall as very porous. That is, there are spaces in the cellulose fibers and in the pectin between cell walls that are large enough to allow relatively unimpeded flow of ions and water into the root. This flow continues until it reaches the endodermis. Flow of ions via this route is termed **apoplastic** movement (Fig. 13–2).

The endodermis is composed of cells that have cell walls impregnated with a waxy substance called suberin. Suberin provides an impenetrable barrier for both water and ions, so that transit through the cell wall and intercellular spaces is no longer possible and access to the xylem is blocked. The ions must pass through the plasma membrane (or plasmalemma) at this point or at some place further out in the cortex of the root. If an ion passes the plasma membrane further out in the root or at the endodermis, it can move to the xylem by going from cell to cell through the cytoplasm. This transit route is termed **symplastic** (see Fig. 13–2). Regardless of their route, once the ions are

key learning concepts

After reading this chapter, you should be able to:

- Explain how nutritional elements are taken into the plant through the roots.
- List the fourteen essential mineral nutritional elements and describe the roles they play in plant growth and development.
- Describe the nitrogen cycle and understand its importance to all life.
- Discuss the concepts of nutrient deficiency and remobilization and recognize the deficiency symptoms for nitrogen, phosphorus, potassium, calcium, magnesium, and iron as well, ways to correct the deficiencies.

Table 13–1
SUMMARY OF MINERAL ELEMENTS THE FORM/S FOUND IN PLANTS

Element	Typical Form/s in Plants
Nickel	Ni^{2+}
Molybdenum	Mo^{3+}
Copper	Cu^{2+}
Zinc	Zn^{2+}
Manganese	Mn^{2+}
Iron	Fe^{2+}, Fe^{3+}, heme
Boron	BO_3^{3-}
Chlorine	Cl^-
Sulfur	SO_4^{2-}, SO_3^{2-}, SH^-
Phosphorus	PO_4^{3-}, $P_2O_7^{4-}$
Magnesium	Mg^{2+}
Calcium	Ca^{2+}
Potassium	K^+
Nitrogen	NO_3^-, NO_2^-, NH_4^+

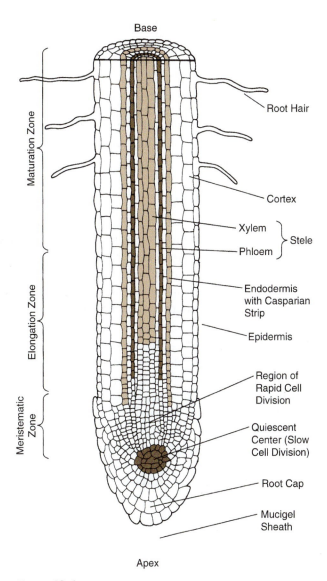

Figure 13–1

Longitudinal view of a growing plant root. The dotted lines for xylem and phloem indicate the regions of the root where differentiation occurs. Note that the phloem, which delivers nutrients to the growing root, extends farther down than the xylem, which transports nutrients from the soil solution to other plant tissues. Casparian strips are the waxy deposits in the cell walls that prevent movement of water and ions into the xylem via the apoplast (Fig. 13–2). The formation of endodermis complete with Casparian strips corresponds to the point where differentiation of xylem is complete.

Source: K. Esau, *Plant Anatomy* (New York: John Wiley & Sons, 1953). Reprinted with permission of John Wiley and Sons, Inc.

past the endodermis, they can move to the xylem vessels relatively easily.

The major control point for movement of ions is at the plasma membrane. The plasma membrane is composed of a lipid bilayer, that is, a double layer of lipid molecules. The important concept to remember is that this lipid bilayer is essentially impermeable to ions. So how do ions get through the plasma membrane? The answer is that proteins are embedded in the membrane (Fig. 13–3), and these proteins serve as channels or carriers for the ions. The exact mechanisms by which ions are transported through membranes are beyond the scope of this discussion. But the transport proteins have specificity for nutrient elements. The proteins can also be controlled independently so ion passage through the plasma membrane is selective. What this means is that, depending on the needs of the plant, nutrients can enter the plant at different rates. Interactions among nutritional elements is also important. Some elements interact synergistically with some elements and antagonistically with other elements. For example, K uptake can be enhanced by adding Ca to the soil. On the other hand, Ca and Mg are antagonistic and an excess in one can lead to a deficiency of the other in a plant. Likewise, phosphorus can interfere with the uptake of iron by a plant.

Before concluding this section, the growth and structural features of roots that allow for absorption of nutrients in the soil must be considered. First, it is important to recognize that roots are constantly growing during the growing season. One can think of roots as exploring the soil; that is, as nutrients are depleted at one area of the soil, it is necessary for the plant to search for ions in new locations. Second, it is important to emphasize the role of root hairs (Fig. 13–1). Root hairs are the site of much of the active ion uptake

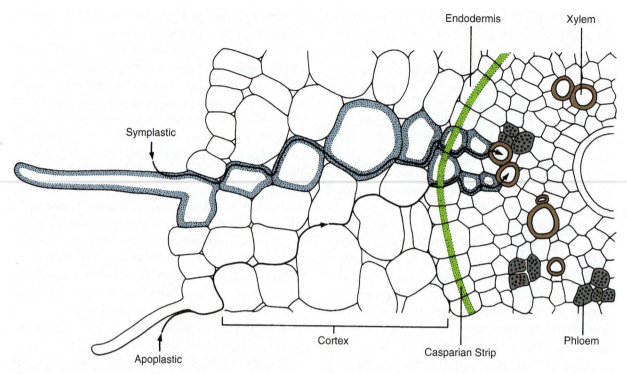

Figure 13–2

Cross section of a root showing the two alternate pathways for the entry of ions into the root tissue. The presence of Casparian strips in the endodermis prevents apoplastic movement of ions into the vascular tissue. Thus, at some point, ions must cross the plasma membrane of root cells and move through the cytoplasm.

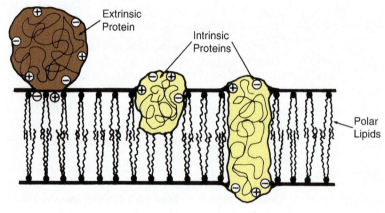

Figure 13–3

General features of proteins associated with the plasma membrane. Surface charges on the proteins allow them to be associated with the surface of the membrane or be partially or completely embedded in the membrane. Proteins that span the entire membrane can act as channels or carriers of ions.

because of their enormous surface area. These structures, which emerge about 1 mm behind the root tip, are extensions of epidermal cells that increase the surface area of the growing root many, many times. As the root grows, these hairs are eventually sloughed off. Thus, continued root growth is necessary to promote continuous root-hair development and proper nutrient uptake.

ELEMENTS THAT PLANTS NEED FOR GROWTH AND REPRODUCTION

Seventeen chemical elements are known to be essential for the growth of most plants, and a few others are used by some plants under certain conditions. An element is considered essential if the plant needs it to properly

complete its lifecycle. The **essential elements** are carbon (C), hydrogen (H), oxygen (O), nitrogen (N), phosphorus (P), potassium (K), calcium (Ca), magnesium (Mg), sulfur (S), iron (Fe), manganese (Mn), molybdenum (Mo), copper (Cu), boron (B), zinc (Zn), nickel (Ni), and chlorine (Cl). Some halophytes (plants that require salts) have been shown to need sodium (Na), and some microorganisms that fix nitrogen symbiotically or nonsymbiotically require cobalt (Co). A mnemonic (memory) device for remembering the elements is: C HOPKNS CaFe Mg B Mn CuZn Mo, Ni, Cl (C Hopkin's café managed by my cousins Mo, Nigel, and Clyde).

Plants are mainly composed of the elements carbon (C), hydrogen (H), and oxygen (O), which come from the air and water the plant takes in. In fact, these three elements comprise greater than 95 percent of the dry weight of a plant. The remaining elements are called the mineral nutrients and primarily come from the soil. Relative to the amount of nitrogen (the greatest mineral nutrient in plant dry weight), the amount of oxygen in a typical plant is thirty times greater, carbon is forty times greater, and hydrogen is sixty times greater. Mineral nutrients are divided into groups according to the quantity plants use.

The primary **macronutrients**—mineral nutrients used in largest amounts—are nitrogen, phosphorus, and potassium; a secondary level of macronutrients—used in lesser amounts than primary—are calcium, magnesium, and sulfur; and the remaining used in the smallest amounts are the **micronutrients**. Micronutrients are sometimes referred to as trace elements. Understanding the interactions among nutritional elements is also important. Some elements interact antagonistically with other elements. As noted earlier, K uptake can be enhanced by adding Ca to the soil. Ca and Mg are antagonistic, an excess in one can lead to a deficiency of the other in a plant. It is important to understand what is called "the law of the minimum" which means that even if all other elements are present in sufficient quantities, if any one element is deficient, no matter how much or little is needed by the plant, then the plant will not grow and develop properly. Think of it as the "weakest link" in the nutrient chain. Table 13–2 lists all the mineral elements and some of their functions in plants.

Table 13–2
MINERAL NUTRIENTS ESSENTIAL FOR PLANT GROWTH AND SOME EXAMPLES OF THEIR ROLES

	Nutrient Element	Function in Plants
PRIMARY NUTRIENTS	Nitrogen (N)	Synthesis of amino acids, proteins, chlorophyll, nucleic acids.
	Phosphorus (P)	Used in proteins, nucleic acids, metabolic transfer processes, ATP, ADP, photosynthesis, and respiration. Component of phospholipids.
	Potassium (K)	Sugar and starch formation, synthesis of proteins. Catalyst for enzyme reactions, stomate activity, growth of meristematic tissue. Opening and closing of stomates.
SECONDARY NUTRIENTS	Calcium (Ca)	Cell walls, cell growth and division; nitrogen assimilation. Cofactor for some enzymes.
	Magnesium (Mg)	Essential in formation of chlorophyll, amino acids, and vitamins. Essential in formation of fats and sugars.
	Sulfur (S)	Essential ingredient in amino acids and vitamins. Flavor in vegetables in the cabbage family and onions.
MICRONUTRIENTS	Boron (B)	Required in nucleic acid biosynthesis and in maintaining membrane structure as well as many other metabolic functions.
	Copper (Cu)	Constituent in enzymes, chlorophyll synthesis, catalyst for respiration, carbohydrate and protein metabolism.
	Chlorine (Cl)	Required for growth and development and a photosynthetic reaction.
	Iron (Fe)	Required for chlorophyll biosynthesis. A component of cytochromes and some reactions involved in electron transfer.
	Manganese (Mn)	Chlorophyll synthesis. Required for the catalytic activity of several enzymes. Also required in one of the key photosynthetic reactions.
	Molybdenum (Mo)	Essential in some enzyme systems that reduce nitrogen. Protein synthesis.
	Nickel (Ni)	Enzyme function.
	Zinc (Zn)	Required for the catalytic activity of several enzymes. Legumes need Zn for seed production.

Mobility within the plant is another important factor in nutrition. Mobility means the plant can translocate the element from one area to another, even after it has been incorporated into metabolites. As a result, deficiencies show up on lower leaves and tissues first as the element is relocated to younger, growing areas of the plant. Immobile elements cannot be easily "scavenged" by the plant. N, P, K, and Mg are the most mobile of the elements; Fe is one of the least mobile. For more details on mobility see the section on remobilization near the end of this chapter.

In the rest of this chapter, you will learn about the mineral elements that most often limit growth in plant growing systems—nitrogen (N), phosphorus (P), potassium (K), magnesium (Mg), calcium (Ca), and iron (Fe).

Nitrogen

Plants require large amounts of nitrogen because N is found in all proteins, it is a part of DNA and RNA, ATP, NADPH and NADH, as well as many other metabolites. Nitrogen is also the element most commonly deficient in plant growing systems. Nitrogen deficiency is a problem not only because of the great quantity the plant requires, but because of its relative low availability in the soil. Nitrogen exists in several forms in nature—as a gas (N_2, approximately 21 percent of the atmosphere is N_2), in organic compounds, and in the inorganic forms of ammonium (NH_4^+) and nitrate (NO_3^-). The only two forms of N that plants can absorb and utilize are ammonium (NH_4^+) and nitrate (NO_3^-). The unavailabe forms are made available to plants through a series of chemical reactions called the **nitrogen cycle** (Fig. 13–4) which like the carbon cycle is critical to life on earth. An explanation of the cycle follows.

Atmospheric N must be converted to nitrate or ammonium before most plants can use it. This occurs in two ways. The first is the conversion of molecular N_2 in the atmosphere to nitrate during lightning strikes. This source of N is small and is highly variable between locations and years. The other method is called **nitrogen fixation**, and is carried out only in some plants. This process is the conversion of N_2 gas from the atmosphere into ammonia (NH_3) by some types of bacteria. The ammonia is quickly converted to ammonium. The N-fixing bacteria of greatest agricultural importance are named rhizobia of which there are several

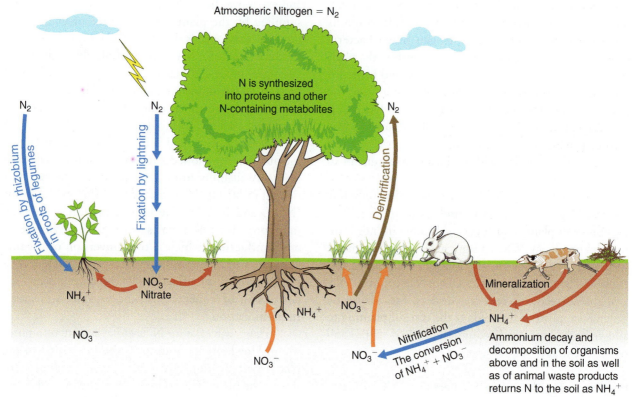

Simplified Nitrogen Cycle

Figure 13–4

The nitrogen cycle is the transfer of nitrogen between air, soil, and organisms.

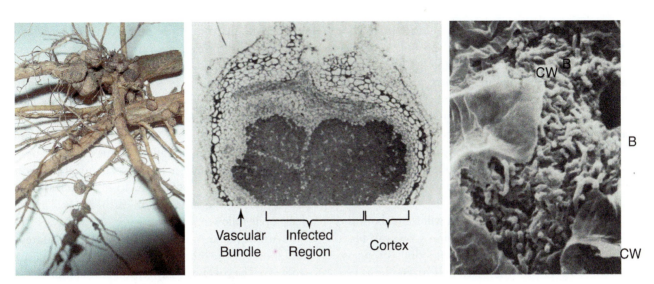

Vascular Bundle   Infected Region   Cortex

Figure 13–5
Left: *Mature nodules on the root of a soybean plant.* Center: *Cross section through a mature nodule. Note the location of vascular bundles in the cortex and the connection of these bundles to the root, which was located at the top of the section.* Right: *Scanning electron micrograph of an infected cell of a soybean nodule showing nitrogen-fixing bacteria. CW = cell wall; B = bacteria.* Source: Left: Margaret McMahon, The Ohio State University.

genera and many species. These bacteria infect legumes such as the vegetable crops peas and beans and the agronomic crops soybean and alfalfa (also legume trees such as the locust). They enter the root from growths on the root called nodules (Fig. 13–5, left).

When one slices open a root nodule, several different tissues can be seen (Fig. 13–5, center). The darker interior region of the nodule contains cells that house millions of nitrogen-fixing bacteria (Fig. 13–5, right). In a symbiotic relationship, the plant supplies the bacteria with carbohydrates, and in return the bacteria supply the plant with ammonium. For the growth of most legumes, additional nitrogen fertilizer is not needed or needed only in small amounts. Some nitrogen fixing legumes are grown as cover crops and plowed under or killed with herbicides to return N to the soil.

In general, a species of rhizobia infects only one species of legume. Thus, pea rhizobia do not form nodules on bean plants and soybean rhizobia do not form nodules on alfalfa. When a legume is first grown in the field, it is very important to supply an appropriate rhizobial inoculant, and these inoculants are commercially available at feed and fertilizer dealers. Inoculants are generally applied to the seeds at the time of planting to allow the bacteria to become established in the soil before the first root hairs appear.

Manufacturing Ammonia for Fertilizer   The chemistry of biological nitrogen fixation is similar to what occurs in the industrial manufacture of ammonia

for fertilizers and other uses. In manufacturing, the conversion of N_2 to NH_3 occurs at a very high temperature and at great pressure in the presence of a catalyst (Fig. 13–6). The process is called the **Haber–Bosch** process. Methane is used as a source of energy and CO_2 is a by-product. The process is quite energy intensive and expensive. In the biological process, carbohydrates supplied by the plant are the source of energy; the product, ammonia, is the same; and CO_2 is again a by-product (Fig. 13–7). The remarkable thing about the biological process is that it occurs at 1 atmosphere of pressure and at ambient temperature. Thus, the catalyst in the bacteria is much more efficient than anything chemists have been able to devise.

Most N in the soil is in organic material (as the metabolites mentioned earlier) and not available to plants. Availability to plants depends on the transformation of organic N to the mineral (inorganic) forms (NO_3^- and NH_4^+). The transformation is done by microorganisms and starts with **mineralization** or **ammonifaction**, which is the conversion of organic N to ammonium. Ammonium is then converted to nitrate through **nitrification**. The reverse process, the conversion of mineral N to the organic N is called **immobilization**.

Plants can take up N as either nitrate or ammonium. However, ammonium is the form used in the plant so nitrate must be converted to ammonium once it is in the plant. The ammonium is then rapidly converted into N-containing metabolites. If there is more nitrate taken up than can be immediately converted by

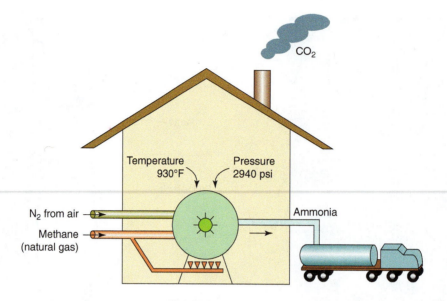

Figure 13–6

Some features of the manufacture of ammonia. Note the very high pressure and temperature required for the conversion of N_2 gas to ammonia. Other common nitrogenous fertilizers—urea, ammonium nitrate, ammonium sulfate—can be synthesized from the anhydrous ammonia. Source: John Streeter, The Ohio State University.

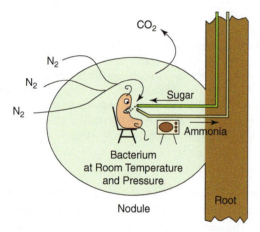

Figure 13–7

The conversion of N_2 gas to ammonia in legume nodules. In this case, sugar from the plant provides the energy, but much of the chemistry is the same as the industrial synthesis of ammonia. Source: John Streeter, The Ohio State University.

Figure 13–8

Nitrogen deficiency. Note the yellow older leaves and green new growth.
Source: Margaret McMahon, The Ohio State University.

the plant, the plant can store the nitrate in the vacuole. Nitrogen can be supplied to plants as ammonia, urea, ammonium, and nitrate. Ammonia and urea have to be converted to ammonium in the soil before plants can take it up. Also, remember that NO_3^- is a cation and can be easily leached from the soil by precipitation or overirrigation, which can contribute to N deficiency.

Deficiency Symptoms Nitrogen is one of the mobile elements so the first symptom is an overall yellowing (chlorosis) of lower leaves (Fig. 13–8).

The symptoms can spread to younger leaves and eventually growth is stunted if the problem is not corrected. Nitrogen can be supplied as: organic material (although its availability is delayed until it can be mineralized) and as ammonia/ urea (has to be converted to ammonium), ammonium, and nitrate fertilizers.

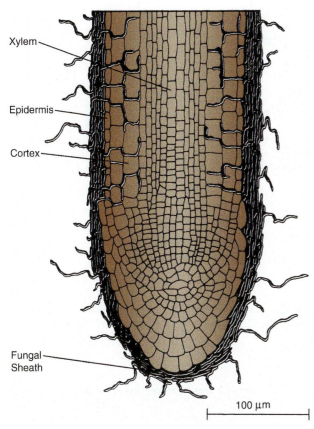

Xylem
Epidermis
Cortex
Fungal Sheath

100 µm

Figure 13–9
A plant root infected with an ectotrophic mycorrhizal fungus. With this type of fungus, the mycelium forms a sheath around the root and also enters the intercellular spaces in the root cortex. Source: L. Taiz and E. Zeiger, *Plant Physiology*, 2nd ed. (Sunderland, MA: Sinauer Assoc., 1991).

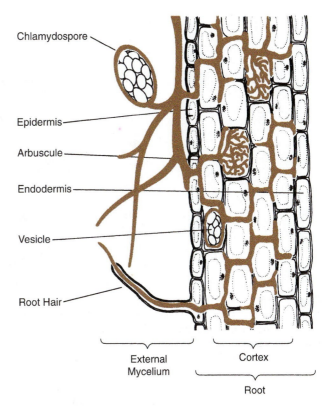

Chlamydospore
Epidermis
Arbuscule
Endodermis
Vesicle
Root Hair

External Mycelium   Cortex

Root

Figure 13–10
A plant root infected with a vesicular-arbuscular mycorrhizal fungus. With this type of fungus, a sheath around the root is not formed, but the fungal mycelium actually enters the cortical cells. The nomenclature of these fungi is based on the formation of sacklike structures called vesicles and branched structures called arbuscles in the root cells.
Source: L. Taiz and E. Zeiger, *Plant Physiology*, 2nd ed. (Sunderland, MA: Sinauer Assoc., 1991).

Phosphorus

Phosphorus plays many vital roles in the plant cell. One role is the presence of this element in energy-rich compounds such as ATP and NADPH, which were discussed in more detail in Chapter 11. A second critical role is as part of the phospholipids that make up membranes. A third role is as a component of DNA, RNA.

After nitrogen, phosphorus is generally the second most limiting element for plant growth because much of the P in the soil is tied up in organic matter and because many phosphorus compounds are only slightly soluble. Soil pH below 6 or above 8 inhibits the uptake of P. Achieving a sufficient supply of P for plant growth and economic yield is often a challenge for the growing plant. Managing soil pH to make P available and/or fertilizing with P is often required.

Mycorrhizal fungi help plants take up P and can be used as part of a fertility system. One type of

mycorrhizal fungi form a sheath around the root surface, and extensions of the hyphae grow between the cells of the root cortex (Fig. 13–9). The hyphae of a second type of mycorrhizal fungus actually enter the cells of the root cortex (Fig. 13–10). In both cases, the function of the fungus is much like that of root hairs. That is, the fungal mycelium allows for exploration or a much larger volume of soil, and the fungus associates with the roots even after the root hairs are gone. The fungus can absorb P from the soil and carry it back to the plant in exchange for carbohydrates.

These fungi associate with the roots of a wide variety of plants, and new mycorrhizal/plant associations are continuously being discovered. Mycorrhizae are available commercially for many plants.

Deficiency Symptoms Phosphorus is a mobile element so deficiencies show up first in lower leaves which are characteristically dark green and red or

Figure 13–11
Deficiency symptoms for phosphorus (tomato). Note the purple coloration of the foliage. Source: Robert McMahon, The Ohio State University, Agricultural Technical Institute.

Figure 13–12
Deficiency symptoms for potassium. Source: Robert McMahon, The Ohio State University, Agricultural Technical Institute.

purplish in color (Fig. 13–11). If not corrected, the plant becomes spindly. Phosphorus can be supplied as a fertilizer in several types of formulations. Supplying the right kind of mycorrhiza can make P more available to plants.

Potassium

Potassium is highly mobile in the plant. Potassium is critical for the osmotic control of stomate opening and closing and thus it is important in the processes of photosynthesis and transpiration. Potassium is also required for the activation of many enzymes. Although addition of K in fertilizer is usually required to provide maximum crop yields, K^+ is retained in the soil because of its positive charge and is not easily leached. Also, sources of K on the earth are more plentiful than are sources of N and P. Potassium fertilizers are also cheaper to produce because the mineral sources of K can be used directly.

Deficiency Symptoms Lower leaves of broad leaf (dicot) plants first show chlorosis as well as some necrosis (browning) starting at the margins of the leaf (Fig. 13–12). In monocots, the symptoms usually start as tip chlorosis that spreads along the margins. If uncorrected the symptoms will involve the entire leaf as well as younger leaves of both broad and monocot plants. Potassium is available in several different fertilizer formulations, including the direct application of mineral sources.

Calcium

Calcium has several structural roles in the plant including giving pectin a stable structure. Pectin is the polymer that acts as a glue between the walls of adjacent cells. Although not generally a nutritional problem for most plants. Deficiency can be a problem if the plant cannot transport Ca to rapidly growing tissues fast enough resulting in tissues collapsing and becoming soft and rotted as is seen in tipburn of lettuce. Another manifestation occurs in tomato plants, where there is often insufficient transport of calcium into the rapidly growing fruit. As a result, the fruit develops a brown patch on the bottom called blossom end rot and is unmarketable. Calcium deficiency has also been implicated in improper development of poinsettia bracts (the part of the plant that gives it its showy flowering characteristics). Often, foliar applications of calcium as $Ca(OH)_2$ or $Ca(NO_3)_2$ can prevent the problem. Applications are especially effective during times of rapid growth or cell expansion, especially when they follow periods of slow growth such as cool or cloudy conditions.

Deficiency Symptoms Along with those described above, the top of the plant can develop a twisted, distorted type of growth (Fig. 13–13).

Magnesium

Magnesium plays several roles in plants, but two stand out. First, Mg^{++} is an essential component of chlorophyll. The second critical role of magnesium is in energy

Figure 13–13
Deficiency symptoms for calcium (pansy). Note the deformed growth. Source: Robert McMahon, The Ohio State University, Agricultural Technical Institute.

Figure 13–14
Deficiency symptoms for magnesium (several species). Note the interveinal chlorosis in the lower leaves. Source: Robert McMahon, The Ohio State University, Agricultural Technical Institute.

transfer as a necessary component of adenosine triphosphate (ATP) mediated reactions. The reactive form in ATP is the magnesium salt. Thus, although we often write:

$$\text{substrate} + \text{ATP} \rightarrow \text{product} + \text{ADP} + PO_4^{-3}$$

The biochemically correct way to write the reaction is:

$$\text{substrate} + \text{MgATP} \rightarrow \text{product} + \text{MgADP} + PO_4^{-3}$$

Thus Mg is critical for energy transfer in plants.

Deficiency Symptoms   Magnesium deficiency is characterized by interveinal chlorosis of the lower leaves (Fig. 13–14). The interveinal pattern distinguishes it from N deficiency. Magnesium deficiencies are often treated with applications of $MgSO_4$ (epsom salts) as part of a fertilizer mix or as a foliar application.

MINOR ELEMENTS

An in-depth look at the other essential nutrients is not included in this book. However, the roles of these elements are listed in Table 13–2 on page 254. Most soils supply adequate amounts of these elements, although inclusion of them in nutrient solutions in greenhouse growth systems is often required.

The minor element that can be a deficiency problem in soil is iron, often because its availability is very

Figure 13–15
Deficiency symptoms for iron (pansy). Note the interveinal chlorosis in the upper leaves. Source: Robert McMahon, The Ohio State University, Agricultural Technical Institute.

pH dependent. Iron is important in several aspects of photosynthesis and energy transfer.

Deficiency Symptoms   Iron is one of the immobile elements. Interveinal chlorosis of the upper foliage is an indication of deficiency (compare to interveinal chlorosis of lower foliage for Mg) (Fig. 13–15). It can be corrected with a foliar application of an $FeSO_4$ or an iron chelate but the spray has to be applied directly to the affected foliage. Adjusting soil pH is often the best way to prevent or correct soil-related iron deficiency.

NUTRIENT DEFICIENCY CONSIDERATIONS

Deficiency symptoms were included in the discussions above of several nutritional elements to help you when you are diagnosing plant problems. However, reliance on typical nutrient deficiencies can be misleading because deficiency symptoms vary slightly from plant to plant, deficiency of more than a single element may be involved, and many deficiency symptoms are similar to symptoms of plant disease or other disorders. Thus, to only examine plant leaves and decide that the plant needs more nitrogen without doing further tests such as soil and leaf analysis can make the problem worse. For example, if a plant shows N deficiency symptoms, that might be the result of too little N in the soil but it might also be the result of damaged roots that cannot take up the N or something interfering with N metabolism in the plant. In this case, a soil test would indicate adequate levels but the leaf analysis would show a deficiency and adding more N to the soil would not correct the problem. In fact, adding more N could create a toxicity effect or, more likely, create a pollution problem.

A good strategy if nutrient deficiency is suspected is to give a small amount of that nutrient but at the same time perform further tests to confirm the diagnosis. Do not apply more of the nutrient until a deficiency in the soil is confirmed. To perform soil and leaf analysis, samples of the soil or medium in which the plant is growing and plant tissue are sent to a laboratory for analysis. Many such laboratories are maintained by agricultural supply companies and can be found through the local agricultural extension agent. These laboratories maintain lists of normal concentrations of elements in a wide variety of plants. Generally, the testing lab supplies an analysis of all major and most minor elements and provides warning if the concentration of one or more elements is outside the normal range for that plant. Although some cost is involved, this approach is clearly the most reliable way to establish the cause of a nutrient deficiency.

Although some deficiency symptoms are included in this text, there are several informative websites that show pictures of deficiencies of elements on various horticultural and agronomic crops. Two such websites are listed in this chapter's Further Exploration, and you are urged to visit them to see what information they provide and to witness the difficulty associated with visual analysis of deficiencies. Other sites continue to appear and can be found through web searches.

REMOBILIZATION

Reproduction is so important in the survival of a species that mechanisms have evolved for plants to preserve precious nutrients for future generations. In practical terms, this adaptation means that seeds of plants are usually a rich source of nutrients relative to other plant tissues, especially in annual plants. It also means that the plant will allocate resources to the seeds from other tissues and organs in a process called **remobilization**.

Nitrogen, phosphorus, and potassium are especially important in seed development. During the growing season, the plant expends considerable energy importing these nutrients from the soil solution. All three of these nutrients are relatively mobile, and large quantities of these elements are moved from vegetative tissues to reproductive tissues during reproductive development and the onset of senescence in annual and deciduous perennials.

An illustration of N, P, and K remobilization is shown in Fig. 13–16. Note that the vertical scale on this figure is percentage, with 100 percent being shown near the end of dry-matter accumulation. Thus, the actual amounts of nutrients translocated from vegetative to reproductive tissues are not shown. However, this presentation emphasizes the relative depletion of nutrients from vegetative tissues during reproductive development. Note that accumulation of N, P, and K continues during early reproductive development but that, at roughly eighty-five to ninety days after emergence, nutrient accumulation stops. Beyond this time, the nutrient content of leaf blades, stems, and petioles starts to decline. Between 85 and 120 days after emergence, a massive expansion of the proportion of nutrients occurs in the beans. Some of the nutrients are lost in fallen leaf blades and petioles, but by the time of leaf drop, much of the N, P, and K has already been withdrawn and remobilized to the seeds.

The same process occurs in all plants that undergo annual senescence of some vegetative tissues, although for some plants, the destination for nutrients is the roots and not the seeds. Thus, by harvest time, the N, P, and K content of corn stalks or grape leaves or apple leaves is very low—too low for these organs to be functional. They are largely composed of cellulose and other polymers of C, H, and O. As such, these expendable plant tissues provide organic material when returned to the soil, but they do not constitute a supply of N, P, and K that is adequate to grow the subsequent crop.

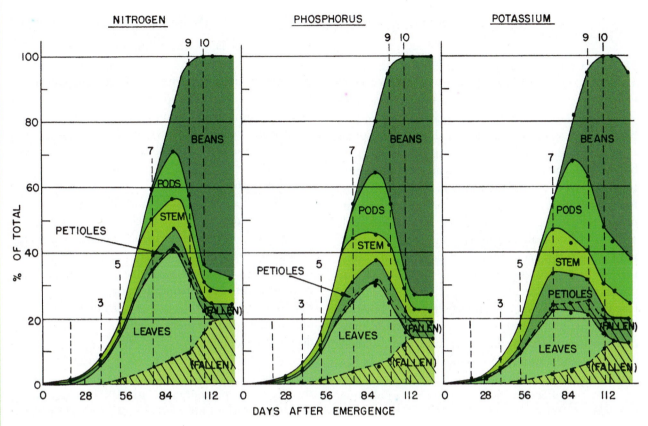

Figure 13–16

Distribution of N, P, and K in soybean plants during vegetative growth and reproductive development. "Fallen" refers to leaf blades and petioles that have senesced and fallen from the plant. Note that the vertical axis of the figure is in units of percent, so that actual amounts of nutrients are not shown, only the relative distribution of elements in the plants.

Source: *Agronomy Journal* 63 (1971): 408. Madison, Wisconsin American Society of Agronomy.

SUMMARY AND REVIEW

Nutritional elements move into the plant root as charged ions through a complex system that involves free movement (apoplastic) and selective transport (symplastic). The plasma membrane is the sight of most selective control. Once inside the root, ions are transported relatively easily through the xylem to other parts of the plant.

The absorption of minerals into the root depends on the growth of the root and its structural features. One can think of growing roots as exploring the soil to search for ions in new locations. The root hairs are the main site of nutrient absorption and in most healthy roots, they have an enormous amount of surface area.

The essential elements are carbon (C), hydrogen (H), oxygen (O), nitrogen (N), phosphorus (P), potassium (K), calcium (Ca), magnesium (Mg), sulfur (S), iron (Fe), manganese (Mn), molybdenum (Mo), copper (Cu), boron (B), zinc (Zn), nickel (Ni), and chlorine (Cl). The first three come from air and water.

The rest, which are called mineral nutrients, most often come from the soil. The mineral nutrients are categorized by amount required as macronutrients and micronutrients. All are essential, meaning that even the micronutrient that is needed in the smallest amounts will cause the plant to grow improperly if it is deficient.

Mobility is another classification of the mineral nutrients. Mobile elements can be "scavenged" from older tissue to support the growth of new vegetative and reproductive tissues.

Nitrogen is the element that most often limits plant growth. It is a component of proteins, DNA, RNA, and other metabolites. Nitrogen is not available for uptake by plants as N_2 or in its organic form; it has to be converted to NO_3^- or NH_4^+. Much of the N in the soil is in an organic form which is converted to a usable form during decomposition. Some plants develop a relationship with rhizobia bacteria. These bacteria form nodules on the plant root where they

convert N_2 to ammonium in a process called nitrogen fixation. The plant can use the ammonium and in turn, the bacteria can feed on the carbohydrates produced by the plants. The nitrogen cycle is the interconversion of all forms of N by physical and biological processes and is necessary for all life.

Most crop plants have a need for N that is greater than the supply in the soil and thus require N fertilization. N fertilizers can be applied as ammonia, urea, NH_4^+, and NO_3^-. Ammonia and urea have to be converted to ammonium for uptake. NO_3^- is easily taken up by plants but it is also the form that is most easily leached from the soil into the groundwater. The most common deficiency symptom is chlorotic lower foliage. Nitrogen can be supplied to plants through many types of material both inorganic and organic in origin.

Phosphorus is the second most limiting element for plant growth. The role of phosphorus in plant metabolism is in energy-rich compounds such as ATP, as a critical element in the structure and function of membranes and as part of DNA and RNA. Although plentiful in most soils, it is in unavailable or only slightly available forms. Soil pH influences P availability, with a pH of 6 to 8 being the ideal. Phosphorus availability can often be improved by adjusting soil pH. Mycorrhizal fungi can play a role in helping plants obtain P. These fungi are associated with many different plants and form a sheath around the plant root or penetrate the root. Hyphae from the fungi spread out and extract P from areas distant from the root. In a sense, they act like root hairs. Phosphorus deficiency is indicated by a dark green and red/purplish coloration to lower leaves. Phosphorus is usually supplied to plants in its inorganic phosphate form.

Potassium is critical for stomate opening and for enzyme activity. Potassium fertilizer is generally needed in less quantity than N or P because it is retained in the soil in a readily available form. Deficiency symptoms are chlorosis and necrosis of the margins of the leaves of dicots. Monocot symptoms start with tip then marginal chlorosis of the leaves.

Calcium is important in structural stability. Though not usually deficient, deficiencies can occur if the plant is not able to move Ca into new growth fast enough. Deficiency is indicated by tissue collapse as is seen in tipburn of lettuce or blossom end rot of tomatoes. Foliar sprays of $Ca(OH)_2$ or $Ca(NO3)_2$ applied during periods of rapid growth can help prevent the problem.

Magnesium is important as a part of the chlorophyll molecule and in energy transfer processes. Deficiency symptom is an interveinal chlorosis of lower leaves. Magnesium can be supplied by adding $MgSO_4$ to the fertilizer or as a foliar spray.

Iron is involved in photosynthesis and other energy transfer reactions. Often deficiency occurs not because of a low amount in the soil but because soil pH is such that uptake is inhibited. Adjusting soil pH often corrects the problem. Foliar applications of iron can also correct the problem but the spray has to reach the affected leaves.

For all suspected deficiencies, it is best to confirm visual symptoms with soil and tissue tests because visual symptoms can be misleading. Different problems can cause symptoms that appear similar. Also, soil and tissue testing are necessary to determine how much of an element is available and how much is actually being taken up and incorporated into plant tissues.

Remobilization is the process plants use to reallocate nutrients from one part of the plant to another. Usually the allocation is from old tissue to reproductive tissue. Some elements are more readily mobilized than others. N, P, and K are readily mobile. Others, such as iron, are relatively immobile.

KNOWLEDGE CHECK

1. In what general chemical form do mineral nutrients move into plants?
2. What is the term for the way mineral elements move between cell walls and into the plant up to the endodermis?
3. What is the term for mineral transport through the plasmalemma and endodermis?
4. What role do proteins play in moving mineral elements into the plant?
5. Why is active root growth and root hair development important for mineral uptake?
6. List the seventeen essential nutritional elements for plants.
7. What is the difference between macro- and micronutrients?
8. What is meant by mobile and immobile nutritional elements?
9. Where do deficiency symptoms first occur with mobile elements? With immobile elements?
10. What does the law of the minimum mean in terms of plant nutrition?
11. What two forms of N are available to plants?

12. Describe the nitrogen cycle and its importance to life.
13. What is nitrogen fixation? Where does it occur in plants? What organism does the fixing?
14. What economically important group of plants is associated with nitrogen fixation?
15. What is the Haber–Bosch process and why is it agriculturally important?
16. Define or describe the following: mineralization, ammonification, nitrification, and immobilization in terms of the nitrogen cycle.
17. Describe a role of each of the following in plant growth and development: nitrogen, phosphorus, potassium, calcium, magnesium, and iron.
18. What is the first symptom of nitrogen deficiency?
19. Why is managing soil pH important when managing phosphorus fertility?
20. What are mycorrhiza and what is their importance in phosphorus fertility?
21. What is the symptom of phosphorus deficiency?
22. Why is managing potassium fertility usually less of a problem than managing N and P?
23. What are the symptoms of potassium deficiency?

24. What are the symptoms of calcium defiency in plants? On tomato fruit?
25. When would foliar applications of calcium be most effective?
26. Describe how deficiency symptoms for magnesium differ from nitrogen.
27. Describe how deficiency symptoms for magnesium differ from iron.
28. What is often the best way to prevent or correct soil-related iron deficiency?
29. Why is it best to not rely solely on visual symptoms when diagnosing what appears to be a nutritional problem?
30. What are other tests you should do to confirm the initial diagnosis?
31. What is remobilization of minerals in plants?
32. How does remobilization explain why plant residues such as leaves, cornstalks, and other plant materials left after completion of a lifecycle are usually not good sources of plant mineral nutrients, though they can be good sources of organic material?

FOOD FOR THOUGHT

1. What do you think some consequences would be if you saw the lower leaves of a tomato crop turning yellow and you added a fertilizer high in NO_3^- to green them up only to find out that there was already enough N in the soil; that the problem was really the roots had been damaged during an earlier rainy period when the field was flooded?
2. What would you tell someone who is appalled at the idea that you want to infect your soybean crop with a bacteria (rhizobium) or your corn crop with a fungus (mycorrhiza) and wants to report you to the Center for Disease Control?
3. A plant "invests" considerable energy and resources into creating the proteins that allow for nutritional elements to be taken in selectively. Why do you think that selectivity is important enough to warrant that kind of investment by the plant?

FURTHER EXPLORATION

BARKER, A. V., AND D. J. PILBEAM. 2006. *Handbook of plant nutrition.* Boca Raton, FL: CRC Press.
EPSTEIN, E., AND A. J. BLOOM. 2005. *Mineral nutrition of plants: Principles and perspectives,* 2nd ed. Sunderland, MA: Sinaur Associates, Inc.
MARSCHNER, H. 1995. *Mineral nutrition of higher plants,* 2nd ed. London: Academic Press.

Websites for nutrient deficiency symptoms:
"Color Pictures of Mineral Deficiencies in Tomatoes," http://www.luminet.net/~wenonah/min-def/tomatoes.htm
NUTDEF Plant Nutrient Deficiency Database, http://hort.ufl.edu/nutdef/
http://agri.atu.edu/people/Hodgson/FieldCrops/Mirror/Nutrient%20Def.htm

unit three
Plant Production and Utilization Systems

✿14
Soil, Water, and Fertility Management

key learning concepts

After studying this chapter, you should be able to:

- Explain why and how land is prepared for growing plants.
- Describe how improper soil handling degrades soil and how improper handling improves soil.
- Explain the practices that improve degraded soil and prevent degradation and conserve soil.
- Discuss the basic principles and components of irrigation and drainage.
- Describe how plant nutrition is managed through fertility practices.

In earlier chapters you saw why soil, water, and nutrients are required for plant growth. Now you will learn how they can be managed to improve and manipulate the growth of plants using ecologically sound practices. Indeed, soil, water, and fertility management cannot be separated into distinct components. They are very tightly integrated and what you do in one area can dramatically impact the other two areas. When you choose the practices you will use with soil, water, and fertility management, you will have to consider these close relationships.

LAND PREPARATION

The first step in growing plants is preparing the soil where the plants are to be grown. Major purposes of land preparation are to:

1. Level the land where needed.
2. Incorporate crop residues, green manure, and cover crops.
3. Improve and maintain the **tilth** of the soil (tilth is a subjective term for the physical condition of the soil with respect to its capability to provide a good environment—aeration and porosity—for optimizing crop production)
4. Help control weeds, diseases, and insects.
5. Help control erosion where needed.

In general, **tillage** is defined as the mechanical manipulation of soil to provide a favorable environment for crop growth. Soil moisture condition at the time of tilling is a factor in the effectiveness of tillage. Unfavorable moisture conditions—too dry, too wet—will result in ineffective tillage and will damage soil structure. Effective tilling improves the moisture holding characteristics of the soil. Ideally, soils should hold enough water to maintain the plant for several days between irrigations yet allow enough air to reach the roots for sufficient root respiration. The "ideal" soil is often considered to consist of 50 percent solid and 50 percent open or pore space. Of the pore space, 50 percent is small enough to hold water by capillary force and 50 percent is large enough to drain by gravity.

Tillage is done with a wide variety of equipment and for various purposes. The soil should provide an environment conducive to rapid germination of seeds and good plant growth. For most crop plants, this means that the surface soil is loose and free of clods (Fig. 14–1). The subsoil is permeable to air and water

Figure 14–2
A rice field ready to be planted. The levees follow contour lines to allow uniform water depth within each paddy.
Source: USDA Natural Resource Conservation Service.

Figure 14–1
A well-prepared raised field that might be used for small seeded vegetable crops such as lettuce or carrots (above). It should be free from unwanted crop residues or excessive cloddiness (below).

Figure 14–3
Heavy equipment such as tractors and implements can contribute to soil compaction especially when the soil is moist and the tires run over the same area several times.
Source: USDA-NRCS, *http://photogallery.nrcs.usda.gov/*

and has adequate drainage and aeration. It should not be water-logged or anaerobic (without oxygen).

Land Leveling

Land is leveled to permit water to flow and spread evenly over the soil surface without causing erosion. In considering the land's suitability for leveling, the land's productive capacity and the method of irrigation to be used are evaluated.

Irrigated land generally benefits from being level, especially if flood (Fig. 14–2) or furrow irrigation is used and row crops are grown. Land leveling can also be used to remove excess surface water from depressions that result in poorly drained fields. The use of GPS and laser technology is common when precision

leveling is required. Heavy equipment may be necessary for leveling when the terrain is rough or soils are heavy (Fig. 14–3). Timing is important. Land should not be leveled when saturated or near saturation because leveling of wet soil subjects it to compaction.

Plowing

Often, the first step in land preparation after any necessary leveling is to plow the land. Plows invert the soil and bury the residues from the previous vegetation, but they often leave the soil in large linear lumps that must be reduced in size. When large amounts of residues are left on the field, they are often chopped with a disk or a rotary stalk cutter before plowing.

Figure 14–4
A moldboard plow is used to turn the soil over and bury in surface vegetation.

Figure 14–5
Chisel plows dig a channel or trough through the soil.
Source: USDA-NRCS, *http://photogallery.nrcs.usda.gov/*

A farmer has the choice between two plow types, the moldboard or the disk plow, each adapted to certain soil characteristics. Moldboard plows range in size from a single moldboard, or bottom, to a gang (group) of plows that turns twelve furrows (twelve bottoms) simultaneously. Each moldboard shears and inverts a slice of so as it moves along, leaving the top of the slice at the bottom of the furrow (Fig. 14–4). Moldboard plows are used when the soil is sufficiently moist to allow the plow to pass through easily but not so wet as to cause the furrow slice to stick to the face of the moldboard and create a drag that requires excessive tractor power and causes poor results. The ideal moisture content for plowing loam soils is slightly less than field capacity.

Two-way reversible plows (flip-over plows) are used to eliminate "dead" furrows (unfilled furrows). They are also used in hilly areas for contour plowing to reduce erosion and in irrigated areas where dead furrows hinder irrigation. These plows always throw the furrow slice in the same direction regardless of the direction the plow travels.

Chisel plows are primary tillage tools consisting of curved shanks spaced widely along a tool bar (Fig. 14–5). The shanks dig through the ground. They are normally operated at depths similar to those attained with moldboard plows, but they shatter rather than invert the soil in the plow layer. Chiseling can be done more quickly than moldboard plowing, disturbs the soil less, and leaves significant quantities of crop residue on the soil surface to provide erosion control. For these reasons, chisel plows are rapidly replacing moldboard plows as the primary tillage tool on many farms, particularly in the Midwest.

The disk plow (not to be confused with the disk harrow) consists of a series of large concave disks 60 to

Figure 14–6
The disk plow accomplishes essentially the same objective as the moldboard plow. However, the disk plow can operate when soil conditions are either too wet or too dry for the moldboard plow. The disks rotate as the plow moves forward, rolling the furrow slice over.
Source: USDA-NRCS, *http://photogallery.nrcs.usda.gov/*

75 cm (24 to 30 in.) in diameter that are set at an angle to the forward movement and cut into the soil while rotating as the plow moves forward (Fig. 14–6). There can be three to ten or more disks on a plow. Soil moisture conditions are less critical for disk plow operation.

Disking

Disk harrows which are smaller in diameter than disk plows are used to reduce the size of larger soil clods by fracturing them with cleavage and pressure. Disking generally follows plowing, but under some conditions disking can substitute reasonably well for plowing.

Figure 14–7
A disk harrow cutting and covering crop residue near prior seedbed preparation. Source: Allis-Chalmers.

Figure 14–8
A spring-tooth harrow is sometimes used in seedbed preparation because it is effective in breaking up soil crusts, reducing clod sizes, and destroying small weeds. Source: Allis-Chalmers.

If the soil is in good tilth, satisfactory field preparation can be accomplished by disking alone.

The depth of penetration is regulated by adjusting the angle of the gangs. A special-purpose disk, called a stubble disk, has semicircular notches cut around the periphery of the disk blade (Fig. 14–7). The notches help cut crop residues more effectively.

Harrowing

The function of the harrow is to reduce further the size of soil clods left after disking, to smooth the soil surface, and to do small-scale leveling. Harrowing also destroys small weeds. This operation generally follows disking. Frequently, farmers attach a harrow behind the disk and do both operations simultaneously. This is a final touch to field preparation, unless beds are to be formed for irrigated row crops.

A wide variety of harrows are used. The principal types are: (1) spike-tooth; (2) spring-tooth (Fig. 14–8); (3) chain or drag; and (4) cultipackers, packers, mulchers, and corrugated rollers. The fourth group crushes clods by applying pressure and tends to break up hard, dry clods better than drag-type harrows. They also pack the soil slightly, reducing large air spaces. Harrowing is becoming less common and necessary, particularly for agronomic crops, as new planters become more capable of operating in uneven fields.

Listing and Ridging

In some areas, row crops are planted on ridges formed by listers. A lister is a plow equipped with two moldboards that cuts a furrow slice two ways—half to the right and half to the left. This forms a ridge of soil commonly about 20 to 25 cm (8 to 10 in.) high and of variable width at the base. Listers can be equipped with attachments to list, plant, and fertilize in one operation. Some farmers flatten the tops of the ridges with a roller, drag, or bed shaper before planting. For some crops at the same time the rows are listed, additional attachments lay irrigation lines and plastic or paper mulch on top of the beds (Fig. 14–9).

Cultivation

Cultivation is the tillage between seedling emergence and crop harvest. The main reason for cultivating is to control weeds, but other benefits are improved water

Figure 14–9
Irrigation lines under plastic mulch research plots. The irrigation lines can be seen as faint stripes under the mulch. The raised objects in the beds are tensiometers to measure water content of the soil at various levels beneath the mulch. Source: Dennis Decoteau, The Pennsylvania State University.

Figure 14–10
A multiple-row crop cultivator loosening soil and destroying weeds in this corn field. The enclosed driver cabs in modern tractors are air conditioned and many have GPS/GIS and other electronic devices to increase production efficiency and reduce environmental impact.
Source: USDA-NRCS, *http://photogallery .nrcs.usda.gov/*

infiltration and soil aeration on soils that crust, the conservation of soil moisture, loosening compacted soils, and in some cases help with insect control. Row-crop cultivators (Fig. 14–10), field cultivators, rotary hoes, and rototillers are types of cultivators as are hand-held hoes.

Deep Tillage

Some farmers use deep tillage to improve problem soils especially when hardpans are present. Extra heavy equipment is used for deep tillage when the soil is dry. Different types include the slip plow (Fig. 14–11) and ripper or deep chisel (Fig. 14–12). The latter consists of one or several shanks that penetrate the soil from 60 to 120 cm (2 to 4 ft). It shatters hardpans best when the

Figure 14–11
The V-shaped blade of this slip plow is pulled horizontally well below the soil surface to break up hard soil layers.
Source: USDA Soil Conservation Service.

Figure 14–12
An example of a deep chisel (ripper) used to break deep compacted hardpans. This procedure is used only when justified because the high amount of energy required makes this an expensive operation. Source: William E. Wildman.

soil is dry. Deep tillage is expensive, and sometimes it does not materially increase crop yields. In established orchards, it can damage trees by severely cutting the roots. When deep tilling, the operator also has to be aware of drainage tile and any utility lines such as gas or electric that may be buried in the area.

Although the above practices were developed for field crops, the principles are applicable for most growing situations. Growers of vegetables and small fruits, landscapers, golf course superintendents and athletic field managers, nursery crop producers, all depend on working the land before growing their respective commodities. In these cases, the equipment they use is often modified for use on smaller areas or for very precise results, but the principles remain the same.

SOIL FUMIGATION AND PASTEURIZATION

For some crops, the soil or media must be fumigated before preparation. These usually are high-value crops where the potential for pest damage is severe enough to justify treatment. Fumigation is normally too expensive to use extensively on most row or field crops. Certain chemicals are used to fumigate soil and destroy harmful bacteria, fungi, and nematodes as well as many weed seeds. A widely used soil fumigant had been methyl bromide (CH_3Br). This toxic gas is colorless and odorless, and is usually mixed with chloropicrin (tear gas). Chloropicrin is used to indicate the presence of the more toxic CH_3Br gas because it is nontoxic in small amounts, but it causes discomfort to the eyes. Chloropicrin also has some effect as a fumigant.

However, CH3Br has been shown to negatively affect the ozone layer that protects the earth from harmful solar ultraviolet radiation. As a result, after 2005, CH3Br cannot be used except for critical use exemptions. These exemptions are crop and location dependent. In general, herbaceous fruits, vegetables, and ornamentals are the most commonly exempted plants. A complete list is available using the keywords: "methyl bromide use" with an Internet search engine. At the time this text was revised, no suitable substitute was available for CH3Br. With many crops, including vegetables and herbaceous fruits, grafting a disease resistant rootstock to the crop is now being used as a way to overcome the loss of methyl bromide.

Nursery and greenhouse managers often prefer to use steam instead of chemicals to pasteurize their soil. A closed container with some means of admitting steam is used. The temperature at the center of the soil mass is brought to above 71°C (160°F) for thirty minutes to destroy disease-causing organisms.

SOIL DEGRADATION

Natural and farmed soils are not static entities; their properties change constantly, but the quality of the soil is usually maintained when these changes occur in response to the slow changes that normally occur in the natural environment. When conditions change abruptly, however, resulting effects on the soil can lead to severe degradation and loss of productivity. Soil management along with some irrigation and fertility practices in the past have been major contributors of rapid soil degradation including accelerated erosion, loss of organic matter, compaction, waterlogging, and salinization.

All upland soils experience small episodes of erosion yearly, but the severity of that erosion is usually minimized by the protection of permanent vegetative cover. When that cover is removed, the soil surface is exposed directly to the erosive forces of wind, rainfall, and runoff. Accelerated erosion removes soil faster than new soil can be formed, reducing the depth of the productive topsoil. If erosion continues unabated, eventually all of the rich A horizon can be lost, forcing crops to root wholly in the undesirable B horizon. Even if the entire A horizon is not lost, erosion preferentially removes smaller particles (including organic matter) first, degrading topsoil productivity long before it is lost completely. Erosion can be accelerated by practices that remove vegetative cover, including overgrazing, intensive and repeated tillage, and deforestation, all common worldwide practices in food production.

Loss of soil organic matter is another facet of soil degradation aggravated by these practices and others. The amount of organic matter in a soil profile is dictated by the balance between additions (mainly plant residues) and losses (mainly as carbon dioxide during respiration). Additions are regulated by the amount of residue the ecosystem supplies on a continuing basis, while losses are limited by the usually low concentrations of oxygen in the soil air. Practices such as overgrazing, crop residue removal, and forest clearing obviously reduce the quantity of organic material available for addition to the soil, but practices such as drainage and tillage have effects that can be even more damaging than these. Drainage removes excess water from the soil, increasing the air (and oxygen) concentration in the soil. Tillage stirs the soil, increasing pore space (though only temporarily) and increasing contact between the soil solids and the atmosphere. Both practices provide aerobic microorganisms with extra oxygen, which they use to respire more carbon than they would under undisturbed conditions. The resulting loss of organic matter may eventually lead to deterioration of soil structure and loss of water-holding capacity, among other effects. Unfortunately, most cultivated soils worldwide have suffered significant losses of organic matter since they were brought into production.

Soil Compaction

The use and size of farm equipment have increased over time and it is common to see heavy equipment make many trips across the fields. Trucks heavily loaded with harvested crops travel over the land (Fig. 14–13). Any of these can seriously compact soil and the effect is

Figure 14–13

Trucks hauling produce out of a field can severely compact the soil, especially if the soil is wet.

Source: USDA-NRCS, *http://photogallery.nrcs.usda.gov/*

worsened when the soil is wet. Grazing animals can also compact soils when pasture soils are wet or many animals graze in a small, confined area.

Soil compaction is a serious problem, directly related to land preparation, tillage, harvesting operations, and herd management). Soil compaction impedes root penetration and thereby limits the volume of soil from which plant roots can extract nutrients and water. This can result in an inadequate water supply and nutrient deficiencies, particularly of phosphorus. The reduction in porosity can also result in a greater accumulation of carbon dioxide in the soil.

Soil compaction is difficult to define precisely and often means different conditions to different people. To growers, a compacted soil is best described as one with abnormally high bulk density (solid mass) with very small pore spaces. Soil is not compacted all at once. The problem develops slowly, worsening each time heavy equipment is used and soil particles are pushed closer together creating a compacted layer in the soil profile (Fig. 14–14). Often years pass before the problem reveals itself in declining crop yield or quality. Although research has found no

easy or simple solution to the problem, the problem when recognized can be alleviated somewhat by proper soil management and organic matter additions but it may take years for any noticeable mitigation to occur.

Whenever a soil does becomes more compacted, the process is accompanied, among other changes, by an increase in **Bulk density**. In simple terms, an increase in bulk density means that the soil has become denser through a loss of pore space (Figs. 14–15 through 14–17).

A simple way to determine the bulk density of a soil or artificial media is to take a measured volume of air-dried soil (in cubic centimeters) or media (A) and weigh it (in grams) (B).

$$\text{Bulk density} = \text{Mass of dry solids } (B^2) \div \text{Volume of media (A)}$$

Soil compaction is especially serious for fresh market vegetable farmers. Wholesale market prices for fresh vegetables frequently change dramatically from day to day. Thus, growers may harvest the crop even though the soil moisture is high and the danger of soil

Soil surface

Compacted layer

Figure 14–14

The upper portion of a soil profile showing a compacted layer about 25 to 30 cm (10 to 12 in.) below the soil surface. This layer was caused by repeated traffic over the field when soil was wet.

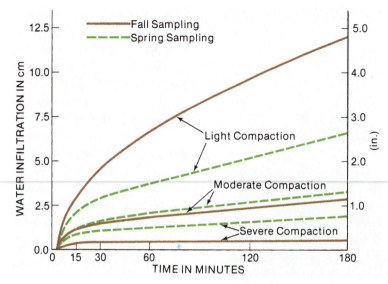

Figure 14–15
Water infiltration rates were measured on each of three plots mechanically compacted to three densities. Tests were made in the fall after compaction and then repeated the following spring on the lightly, moderately, and severely compacted plots. Obviously, severe compaction decreases the water infiltration rate.

Figure 14–16
Soil samples taken from the top 10 cm (4 in.) of adjacent test plots of a noncompacted (left) and compacted (right) soil. The photographs are magnified 40 times. Note the greater porosity in noncompacted soil samples.

Figure 14–17

Soil samples from noncompacted and compacted fields were air-dried and screened to simulate tillage. Water was allowed to infiltrate into each for four hours. Note that much more water infiltrated into the noncompacted soil, showing that the ill effects of previous compaction still remained.

compaction great if the value of the crop is high that day. Farmers in irrigated regions sometimes irrigate up to the time of harvest to keep plants fresh and turgid, and then quickly harvest the crop, resulting in the use of heavy equipment in the wet fields.

Soil compaction is easier to prevent than to correct. Good soil management to prevent compaction includes using tillage implements of the lightest possible weight at the proper soil moisture content, avoiding unnecessary and expensive tillage, keeping equipment off the land when the soil is wet, and leaving green manure crops or crop residues on the soil surface.

Soil crusting, similar to soil compaction, appears as a dense, hard layer of varying thickness on the soil surface. Soil crusts result from the action of raindrops or irrigation, which break apart surface aggregates. Crusting is particularly severe when the amount of water applied exceeds the infiltration rate. Soil crusts restrict seedling emergence, gas exchange, and water infiltration (Fig. 14–18).

Salinization

Salinization is related to the introduction of cropping and irrigation practices into arid and semiarid regions and saltwater coastal regions. Many arid-region soils exist under conditions where only a few millimeters of water may percolate below the root zone each year. Adding excess water with irrigation can increase this depth significantly. The excess water can dissolve, transport, and concentrate soluble salts that would have remained undisturbed under natural conditions. Saltwater may be released from the soil on the lower reaches of hillsides, forming saline seeps that eventually make the soil below the seep too salty to support plant life. Saltwater can also accumulate at the surface of the production field itself through capillary rise, eventually rendering it too salty for further use. In many instances, the salts are compounds of sodium, which cause the soil to disperse, destroying its structure. Dispersed soils

Figure 14–18

Soil that has become crusted. Source: USDA-NRCS, *http://photogallery.nrcs.usda.gov/*

do not transmit water to any degree and are prone to becoming almost permanently waterlogged and useless for further production.

Saltwater intrusion is a serious issue along saltwater coasts. Water pumped from the wells near the coast for irrigation and other uses lowers the water table. That allows salt water to move into the irrigation wells. The irrigation water becomes too salty to be used for growing plants and a new source has to be found.

SOIL IMPROVEMENT AND CONSERVATION

Improvement

Often, the best strategy for improving a degraded soil is just to leave it alone for several years. Significant soil improvement can often be accomplished by establishing a permanent cover crop that includes a deeply rooting species, such as a mixed grass-alfalfa cover crop, which ensures sufficient fertility to get improvement started, and then simply watching the cover crop grow. After a few years, the crop will provide fresh organic material throughout the profile, provide an environment for the reestablishment of soil organisms, and aid in rebuilding soil structure while protecting the soil from erosion. One of the major benefits of using a living cover crop for remediation is that roots eventually penetrate deeply into the soil and dry it to deeper depths. This promotes repeated shrinking and swelling of the soil mass, which aids in the formation of pore space and structure deep in the profile.

If establishing a permanent cover over a long period of time is not feasible, adding large quantities of organic material can often help improve a soil, particularly if the degradation is not too deep or severe. Repeated additions of organic matter provide a constant supply of food and raw materials to soil organisms, allowing populations to flourish and create compounds needed to promote soil aggregation. It takes much more organic material than most people realize to improve a soil this way, but it is an effective method if it is practiced repeatedly.

On larger areas, such as corn, wheat, or soybean fields, no-tillage crop production may provide a way to improve the soil. In such systems, crops are planted into the undisturbed remains of the previous year's crop by dropping the seeds immediately into the shallow furrow created by a chisel plow, with no preplant soil disturbance. Crop residues protect the soil from erosion, and lack of tillage allows soil organisms to flourish. Repeated use of the practice improves the

structure, organic matter content, and infiltration capacity of the surface soil after a few years. Improvement slowly moves downward through the profile.

Carbon: Nitrogen Ratio   As described above, improving the physical condition of the soil by incorporating crop residues, green manure crops, or other organic matter is a beneficial practice, but one must be aware of the resulting effect it has on the carbon:nitrogen (C:N) ratio. Incorporation of the straw residue having a high C:N ratio (50:1) will result in a change in the soil's C:N ratio. Soil organisms, primarily fungi and bacteria, rapidly multiply because of the large source of added carbon food (Fig. 5–14). However, these organisms also require large amounts of nitrogen for their own growth. Consequently, they tie up the soil nitrogen, causing a temporary nitrogen deficiency for the growing crop. During this time, nitrogen fertilizer should be applied to prevent nitrogen deficiency in the plants.

Salinity   Salt-affected soils require different approaches for improvement. Excess salts must be removed before other practices can be employed. Desalinization often involves installing drainage improvements deep in the profile and flushing the soil with fresh water. Salts are removed in the drainage tailwater. If the soil is sodic, additional measures must be taken to remove excess sodium from the cation exchange sites before the structure can be restored, usually by incorporating large quantities of a calcium salt such as gypsum into the soil before it is flushed. Calcium is bound preferentially over sodium to the exchange sites, so when the gypsum dissolves, its calcium displaces sodium from the sites and prevents its readsorption, forcing it to leach and leave the soil in the tailwater.

Reverse osmosis is a process that is used to desalinate water. It is an expensive process usually reserved for high value crops such as ornamentals and some vegetables and fruits. The salty brine that is a byproduct of the process can create a disposal problem. However, when dried the brine is sometimes sold as salt for use in highway snow removal.

Erosion

Recent soil surveys in several countries show that vast areas of productive land have been damaged beyond recovery. Erosion continues to be a critical concern in almost every agricultural region of the world. Erosion is particularly severe where intense torrential rains are frequent. Unfortunately, the need for food in some nations has overshadowed the danger of uncontrolled erosion. This is particularly true in East Africa, the Yellow River basin in China, Eastern Europe, Latin

America, and parts of Australia, India, and the United States—all plagued with serious and widespread erosion. In the United States, erosion has damaged nearly 110 million hectares (272 million acres). The soil erosion problem in these nations is not over the entire nation, but it is usually severe in hilly high rainfall areas and/or in the semiarid regions and especially in those areas where irrigation is not used.

Factors Affecting Erosion   An important factor in erosion control is the amount of plant cover. Land covered with sod or trees loses little, if any, soil, while barren land can quickly lose considerable topsoil. The intensity, duration, and distribution of rainfall are also factors. A torrential rain of short duration on land with little plant cover causes severe soil losses while a gentle, evenly distributed rain causes less. It does not take long for water to erode and remove the fertile topsoil and form gullies (Figs. 14–19 and 14–20). Topography of the land is another factor. Level land is less likely to erode from water than sloping land but it can be more vulnerable to wind erosion (Fig. 14–21). The soil's physical properties affect erosion. Deep permeable soils that quickly absorb water are less likely to erode than shallow, slowly permeable soils.

Gently sloping land can be cropped if proper erosion controls are used, but row crops that require tillage for weed control should never be planted in rows that run up and down steep hills. Row crops on gentle slopes require contoured rows. Sod crops or crops planted by broadcast methods should be used to reduce hillside erosion losses. At times when the demand for food crops (at home and abroad) is high, economic pressure is put on the farmer to seed land to grain or

Figure 14–20
The beginning of a deep gully. Special care and treatment will be required to conserve this land if the erosion is not stopped. Source: USDA-NRCS, *http://photogallery.nrcs.usda.gov/*

other row crops when the soil should remain as grassland. This practice contributes to erosion.

Wind erodes land by removing topsoil just as water does. Unlike with water, however, level land is vulnerable to wind erosion. As with water erosion, the best protection against wind erosion is to provide vegetative cover for the land during periods of high winds. Tillage methods such as stubble mulching have been helpful. Leaving the soil surface rough or cloddy reduces wind velocity at the soil surface, and windbreaks are helpful. These vary in size from tall trees to hedges planted close together perpendicular to the prevailing wind.

Figure 14–19
Rill erosion has begun in field, a forerunner of severe gully formation if not checked soon.
Source: USDA-NRCS, *http://photogallery.nrcs.usda.gov/*

Figure 14–21
Wind erosion occurs most readily on flat land without vegetation. Source: USDA-NRCS, *http://photogallery.nrcs.usda.gov/*

Conservation

Soil conservation is the preservation and extension of the life of soil by using land wisely to keep it in its most productive state for the present and future generations. One of the complicating factors in soil conservation is our dependence on the soil for food and fiber. The land must be used but at the same time saved for future use. To do both takes wise land management.

George Washington and Patrick Henry were among the earliest American land conservationists. In his final message to Congress as president in 1796, Washington urged the creation of a board of agriculture. More than a half-century later, in 1862, Abraham Lincoln established the Department of Agriculture. Little interest in soil conservation was felt for the next seventy-five years because of the availability of new, western lands. During the 1920s, a soil surveyor, H. H. Bennett, called attention to the waste and depletion of America's greatest natural resource—land. Finally, in 1929, Congress established ten soil conservation experiment stations, and assigned personnel to study and gather information on erosion control measures.

The Natural Resources Conservation Service (NRCS) was established in 1933 as the Soil Conservation Service in the Department of Interior as one means of helping the United States recover from the Great Depression of 1929 to 1935. The need for immediate soil conservation became evident on Black Sunday, April 14, 1935, because that day the most severe dust storm in U.S. history completely blotted out the noonday sun (Fig. 14–22). During the summer and fall of 1935, the skies over Washington, DC, and New York were darkened with topsoil blown from Texas, Oklahoma, and other prairie states where land tillage and drought had made the land exceptionally vulnerable to erosion.

Methods of Conservation   The appropriate method of soil conservation depends on the topography, soil type, cropping and livestock system, and climate. The Natural Resources Conservation Service (NRCS) provides growers with valuable information and assistance regarding erosion control. A professional conservationist will survey and classify the soil into one of eight broad land-capability classes according to its best use with least erosion (Table 14–1). The important consideration is that each parcel of land is managed according to its needs. This means that land not suitable for any type of agriculture, even though unaffected by erosion, should be left for wildlife and recreation; forest land should be used to produce trees, range and

Figure 14–22
Two consecutive years of drought, followed by high winds, blew immeasurable amounts of fertile topsoil from areas of Texas and Oklahoma into the Atlantic Ocean. This catastrophe created what came to be known as the dust bowl on Black Sunday, April 14, 1935. The late afternoon sun, a circular ball above the auto, is barely visible through the dust cloud.
Source: Library of Congress.

grassland should be used to produce forage, and land suitable for cultivation should be reserved for crop production.

Each kind of farming needs its own special conservation practices, but even with careful land management, additional measures are often necessary to improve land use. In addition to the conservation tillage methods described earlier in this chapter, the following subsections describe ways to conserve soil.

Grass Waterways   These are strips of land of varying width permanently seeded to a grass sod (Fig. 14–23). They conduct water to drainage outlets and control runoff from sloping land with cultivated crops. Waterways are used with contours or terraces that drain into them.

Figure 14–23
Waterways are grassy areas left in fields to control water run-off from cultivated fields.
Source: USDA-NRCS, http://photogallery.nrcs.usda.gov/

Table 14–1
LAND CAPABILITY CLASSES

Class	Use	Conservation Practices Needed
I	Few limitations. Suitable for wide range of plants. Can be used for row crops, pasture, range, woodland, and wildlife. Not subject to overflow.	Needs ordinary management practices—fertilizer, lime, cover, or green-manure crops, conservation of crop residue, animal manures, and crop rotations.
II	Some limitations. Choice of crop plants reduced. With proper land management, land can be used for cultivated crops, pasture, range, woodland, or wildlife.	Limitations few and easy to apply. Problems may include gentle slopes, moderate susceptibility to wind or water erosion, less than ideal soil depth, slight salinity. May require special conservation practices, water-control devices, or tillage methods, terraces, strip cropping, contour tillage, special crop rotations, and cover crops.
III	Severe limitations reduce choice of plants and/or require special conservation practices. May be used for cultivated crops, pasture, range, woodland, or wildlife.	May require drainage and cropping systems that improve soil structure. Organic matter additions might be needed. In irrigated areas, soils may have high water table, high salinity, or sodic accumulations. Soils may be slowly permeable.
IV	Severe limitations that reduce choice of plants. Requires very special management. Limited use for cultivated crop but can be used for pasture, range, woodland, or wildlife.	Limited cultivated crops because of steep slopes, susceptibility to wind or water erosion, effects of past erosion, shallow soils, overflows, poor drainage, salinity, adverse climate. May be suitable for orchards and ornamental trees and shrubs. Special practices needed to prevent soil blowing and to conserve moisture.
V	Land limited in use—generally not suitable for cultivation. Little or no erosion hazard but has other limitations. Use limited to pasture, range, woodland, or wildlife.	May be nearly level but has excessive wetness, frequent overflow, rocks, or climate variations. Cultivation of common crop not feasible but pastures can be improved and benefits from proper management can be expected.
VI	Severe limitations make the land unsuitable for cultivation. Restricted to pasture, range, woodland, or wildlife.	Pastures can be improved by seeding, liming, fertilizing, water control with contour furrows, drainage ditches, etc. Have severe limitations that cannot be corrected, thus not suitable for cultivated crops. Some soils can be used for crops such as sodded orchards, berries, etc. Pastures can be improved by seeding, liming, fertilizing, water control with contour furrows, drainage ditches, etc. Have severe limitations that cannot be corrected, thus not suitable for cultivated crops. Some soils can be used for crops such as sodded orchards, berries, etc.
VII	Severe limitations make land unsuitable for cultivation. Use limited to grazing, woodland, or wildlife.	Physical condition of soils prevents range or pasture improvement practices. Restrictions are more severe than those of Class VI. Can be used for grazing. May be possible to seed some areas.
VIII	Limitations preclude use for commercial plant production. Use restricted to recreation, wildlife, or water supply.	Cannot be expected to yield any significant return from crops, grasses, trees, but benefits from wildlife use and watershed protection or recreation are possible. Class VIII includes badlands, rock outcrops, sand beaches, river wash, mine tailings, etc.

Source: USDA Handbook 210, 1973. A detailed description is available at http://soils.usda.gov

Figure 14–24
Contour tillage curves around or along a slope, not up and down it. The furrows act as small dams to slow the flow of water rather than conduct it.
Source: USDA-NRCS, *http://photogallery.nrcs.usda.gov/*

Figure 14–25
Contour strip-cropping is useful for conserving water as well as reducing soil losses by wind erosion. In this case a cereal grain crop has been alternately planted with a hay crop.
Source: Margaret McMahon, The Ohio State University.

Contour Tillage One easy cropping practice that reduces losses of topsoil is to till the land on the contour (level elevation) instead of up and down the hill (Fig. 14–24). The land is plowed and the crop rows planted and cultivated around the slope, always at the same elevation from end to end. The rows are curved and sometimes come together in points. The ridges left by the tillage tools form small dikes to catch water, allowing more time for it to percolate into the soil instead of running down the hill.

Contour Strip-Cropping This effective practice is used to conserve both soil and water. Soil conservation is enhanced by alternating strips of solid-planted crops with row crops; for instance, strips of grain or hay crops can alternate with corn or sugar beets (Fig. 14–25). The strips always run on the contour. In some cases this practice reduces erosion by more than half of what it would be if either crop were planted alone.

Terraces Terraces are used on long gentle slopes to decrease runoff and to increase water infiltration. On gently rolling land, terraces are low broad mounds that follow the contour and retain water that would otherwise run down the slope. The terraces are constructed with a slight grade so excess water flows slowly to an outlet, often a grass waterway. Terraces are also used in some places on steep slopes (Fig. 14–26). They have been used for centuries in the Andes, China, and many other parts of the world where insufficient flat arable land is available. In Thailand, rice is grown on

Figure 14–26
Terrace cultivation a field in Iowa.
Source: USDA-NRCS, *http://photogallery.nrcs.usda.gov/*

steep, terraced hillsides. Many landscape and golf course designs incorporate terracing for not only soil conservation but also for aesthetic appeal.

Low- and No-till Farming A more recent conservation practice that has become very widespread is no- or low-till farming. Instead of plowing, disking, and otherwise cultivating the soil to remove old vegetation, herbicides are used to kill the vegetation which is left in place. Seeds or transplants are planted in trenches made by a shallow chisel plow within the vegetation residue (Fig. 14–27). The residue serves as mulch. Bare soil is not exposed so it is less vulnerable to erosion than with plowing. The subterranean ecosystem is barely disturbed so it functions more naturally. Genetically engineered

Figure 14–27

No-till fields are not plowed or otherwise cultivated before planting. In some cases a small chisel creates a shallow trough in which the seeds are immediately planted. Old vegetation remains on top of the soil but has been killed by application of an herbicide. Source: USDA-NRCS, *http://photogallery.nrcs.usda.gov/*

herbicide resistant crops are often part of a low- or no-till farming operation to allow herbicide applications to control weeds that emerge after the crop is planted.

Artificial Soils

The extreme example of "land" preparation for growing plants is the creation of growing mixes or media that contain little or no soil. Several products are available for growers that are prepared and marketed under descriptions such as "artificial media," "potting soils," "soilless mixes," "container mixes," and so on. These can be bought premixed or custom-blended by a grower. They are prepared for very specific uses, but most are designed to facilitate rapid seedling emergence, rooting, and early plant development, and allow for relatively easy management of water and fertilization.

Some artificial soils are used on golf courses and athletic fields and at construction sites. Artificial soils on golf courses and athletic fields have to have characteristics that not only support good plant growth but also meet the requirements of the athletes. For example the way a golf ball bounces when hit onto a green or the footing required by a football player influences the choice of soil components for a green or football field. Construction site soils must provide the proper density to support buildings while allowing for plant growth.

The factors that influence soilless mix characteristics are the same as those that influence native soils. Particle size and structure determine porosity and water-holding characteristics. Handling influences compaction and particle degradation. Media mixing and pot-filling machines are specifically designed to avoid destroying particle size or compacting the mix as it is blended then put into pots.

The media can contain a blend of many different components, such as organic soil, sand, peat moss, and expanded vermiculite. Other commonly found ingredients include coir fiber from coconut husks; bark; sawdust; perlite (a special kind of sand that is heated to extreme temperatures until it pops like popcorn); and calcined clay, which contains small ("kitty-litter" size) particles of kiln-fired clay (Fig. 14–28). A recent addition to the list is parboiled rice hulls (PBH), a waste product from rice production. Each component gives the mix a certain characteristic. For example, a typical mix may contain sand for large pore spaces and good drainage, vermiculite to hold water and some nutrients, and sphagnum peat moss for its nutrient-holding

A

B

C

Figure 14–28

Some examples of material that can be used in artificial media: (A) perlite, (B) calcined clay, (C) peat moss.
Source: Margaret McMahon, The Ohio State University.

capacity and to give the mix an acidic pH. The addition of a wetting agent is often needed to allow the mix to take up water when the ingredients have been dry when first added. Peat moss is very good at holding water when it is already wet, but it repels water when it gets dry. (If you have ever tried to moisten a dry pot containing peat moss, you know what we mean.) Many mixes also contain a small amount of a complete fertilizer called a starter charge, which provides nutrients during the early stages of growth when the plant needs fertilizer but is not taking up enough water to allow an application of liquid fertilizer.

WATER MANAGEMENT

Water management is a very important part of successfully growing plants. Even in areas where growers natural precipitation is sufficient for plant growth and irrigation is not needed, the growers have to be aware of daily precipitation amounts as those affect other management practices such as fertilization, pest control, and equipment use for planting, cultivating, and harvest. In areas where irrigation is used, determining when to apply water, how to apply it, and how often to apply it requires considerable understanding of the principles of good water management. Drainage systems can be installed in areas where too much water in the soil impedes the growing of plants but care must be taken to preserve the wetland areas that are important parts of many well-functioning ecosystems.

Irrigation

When precipitation is inadequate to meet the needs of plants, irrigation is necessary. Farmers have irrigated crops for over 4,000 years. Early civilizations began alongside the rivers in arid or semiarid regions. Records indicate that crops were irrigated along the Nile, Ganges, Tigris, and Euphrates rivers as early as 2600 BCE. It has been suggested that crop irrigation contributed to the founding of the great civilizations in these areas. Irrigation canals over 1,000 years old have been found along the Gila River in Arizona. Today, irrigation canals are still used along the Tigris in Iraq (Fig. 14–29). In early times, water flow was controlled by humans building and removing small dams at the head of the canals and pumping water by treadmill or water screw. Today, the water flow is controlled mechanically and often by environmental sensors feeding data into computers.

Water quality is as important as having an adequate water source for irrigation systems. Water used for irrigation almost always contains measurable quantities of dissolved substances that, in general, are

Figure 14–29
Irrigation ditch currently in use in Iraq.
Source: Margaret McMahon, The Ohio State University.

soluble salts. These include small but important amounts of dissolved salts originating from the weathering of parent rocks and salts and chemicals carried into the water by runoff and leaching. The value of water for irrigating crops is determined by the amount and kind of salts and other chemicals present.

pH is important in the water supply because it can affect soil pH and nutrient availability. (A more complete discussion of pH and fertility is found later in the chapter.) Irrigating with a high pH water can raise soil pH, likewise irrigating with low pH water can lower pH.

Alkalinity is the ability of the water to resist a change in pH. It is determined by the amount of carbonates present in the water. The more carbonates, the greater the resistance (buffering) capacity of the water. This can be important if a grower needs to change pH. Water with a high alkalinity would need much more acid added to it to lower the pH than water with low alkalinity.

Saline Water and Salinity

If irrigation water contains soluble salts in sufficient quantities to accumulate within the root zone and interfere with crop yields, a salinity problem arises (Fig. 14–30). The concentration of salts in the water can increase soil salinity to the point where the osmotic pressure leaves plants unable to extract sufficient water for growth. The plants show much the same symptoms they would in a drought—wilting, reduced growth, and, in some plants, a color change from bright green to bluish green. The wide range of salt tolerance among agricultural crops permits the use of some saline water for irrigation—and water that is too saline for one crop can be used for a more salt-tolerant crop.

Figure 14–30
Excessive soluble salts in soils and from applied irrigation water may accumulate in lower areas of a field or where drainage is poor. Source: USDA-NRCS, *http://photogallery.nrcs.usda.gov/*

Figure 14–31
Flood irrigation of a field.
Source: USDA-NRCS, *http://photogallery.nrcs.usda.gov/*

Toxicity   High concentrations of certain salts in water are toxic to some plants. The problem occurs when certain specific ions such as boron, chloride, or sodium are taken up by plants from the soil solution in sufficient quantities to reduce yield, stop growth, or reduce the aesthetic appeal of the plant.

Various other situations related to salt concentrations in water can arise. Calcium, magnesium carbonates, or bicarbonates deposit on the leaves and fruits of orchard crops, grapes, and vegetable crops after sprinkle irrigation. They are also often seen as a white crust or stain on the sides or tops of clay pots. These white deposits, while not particularly harmful, do tend to lower quality and value by creating an unsightly product.

Methods of Application

Selection of the proper water distribution system can save expensive labor and assure better crop yields as well as saving water. The method of application is important, especially if the cost of water is high. Some factors that determine the method and type of system used are: (1) climate, (2) type of crop, (3) cost of water, (4) slope of field, (5) physical properties of soil, (6) water quality, (7) water availability, (8) and drainage capability.

Border or Flood Method   Flood irrigation is used where the topography is flat and level. This method is often used for drilled or broadcast crops, such as hay, pasture, and small cereal grains. Orchards and vineyards are also sometimes flood irrigated (Fig. 14–31).

The land must be graded and leveled with a slight downslope for flood irrigation. The amount of grading needed depends on the topography, cropping system, and cost of grading. Permanent or temporary levees are constructed running downslope, with border ridges that divide the field into strips or checks.

Water from an irrigation pump or canal is turned into the supply, or head, ditch at the higher end of the field. It is released or siphoned into one or more checks and allowed to flow slowly downslope, spreading evenly and uniformly over each entire check as it advances toward the lower end. Designing and operating such a system efficiently requires considerable experience, skill, and knowledge.

Furrow Irrigation   Furrow irrigation is a modification of flooding—water is confined to furrows rather than wide checks. The plants are grown in the raised beds created between the furrows (Fig. 14–32). Water is

Figure 14–32
Furrow irrigation. Source: USDA-NRCS, *http://photogallery.nrcs.usda.gov/*

used more efficiently with furrows than with flooding because the entire surface is not wetted, thus reducing evaporation losses. Furrow irrigation is frequently used for vegetable crops, orchards, and vineyards. The length of furrow varies but the longer the furrows the greater the loss of water because of deep percolation and excessive soil erosion at the head of the field.

The depth of the furrow should be such that the water can flow in the furrow for sufficient time to allow it to percolate into and through the raised bed, although not directly wetting the surface of the bed.

A variation of flood irrigation called ebb and flood or ebb and flow is used in many greenhouses. Plants are grown on benches or floors designed to fill and drain mechanically (Figs. 14–33 and 14–34). The process can be controlled manually or by a computer. The benches or floors flood rapidly (≈5 minutes) and drain equally as fast. The water level reaches approximately 0.75 in. up the pot and is held there for 5 to 10 minutes. The pots absorb the water through capillary action. If the pots are dry when the beds are flooded, capillarity may not be established with the water in the bench. In that case, the pots have to be irrigated with a hose until there is enough moisture in the pot to establish capillarity.

Liquid fertilizers can be applied through the system too. The water and nutrient solutions can be recycled. When recycling is practiced, caution must be taken to prevent the spread of disease. Methods of sanitation include treatment with hydrogen peroxide, ultraviolet light, or copper ionization. Also, fertilizer concentrations must be closely monitored and adjustments made for losses due to plant uptake.

Figure 14–34
Ebb and flood floor.
Source: Margaret McMahon, The Ohio State University.

Sprinkler Irrigation   Sprinklers are often used when flood or furrow irrigation is impractical. There are many types of systems, but they all depend on pipes or hose to get the water to a field and nozzles to deliver the water. Sprinklers can have some advantages over other irrigation methods; for example, through nozzles with different flow rates and spray patterns, sprinklers are adaptable to high or low soil permeability.

Through uniform wetting of the surface area by sprinklers, seed germination is more uniform; less total water may be used, and sprinklers are effective in washing salts away from salt-sensitive crops. Additionally, the surface need not be level; a properly designed system offers a method of adequately and accurately applying water, even on sloping land with grades up to 3 percent. Sprinklers are sometimes also used for frost control. The systems have to be designed to distribute water evenly from the input end to the farthest point. This requires keeping the flow rate and pressure uniform throughout the system.

In some situations, sprinkler systems may reduce labor need compared to flood systems, whereas, systems that are portable and moved frequently may increase labor. In general, equipment and energy costs are higher than with flood or furrow methods.

There are different types of sprinkler systems. One type is portable and has a main line located at the edge of the field with laterals usually set perpendicular to the main line either parallel along or perpendicular across the crop rows. The sprinkler nozzles are on top of short lengths of pipe (risers) spaced along the laterals. The system can be moved from field to field as needed. This system is used on many crops and is particularly useful for germinating small-seed crops, especially if the seedbed is

Figure 14–33
Ebb and flood bench. Source: Margaret McMahon, The Ohio State University.

Figure 14–35
Greenhouse sprinkler irrigation.
Source: Margaret McMahon, The Ohio State University.

rough. Greenhouses use a similar system with the lines and risers located down the center of the bench or along the floor where plants are being grown (Fig. 14–35).

Another sprinkler type is permanent and has all lines buried below the soil or permanently lying on the surface. Buried systems may have pop-up sprinklers that raise up when the water is flowing but drop back to the surface or below the surface of the ground when the water is off. The pop-up type is commonly used in landscapes, golf courses, and sports fields (Fig. 14–36). Some above ground systems use lines made of soft plastic that lay flat when not in use allowing them to be driven over by equipment. The permanent systems require a higher investment in equipment while the portable systems require more labor.

The wheel line system was designed to reduce labor by moving pipe across the field with a small gasoline engine. The sprinklers are mounted in the lateral pipe, which serves as the axle. To move to the next position, the gasoline engine turns the axle and propels the entire line across the field, saving labor and time (Fig. 14–37).

A variation of the wheel line is found now in greenhouses. The sprinklers are mounted on a gantry that moves along an overhead track the length of the greenhouse (Fig. 14–38). By varying the nozzle type (Fig. 14–39) and gantry speed, the system can be used to mist, irrigate, and even apply pesticides and chemical growth regulators.

Figure 14–36
Pop up sprinklers. (A) when not irrigating, (B) when irrigation is turned on.
Source: A) Margaret McMahon, The Ohio State University, B) USDA-NRCS, *http:// photogallery.nrcs.usda.gov/*

A

B

Figure 14–37
(A) Wheel line system of sprinkler irrigation. (B) Center pivot. Both systems roll across a field, but the wheel line moves in a straight line while the center pivot pivots around one end. Source: USDA-NRCS, http://photogallery.nrcs.usda.gov/

A

B

Figure 14–38
Irrigation gantry in a greenhouse.
Source: Margaret McMahon, The Ohio State University.

Figure 14–40
Aerial view of fields being irrigated by central pivot systems.
Source: USDA-NRCS, *http://photogallery.nrcs.usda.gov/*

with wheel line and center pivot systems. To reduce evaporation, newer systems have the water emitted as close to the ground or plant canopy as possible (Fig. 14–41).

Drip Irrigation   Drip or trickle irrigation was developed to apply water more precisely than flood or sprinkler systems. These systems wet the soil without

Figure 14–39
Multinozzle sprinkler heads allow a grower to rapidly change the pattern of the water being applied to the plants.
Source: Margaret McMahon, The Ohio State University.

The center pivot system is another labor-saving variation of sprinkler irrigation. The system is used more often in areas where land values are low or the availability of labor is low. The line, mounted on wheels driven by water pressure or electric motors, follows a circular path, pivoting around a fixed central point (Fig. 14–40). One disadvantage of the system is that the circular irrigation pattern leaves the corners in rectangular fields unirrigated.

Depending on the type of nozzle on the sprinkler, a filter may be needed to remove particles that could plug the nozzle. Sprinkler systems are prone to high losses due to evaporation of the water while it is suspended in the air. The highest evaporation occurring from the systems that have the sprinklers high above the ground occurs

Figure 14–41
New wheel line and center pivot systems have been developed where the nozzle is closer to the ground to reduce evaporation before the water reaches the soil.
Source: USDA-NRCS, *http://photogallery.nrcs.usda.gov/*

runoff (Fig. 14–42) and reduce evaporation losses Small amounts of water are allowed to trickle slowly into the soil through a hose that slowly and precisely "leaks" water along its length or through nozzles called emitters that apply water to a small area near the plant or into a pot. Emitters are connected to a small plastic tube that is attached to a larger tube called a lateral line that runs along the area to be irrigated. The lateral lines are connected to main line carrying water. In fields, most of the system can be buried allowing equipment to be driven harmlessly over the irrigation system.

Some advantages of drip irrigation are (1) it does not have to be moved when equipment is being driven over it; (2) there is little interference with other agricultural operations because much of the soil surface is not wetted; (3) there is less fluctuation of soil moisture in the root zone area because of the constant and slow drip application of water; and (4) less water is needed to grow a crop. The area of wetted soil can be as little as

Figure 14–42
Drip irrigation systems deliver the water directly to the soil surface near the plant. Source: USDA-NRCS, *http://photogallery.nrcs.usda.gov/*

10 percent of the total area of newly planted tree crops or up to 60 percent of the area of a mature crop.

The amount of soil wetted depends on the soil's physical properties, the time of application, and the number of emitters used. Some objections are (1) expensive filtration equipment is needed to avoid frequently clogged emitters; (2) water distribution may be uneven on hilly land; (3) salts tend to concentrate on the soil surface and near the wetted area boundary because leaching with excess water does not occur; and (4) the distribution of roots may be restricted to the small volume of wetted soil.

Drip irrigation may not fit the needs of every crop or situation, but it is used by many orchardists, strawberry growers, and for some high-yield field crops. It is the most common form of irrigation in greenhouses and nurseries where crops are grown in containers.

Controlling Irrigation

Irrigation can be controlled manually or by using simple electronic controls such as timers to open and close valves. Timers, however, are not a good means to control irrigation because they do not take into account any variations in environment such as temperature or light conditions. If you have ever seen irrigation sprinklers running during a rainstorm, it is probably because they are controlled by timers and not tensiometers, which measure soil moisture (Fig. 14–43).

Irrigation is much more effective when controlled by computers using sensors that measure such things as wind speed and duration, relativity humidity, light intensity and accumulation, soil moisture, evaporation

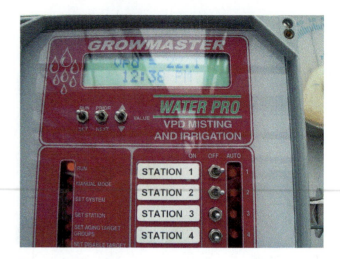

Figure 14–44
Connecting a vapor pressure deficit sensor to an environmental control computer is an effective way to control misting and irrigation. Source: Margaret McMahon, The Ohio State University.

potential, and even plant growth and development (Fig. 14–44). The data from the sensors is analyzed by a computer using irrigation software to precisely control water applications. In addition to providing the water needed for growth, proper timing of irrigation can be an effective means to control pests and pathogens and regulate plant growth.

Drainage

Many areas of land would be agriculturally productive except that they drain poorly and the soil is saturated during some or most of the growing season. In the past, extensive drainage systems were installed in these fields. The systems consisted of clay tile or perforated plastic pipe laid buried underground in a pattern that would move the water from the field to a ditch or creek. Today most of those fields continue to be productive.

However, with time we have come to realize that the wetlands that were drained for these fields played an important part in the local ecosystem by serving as traps or filters for sediment and pollutants including excess plant nutrients. They can serve as reservoirs or "sponges" for water that could be used by plants during dry periods and to mitigate flooding during rainy periods. They also provide habitat for many ecologically important wetland plants, and for permanent and migratory species of birds and animals. Currently, there is considerable effort being made to restore some of those wetlands (Fig. 14–45). New drainage systems are designed to retain the wetland's function in the ecosystem.

Figure 14–43
Tensiometers measure soil moisture and can be used to schedule irrigation.
Source: USDA Image Gallery, http://www.ars.usda.gov/is/graphics/photos/

Figure 14–45
Restored wetland. Source: Margaret McMahon, The Ohio State University.

Figure 14–46
Lime being applied to a field as a powder.
Source: USDA-NRCS, *http://photogallery.nrcs.usda.gov/*

Perhaps most interesting is a developing practice that combines wetland management with drainage and irrigation management in an intergrated manner. An increasing number of farm fields combine a type of sub irrigation with wetland management. These fields have a drainage system that flows into a wetland pond. The pond keeps the water table high to maintain the wetland. During periods of drought, some of the water from the pond is pumped back into the field through the drainage pipes, thus irrigating the crop from below the surface of the ground.

FERTILITY MANAGEMENT

Fertilizers

Use of fertilizers has probably increased crop yields and reduced hunger more than any other single agricultural practice. In the United States, from 1970 to 1985 alone, the annual consumption of chemical fertilizers increased about 5 percent per year. In addition to supplying nutrients to crops to increase yields, fertilizers can pollute the environment and cause marked changes in soil characteristics, some beneficial, some not. These secondary influences play an important role in the choice of fertilizer.

As you learned in Chapter 13, sixteen chemical elements are known to be essential for the growth of most plants, and a few others are used by some plants under certain conditions. Essential means that without the element the plant cannot grow properly and will ultimately die. The essential elements are carbon (C), hydrogen (H), oxygen (O), nitrogen (N), phosphorus (P), potassium (K), calcium (Ca), magnesium (Mg), sulfur (S), iron (Fe), manganese (Mn), molybdenum

(Mo), copper (Cu), boron (B), zinc (Zn), and chlorine (Cl). Some plants have been shown to benefit from sodium (Na), and some microorganisms that fix nitrogen symbiotically or nonsymbiotically require cobalt, but these are not usually considered essential.

Mineral nutrients are divided into groups according to the quantity plants use. The primary **macronutrients**—mineral nutrients needed in largest amounts—are nitrogen, phosphorus, potassium, calcium, magnesium, and sulfur. The remaining are needed in much smaller amounts and are called **micronutrients**. Micronutrients are sometimes referred to as **trace elements**.

Mineral nutrients are supplied to the soil by applying crop residues, animal manures, chemical fertilizers, or naturally occurring minerals. Other sources are the atmosphere, irrigation water, rainfall, and the elemental nutrients of the soil itself that can enter the water solution. The actual source of the nutrient (organic or inorganic) is unimportant to the plant as long as the mineral nutrients are available in sufficient quantity and can be easily assimilated. Nutrients are lost from the soil when any part of a crop is removed. Returning the crop residue to the soil does not replace all of the removed nutrients. To maintain or improve the soil's fertility, nutrients must be added in one form or another in amounts equal to or greater than those removed by crop harvest (Figs. 14–46 and 14–47). Generally, commercial fertilizers are easier to apply and manage than manures and crop residues, but the latter two should not be disregarded. They are especially beneficial in adding organic matter to help improve soil structure. Using manure as a soil additive also reduces its impact as a pollutant (Fig. 14–48).

Figure 14–47
Applying a dry fertilizer to a field before plowing.
Source: Robert Mullen, The Ohio State University.

Figure 14–48
Manure spread on fields provides nutrients to plants and reduces the polluting effects the manure can have on the environment. Source: USDA-NRCS, http://photogallery.nrcs.usda.gov/

For most field crops fertilizer is added at the beginning of the growing season. An additional application may be made later in the season. For more intensively managed crops such as vegetables that are harvested throughout the growing season, fertilization may be done several times during the growing season. For very intensively managed plants such as in ornamental nurseries and greenhouses, fertilization may be done as frequently as daily. In general, the more frequent the application of fertilizer, the lower the concentration applied at each application.

Fertilizer must be applied correctly, in the right amount, and with regard to the surrounding ecosystem. Overapplication and misuse of commercial fertilizers has resulted in the nutrients infiltrating ground

and surface water in some areas. Nitrates in ground water have become a serious issue with human health in some areas, while phosphorus has been associated with surface water problems such as the Dead Zone at the mouth of the Mississippi River and a similar problem in Lake Erie. Fertility management programs have to consider the fate of unused nutrients.

Types of Fertilizers

A **complete fertilizer** contains the three primary nutrients: nitrogen, phosphorus, and potassium. It may also contain some secondary or micronutrients. Each bag of fertilizer must carry a label stating the analysis of its contents (Fig. 14–49) regardless of the source of the nutrients. This analysis is represented by three figures prominently displayed on the bag; for example, 5–10–5. The first figure is the percentage nitrogen, the second figure represents phosphorus, and the third figure is potassium. Other nutrients would be listed by percentage elsewhere on the label. Any remaining percentage consists of filler, which most often is dolomitic limestone or gypsum, but other materials, even sand, can be used.

Although the label indicates percent N, P, and K, only N is in the elemental form. Phosphorus is in the form of phosphate (P_2O_5) and K is potash (K_2O). To get the percent of elemental, P multiply the percent, by 0.44 (the percent of P in P_2O_5) and the percent K by 0.83 (the percent K in K_2O).

While the label on the bag states the percentage of each primary nutrient, it may or may not indicate

Figure 14–49
Any material that is sold as a fertilizer has to have a label stating the percent of nitrogen, phosphorus, and potassium (N-P-K) in the fertilizer. This bag has 15% N, 16% P, and 17%K.
Source: Margaret McMahon, The Ohio State University.

the compounds used to make up the fertilizer. For example, the nitrogen in the fertilizer might be supplied as urea, ammonium nitrate, ammonium sulfate, or calcium nitrate, or a combination of these. The formulation is important because it informs the user of what compounds are used and their chemical form. As you learned in Chapter 13, nutrients can influence plant growth. By knowing the formulation, a grower can manipulate plant growth by manipulating the type of fertilizer applied to the plants. For example, nitrogen in the nitrate form promotes a slower, less succulent type of growth than ammonium.

Fertilizer recommendations are often given as a ratio. A ratio differs from an analysis because it expresses the amount of one nutrient in relation to the other. For example, the ratio of nutrients in 5–10–5 fertilizer is 1:2:1. Thus, if the recommendation was for a 1:2:1 fertilizer, a product labeled 5–10–5 or 10–20–10, or 15–30–15 would be equally acceptable as long as the rate of application was adjusted accordingly.

The fertilizer's physical properties are worthy of consideration because a lumpy or caked fertilizer is difficult to apply evenly. Some constituents, such as ammonium nitrate (NH_4NO_3), calcium nitrate [$Ca(NO_3)_2$], sodium nitrate ($NaNO_3$), and urea [$CO(NH_2)_2$], absorb water from the air (i.e., they are hygroscopic) and thus must be protected from moisture. They are packaged in moisture-proof bags, and often a conditioning material is added to decrease moisture absorption and caking. Another issue with NH_4NO_3 is its explosive (flashpoint) properties. As a result of the bombing of the Federal Building in Oklahoma City, sellers and purchasers of NH_4NO_3 in the United States now have to be registered with Homeland Security to be in compliance with the **Secure Handling of Ammonium Nitrate Act** of 2007. A similar law exists as the **Explosives Act** in Canada as well as Afghanistan.

Plant Requirements   Each plant or crop has its own nutritional requirement. While most of these are very similar, some plants have particular needs or sensitivities. You have to be aware of the needs of your plants. For many crops, fertility recommendations are available.

Any fertilization program requires regular and consistent **soil or growing media testing** along with **tissue testing** of the plant to know what nutrients are needed. Soil/media testing allows a grower to monitor the fertility of the soil or growing media and adjust fertilizer formulations or application frequency if necessary. The results will usually tell the grower if each nutrient is above, within, or below its recommended

rate. If a nutrient is below the recommended rate, the report often gives the amount of a corrective material that should be added to correct the deficiency. For field crops, this is often in pounds per acre or kilograms per hectare. Tests should be made, at a minimum, at the beginning of each crop or growing season. More frequent testing is recommended if fertilization is done frequently, as with golf course and greenhouse or nursery management, or if problems develop. Fertility maps of fields can be created and incorporated into a global positioning system (GPS), thus enabling precision applications of appropriate fertilizers.

Tissue testing allows the grower to know what nutritional elements are actually in the plant. In some cases, a deficiency or toxicity is not the result of how much or how little a nutritional element is in the soil but how much the plant is taking up. For example, a plant with a diseased or damaged root system might exhibit a nutritional deficiency even though there is a sufficient amount of that element in the soil. Adding additional fertilizer will not help the problem and may even make it worse or create a potential for nutrient runoff and environmental damage.

pH soil/media   pH is a critical component of fertility management. The availability of certain elements is regulated by the pH of the soil (see Fig. 5–11). In certain acid soils, aluminum solubility increases to toxic levels at low pH values, but its availability decreases at high pH values. Iron and zinc become less available to plants as the pH increases, but molybdenum is more available at higher pH levels. Phosphorus is more available at a soil pH of about 6.5 to 7.0 than at either higher or lower values. When necessary or desirable, soil pH can be adjusted by adding certain materials. The pH of an acid soil can be increased by adding basic amendments or fertilizers. Most often, calcium carbonate ($CaCO_3$), or agricultural lime, is used. This finely ground limestone is spread evenly over the surface and tilled into the soil. The CO_3^{-2} ions in the calcium carbonate neutralize acidity by combining with H+ ions to produce carbon dioxide and water. Estimating the amount of lime required involves several factors that include the soil pH, texture, organic matter content, structure, crops to be grown (see Table 5–3 on page 71), and the fineness of grind (mesh).

Residually basic (alkaline) fertilizers can also make the soil more alkaline. These fertilizers include sodium nitrate, potassium nitrate, and calcium nitrate.

When the soil pH is neutral or alkaline, the pH often has to be lowered. Lowering the soil pH can be performed by adding acid-forming chemicals such as

aluminum sulfate, ferrous sulfate, and elemental sulfur. These sulfur-containing compounds lead to the formation of sulfuric acid and acidify the soil. Alternatively, though less effective, acidic organic materials—such as peat and decomposed plant materials—can also be mixed with the soil to lower pH. The prolonged use of chemical fertilizers that are residually acid such as ammonium sulfate, ammonium nitrate, and ferrous sulfate tend to make the soil acid. Calcium and magnesium applied as carbonate salts not only increase their availability, but also decrease soil acidity, raising the soil pH. Knowing the pH of a field allows a grower to apply the proper type of fertilizer. For intensively cultivated plants such as in nurseries and greenhouses, soil and water pH is closely controlled along with fertility. The pH is deliberately raised or lowered to get the desired response from the plants.

Cation Exchange Capacity (CEC)   The CEC of the soil/media is another important factor in fertility management. CEC determines how well the positively charged nutrient ions will be held and made available to the plant. Plants growing in soil or media with a low CEC would require a higher rate or frequency of fertilization than those growing in a high CEC situation. The addition of clay or organic material to a soil can increase CEC. Inert materials such as sand have little or no ability to hold cations. In some cases, such as hydroponic systems, a growing media with little or no CEC is desired. The nutrients are supplied by a constant or near constant flow of fertilizer solution.

Another important fertility management issue with CEC is the potential for anions, especially NO_3^-, to leach out of the soil and into the ground water or surface water at elevations lower than the field. This is a serious concern as nitrates can cause physiological problems in humans, the most notable being blue baby syndrome where excessive nitrates interfere with the ability of hemoglobin to carry oxygen to the cells in an infant's body. There are some areas of the United States where NO_3^- levels in the water supply has risen as the result of agricultural runoff that the water to the level where the water is not fit for human consumption. The Environmental Protection Agency (EPA) has set the Maximum Contaminant Level (MCL) of nitrate as nitrogen at 10 mg/L (or 10 parts per million) for the safety of drinking water.

FERTILIZER APPLICATION METHODS

Nutrients can be supplied to plants in many ways and in many forms such as a gas injected into the soil (N from ammonia is applied this way), solids, and liquids.

Figure 14–50
Liquid fertilizer being applied to a field through the furrow irrigation system. Source: USDA-NRCS, *http://photogallery.nrcs.usda.gov/*

The solids can be organic matter, powders and dusts such as lime, or manufactured granules and pellets). The granules and pellets can be made to release the nutrients almost immediately or slowly over several weeks (slow release). Liquids include manure slurries and fertilizers dissolved in water (Figs. 14–48, 14–50, 14–51). Some nutrients such as iron are not very soluble in water. These may be chelated to improve solubility. The **chelate** is a soluble molecule attached to the nutrient. When the chelate goes into solution, it takes the nutrient with it. The concentration of fertilizer dissolved in the liquids can be varied depending on the manure source and/or by the amount of the manure or fertilizer added to the water.

Fertigation is the application of soluble fertilizer in the irrigation water. It is most often used in areas of intensive cultivation such as landscapes, sports and recreational fields, nurseries, and greenhouses. Because of the high frequency of application (with nearly every irrigation), fertilizer rates are relatively low and are usually expressed in parts per million (ppm) of the dilute fertilizer solution. A typical fertigation dose of a complete (N–P–K) 2:1:2 fertilizer is 200–100–200 ppm.

Fertilizer injectors are usually used in fertigation programs (Fig. 14–51). The injector allows a concentrated fertilizer solution to be prepared to save space, then the injector measures a precise amount of that concentrate into the irrigation water at a constant and reliable rate. Typical dilutions for injectors are 1:16, 1:50, 100, and 1:200. Thus, when a 1:100 injector is used, the concentration in the fertilizer tank is 100 times the recommended rate. For example, 250 ppm would be mixed at 25,000 ppm. The injector injects 1 part of the concentrate into 100 parts of

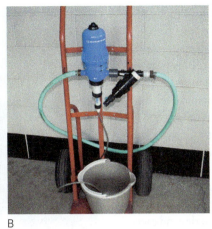

A   B   C

Figure 14–51

Different types of fertilizer injectors. Some require electricity to run (A). Some run from the force of water passing through the injector (B,C). Source: Margaret McMahon, The Ohio State University.

water so the rate coming out the end of the hose or irrigation line is 250 ppm. It also means that it would take a 10,000-gallon tank to hold the same amount of unconcentrated fertilizer that can be held in a 100-gallon tank of concentrated fertilizer. The space-savings is well worth the investment in an injector and the small amount of additional plumbing required. Although it might seem to make sense to have injectors with higher ratios to be even more space-efficient, at concentrations higher than 1:200, the fertilizer solution usually becomes saturated so that not all of the fertilizer can dissolve.

SUMMARY AND REVIEW

Land preparation is the first step in growing plants. Leveling, incorporation of existing plant material back into the soil, improving tilth, control of weeds, insects, and diseases, and control erosion are the purpose of land preparation. Tillage is the mechanical manipulation of soil to provide a favorable environment for plant growth. Plowing, disking, harrowing, and cultivating are types of tillage. Some soil and growing media require fumigation or pasteurization to destroy pathogens before plants can be grown.

Soil degradation can occur when poor soil management practices are used. Removal of the vegetative cover by intensive and repeated tillage has lead to erosion and loss of organic matter in the soil. Soil compaction occurs when heavy equipment is run over the field for tillage, planting, and harvest especially when the soil is wet. Compaction is the loss of pore space, especially the large, drainage pores resulting in poor water infiltration through the soil. Compacted soils also impede root penetration which limits the area from which the root can extract water and mineral nutrients. The amount of pore space compared to solids in the soil can be determined by measuring the soil's bulk density. Salinization is the introduction of excess soluble salts into the soil making the soil too salty to support plant growth.

Soil improvement can often occur naturally by leaving the soil alone for several years and establishing a permanent cover crop that includes a deep-rooted species such as alfalfa. Adding large quantities of organic material or practicing no-till farming are other ways to improve the soil. When organic material is added the carbon:nitrogen ratio can be altered. A material that has a high C:N (50:1 or greater) may cause a temporary drop in soil N while soil organisms feed on the carbon-based food and require N for their own use. During this time supplemental N may have to be added for the plants growing there. Salinization can be mitigated by flushing the land with large amounts of water. To do this, drainage systems may be required to take the salty solution away.

Erosion is a major factor in soil degradation. It occurs when land is tilled in ways that leave it vulnerable to erosive forces such as wind and water. Removal of vegetation or creating planting rows that run up and down slopes are two common tillage practices leading to erosion.

Soil conservation is preservation of soil by using the soil wisely. Conservation practices include the use

of grassy greenways, contour tilling, strip-contour farming, terracing, and no- or low-till farming.

Artificial soil is made with several components blended together to give the media the characteristics for good plant growth. The factors that influence the growing characteristics of these mixes is the same as those that influence the characteristics of native soils.

Water management is the application and removal of water as needed for good plant growth. Irrigation is one of the oldest agricultural practices with evidence that it was practiced several thousand years ago. One of the first considerations of irrigating is the quality of the water to be used. The pH, alkalinity, and soluble salt content of the water can have a profound effect on plants so these qualities must be known and adjusted if necessary.

The need for irrigation and type of system needed is determined by several factors including climate, type of crop, slope and physical characteristics of the soil, water availability and quality, and drainage capability. Types of irrigation systems are border/flood (including ebb and flood in greenhouses), furrow, sprinkler, and drip. Each has advantages and disadvantages and some are good for some situations while others work better in other situations. Irrigation systems are controlled by many different methods from manual to computer. The best systems operate by sensing an appropriate environmental condition such as soil moisture, evaporation potential, light intensity and accumulation, or condition of the plant.

In the past, extensive drainage systems were installed in wet areas to create what are now large areas of agriculturally productive land. The systems consisted of clay tiles or perforated pipe buried in the field to conduct water to a ditch or creek. Although the land is very productive agriculturally, we have come to realize the importance wetlands have in the local ecosystem, where they act as filters, reservoirs, and wildlife habitat. In many areas, wetlands have been partially restored to improve the ecosystem. Drainage systems continue to be installed but with regard to the impact they have on wetlands and include regions set aside to remain wetlands.

Fertility management is the application of nutrient elements not provided by the soil in which plants are growing. It requires careful monitoring of the soil and adding only those elements that are needed and only in the amounts that are needed. Underapplication can result in deficiency problems with the crop. Overapplication can result in toxicity and environmental pollution. Nutrients can be added in many different forms and in many different ways. They come from crop residues, animal manures, chemical fertilizers, or naturally occurring minerals. They can be applied as solids or liquid. Fertilizer is added at the beginning of the growing season for most field crops, at the beginning and occasionally during the growing season for vegetables, and as often as daily for intensively managed greenhouse and nursery crops. The concentration of the fertilizer for the frequent applications is much lower than for the less frequent applications.

A complete fertilizer contains nitrogen, phosphorus, and potassium. Incomplete fertilizer lacks one or two of these elements. Each bag of fertilizer has to give the percent of N, P, and K on the label, even if the percent is zero. Other nutrients may also be provided by the fertilizer and these may be listed too.

Soil/media testing and plant tissue testing is required to know what nutrients are needed. Soil pH affects nutrient availabilty and is an important factor in fertility management. pH is adjusted or the fertilizer rate is adjusted to make sure the nutrients are available to the plant. The cation exchange capacity (CEC) of soil and media is another factor in fertility management. A soil or media with a high CEC usually requires less application of cation nutrients because they can be held in reserve by the soil or media particles.

Fertilizers can be applied as gas injections into the field, as solids, and as liquids. They can release the nutrient elements immediately or over a long period of time (slow-release). Fertigation is the practice of applying fertilizer in the irrigation water. Fertigation usually requires the use of a fertilizer injector that injects the concentrated fertilizer solution into the water at a very precise rate.

KNOWLEDGE CHECK

1. What are two reasons for land preparation?
2. What is tillage and why is soil moisture an important factor in tillage?
3. How much water should a soil ideally hold?
4. What is the reason for using a plow, a disk, a harrow, a cultivator?
5. What is the purpose of ridging or listing a field?
6. Why are some soils fumigated or pasteurized?

7. Why is methyl bromide use as a soil fumigant highly regulated?

8. What is being done to some plants to overcome the loss of methyl bromide?

9. Why can loss of vegetation cover lead to soil degradation?

10. How do tillage and drainage contribute to the loss of organic matter from the soil?

11. What is soil compaction and how does it occur in field where plants are produced?

12. What happens to soil bulk density in compaction? How can bulk density be measured?

13. What are soil salinization and saltwater intrusion?

14. How does a permanent cover crop help improve degraded soil?

15. What is no-till production and how does it help improve soil?

16. Why is it important to understand C:N when adding organic material to soil?

17. Why are row crops more likely to contribute to soil erosion than crops such as grasses that are planted by broadcast seeding?

18. What is Black Sunday?

19. How does each of the following help to control erosion? Grassy waterway, contour tillage, contourstrip-cropping, terracing.

20. What is an artificial soil? What factors influence their characteristics?

21. Why are water pH, alkalinity, and salinity important factors in irrigation?

22. Describe border, flood, and furrow methods of irrigation.

23. What is ebb and flood irrigation and where is it used?

24. Briefly describe each of the following as it relates to sprinkler irrigation: riser, pop-up, wheel line, gantry, center pivot.

25. Why is water use more efficient with drip irrigation compared to other systems.

26. Why are irrigation systems controlled by sensors that detect soil moisture, evaporation potential, or light better than systems controlled by timers.

27. Why was drainage installed in fields? What problem do we now realize it created?

28. Explain why replacing plant nutrients is needed when crops are harvested and removed from a field.

29. Besides nutritional elements, what do manures and crop residues add to soil?

30. When is fertilizing usually done for field crops? For vegetables? For intensely managed crops?

31. What is the difference between a complete and incomplete fertilizer?

32. Why is it important to know the materials that make up a fertilizer formulation (for example the amount of nitrogen as nitrate and ammonium)?

33. What is the Secure Handling of Ammonium Nitrate Act? How does it affect those who want to use ammonium nitrate as a fertilizer?

34. Why are both soil and tissue testing important in a fertility management program?

35. How can soil pH be increased? How can it be decreased?

36. Why is cation exchange capacity (CEC) important in fertility management?

37. What is fertigation?

38. What is a fertilizer injector? Why does using one reduce the size of the container needed to hold the fertilizer solution?

FOOD FOR THOUGHT

1. A low-lying area of a large field is often flooded and stays wet even though much of the rest of the field can get very dry during the summer. You think you could probably increase productivity of the wet area if you drained it. What things would you consider as you decide whether or not to drain it.

2. When no-till farming was first introduced there was considerable resistance by farmers to implement it, in part because the no-till fields did not look "tidy" and well-kept like tilled fields looked. Today however "untidy" no-till is a very common practice. What do you think made so many farmers change their minds?

3. You are a cost-conscious homeowner who is also environmentally responsible and you do not want to waste resources such as water and electricity. You are about to install an irrigation system in your landscape and you are considering what kind of irrigation control system to use. You have the

choice between an inexpensive timer and a more costly evaporation potential sensor. Which would you choose and why would you choose it?

4. Hydroponic growers usually use a well-drained growing media that is inert (no or very low CEC)

and they flood the media constantly with fertigation water. How do you think their fertigation practices would be affected if they switched to a media that, while still well-drained, had a high CEC?

FURTHER EXPLORATION

HAVLIN, J., S. TISDALE, W. NELSON, AND J. BEATON. 2004. *Soil fertility and fertilizers*, 7th ed. Upper Saddle River, NJ: Pearson Education.

MADOFF, F., AND H. VANES. 2000. 2nd ed. Waldorf, MD: Sustainable Agriculture Network SARE Outreach Publications.

REED, W. 1996. *Water, media, and nutrition for greenhouse crops.* Chicago: Ball Publishing.

*15

Integrated Management of Weeds, Insects, Disease, and Other Pests

MICHAEL BOEHM, JERRON SCHMOLL, AND DAVID SHETLAR

key learning concepts

After reading this chapter, you should be able to:

- Explain the foundational concepts of weed science, entomology, and plant pathology.
- Discuss five major strategies for managing weeds, insects, and diseases and how they can be combined to develop an integrated pest management (IPHM) program.

Since the beginning of agriculture, farmers have had to develop ways to manage weeds, insect pests, and diseases. Even today, with all the scientific research on crop protection, it is estimated that insects, diseases, weeds, and animal pests eliminate half the food produced in the world during the growing, transporting, and storing of crops. In the tropics, where heat and humidity favor the development of many diseases, two-thirds of some crops are lost, thereby aggravating the world hunger problem. Although not directly related to the world's supply of safe food, similar losses occur in timber, ornamental, floricultural, and turfgrass production systems. In economic terms, annual losses in food, fiber, and ornamental production systems caused by weeds, insects, and diseases are estimated in the hundreds of billions of dollars. Because of the significant impact of weeds, insects, and diseases—on both human and animal welfare as well as on the global, regional, and local economies—it is important for those interested in growing plants to develop a firm understanding of weed science, **entomology** (study of insects), and **plant pathology** (study of plant disease) and to learn how to eradicate, manage, or otherwise minimize plant losses caused by these important plant pests.

INTEGRATED PLANT HEALTH MANAGEMENT (IPHM)

Whether you are managing a weed, insect pest, or disease-causing organism (also called a **pathogen**), most specialists interested in plant health recommend the use of a multipronged approach or strategy—commonly referred to as an **integrated plant health management** (**IPHM**) approach. Integrated plant health management programs rely on the use of several methods rather than on a single means for developing an effective and sustainable pest management program (Fig. 15–1). Although sometimes called different names by weed scientists, entomologists, and

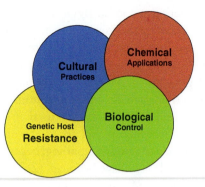

Figure 15–1

Four of the five categories of pest/disease management approaches available to plant production specialists for use in developing effective integrated plant health management programs. In an ideal situation, plant production specialists would have a pool of equally effective pest/disease management tools from which to choose.

Source: Michael Boehm, The Ohio State University.

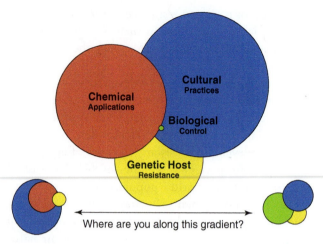

Figure 15–2

Regardless of the cropping system being managed, plant production specialists need to consider the complexities of their system to develop an IPHM strategy that best meets their needs. Source: Michael Boehm, The Ohio State University.

plant pathologists, the methods for managing plant pests fall into five distinct categories:

1. ***Genetic host resistance***—use of genetically resistant plants to minimize or avoid losses caused by insect pests and/or pathogens.
2. *Cultural practices*—use of agronomic or horticultural practices that favor plant development and minimize pest or pathogen activity.
3. *Chemical applications*—use of pesticides such as herbicides (weeds), insecticides (insect pests), miticides (mites), antibiotics (bacteria), fungicides (fungi), and nematicides (nematodes) to suppress or inhibit pest/pathogen activity.
4. ***Biological control***—the use of beneficial or antagonistic organisms that kill or otherwise suppress plant pests or pathogens.
5. *Government regulatory measures*—the use of quarantines and pest-eradication programs to stop the introduction or spread of deleterious plant pests and/or pathogens.

How plant production specialists integrate or mix and match these individual pest/disease management approaches to develop an effective IPHM strategy depends on many different considerations, such as personal preference, public perception, availability of effective options or tools in each management category, and profit margin. Each manager, regardless of cropping system, must consider the complexities of the unique system being managed to develop an IPHM strategy that best meets his or her needs (Fig. 15–2). The use of IPHM strategies is often perceived to be more environmentally sound compared to those that

rely heavily on the use of pesticides. In response to recently mandated and self-imposed restrictions on pesticide use (based primarily on concerns over human, animal, and environmental health), many farmers, growers, and professional plant production specialists have either been forced or have made a conscious effort to rely less on the use of pesticides and more on IPHM strategies. In some cases, the shift from a reliance on chemical control has been intensified by the development of pest populations resistant to previously effective pesticides.

Genetic Host Resistance

Genetic host resistance involves the use of genetically resistant plants to avoid or minimize plant losses caused by insect pests and/or pathogens. The use of genetically resistant plants is often recommended by entomologists and plant pathologists as the first line of defense against avoiding or minimizing plant damage caused by insects and pathogens. In some cropping systems, such as large-acreage field or row-crop agriculture (corn, soybeans, cereal crops, rice, cotton, etc.), the use of genetically resistant plants may be the only cost-effective means for managing a particular pest or disease, as is the case of managing plant diseases caused by viruses. The development of resistant plant types also reduces the need for costly and sometimes dangerous pesticides. Although genetic resistance should be considered when dealing with all insect pests and diseases, it is especially useful when dealing with annual cropping systems where new seed is sown each season, thereby providing an opportunity to introduce new cultivars or varieties

with insect or pathogen resistance. Although important in perennial cropping systems such as orchards, timber production, and turfgrass swards, once the initial crop is planted, the introduction of resistant lines is limited due to the long-term nature of these crops.

In nature, plants susceptible to a prevalent disease or insect pest tend to disappear, while resistant types remain. The work of geneticists and plant breeders in purposely developing resistant cultivars of agriculturally important species has been one of their most useful pursuits, and the world's population owes much to them. The development of rust-resistant wheat cultivars, for example, has added food for untold millions of people. Genetically resistant plants may occur naturally, be derived through classical plant-breeding programs, or be derived through the use of molecular biology techniques. Plants derived from the latter are commonly referred to as **genetically modified organisms (GMOs)**. Two examples of GMOs that are widely used as part of pest management programs include the use of Roundup® resistant cotton, corn, and soybeans, and Bt corn that has been modified to produce a protein that is toxic to several species of Lepidoptera (a more detailed discussion of these crops is found in Chapter 9). Also of recent significance in terms of human health are low linolenic soybean varieties. New pest management GMOs expected to soon be on the market include dicamba-resistant soybeans and highly "stacked" corn hybrids containing multiple herbicide and insect-resistance traits.

Many cultivars or varieties have been developed through traditional breeding and selection as well as genetic engineering that maintain yields in spite of heavy disease or insect pest pressure. (The ability of a plant to maintain its yield in spite of heavy disease or pest pressure is known as **tolerance**.) Ideally, all insect pests and diseases would be managed in this manner. Also of recent significance in terms of human health are low linolenic soybean varieties.

Cultural Practices

Cultural practices in pest management involve the use of agronomic or horticultural practices that favor plant development and minimize pest or pathogen activity. The second line of defense available to those managing plant pests is the use of cultural management practices that favor plant growth over pest or pathogen activity. As discussed in earlier chapters, all plants require light, water, and nutrients to grow. Knowing how to provide the correct balance of these essentials to maximize plant growth and yield—whether measured in bushels per acre or aesthetics—is the responsibility of the professional plant production specialist. By providing growing conditions that favor plant growth over insect and pathogen development or activity, it is possible to avoid or minimize losses caused by these pests. Plant pathologists and entomologists have developed a simple model called the disease/pest triangle to illustrate this concept (Fig. 15–3). Several cultural practices can be used to change the environment in which plants are grown and they can severely influence pest and pathogen activity. These practices include (1) tillage practices, (2) water management, (3) fertility, (4) crop rotation, and (5) sanitation (cleaning or removal) of equipment and diseased or infested plant material.

In field agriculture, tillage—both plowing and cultivation—is widely used to reduce soil compaction, increase water infiltration, increase the rate of crop residue decomposition, control emerged weeds, and bury survival structures of plant pathogens. Cultural practices that focus on water management—drainage, irrigation, and other specialized practices such as syringing on golf courses—is another group of cultural practices that can often be used to manage important pests and diseases. For example, management of *Phytophthora* and *Pythium,* organisms that grow best in poorly drained soils, is directly related to avoidance of saturated soils and flooding. The use of a balanced fertility program (discussed in Chapter 6) is another important requirement for growing healthy plants. When plants are either over- or

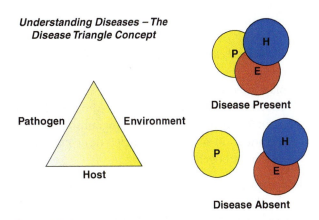

Figure 15–3

The disease/pest triangle concept. A susceptible host (H), a virulent pathogen (P), and environmental conditions (E) that favor the pathogen must be present in the right mix to yield disease. If any one of these three components is missing or minimized, disease will not occur. Genetic host resistance, cultural practices, chemical applications, and biological control are tools used by professional plant production specialists to favor plant growing conditions and minimize pathogen activity or development.

Source: Michael Boehm, The Ohio State University.

underfertilized, however, they may become more or less susceptible to pathogen and insect attack. Knowing the diseases and insect pests for which a particular crop is predisposed is important and is often linked directly to the fertility status of the crop. Crop rotation is a simple but often effective means of controlling many pests. If the same annual crop is grown year after year on a given plot of land, a particularly serious pest may keep increasing year after year, overwintering in crop residues, until it reaches such overwhelming populations that the crop cannot be produced on that piece of land. But by rotating the crop in a one- to several-year rotation with other nonsusceptible crops, the pest population, lacking a crop host for a long period, may decline significantly. Cultural practices designed to achieve or maintain disease- or pest-free conditions on the farm or in the greenhouse or nursery play a key role in reducing plant losses by minimizing or eliminating sources of insect pests, weed seed, or pathogen inoculum. Practices such as the removal and destruction of infected plants or infested soil or potting mix, the use of certified pathogen-free or weed-free seed, and the use of clean tools and equipment is critical to reducing reservoirs and the spread of weeds, insects, and pathogens.

Chemical Applications

The use of pesticides such as herbicides (weeds), insecticides (insect pests), miticides (mites), antibiotics (bacteria), fungicides (fungi and fungal-like organisms), and nematicides (nematodes) to suppress or inhibit pest/pathogen activity are included in this category. In an ideal world, plant pests and diseases could be managed effectively using genetically resistant plants and cultural management practices. Although some diseases and pests can be managed effectively using these two approaches, many still require the use of pesticides to achieve commercially acceptable disease or pest thresholds (Fig. 15–4). Thus, the third line of defense available to those managing plant pests is the use of pesticides to either kill or, as is more often the case, suppress plant pests and pathogens. Because the use of pesticides increases the cost of production and also causes concerns over the potential dangers of pesticides to humans, the environment, food products, farm and domestic animals, wildlife, beneficial insects, and the atmosphere, they are often considered the least desirable method of managing harmful insects and plant diseases (Fig. 15–5). As a result of these concerns, pesticide manufacturers, governmental regulatory agencies, and growers' groups have worked hard over the several decades to reduce the use of broad-spectrum pesticides in favor of the development and use of pest/pathogen-specific compounds. Detailed information about the current status and use of herbicides, insecticides, antibiotics, fungicides, and nematicides is provided later in this chapter.

Biological Control

Biological control (sometimes referred to as biocontrol or biorational pest management) involves the use of beneficial or antagonistic organisms that kill or otherwise suppress plant pests or pathogens. The use of biological control is considered advantageous and environmentally

Figure 15–4

(A) Aerial spraying with a low toxicity pesticide. (B) Certain insect and disease pests are controlled in commercial orchards by insecticides and fungicides applied with high-efficiency power sprayers.

Source: USDA Agricultural Research Service Image Gallery, http://www.ars.usda.gov/is/graphics/photos/

A

B

Figure 15–5

When spraying pesticides to control insects workers should wear protective masks and clothing.

Source: Margaret McMahon, The Ohio State University.

sound because it provides an environment-friendly alternative to the use of pesticides. However, few biocontrol products provide consistent and commercially acceptable levels of pest or disease control.

Within all eco- and plant production systems, a balance usually develops among organisms, both plant and animal. Certain organisms are antagonistic to others and retard their expansion. Environmental or human-induced changes (such as those used by plant production specialists in the course of growing plants) that upset this balance by eliminating one of the organisms can lead to

explosive proliferation of the others and to subsequent attacks on vulnerable crops. In a similar manner, if an organism is introduced into an environment, but its antagonizing organism is not, the results can be devastating to a vulnerable crop plant. Biological control in such a situation would consist of introducing the antagonizing organism of the pest into the area, thus bringing the pest under control again. Biocontrol organisms can kill or suppress pathogens and pests by (1) parasitizing the pest or pathogen directly; (2) competing with the pest or pathogen for space or nutrients; (3) producing toxins that kill or make the pest or pathogen sick; or (4) inducing a physiological or biochemical change in the host plant, thereby making it less susceptible to pest or pathogen attack. Specific examples of how biological control can be used to manage weed and insect pests and plant pathogens will be discussed later in this chapter. Several examples of how biological control has been used successfully to suppress important insect pests and disease include the use of parasitic wraps (introduced from Asia to control *Olive parlatoria* scale in California olive orchards) and the use of compost-amended potting mixes to suppress *Pythium* damping-off and root rot in greenhouses and nurseries (Fig. 15–6).

Government Regulatory Measures

The fifth and final means for managing plant pests and diseases is through the use of intensive inspection programs and quarantines. Although used by growers on a small-scale (i.e., the inspection and isolation of newly

A   B

Figure 15–6

The use of compost-amended potting mixes by the floricultural and nursery industry for suppressing damping-off and root rot caused by Pythium *and* Rhizoctonia *species. (A) The potting mix in both pots was inoculated with* Pythium ultimum. *The potting mix on the left (CP) was comprised of sphagnum peat and perlite and was severely stunted with extremely rotted roots. The potting mix on the right was the same as that on the left but was amended with composted pine bark (CPB) that contained beneficial biological control organisms (bacteria and fungi). (B) The top row shows the roots of five plants grown in CP. The bottom row shows roots from plants grown in CPB. The middle row is another type of potting mix without composted pine bark.* Source: Michael Boehm, The Ohio State University.

received plant materials prior to introduction), the use of quarantines is often implemented and regulated by local, state, regional, national, and international agencies and authorities. Whether dealing with people, animals, or plants, a shared concept in preventive medicine is that when pathogens and insect pests evolve with their hosts, the host will develop resistance or tolerance against the pathogen or pest. The result of such coevolution is that the pathogen or pest is kept in check. A practical corollary to this concept is that when an exotic pathogen or pest is introduced into a population that has never been exposed to or that did not coevolve with it (such populations are said to be naive), serious losses are likely to take place because of a lack of resistance in the host. The ash species in the United States are rapidly being destroyed by the recently introduced Emerald Ash Borer. However, the ash trees in Asia, where the insect originated, tolerate the insect and do not decline and die when infected. German soldiers hired by the British during the Revolutionary War brought the Hessian fly to North America with them in their straw bedding. This pest later moved westward through the United States, devastating wheat fields and ultimately causing far more problems for the new country than the soldiers did. Other historical examples of the devastating impact of exotic pests or pathogens on our native plant populations include the introduction of chestnut blight and Dutch elm disease fungi to North America in the early 1900s.

Strict government inspections and quarantines of imported plants, plant products, and soil have kept such pests out of many countries and many areas. However, modern jet travel, taking people and their belongings swiftly from country to country, greatly increases the possibility of introducing pests dangerous to people, plants, and animals. In addition to the passive or accidental means for moving pathogens and pests, the 2001 anthrax attacks in the United States highlight the potential for the intentional release of exotic pests and pathogens as agents of bioterrorism. Recent outbreaks of sudden oak death, Asian soybean rust as well as the Emerald ash borer emphasize the continued need for strict government monitoring and enhanced reporting and response programs for key pests and diseases. The National Plant Diagnostic Network (NPDN), implemented following the events of September 11, 2001, is an excellent example of just such an effort. The NPDN was designed and implemented to enhance agricultural security within the United States by linking plant disease and pest specialists at land grant universities and government agencies so that high-consequence pests and pathogens could be quickly identified and an appropriate response coordinated. The identification and subsequent quarantining of imported geraniums infected with the bacterial pathogen *Ralstonia solanacearum* race 3 biovar II is an example of this effort. This biotype is not naturally found in the United States and could pose a threat to U.S. tomato and potato production.

In some cases, certain plants or plant products are forbidden entry into the country or areas of the country, and agricultural inspectors check luggage for such outlawed products. For example, mangoes, guavas, and passion fruit from Hawaii are not permitted entry into the U.S. mainland because they may be carrying the Mediterranean fruit fly, which is widespread in Hawaii but absent on the mainland. If introduced into California, for example, the fly could devastate the state's huge fruit industry. However, such plant products can be imported commercially if they are properly fumigated to kill the pests before shipping. Certain kinds of living plant material can be imported into the United States under permit or if it is quarantined after entry for two growing seasons to reveal any diseases. Travelers coming into the United States should not attempt to bring or send in agricultural materials such as fruits, vegetables, plants, bulbs, seeds, or cuttings unless advance arrangements have been made and a permit obtained. Any such plant materials being carried or imported through commercial channels must be reported to agricultural quarantine or customs officials upon arrival. The movement of certain crops among the contiguous forty-eight states is also prohibited or closely monitored.

Government eradication programs are conducted when a serious insect or disease pest breaks out. Often the trouble is eliminated before it has a chance to spread. Such programs require highly trained personnel who know the potential insect and disease problems and can recognize the pathogens and the symptoms of their activities. The dangerous Mediterranean fruit fly has been introduced accidentally into Florida on several occasions. It has been eradicated each time, but always at considerable expense. In California, too, discoveries of the Mediterranean fruit fly, probably brought in on illegal tourist importations of fruit from Hawaii, have also caused costly eradication procedures several times.

Lifecycle   A very important concept in controlling all pests is **lifecycle**. Lifecycle is the stages an organism goes through from fertilization to death. For plants the stages can be fertilization, seed development, germination, juvenile (strictly vegetative), vegetative, reproductive, senescence, and death. For insects the stages can

be egg, larva, pupa, adult. Development through these stages is also called **metamorphosis**. A fungus can have sexual, asexual, or both stages during its lifecycle, depending on the type of spores it produces. Regardless of the type, typically when the fungus spores mature, they leave the parent and germinate on a host. The germination results in the development of several filamentous structures, collectively called mycelia. Some of these structures (appresoria) are capable of penetrating plants cells and absorbing liquid and nutrients from the cells. Pathogenic fungi are the appressoria-producing type. Knowing the host plant lifecycle is important because, depending on the plant and pest, some of the lifecycle stages may be more susceptible to infestations than others. For weed and disease pests, some lifecycle stages are easier to control or to cause problems than others. For example, some weeds are easier to control as germinating seedlings while for others it is a larger plant. The larva of a leaf miner fly causes the damage, but the adult is often easier to control.

Threshold Level   Another general pest management concept that is important is **threshold level**. This refers to a level of infestation or infection below which the economic value of a plant or crop is not compromised. The level depends on the crop and its use. For example, some insect damage on a wheat plant would be acceptable as long as it did not lower the grain yield. However, for leaf lettuce or an ornamental foliage plant little or no damage could be tolerated.

WEEDS

A **weed** can be defined as a plant out of place, growing where it is not wanted, that interferes with human activities or attempts to grow other plants. A weed can be any plant that competes with a desired plant for water, nutrients, or sunlight; interferes with harvesting a crop; reduces crop quality; reduces the aesthetic appeal of a crop; harbors undesirable insects, pathogens, or nematodes; or competes with the crop for pollinating insects. For the purpose of organization, the study of weeds is generally divided into two broad components: (1) weed biology and ecology and (2) weed control and management. However, all effective weed control strategies involve careful consideration and integration of both components.

Weed Biology and Ecology

Weed biology is the study of the classification of an individual weed species, its competitiveness, growth and reproductive capabilities, and genetics. Weed ecology is the study of the interaction of a weed species with its physical and biological environment. The physical components of the environment includes soil structure, soil acidity, soil fertility, and the quantity of rainfall and sunlight available. The biological components of the environment include the dynamics involved in regulating the interaction of plants, animals (such as insects and herbivorous mammals), and microorganisms (such as bacteria, viruses, and fungi). Understanding the biology and ecology of a weed species is fundamental to the development of effective weed management strategies.

Classification

Plants can be classified in many different ways. Plants may be referred to by their habitat, for instance, tropical versus desert plants, or aquatic versus terrestrial plants. Broad classification of plants as either monocots or dicots is sometimes helpful when making general comparisons among plant species. Weed scientists have found that classification by life cycle (i.e., annuals, biennials, or perennials) is particularly useful because understanding weeds' life cycles is a key component of weed management. The term *invasive weed species* has recently been added to the vocabulary of weed scientists and is used to classify species that are particularly aggressive, regardless of their life cycle.

Annuals   Annual plants reproduce by seed only and complete their life cycle within one year. Annuals may be classified as either summer annuals or winter annuals. Winter annuals may be further subclassified as either obligate or facultative winter annuals. Summer annuals normally germinate in the spring or early summer, grow vegetatively throughout the remainder of the summer, and then set seed and die in the fall or early winter. The seeds of summer annuals typically lie dormant all winter and then germinate the following spring. Summer annual weeds tend to be troublesome in summer annual crops such as corn, soybeans, cotton, rice, peanut, and many vegetable crops (Fig. 15–7). Obligate winter annual weeds germinate in the fall, overwinter as immature plants, and then set seed and die in the spring or early summer. Facultative winter annuals normally behave as obligate winter annuals, but under the correct environmental conditions, they may follow the summer annual life cycle. The seeds of obligate winter and fall-germinating facultative annual weeds tend to lie dormant in the soil through the summer, and then the cycle continues with seed germination in the fall. Like summer annuals, winter annuals tend to be most problematic in crops that share the winter annual life cycle, such as fall-planted nursery crops and small grains.

Figure 15–7

Foxtail (Setaria spp.), a summer annual weed, competing with soybean (Glycine max), a summer annual crop.

Source: Kent Harrison, The Ohio State University.

Figure 15–9

Plantains (Plantago spp.) have a simple perennial life cycle.

Source: Kent Harrison, The Ohio State University.

Biennials Biennial weeds live for two years and reproduce only by seed. In the first year of the life cycle, biennial plants grow vegetatively, producing a fleshy tap root and a rosette (a loose whorl of leaves attached to a highly compressed stem; (Fig. 15–8). The tap root and rosette overwinter, and exposure to the cold vernalizes the plant and induces production of an elongated flower stalk the second year. After seed are produced, the plant usually dies. Biennial weeds can cause losses in pastures, orchards, and agronomic crops.

Perennials Perennial weeds live for an indefinite period of time and can be classified as simple perennials, herbaceous (nonwoody) creeping perennials, or woody perennials. Simple perennials (Fig. 15–9). The herbaceous (nonwoody) shoot tissue normally dies back during the winter, and plants regrow in spring from either seed or, more frequently, the vegetative propagules that overwinter in soil. Over time, one individual creeping

perennial seedling can grow into a large patch with an extensive root system and many shoots, which makes these species particularly difficult to control. These species tend to become established in crops that are grown under some form of reduced tillage, as well as in perennial crops such as pastures and hayfields. Once established, however, they can also persist in tilled environments for a long period of time because the tillage operations may fragment and spread the vegetative propagules to other areas of the field (Fig. 15–10).

Woody perennial weeds are usually vine or shrub plants that regrow each year from persistent woody stems. These weeds are sometimes encountered in crop fields in reduced tillage systems; however, they are most troublesome in low maintenance areas, such as pastures, forests, and rights-of-way.

Invasive Weed Species The U.S. National Invasive Species Council, established in 1999, defines an invasive species as "a species that is (1) nonnative (or alien) to the

A   B

Figure 15–8

(A) Wild parsnip (Pastinaca sativa), young first-year rosette. (B) Second-year flower stalk. Source: Kent Harrison, The Ohio State University.

Figure 15–10

Examples of vegetative propagules. (A) Johnsongrass (Sorghum halepense) rhizomes. (B) White clover (Trifolium repens) stolons. (C) Yellow nutsedge (Cyperus esculentus) tubers. (D) Wild garlic (Allium vineale) bulbs. (E) Honeyvine milkweed (Ampelamus albidus) creeping roots.

Source: Kent Harrison, The Ohio State University.

ecosystem under consideration and (2) whose introduction causes or is likely to cause economic or environmental harm or harm to human health. Invasive species can be plants, animals, and other organisms (e.g., microbes). Human actions are the primary means of invasive species introductions." Some weed scientists argue that some particularly aggressive and problematic native plants should also be considered invasive species. Regardless of origin or terminology, the hallmark of invasive weeds is their lack of natural enemies and the ability to infest new environments rapidly. These species are capable of aggressive growth under a variety of environmental conditions, which allows them to compete successfully with a wide range of native and cultivated plants. In North America, garlic mustard (*Alliaria petiolata*, a biennial) and Japanese honeysuckle (*Lonicera japonica*, a woody perennial) are examples of exotic (nonnative) invasive weed species that were introduced for use as ornamental garden. Kudzu (*Pueraria lobata*) was introduced to control erosion.

Competition

Competition for sunlight, water, and mineral nutrients is often the first factor that comes to mind when one considers the negative effects of a weed on a crop. Indeed the time that a weed spends in contact with a crop plant can have a dramatic effect on yield. Fig. 15–11 illustrates the effects of competition on crop yield.

The dashed line in Figure 15–11 shows what happens when weeds emerge with the crop and are allowed to compete for a period of time before they are removed. The weeds have very little effect on crop yield during the first few weeks of crop growth primarily because there is sufficient light, water, and nutrients available for all plants in the field. As time passes and the crops and weeds grow larger, competition starts to have an increasingly negative impact on crop yields, which is indicated by the steep slope in the middle of

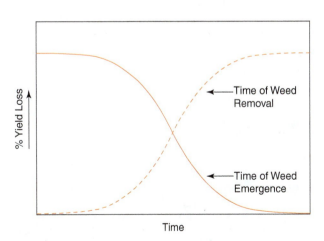

Figure 15–11

Effects of weed competition on crop yields. Source: Jerron Schmoll, Pioneer, Inc.

Figure 15–12

The corn plot to the left had weeds removed early in the growing season; while in the plot to the right, the weeds were allowed to persist until much later in the growing season, which greatly restricted the growth of the corn crop.

Source: Jerron Schmoll, Pioneer, Inc.

Figure 15–13

Giant ragweed in the background overtook the field corn planted there, completely eliminating any harvest from that section of the field. Source: Jerron Schmoll, Pioneer, Inc.

the curve. At the top of the curve, weed competition is maximized and crop yields may be driven to zero by some of the most competitive weed species. Figure 15–12 provides a further demonstration.

The solid line in Fig. 15–11 shows that weeds emerging early in the season have a much greater potential impact on crop yield than those that emerge later in the season. Crop yields can be reduced drastically by a large population of weeds that emerges with the crop (Fig. 15–13), but with later emerging weeds, the crop can use its head start to grow rapidly and shade out the weeds. Once the crop canopy closes and the entire soil surface is shaded, weeds have a significantly reduced capacity to affect crop yield. Decreasing the time to canopy closure by maintaining a healthy, actively growing crop is an important goal for many crop producers.

Reproduction

Annual weeds rely on high seed production, seed longevity, seed dormancy, or a combination of these three factors for survival. Table 15–1 shows the variation in the number of seed produced per plant for a few common weeds and lists an estimate of the length of time that seeds from that plant may remain viable in the soil. The high numbers produced and the longevity of many common annual weed species allows a **seed bank** to build up in the soil. This seed bank serves as the primary source of seed for annual weed infestations in crop fields. While these species can be controlled effectively by tillage performed when seedlings are still small, tillage can also serve to bury ungerminated seeds deeper in the soil, where the cool, relatively stable temperatures, protection from seed-eating organisms; and lack of light provide an ideal environment for long-term survival of the seed.

Table 15–1
SEED PRODUCTION AND LONGEVITY

Species	Life Cycle	Seeds/Plant	Longevity (years)
Ambrosia trifida (giant ragweed)	Summer annual	5,000	21
Capsella bursa-pastoris (shepherd's-purse)	Winter annual	38,500	35
Digitaria sanguinalis (large crabgrass)	Summer annual	150,000	50
Amaranthus retroflexus (redroot pigweed)	Summer annual	230,000	40
Chenopodium album (common lamb's quarters)	Summer annual	500,000	39
Cirsium arvense (Canada thistle)	Creeping perennial	5,300	25

Source: E. E. Regnier, "Teaching Seed Bank Ecology in an Intergraduate Laboratory Exercise," *Weed Technology*: 9 (1995): 5–16.

Seed dormancy also helps to ensure the long-term survival of many weed species. Most of the seeds that are produced by annual plants will germinate after exposure to a prolonged cool, moist period (summer annuals) or a prolonged warm period (winter annuals). A small percentage of seeds will not germinate, however, even after this primary dormancy is broken and the seeds are exposed to conditions that are favorable for growth. The seeds that do not germinate are considered to be in a state of secondary (or conditional) dormancy and will not grow until the proper conditions are present to break this secondary dormancy. These conditions might include exposure to an additional cold (or warm) period, decay of the seed coat, exposure to light, leaching of growth-inhibiting hormones from the seed coat, or some combination of these factors. Thus, the seed production from a single year's annual weed infestation can provide enough seed to infest a field for many years. This predicament has led many growers to adopt the goal of zero seed return for annual weed species.

Weedy species may also produce seed that differ significantly in size. Larger seeded weeds may produce relatively low numbers of seeds, but these species can rely on the large reserves of energy stored in their cotyledons to emerge from considerable depths in the soil profile. Smaller seeded weeds may produce many more seeds per plant but the seed can germinate only from shallow depths in soil. Both strategies can be equally successful.

Biennial weeds also rely on seed longevity and seed dormancy for long-term survival. One additional survival mechanism that many biennials possess is that they have the capacity to behave as short-lived perennials. If the elongated flower stalk is removed, perhaps by mowing or grazing animals, the plant may overwinter a second year and produce seed in the third year.

Unlike annuals and biennials, perennials do not rely as heavily on seed production for establishment. While perennials do sometimes establish by seed, they are also capable of spreading via vegetative **propagule** fragments, which can easily be carried from one area to another by tillage equipment or other transport mechanisms (Fig. 15–10). Note that carbohydrate storage is at its maximum in the winter months, which provides the plant with sufficient energy to produce new shoots aggressively in the spring (Fig. 15–14). Carbohydrate levels are at their minimum levels in the spring and early summer months because they have been expended to support early spring shoot growth. At this

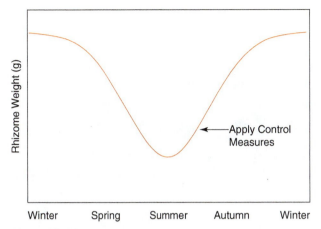

Figure 15–14

Seasonal carbohydrate distribution in rhizomatous creeping perennials. Source: Jerron Schmoll, Pioneer, Inc.

minimum point, which usually corresponds to the mid- to late-vegetative stage, depending on species, the plant shoots have become large enough that they start to export sugars downward to replenish the vegetative propagules, which accumulate adequate levels of stored energy for winter survival and growth the following spring. Application of systemic herbicides (herbicides that are mobile within the plant) when the carbohydrate flow starts to move back to the vegetative propagules can be very effective in controlling creeping perennials. The herbicide moves with the carbohydrate flow into the below-ground structures, thereby providing whole plant control. It is also possible to deplete the carbohydrate supply of vegetative propagules gradually and eliminate perennials by repeated tillage, mowing, or with other methods of shoot removal.

Genetics

The study of weed genetics has gained more attention as the number of reported cases of herbicide resistance has steadily increased. Understanding the basic genetic concepts that drive herbicide resistance is key to maintaining the long-term utility of commonly used herbicides. Since the first reported case of herbicide resistance in common groundsel (*Senecio vulgaris*) in 1968, the number of herbicide resistance cases has increased to 187 species and 323 biotypes. A **biotype** refers to a subpopulation of plants within a species that differs only slightly from the norm of the general population. If a biotype of a weed species, which has some level of resistance to one particular herbicide (or to a group of herbicides that have a similar mode of

action), survives and reproduces, some of the pollen or seed resulting from that plant may carry the gene that imparts resistance. If a herbicide with the same mode of action is applied again in subsequent years, plants that carry the gene for resistance will survive and reproduce. If this cycle repeats, a population of herbicide-resistant biotypes may become prevalent within a relatively short period of time. Integration of several weed-management strategies, rather than strict reliance on chemical control, is a useful tool for controlling these especially problematic weeds. The same strategy applies for preventing resistance to managing other pests.

Genetic variability within a weed species is not restricted to physiological differences. Many weed species also vary significantly in vegetative features such as leaf and seed morphology. For instance, in giant ragweed (*Ambrosia trifida*), the size and shape of the seeds produced varies dramatically from one individual to the next. Even individuals from closely related species may differ so much in their morphology that accurate identification becomes challenging. The pigweeds are a good example because there are several different species of pigweeds (*Amaranthus* spp.) that can be differentiated only by slight differences in seed-head shape and flower structure. Morphological variation likely plays a significant role in the adaptive survival and evolution of a species. For instance, a seed-boring insect might prefer to lay its eggs on the smaller seeds of a given species, while a seed-eating rodent might prefer to forage for larger seeds of the same species. Therefore, if variability exists in the size of seeds produced in a given year, the weed species may prove to be less susceptible to one particular predator, which increases the chance that some of the seed will survive until the following year. This type of interaction among species is broadly described as the science of ecology.

Ecology

Ecology is the study of a species in both its physical and biological environment. Many weed species have special adaptations that allow them to compete effectively in specific physical environments. For instance, field horsetail (*Equisetum arvense*), reed canarygrass (*Phalaris arundinacea*), and the sedges (*Cyperus* spp.) thrive in damp areas. Red sorrel (*Rumex acetosella*) favors acidic soil conditions. Goosegrass (*Eleusine indica*) tolerates highly compacted soils. These adaptations give each species a distinct advantage over other species under specific environmental conditions. At the same time, these adaptations provide land managers with a powerful control strategy: removing the set of conditions to which a particular species is adapted will reduce the competitive ability of that species. For instance, sedges are rarely a problem in a field with a properly installed tile drainage system.

The study of the biological environment as it relates to a weedy species is a rich field of study. Numerous organisms use weeds as a food source, for shelter, and for nesting habitat. For instance, quackgrass (*Elytrigia repens*), while a troublesome weed in agronomic and vegetable crops, is a nutritious food source in pastures for grazing animals. Insect larvae such as the ragweed fruit fly (*Euaresta festiva*) use giant ragweed (*Ambrosia trifida*) seeds as a food source. Foxtail (*Setaria* spp.) seeds are eaten by many species of birds, including the mourning dove (*Zenaida macroura*). A wide variety of birds; including ring-necked pheasants (*Phasianus colchicus*), meadowlarks (*Sturnella* spp.), and sparrows (*Ammodramus* spp., *Chondestes grammacus,* and *Passerculus sandwichensis*), use grasslands (including weeds) as nesting habitat and foraging grounds.

The study of weed ecology may provide information that increases our capacity to use biological methods to control weeds. For instance, researchers have been studying the potential to reduce the number of weed seeds that enter the seed bank by optimizing habitats for seed-feeding insects, rodents, and birds. While practical recommendations are not yet available, researchers have determined that various seed feeders can consume a large number of weed seeds within days after the seeds are deposited onto the soil surface. Integration of practices that increase the populations of seed feeders may become an important nonchemical weed-control method available to crop producers. As a result of this study and others like it, it is very likely that future weed management strategies will have to include the effect that weed control practices could have on all other ecosystem inhabitants.

WEED CONTROL AND MANAGEMENT

Weed control and management is the second broad component of weed science. The understanding of weed biology and ecology from the previous section,

allows a more robust treatment of the preventative (regulatory), cultural, chemical, biological, and measures.

Cultural Weed Control

Cultural weed control describes any method of crop production that enhances the capacity of a crop to compete with weeds. Modification of crop planting date, crop row spacing, crop rotation, flooding, the use of cover crops, mechanical weed control, use of mulches, and fire can all be used as components of a cultural weed-control program.

While early planting dates are normally recommended to provide the crop with a longer growing season and thus higher potential yields, early planting also increases the length of time that the crop must compete with weeds. Many summer annual weeds also emerge early in the season, so planting a crop a few weeks later may allow the grower an opportunity to implement an additional weed-control measure such as disking or harrowing or chemical application to remove these weeds prior to planting.

Modification of crop row spacing is a widespread practice that directly affects weed competitiveness. The goal in optimizing row spacing is to maximize crop yield potential while reducing the time required for the plant canopy to close, thereby effectively shading the soil surface and reducing the light available to late-emerging weeds. A tremendous amount of research has been completed on the subject of optimum plant population and row spacing, and local recommendations can be obtained for most economically important crops.

Crop rotation can be an effective weed-control strategy as management practices such as planting and harvest dates, fertilizer regimes, tillage practice and herbicide usage vary dramatically between different crops. These differences allow different weed control measures to be utilized with each crop that is grown. As an example, because the weeds that are most prevalent in a crop often have the same life cycle as the crop, a perennial-annual crop rotation, such as a short-term hay crop with a row crop, can serve to disrupt weed life cycles and prevent the establishment of large weed populations.

Flooding is a cultural weed-control practice that can be used in rice. Flooding allows the grower to regulate the germination of annual weed seeds by indirectly managing the amount of oxygen available in the soil. Flooded soils typically have very low oxygen content, which is required by most weed species for germination. Water management is an important method available for controlling red rice, which is closely related to cereal rice, in the southeastern United States.

The use of cover crops is another effective cultural weed-control measure. Cover crops such as cereal rye (*Secale cereale*) and hairy vetch (*Vicia villosa*) can suppress germination and compete effectively with existing weed species. These crops are often planted in the fall and killed in the spring prior to the establishment of the crop intended for harvest. Because they provide winter cover, this practice can also be an effective method of reducing soil erosion. Historically, legume cover crops were widely used as an important nitrogen source as well as for weed control. In addition to the effects of competition, **allelopathy** (the production of chemicals by plants that are toxic to other plants) may play a role in the suppressive effect of cover crops on weeds. Leachates from crops such as cereal rye have been shown to inhibit germination and growth of susceptible weed species in laboratory and greenhouse experiments. Under field conditions, however, it has been difficult to ascertain the role of allelopathy in weed control because separating the effects of competition from allelopathic chemical inhibition is challenging.

Mechanical weed control is the term used to describe any physical weed-control measure. Prior to the 1940s, weed control in agronomic crops was accomplished primarily by hand removal or by some form of tillage. Even today in gardens, flower beds, and small crop fields, the hand removal and hoeing is often the principal means of eliminating weeds. With the advent of mechanized agriculture, tools were developed that accomplished the work of the hand hoe on a much larger scale. A variety of rotary hoes, cultivators, disks, and harrows are available to today's producers to accomplish the work of weed control in crop fields. In some areas, cultivating between the rows can remove most of the weeds, and only a small band of herbicides applied over the crop row is required for adequate weed control. While tillage is a highly effective control strategy for many weed species, concerns about soil erosion and compaction have reduced the utilization of this control measure in recent decades.

Mowing is another important method of weed control in many situations. Home lawns, roadsides, pastures, waterways, fallow ground, and hay fields are examples of areas in which mowing is the principal, if not the sole, source of weed control. Mowing at the correct time accomplishes two goals: (1) it eliminates the production of seed, which is of particular importance in the management of annual crops; and (2) if timed correctly, it depletes the stores of carbohydrates in the roots and vegetative propagules of weedy plants, which is of importance in perennial crops.

Mulching is a common method of weed suppression in flower beds, vegetable crops, and nurseries. Many different types of mulch are available, and essentially any substance that can be used to cover the ground and smother weeds can be considered mulch. Wood chips, sawdust, gravel, newspaper, and straw are just a few of the materials that can be used. While these mulches can effectively suppress the germinating seeds of many annual weeds that may be present in the seed bank, they are often less effective against creeping perennials, whose shoots can grow through the mulch layer. Another issue is that over time, as organic mulches decay, they can serve as a fertile seedbed for annual weed seeds; thus, new layers of organic mulch must be put down every year to maintain weed control. Plastic and fabric mulches have been developed to suppress the more aggressive creeping perennials species and they also prevent the establishment of newly deposited annual seed. However, plastic mulches often significantly alter soil temperatures and restrict the amount of rainfall that penetrates the ground, which can complicate other aspects of crop management in exchange for weed control.

Fire has been tried as a weed-control measure in almost every environment that humans have attempted to manage. Controlled burning has been of considerable value in removing unwanted brush from forest areas, in prairie management and restoration, warm-season grassland management, and rangeland management. While fire is not widely used in row crop production, specialized flame weeders are sometimes used to burn off small annual weeds in organic vegetable production.

Biological Weed Control

Biological weed control involves introduction of a living organism, often a natural pest or disease, to control a weed species. There are two approaches to biological weed control: the classical (or inoculative) approach and the inundative (or augmentative) approach. The classical approach involves a one-time introduction of an organism to manage weeds. In a successful introduction, the organism must establish itself in the ecosystem and reproduce, thereby providing long-term weed management. The inundative approach involves application of the biological organism on an as-needed basis. The organism is expected to provide short-term control of the target weed, but it is not expected to establish itself in the environment.

The control of prickly pear cactus (*Optunia* spp.) with a moth (*Cactoblastis cactorum*) from Argentina is example of a classical biological weed-control effort. The cactus was introduced in the mid-1800s to Australia as an ornamental, but it escaped and spread rapidly throughout Australia, covering 60 million acres by 1925. It was eventually discovered that the *Cactoblastis* moth larvae fed exclusively on the cactus, so the moth was released in Australia. The feeding damage caused by the moth allowed a fungus to establish in the cactus, which enhanced control. The moth populations grew quickly and reduced the cactus populations, to the extent that the moths ran out of food. This allowed a resurgence in the cactus population, which was soon followed by an increase in the moth population. This pattern is typical of classical biological weed control: a cyclic balance between the two species is often established (Fig. 15–15). Therefore, weed control in this system is typically neither instantaneous nor complete; however, the system can still be very cost-effective if the controlling species establishes itself successfully in the environment. Other successes have followed, often involving an interaction between a weed and an insect. The approach has been most successful against single invasive species in relatively undisturbed, low-management areas, such as rangelands.

Some commercial successes have also been achieved with the inundative approach to biological weed control. *Colletotrichum gloeosporioides* is a fungus that was sold for several years under the trade name Collego®, which was labeled for control of northern jointvetch (*Aeschynomene virginica*) in rice and soybeans.

Chemical Weed Control

Sporadic research with chemical weed control started about 1910 in both Europe and the United States, but the basis for modern chemical weed control was

Figure 15–15
USDA scientist studying the effects of the cactus moth on invasive Opuntia cactus in the United States.
Source: USDA Agricultural Research Service Image Gallery, http://www.ars.usda.gov/is/graphics/photos/

laid out with the initial studies of auxins and plant hormone physiology in the 1930s, culminating with the discovery of the herbicidal properties of 2,4-dichlorophenoxyacetic acid (2,4-D) in 1942. Since that time, many more herbicides have been discovered and introduced into the market. Chemical weed control helped revolutionize crop production in the second half of the twentieth century, when crop producers integrated the use of herbicides with improved crop varieties, mechanized farm equipment, and advances in crop production practices to reduce labor costs and increase yields.

Herbicides have several characteristics that are important to understand so that they can be used effectively. These characteristics are the selectivity of the product, the timing of its application, its use rate, and its mode of action. Recently, crops that have been bred or genetically engineered for herbicide resistance have been introduced into the market. These products have greatly expanded the utility of several commonly used herbicides.

Selectivity   One of the key characteristics of a successful herbicide is selectivity. Herbicides are often broadly classified as grass killers, broad-leaf killers, or nonselective (kills both grasses and broad leaves), although some grass herbicides possess a certain degree of broad-leaf activity and some broad-leaf herbicides control certain grasses. The development of herbicide-resistant crops represents the latest important advance in herbicide selectivity.

Herbicide Application Timing   Herbicides can be classified according to timing. Burn-down, preplant, preemergence, and postemergence applications are the four basic classifications. Herbicides that are applied to remove existing weeds in the field and thus prepare a clean seedbed for planting are considered **burn-down herbicides**. **Preplant herbicides** are applied before the crop is planted. These herbicides often require incorporation into the soil, either by mechanical means or by rainfall. **Preemergence herbicides** are applied to soil after crop planting but prior to the emergence of the weeds, the crops, or both. Many preemergence herbicides also require rainfall for maximum efficacy. **Postemergence herbicides** are applied directly to foliage after weed emergence. Because herbicidal chemicals are often toxic to a broad number of plants, it is advised that the equipment used for the application of herbicides not be used for other purposes to avoid the possibility of accidentally harming plants.

Herbicide-application technology has evolved significantly in recent years. Utilization of satellite positioning technology in conjunction with advanced application controls has allowed growers to apply herbicides with pinpoint accuracy and to vary the application rate across the field based on the history of weed pressure or on visual observations of high-density patches of weeds. This technology has been significant in allowing growers to utilize herbicides more judiciously.

Herbicide Use Rates   Utilization of the proper herbicide rate is key to effective weed control. In general, smaller weeds of a given species can be controlled with a much lower use rate than larger weeds of the same species. Growers typically favor low use rates because it lowers production costs while simultaneously reducing any potential negative environmental impacts from leaching or runoff. However, the difficulty with using low rates is that larger weeds in the field may survive the herbicide application, mature, and produce seed. Once this seed enters the seed bank, the weed will likely be present in subsequent years, requiring more herbicide applications in the future. Therefore, an optimum use rate decision requires that the grower have a thorough knowledge of the size and species of weeds present in the field.

Herbicide Mode of Action   Herbicides can be classified into groups based on the plant biochemical processes with which they interfere. **Mode of action** is the term commonly used when discussing the action of herbicides on these biochemical processes. Table 15–2 lists nine of the most common modes of action. Many weed scientists will further divide these modes of action into more precise groups based on the specific enzymatic or cellular mode of action of a given herbicide, however such a discussion is beyond the scope of this text. For the purpose of basic weed management, these categories are sufficiently robust. Each of these modes of action contains one or more groups of chemical families, and each chemical family contains one or more commonly used herbicides, examples of which are provided. Understanding the mode of action of an herbicide is important for three main reasons. First, knowing how the herbicide kills the plant allows a producer to predict how a particular herbicide will perform under different growing conditions. For instance, many herbicides work much more quickly on actively growing plants; thus, if such a herbicide is applied during cool conditions when plants are growing slowly, it may take much longer for injury symptoms to be observed on the target plant. Second, understanding mode of action helps a grower identify and assess the extent of injury symptoms on desired plants and on nontarget plants. Third, knowledge of mode of action is an important asset in

Table 15–2
HERBICIDAL MODES OF ACTION

General Mode-of-Action Categories	Chemical Family Examples	Specific Herbicide Examples
Amino acid synthesis inhibitors	Glycine derivatives, imidazolinones, sulfonamides, sulfonylureas	Glyphosate, imazethapyr, cloransulam-methyl, chlorimuron
Ammonia assimilation inhibitors	Phosphonic acids	Glufosinate
Auxin transport inhibitors	Semicarbazones	Diflufenzopyr-Na
Cell membrane disruptors	Bipyridiliums, diphenylethers, triazolinones, thiadiazoles, N-phenylphthalimides	Paraquat, lactofen, sulfentrazone, fluthiacet, flumiclorac
Growth regulator (auxinic) herbicides	Phenoxy acids, benzoic acids, 2-pyridine carboxylic acids	2,4-D, dicamba, clopyralid
Lipid biosynthesis inhibitors	Aryloxyphenoxypropionoates, cyclohexanediones, thiocarbamates	Fenoxaprop, sethoxydim, butylate
Meristem mitotic inhibitors	Benzamides, chloroacetamides, dinitroanilines, oxyacetamides	Pronamide, acetochlor, pendimethalin, flufenacet
Photosynthesis inhibitors	Benzothiadiazole, nitriles, triazines, triazinones, uracils, ureas	Bentazon, bromoxynil, atrazine, metribuzin, terbacil, diuron
Pigment synthesis inhibitors	Isoxazolidinones, isoxazoles, callistemone	Clomazone, isoxaflutole, mesotrione

Source: Jerron Schmoll, Pioneer, Inc.

helping producers manage herbicide-resistant weeds. Repeated applications of herbicides with the same mode of action increase the probability that a herbicide-resistant biotype of a weed species will be selected. Rotating herbicidal modes of action reduces the probability that a herbicide-resistant biotype will be selected for a given field. However, a few species of weeds such as rigid ryegrass (*Lolium rigidum*) and horseweed (*Conyza canadensis*) have developed resistance to multiple modes of action. Therefore, while rotating modes of action is important, relying strictly on herbicides to manage weeds is not a viable long-term weed-control strategy for some species.

Herbicide-Resistant Crops   One of the latest advances in weed management has been the development of crops that are resistant to broad-spectrum, commonly used herbicides. These crops have been developed using both conventional plant-breeding methods and genetic engineering technology. Since the release of the first herbicide-resistant crop in 1991, the acreage planted with herbicide-resistant crops has grown dramatically. The level of weed control obtained with herbicide-resistant crops can be quite impressive, and this has fueled the rapid adoption of these crops by many producers throughout the world. However, some countries have restricted or banned the use of herbicide-resistant crops or stipulated that the grain from these crops be used only for animal feed. One of the factors that has hindered the utilization of herbicide-resistant crops has been opposition from groups who point out that the

long-term effects of releasing these new crop biotypes into the environment are unknown. Proponents of herbicide-resistant crops point out the many potential benefits from the utilization of these crops, including the potential to reduce herbicide use, the opportunity to utilize less toxic herbicides, and the opportunity to increase crop yields to feed a growing population. Many groups of academic and industry scientists are currently working to answer questions pertinent to both sides of the debate, but this issue seems unlikely to be fully resolved, at least in the near future.

One of the first lines of defense against weeds is to prevent their initial establishment. Many different preventive measures are effective at keeping weeds from entering a field. Cleaning planting, tillage, and harvesting equipment when moving from one area to the next is essential to reducing the spread of weed seed and vegetative propagules. Using weed-free crop seed is another method of preventing weed establishment. Seed companies must comply with rigorous quality standards restricting weed seed content of planting seed, thus making the spread of weeds through contaminated seed much less problematic than in the past. Preventing seed from moving into a field from surrounding areas such as roadsides, ditches, and fencerows helps to control weed spread into a field. These areas should be regularly mowed, sprayed, or otherwise managed to keep weeds from spreading. Preventing weed reproduction in crop fields is another important preventive weed control measure, particularly for the control of annual weed species. Finally, many US federal, state, and local laws

require landowners to control certain aggressive weed species on their property.

Weed Control in Organic Crop Production Systems

Organic producers grow crops without the use of synthetic herbicides and fertilizers; therefore, successful weed control in these systems requires more careful integration of preventive, mechanical, and cultural weed-control methods than does conventional crop production, in which chemical weed control plays a major role. Because most of the research in recent decades has focused on chemical weed control, the most effective and economical weed-control strategies in annual crop production systems will almost invariably include chemical weed control. The result is that organic producers will normally have higher weed-management costs for less effective overall weed control than their counterparts in conventional production. Increasing interest in organic production has spurred a corresponding increase in weed-management research in organic systems, which should provide producers with better weed-control strategies in the future. In the meantime, organic crop producers must continue to receive premiums for their crops to compensate them for the increased production costs. If consumer demand for organically grown produce continues to increase, so will the need for more effective weed-control strategies in organic systems.

INSECTS AND MITES

Plants and products are the only sources of food for a host of insects and mites. When these plants are grown as vast monocultures (e.g., agricultural crops, forestry, lawns) or their products are stored in large quantities, we often enter into a fierce competition with these insects and mites for these resources. This competition has gone on since humans first cultivated plants as agricultural crops. Insects and mites also pose secondary hazards because they spread diseases among plants and animals.

The conflict between agricultural production and insect and mite pests took a major turn shortly after World War II. During this period synthetic organic pesticides were discovered and developed. These new chemicals were cheap to produce and they had an amazing ability to kill insects and mites. However, as with weeds, many pests became resistant to the pesticides that bombarded them constantly and many of these chemicals had a very negative impact on the environment. As a result there has been widespread adoption of IPHM strategies for insect and mite control.

It is important to remember that when controlling insect pests, many insects provide essential or useful benefits to plants agriculture. Certain harmless insect species prey on other insect species that destroy plants or crops. Many of these beneficial insects are available commercially. The honeybee and certain other insects do a tremendous service in pollinating fruit trees and other crops, such as alfalfa. Honeybees also produce useful products, like honey and beeswax. When controlling insect and mite pests, it is extremely important to consider how those control measures will affect the beneficial insects and mites in the area.

With that consideration in mind, we still have to control innumerable kinds of insects and mites that, if left uncontrolled, would soon reduce the world's food supply to a shambles. These measures are the same as listed at the beginning of the chapter. Special considerations for insects and mites are listed below.

1. *Genetic host resistance.* Plant species and even some cultivars within plants, are resistant to attack by some insects. These plants may have a morphology that discourages insects from settling on them (trichomes that are thick or spiny) or contain a phytochemical that deters some insects.
2. *Biological control.* There are several types of biological controls for insects. These include parasites, predators, and disease-causing pathogens. The praying mantis and several lady bird beetle species, both larvae and adults, feed on aphids (Fig. 15–16). The larvae of the green lacewing feeds on mealybugs,

Figure 15–16

Lady bird beetle devouring a pea aphid. Source: USDA Agricultural Research Service Image Gallery, http://www.ars.usda.gov/is/graphics/photos/

Figure 15–17
Green lacewing larva attacking whitefly nymphs.
Source: USDA Agricultural Research Service Image Gallery, *http://www.ars.usda.gov/is/graphics/photos/*

scale, whitefly, and aphids (Fig. 15–17). Predatory wasps attack several inspect species (Fig. 15–18). Some can be kept in a balance with the pest for a long time, however, most insect pathogens in use today are not capable of perpetuating themselves in the field and must be applied—usually as a spray—each time they are needed.

3. *Cultural control.* An important cultural control for controlling some insects is time of planting. By planting at the right time, it is sometimes possible to have the crop's susceptible stage not matched up with the damaging stage of the insect. For example, planting wheat after the "fly free date" for Hessian fly (*Mayetiola destructor*) allows the wheat to escape infestation.

4. *Physical or mechanical control.* Leaving fields fallow during critical periods in an insect's life cycle, screening of seedbeds or ventilators on greenhouses (exclusion), disking soil between tree and vine rows, and burying or chipping prunings are examples of physical practices that control specific insects. Hand picking insects, syringing plants with water to dislodge insects, and picking up and destroying fallen fruit are examples of physical controls useful in small-scale plantings.

Insects and Their Relatives

Insects and their near-relatives are placed in the phylum Arthropoda. This phylum contains the Arachnida (spiders, mites, ticks, and scorpions), Crustacea (crabs,

shrimp, isopods, etc.), Chilopoda (centipedes), Decapoda (millipedes), and the Insecta. The class Insecta is usually broken down into twenty-five or more orders that contain more than 1 million species: over 80 percent of all the species of animals known on the earth! Fortunately, most insects are of no economic importance; most estimate that less than 5 percent of the insect species have any direct importance to humans, and only about 1 percent regularly reach pest status. The insect orders with the most species are:

Coleoptera—beetles, weevils.

Lepidoptera—butterflies, moths, caterpillars.

Hymenoptera—bees, wasps (including many insect parasites), ants.

Diptera—true flies.

Heteroptera—true bugs and buglike insects (scales, whiteflies, leafhoppers, etc.).

Most entomologists emphasize insect metamorphosis (development) because this knowledge often assists in developing control strategies. Three main types of metamorphosis are generally recognized: no metamorphosis (ametabolous), born fully developed; gradual or incomplete metamorphosis (paurometabolous), some lifecycle stages are not present; and complete metamorphosis (holometabolous). These differences are illustrated in Fig. 15–19.

Figure 15–18
Predatory wasp preparing to lay an egg in a tarnished plant bug nymph. Source: USDA Agricultural Research Service Image Gallery, *http://www.ars.usda.gov/is/graphics/photos/*

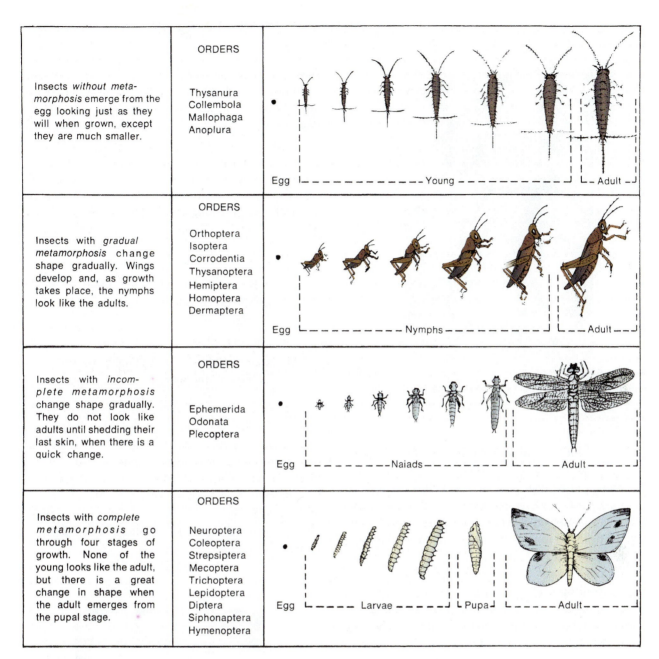

Insects *without metamorphosis* emerge from the egg looking just as they will when grown, except they are much smaller.	ORDERS Thysanura Collembola Mallophaga Anoplura	Egg — Young — Adult
Insects with *gradual metamorphosis* change shape gradually. Wings develop and, as growth takes place, the nymphs look like the adults.	ORDERS Orthoptera Isoptera Corrodentia Thysanoptera Hemiptera Homoptera Dermaptera	Egg — Nymphs — Adult
Insects with *incomplete metamorphosis* change shape gradually. They do not look like adults until shedding their last skin, when there is a quick change.	ORDERS Ephemerida Odonata Plecoptera	Egg — Naiads — Adult
Insects with *complete metamorphosis* go through four stages of growth. None of the young looks like the adult, but there is a great change in shape when the adult emerges from the pupal stage.	ORDERS Neuroptera Coleoptera Strepsiptera Mecoptera Trichoptera Lepidoptera Diptera Siphonaptera Hymenoptera	Egg — Larvae — Pupa — Adult

Figure 15–19

Four types of insect metamorphosis, from egg to adult.

TO TELL THE DIFFERENCE AMONG INSECTS, MITES, SPIDERS, AND TICKS, CONSULT THE FOLLOWING LISTS.

Insects

The adults are characterized by:

1. Three definite body regions—head, thorax, and abdomen.
2. Three pairs of jointed legs.
3. One pair of feelers or antennas.
4. Eyes that are usually compound.
5. One or two pairs of wings.

Mites, Spiders, and Ticks

These are characterized by:

1. Two main body regions—the cephalothorax (head and thorax fused together) and the abdomen.
2. Four pairs of jointed legs.
3. A lack of antennas or wings.
4. Simple eyes.

The mode of feeding is also emphasized by entomologists. Insects all have mandibulate mouthparts, but these may be chewing or modified into sucking forms. Chewing insects (like caterpillars, beetles, and their larvae) devour plant foliage, shoots, flowers, and fruit, or they may burrow into the plants (borers) (Fig. 15–20). Sucking insects (and mites) consume plant juices or cell contents through their mouthparts that pierce the interior of the plant (Figs. 15–21 and 15–22). Some insects such as thrips have mouthparts that both chew and suck (rasping-sucking). Many sucking pests inject saliva that can clog the plant's vascular bundles, thereby causing parts of the plant to wilt or die. Other sucking pests commonly pick up plant diseases and transmit them when the insect moves to another plant (Fig. 15–23). The mode of feeding is also important in determining what kind of pesticide should be used and what may be the most efficient method of application. Insecticies that work by entering the body through ingestion are effective when applied to the surface of a plant then readily ingested by foliage eating insects, but a systemic (taken up by the

A   B

Figure 15–20

(A) Gypsy moth chewing on leaves. (B) Stem borer.

Source: USDA Agricultural Research Service Image Gallery, *http://www.ars.usda.gov/is/graphics/ photos/*

Figure 15–21

Aphids are sucking insects.

Source: Luis Cañas, The Ohio State University

Figure 15–22

Spider mite on rose leaf under scanning electron microscope.

Source: USDA Agricultural Research Service Image Gallery, *http://www.ars.usda .gov/is/graphics/photos/*

Figure 15–23
(A) The silverleaf whitefly (Bemisia argentifolii) is a vector for viruses. (B) Effects of silverleaf whitefly on a melon leaf in the foreground.
Source: USDA Agricultural Research Service Image Gallery, *http://www.ars.usda.gov/is/graphics/photos/*

A   B

plant) insecticide most likely would be more effective against a sucking pest.

Insect and Mite Pest Management

As has been previously described in this chapter, insect and mite pest management generally relies on the integration of cultural, biological, and chemical controls using the IPHM approach. Pest managers rarely use the term *pest control* (or *pest eradication*) because we have never been able to control insect and mite pests permanently. At best, we can manage them to keep them at or below levels that allow us to obtain a reasonable return on our investment when producing a crop. By using a mix of cultural and chemical controls to conserve effective biological controls, most crops can be coaxed into providing adequate returns.

While genetic, cultural, mechanical, and biological controls are important in managing insect and mite pests, pesticides still dominate. Their characteristics will be discussed more extensively in the following subsections.

Pesticide Modes of Entry   Each pesticide has one or more ways in which it can get into an insect or mite. These modes of entry are important to understand in terms of making the best application for control of each type of pest.

Stomach poison refers to materials that enter the insect by mouth and kill by being absorbed into the body through the digestive tract. For most chewing insects, pesticide residues should be on the plant surface(s) being ingested. Some insecticides can be absorbed into the tissues of plants (translaminar systemic) and/or even transported through the vascular system of the plant (translocated systemic). Systemic pesticides are most useful for control of sucking pests, borers, and leafminers. Because borers and leafminers feed within

the plant tissues, systemics are often the only way to reach these pests. Great care must be taken when using systemic insecticides because the residues can remain in the plant tissues eaten by humans. Such pesticides usually have a preharvest interval period, which is the time needed to degrade the pesticide residue below active levels, written on the label.

Contact poison refers to materials that enter the insect or mite body directly through the pest's cuticle (outer covering on the body). Some contact poisons actually have to be applied to the pest, while others have sufficient residual action that the residues are picked up as the pest walks across treated surfaces and sufficient material is absorbed to kill the pest.

Inhalants are applied as fumigants and inhaled as a vapor or dust by the insect or mite, and the material is then absorbed through the respiratory system. Fumigants are generally used in enclosed spaces (e.g., greenhouses and grain storage bins), but some can also be used in the soil if the surface is sealed with water or an impermeable covering (e.g., plastic sheeting or tarp).

Pesticide Modes of Action   Each pesticide group tends to have a unique way (mode of action) of disrupting vital body functions of insects and mites. While the list of modes of actions discussed in the following subsections is not complete, it covers many of the most common.

Neural Disruption   Nervous systems are very complex, and pesticides can interfere with their normal function in numerous ways. The two major neural systems affected are sodium pump channels and neural transmitter systems.

The sodium pump is the system used by nerve cells to transmit an electrical charge from one end of the nerve to the other end. As the electrical impulse moves, ions move in and out of the cell membrane, then the

sodium pump system resets the electrical potential. Pesticides such as pyrethroids and chlorinated hydrocarbons interfere with this system.

Neural transmitter systems are located in the nerve synapsis—the microscopic gap between two nerves. In this gap, a neural transmitter compound is released by the sending nerve to stimulate the receiving nerve. This neural transmitter is rapidly degraded so that the receiving nerve doesn't continue to be stimulated. Organophosphate and carbamate insecticides block the neural transmitter–degrading chemical, thereby causing the receiving nerve to continue to fire uncontrollably.

Neonicotinyl insecticides block the receiving nerve at the synapsis, which means that the sending nerve sends the signal, but the receiving nerve doesn't fire. The result is that the insects stop their normal behavior or simply sit and do nothing.

Cellular Metabolic Disrupters   Various chemicals can disrupt different metabolic pathways of a cell. Arsenicals are examples of this group. Some interfere with the energy pathways, some disrupt oxygen utilization, and still others interfere with gene functions. Most chemicals in this group are general toxins, with similar actions in all cells, whether insect, animal, or plant. Many of these chemicals can also be slow acting, so when they are used in baits, affected colonizing insects such as termites and ants can eliminate the entire colony.

Cell Membrane Disruption   In the past, many maintained that mineral oils (e.g., dormant oils, horticultural oils), crop oils (e.g., soybean oil), and insecticidal soaps (i.e., fatty acid salts) have killed insects and mites through suffocation and were considered to be suffocants. The concept was that oils and soaps coat the insects and/or clog the spiracles (breathing holes). In reality, these materials disrupt the lipoprotein matrix of cells, causing the cells to lose their contents and cease functioning. Even though insect cuticle is considered to be an impermeable structure, in fact, it is filled with microtubules that lead to the underlying cells. When oils or soaps contact insects, they penetrate into the underlying cells, causing them to be destroyed. The same process occurs in insect egg cells, a common target of dormant oils. The protein toxin of *Bacillus thuringiensis* (Bt), often considered to be a biological control, actually attaches to the cell wall of insect gut cells, poking a hole in them. The dead cells then allow the leakage of cell bacteria into the insect's body cavity, eventually causing death.

Hormone Disruption   Insects use various hormones to control their growth, development, and exoskeleton formation. Various pesticides can either mimic insect hormones or disrupt the normal function of hormones resulting in abnormal lifecyle stages. These chemicals are also called insect growth regulators (IGR). Juvenile-hormone mimics cause insects to remain in a juvenile state. Chitin inhibitors disrupt the normal formation of chitin: it may not harden properly or molting is disrupted. Molt accelerators cause the insects to molt before the proper time. Many of these chemicals have very specific targets, making them safe to use with predatory insects and mites.

Pheromones   Pheromones are the chemicals that insects normally use to communicate with each other. Most pheromones are used in insect traps to monitor activity, but some are used to disrupt mating. In essence, permeating the air with a sex pheromone causes males to have great difficulty locating unmated females.

Dessicants   Dessicants are chemicals or materials that can cause insects to lose water faster than they can replace it. Physical desiccants include diatomaceous earth, which makes microcuts in insect cuticle, thereby reducing the waterproofing properties of an intact cuticle. Other compounds, like silica aerogel, can pull water through insect cuticle when the dust adheres to the exoskeleton. Obviously, desiccants work best in dry environments, like house or building crawl spaces and wall voids.

Insecticides Classified by Chemistry

A great many chemicals have been used over the years for controlling insects. Some were found to be so potentially harmful to humans and other animal life, including beneficial insects, and to the environment that their use is no longer permitted. Use of all insecticides is tightly controlled in most countries. In the United States, insecticides must be approved and registered by the Environmental Protection Agency (EPA), and by state agencies also, before they can be used. Even then there are usually heavy restrictions that confine use to certain plants at certain times and at certain concentrations. Residues on food or feed crops exceeding fixed tolerances subject the product to seizure and destruction. Although no pesticide tolerances have been established for ornamental or forest crops, use of these chemicals on such commodities is nevertheless tightly regulated.

Insecticides can be classified as follows:

1. *Inorganic compounds.* These compounds include arsenic, fluorine, phosphorus, and sulfur compounds. These use of these has diminished greatly over time.
2. *Organic compounds.*
   a. *Plant derivatives.* These derivatives include materials such as pyrethrum (from the dried and powdered flowers of *Chrysanthemum cinerariaefolium* which is a safe and effective insecticide; rotenone (from the roots of several plants in the pea family); nicotine from *Nicotiana* spp.; and an extract from the neem tree. Many others could possibly be developed for commercial use if there were sufficient demand.
   b. *Synthetic organic chemicals.* After World War II, an entirely new concept in insect control emerged with the development of synthetic organic insecticides.
      i. *Chlorinated hydrocarbons.* DDT, first discovered in Germany in 1874, was found to have insecticidal properties. During and after World War II, it was used to control mosquitoes, flies, fleas, and many agricultural insects. However, the buildup of DDT in the world's ecosystems to levels many scientists considered dangerous led to banning its use in several countries, including the United States. Other chlorinated hydrocarbons, closely related to DDT, have also been very effective in insect control. They are not easily biodegraded, however, tending to build up in plants and in the soil and to be transmitted into fish, fish-eating birds, meat, and milk products.
      ii. *Organic phosphates.* Compounds in this group were developed near the end of World War II and were found to have good insecticidal properties. They decompose more rapidly than the chlorinated hydrocarbons, but some of the materials can be very toxic to mammals and must be used with great care.
      iii. *Carbamates.* Carbaryl was the first chemical in this group to be widely used. It has low mammalian toxicity and its residual action is short-lived. It is effective against a wide range of both sucking and chewing insects, acting as a contact insecticide. Some carbonates are **systemic** and highly toxic.
      iv. *Pyrethroids.* This is a relatively new class of insecticides. They mimic properties of pyrethrins such as low mammalian toxicity. They kill a broad spectrum of insects and have a much longer residual life than pyrethrins.
      v. *Neonicotinyls.* The neonicotinyls form a recent class of insecticides that block the synoptic transmission of insect nerves. Affected insects do not behave normally; they stop feeding, thus causing the insects to be more vulnerable to predators, parasites, and diseases. Unlike the other neurotoxins, these compounds have very low toxicity to vertebrates.
   c. *Spray oils.* Long used to control scale insects and mites on fruit trees and ornamentals, spray oils are prepared by the distillation and chemical refining of mineral oils. A distillation range is chosen to give an oil fraction that is relatively nontoxic to plant tissue, yet lethal to insects. Spray oils are treated with hot sulfuric acid to remove many of the unsaturated molecules in the oil that cause plant injury. Botanical oils have similar properties.
   d. *Microbial insecticides.* Some kinds of insects are susceptible to certain toxins produced by bacteria, fungi, and viruses. Often these pathogens cause the spectacular disappearance of insects. Such natural biological control of insects has been encouraged and used commercially. For example, *Heliothis* virus is used commercially to kill the tobacco budworm and the cotton bollworm. It infects only these two insect species. Several species of caterpillars (*Lepidoptera*) can be controlled by the toxins produced by the bacterium *Bacillus thuringiensis*. The toxins are applied in dust suspensions or sprays to foliage, which is then ingested by the caterpillars. These bacteria do not sustain themselves naturally and must be reapplied each time caterpillar control is needed. Bt corn and other crops have been genetically engineered to produce the microbial toxin that affects pests.

Important Insect Pests of Agricultural Crops and Plant Products

There are innumerable insect pests of food crops, but certain pests are outstanding for the havoc they have wreaked over the years. Several species are discussed in the following subsections. When infestation of these or other pests occur, it is best to consult a local agricultural extension agent, pest control consultant, or garden supply center. Insecticide recommendations are continually changing.

Figure 15–24
Codling moth larva in an apple. Source: USDA Agricultural Research Service Image Gallery, *http://www.ars.usda.gov/is/graphics/photos/*

Figure 15–25
Tarnished plant bug. Source: USDA Agricultural Research Service Image Gallery, *http://www.ars.usda.gov/is/graphics/photos/*

Corn Ear Worm (*Heliothis zea*) This insect, with three to five generations per year, occurs all over the world wherever corn is grown. Caterpillars feed on the corn silks and kernels, making the ears wormy. The worm also feeds on the other crops—beans, cotton, lettuce, tomato, alfalfa, clover, peanuts, and tobacco. It is best controlled by insecticides.

Codling Moth (*Laspeyresia pomonella*) This insect, with two to three generations per year, attacks apples and pears wherever they are grown throughout the world. It is also a pest for walnuts. The larvae tunnel into the fruits, making them wormy (Fig. 15–24). Unless insecticides are used, up to 90 percent of the fruits can become affected.

Peach Twig Borer (*Anarsia lineatella*) This species occurs all through the peach-producing areas of the United States. It also attacks most other stone fruits. Overwintering larvae bore into buds and shoots as they start to grow in the spring. There are two to three broods a year, the later ones feeding directly on the fruit.

Lygus Bugs (*Lygus spp.*) These insects principally attack alfalfa, Ladino clover, sugar beets, safflower, beans, cotton, and carrots. *L. lineolaris* is commonly called the tarnished plant bug (Fig. 15–25). The sucking mouthparts are inserted into buds, flowers, and young fruits, so damaging the crops that they may be unmarketable. In some seed crops, such as alfalfa, no seeds may develop because of blasting of the flower buds by the lygus bugs.

San Jose Scale (*Quadraspidiotus perniciosus*) This scale insect is well established throughout North America, Europe, and Asia, attacking most fruit crops and many ornamental trees and shrubs. Heavy infestations of scale

can reduce tree vigor and cause death unless they are controlled. Scale spots on fruits reduce their market quality.

Green Peach Aphid (*Myzus persicae*) This insect is found throughout the world. It is a sucking pest on many vegetable crops, all stone fruits, and many ornamentals (see Fig. 15–22 on page 315). While its feeding reduces plant vigor, the chief source of damage is its transmission of viral diseases. Viral particles in its salivary fluid are injected into the host plant. Viruses known to be transmitted by the green peach aphid are sugar beet yellows, potato leaf roll, bean mosaic, lettuce mosaic, and cucumber mosaic.

Egyptian Alfalfa Weevil (*Hypera brunneipennis*) This is a very destructive pest for alfalfa grown as a hay crop. Principal damage is done by the larvae, which feed on shoot tips, buds, and leaves.

Spider Mites (*Bryobia praetiosa, Panonychus ulmi, and Tetranychus urticae*) These pests (which are not true insects) are widely distributed and feed on many species of host plants, including a wide array of vegetable and field crops, greenhouse and nursery plants, fruits, nuts, and ornamentals (Fig. 15–23 on page 316). A typical webbing sometimes appears as the mite population increases, with a yellow stippling on leaf surfaces as defoliation begins, resulting from removal of plant fluids by the mites' piercing and sucking mouthparts. Six to ten generations per year can occur. Spider mites have developed resistance to many miticides.

Cereal Wireworms (*Agriotes spp.*) These are the main insect pests of cereals such as wheat, barley, oats, and rye, especially in the northern growing regions. They are the larval stage of so-called click beetles, which themselves do no harm. Their life cycle spans five

years, most of which is spent in the larval wireworm stage in the soil.

Colorado Potato Beetle (*Leptinotarsa decemlineata*) This well-known insect, with chewing mouthparts, occurs from Colorado to the eastern United States in all potato-growing areas (Fig. 15–26). It is established on the European continent but has been eradicated from the British Isles. Both adult and larval stages feed on the foliage of the potato plant, completely denuding it. It seldom feeds on other plants. In both stages the insect is large and easily seen.

Cotton Boll Weevil (*Anthonomus grandis*) This insect attacks cotton plants in the United States, Mexico (where it originated), and Central America. The larvae develop inside the flower or boll, arising from eggs deposited by the female (Fig. 15–27). They feed on the developing floral parts and fibers. Many generations occur in a single season. Losses are heavy unless insecticides are used. Probably no insect, other than perhaps the codling moth, has had such an impact on agriculture and probably no other insect has received more study.

Sugarcane Shoot Borer (*Diatraea saccharalis*) This is the most important insect pest of sugarcane in the Caribbean. Caterpillars feed on the leaves, then enter the stalk and bore into the center, killing the central growing shoot. Later generations bore into side shoots. Such mechanical injury causes the stalks to break and fall over. Yield and quality of extracted juice drops. Control attempts include resistant cultivars, insecticides, and the release of parasites of the borer; none is very effective.

Mediterranean Fruit Fly (*Ceratitis capitata*) This notorious insect is of the greatest importance on all fruit species in the Mediterranean countries, south and

Figure 15–27
Boll weevil on cotton boll. Source: USDA Agricultural Research Service Image Gallery, *http://www.ars.usda.gov/is/graphics/photos/*

central Africa (where it originated), western Australia, the west coast of South America, and throughout Central America. It has been kept out of the United States (except for Hawaii) by strict government inspection, quarantine, and eradication measures. The fly attacks peaches, apricots, apples, citrus, bananas, and many other fruits and vegetables. The fly is slightly smaller than the common housefly and is mostly yellow and brown in color, with black markings (Fig. 15–28). It is controlled mainly by poisonous sprays containing attractants and by the release of sterile male flies. The Oriental (*Dacus dorsalis*) and Mexican (*Anastrepha ludens*) fruit flies can also be devastating to fruit crops.

Grape Berry Moth (*Endopiza viteana*) This is the principal insect pest on grapes in Europe, eastern North America, North Africa, and Japan. The larvae feed on developing fruit. Two or three generations can

Figure 15–26
Colorado potato beetle. Source: USDA Agricultural Research Service Image Gallery, *http://www.ars.usda.gov/is/graphics/photos/*

Figure 15–28
Mediterrean fruit fly. Source: USDA Agricultural Research Service Image Gallery, *http://www.ars.usda.gov/is/graphics/photos/*

occur during the season. Insecticide spray is the only control measure.

Insects Attacking Dried Fruits   Several fruits are preserved by drying, either outdoors on trays in the sun or in forced hot-air dehydrators. Important dried fruits are raisins, prunes, dates, figs, apples, peaches, apricots, and pears. All these dried fruits are food for various insects, which are best controlled by fumigation treatments. Several types of insects feed on dried fruits:

Beetles: dried fruit beetle, saw-toothed grain beetle, small darkling beetle, hairy fungus beetle, corn sap beetle, pineapple beetle, and date stone beetle.

Moths: raisin moth, Indian meal moth, almond moth, dried fruit moth, navel orange worm, dried prune moth, and dusky raisin moth.

Flies: vinegar fly *(Drosophila),* soldier fly, blowfly, housefly.

Insects Attacking Stored Grains   Conservative estimates maintain that insects destroy at least 5 percent of the world's production of cereal grains, amounting to about 15 million tons annually. Important insects that feed on stored grains are the sawtoothed grain beetle, lesser grain beetle (Fig. 15–29), flat grain beetle, red flour beetle, foreign grain beetle, larger black flour beetle, Angoumois grain moth, hairy fungus beetle, granary weevil, and the rice weevil.

Control measures include prompt harvesting, drying the grain with heated air to a low moisture content (11 to 13 percent), and storage in tight, insect-free bins raised above ground. After two to six weeks, the grain is fumigated.

Figure 15–29
Lesser grain beetle on wheat grain. Source: USDA Agricultural Research Service Image Gallery, *http://www.ars.usda.gov/is/graphics/photos/*

VERTEBRATES

Rodents, particularly Norway rats, roof rats, and house mice, cause great losses to food crops. Such losses occur mainly in stored grains and other food products in open storage, although rats also feed on unharvested fruits and vegetables. Sugarcane fields in Hawaii, for example, are often invaded by rats. Several million metric tons (MT) of grain are lost annually to rats.

The strategy in rodent control is, first, to remove all food and water available to them from the areas they inhabit, and second, to place bait traps containing an anticoagulant rodenticide such as Warfarin in their runways. Control should be done around granaries just before harvest begins. Some rat species, however, show increasing resistance to rodenticides, and in the future the chemicals may not be effective. Chemical sterilants, acting as oral contraceptives, reduce rodent populations.

Young fruit trees and fall-planted seeds in nurseries are often damaged by deer, mice, gophers, squirrels, and rabbits. Effective repellents of rodents and birds have been developed as coatings of forest seeds sown in logged- and burned-over areas.

Certain birds—particularly crows, ducks, geese, starlings, blackbirds, ravens, magpies, and scrub jays—are a major menace to grain crops and many fruit and nut crops, such as cherries, grapes, prunes, plums, strawberries, almonds, pecans, walnuts, and pistachios. Blackbirds, ducks, and starlings, in particular, can decimate grain crops such as wheat and field and sweet corn just ready for harvest. They can also cause considerable losses in peanut crops. Canadian geese cause extensive damage to golf course greens and fairways. The use of nonlethal chemical repellents is one type of bird control. Scare tactics—carbide explosives, shell crackers, and amplified recordings of bird distress calls, and dogs—are also used with varying degrees of success. Research on chemical reproduction inhibitors may eventually provide the best method of controlling depredating bird populations.

PLANT DISEASES

All species of native and cultivated plants are susceptible to disease and prone to injury. **Disease** is defined as suboptimal plant growth brought about by a continuous irritant such as a pathogen (an organism or entity capable of inciting disease) or via chronic exposure to less than ideal growing conditions. In contrast, injury is loss of plant vigor resulting from an instantaneous event such as a lightning strike, hail damage, chemical burn, or mechanical damage. Because of the instantaneous

and cause-and-effect nature of injuries (i.e., the plants were pruned yesterday and they are dying today), they are often easy to diagnose. In the case of disease, the source of continual irritation may be abiotic (nonliving) or biotic (caused by a pathogen). Abiotic diseases are also referred to as noninfectious diseases because they do not spread from plant to plant. In lay terms, they are not contagious.

Abiotic diseases are very common and should be considered the likely suspect when attempting to diagnose the cause of decreased plant vigor or death. This approach is very important when working with intensively managed cropping systems in which a high degree of manipulation or handling takes place. Examples of abiotic plant diseases include damage caused by chronic exposure to air pollutants such as nitrogen dioxide (NO_2—automobile exhaust), sulfur dioxide (SO_2—smokestacks of factories), and ground-level ozone (O_3—a by-product of photochemical reactions in the atmosphere); nutritional deficiencies and toxicities; and growth under less than ideal light, moisture, or temperature conditions. The latter emphasizes why it is so important for plant production specialists to know and understand the complexities of their unique cropping system and the environmental growth requirements of the plant species being cultivated.

Biotic diseases are caused by pathogens and are often referred to as infectious diseases because they can move within and spread between plants. Plant pathogens are very similar to those that cause disease in humans and animals and include viruses, bacteria, spiroplasmas, phytoplasmas (formally called mycoplasmalike organisms), fungal-like organisms, fungi, nematodes, and parasitic higher plants. Pathogens may infect all types of plant tissues including leaves, shoots, stems, crowns, roots, tubers, fruit, seeds, and vascular tissue and can cause a wide variety of disease, ranging from root rots and rusts to cankers, blights, and wilts. Most plants are immune to most pathogens; however, all are susceptible to attack by at least one pathogen in each of the groups listed above. Some plants are susceptible to many. For example, wheat, rice, corn, and soybean are each susceptible to more than thirty different fungal pathogens. Similar levels of susceptibility exist in most other crops such as citrus, turfgrass, trees, and the vast diversity of solanaceous, cucurbit, floricultural, and ornamental plant species. Some pathogens like *Rhizoctonia*, *Pythium*, *Fusarium*, and *Sclerotinia* infect a broad range of plant species and yet others only a given species. For an infectious disease to develop, the following three conditions must

exist: a susceptible host, a pathogen capable of inciting disease, and a favorable environment for pathogen development (see Fig. 15–3 on page 298). If any one of these factors is not present, disease will fail to develop. In the case of infectious plant diseases, any practice that favors plant growth and reduces either the amount of pathogen present or its development or activity will result in significantly less disease.

Regardless of the type of pathogen, the development of biotic disease symptoms on a plant requires that the infectious agent must (1) come into contact with a susceptible host (referred to as inoculation); (2) gain entrance or penetrate the host through either a wound or a natural opening (stomata, lenticel, hydathode, nectarthode) or via direct penetration of the host; (3) establish itself within the host; (4) grow and reproduce within or on the host; and ultimately (5) be able to spread to other susceptible plants (referred to as dissemination). Successful pathogens must also (6) be able to survive prolonged periods of unfavorable environmental conditions and the absence of a susceptible plant host. Collectively, these steps are referred to as the disease cycle (Fig. 15–30). If this cycle is disrupted, either naturally or via the concerted efforts of a grower, the disease will be less intense or fail to develop.

Disease Signs and Symptoms

Two terms used often when discussing plant disease and injury are *sign* and *symptom*. The term *sign* is used when the pathogen or part of the pathogen is observed in or on an infected plant. Examples include fungal hyphae or mycelium, spores, fruiting bodies, bacterial cells, virus particles, nematodes, and parasitic higher plants. Although many plant diseases can be diagnosed in the field based on the observation of key diagnostic signs, many require observation by trained specialists under laboratory conditions. *Symptoms* are visual or otherwise detectable reactions or alterations of a plant as the result of disease or injury. Symptoms of disease often change as the disease progresses. Initial symptoms are often invisible to the naked eye or very small and nondescript, and they may be quite different from those observed in the final stages of disease development. Symptoms can generally be placed in the following categories:

1. *Abnormal tissue coloration.* Leaf appearance commonly changes. Leaves may become chlorotic (yellowish) or necrotic (brown) or exhibit purpling, bronzing, and reddening. Mosaic or mottling patterns may appear, especially with virus diseases. Nitrogen-deficient plants often exhibit a generalized

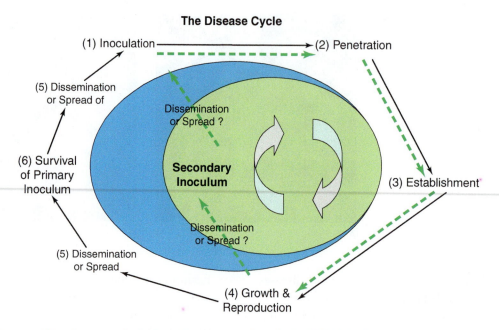

The Disease Cycle

A disease with only one cycle (outbreak) per season is called a MONOCYCLIC or "one-cycled" disease
A disease with multiple cycles (outbreaks) per season is called a POLYCYCLIC or "many-cycled" disease

Figure 15–30

The six steps of a disease cycle: survival, inoculation, penetration, establishment, growth and reproduction, and dissemination or spread. A disease with only one cycle (outbreak) per season is called a monocyclic, or one-cycled, disease. A disease with multiple cycles (outbreaks) per season is called a polycyclic, or many-cycled, disease. Information about a pathogen's disease cycle can provide clues about how to intervene, disrupt, or break its cycle and thus reduce plant losses.

Source: Michael Boehm, The Ohio State University.

chlorosis or yellowing. Iron-deficient leaves often exhibit interveinal chlorosis. Phosphorus-deficient plants are often purple (Fig. 15–31).

2. *Wilting.* Drought stress causes wilt. If a pathogen interferes with the uptake of water by the host plant, part or all of the plant may die. Fungi belonging to the genera *Verticillium* and *Fusarium* and bacteria in the genus *Xanthomonas* are often associated with wilt diseases because they colonize the xylem of plants, leading to a lack of water transport.

3. *Tissue death.* Necrotic (dead) tissue can appear in leaves, stems, or root, either as spots or as entire organs. Decay of soft succulent tissue, as in damping-off in young seedlings, is common. Cankers caused by death of the underlying tissue sometimes appear as sunken, dead tissue on the trunks or limbs of woody plants.

4. *Defoliation.* As the infectious disease progresses, the plant may lose all its leaves and sometimes drop its fruit. Defoliation is a common symptom in sycamore anthracnose (caused by *Gnomonia veneta*) and apple scab (caused by *Venturia inaequalis*) (Fig. 15–32). Both are fungal diseases.

Figure 15–31

Phosphorus deficiency in young corn growing in a drought-stressed field.

Source: Michael Boehm, The Ohio State University.

Figure 15–32
(A) Apple scab causing leaves of a cherry tree to die and drop off.
(B) Apple scab fungus emerging from inside an apple leaf.
Source: USDA Agricultural Research Service Image Gallery, http://www.ars.usda/.

A

B

5. *Abnormal increase in tissue size.* Some diseases increase cell numbers or cell size in the plant tissues, twisting and curling the leaves or forming galls on stems or roots (Fig. 15–33).

6. *Dwarfing.* In some cases, the pathogenic organism reduces cell number or size, stunting parts or all of the host plant.

7. *Replacement of host plant tissue by tissue of the infectious organism.* This development occurs commonly when floral parts or fruits are involved.

Examples include ergot on rye and other cereal crops caused by the fungus *Claviceps purpurea* and corn smut caused by the fungus *Ustilago maydis* (Fig. 15–34).

Figure 15–34
Common smut of corn is caused by the fungus Ustilago maydis. Sweet corn is considered more susceptible to smut than field corn. The disease occurs on leaves, stalks, tassels, and ears. The disease is easily recognized early in the season by the presence of silvery white tumorlike galls that, as the season progresses, mature, rupture, and release large masses of soot-colored smut spores (called teliospores) similar to those shown above. Young galls are considered a delicacy in certain parts of the world. Source: Michael Boehm, The Ohio State University.

Figure 15–33
Galls on a young fruit tree infected by crown gall bacteria (Agrobacterium tumefaciens).

Classification of Infectious Plant Diseases

The pathogens responsible for causing most biotic plant diseases include viruses, bacteria, spiroplasmas, phytoplasmas (formally called mycoplasmalike organisms), fungi, fungal-like organisms, nematodes, and parasitic higher plants. Each will be discussed in the following subsections.

Viruses   Viruses are intracellular (inside cells) pathogenic particles that infect other living organisms. Human diseases caused by viruses include chickenpox, herpes, influenza, rabies, smallpox, and acquired immunodeficiency syndrome (AIDS). Although most of us are familiar with these viruses, the first virus ever described and from which the term was eventually derived was tobacco mosaic virus (TMV).

Virus particles are extremely small and can be seen only with an electron microscope. Because the cell membrane in plants is surrounded by a rigid cell wall that a virus cannot penetrate, plant viruses require a wound for their initial entrance into a plant cell. Wounds in plants occur naturally, such as in the branching of lateral roots, or they may be the result of agronomic, horticultural practices or other mechanical means; fungal, nematode, or parasitic plant infections; or insects. In some cases, the organism creating the wound can also be carrying and can pass or transmit the virus. Organisms that transmit pathogens are called **vectors**. The activity of humans in propagating plants by budding and grafting or by cuttings is one of the chief ways viral diseases are spread. The seedling offspring of a virus-infected plant is usually, but not always, free of the virus, depending on the plant species and the kind of virus. Insect transmission is perhaps the most important means of virus transmission in the field. Insects in the Order Homoptera (such as aphids, planthoppers, leafhoppers, whiteflies, and mealy bugs) that have piercing, sucking mouthparts are the most common and economically important vectors of plant viruses (see Figs. 15–22, page 315 and 15–24, page 319). Some plant viruses can also be transmitted in pollen grains or by seed.

Viruses are obligate parasites; that is, they require a living host to be able to grow and multiply. Once in a wounded cell, the virus particle sheds its protein coat, and the nucleic acid is replicated, leading to the development of new virus particles. Cell-to-cell movement of plant viruses occurs through the cytoplasmic bridges between cells called **plasmodesmata**, and viruses move systemically throughout infected plants via the phloem. Although the details of plant virus replication are complex and beyond the scope of this chapter, the general idea is that plant viruses cause disease in part by causing a reallocation of photosynthates and a disruption of normal cellular processes as they replicate. Many kinds of plants are infected with viruses and show no symptoms. Such infections are referred to as being latent.

Viruses are difficult to classify and, for want of anything better, they are given descriptive (and sometimes colorful) names based on the disease they cause—for example, tobacco mosaic virus, watermelon mosaic, barley yellow dwarf, potato mop top, carnation streak, and tomato spotted wilt (Fig. 15–35). Many of these viruses also infect plants of other species. For example, tobacco mosaic infects tomatoes where it can cause devastating losses. Maize dwarf mosaic infects sorghum, Sudan grass, sugarcane, and johnsongrass in addition to corn.

Once plants are infected, little can be done to free them from the virus. Because different cultivars and

---

KOCH'S POSTULATES: RULES FOR PROOF OF PATHOGENICITY

How do plant pathologists know that an organism isolated from a sick plant is actually the causal agent of a particular disease, especially the first time a new disease or set of symptoms is observed? Dr. Robert Koch, a German bacteriologist and a physician, asked this question in 1876 when studying anthrax in cattle. He developed a simple four-step process, commonly referred to today as Koch's postulates, to answer this question. Koch's postulates, or rules for proof of pathogenicity, have since been adopted as the principle means of confirming that an organism isolated from a sick or symptomatic plant, animal, or human is the pathogen responsible for a given disease. In simple terms, Koch's postulates state that (1) the microorganism in question must be present in every case of the disease; (2) the microorganism must be isolated from the diseased host and grown in pure culture; (3) the specific disease must be reproduced when a pure culture of the microorganism is inoculated into a healthy susceptible host; and (4) the introduced microorganism must be recovered from the experimentally infected host. If one can prove Koch's postulates, it is then possible to say conclusively that a particular organism is the causal agent of the disease in question.

A

B

Figure 15–35

(A) Healthy barley and barley infected with Barley Yellow Dwarf virus. (B) The tomato plant on the right is infected with Tobacco Mosaic Virus, note the puckered leaves with yellow mottling.

Source: USDA Agricultural Research Service Image Gallery, http://www.ars.usda.gov/is/graphics/photos/

species show different degrees of resistance to some viruses, resistant types should be planted whenever they are available. Recent advances in plant cell molecular biology and virology have lead to the development of genetically modified plants with superior resistance to some viruses. For many crops, the best management approach is the planting of stock that has been propagated from known virus-clean or certified sources. Another successful way to eliminate viruses, particularly from herbaceous plants, is to use meristematic tip-culturing techniques and tissue culturing to develop virus-free callus tissue that can then be used to generate new virus-free clones of the original plant. This procedure is based on the fact that virus is usually not present in the actively growing shoot tip of an infected plant.

Numerous cultural practices can be used to reduce plant losses due to virus infection, including (1) scouting and the removal of symptomatic plants or known alternative weed or volunteer plants that may serve as a reservoir for a given virus; (2) the use of clean or sanitized tools and equipment; (3) hand washing; and (4) the use of disposable overgarments. Rotations with nonhost crops as well as isolation of newly received plant material prior to its introduction into the rest of a production system can also minimize perpetuating the pathogen. Some viruses are permanently inactivated by prolonged exposure of infected tissue to relatively high temperatures—for example, twenty to thirty days at 38°C (100°F). This procedure, called heat therapy, frees individual plants or cuttings of the virus. There are no chemical sprays or biological control approaches to eradicate viruses, although insecticides and biocontrol products

can be used to control insect vectors. Management of insect vector populations in the field can be difficult to impossible unless it is coordinated on a regional basis, but it may be highly effective in closed production systems such as greenhouses or interiorscapes.

Bacteria Bacteria are microscopic, single-celled prokaryotic organisms. They occur singly or in colonies of cells. Bacteria are classified into four groups based on cell shape: (1) the spherical cocci, (2) the rod-shaped bacilli, (3) the spiral-shaped spirilli, and (4) a small group of filamentous forms—the actinomycetes. Only bacilli and actinomycetes are known to cause diseases in plants. Some types of bacilli and spirilli are motile—they have whiplike flagella that propel them through films of water. Bacteria multiply at alarming rates under suitable environmental conditions. Bacteria are divided into two groups based on their reaction when subjected to a relatively simple staining procedure called the Gram stain: Gram-negative bacteria stain red or pink, and Gram-positive bacteria stain purple. The difference in color is directly related to the chemical composition and structure of the cell wall.

Although considered structurally simple, bacteria are extremely diverse from a metabolic standpoint and are found almost everywhere on earth in vast numbers—from living in jet fuel and on the rims of volcanoes to thriving in hydrothermal vents deep on the ocean floor. Beneficial bacteria are involved in diverse processes such as digestion in animals, nitrogen fixation in the roots of certain legumes, the decomposition of animal and plant remains, and sewage disposal systems.

Pathogenic bacteria, on the other hand, cause severe and often fatal diseases in humans, animals, and plants. The first bacterial disease ever discovered was anthrax (*Bacillus anthracis*) of cattle and sheep in 1876. The discovery of anthrax in cattle was immediately followed by the discovery of fire blight (*Erwinia amylovora*) in pear and apple (Fig. 15–36). Table 15–3 gives some examples of important bacterial diseases of plants.

The taxonomy of plant-pathogenic bacteria is currently in flux based on recent advances in the classification of bacteria. Most plant-pathogenic bacteria belong to the following genera: *Erwinia, Pectobacterium, Pantoea, Agrobacterium, Rhizobium, Pseudomonas, Ralstonia, Burkholderia, Acidovorax, Xanthomonas, Clavibacter, Streptomyces,* and *Xylella*. Plant-pathogenic

Figure 15–36

Fireblight on apples. Source: USDA Agricultural Research Service Image Gallery, http://www.ars.usda.gov/is/graphics/photos/.

Table 15–3
SOME IMPORTANT PLANT DISEASES CAUSED BY BACTERIA

Common Name of Disease	Pathogen	Hosts Attacked	Integrated Disease Management Measures
Crown gall	*Agrobacterium tumefaciens*	Woody ornamentals and tree fruits	Use of pathogen-free nursery stock; avoid planting in infested soils; avoid wounding; treat graft junctions and rootstock with biocontrol bacteria (Note: *A. tumefaciens* is an important means of transferring genetic material in genetic engineering)
Bacterial wilt of cucurbits	*Erwinia tracheiphila*	All cucurbits (cucumber, pumpkin, squash, and muskmelon)	Resistant varieties; control of insect vectors (spotted and striped cucumber beetles)
Stewart's wilt of corn	*Pantoea stewartii* (formerly *Erwinia stewartii*)	Corn (sweet and field hybrids)	Resistant varieties
Fire blight	*Erwinia amylovora*	Pome fruits (apple, pear, quince, and some ornamentals such as pyracantha)	Moderately resistant varieties; prune out diseased tissues; maintain a balanced fertility program; chemical applications—copper sulfate, Bordeaux mixture, streptomycin or oxytetracycline; biological control—blight ban
Common blight of beans	*Xanthomonas phaseoli*	Beans (field or dry, garden or snap, lima, etc.)	Pathogen-free seed; sanitation; disposal of crop residue; three-year crop rotations
Bacterial canker	*Pseudomonas syringe*	Almonds, apricots, avocado, cherry, peach, plum	Resistant cultivars; use only pathogen-free budwood; prune out diseased tissues; chemical applications—copper, Bordeaux mixture
Citrus canker	*Xanthomonas axonopodis*	Citrus crops (oranges, grapefruits, lemons, limes, etc.)	Quarantines to exclude and restrict movement of pathogen and/or infected tissue; destruction of all trees in a location when disease is detected; chemical applications with copper
Soft rot of vegetables	*Pectobacterium carotovora* (formerly *Erwinia carotovora*)	Fleshy storage tissues of vegetables and ornamentals (potatoes, carrots, cucumber, lettuce, cabbage, onions, iris, etc.)—in the field, transit, and storage	Sanitary and cultural practices; avoid mechanical damage during harvest and transit; good ventilation during storage; storage under cool, dry conditions

bacteria cause many different kinds of symptoms, including galls and overgrowths, wilts, leaf spots, specks and blights, soft rots, and scabs and cankers (Table 15–3). In contrast to viruses, which are intracellular pathogens, bacteria are considered intercellular (between cells) pathogens. The means by which plant-pathogenic bacteria cause disease are as varied as the types of symptoms they cause. Some plant-pathogenic bacteria produce toxins that lead to cell death or enzymes that break down key structural components of plant cells, such as the case of the soft-rotters that produce pectinase that degrades the middle lamella. Still others colonize the xylem and restrict water movement, resulting in wilt. Some even have the ability to modify or transform their host genetically and bring about the formation of galls and overgrowths such as the case of *Agrobacterium tumefaciens* (causal agent of crown gall).

Bacteria that cause plant diseases are spread in many ways—they can be splashed about by rains or carried by the wind, birds, or insects. People can unwittingly spread bacterial diseases, for instance, by pruning infected orchard trees during the rainy season. Water facilitates the entrance of bacteria carried on pruning tools into the pruning cuts. Propagation with bacteria-infected plant material is a major way pathogenic bacteria are moved over great distances. However bacterial pathogens are disseminated, they require a wound or natural opening to penetrate a plant host. As previously described, the means by which bacterial pathogens cause disease are varied. All bacterial plant pathogens replicate by means of binary fission. They overwinter or survive unfavorable environmental periods or the absence of a susceptible host by going dormant in infected tissue, infested soil or water, or an insect vector.

Bacterial diseases in plants are difficult to control. Integrated management measures for bacterial plant pathogens include using resistant varieties, cultivars or hybrids, bacteria-free seed or propagation materials, and sanitation and cultural practices that either eliminate or reduce sources of bacterial contamination; preventing surface wounds that permit the entrance of bacteria into the inner tissues; and propagating only bacteria-free nursery stock. Prolonged exposure to dry air, heat, and sunlight sometimes kills bacteria in plant material. Applications of copper-containing compounds or Bordeaux mixture (copper sulfate and lime) as well as the antibiotics streptomycin and/or oxytetracycline may also help kill or suppress plant-pathogenic bacteria. Similarly, the use of antagonistic or biological control products such as Blight Ban and Agrosin K84 may also be effective for managing bacterial diseases of plants.

Spiroplasmas and Phytoplasmas   **Spiroplasmas** and **phytoplasmas** are prokaryotic organisms that lack rigid cell walls and infect plants. Spiroplasmas are helical in shape (resembling a corkscrew), whereas phytoplasmas are round or ovoid. These organisms resemble bacteria in some ways but are very small (perhaps even submicroscopic) and are wall-less. Phytoplasmas were once called mycoplasmalike organisms because they resemble mycoplasmas—organisms that cause disease in humans and animals. About 200 different plant diseases have been shown to be caused by spiroplasmas and phytoplasmas. Some of the diseases they cause are aster yellows, western-X of peaches, cherry buckskin, pear decline, mulberry dwarf disease, corn stunt, and stubborn disease of citrus. As with viruses, a disease caused by phytoplasmas is named after the plant on which is was first studied, but it can also occur on many other plants. For example, aster yellows also affects other ornamentals—gladiolus and phlox, for example—and tomato, spinach, onion, lettuce, celery, carrots, strawberry, and many weeds. Spiroplasma- or phytoplasma-infected plants exhibit a wide variety of symptoms, similar to those typically associated with viral infections, including generalized stunting and decline, yellowing, excessive proliferation of shoots, sterility of flowers, the development of green flowers, dieback, and plant death. In fact, until the discovery that these organisms caused disease in plants in 1967, most diseases now known to be caused by spiroplasmas or phytoplasmas were believed to be caused by viruses. Unlike most bacterial pathogens, spiroplasmas and phytoplasmas live in the phloem and are vectored by sucking insects such as leafhoppers, planthoppers, and psyllids. The pathogens can then be spread to other plants when the insect feeds on them. One obvious method of controlling the spread of these diseases is an effective spray program that eliminates the insect vectors. As discussed earlier in this chapter, this type of program is often difficult or impossible in the field but may work well in enclosed production systems like greenhouses. The antibiotic tetracycline may be slightly effective against these pathogens.

Fungi and Fungal-Like Organisms   Collectively, fungi and fungal-like organisms (FLOs) cause more plant diseases than any other group of plant-pathogenic organisms. Over 8,000 species of fungi and FLOs have been shown to cause disease in plants. Fungal-like organisms are organisms like *Pythium* and *Phytophthora* and those that cause downy mildew. Until recently, they were considered fungi but, because of changes in fungal taxonomy, are now in the Kingdom Chromista (also called Stramenopila). Fungi and FLOs are heterotrophic

eukaryotic organisms that lack chlorophyll and thus the ability to photosynthesize. Fungi and FLOs obtain nutrients by absorbing them through tiny threadlike filaments called **hyphae** that branch in all directions throughout a substrate. A collection of hyphae is referred to as mycelium (pl., mycelia). The hyphae are filled with protoplasm containing nuclei. Mycelia are the key diagnostic sign associated with diseases caused by fungi and FLOs. Most of us have seen mycelium growing on old bread or rotten fruit or vegetables and may have referred to these organisms collectively as molds or mildew.

Fungi and FLOs, like bacteria, can be beneficial as well as pathogenic. Beneficial fungi participate in biological cycles, decaying dead animal and plant materials and thus converting them into plant nutrients that are absorbed by living plants. **Mycorrhiza**, as you learned earlier in this book, are fungi that form a symbiotic relationship on the roots of many plants. Other beneficial fungi, such as those belonging to the genus *Trichoderma*, are effective biocontrol agents of plant pathogenic fungi, while others, like *Arthrobotrys dactyloides*, have been shown to trap and parasitize plant pathogenic nematodes.

Fungi and FLOs (indeed all pathogens) can be grouped into the following four categories based on their preference for surviving on dead or decaying organic matter versus living tissue:

1. *Obligate saprophytes*—always saprophytes. These organisms can only survive or are obliged to gain nourishment by colonizing dead or decaying organic matter. They are *not* parasites.
2. *Obligate parasites*—always parasites. They can grow only as parasites on or in a living host. They cannot survive as saprophytes or be cultured in the laboratory. This group of pathogens have a vested interest in prolonging the life of their host to increase their own viability. All viruses, downy mildews, powdery mildews, rusts, and smuts are obligate parasites.
3. *Facultative parasites*—usually survive as saprophytes but have the ability to parasitize and cause disease under certain conditions. Examples include *Pythium* species and many bacterial pathogens.
4. *Facultative saprophytes*—usually survive as parasites but have the ability to live on dead and decaying organic matter under the right conditions. Examples include *Phytophthora* and *Botrytis* species.

Some fungi and FLOs can live on only one host species, while others develop on many different kinds. Most plant diseases are caused by fungi, and the food loss to fungal diseases is staggering (Fig. 15–37). Some of the world's great famines can be blamed on pathogenic

Figure 15–37
Fungal diseases can cause considerable damage to many crops. Shown here is a strawberry infected with Botrytis cinerea. *Note the filamentous mycelia.* Source: USDA Agricultural Research Service Image Gallery, *http://www.ars.usda.gov/is/graphics/photos/*

fungi. Wheat crops of the Middle Ages were ruined when the grains became infected with a dark, dusty powder now known to be the spores of the fungus called bunt or stinking smut (*Tilletia* spp.) The potato blight in Ireland and northern Europe, rampant during two successive seasons (1845–1846 and 1846–1847), was caused by the FLO *Phytophthora infestans*. It resulted in the death of more than 1 million people by starvation and caused mass migration from Ireland. In the 1870s, an epidemic of downy mildew, caused by the fungus *Plasmopara viticola*, struck the grape vineyards of central Europe, causing great losses to grape growers and wine makers. In the United States alone, hundreds of millions of bushels of wheat have been lost in epidemic years to stem rust (*Puccinia graminis tritici*) (Fig. 15–38). Managing Dollar spot (Fig. 15–39) on

Figure 15–38
Wheat rust (Puccinia graminis) on a wheat stem. Source: USDA Agricultural Research Service Image Gallery, *http://www.ars.usda.gov/is/graphics/photos/*

Figure 15–39
Dollar spot (Sclerotinia homoeocarpa) *on turf.* Source: USDA
Agricultural Research Service Image Gallery, *http://www.ars.usda.gov/is/*
graphics/photos/

turf can be very expensive for golf course superintendents, sports field managers, and landscapers.

Because of the sheer number of plant diseases caused by fungi and the huge diversity in how plant-pathogenic fungi cause disease, it is impossible and beyond the scope of this text to provide details about specific disease cycles and integrated fungal disease management strategies. For those interested in such information, an introductory level course or workshop in plant pathology is recommended. But like other groups of plant pathogens, fungal pathogens have developed ways to survive periods of unfavorable environmental conditions or in the absence of a susceptible host, spread, infect, and grow and reproduce on and within plants. The steps involved in a fungal or FLOs disease cycle are identical to those described previously for bacterial pathogens (see Fig. 15–30 on page 323). One important difference between fungi and FLOs, and bacteria and viruses is that fungi and FLOs not only penetrate a host via a wound or natural opening, but they can also penetrate actively via the production of specialized hyphal structures called appresoria (sing. appresorium). Appresoria are swollen tips of hyphae that allow the fungus to penetrate plant tissues directly through mechanical and enzymatic activity. Once inside the plant, the fungus can extract liquid and nutrients from the plant cells causing them to weaken and even die. The ability of fungi and FLOs to penetrate healthy plants is undoubtedly responsible for their place collectively as the most important group of plant pathogens.

More options are generally available to professional plant production specialists and growers to manage

fungal and FLO diseases compared to viral and bacterial diseases. One of the most satisfactory methods of dealing with fungus diseases is strict sanitation to eliminate the pathogenic organism, starting with the initial stages of propagation and growth of the potential host plants. Proper irrigation practices can help reduce fungus infections, as many fungi are favored by free moisture or high relativity. In greenhouses, controlling relative humidity by proper irrigation and venting to make the environment unfavorable for some fungi and FLO's can be very effective.

In many of the major crops, cultivars resistant to prevailing diseases are available, and more are continually being developed by plant breeders. As has been discussed with other types of diseases, the use of genetically resistant plants should be the first line of defense for diseases caused by fungi and FLOs, if they are available. Several examples of cultivars genetically resistant to fungal and FLO disease are notable. Certain hybrid potato cultivars are resistant to late blight (*Phytophthora infestans*). Soybean cultivars resistant to downy mildew (*Peronospora manshurica*) have been developed. In the United States, apple cultivars have been developed that show high resistance or immunity to apple scab (*Venturia inaequalis*)l. In the cereal crops (oats, wheat, rye, barley), powdery mildew (*Erysiphe graminis*) can be controlled only by the use of resistant cultivars. Tomatoes can be grown in *Fusarium*-infested soils if *Fusarium*-resistant cultivars are planted. The importance of resistant cultivars in controlling fungus diseases is shown in Table 15–4.

Although the use of resistant cultivars and eradication of the pathogen through the use of cultural practices are the most satisfactory ways of dealing with diseases caused by fungi and FLOs, these measures are not possible in many instances. Often the disease appears and its development must be slowed or stopped by whatever means are available. The use of fungicides are relied on heavily for certain fungal and FLO problems and may be essential where there is a demand for plant health during environmental periods that favor pathogen growth. Fungicides are typically more effective when applied prior to the onset of disease symptoms (referred to as preventive applications). Some fungicides are effective when applied after the onset of symptoms and are said to have curative activity. In either case, fungicides must be delivered to the area of the plant where the pathogen is active to be effective. Many different types and chemical classes of fungicides are currently available. Numerous online extension-outreach and

Table 15–4
SOME IMPORTANT PLANT DISEASES CAUSED BY FUNGI AND FUNGAL-LIKE ORGANISMS

Common Name of Disease	Pathogen	Hosts Attacked	Integrated Disease Management Measures
Stem rust of wheat	*Puccinia graminis tritici*	Wheat	Resistant varieties
Corn smut	*Ustilago maydis*	Corn	Resistant varieties
Fusarium wilt	*Fusarium oxysporum*	Tomato, pea, celery, banana, cotton, watermelon, ornamentals	Resistant varieties
Powdery mildew	Many different species	All hosts	Resistant varieties; fungicide applications
Dollar spot	*Sclerotinia homoeocarpa*	Many turfgrass species	Moderately resistant species or cultivars; avoid excessive nitrogen fertility and prolonged leaf wetness; adequate soil moisture; fungicide applications
Rice blast	*Magnaporthe grisea*	Rice	Resistant cultivars; avoid excessive nitrogen; fungicide applications
Apple scab	*Venturia inaequalis*	Apples and crab apples	Resistant varieties; fungicide applications
Peach leaf curl	*Taphina deformans*	Peaches and nectarines	Fungicide applications
Verticillium wilt	*Verticillium albo-atrum* and *V. dahliae*	Potatoes and a wide range of woody fruit and ornamental species; many herbaceous plants	Resistant varieties; remove infected plant material; avoid infested soils; soil fumigation; fungicide applications
Fusarium head blight	*Fusarium graminearum*	Wheat and barley	Moderately resistant varieties; crop rotation; fungicide applications have limited effectiveness; biological control
Gray mold, botrytis	*Botrytis cinerea*	Leaves, flowers, and fruit of many herbaceous plants, grapes (when infected at the right time, *Botrytis* gives the wine made from infected grapes a desirable sweet taste)	Resistant varieties, good air circulation around the plants, lowering relative humidity in greenhouses
Downy mildew of grape	*Plasmopara viticola*	Grape	Resistant varieties; fungicide applications
Late blight of potato	*Phytophthora infestans*	Potato and tomato	Resistant cultivars; destroy all culled potatoes; crop rotation; fungicide applications
Phytophthora root rot of soybean	*Phytophthora sojae*	Soybean	Resistant varieties; fungicide applications
Damping off	*Pythium* and *Rhizoctonia* Species	Seedlings of most plant species	Fungicide seed treatments; biological control—use compost-amended mixes and antagonistic microorganisms
Soybean rust	*Phakopsora pachyrhizi*	Soybean and many alternative hosts	Quarantines to exclude/restrict movement of pathogen; destruction of infected plant material; fungicide applications

Typical root symptoms indicating nematode attack are root knots or galls (see Fig. 15–40), root and tuber lesions

agrichemical company resources provide specific fungicide recommendations for nearly every major cropping system and pathogen. Always read and follow label recommendations when applying fungicides and pesticides.

OTHER PLANT PESTS

Nematodes

Nematodes are simple, microscopic, multi-cellular animals—typically containing 1,000 cells or less. They are wormlike in appearance but are taxonomically distinct from earthworms, wireworms, or flatworms. They are microscopic soft-bodied (no skeleton), nonsegmented roundworms. The basic body plan of a nematode is a tube within a tube. Most nematodes are not pathogens but rather saprophytes. Some are serious human, animal, and plant pathogens. Those that attack animals or humans do not attack plants, and vice versa. Heartworm in dogs and cats and elephantiasis are examples of nematode diseases in animals and people, respectively. Plant parasitic nematodes may attack the roots, stem, foliage, and flowers of plants. All plant parasitic nematodes have piercing mouthparts called **stylets**. The presence of a stylet is the key diagnostic sign differentiating plant parasitic nematodes from all other types of nematodes. In other words, if a nematode has a stylet, it is a plant pathogen. The bacterial-feeding nematode, *Caenorhabditis elegans,* is one of the best-understood animals on earth. It was the first animal to have its genome completely sequenced. The study of *C. elegans* has led to many new insights into animal development, neurobiology, and behavior.

Several genera and species of nematodes are highly damaging to a great range of hosts, including foliage plants, vegetable crops, fruit and nut trees, turfgrass, and forest trees. Some of the most damaging nematode species are root knot (*Meloidogyne* spp.) (Fig. 15–40), cyst (*Heterodera* spp.), root lesion (*Pratylenchus* spp.), spiral (*Helicotylenchus* spp.), burrowing (*Radopholus similis*), bulb and stem (*Ditylenchus dipsaci*), reniform (*Rotylenchulus reniformis*), dagger (*Xiphinema* spp.), and bud and leaf (*Aphelenchoides* spp.).

Parasitic nematodes are readily spread by any physical means that can move soil particles about—equipment, tools, shoes, birds, insects, dust, wind, and water. In addition, the movement of nematode-infested plants or plant parts spreads the parasites.

Along with the usual IPHM methods of controlling pests, nematode damage can be controlled in some

Figure 15–40
Southern root knot nematode (Meloidogyne incognita) *on pepper.* Source: USDA Agricultural Research Service Image Gallery, *http://www.ars.usda.gov/is/graphics/photos/.*

cases by using resistant rootstock, for example, a root-knot nematode-resistant peach rootstock called "Nemagard," is available, thus permitting peach production even on infested soils. Heat treating the soil or fumigating the soil before planting can reduce nematode populations. Methyl bromide had been the most widely used fumigant, but its use is now severely restricted because of its polluting effect in the atmosphere. To date, there is no satisfactory substitute for methyl bromide. In certain cases nematicides can be used, but they are very poisonous and must be used carefully. Most of these kinds of materials will injure or kill plants if they are applied too close to their root zone. Many nematicides have been taken off the market, which, along with the loss of methyl brominde, has resulted in a greater emphasis on the development of alternative IPHM practices. Several organisms, including bacteria, fungi, as well as other nematodes, will attack harmful nematodes.

Parasitic Higher Plants

Some higher plant forms live on the surface of or parasitize other plants and often cause harmful reactions in their hosts. These **higher plant parasites** can be placed in three groups: epiphytes, hemiparasites, and true parasites. The **epiphytes** do little or no harm to their host plants, using them merely for physical support and protection. Examples are Spanish moss and epiphytic orchids, which in their native habitat commonly grow on tree limbs.

The hemiparasites, sometimes called water parasites, do injure their host plants, absorbing water and mineral nutrients from them. However, they possess chlorophyll and can manufacture their own

carbohydrates by photosynthesis. Witchweed (*Striga asiatica*) is a hemiparasitic seed plant that severely damages sugarcane, corn, sorghum, many other grasses, and some broad-leaved plants. It attaches itself to the host's roots and utilizes most of the host's water and mineral nutrients, causing it to wilt, yellow, stunt, and die. The best control is to plant a crop, such as Sudan grass, which stimulates the witchweed seed to germinate, and then plow under the entire field. Crops should be rotated, and susceptible crops should not be planted.

Mistletoe (*Phoradendron* spp.), another member of the hemiparasitic group, attacks many broad-leaved trees such as Modesto ash, silver maple, honeylocust, hackberry, cottonwood, walnut, oak, birch, and some conifers (Fig. 15–41). The seeds germinate on the limbs of susceptible hosts, forming an attachment disk on the bark. The sticky berries are disseminated throughout the tree and from tree to tree by birds and wind. Although not very effective, the usual control is to cut out the mistletoe branches deep into the tree under the point of attachment. No good herbicidal control has been developed; however, a dormant-season application of ethephon, an ethylene-releasing material, is a possible control on some plants. The best control is to plant only tree species resistant to mistletoe attacks.

True parasites lack chlorophyll and depend on their hosts for all nourishment—carbohydrates, minerals, and water. Examples of this group are the dwarf mistletoe (*Arceuthobium* spp.) and dodder (*Cuscuta* spp.). Broomrape (*Orobanche* spp.) is a serious parasitic pest in Europe and has caused extensive damage to tomatoes in California. Dwarf mistletoe attacks many coniferous species in the western United States, reducing tree vigor and lowering lumber quality. The sticky seeds (not the fruits) are forcibly ejected and can travel up to about 19 m (60 ft). This is the principal means of dissemination. Birds are known to carry the sticky seeds on their feathers; wind plays a very minimal role in dissemination. The best control is removal of infected trees.

Dodder has many species, but about six cause the major damage, attacking crops such as alfalfa, lespedeza, clover, flax, sugar beets, some vegetable crops, and ornamentals. Dodder seriously reduces yields and quality of crops. Strict regulations prohibit the sale of crop seed contaminated by dodder seed. Great effort should be taken to avoid planting seed that has dodder seed mixed with it. Patches of dodder in field crops or along fences or ditch banks should be eradicated by burning or with herbicides.

Practical Tips for Diagnosing Plant Disorders

Proper diagnosis is a critical step in the management of plant disorders. Without a solid diagnosis, it is impossible to suggest an adequate management approach. The more you know, the better equipped you will be to take corrective action. In the case of plant disease diagnosis, the more you know about the host, environment, and biotic and abiotic factors that cause disease (the disease triangle), the greater chance you have of making a correct diagnosis. The following five-step approach is just one of many approaches available for diagnosing plant disorders:

1. *Define the problem.* In other words, determine that a problem actually exists. Start by correctly identifying the host plant and by being familiar with its normal or healthy state and characteristics. Make sure to take seasonal effects into account. After all, a maple tree without leaves in December in Ohio should not trigger an alarm. A defoliated maple tree in June is another story. Know your hosts and how they change with the seasons. Only then can you determine that a problem exists.

2. *Examine the entire plant community.* Don't jump into examining the affected individual plant or area. Take stock of the entire plant community. Note light intensities, wind direction, slope of the land, air movement, and so on. Take time to

Figure 15–41

Mistletoe on oak trees that are dormant for the winter.

Source: Margaret McMahon, The Ohio State University.

develop an overview of the situation at hand because this approach often provides many valuable clues about the situation. Then focus your attention on the affected plant(s) or area. Even then, however, look at the entire plant first before jumping directly to any signs or symptoms that might be present. Check out the leaves, stems, roots, and fruits or flowers if appropriate.

3. *Look for patterns.* Is only a single plant affected? Is the potential disorder restricted to a certain area or a single species? Are the symptoms randomly distributed, or can you see any distinct patterns or clear lines of demarcation between healthy and affected plants? Random patterns are often indicative of disorders caused by pathogens or insects, whereas uniform damage such as streaks or lines or damage over a large area is indicative of an abiotic (chemical, physical, or mechanical) culprit.

4. *Consider how the damage developed over time.* Did the damage appear suddenly or over time? Has the damage spread or stayed in the same location? Progressive development and spread over time often indicates damage caused by pathogens and insects. In contrast, damage that does not spread and which occurs suddenly is typically caused by an abiotic factor.

5. *Ask questions, gather information, and determine causes of plant damage.* Gather as many clues (i.e., as much information) as possible about the crop. Determine cultivar or variety, age of the stand (especially important when dealing with some perennial crops), recent fertilizer or pesticide applications, cultural practices implemented, recent weather trends, irrigation practices, the history of the site or stand, and how the damage progressed over time. Collect as much information as possible to help you develop a solid mental picture of what led up to the damage you are observing. Look for evidence of pathogen or insect activity. Specifically, look for key diagnostic signs or symptoms indicative of plant pathogens or insects. After you have gathered sufficient background information and nothing strikes you as being obvious, such as a chemical misapplication or hail damage, and you have eliminated the possibility of pathogens and insect pests, retrace your steps and focus your diagnosis on abiotic factors. You may need to enlist the services of a plant pest or disease diagnosis laboratory or clinician to help narrow the range of probable causes. Consider contacting and coordinating with the clinician before sending samples. Whenever possible, include photographs or digital images to aid the diagnostician in the task.

The Safe Use of Agricultural Chemicals— Herbicides, Insecticides, Fungicides, Miticides, and Nematicides

The application of agricultural chemicals to food-producing plants must not create a health or environmental hazard. Many countries have elaborate procedures for determining whether agricultural chemicals are reasonably safe before they can be registered for sale to growers of agricultural crops. In the United States, the Environmental Protection Agency (EPA)[1] is responsible for determining the safety of agricultural chemicals; state and local government agencies can also add their own safety requirements.

Before the EPA grants approval for the sale of agricultural chemicals, exhaustive tests are conducted to show:

1. That the product, at the recommended application rate, has low toxicity levels (both acute and chronic) as determined by experiments with test animals.
2. An absence of residues in food or feed crops—or, if there is a detectable residue, that it is no more than the tolerance level established as safe.
3. The fate of residues and breakdown products in the environment—in the soil, runoff water, groundwater, or wildlife.
4. Whether the product affects the environment by inducing changes in the natural populations of higher plants and animals or of microorganisms.

It is estimated that an agricultural chemical company developing a new pesticide spends $50 million and that six to ten years of research are required to develop information sufficient to satisfy EPA standards.

When the EPA registers a pesticide for use, the label lists very specific restrictions on the product. It is registered for use on a certain crop or crops, to be applied at specific times and at specific concentrations.

---

[1]"The Environmental Protection Agency is charged by the United States Congress to protect the nation's land, air, and water systems. Under a mandate of national environmental laws focused on air and water quality, soil waste management, and the control of toxic substances, pesticides, noise and radiation, the Agency strives to formulate and implement actions which lead to a compatible balance between human activities and the ability of natural systems to support and nurture life." *Source: EPA Journal.*

The reentry interval (REI) specifies the amount of time that must pass after chemical application and before reentry in the area of application is allowed. While a certain herbicide, for example, may be known to control a given weed species, it may not be permissible to use the herbicide to control such weeds if they are growing in a crop for which the herbicide is not registered. When using pesticides, including herbicides, the warnings given in the box below should be carefully read and followed.

The term LD_{50} may be seen on labels of agricultural chemicals. LD means "lethal dose," and refers to the chemical's toxicity. Oral LD_{50} is the dose that will kill 50 percent of test animals ingesting the chemical by mouth. LD_{50} is expressed in milligrams of the chemical per kilogram of body weight of the test animal. The higher the LD_{50} value, the safer the chemical. According to EPA toxicology guidelines a chemical with an oral LD_{50} of 50 or less must be labeled "DANGER—POISON (FATAL)"; one with an LD_{50} between 50 and 500 is labeled "WARNING (MAY BE FATAL)"; an LD_{50} from 500 to 5,000, "CAUTION", and one with an LD_{50} over 5,000 is also labeled "CAUTION." All labels must also state "KEEP OUT OF REACH OF CHILDREN." Other toxicity categories that may be listed on the label or Material Data Safety Sheet (MSDS) include dermal (transmission through skin), mutagen, and carcinogen.

PESTICIDE EFFECTS ON THE ENVIRONMENT

Insecticides, miticides, fungicides, nematicides, and herbicides can be thought of as necessary evils. Without these chemicals, large-scale agriculture and our standard of living today would not exist. Too little food would be produced to feed the world's 6 billion people, and mass starvation would result.

However, chemical applications in agriculture do not always do just what they are supposed to do and nothing else. This problem is recognized more and more now, and greater and greater precautions are being taken to avoid unwanted side effects from these chemicals. Government regulations on pesticides have become steadily tighter. Pesticides that are chemically stable and persist in ecosystems are the ones largely responsible for environmental contamination. Persistent chemicals such as DDT, DDD, dieldrin,

aldrin, chlordane, BHC, and heptaclor have been replaced with the low-persistence organophosphates (malathion, diazinon) and the carbamates (carbaryl and methomyl), which break down rapidly and are not taken up in food chains. There is the risk, however, that large-scale applications of pesticides followed by irrigation or heavy rains, can result in leaching of the chemicals into the underground water supply and eventually into drinking water. In fact, traces of several pesticides have been found in several underground water supplies in many states in the United States. Some of this is attributable to runoff from pesticide and herbicide applications to plants in urban and suburban areas. Depending on location, these sources may contribute more than agriculture in the area.

It was pointed out earlier in this chapter that many insect pests are held in check very well by their own natural enemies—often other insect predators. Reducing populations of one serious primary insect pest with insecticides may, at the same time, so reduce the numbers of insect predators feeding on a secondary insect that the secondary pest increases explosively. For example, insecticidal control of codling moth, pear thrips, and pear psylla in pear orchards may be followed by large increases in spider mite populations because the spray applications have reduced other predatory mites that feed on the spider mite.

Production of several major food crops relies on pollination of the flowers by honeybees and other bees. Many nut and fruit tree species—almonds, apples, plums, and sweet cherries—as well as certain vegetables, such as muskmelons and honeydew melons, and forage crops such as seed alfalfa and seed clover require thorough working of the flowers by bees during bloom. Elimination of bees by haphazard insecticidal applications and drift cannot be tolerated. Some states impose strict legal requirements wherever honeybees could be involved.

Insecticides should not be applied in areas where bees are working, particularly with chemicals highly toxic to bees such as diazinon, Guthion, malathion, parathion, and carbaryl. Some insecticides (Aramite, ethion, methoxychlor, Omite, pyrethrins, rotenone, and Tedion) are relatively nontoxic to honey bees.

The state of California requires that beekeepers post their names and telephone numbers on all hives. Anyone planning to apply pesticides in the vicinity must inform the beekeepers of upcoming spray applications. If potentially hazardous insecticides are to be used, spray applicators must notify all beekeepers

WARNINGS ON THE USE OF PESTICIDE CHEMICALS AND SUGGESTIONS FOR THEIR PROPER USE

Pesticides are poisonous and should always be used with caution. The following suggestions for using and handling pesticides help minimize the likelihood of injury from exposure to such chemicals to humans, animals, and crops other than the pest species to be destroyed:

1. *Always* read and exactly follow all precautionary directions on container labels before using sprays or dusts. Read all warnings and cautions before opening the container. Repeat this process every time you use the pesticide—regardless of how often you use it or how familiar you think you are with the directions. Apply materials only in amounts and at times specified. *Remember:* The label is the law and violation of its directions is a criminal act subject to prosecution. The specifics of many of the following procedures (for example, what protective clothing has to be worn, disposal procedures, etc.) for a chemical will be given on the label.

2. Keep sprays and dusts out of reach of children unauthorized persons, pets, and livestock. Store all pesticides outside the house in a locked cabinet or shed and away from food and feed.

3. Always store sprays and dusts in their original labeled containers and keep them tightly closed. Never store them in anything but the original container.

4. Never smoke, eat, or chew anything while applying chemicals.

5. Avoid inhaling sprays or dusts. Wear protective clothing as directed on the label.

6. Remove contaminated clothing immediately and wash the contaminated skin thoroughly if pesticides accidentally contact skin or clothing.

7. Always bathe and change into clean clothing after spraying or dusting. If this is not possible, wash hands and face thoroughly and change clothes. Wash clothing after applying pesticides, never reuse before laundering. Launder this clothing separately from the family wash.

8. Cover food and water sources when treating around livestock or pet areas. Do not contaminate fishponds, streams, or lakes.

9. Always dispose of empty containers so that they pose no hazard to humans, animals, valuable plants or wildlife. Never burn pesticide containers, especially aerosol cans.

10. Read label directions and follow recommendations to keep residues on edible portions of plants within the limits permitted by law.

11. Call a physician or get the patient to a hospital immediately if symptoms of illness occur during or shortly after dusting or spraying. Be sure to take the container or the label of the pesticide used to the physician. If not possible to take label or container, take the EPA number as this will allow identification of the material used.

12. Do not use the mouth to siphon liquids from containers or to blow out clogged lines, nozzles, and so on.

13. Do not spray with leaking hoses or connections.

14. Do not work in the drift of a spray or dust.

15. Confine chemicals to the property being treated and avoid drift by stopping treatment if the weather conditions are not favorable.

16. Protect all non-target areas.

17. Do not use household preparations of pesticides on plants because they contain solvents that can injure plants.

18. Obey the **Restricted Entry Interval (REI)** given on the label and do not work in the area until the time limit is up or other conditions listed on the label have been met after spraying. Be sure the sprayed area is posted with time of reentry as well as the chemical used.

19. Record all pesticide use as directed by law.

Source: Adapted from the Division of Agricultural Sciences, University of California.

---

within a one-mile radius and allow them forty-eight hours to move their hives.

Effects on Wildlife

There is no doubt that pesticides have harmed wildlife, even though such effects may be difficult to document. Some reported losses of fish and fish-eating birds have resulted from the improper or illegal use of pesticides and sometimes from legal use. Certain pesticides most lethal to wildlife, such as DDT, have been withdrawn from general use in the United States.

Some of the adverse effects of pesticides on wildlife have been indirect. For example, in some areas the pheasant population has declined when the weed cover, which had offered protection and nesting places, was cleared out along ditch banks, fencerows, and fallow lands by herbicides. Pesticide labels will give instructions for handling the chemical to protect wildlife and aspects of the environment.

SUMMARY AND REVIEW

The management of weeds, insects and their relatives, and disease pathogens is important to anyone growing plants for a living. In the past, we tried to rely mainly on chemicals to control these pests, but today we realize that we have to take an integrated plant health management (IPHM) approach and use many different methods. In general, these methods are genetic host resistance, chemical applications, cultural practices, biological controls, and government regulatory measures. Understanding the lifecycles of plants and their pests is fundamental to an effective pest control program. Threshold level is the maximum level at which a pest can be tolerated. Most of the time, pest management means keeping the pest population at or below the threshold level, not at zero. Using these methods and concepts in the proper way at the proper time results in effective and sustainable management of the pests and at the same time protects the environment, workers who grow and handle the plants, and consumers who eat or use the plants.

Weed biology and ecology is the study of weeds and their interactions in the physical and biological environment. Classifying weeds by their life cycles (annual, biennial, and perennial) is a useful component in weed management, as is the broader classification of monocot or dicot. Invasive weeds are those species that lack natural enemies and have the ability to infest an area rapidly. Most weeds harm cultivated plants by competing more successfully for resources and reducing the growth or yield of the desired plant. Weed control usually involves removing the weed before it can effectively compete for resources. Annual and biennial weeds rely heavily on seed reproduction. The seed bank is an important factor in the infestation of annual weeds in a field. Biennials also rely on their seed bank. Perennials rely somewhat on a seed bank, but they are more likely to persist year after year by the spread of propagules. Herbicide resistance is becoming more widespread in many weed species, especially if the herbicides with the same mode of action are continuously used. Weed species often thrive better in one set of environmental conditions than others. Knowing these conditions can help a grower plan an effective control strategy based on altering the environment or using biological controls. Weed management includes preventive, mechanical, cultural, biological, and chemical (herbicide) control. Herbicides are classified by their time of application and their mode of action.

Insects and mites present a serious threat to cultivated plants. After the second World War, powerful chemical insecticides were developed that were very effective but also posed a serious environmental threat. Insect resistance to these chemicals reduced their effectiveness. In addition, cultivated and neighboring wild plants depend on or are benefitted by several insects and mites. Populations of these beneficials have to be considered in an insect and mite management program. As a result, insect and mite management programs now usually incorporate the principles of IPHM. Insects and mites are both in the phylum Arthropoda, but have very distinct characteristics, especially as adults. Insects and mites are classified by their feeding habits which can be classified generally as chewing, sucking or both (rasping-sucking). Many sucking insects inject saliva into the plant's vascular tissue that causes parts of the plant to wilt or die. Others transmit diseases from one plant to another.

Although other IPHM practices are commonly used, pesticide use is still very important in controlling insects and mites. Pesticides are classified by their entry into the pest and include stomach, systemic, and contact poisons, as well as inhalants. Another pesticide classification is by mode-of-action. These include the neural disrupters, cellular metabolic disrupters, cell membrane disrupters, hormone disrupters (insect growth regulators), pheromones, and dessicants. The third classification of insecticides is by chemistry. These classifications include inorganic compounds, organic compounds both natural (plant derivatives) and synthetic, oils, and microbial. Several insect and mite pest species are very important on a wide regional or even global scale.

Several vertebrates, most notably rodents, as well as deer, squirrels, gophers, rabbits, and certain birds can be pests in cultivated plants.

Plant disease can be biotic (caused by a living organism) or abiotic (caused by environmental factors). Biotic diseases are considered to be infectious because they can be spread from plant to plant. Abiotic diseases are noninfectious. Disease symptoms include abnormal tissue coloration, wilting, tissue death, defoliation, abnormal increase in tissue size, dwarfing, and replacement of host tissue by tissue of the infectious organism. Pathogens responsible for causing most biotic diseases include viruses, bacteria, spiroplasmas, phytoplasmas, fungi, fungal-like organisms, nematodes, and parasitic higher plants.

Viruses require a host to grow and multiply. They are difficult to control once a plant has become infected. Prevention is the best control measure. Preventive methods include purchasing material that is pathogen-free, using resistant varieties and cultivars, eradicating infected plants, sanitation of the facility and workers, and managing insect vectors. Pathogenic bacteria interact with the plant host and are spread in many different ways. Symptoms include galls and growths, wilts, spots, specks, blights, soft rots, scabs, and cankers. They are spread by splashing water, animals and insects, and people working with plants. Like viruses, bacteria are also difficult to control once plants become infected, so the same methods used to prevent viral infections can be used for bacterial infections. Not all bacteria are harmful to plants; some can be beneficial. Spiroplasmas and phytoplasmas resemble bacteria but are smaller and lack cell walls. They live in phloem and are vectored by sucking insects. Controlling the insects is the most effective way to control these pathogens.

Although most fungi and FLOs are nonpathogenic and some are even beneficial, pathogenic fungi and FLOs are the cause of most plant diseases. One reason for this is their ability to penetrate a healthy plant without the aid of a wound or natural opening. These pathogens can be controlled in part with the use of chemicals, but the other control mechanisms of an integrated plant health management program are equally important and are often more effective, with sanitation and host genetic resistance (when available) being two of the most effective.

Nematodes are microscopic multicellular organisms. The plant parasitic nematodes are distinguished by their stylets used for feeding. Genetic host resistance, particularly using resistant rootstock, is an important method of control of some nematodes, as is the government regulation that forbids the importing of soil or plants with soil on their roots into the United States. Other control measures include chemical fumigation or steam pasteurization of the soil, and the use of some nematicides. Higher parasitic plants include the hemiparasites, such as witchweed and mistletoe, and true parasites, such as dwarf mistletoe and dodder. Hemiparasites possess chlorophyll and can photosynthesize, but they take water and mineral elements from their host. True parasites lack chlorophyll and depend on the host for all nourishment. It is difficult to kill these pests with chemicals; genetic host resistance and cultural methods such as physical eradication of host and pest work best.

There are several practical ways to diagnose plant disorders. These include defining the problem, examining the entire plant community, looking for patterns in signs and symptoms and considering the timeframe for their development, gathering as much other information as possible, then, using a diagnostic clinic if necessary, determine the cause.

The use of herbicides, insecticides, fungicides, and other pest-management chemicals is highly regulated by the government. Their safe use is determined by the EPA and by other state and local government agencies. Legal use of these chemicals requires compliance with the label instructions. In addition to keeping the use legal, compliance helps to ensure that those handling the chemicals (and the plants treated with them) and the environment are not harmed. LD_{50} refers to the chemical's toxicity. The higher the LD_{50}, the lower the toxicity. There are several suggestions for safe pesticide handling given in this book. Details for many of these are also found on the pesticide label.

KNOWLEDGE CHECK

1. What are the methods used in integrated plant health management?

2. What is tolerance in terms of genetic host resistance?

3. Why is knowing the lifecycle of a pest and host plant necessary to managing the pest?

4. What does threshold level mean in pest management?

5. List three things a weed can do to impact a cultivated plant.

6. What is an invasive weed species?

7. Why does tillage often help to control annual weeds but spread creeping perennials?

8. Give two examples of invasive weeds that were deliberately introduced into the United States (though not for the purpose of being weeds).

9. Do weeds that emerge early or late in a crop's development usually cause the most loss?

10. Why is maintaining a healthy, actively growing crop that quickly closes its canopy a weed management strategy for many growers?

11. What is a seed bank and why is it important in controlling annual weeds?

12. Why is applying a systemic herbicide to perennial weeds when carbohydrates are flowing to the underground structures usually very effective at controlling those weeds?

13. How does herbicide resistance (or resistance to any pesticide) occur? How can it be prevented?

14. Why do you want to rotate crops of different life-cycles to help control weeds?

15. What is allelopathy and how might it help make cover crops effective as a weed control strategy?

15. How does mowing help control weeds?

17. What are the four classifications of herbicides based on timing of application?

18. What is meant by exclusion in insect control?

19. What are three different ways pesticides can enter an insect or mite?

20. How do insect growth regulators work?

21. What advantage do neonicotinyls have over other neurotoxins?

23. What are microbial insecticides?

24. To what type of commodity do rodents cause the most damage?

25. What is the difference between a disease and an injury?

26. What is the difference between a biotic and abiotic disease?

27. List four types of plant pathogens.

28. What three conditions must be met to have an infectious disease develop?

29. What are the six steps of the disease cycle?

30. List four categories of plant disease symptoms.

31. What are the four steps of Koch's Postulates and what do they show conclusively about a disease's cause?

32. List two ways that viruses can be controlled.

33. List three symptoms of bacteria infection in plants.

34. List three ways bacteria can be spread.

35. What makes fungal and FLO pathogens different from how other pathogens enter plants?

36. At what time of disease development is the use of fungicides most effect?

37. What are nematodes? How does a pathogenic nematode differ from a nonpathogenic nematode?

38. What are the two classes of parasitic higher plants that are detrimental to their hosts?

39. What are the five steps for diagnosing plant disorders?

40. What is LD_{50}? Is a low number indicative of high or low toxicity?

41. Give three reasons why the safe handling of pesticides is important.

FOOD FOR THOUGHT

1. Describe the things you might consider if you were setting threshold limits of insect damage on the leaves of spinach, potato tubers, the stems of asparagus, tomato fruit, and Easter lily flowers.

2. What would you say to someone who wants to know why the herbicide he used killed the grass in his lawn, but left the dandelions and thistles?

3. As a grower of high quality organic vegetables for which you get a premium price, what methods do you think you would use to control weeds? What methods would you use to control diseases?

FURTHER EXPLORATION

AGRIOS, G. N. 2005. *Plant pathology*, 5th ed. New York: Elsevier Academic Press.

HOROWITZ, A. R., AND I. ISHAAYA (eds.). 2004. *Insect pest management: Field and protected crops.* New York: Springer.

LIEBMAN, M., C. L. MOHLER, AND C. P. STAVER. 2001. *Ecological management of agricultural weeds.* Cambridge: Cambridge University Press.

PEDIGO, L.P. AND M.E. RICE, 2005. *Entomology and Pest Management.* Upper Saddle River, NJ: Prentice Hall

ROSS, M. A., AND C. A. LEMBI. 2008. *Applied weed science: Including the ecology and management of INVASIVE plants,* 3rd ed. Upper Saddle River, NJ: Prentice Hall.

The following websites provide information about integrated plant health management as well as information about specific weeds, insect and mite pests, and diseases. In addition to these online resources, factsheets and the like are also available on the Internet from other institutions, organizations, and

universities. Search using keywords such as name of the pest or integrated plant health management (IPHM) or its older name, integrated pest management (IPM).

International Survey of Herbicide Resistant Weeds,
http://www.weedscience.com/

The American Phytopathological Society,
http://www.apsnet.org/

The Entomological Society of America,
http://www.entsoc.org

USDA Agricultural Research Service,
http://www.ars.usda.gov

*16

General Considerations for Production, Harvest, Postharvest Handling, and Marketing

Margaret McMahon

In the following chapters, you will learn the principles for producing or utilizing several different plant commodities. However, there are some principles and practices that are similar for all or nearly all commodities. These include analyzing the site where the plants will be grown, choosing a production/growing system, harvest and preservation techniques, marketing systems, and transport. This chapter describes those common principles and practices. A distinction between agronomic (field) and horticulture crops should be made at this point. In general, horticulture crops are those crops that require more intense cultivation while growing and greater care during harvest than field crops. They often do not need as much processing as agronomic crops to make them edible, though they can be processed to enhance their salability or value. The distinction is not always clear. Turfgrass is considered by some to be a horticulture crop and by others to be agronomic. The distinction may also depend on how the product is used. For example, fresh market tomatoes are considered to be a horticulture crop whereas processing tomatoes are often included with agronomic crops. The same would be true for food crops harvested at an immature state, such as sweet corn, green beans and peas (i.e., horticulture) and when harvested at a mature ("dry") state (i.e., agronomic). Regardless of classification, starting or renovating a plant growing business requires many decisions. Analyzing the area where the plants are to be grown and the type of production/growing system to use are two of the most important considerations to be made by anyone growing plants.

SITE ANALYSIS OF A PLANT GROWING OPERATION

Successfully growing plants for any reason requires that before starting you carefully analyze the site where you intend to grow the plants. It is preferable to analyze the site before acquiring it.

key learning concepts

After reading this chapter, you should be able to:

- Discuss the factors to consider when doing a site analysis of the area where plants will be grown.
- Explain the differences among traditional, organic, and sustainable production practices.
- Describe how environmental factor management applies to growing plants.
- Discuss the basic principles of harvesting.
- Describe how quality changes after harvest.
- Identify strategies to maintain quality after harvest.
- Explain how the production of crops is linked with consumption through marketing and transport.

Sufficient Area

In addition to immediate production plans, enough area should be available to allow for future expansion. Sufficient area will likely be required for other uses too. For example, purchasing enough land for storage facilities, nongrowing work areas such as packing and processing, lunch- and restrooms for employees, coolers and driers, and the like would apply in many situations. A nearby rural–urban interface could impact the ability to expand in the future.

Water Supply

An adequate supply of good quality water must be available through the growing season either from natural precipitation or, if irrigation is to be used, from surface, subsurface, or municipal sources. If irrigation is needed, the type of irrigation system to be used and the cost of its installation and maintenance has to also be considered.

Physical Features

The physical features, or topography, of the property is important. Is it flat? Sloped? Rolling? Is it composed of steep hills? Is it subject to flooding? These and other questions about the topography of a production area need to be addressed. Topography influences how the soil is tilled and what precautions must be taken to avoid erosion and other sources of soil degradation. Flat fields with poorly drained soil conditions may warrant use of surface drainage (ditches) and subsurface drainage (tiling). Microclimates of warm and cold zones formed by changes in elevations and other geographic or architectural features have to be considered.

Soil Type

Soil type is important because it determines the nutrients the soil has or can hold as well as its ability to hold water. It also determines how the root system will develop. There may be several different soil types in a single location. You have to know where each is and its characteristics because you will most likely have to use different growing strategies for each type.

Climate

Another consideration is the climate of the area. You may have to provide protection from heat or sun in hot regions. Protection during the winter in cold climates may be needed if you are growing perennials such as strawberries or landscape plants. Annual drought and flood patterns have to be considered as well as the potential for hail. Prevailing wind direction influences many growing practices such as row orientation, greenhouse structure orientation, where to place windbreaks, and the like. The possibility of extreme weather events such as 100- and 500-year floods, tornadoes and hurricanes, and extreme summer highs and winter lows is another very important consideration.

Economic and Social Factors

The site analysis also includes economic factors, with probably the biggest initial expense being land cost. For many crops, such as in a nursery or orchard the plants have to be grown for several years before they begin to generate a return on investment dollars so ways to finance the operation during that time have to be considered.

Labor is another very important consideration when looking at economics, with the availability of reliable labor often being the major issue. The wages and benefits you will have to offer to be competitive for labor in the region will be a significant part of the budget. In general, labor accounts for a third of the costs of business.

Transportation

Transportation is another factor when analyzing a site. You have to consider the equipment you will be using and how you will move it from one area to another both off- and on-site. Equipment access to all parts of the operation is essential and the on-site roadways and walkways necessary for access must be factored into the amount of area available for production. If you are selling a product, you have to consider how you will move it to the market. In the case of many crops, trucks, along with rail and waterways, play an important role in transporting the product to distant locations. Having access to highways, railroad yards, and docks is important. Other crops, especially those that are perishable (short shelf-life), are shipped by air so a nearby airport may be beneficial.

Parking for those workers that drive to work has to be provided. If the workers do not drive (for example, Amish and migrant workers), transportation has to be made available. For many producers, local retail markets are important—and purchasing or leasing trucks large enough to transport the product to those markets may be necessary. If a retail business is part of the on-site operation, allowing enough space for the sales facility and customer parking is crucial. In addition, a retail operation must be located close enough to customers for them to be willing to travel there.

OVERVIEW OF PRODUCTION SYSTEMS

Once a site is analyzed, then a strategy for growing the plants has to be developed. As you have been learning throughout this book, growing plants either for commercial or personal use and enjoyment requires managing the ecosystem that surrounds the plants. These managed ecosystems are also called production or growing systems when applied to commercial production. There are several approaches to managing ecosystems.

Conventional or Traditional Production Systems

Until the late 1900s, the conventional practices for growing plants in developed countries relied heavily on high-resource input. One of the resources most heavily used was nonrenewable energy in the form of fossil fuel. Other major inputs included synthetic fertilizer and equipment for soil cultivation, fertilizer and pesticide applications, and harvesting, as well as irrigation, drainage, and chemicals other than fertilizers. The benefits of these inputs were a reduction in labor and increased productivity per acre/hectare. However, over time, hidden costs, especially to the environment, were discovered. Soil degradation and erosion and nutrient/chemical runoff and leaching had a detrimental impact on the environment. For example, hypoxia in Gulf of Mexico has been linked to excessive application of fertilizers by farmers and homeowners in the Midwest and along the Mississippi River. The loss of wetlands and the benefits they provide was also costly. It should be noted that the value of a crop in part determines inputs used. For example, fungicides are regularly used during production of high-value vegetables, but in grain production the routine use of fungicides is cost prohibitive.

As a result, conventional high-input farming is still commonly practiced, but now more ecologically sound practices are usually a part of it. As you learned in earlier chapters, those practices included strip farming, contour farming, the use of greenways and waterways, and no- or low-till farming. Alternatives to synthetic fertilizers, as well as chemicals to control insects, diseases, and weeds, are used whenever possible by environmentally conscious growers. When chemicals have to be used, their selection is based on not only their efficacy and cost, but also having low toxicity and low potential to harm the environment or cause other damage. The widespread adoption of transgenic crops that contain proteins toxic to select insect pests has reduced the use of insecticides that often exposes human applicators to highly toxic chemicals and damages the environment by killing nontarget species. Similarly, the use of herbicide-resistant crops has allowed the use of less-toxic chemicals to maintain weed control in those fields. On a broader scale, too, fungicides and insecticides are being applied to seed prior to planting. That minimizes broadcast application of chemical pesticides above ground after crop emergence.

The use of computers and global positioning systems (GPS) and geographic information systems (GIS) have been in use for several years in traditional production systems. Computer analysis of a field, along with GPS and other sensors, enables the precise application of chemicals and fertilizers only in the areas where and when needed, greatly reducing the risk of runoff and leaching. Sensors that measure yield during harvest allow machinery to run more efficiently and conserve fuel (Fig. 16–1).

The preservation and restoration of natural and creation of new wetlands are practices that are being incorporated into the conventional production and utilization systems for plants (Fig. 16–2). These wetlands provide habitat for many plant and animal species, some of which are endangered. Wetlands can also provide a source of water for the cultivated plants during times of drought.

Figure 16–1

Yield sensors connected to onboard computers automatically adjust the speed of a harvester to increase harvest efficiency.

Source: USDA Agricultural Research Service Image Gallery, http://www.ars.usda.gov/is/graphics/photos/

Figure 16–2
Restored wetlands on a farm. Source: USDA Agricultural Research Service Image Gallery, http://www.ars.usda.gov/is/graphics/photos/

Other high-input systems include greenhouses, nurseries, golf courses, and athletic fields. In most cases, these systems now use several of the aforementioned techniques to reduce environmental impact. Where irrigation is used, recycling the irrigation water is becoming common. Runoff is being captured for recycling on-site or being filtered through greenways to reduce the risk of fertilizer and other chemical contamination in the environment.

Organic Growing Systems

Organic growing systems are being used to reduce dependence on nonrenewable resources, chemicals, and other inputs. This practice is ecologically based and relies on naturally derived nutrient sources and the use of cultural practices and biological agents to control pests and diseases. Fertilizers not directly from plant and animal sources and many chemicals, especially those derived from fossil fuels cannot be used. The USDA strictly regulates the standards for organic production. According to its USDA's **National Organic Program (NOP)** standards:

> A producer of an organic crop must manage soil fertility, including tillage and cultivation practices, in a manner that maintains or improves the physical, chemical, and biological condition of the soil and minimizes soil erosion. The producer must manage crop nutrients and soil fertility through rotations, cover crops, and the application of plant and animal materials. The producer must manage plant and animal materials to maintain or improve soil organic matter content in a manner that does not contribute to contamination of crops, soil, or water by plant nutrients, pathogenic organisms, heavy metals, or residues of prohibited substances.

To become a certified organic grower, stringent requirements set by the USDA must be met. After certification, a crop produced at this site can be labeled "organic."

Because of the difficulty and expense of becoming certified, some growers use organic practices but do not become certified and cannot label their products as organic. However, they can describe their production methods as being organic based.

Sustainable Production Systems

More recently another way of looking at a growing system is its sustainability. In broad terms, sustainability is defined as "meeting the needs of the present without compromising the ability of future generations to meet their needs." Sustainability is not new. Soil conservation and other environmental stewardship practices mentioned earlier in this book are sustainable practices. One of the oldest and most sustainable crop systems is "slash and burn" agriculture, where small area of forest is burned to release nutrients into the soil. Crops are produced for a few years, then the land is left alone (**fallow**) to return to forest for several decades or longer, then the process is repeated. Nonrenewable inputs are low. However, this system causes environmental degradation when demand for the crops increases and land must be re-used before the forest regrows. Without reforestation, soil erosion and nutrient loss occur from the bare soil decreasing crop productivity. Native species are often replaced by invasive species that may further disrupt the ecosystem. Restoring the ecosystem requires enormous inputs of energy and resources.

The current concept of sustainability takes a broad look at the production system and its long-term effects and includes many aspects beyond soil conservation and organic standards. For example, how the labor force is treated and how local communities are served, along with generating a profitable income for the producer, are part of sustainability. The USDA's **Sustainable Agriculture Research and Education (SARE)** Program lists the primary goals of sustainable agriculture as:

- Providing a more profitable farm income.
- Promoting environmental stewardship, including:
  - Protecting and improving soil quality.
  - Reducing dependence on nonrenewable resources, such as fuel and synthetic fertilizers and pesticides.
  - Minimizing adverse impacts on safety, wildlife, water quality, and other environmental resources.
- Promoting stable, prosperous farm families and communities.

The Scientific Certification Systems (SCS) and the Leonardo Academy are in the process of establishing criteria for certification in sustainability. Currently, the eight key elements of sustainability identified by SCS are:

- Sustainable crop production
- Ecosystem management and protection
- Resource conservation and energy efficiency
- Integrated waste management
- Fair labor practices
- Community benefits
- Product quality
- Product safety and purity

SCS has defined sustainable crop production as being "practices that: (1) build and maintain a healthy agroecosystem, based on healthy soil structure and functioning; (2) preferentially employ biological, mechanical, and cultural methods to control pest and disease vectors; (3) minimize agrochemical inputs, favoring the use of reduced risk or NOP permitted agrochemical options; and (4) phase-out those agrochemical inputs that pose significant acute and chronic risks to human health or ecotoxic risks to the environment."

In many cases, growers may have implemented several sustainable practices over the years without realizing that those practices fall in the category of sustainability. Those growers may be able to become nationally certified sustainable when certification is available by implementing only a few more sustainable practices. In addition to a national certification, there are several organizations such as the Environmental Protection Agency, Veriflora, MPS, and the Audubon Society that already offer sustainable certification in specific areas such as irrigation and water-use efficiency, floral, fruit, and vegetable production, and wildlife habit preservation on golf courses.

Many growers do not use one type of system exclusively. These growers use a blend of practices from all three systems to produce plants in a way that works for them. An exception would be those organizations that are certified organic or sustainable, which have to comply with prescribed laws, regulations, and criteria.

MANAGING ENVIRONMENTAL FACTORS

The management of environmental factors is discussed in detail elsewhere in the book, but a summary follows.

Soil, Water, and Nutrients

Soil, water, and nutrient management includes both long- and short-term practices. Initially fields may have to be leveled, irrigation and drainage systems installed, and fertilization equipment purchased and, in some cases, permanently installed. Frequent decisions have to be made regarding the application of water and fertilizers. The type of plants being grown, their stage of development, and a targeted harvest or ready-to-use date determine how water and nutrients are managed. (See Chapter 14 for details of soil, water, and nutrient management.)

Temperature

For field crops, temperature management is limited to a few practices such as spraying plants with water during a frost period to protect them from freezing. As the water freezes, it gives up heat that is absorbed by the plant and keeps ice from forming in the cells. Instead, temperatures are frequently used to predict events such as harvest date. The concept of **growing degree days (GDD)** is used for many crops to predict when they will be at a harvestable stage (field crops). Growing degree days are calculated as the average daily (24 hr.) temperature above a baseline (usually 50°F). Each degree above the baseline is considered a degree day. For example: a day that has an average temperature of 65°F would be the equivalent of 15 growing degree days. Days with average temperatures below 50°F would count as zero. For many crops, we know how many GDDs are required to reach harvest so we can predict, based on daily temperature recordings, when the crop will be ready to harvest when the time comes near. For example: if corn requires 1,500 GDDs and there are already 1,250 GDDs accumulated, only 250 more are needed. It would take only 10 GDDs for the corn to be ready to harvest if the next 10 GDDs averaged 75°F.

Temperature management is a major factor in greenhouse production. Temperature is controlled by the use of heating and cooling systems. As with other environmental factors, the type of plants being grown, their stage of development, and a targeted harvest or ready-use-date determine how temperatures are managed. Some crops, such as Easter lilies, use a modified system of growing degree days to precisely time the crop to be ready for the Easter market window (Fig. 16–3). (Chapter 24 explains temperature management in greenhouses in greater detail.)

Light

Although most light manipulation is done in greenhouses, some principles are used in outdoor crops. Row orientation and plant spacing are the most common

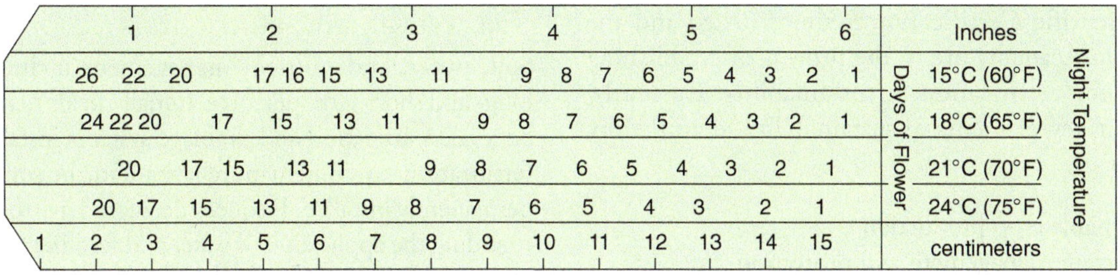

																			Inches	
		1			2		3		4		5			6						
	26	22	20		17	16	15	13	11		9	8	7	6	5	4	3	2	1	15°C (60°F)
		24	22	20		17		15		13	11		9	8	7	6	5	4	3	2	1	18°C (65°F)
			20		17	15		13	11		9	8	7	6	5	4	3	2	1	21°C (70°F)
		20	17		15		13	11	9	8	7		6	5	4	3		2	1	24°C (75°F)
1	2	3	4	5	6	7	8	9	10	11	12	13	14	15						centimeters

(Side labels: "Days of Flower", "Night Temperature")

Figure 16–3

This device measures Easter lily buds to determine what temperatures to use to have the buds open (a temperature-controlled process) for Easter. Source: Margaret McMahon, The Ohio State University.

ways to manipulate light outdoors. Row orientation can influence how the light penetrates the rows as the sun moves across the sky. The influence varies with latitude and season. In general for temperate regions at and below 40° north, east–west row orientations are better in the winter and lower in the summer, while above 40° north, north–west orientations are better. The space each plant is given greatly influences the amount of light it receives. Seeding/planting density and row spacing are carefully measured to provide enough space for good growth and development of the crop. These measurements vary by crop and location. Planting dates for many crops depend on there being enough accumulated solar energy after the planting date for the plant to develop to maturity.

Even the most efficient artificial light sources convert only a fraction of electrical energy to photosynthetically active radiation. Thus, it is expensive to achieve light intensities similar to sunlight and almost impossible on a large scale. However, **high-intensity discharge (HID)** lights are used in greenhouses to promote growth of high-value crops.

On the other hand, some crops, particularly ornamentals, can be injured by high light intensities and shade is used to prevent this type of injury. Shade can come in the form of wooden laths or screens covering plants out in the open. In a greenhouse, a paint-like shading compound applied to the glass or a shade cloth installed over or inside the greenhouse are used to reduce light.

Some plants are photoperiodic short- or long-day plants. For field crops such as the short-day soybean or long-day spinach, growers have to be aware of how daylength (photoperiod) changes during the growing season will affect the development of the crop. In greenhouses, photoperiods are manipulated with lights and blackout cloth to time the development of photoperiodic crops to be ready for specific dates.

HARVEST

Harvest is final step of production for many plants. The functions involved in the harvesting of plant products are varied and highly dependent on the particular commodity. These functions can include digging, picking, cutting, lifting, separation, grading or sorting, cleaning, conveyance, loading, and so on (Fig. 16–4). Such functions depend on various factors related to the crop, which could include field conditions at the time of harvest, crop maturity, crop use (fresh market or as a processed product), plant part or portions involved, and crop worth. During harvest, care must be taken to not damage the product so much that it loses value. Harvesting was originally done by hand and today many fresh market vegetables and fruit and nearly all cut floral crops are still picked by hand (Fig. 16–5). However, the harvesting of most other crops has been mechanized.

Mechanization of crop harvesting began with the cereal grains. The forerunner of the grain combine was the hand sickle and flail used as far back as 5,000 years ago.

Figure 16–4

Mechanized cranberry harvest Source: USDA Agricultural Research Service Image Gallery, *http://www.ars.usda.gov/is/graphics/photos/*

Figure 16–5
Hand harvesting cut flowers.
Source: Margaret McMahon, The Ohio State University.

Figure 16–6
Cotton harvesting. Source: USDA Agricultural Research Service Image Gallery, *http://www.ars.usda.gov/is/graphics/photos/*

The mechanical reaper, threshing (the removal of the inedible chaff from the grain) machine, and combine (which combines reaping and threshing) followed. Other examples of significant machinery development were the cotton picker (Fig. 16–6) and cotton gin which separates the seeds from the cotton fibers. Harvest mechanization, whether for grain, potatoes, sugar beets, carrots, or processing tomatoes (Fig. 16–7) has greatly reduced the level of human labor previously needed in the harvest of these crops. However, mechanization has dramatically increased the need for non-renewable fuel.

Generally, harvest mechanization was developed for crops intended for processing long before crops grown for fresh market purposes. For example, tomatoes for processing are machine-harvested while fresh market tomatoes along with table grapes, strawberries, and oranges are still often harvested by hand to limit physical injury. A mechanical picker has been developed for sweet corn that removes the ear from the stalk without injury (Fig. 16–8), though much of the sweet corn crop is still picked by hand. Similarly, mechanical harvesters are used for leaf lettuce but much of each crop is also picked by hand (Fig. 16–9). Mechanical shakers are used to harvest nuts and some fresh market fruit such as cherries from trees. By shaking the tree, the fruit is dislodged and falls onto a large platform or tray which is heavily padded when used for fruit that bruises easily. Potatoes for both processing and fresh market use are dug, cleaned, and sorted by machines. Although most cut flowers are harvested by hand there are machines that aid harvest

A

B

C

Figure 16–7
(A) Barley harvester. (B) Corn picker. (C) Carrot harvester.
Source: USDA Agricultural Research Service Image Gallery, *http://www.ars.usda.gov/is/graphics/photos/*

Figure 16–8

Mechanical sweet corn harvester. The ears are pulled from the stalk as the stalk moves the harvester.

Source: Margaret McMahon, The Ohio State University.

by stripping leaves, trimming stems, and sorting flowers such as roses and carnations by size or quality (Fig. 16–10).

Technology incorporates computers and electronic sensors in harvesting equipment to greatly improve harvest efficiency. Global positioning systems allow field equipment to be controlled precisely as it moves through a field (Fig. 16–11). Yield sensors monitor the instantaneous yield and slow the harvester down if the yield is so high that loss is occurring because grain is being left in the field (Fig. 16–1 on page 343). Conversely, if the yield is low, the harvester speed is increased to save time and reduce soil compaction. Moisture sensors measure and record the moisture content of grain as it is harvested. Harvest can be stopped if the grain is too moist, thus preventing a loss of quality and value of the grain or the need to run driers unnecessarily. Sensors can be used to sort a product, tomatoes and oranges for example, by color or appearance. Other sensors sort

Figure 16–9

A. Mechanized leaf lettuce harvest.
B. Hand harvesting leaf lettuce.

Source: USDA Agricultural Research Service Image Gallery, http://www.ars.usda.gov/is/graphics/photos/

A

B

Figure 16–10

This machine strips the leaves from cut flower stems.

Source: Margaret McMahon, The Ohio State University.

Figure 16–11

This tractor is equipped with a GPS device that allows precise management of cultivation techniques such as fertilizer applications. Source: USDA Agricultural Research Service Image Gallery, http://www.ars.usda.gov/is/graphics/photos/

A   B   C

Figure 16–12
(A) Machine that replaces empty cell with another seedling. (B) Tray of seedlings before mechanical replacement.
(C) A tray that has had any empty cells refilled by machine. Source: Margaret McMahon, The Ohio State University.

by weight or volume. In greenhouses, transplanting machines using electronic sensors automatically find and replace plants missing from packs before they are shipped (Fig. 16–12).

POST-HARVEST HANDLING

The ability to store food for later consumption was a critical necessity in the development of human societies. This ability enabled people to move beyond the limited area of the earth where food is continuously available. Life in North America and Europe would have been impossible without the ability to store food for the winter. Transportation is another aspect of post-harvest handling of crops that has gained in importance as people moved from the country, where they produced their own food, to an urban existence, where food comes from the store. Today, locally produced food is supplemented with crops from other regions and countries. The urban lifestyle has also led to demand for nonfood crops. Cut flowers, garden plants, and even turf require specific attention to post-harvest handling if they are to reach the consumer in good condition. Although we may focus on the growing of crops to the point of harvest, the ability to get them from the production field or greenhouse to the consumer is equally critical and presents a distinct set of challenges. Crop losses after harvest remain unacceptably high, and limiting post-harvest deterioration would make a major contribution to improving the world's food supply. How plants are handled after production is often the responsibility of or a source of concern for the grower.

Post-Harvest Deterioration

Crops that are harvested represent all parts of the plant at many different developmental stages (Table 16–1). Crops may be actively growing parts of the plant, such as asparagus stems or spinach leaves; they can be dormant structures with natural longevity, such as seeds, bulbs, or tubers; or they may be terminal structures, such as fruits and flowers that are naturally programmed to senesce and die. Deterioration can occur for many different reasons: physical, chemical, or biological. Biological deterioration may involve physiological processes in the crop itself or pests and pathogens that may be the same as those in the growing crop, but the post-harvest environment can be conducive to a whole new set of pests too.

Physical deterioration often begins at harvest, which for most crops involves separation of the part of the plant that will be marketed and consumed. Damage can occur even when the whole plant is harvested. Vegetables such as radishes and carrots may be uprooted, and ornamental plants may be dug from the field, resulting in breakage and loss of much of the root system. Container-grown plants may suffer the least damage through harvest, but there is still likely to be accidental breakage and loss of plant parts as containers are lifted from the growing area to begin their journey to the consumer.

Many fruits and seeds are harvested by taking advantage of a natural separation between them and the parent plant. This natural separation occurs at abscission zones that develop in fruit pedicels, ovary walls, and placentas. The cell walls in these zones weaken as the structure matures, allowing for easy

Table 16–1

EXAMPLES OF THE PLANT PARTS THAT ARE HARVESTED AND THE STAGE OF DEVELOPMENT AT WHICH THE PART IS HARVESTED FOR SEVERAL CROPS

Plant Part	Stage of Development	Examples
Whole plant	Immature	Radish, carrot, bedding plants, turfgrass
Whole plant	Immature	Silage (corn, soghum)
Whole plant	Mature	Ornamentals (trees, shrubs, potted plants)
Leaves	Immature	Brussels sprouts, lettuce
Leaves	Mature	Collard greens, spinach, rhubarb
Stems	Immature	Asparagus
Stems	Mature	Cut-foliage ornamentals
Roots	Immature	Radish (includes hypocotyls)
Roots	Mature	Rutabaga, carrot
Flowers	Immature	Broccoli, cauliflower, artichokes
Flowers	Mature	Cut flowers
Fruits	Immature	Green beans, okra
Fruits	Mature	Tomatoes, squashes, apples, etc.
Seeds	Immature	Green peas, sweet corn*
Seeds	Mature	Pulses, grains*

*Cereals are botanically fruits with a thin, dry ovary wall.
Source: Michael Knee.

separation and a wound area that heals naturally. Abscission can be promoted with growth regulators to facilitate harvesting of fruits such as apples, tomatoes, and coffee. On the other hand, it may be desirable to retard abscission to avoid loss of crop.

Many grain crops were selected long ago for seed retention in the ear or head so that the whole stem could be harvested and the grain separated by threshing. Although dry seeds and grains are much more robust than most other crops, threshing is of necessity a force-ful process and causes damage. Grain damage or break-age may lead to insect and diseases infestations and result in reduction in quality and value. Other fruits and seeds are harvested in an immature state when the abscission zone is not fully developed and the support-ing or enclosing structure must be cut or broken, which results in wound areas. These wounds also occur on ornamental and edible stems, foliage, and flowers that are often harvested by cutting the stem or petiole.

Physical damage at harvest time may be unavoid-able, whereas avoidable damage tends to occur during subsequent handling and storage. Depending on the commodity and the handling procedure, damage can include abrasions, cuts, cracks, and bruises. These defects occur through contact between one item of pro-duce and another, and between produce and the edges or surfaces of containers and handling equipment. The kinetic energy of a moving crop item tends to cause damage whenever it experiences sudden deceleration by contact with something else. The damage is directly proportional to momentum (the product of mass and velocity) and inversely proportional to the surface area of contact. Wide contact areas cause less damage than narrow edges or points for a given energy of impact. Of course, the physical properties of the crop itself have a major influence on susceptibility to damage. These physical properties include its size, shape, structure, and the fundamental physical properties of its tissues, including stiffness, hardness, elasticity, and viscosity. Many crops can absorb a certain amount of pressure if the load is applied slowly rather than when it is sudden. However, eventually, the pressure causes a bruise, crack, or hole to form in the product.

Much of the avoidable physical damage occurs as crops are being moved along conveyors for packing, sorting, and storage. For delicate crops, such as straw-berries or cut flowers, this damage can be minimized by harvesting directly into containers in which they remain until the point of sale. Many cut flowers and potted ornamentals are placed in sleeves to protect them during shipping (Fig. 16–13). The sleeves should allow good air exchange to prevent the buildup of eth-ylene inside the sleeve. Damage can also occur in pack-aged produce through vibration or sudden changes of direction during transport. Good design of packaging, handling, and transport systems can minimize mechan-ical damage, but it remains a major cause of crop loss after harvest. Some of the loss from physical damage occurs indirectly because cuts and bruises provide entry points for disease organisms.

Figure 16–13
Plastic sleeves are placed around these pots of poinsettias to protect them from damage while being transported. The sleeves are perforated to allow air to circulate around the plant. Source: Margaret McMahon, The Ohio State University.

Figure 16–14
Sprouting while in storage is a problem with many tubers such as this potato. Source: USDA Agricultural Research Service Image Gallery, *http://www.ars.usda.gov/is/graphics/photos/*

Postharvest handling is greatly simplified if the harvested part is a mature structure that is naturally involved in long-term survival of the plant. Examples include dry grains, onions, and potatoes. However, developmental changes can still decrease the value of the crop. Seeds lose the ability to germinate over time because of subtle changes in cell structure and genetic material. This loss is of less consequence if the seed is to be used for food than if it is required for propagation. Even in a food crop, deterioration may be associated with the development of off-flavors that render it unpalatable. Another cause of loss in perennating structures such as seeds, bulbs, and tubers is germination or sprouting. Germination of seeds is usually avoided by maintaining low temperatures and dry conditions. Sprouting in bulbs and tubers may be controlled with chemical inhibitors, as with potatoes (Fig. 16–14) and onions, or by physical removal of the growing point, as with carrots and rutabagas. If the bulbs or tubers are required for regrowth, natural sprouting may simply be tolerated or manipulated by adjusting the temperature and/or light regime during storage.

Immature structures, such as green peas, sweet corn, and green onions, are among the most perishable crops. Other highly perishable items include immature stems, such as asparagus, and flowers. All of these products are harvested when they are importing food from the rest of the plant to sustain rapid growth. When this process is interrupted, their metabolism remains rapid; but without the constant supply of nutrients, spoilage tends to be rapid. Spoilage often involves yellowing and premature senescence. Breakdown of chlorophyll and abscission of leaves are common responses of green plant material of all kinds when held in the dark or low-light conditions. These problems apply as much to cabbage and lettuce on the supermarket shelf as to ornamental foliage plants during transit from Florida or to evergreen azaleas in cold storage for vernalization.

Mature fruits and flowers are not importing food as rapidly as at immature stage, but they would not naturally persist much beyond the time of pollination in flowers or seed maturation in fruits. Many fruits and flowers show a climacteric pattern of development, which involves a burst of respiration and other metabolic changes that are stimulated by ethylene synthesis in the fruit, which in turn hastens ripening. The ethylene produced by these plants can hasten ripening in ethylene-sensitive produce stored nearby. Controlling ethylene is an important factor in postharvest handling.

Although insects and other field pests may persist and cause damage on fresh plant produce after harvest, they present few special postharvest problems for these crops. However, insects are a major problem for dry seed crops, and several species of beetles are specifically associated with storage of cereals and pulses (Fig. 16–15). Rodents are a problem in open storage systems for seed crops (Fig. 16–16) and in perennial ornamentals held in plastic houses for the winter. Fungi attack crops of all kinds, and the postharvest environment is often conducive to fungal infestations. Infection of grain by various preharvest and storage molds may result in mycotoxins. Some of these fungi produce mycotoxins such aflatoxin and vomitoxin, which can be toxic to humans and animals. The fungal species involved are often opportunistic pathogens that do not cause major problems in the field. Frequently they are saprophytes

Figure 16–15

Insects such as these grain borers are a problem with stored grains. Source: USDA Agricultural Research Service Image Gallery, *http://www.ars.usda.gov/is/graphics/photos/*

Figure 16–16

Rodents are a serious problem in open grain storage structures such as this one. Source: USDA Agricultural Research Service Image Gallery, *http://www.ars.usda.gov/is/graphics/photos/*

that colonize on dead plants, persist on equipment and containers, and attack the crop through wounds or other damage occurring during handling and storage. Whereas growing crops tend to be attacked by specific fungal pathogens, after harvest a very wide range of crops can be attacked by a few species of *Botrytis* and *Penicillium. Penicillium* can be especially devastating because some species generate ethylene. A few postharvest bacterial diseases cause problems, particularly on vegetables, flowers, and herbaceous ornamentals.

Quality

After harvest, crop quality generally declines and retarding loss of quality becomes a major objective. Quality can be defined as "fitness for purpose"; clearly, the quality standards required for soybeans as animal feed are different from those required for the manufacture

of tofu. For most crops, market quality standards are established by the USDA and other bodies such as the American Association of Nurserymen. These organizations set minimum standards for size, shape, color, and freedom from damage. However, consumers have an interest in more subtle aspects of quality that are not easy to identify in a market environment. Harvested crops are inspected to make sure they meet quality standards (Fig. 16–17).

Nutritional quality can be a concern for consumers, and federal law requires food packaging to carry nutritional information (Fig. 16–18). Similar information is recommended but not required for fresh produce. The nutritional content of a particular kind of crop at harvest can be highly variable, and it can change after harvest, particularly for critical components such as vitamins A and C. So when values are presented for nutritional content, they have to be generic and represent minimal expectation for the crop at the end of the marketing chain.

Safety is another aspect of crop quality that is very important. Samples of crops passing through the markets are analyzed to ensure that pesticide residues fall below accepted levels, and retailers sometimes give assurances of the absence of detectable residues. As a result, cases of pesticide poisoning from commercially produced food are almost unknown. However, there have been many instances of food poisoning from microbial contamination of plant foods. As with pesticides, samples are routinely screened for contamination,

Figure 16–17

Inspectors such as this man inspect harvests to make sure the product meets quality standards. Source: USDA Agricultural Research Service Image Gallery, *http://www.ars.usda.gov/is/graphics/photos/*

Nutrition Facts

Serving Size: 1 oz (28g)

Amount Per Serving	
Calories 78	Calories from Fat 3

	% Daily Value*
Total Fat 0.34 g	1%
Saturated Fat 0.05 g	0%
Trans Fat	
Cholesterol 0 mg	0%
Sodium 151.96 mg	6%
Potassium 34.02 mg	1%
Total Carbohydrate 15.79 g	5%
Dietary Fiber 0.62 g	2%
Sugars	
Sugar Alcohols	
Protein 2.58 g	
Vitamin A 0 IU	0%
Vitamin C 0 mg	0%
Calcium 24.38 mg	2%
Iron 0.4 mg	2%

Figure 16–18

A label that carries nutritional information is required to appear on every packaged food Source: USDA Agricultural Research Service Image Gallery, http://www.ars.usda.gov/is/graphics/photos/

Figure 16–19

Food producers often use human testers to evaluate the product for specific sensory attributes. Source: USDA Agricultural Research Service Image Gallery, http://www.ars.usda.gov/is/graphics/photos/.

but harmful organisms do get into the food supply as happened in the United States in 2006 with a serious outbreak of *Escherichia coli* (*E. coli*) on spinach and again in 2008 on tomatoes, as well as *Salmonella* on tomatoes and peanuts in 2008. The bulk handling and wide distribution methods involved in modern crop marketing are also vulnerable to the spread of waterborne organisms such as coliform bacteria. Emphasis on microbial food safety has become a very important focus of food producers and processors.

All of the quality attributes considered so far are objective: they are capable of being observed or measured, even if this is not commonly done. However, consumer satisfaction also depends on subjective quality attributes. This kind of quality is sometimes called hedonic, which is a word related to *hedonism,* the pursuit of pleasure. Depending on the crop, these attributes can include the aesthetic appreciation of the form and color, the scent or aroma, and eating quality. Eating quality is a complex blend of all the senses, from the visual impression as we are about to bite the item, the texture at the first bite and on subsequent mastication, the taste on the tongue, the flavor perceived as

aromatic compounds move from the nose to the mouth, and even the sounds made during eating. Maintaining subjective quality attributes from harvest to the consumer is as important as maintaining the objective attributes (Fig. 16–19).

Postharvest Treatments

Specific measures must be taken to maintain crop quality from the time of harvest onward. Handling procedures should minimize mechanical injury. Other procedures are commonly applied to prevent physiological deterioration and attack by pests or pathogens. For some crops, a curing period during which they are held at high temperature and low humidity is advantageous to allow wounds incurred in harvesting to heal. Examples of such crops include potatoes and onions. Low moisture content is critical for successful storage of dry seed and grain crops. If it is higher than about 13 percent, hot air is forced through the grain to dry it.

Chemical treatments are used to prevent various kinds of deterioration on a wide variety of crops. Fumigants are used on grain crops to eradicate pests; fungicides are used on many fruits and vegetables; and sprouting inhibitors are applied to vegetables; such as potatoes and onions. Although these chemicals were very helpful in minimizing crop losses, public opinion increasingly opposes their use and many have been withdrawn. Irradiation has been investigated and advocated as a replacement for many of the chemical treatments, but it has not been generally approved or widely used. Currently, heat treatment at temperatures above 45°C for a few seconds or minutes is being investigated for insect and fungus control on many fruits and vegetables.

Edible oils or waxes are still applied to many fruits and vegetables to improve appearance and reduce water loss. Application is usually preceded by thorough cleaning with detergent; however, deterioration may still occur after storage if crops are not to be marketed immediately.

STORAGE PRINCIPLES

Simple Storage

In primitive communities, crops were, and still are, stored under a cover of leaves or earth in pits or heaps on the ground. Such simple storage systems have been used for grains, root vegetables, and fruits such as apples. The system is best suited to root vegetables and can be used for low-cost storage of such crops, particularly when intended for animal feed. Losses of grain crops from pests and dampness in simple storage can be reduced by raising them off the ground and providing a waterproof cover. This kind of storage can still be seen occasionally on North American farms today, but most grain is stored in bins or a type of silo called grain elevators (Fig. 16–20). Silage is stored in silos that are air-tight.

Cooling

Cooling generally slows down respiration and other metabolic processes in crops and in the pests and pathogens that affect them, which means that most kinds of deterioration are delayed. As a general rule, the rate of biological processes doubles for every 10°C rise in temperature. Therefore, the storage life of many crops is increased by reducing the temperature. For example, sweet corn might last two days at 20°C and eight days at 0°C. For most temperate crops, the lower limit of storage temperature is set by their freezing point, which is generally just below 0°C. However, crops of tropical and subtropical origin are injured by temperatures below 10°C. So avocados, bananas, tomatoes, and potatoes must be held above a threshold temperature in the range of 5°C to 10°C which decreases the amount of time they can be stored.

Many fruits, vegetables, and cut flowers can get very warm when they are growing. This heat is called "field heat." It is critical that the field heat be removed quickly after harvest. Field heat can be reduced if crops are harvested in the early morning when temperatures are naturally low. However, many crops cannot be harvested during the cool part of the day. In this case, mobile coolers are taken into the field to start cooling the crop as soon as it is harvested. Most coolers rely on convective removal of heat by air or water (hydrocooling). Low-bulk items can be cooled by placing them in a refrigerated room (Fig. 16–21), but room cooling is generally too slow for packaged crops. Forced-air cooling is used for delicate, low-bulk crops such as leafy vegetables, berry fruits, and flowers (Fig. 16–22). Packages of product are stacked against a partition in a refrigerated room or container. Powerful fans behind the partition draw cold air through tubes or vents that match perforations in the packages,

Figure 16–20

Extended storage of dried seed grains is commonly made in large "grain elevators," as pictured here. These facilities often have the capability to dry seed further if it is necessary to lower the moisture content. Source: USDA Agricultural Research Service Image Gallery, *http://www.ars.usda.gov/is/graphics/photos/*

Figure 16–21

An example of a large-capacity cold-storage room widely used to store many fruits and vegetables. These rooms are not as effective as other methods in the initial cooling of the product, but they are useful for maintaining the temperature and storage of products already cooled.

Source: Margaret McMahon, The Ohio State University.

Figure 16–23
Broccoli packed in ice for shipping. Source: Dennis Decoteau, The Pennsylvania State University.

Figure 16–22
Air that has been cooled is drawn into this forced-air cooler through circular vents (upper right of photo) in the wall. After the cold air flows through the produce, it exhausts through holes in the opposite wall. This type of cooler is used extensively for strawberries, but in this case, cauliflower is being cooled. Cut flowers are also cooled by this method. Source: Western Grower and Shipper.

Figure 16–24
A vacuum cooler can cool lettuce from 30°C to 1°C (86°F to 34°F) in about thirty minutes. While the cooler was originally designed for lettuce, other crops, such as the cauliflower shown here, are now cooled in this manner. Source: Western Grower and Shipper.

drawing the cooled air through the packages. Air cooling avoids the spread of pathogens, which is one of the hazards of water used for cooling. Air has a low thermal capacity, however, and air cooling is slow for bulky products. This delay may not matter if the food products are not highly perishable, but water provides much better cooling for crops such as sweet corn and asparagus. Typically, packages of the crop pass slowly under a falling curtain of refrigerated water. The water is usually recirculated to conserve energy. It must be treated with chlorine or ozone to prevent the carryover of fungal and bacterial pathogens. In some cases, produce is cooled by packing it in ice (Fig. 16–23).

Vacuum cooling utilizes latent heat to lower temperature (Fig. 16–24). Crops are sealed into a large steel container and the pressure is rapidly reduced by a pump. The temperature of the produce is reduced as water evaporates from its surface. This process can cause wilting of leafy crops, and it is common practice to spray crops with water before vacuum cooling to minimize this problem.

Moisture

Moisture is a critical storage variable. The importance of dry storage for seed crops has already been mentioned. Fresh produce must be kept close to fully hydrated. Generally a loss of 5 percent of fresh weight results in wilting or shriveling, and the crop becomes unsalable. Refrigeration tends to cause drying because water condenses from the air onto the surface of the cooler, reducing humidity in the cooler. Humidifiers can be installed to replenish the water in the atmosphere, but the proper design of storage units is critical to moisture retention. However, high humidity can allow fungal spores to germinate and invade plant tissues. An optimum humidity for a crop balances the

risk of infection against the tendency to lose moisture from the crop.

Dehydration

Dehydration is a way to preserve many crops that are not damaged when water is removed. Several preservation techniques involve dehydration to preserve the product.

Preservation by Solar Drying   The drying of vegetables, fruits, and berries by solar radiation was probably the first attempt at food preservation, and its use and popularity remain high. Dates and figs were grown and sundried by early civilizations in the eastern Mediterranean area. Preservation by drying succeeds because decay-causing organisms usually do not grow at moisture contents below 10 to 15 percent.

Some advantages of dehydration over other methods are a reduction in weight and volume, thus reducing handling and storage costs. Storage generally requires minimal or no refrigeration. Solar drying is also less expensive than other methods of dehydration.

A climate having high temperature, clear days, and low humidity is ideal for solar dehydration of various commodities. Crops preserved by solar dehydration include apples, apricots, currants, grapes, peaches, figs, dates, pears, and plums (Fig. 16–25). Sundried tomatoes have become very popular in the United States.

Forced Hot Air Dehydration   The popularity of dehydrated foods, particularly with various fruits, has stimulated the development of efficient methods for rapid

dehydration with forced hot air. This method is more appropriate when natural solar drying is not practical.

Air heated to 60°C to 70°C (140°F to 158°F) in the drier and circulates around the product. The time for dehydration varies from a few hours to as much as thirty-six hours, depending on the kind and size of the product.

Large volumes of onions, potatoes, mushrooms, and other commodities are dehydrated. The use of ready-mixed vegetables for soups and other uses has increased the use of dehydrated celery, carrots, peppers, tomatoes, peas, parsley, chives, and other vegetables and herbs.

Dehydration by Freezing   Freeze-drying involves removing water by sublimation of ice at temperatures below the freezing point. **Sublimation** is the process by which a solid (ice) passes directly to a vapor without going through the liquid phase. Heat is absorbed in this process. Upon rehydration, the quality of the product is equivalent to that of food preserved by freezing. While freeze-dried products are expensive, they are valuable when weight reduction is necessary. These products are used in special markets where convenience of preparation is desired, such as in backpacking or space travel.

There are several other methods of preservation in addition to controlling moisture.

Modified and Controlled Atmosphere Storage   For certain crops such as apples, additional control of deterioration can be achieved by lowering the oxygen, and thereby respiration, in the atmosphere in which the crop is stored (Fig. 16–26). This change is usually done by exchanging nitrogen for oxygen in the sealed storage

Figure 16–25
The typically brilliant cloudless summer days in the San Joaquin Valley in California are ideal for drying grapes to make raisins. The grape berries are hand-harvested and laid on heavy brown paper to dry. The use of a continuous roll of paper permits some mechanization of the removal of the raisins after drying and before they are taken to the packing house for processing. A rain during this period can be a catastrophe to raisin producers.

Figure 16–26
Modified or controlled atmosphere storage units such as those being tested here preserve produce by not only cooling but also by removing some of the oxygen from the air and replacing it with nitrogen gas. Source: USDA Agricultural Research Service Image Gallery, http://www.ars.usda.gov/is/graphics/photos/

room. Crops need only a minimal level of oxygen, usually around 1 to 2 percent, to remain healthy. At that low level of oxygen, pathogens do not thrive. Another type of modified atmosphere storage is the use of special polymeric bags to package produce such as cut lettuce. Injecting the bags with an atmosphere that is high in carbon dioxide and relatively low in oxygen in conjunction with refrigeration, increased fresh produce storage times by several days. Ideally, the polymer selectively controls the movement of oxygen and carbon dioxide through the bag's surface allowing the oxygen concentration to remain low but still meet the respiration needs of the product.

Canning

The purpose of canning is to destroy spoilage organisms by heat, although complete sterilization is not mandatory. The product is placed into airtight containers that are sealed. The major containers are cans made from tin-plated sheet steel or glass bottles. Plastic- or foil-coated paper containers are also used, especially to reduce weight. The filled containers are closed and then sterilized, usually by heat and pressure in a pressure cooker. The characteristic "pinging" heard in kitchens where glass canning jars are cooling is the result of a decrease of air pressure in the closed container as the contents cool and the lid is pulled inward by the partial vacuum. The inward pull insures that the lid is sealed.

Processing with Sugar

Some fruit products are processed with high concentrations of sugar for the production of jams, jellies, marmalades, crystallized fruits, and the like. Sugar increases the osmotic pressure to levels where water is unavailable for microbial activity, thereby reducing spoilage possibilities.

Processing with Salt

The processing of vegetables with brine has changed little over the years. This process preserves cucumbers and gherkins as pickles; cabbage as sauerkraut; and onions, beets, beans, peppers, and olives. Fresh cucumbers or cabbage are allowed to ferment anaerobically in a salt solution concentrated enough to prevent the activity of spoilage organisms but not concentrated enough to destroy the bacteria producing lactic acid. Lactic acid lowers the pH, thus helping to prevent the growth of organisms.

Ethylene

The plant hormone ethylene tends to accumulate in the atmosphere during storage, handling, and distribution of produce. Although ethylene is used to ripen bananas and tomatoes, its effect on other produce is not so desirable. Even at very low levels, from a fraction of a part per million upward, ethylene promotes yellowing and abscission of leaves, distorted growth, flower senescence, and premature ripening in many fruits and vegetables. Ethylene comes from many sources such as climacteric fruit, senescing and decomposing plant tissues as well as internal combustion engines used on forklifts and other equipment. Many produce storage facilities only use electric equipment for this reason. Ethylene does not escape readily from storage facilities because they are generally enclosed spaces. Although equipment has been developed to remove ethylene from the atmosphere, the problem can often be minimized by removing the source, whether artificial or natural. For example, climacteric fruits produce large amounts of ethylene as they ripen and they should be kept away from most other kinds of plant material. Many cut flowers are particularly responsive to ethylene, and for many years they were protected by treatment with silver thiosulfate. A less toxic treatment (1-methylcyclopropene or MCP) has replaced silver thiosulfate.

MARKETING OF AGRICULTURAL PRODUCTS

Primitive societies had little marketing. Individuals or family groups gathered or produced their own food and other necessities. As a barter system and, later, money exchange developed, the variety of products available for exchange widened. Hunters could barter their game with growers of grain, vegetables, and fruits, enabling each to obtain and exchange products. Individuals not directly producing food could buy or trade for food, thus allowing them to participate in their societies in numerous and varied work activities other than agriculture and food production.

Today, specialization has progressed beyond the self-reliant local community. Producers tend to concentrate on growing the products best adapted to their particular areas, soils, climate, then distribute the product regionally, nationally, or even globally rather than locally. However, as our understanding of sustainability increases, this trend is reversing somewhat and locally grown is becoming an attractive marketing strategy, creating a resurgence in farmers' markets, roadside outlets, pick-your-own operations, and community supported agriculture (CSA). In a CSA system, consumers contract with a farmer who delivers on a regular basis (usually weekly) a container of whatever is in season. Indications are that the general public will continue to

support these local forms of marketing, even though they will account for a relatively small portion of the total distribution.

Marketing other than directly to a consumer usually involves a series of intermediates. Intermediates are prevalent in countries with highly developed systems of agricultural marketing that includes imports and exports in addition to the domestic markets. Some intermediates buy the product and further process it, some transport or store it, and others distribute it as wholesalers or retailers. A multitude of intermediate functions occur in this process and are essential in the orderly distribution of most agricultural commodities.

The Process of Marketing

Marketing plays a key role in the distribution of a product. The process is better understood by discussion of its major components, which are assembly, distribution, exchange, financing, storage, and transportation.

Assembly   Assembly is the concentration of smaller quantities of a commodity into a central place for the necessary handling, processing, or other procedures that must occur before transfer to market. Concentration at convenient locations gives prospective buyers the opportunity to examine products and arrange their purchase. Not all agricultural products need to be assembled and, in fact, a considerable volume is sold directly to processors, wholesalers, or retailers.

Systems have been developed to distribute products from places of assembly to places of consumption. The extensiveness of the distribution system is the foundation of marketing in the industrialized countries. The system must be able to adjust the supply of commodities to market demands quickly and easily as either supply or demand changes. Food sale chains and

institutional and fast-food suppliers have captured a large share of food distribution in developed nations. In North America and western Europe, management of the retail distribution of food has become concentrated in relatively few hands. In the United States, large food chains retail food through thousands of stores. Fast-food restaurant corporations prepare a significant proportion of the total number of meals consumed nationally each day in the United States. The same sort of food retailing has occurred to a large degree in Canada, England, and other European countries, and the trend is expanding to other areas of the world.

Exchange   Before any buying or selling can occur, the buyer and seller must meet or communicate. The exchange process involves two phases: contacting possible buyers and sellers, and negotiating an agreeable price. If agreement is reached, a sale is consummated. In some larger markets, sales are negotiated for a fee by intermediates called brokers. They do not take possession of the goods but facilitate the transaction. This setup enables buyers and sellers, who may be separated by great distances and thus unable to physically meet, to arrive at an exchange. Because brokers often represent the producers of many different products, buyers can obtain many different products through a single broker.

Grade standards developed by the USDA or individual state departments of agriculture are helpful because they allow the buyer to know about the quality of the product without actually seeing it. Specialized market reporting and news carried by the press, radio, and television also provide helpful information.

Many producers sell directly to the public through their own retail stores. These can be retail garden centers, produce stands, farmers markets, or other enterprises. Golf courses and sports fields (Fig. 16–27) are a special

Figure 16–27
Golf courses, sports fields, and private botanical gardens are a unique type of agriculture retail exchange where golfers, athletes, and spectators pay to use or enjoy the plants. Source: Margaret McMahon, The Ohio State University.

type of retail exchange where golfers pay for the use of the turfgrass or spectators pay to watch their teams play on turfgrass. Managers of these facilities have to be skilled in not only growing plants but communicating with the public.

Financing   Financing is essential in all marketing processes. Producers, wholesalers, processors, retailers, and other intermediaries must either have the capital or credit to produce, hold, and handle a commodity until it is sold and payment is received. Market financing entails a certain amount of risk. A major risk is a falling market; another is rapid price fluctuation. Prices for agricultural commodities generally are less stable than those for manufactured goods because supply and demand are subject to greater and less controllable fluctuations. Deterioration of quality entails another financial risk. A buyer must be aware that the quality is the most important attribute of most agricultural commodities and that the quality of many of these commodities depreciates rapidly.

Commodity Markets and Exchanges   Commodity markets transfer ownership of products from producers to processors, manufacturers, or consumers and determine the price for the exchange. Commodities often handled by this type of market are foodstuffs and non-food raw materials. Some examples are cereal grains, oil seeds, tea, coffee, rubber, and tobacco. The commodity market is concerned with trading a given amount of a certain grade of commodity for present delivery. The goods need not be on hand at the time but can be in transit or stored. Terms are arranged, payment is made, and title transferred.

A commodity exchange, or futures market, differs from a commodity market because the exchange involves the purchase or sale of a contract to deliver a certain amount of a given grade of commodity on an agreed date in the future. This arrangement allows each party to hedge against a future market supply and demand situation. Hedging means that one party covers his or her position in the cash market by taking an opposite position in the futures market. This setup may or may not result in a profit, but it provides a means to avoid excessive losses. The Chicago Board of Trade (CBOT) is an example of a commodity exchange handling mainly agricultural commodities.

Governmental Marketing Services

Most governments participate in the regulation of marketing to some degree. Some go to the extreme of setting prices, and regulating production and distribution of products. For example, the price of bread is regulated in Egypt; that of milk in Denmark. In the United States, California and some other states regulate milk prices, and the production of tobacco and other crops is regulated through acreage allotments. There are many examples of governmental regulations, which often arise during wars or other emergency periods.

Services provided by the federal government are intended for the protection of public health; establishment of standards and grades; and enforcement of weights, measurements, and other regulatory needs. Federal and state governments cooperate to supply production statistics and estimates for various crops, their condition, and other important market characteristics. For some commodities, information on shipments and market prices is provided on a daily basis.

Grade standards are degrees of quality, each established by definition. The standards define the color, size, and freedom from defects or any other attributes that pertain to quality. These standards are published and maintained by the USDA, although some states also establish grade standards. Regulatory departments of the USDA or state may provide inspection and grade determination services. These official standards permit marketing, even over long distances, without the buyer's actual inspection of the commodity. For example, lettuce from Arizona or Florida oranges are sold to New York buyers who order these commodities by well-defined grade standards, knowing before arrival that the desired quality has been guaranteed for delivery. Not all commodities have grade standards established. In general, though, quality is recognized as having the attributes for which the crop is grown and little or no damage or blemishes.

Agricultural Cooperatives

Producers may join together to form businesses known as cooperatives (co-ops), which differ from other businesses because the farmers are owners as well as customers. Co-op members provide capital, elect officers, and may employ a manager. Farmers use the co-ops to market their products, to procure supplies (fuel, fertilizers, containers, etc.), and to obtain services (insurance, credit, etc.). Co-ops generally do not operate as a profit-making business but function to lower member costs through large volume buying discounts. Through collective marketing and having a large presence in the marketplace, co-ops can often strengthen their selling abilities.

Marketing Boards

Marketing boards were first developed in England and some of the commonwealth nations. Their primary

objective was stabilizing producer prices. Examples of marketing boards established in some countries are the National Coffee Board of Ethiopia, the Sri Lanka Tea Propaganda Board, the Canadian Wheat Board, the Australian Federal Marketing Board.

On a multinational level, the European Community sets common policies on agriculture including subsidies or land stewardship payments, guaranteed minimum prices, and import tariffs and quotas for its member nations.

In the United States, a similar program, called a marketing order, exists. Used originally for fruits and vegetables in California and for milk in many milk-producing states, marketing orders are agreements among producers to solve a specific problem or to achieve a goal for the common good of the members. Under the authority of a marketing order, either state or federal, an elected administrative board can levy assessments to collect funds to support advisory services, promotional activities, or research. Marketing orders cannot control supply.

Direct Marketing

Many producers sell directly to customers. The sales may be wholesaler or retail. Wholesale means the buyer then sells the product to another customer. Retail is direct sales to the end consumer. Examples of retail sales are pick-your-own enterprises and garden centers. Retail usually requires the producer to advertise their products. The advertising can be on-site (Fig. 16–28) or through some broader reaching medium such as newspapers, fliers, radio and television, or the internet. An interesting development in direct marketing for farmers is the concept of **agritourism**. To supplement their regular income, some farmers as well as fruit, vegetable, and greenhouse growers now provide

Figure 16–28
A retail greenhouse advertisement along a busy highway.
Source: Margaret McMahon, The Ohio State University.

the public with some form of entertainment during the year. The entertainment can be on-going (daily or weekly hayrides and corn mazes) or seasonal (pumpkin toss at Halloween). "Tourists" (they may be local residents or from more distant locations) to the production site pay the grower to participate in the activities.

Internet Marketing

For some agricultural commodities, the Internet has become an important means of marketing.

The Internet is increasingly important in the production and marketing of plants and plant products through its business-to-business (B2B) and business-to-consumer (B2C) roles. On the marketing side, these exchanges occur at three levels.

1. *B2B exchange of raw plant commodities throughout the marketing chain.* The Internet's role in exchanging raw plant commodities is the least visible and, perhaps, the most important of the three. Much of this exchange takes place in closed network environments; but it is here that prices are set, the availability of most crops is determined by buyers, and exchange deals are developed.

2. *B2C direct sales of some plants/plant products to end consumers.* Direct sales range from food crops to flowers and other green products. Indeed, florist services were among the early successes in e-commerce. These exchanges offer their own special set of concerns. Sales of many plant products and foods to consumers involve complex signaling of organoleptic traits (smell, color, texture) that do not transfer easily in a digital environment. Nevertheless, this form of B2C marketing continues to grow steadily.

3. *Aiding information search for buyers and sellers.* In all types of marketing, the primary role of the Internet is as a provider of information about products and sellers in an attempt to enhance the marketing process. Industry research estimates that for every $1 consumers actually spend, they buy another $6 of product at stores based on their online research. This percentage is likely higher for food and agricultural products given the relatively low adoption rate for B2C at present. This also works well for the rising consumer interest in connecting with food producers, with the Internet used as a method to tell the farmer's story of how food is produced. This story helps to reestablish the connection between agricultural producers and the consumers of their products (see Chapter 1).

As adoption of Internet technologies continues to grow, new marketing applications will evolve, permitting enhanced integration of computer and network technologies with the production side of the industries.

Storage

Storage must be considered part of the marketing system, and certainly is necessary at various stages in the marketing sequence. Producers often find it necessary to store their products; buyers, whether processors, wholesalers, or retailers, operate storage facilities to control the supply and time of resale of the commodity. It would be difficult for a large apple grower to stay in business without the ability to hold a proportion of the crop for up to twelve months. On the other hand, a flower grower may use storage only to hold a very perishable crop overnight for shipment on the following morning.

Processing plants operate storage facilities to hold raw product stocks to operate the plant when incoming supplies are low as well as to hold the processed product until it is distributed.

TRANSPORTATION

An essential part of marketing is rapid, dependable transportation. The products are generally transported from their origin or assembly areas to various markets or processing facilities by the producer or the buyer, or by transport firms contracted to provide such services. Who provides the transportation and how it is performed varies considerably.

Transporting Commodities

Truck Transport   Trucks now haul the major portion of all perishable goods in the United States (Fig. 16–29). They offer dock-to-dock service, an advantage over trains. However, as fuel prices increase, an increase in rail transport over trucks has started to occur. Most trucks that transport perishables have mechanically operated heating and refrigeration systems.

Rail Transport   Rail transport of nonperishable commodities such as grains began during the Civil War era. Perishable commodities began to be shipped later with the invention of refrigerated rail cars. Temperature is regulated by thermostats and cooled air is circulated through the cars. Often, the system provides heated air when needed, too.

Another adaptation is a system of truck-train transportation called the piggyback method, in which

Figure 16–29

Trucks account for a substantial amount of transportation of all forms of goods and particularly for perishables in the United States. Trucks can be equipped to provide refrigeration or heat as needed as well as advertisement.
Source: Margaret McMahon, The Ohio State University.

a refrigerated truck-trailer is carried on a railroad flatcar. At destination, the trailer is removed from the flatcar and attached to a tractor for further transportation to the final destination. These piggyback units combine the truck's flexibility of delivery with the train's reduced cost of transport. With rising fuel costs, the use of railways to move produce has been increasing.

Air Transport   Transportation by air, except for small packages and mail, was not important until after World War II. The development in the late 1960s of large-capacity, jet-engine-powered aircraft made such transport feasible.

Air transport has opened up previously inaccessible markets. High-value, perishable crops such as cut flowers, strawberries, pineapples, and various fruits and off-season vegetables are shipped around the globe via this transportation method – often in just hours from one continent to another (Fig. 16–30. The use of containers especially fitted to aircraft contributes to handling efficiencies (Fig. 16–31).

Utilizing this method for perishables requires other components, including precooling, short-term refrigerated storage at departure and arrival destinations, and effective scheduling with unloading and distribution to final destinations. Any delays negate the benefits of the rapid transport provided by aircraft of perishable produce.

Waterway Transport   Most waterway transport is done by ships designed to efficiently handle and transport

Figure 16–30

Large quantities of flowers and fresh fruits and vegetables are carried throughout the world by air cargo and regularly scheduled passenger aircraft. These flowers at a large flower auction near the Amsterdam airport will be in New York City, Atlanta, and other major cities around the world in just a few hours.

Source: Margaret McMahon, The Ohio State University.

Figure 16–31

At the production facility or central collection point, plants that are to be shipped by air are packed into specially shaped cargo holders as seen in the picture. These containers being filled at a large greenhouse are specially designed to fit tightly in the cargo hold of a plane.

Source: Margaret McMahon, The Ohio State University.

containerized freight. The containers can be equipped to heat or refrigerate perishable produce. They are transported to docks for shipboard loading and unloading. However, in some parts of the world, canals and small watercraft contribute significantly to shipping plants and plant products. For example, in The Netherlands small boats are used by some flower growers to ship their product to local and regional markets as well as being used as retail floral shops.

Transportation Challenges

Modern methods of transportation allow for a tremendous and often rapid exchange of products throughout the world and provide access to many new and distant markets. But it has allowed for the exchange of plant pests and pathogens at the same time. The Emerald Ash Borer, which is wreaking havoc in the lower Great Lakes area, is an example of a recent pest introduced by transportation. The beetle is believed to have come into Detroit on shipping pallets from China. Often these entities are introduced into areas where they have no natural enemies and they create enormous problems as the populations expand unchecked. Increasing fuel costs are impacting all forms of transportation. The challenges of sustainability will undoubtedly generate significant changes in the way plants and plant products are moved from producer to consumer.

SUMMARY AND REVIEW

Establishing a plant growing business requires a site analysis of the area where the plants are to be grown. The analysis includes water supply, physical features, soil, climate, economic and factors such as costs of land, labor, and equipment. Labor availability and transportation factors both on- and off-site also have to evaluated.

Plant growing systems fall into three general practices: (1)conventional or traditional (high input of resources, especially nonrenewable resources); (2) organic (reduced dependency on nonrenewable resources); and (3) sustainable (using practices that allow present needs to be met without compromising the ability to do so in the future). While these are separate categories, they are not mutually exclusive. Most plant growing operations can and do use practices from all categories. An exception would be organically certified operations which have to comply with a very strict set of regulations set forth by the United States Department of Agriculture.

Crop harvesting can be done mechanically or by hand. Mechanical harvesting is done by machines that have been engineered to efficiently collect the part of the plant being harvested and separating it from the

rest of the plant. At first, harvesters could be used only on uniformly ripening, durable crops that could be harvested in bulk, such as grains, hay, and cotton. Now harvesters are available for a wide array of crops. Other crops were too fragile, did not ripen uniformly, or were too difficult to harvest in bulk to be harvested by machines. Now, with the development of uniformly ripening species and advances in technology such as computers, sophisticated sensors, and robotics, more and more crops are being harvested mechanically. Hand harvesting can be slow and labor intensive, but some crops still require the greater care that it provides. In many parts of the world where labor is inexpensive, it is still the most common method of harvest.

The problems of crop preservation after harvest and the solutions to those problems are influenced by the type of structure that is harvested and its stage of development. Crops are subject to several kinds of deterioration after harvest. Mechanical injury during harvest and subsequent handling is a major cause of loss. Crops also undergo developmental and metabolic changes that may render them unfit for consumption. Insect pests are mainly a problem in the storage of dry crops and some ornamentals. Fungal and, to a lesser extent, bacterial diseases affect many kinds of crops in storage, and the pathogens are often distinct from those causing problems in the field.

While market standards for crops emphasize visible aspects such as size, color, and shape, consumer quality is determined by other factors such as safety, nutritional quality, and the subjective experience of consumption. Various chemicals have been used to prevent crop deterioration and maintain quality, but their use is declining. Refrigeration is the main technology employed for preservation of fresh produce. The moisture level during storage may need to be high to maintain full hydration in the crop or low to minimize disease. A compromise between these two extremes is often necessary. The storage life of some crops can be extended with atmospheres containing high nitrogen or low oxygen concentrations. Ethylene is produced by several sources including climacteric fruit in the postharvest environment and can cause damage to a wide variety of crops.

Marketing and distribution of crops involves several steps, including assembly of the product, transportation and distribution, storage, and exchange and financing. Some commodities have marketing boards that can levy assessments or collect funds to provide advisory services, develop promotional activities, and fund research. Internet marketing has become an important sales option for many commodities.

Methods of transporting plant commodities include truck, rail, a combination of truck and rail, air, and water. If the product is perishable, refrigeration and heating systems are used during transit to keep the product in the best possible condition.

KNOWLEDGE CHECK

1. List the six major categories that are considered when doing a site analysis for a plant growing operation.
2. Give a factor you have to consider for each of the above six categories.
3. Briefly describe each of the following systems of growing practices: conventional, organic, and sustainable.
4. Can a conventional grower use practices from organic or sustainable systems? If so, list two of these practices.
5. What agency certifies organic farms?
6. Why do some organic growers choose to label their crops as being organic-based rather than as being organic?
7. What is one of the oldest forms of sustainable crop growing? Under what conditions does it lose its sustainable characteristics?
8. What are two areas in which sustainable certification is available for plant growing systems?

9. What factors determine how water, nutrients, and temperature are managed?
10. What is a "growing degree day"? How many growing degree days are there if the base is 50°F and the average temperature for one day is 78°F?
11. How are growing degree days used to predict harvest?
12. Why are high-intensity discharge lamps only used on high-value crops?
13. How are plants protected from high light intensities in a greenhouse?
14. What does a combine do in relation to grain harvest?
15. Why is hand-picking for some high-value crops preferred to mechanical harvesting?
16. Although the flowers are cut by hand, how does mechanization aid in the harvest of cut flowers?
17. How do yield sensors on harvesters help make harvest more cost efficient?

18. Would a cut tulip flower or a tulip bulb most likely have the shortest postharvest life?

19. List two ways damage can occur to plants during the harvesting operation.

20. Explain why a striking an object quickly is more likely to damage a soft fruit than striking the same object more slowly.

21. What is the reason many cut flowers and potted plants are sleeved before shipping?

22. What is a major problem with storage of plant structures such as bulbs and tubers? How can it be controlled?

23. Why are immature structures such as green peas, sweet corn, and green onions more perishable than the same structures when they are mature?

24. What is a climacteric fruit and why should they not be stored with ethylene sensitive plants and produce?

25. Why is it often a different set of pathogens that attack harvested plants compared to what attacks the same plant when it is growing?

26. List three characteristics that are often used to measure quality of plants or produce.

27. Explain why there is a Federal law that requires the packaged food to carry nutritional information but fresh produce does not have a similar law.

28. What are two recent examples of microbial contamination in vegetables that occurred in spite of routine screening of produce for contamination?

29. What is hedonic quality?

30. What is the purpose of "curing" some crops after they are harvested?

31. What types of crops are cooled by forced air? What types are hydrocooled?

32. What are two ways of drying plant material for storage?

33. Why does drying work as a preservation method?

34. What is the principle behind modified and controlled atmosphere storage?

35. How are spoilage organisms controlled by canning, sugar, and salt?

36. Why can it be said that ethylene is both a blessing and a curse when it comes to handling produce?

37. List the major components of marketing.

38. What are agriculture cooperatives and marketing boards?

39. What problem for ecosystems has occurred with the rapid exchange of products between many distant markets?

FOOD FOR THOUGHT

1. Many cities are turning inner city vacant lots into community gardens. These gardens have been shown to have a very positive effect on what is often a marginalized section of society. You have been hired by a city close to where you grew up to analyze a group of these properties for growing plants. What are some plant-growing related problems you think your analysis might reveal?

2. Of the three main plant growing systems (traditional/conventional, organic, sustainable) which would you choose to use as a professional or amateur grower? Give the reasons you would choose it. What justification would you use for your reasons if someone who chose another system challenged them?

3. As a grower of five acres of fresh market tomatoes (the tomatoes you buy in a produce department), you have a choice between two types to grow. One type stores and ships well but tastes somewhat bland. The other tastes great but does not store or ship very well. Your main market is a grocery chain in a city about seventy-five miles away and you deliver the tomatoes by your own truck to the chain's local distribution center on the far side of town (about 100 miles from you).

   1. Which type would you grow and why?

   2. Would your choice be different if you sold the tomatoes from a roadside stand beside the field? If so, why?

FURTHER EXPLORATION

FOOD AND AGRICULTURE ORGANIZATION OF THE UNITED NATIONS (FAO). 1989. *Prevention of post-harvest food losses: Fruits, vegetables and root crops. A training manua*l. New York. Available online at: http://www.fao.org/docrep/T0073E/T0073E00.htm

KADER, A.A., AND R.S. ROLLE. 2004. *The role of postharvest management in assuring the quality and safety of horticultural produce.* New York. Food and Agriculture Organization of the United Nations (FAO). Available online at http://www.fao.org/docrep/007/y5431e/y5431e00.htm

RICKETTTS, C., AND O. RAWLINS. 1999. *Introduction to agribusiness.* Albany, NY: Delmar.

SUSTAINABLE AGRICULTURE RESEARCH AND EDUCATION. 2003. *Building a sustainable business.* Washington, DC. USDA Available online at http://www.sare.org/publications/business.htm

The following USDA websites provide additional information relevant to the topics discussed in this chapter:

1. USDA Agriculture Marketing Service (AMS), http://www.ams.usda.gov/
2. USDA National Agricultural Statistics Service (NASS), http://www.nass.usda.gov/.
3. USDA National Organic Program (NOP), http://www.ams.usda.gov/AMSv1.0/nop

✳17

Field Crops Grown for Food, Fiber, and Fuel

PETER THOMISON

key learning concepts

After reading this chapter you should be able to:

- Discuss the cultural practices common to nearly all field crops and the reasons behind those practices.
- List the major field crops grown for food, fiber, fuel, and other industrial uses.
- Describe the specific cultural practices used for growing many of those crops.

INTRODUCTION

Humans depend on field crops for much of their food, grains for animal feed, clothing, rope, biofuels, and for many other industrial uses. Table 17–1 lists the world's major food field crops and the principal uses of each. Field crops for forage will be covered in the following chapter. As you learned in Chapter 14, 15, and 16, successful crop production requires an understanding of the various management practices and environmental conditions affecting crop performance. Planting date, seeding rates, variety selection, tillage, fertilization, and pest control all influence crop performance. A crop's response to a given cultural practice is often influenced by one or more other practices. The keys to developing a successful production system are (1) recognizing and understanding the types of interactions that occur among production factors, as well as various yield limiting factors; and (2) developing management systems that maximize the beneficial aspect of each interaction. Knowledge of crop growth and development is also essential to using cultural practices more efficiently to obtain higher yields and profits. The following sections discuss the general aspects of field crop production, usually using corn or other grains as the example. Most aspects of the general discussion will apply for other field crops as well. Following the general discussion are brief descriptions of the specific production considerations for several crops.

CROPPING SEQUENCES

One of the most important aspects of growing field crops is the order in which crops follow one another (sequence) either during the same growing season or through consecutive seasons. Field crops are produced using different cropping sequences based on agronomic and economic considerations. Some of these cropping sequences are complicated and require a high level of expertise and management to be successful, whereas others are fairly simple. The most popular cropping sequence in the US corn belt is the corn–soybean rotation, which involves alternating crops of corn (a grass) and soybean (a nitrogen-fixing legume). Crop rotations may also include a small grain, usually wheat, or forage crops. The corn–soybean (or other

Table 17–1

FIELD CROPS GROWN FOR FOOD (HUMAN AND/OR ANIMAL), FIBER, AND FUEL

Crop	Food	Fiber	Fuel and Other Industrial Uses
Agave	X	X	
Barley	X		
Beans	X		
Castor bean			X
Coconut	X		
Corn	X	X	X
Cotton	X	X	
Flax	X	X	
Hemp	X	X	
Jute		X	
Kenaf	X	X	
Oats	X		
Olive	X		
Palm oil	X		
Peanut	X		X
Ramie		X	
Rape	X		
Rice	X		
Rye	X		
Safflower	X		
Sesame	X		
Soybean	X		X
Sorghum	X		
Sugar beets	X		
Sugarcane	X		
Sunflower	X		
Wheat	X		
Wild rice	X		

Figure 17–1

Intercropping barley with clover. Source: USDA Agricultural Research Service Image Gallery, *http://www.ars.usda.gov/is/graphics/photos/*

grass-legume combination) rotation offers several advantages over growing either crop continuously. Benefits to growing corn in rotation with soybean include more weed control options, fewer difficult weed problems, less disease and insect buildup, and less nitrogen fertilizer use. Corn grown after soybeans typically yields about 10 percent more than growing continuous corn. No-till cropping systems are more likely to succeed on poorly drained soils if corn follows soybean or meadow rather than corn or a small grain, such as wheat. The yield advantage to growing corn following soybean is often much more pronounced when drought occurs during the growing season.

In contrast to crop rotation, monoculture is growing the same crop in a field continuously over a period of years. In the US corn belt, continuous corn production is the most common monoculture practiced. Farmers have used monoculture to take advantage of favorable markets or government support programs, an environment that strongly favors producing a high-value crop, or to reduce the need for diverse machinery or facilities. The ratoon cropping system, another form of monoculture, is used in sugarcane production. After the first harvest is completed, the crop is allowed to regrow and a second harvest is obtained from the regrowth prior to destroying it in the fall.

Double cropping involves the production of two crops, one following another, during one growing season. The most common double-cropping system in the corn belt is following winter wheat with soybeans. Intercropping consists of simultaneously producing two or more crops in the same field (Fig. 17–1). Intercropping is most advantageous when crops of different characteristics are planted. Examples of intercropping include growing rows of corn and soybeans, or growing cassava and beans or peas in tropical systems. Relay intercropping consists of seeding a second crop into an initial crop prior to harvest of the initial crop. This provides a longer growing season for the second crop. Seeding soybeans into winter wheat prior to wheat harvest in the spring is an example of relay intercropping.

VARIETY SELECTION

Selecting varieties for planting is a key step in designing a successful crop production system. To stay competitive, growers must evaluate and select the most adapted varieties for their locale on an annual basis.

Growers need to choose the highest-yielding varieties that are maturity adapted to their growing region. Maturity, yield potential, and insect and disease resistance are important factors to consider in variety selection for all crops. For the cereal crops (including wheat, barley, oats, rye, spelt, and triticale), winter hardiness, straw stiffness, and nonshattering characteristics are also important cultivar selection criteria. High-quality seed is also important to ensure varietal purity, high germination, and minimum contamination by weed seed and other impurities.

Maturity is a key characteristic in choosing a variety. Varieties that use most of the growing season to mature generally have greater yield potential than those that mature more quickly. However growers often plant varieties of different maturities. This practice may help limit damage to the crop from a period of stress during the growing season and may allow the harvest period to begin earlier. Planting varieties with different maturities also ensures some degree of genetic diversity, which may help buffer the crop from various biotic and abiotic stresses. Most crops are characterized by a maturity designation. Soybeans are assigned to ten different maturity groups from 00 (grown in short season environments, e.g., Canada) to VIII (grown in semitropical environments). Winter and spring designations for small grain crops are associated with their planting date within a region.

In corn, the most common maturity rating system in the United States, is the days-to-maturity system. This system does not reflect actual calendar days between planting and maturity—108-day hybrid, for example, does not actually mature 108 days after planting. A days-to-maturity rating is based on relative differences within a group of hybrids for grain moisture at harvest. A one-day maturity difference between two hybrids is typically equal to a ½ to ¾ percentage point difference in grain moisture. For example, a 108-day hybrid would be, on average, 3 to 4.5 points drier than a fuller season 114-day hybrid if they were planted the same day (6 days multiplied by 0.5 or 0.75). The days-to-maturity–moisture relationship is usually dependable when comparing hybrid maturities within a single seed company. However, because there are no industry standards for the days-to-maturity rating system, grain moisture comparisons of similar hybrid maturities from different seed companies may vary considerably. Days-to-maturity ratings are satisfactory for pre-season hybrid maturity selection when the length of the growing season is usually not an issue. For delayed planting or replanting hybrid selection needs, growers need more absolute descriptions of a hybrid's

maturity requirements to manage the risk of a killing fall frost to late-planted corn.

The growing degree day (GDD) system you learned about in Chapter 16 is more accurate in determining hybrid maturity than the days-to-maturity system because growth of the corn plant is directly related to the accumulation of heat over time rather than the number of calendar days from planting. The GDD system has several other advantages over the days-to-maturity system. The GDD system provides information for choosing hybrids that will mature reliably, given a location and planting date; allows the grower to follow the progress of the crop through the growing season; and aids in planning harvest schedules.

The GDD calculation method most commonly used for corn in the United States is the 86/50°F cutoff method. Growing degree days are calculated as the average daily temperature minus 50:

$$GDD = \frac{T_{max} + T_{min}}{2} - 50$$

If the maximum daily temperature (T_{max}) is greater than 86°F, 86 is used to determine the daily average. Similarly, if the minimum daily temperature (T_{min}) is less than 50°F, 50 is used to determine the daily average. Growing degree days are calculated daily and summed over time to define thermal time for a given period. Each corn hybrid requires a certain number of accumulated GDDs to reach maturity, and seed corn dealers have information on specific hybrids. As with any system, the GDD system has some shortcomings. Under certain delayed planting situations and stress conditions, GDD requirements for maturity may be reduced significantly.

Because weather conditions are unpredictable, the most reliable way to select superior varieties is to consider their performance during the last growing season and previous year over a wide range of locations and climatic conditions. In many countries, including the United States, state universities, and/or government agricultural agencies conduct crop evaluations of private and public varieties to provide farmers with objective, unbiased performance information. Two to three years of data from several test locations is usually adequate when using results of performance trials. Test summaries for more than three years may exclude recently released varieties with better performance potential.

Resistance to lodging is an important trait in the selection of many crops (Fig. 17–2). Lodging is the breaking or collapse of stalks (e.g., for corn, sorghum) or stems (e.g., for soybean, wheat, rice) usually at or

Figure 17–2

Lodging, as shown here, is the collapse of stems, causing the plant to fall over. This condition causes serious crop loss of grain in grain crops such as barley, corn, rice, and wheat because it makes harvesting difficult. In wheat, excessive nitrogen fertilization causes the heads to become too heavy with grain. Lodging is aggravated by strong winds. Lodging is also promoted by stalk rots in corn.

Source: Clay Sneller, The Ohio State University.

near maturity. Lodging causes serious crop loss of grain because it makes harvesting difficult. Strong winds often contribute to the severity of lodging problems. In corn, lodging is promoted by stalk rots and harvest delays. In some wheat varieties, excessive nitrogen fertilization causes the heads to become too heavy with grain and may result in lodging.

The number of hectares, soil type, tillage practices, desired harvest moisture, and pest problems determine the need for traits such as disease resistance, early plant vigor, plant height, and so on. End uses of

the crop should also be considered. For example, with corn, will it be sold directly to an elevator as shelled grain or used on the farm? On-farm capacity to harvest, dry, and store grain is another consideration. Premiums may be available for the contract production of specialty hybrids or cultivars with nutritionally enhanced grain characteristics.

The development of transgenic crops has given farmers new options for pest control and additional traits to consider when selecting hybrids to plant. Since the mid-1990s, transgenic (also referred to as GM or genetically modified) grain and fiber crops have been commercialized and widely produced. Insect or herbicide resistance are the two most common traits placed into transgenic crops. In some cases, both insect and herbicide resistance are engineered into the same plant. Transgenic corn, soybean, cotton, canola, and sugar beets account for most of the hectarage planted to transgenic crops. In the United States as of 2008, transgenic soybean cultivars were planted on more than 92 percent of the soybean hectarage, transgenic cotton cultivars were planted on more than 86 percent of the cotton hectarage, and transgenic corn hybrids were planted on nearly 80 percent of the corn acreage (Fig. 17–3). Transgenic insect-resistant corn hybrids contain a gene from bacteria that produces a protein known as Bt, which is toxic to some insect larva, including European corn borer and corn rootworm which can be very harmful to corn crops. Planting Bt hybrids may eliminate or at least greatly reduce the need for insecticides that are less effective on the pest while being potentially harmful to nontarget beneficial insects. Planting crops tolerant or resistant to herbicides, like Roundup®, offers growers new weed-control techniques that involve fewer

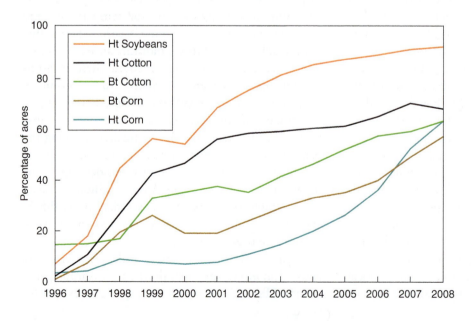

Figure 17–3

Growth of production of genetically engineered crops in the United States. Ht: Herbicide tolerant, Bt: Contains the insect toxin Bt from Bacillus thuringensis.

Source: USDA Agricultural Research Service

Image Gallery, *http://www.ars.usda.gov/is/graphics/photos/*

herbicide applications and use of more environmentally benign chemicals. Consumer protests against GM foods, combined with marketing restrictions in export markets on grain from crops containing the transgenic insect-resistance and herbicide-tolerant traits, have limited adoption of transgenic varieties in some countries. There is also a risk that the efficacy of transgenic crops may also be reduced if targeted insects and weeds develop resistance to Bt and Roundup®, however careful management of the crops can greatly reduce the risk. For example, in a corn and soybean rotation, using a grass herbicide occasionally when growing soybeans can eliminate any grasses that have developed resistance to Roundup®. Likewise, occasionally using a broadleaf herbicide when growing corn can eliminate resistant broadleaves.

PLANTING AND CROP ESTABLISHMENT

The planting operation is a key step in crop establishment. Soil and residue conditions should be evaluated before planting. Soil should be slightly moist and should crumble when squeezed. Planting in excessively wet soil may result in poor seed-soil contact or seed furrows that reopen upon drying as well as compacting the soil. Wet residues may be jammed into the seed slot rather than being cut by the coulter (the blade that makes a vertical cut in the soil where the seeds fall), thus causing poor seed–soil contact and germination. Planting in dry soil may cause penetration problems with the coulter, resulting in too shallow planting, and failure of the seed slot to close. All of these factors affect plant stands, either by reducing stand or causing uneven emergence of the crop and thus result in uneven development.

Many types of equipment are available for planting crops. Row crop planters are used to plant corns and soybeans in row spacings of 0.38 to 1.02 m (15 to 40 in.). Most soybeans and cereal crops are usually broadcast or planted with a seed drill (box drill) and air seeder. Row crop planters provide more effective seed placement that results in more accurate crop plant populations. To achieve this accuracy, the planting units on row crop planters take up more space than drill, thereby limiting the row spacings to no less than 0.38 m (15 in.). Row crop planters are also designed to accommodate attachments for applying liquid or dry fertilizer, insecticides, and herbicides during the planting operation. The use of GPS/GIS on farm equipment now allows very precise seeding, fertilizer and chemical applications, and harvesting (Fig. 17–4). An additional

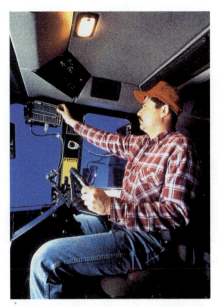

Figure 17–4
This tractor operator is using GPS/GIS as part of a system that monitors harvest as the tractor moves across the field.
Source: USDA Agricultural Research Service Image Gallery, *http://www.ars.usda.gov/is/graphics/photos/*

benefit is that there is less overlap of tire tracks thus reducing soil compaction.

Rice is also planted with grain drills, but in some areas it is seeded into flooded fields using aircraft. Precision planters capable of metering seed at desired spacings are widely used in sugar beet production to achieve uniform emergence. Pelleted seed used in conjunction with these precision planters enhances the uniformity of seed spacing.

Planting Date

In temperate regions, the major environmental factors determining planting date are soil temperatures and moisture. Early plantings in the US corn belt are often associated with excessive soil moisture and low temperatures, which may delay emergence. These environmental conditions are conducive to infection of seed and seedlings by various fungal and bacterial pathogens, which may reduce emergence or cause postemergence seedling blights, further reducing the plant population. In the warmer tropical and subtropical climates, the planting date is often determined by the onset of rains (i.e., the monsoons in India). In such regions, high soil temperatures exceeding 35°C (95°F) and low soil moisture availability may limit germination of cool season crop species.

In the United States and Canada, cereal crops are planted in either the late fall or early to mid-spring months (Table 17–2). Winter and spring types exist for

Table 17–2

SUGGESTED RATES AND DATE OF SEEDING FOR IMPORTANT FIELD CROPS OF THE EASTERN US CORN BELT AT NORMAL PLANTING TIMES

| Crop | Rate to Plant | | Date to Plant | |
	Kg/ha*	Seeds/ha†	Northern	Southern
Barley, winter	100.8–134.4 (90–120)	3.0–3.7 million (1.2–1.5)	9/15–9/25	9/15–10/5
Corn, dent	137.9–20.2 (12–18)	54,000–79,000 (22,000–32,000)	4/28–5/10	4/15–5/10
Corn, pop	6.7–11.2 (6–10)	61,700–84,000 (25,000–34,000)	4/25–5/10	4/15–5/10
Oats, spring	84–112 (75–100)	3.0–3.7 million (1.2–1.5 million)	3/15–4/15	3/5–4/5
Soybeans		272,200–395,500 (110,000–160,000)	5/1–5/20§	4/20–5/10§
Wheat	84–134.4 (75–120)	2.5 million–3.7 million (1 million to 1.5 million)	9/22–10/13	9/28–10/18

*Pounds per acre in parentheses.
†Seeds per acre in parentheses.
§Double cropped soybeans planted as late as early July.
Source: Adapted from *Ohio Agronomy Guide*, 13th ed., Bulletin 472, The Ohio State University Extension, 1995.

most of the cereal crops. Winter types usually require vernalization in order to flower and set seed while spring types usually do not survive the winter so they are not very interchangeable in regard to the season they are planted. Winter wheat, barley, and oats are planted in fall or early winter in areas where the winters are mild enough to allow the small plants to survive. The spring types, including corn which does not have a winter type must be planted in the spring. There is only a limited period when fieldwork can occur throughout much of the US corn belt because of its wet spring weather. This indicates the need to begin planting as soon as field conditions allow. Table 17–3 shows the usual planting range for Ohio. Other regions usually have similar planting windows.

Another consideration in planting date in temperate regions is susceptibility to cold and freeze damage. If the seedlings emerge before the last frost or freeze and are not tolerant of these conditions, the field may have to be replanted. One factor that determines tolerance is where the apical meristem (growing point) of the seedling is at the time of the exposure to cold. Soybeans are more cold tolerant than corn, but sometimes a field of soybeans will have to be replanted when a nearby corn field does not even though both have germinated. The difference is that the apex of the corn plant is still below ground and protected from freezing whereas the soybean apex was above ground and damaged or destroyed.

Plant Population and Seeding Rate

The seeding rate of a crop is determined by the plant population (or plant density) required to optimize yields in a particular environment. For many field crops, a strong relationship exists between plant population and yield potential. Crop yields can be optimized only when

Table 17–3

DAYS AVAILABLE TO PLOW, TILL, OR PLANT IN OHIO, MARCH 21–JUNE 30, 4 OUT OF 5 YEARS (7-DAY WEEK)

Soil Drainage Types	March 21–30	April 1–30	May 1–31	June 1–30
Poorly drained	0 or more	4 or more	8 or more	9 or more
Poorly drained with drainage improvements	1 or more	6 or more	11 or more	11 or more
Well drained	2 or more	8 or more	13 or more	13 or more

Source: Adapted from Nolte et al., Bulletin 605, The Ohio State University Extension, 1976.

Table 17–4
CORN YIELD RESPONSE TO PLANT POPULATION ACROSS VARIOUS ENVIRONMENTAL CONDITIONS*

Yield level*	Number of Environments	Harvest Population (Plants/A)			
		18,000	22,000	26,000	30,000
		Yield (Bu/A)			
<100	5	86	84	83	82
101–120	14	109	114	118	116
121–140	27	125	132	135	135
141–160	46	139	149	155	158
161–180	35	157	168	174	177
>180	14	177	189	199	203

*Lower yield levels usually associated with less favorable growing conditions (drought, late planting, low soil fertility, etc.)

Source: Adapted from *Ohio Agronomy Guide*, 13th ed, Bulletin 472, The Ohio State University Extension (1995).

water, nutrients, and light are not limiting, and temperatures and soil conditions are conducive for growth. The relative availability of water and nutrients are key factors determining the optimal plant population. In production environments with low rainfall and reduced fertility, the optimal crop population is usually lower than in environments associated with high fertility, adequate rainfall, and soils with good moisture holding capacity (Table 17–4). Yields for grain crops generally rise to a maximum as plant density increases and then fall or level off with further increases in plant density.

Plant population can also affect other economically important crop traits. Increases in plant population may increase plant height and reduce stem and stalk strength, thereby increasing the risk of lodging. Increased plant height may be beneficial in some crops such as soybean and cotton because the height of the lowest pod or boll may be increased, thereby facilitating harvest and producing a cleaner crop. The cutter bar of the combine used for harvest does not have to be set as low to the soil surface to remove soybean pods and cotton bolls, thereby minimizing the potential damage and interference of the harvest operation caused by rocks and residue.

Seeding rates vary greatly among crops (Table 17–2). As the plant population increases, tillering and branching are reduced. Grain crops such as wheat can compensate with tillers to some extent for lower-than-optimal plant density. For other crops such as soybeans, which produce branches, specific plant populations are less critical, with crops producing optimal yields across a broader range of plant populations.

An inverse relationship usually exists between plant population and the ear (corn) or seed head (grain). In some crops, this effect may be advantageous because smaller seed heads dry more quickly and may facilitate the harvesting operation. However, when drought conditions limit nutrient and water availability in corn, plant populations above the optimum may result in ears with few or no kernels.

Seed germinability and the vigor and timing of planting are major factors that need to be considered when determining the seeding rate required to achieve a target plant population. When planting is delayed in some crops, including small grains and soybeans, seeding rates are increased to compensate for a reduced period in which tillering, branching, and flowering can occur. In other crops, like corn, the seeding rates may be reduced because plant mortality associated with emergence declines as soils become drier and warmer later in the spring.

The number of plants/hectare at harvest is always less than the number of seeds planted. Planting date, tillage practices, pest problems, chemical injury, planter performance, and seed quality affect final populations obtained in the field. To compensate for these losses, a grower needs to plant more seed than the desired population at harvest. To determine an appropriate seeding rate, use the following formula:

$$\text{Seeding rate} = \frac{\text{Plant population per acre at harvest}}{(\text{Seed germination} \times \text{expected survival})}$$

Seed germination is the percentage germination shown on the seed tag. Most seed corn has a germination rate of 95 percent or higher. Expected survival is the percentage of plants that you expect to survive to become harvestable plants in the fall. Keep in mind that survival rates for corn are often in the range of 85 to 95 percent

but can vary considerably depending on planting conditions and other environmental factors. When early planting is likely to create stressful conditions for corn during emergence (e.g., no-till in early to mid-April), seeding rates 10 to 15 percent higher than the desired harvest population are used to ensure that the target population is achieved. Greater plant mortality usually occurs with early spring plantings in temperate regions.

Plant Spacing Configuration

In many grain crops, increased yields can be expected from reducing row width. At a constant plant population, this increase in yield results from more uniform distribution of plants and reduced competition of plants within the row for sunlight, water, and nutrients. Plants in narrower rows are more crowded and cannot make as efficient use of sunlight as plants in wider rows. Plants also tend to become taller and more subject to lodging. However, narrowing row width may enhance weed control and moisture utilization. With narrow rows, the crop develops a canopy earlier that restricts light to weeds between the rows. Earlier canopy formation may also shade the soil surface earlier and reduce evaporation of soil moisture, thereby conserving moisture for the crop. In some crops, narrower rows in conjunction with high plant population may create a high humidity microclimate conducive for certain diseases such as *Sclerotinia* stem rot of soybean (fungal causal agent – *Sclerotinia sclerotiorum*). Narrowing row width increases dependence on herbicides because postemergence mechanical cultivation may injure the crop. In production environments subject to late season droughts, the benefits of narrowing row width may be limited.

Seeding Depth

Crop seed needs to be planted deeply enough to place each seed in contact with soil moisture and prevent rapid desiccation, yet shallow enough to allow rapid emergence of the shoot. Factors affecting seeding depth include seed size, type of emergence (i.e., **hypogeal** or **epigeal**), soil and seedbed conditions, date of planting, and available moisture supply. A rule of thumb for the planting depth of many crops is four to five times the average seed diameter. The seeding depth of small seeded crop species (such as rape) is shallower than larger seeded species (such as corn). Small seeded species have smaller food reserves compared with larger seeded crop species, which may limit shoot elongation and seed vigor. Increasing seed size within a species usually increases final emergence, with the greatest advantage occurring with the deepest planting.

In crops like soybean that are characterized by epigeal emergence and where the cotyledons are brought above the soil surface during emergence, the seeding depth needs to be shallower compared to a crop like corn that exhibits hypogeal emergence and where the cotyledons remain below ground. When soil moisture is limited, it may be necessary to plant deeper than normal to place seed in moisture. Recommended seeding depths generally are greater for the lighter textured (sandy) soils than heavier textured (clay) soils because of lower water-holding capacity and faster drying rate at a given depth. Deeper planting is feasible in these soils because they offer less physical resistance to the emerging seedling than heavier soils. Early spring temperatures in temperate regions decrease rapidly with depth. Therefore, early plantings are usually shallower than later plantings because temperatures are warmer near the soil surface and soil moisture conditions are better early in the season.

WATER MANAGEMENT: IRRIGATION AND DRAINAGE

In areas with insufficient rainfall during the growing season or sandy soils with poor soil moisture retention, irrigation is essential for crop production. Conversely, tile drainage is essential throughout much of the US corn belt to allow for timely planting and to minimize saturated soil conditions. Some crops like rice are tolerant of saturated soil conditions.

Most agronomic crops in the United States east of the Mississippi are produced under rainfed conditions. In many areas of the western United States, however, where rainfall is erratic and limited, irrigation is required for successful crop production. Some of the major crops produced under irrigation in (mainly) the southwest include corn, sorghum, sunflowers, soybeans, and sugar beets. Most of these crops are irrigated using center pivot or furrow irrigation systems. (See Chapter 14 for a more detailed description of irrigation systems.)

CROP NUTRITION

Most field crops grow best on well-drained, fertile loam soils well supplied with organic matter. When adequately fertilized, however, field crops perform well in a variety of soils. Most do well on a neutral pH or slightly acid (6.0) soil. Sorghum and cotton are more tolerant of alkaline (high pH) soils and saline conditions than are most field crops. Sugar beets are also

Table 17–5
APPROXIMATE AMOUNTS OF PLANT FOODS REMOVED IN SEVERAL MAJOR CROPS (ON A KG/HA BASIS)[*]

Crop	N	P_2O_5	K_2O
Corn at 9.4 Mg/ha (150 bu/a)			
Grain	121 (135)	49 (55)	36 (40)
Stover (leaves and stalks left in field)	210 (235)	22 (25)	143 (160)
Corn silage at 56 Mg/ha (25 T/a)	210 (235)	71 (80)	179 (200)
Oats at 3.6 Mg/ha (100 bu/a)			
Grain	58 (65)	22 (25)	18 (20)
Straw	31 (35)	13 (15)	89 (100)
Sorghum—grain at 8.5 Mg ha (7,600 lbs/a)			
Grain	94 (105)	27 (30)	27 (30)
Stover	71 (80)	45 (50)	205 (230)
Soybean at 3.4 Mg/ha (50 bu/a)	170 (190)	40 (36)	63 (70)
Sugar beets—roots at 56 Mg/ha (25 T/a)	89 (100)	45 (50)	223 (250)
Wheat at 5.0 Mg/ha (75 bu/a)			
Grain	85 (95)	73 (48)	27 (30)
Straw	27 (30)	4 (5)	45 (50)

*Pounds per acre in parentheses.

Source: Adapted from *Ohio Agronomy Guide*, 13th ed., Bulletin 472, The Ohio State University Extension (1995).

more saline-tolerant than other crops, except during the early stages of germination.

In natural ecosystems, most nutrients are cycled within the system. However, for crop production systems to maintain the original fertility level and optimize performance, crops must be fertilized to replace nutrients removed by the grain (Table 17–5) as well as losses that may occur because of erosion, leaching, denitrification, and volatilization. A corn crop that produces 9.4 Mg/hectare (150 bu/acre) of grain removes 121 kg/hectare of nitrogen (N), 49 kg/hectare (55 bu/acre) of phosphate (P_2O_5), and 36 kg/hectare (40 bu/acre) of potash (K_2O). Secondary nutrients and micronutrients are also removed and need to be supplied to crops, but in lesser amounts and less often. Nitrogen requirements for grain crops such as corn may be reduced when they are grown in a crop rotation following a legume like soybean that fixes nitrogen. Soybean can convert (fix) adequate atmospheric nitrogen to produce yields of seventy to eighty bushels per acre if well nodulated. Seed of soybean and other grain legumes should be treated with commercial inoculants containing *Bradyrhizobia* bacteria when planting occurs in a field where these crops have not been grown previously or not grown within the past three years.

PEST MANAGEMENT

Pest and disease management is discussed thoroughly in Chapter 15, so the following is just a brief overview.

Weed Control

Weeds are the major pest control problem in many crops. Several factors need to be considered when developing weed-control programs, including soil type, weeds present, crop rotation, and budget. No single control program effectively handles the various weed problems that arise under different environmental conditions. Most weed-control programs use one or more of the following three strategies—prevention (restricting introduction or expansion of a weed problem), eradication (complete elimination of weed plants or seeds from an area), and reduction of weed plants and seed to a tolerable level.

Several different weed-control methods have been developed. These methods involve cultural practices (rotation, row spacing, and population), mechanical weed control (tillage for seedbed preparation and row cultivation), biological weed control (using natural enemies of weeds), and chemical weed control through the use of herbicides. Most grain producers regard herbicides as the most effective, economical means of controlling weed populations. Several methods of herbicide application are used in producing grain crops, including pre-plant (applying before the crop is planted), preemergence (after the crop is planted but before crop emergence), and postemergence (after crop emergence). The commercialization of several herbicide resistance crops in recent years (including Roundup® Ready soybeans, corn, and cotton) has increased the popularity of postemergence herbicide programs. By

growing herbicide resistant crops, the herbicide to which the crop is resistant can be used to control weeds but not harm the crop.

Insect Control

Despite the number of different pests that feed on crops (Table 17–6), most crops generally exhibit minimal insect problems because the normal activity of most insect populations is below a threshold level; that is, it is too low to inflict significant injury to warrant attention. Economic levels of pest injury do occur, however, and economic losses resulting from sudden outbreaks of pest activity can be prevented if growers are aware of the biology of the various crop pests, monitor fluctuations in pest population activity, and implement timely corrective action. Management of pest populations requires

Table 17–6
MAJOR PEST PROBLEMS IN CROPS GROWN FOR FOOD, FIBER, FUEL, AND OTHER INDUSTRIAL USES

Crop	Diseases	Insect Pests
Agave	Bacterial leaf rot, stem canker	Myriad bugs
Barley	Powdery mildew, stripe mosaic, yellow dwarf, leaf scald, leaf and stem rust, fusarium blight, helminthosporium fungus, smut	Wireworms, cinch bugs, greenbugs, grasshoppers, armyworms, aphids, thrips
Beans	Bacterial brown spot, bacterial blight, halo blight, bean common mosaic virus, anthracnose	Western bean cutworm, grasshoppers, flea beetles, Mexican bean beetle
Castor bean	Fusarium, rhizoctonia, sclerotium, scab, wilt, leaf spot, seedling blight, inflorescence rot, pod rot, rust spot, graymold, crown rot, stem canker, leaf blight, bacterial wilt, angular leaf spots	Castor whitefly, red spider mite, pink bollworm, mung moth
Coconut	Lethal yellowing, phytophthora, coconut heart rot, stem bleeding, bud rot, basal stem rot	Coconut caterpillar, coconut leaf hispid, black headed caterpillar, red palm weevil, rhinoceros weevil, slug caterpillar, termites
Corn	Gray leaf spot, northern leaf blight, southern leaf blight, various ear and kernel rots (including Gibberella, Diplodia, Fusarium, Aspergillus), various stalk rots (including Anthracnose and Gibberella) and maize dwarf mosaic virus	Corn rootworm, European corn borer, stalk borer, cutworm, cinch bugs, slugs, grasshoppers, aphids, flea beetles, stinkbugs
Cotton	Damping-off fungi, fusarium, verticillium, bacterial blight, leaf crumple	Boll weevil, pink bollworm, cotton bollworm (corn earworm), thrips, cotton leafworm, fleahopper, tobacco budworm, banded wing whitefly, lygus bug, cotton aphid, spider mite
Flax	Rust, fusarium, pasmo, stem break and browning, seedling blight and root rot, aster yellows, crinkle	Flax bollworm, grasshoppers, cutworms, army cutworm, Bertha cutworm, beet webworm, aphid, aster leafhopper, tarnished leaf bug
Hemp	Gray mold, sclerotinia, fusarium	European corn borer, grasshoppers, Bertha armyworm, flea beetle, hemp borer
Jute	Stem rot, black band, anthracnose, soft rot, powdery mildew, root knot disease, mosiac, leaf spot	Jute hairy caterpillar, jute semilooper
Kenaf	Anthracnose, damping-off	None
Oats	Rust, smut, yellow dwarf, blast, septoria leaf spot	Grasshoppers, cutworms, aphids, thrips, armyworms, greenbugs
Olive	Verticillium wilt, olive knot	Olive fruit fly
Palm oil	Stem rot	Red-striped weevil
Peanut	Leaf spots, southern blight, collar rot, peg rot, black rot	Cutworms, thrips, leafhoppers, corn earworms, armyworms, lesser cornstalk borers, cadelle beetle, carpet beetle, confused flour beetle, flour and grain mites, rice weevil

(Continued)

Table 17–6 (Continued)

Crop	Diseases	Insect Pests
Rape	Black spot, collar rot, white rot, leaf spot, downy mildew, powdery mildew, clubroot	Cabbage stem flea beetle, turnip flea beetle, rape beetle, pollen beetle, rape stem beetle, cabbage seed weevil, brassica pod midge, cabbage aphid, field slug, cabbage root fly
Rice	Seedling blight, seed rot, brown leaf spot, blast, stem rot	Rice leaf miners, rice water weevils, midges, rice leaf folders, armyworms, leafhoppers, thrips, water scavenger beetles
Rye	Ergot, now mold, leaf and stem rusts, stalk and head rots, blotch, root rots, anthracnose	Grasshoppers, aphids, cutworms, armyworms
Safflower	Rust, phytophtora root rot, verticillium wilt, fusarium wilt, leaf spot, bud rot	Flower thrips, lygus bugs, bean and peach aphids, wireworms, loopers
Sesame	Charcoal rot, alternaria leaf spot, bacterial leaf spot, bacterial blight, phyllody, root wilting	Whitefly, til gallfly
Sorghum	Loose kernel smut, covered kernel smut, head smut, pythium root rot, bacterial spot, crazy top, downy mildew, fusarium stalk rot, leaf blight, milo disease, rhizoctonia stalk rot	Chinch bugs, corn earworm, leaf aphid, corn borer, armyworm, sorghum webworm, southwestern corn borer, sorghum midge, greenbug
Soybean	Phytophthora root rot, pythium rot, rhizoctonia rot, fusarium rot, anthracnose, soybean mosaic, bacterial blight, downy mildew, stem canker, pod and stem blight, soybean rust	Mexican bean beetle, seed corn maggots, seed corn beetles, wireworms, white grubs, thrips, southern corn rootworms, Japanese beetles, grasshoppers, alfalfa hoppers, blister beetles, cabbage loopers, stink bugs, velvet bean caterpillars, spider mites, soybean aphid
Sugar beets	Beet yellow and western yellow virus, curly top, cercospora leaf spot, powdery mildew, downy mildew, rust, mosaic, rhizoctonia	Aphids, leafhoppers, root maggots, wireworms, white grubs, flea beetles, cutworms, sugar beet crown borers, beet web worms, armyworms, alfalfa loopers, grasshoppers
Sugarcane	Mosaic, gumming disease or gummosis, red rot, smut, ratoon stunting disease	Corn aphid, cane leafhopper, moth borer
Sunflower	Rust, downy mildew, head mold, sclerotinia rot	Salt marsh caterpillar, stalk borers, sunflower moth
Wheat	Stem rust, leaf rust, stripe rust, powdery mildew, bunt, loose smut, stinking smut, wheat scab, root rots, crown rots	Hessian fly, wheat stem sawfly, wheat jointworms, strawworms, chinch bugs, aphids, grasshoppers
Wild rice	Bacterial brown spot, bacterial leaf streak, anthracnose, ergot, fungal brown spot, phytophthora, scab, spot blotch, stem rot, stem smut, eye spot, wheat streak mosiac	Wild rice worm, wild rice midge, rice stalk borer, rice water weevil, rice leafminer, wild rice stem maggot

a combination of preventive and timely responsive actions based on the risks associated with various cultural practices and pest activity observations collected through periodic field inspections. Management practices for controlling insect pests include the use of cultural practices (e.g., timing of planting, tillage, crop rotation); seed, soil, and foliar applied insecticides; biological control (reducing pest populations through use of natural predators); and insect resistant crops (varietal/hybrid tolerance to insect pests).

Disease Control

A host of plant diseases limit the yield potential of field crops (Table 17–6). Although some diseases can be controlled by a single practice, such as planting a resistant hybrid, most diseases require a combination of practices to ensure that economic damage is kept to a minimum. Once a disease has been identified, its prevention and/or control depends on understanding its cause(s), the factors that favor the disease, which plant parts are

affected, and when the disease organisms are spread. Sometimes a change in growing practices can reduce a disease (or other pest problem) but in some cases, it can promote it. In the United States gray leaf spot (GLS), caused by the fungus *Cercospora zeae-maydis,* was a relatively minor disease problem of corn until the advent of no-till crop production, However, following widespread adoption of no-till cropping in continuous corn production in the southern United States in the early 1970s, GLS became a serious, and eventually a major foliage disease problem of corn because the pathogen survives in the stubble. Many Midwestern states were affected by a major GLS epidemic in 1995. Annual crop rotation in no-till corn greatly reduces the risk of GLS.

HARVEST AND STORAGE

Most agronomic grain crops are mechanically harvested with combines, which separate the heads, pods or ears from wheat or soybean stems or corn stalks, and shell grain from the pods or cobs in one operation. In the US corn belt, corn and soybeans are harvested in the fall; wheat, barley, and oats are harvested in the late spring or early summer.

For most crops, there is a preferred grain moisture content at which to harvest. The ideal kernel moisture level to harvest corn for dry grain storage is 23 to 24 percent. The yield potential can drop considerably if harvesting is delayed much beyond maturity and if poor stalk quality causes stalk lodging. Stalk and root lodging may slow harvest and contribute to yield losses. The loss of one normal size ear per 30 m (100 ft) of row translates into a loss of more than 67 kg/hectare (1 bu/acre). An average harvest loss of 18 kernels/sq m (2/sq ft) is about 67 kg/hectare (1 bu/acre). According to an Ohio State University agricultural engineering study, most harvest losses occur at the gathering unit, with 80 percent of the machine loss caused by corn that never gets into the combine. There is a significant grade and price penalty for grain with a high moisture content. For that reason moisture sensors are used on harvesters to monitor grain moisture. Grains with a moisture content above 14 percent have to be dried to below 14 percent for long-term storage to limit fungal growth and mycotoxins production.

Following the harvesting of small grains, some farmers harvest the straw residue. Straw, especially oat straw, makes excellent bedding for livestock, and the straw is often harvested for that purpose. Straw is also used as a mulch. Straw harvest has become an important income source, especially near urban areas, partly because of its use as a mulch. Some farmers burn the residue and stubble after harvesting the grain to reduce trash and help control some diseases (e.g., rice). This practice is decreasing, however, because the value of residue in conservation tillage is more widely recognized and air pollution laws banning or limiting such fires are becoming more common.

FIELD CROPS GROWN FOR FOOD, FORAGE, FIBER, AND INDUSTRIAL USES

Food Crops

Worldwide food scarcity in the years ahead is a real possibility unless measures are taken to both control population and increase food production. The need for more efficient crop production and reduction of environmental pollution is clear, and a likely area for success with food crops is increased grain production.

Increased food production must come primarily from crop plants with high caloric output per unit area of land. Generally these crops are high carbohydrate producers, and they include cereal grains, potatoes, cassava, and sugar crops. Cereal grains are a concentrated source of energy, and they are easily processed, stored, and distributed. They directly supply the world's population with about 80 percent of its total food calories. The word *cereal* comes from the name of the goddess Ceres, "giver of grain," who is said to have given wheat to the early civilizations. It soon becomes evident that the true cereals are grasses, members of the family POACEAE. Other families are considered cereals because their seeds are used in ways similar to those of the grasses. The cereals are humans' principal food source because of their adaptability to many climates, soils, and handling methods. They are efficient converters of light energy into food, are hardy, produce many seeds per plant, store well, and are readily processed into many uses, including their conversion into meat through livestock feeding. Table 17–1 on page 367 identified the world's major field crops and the principle uses of each. Several of these crops have multiple end uses. For example, corn (maize) and soybean are used for feed and have many industrial uses, including the production of ethanol and biodiesel. Polymers derived from soybeans and corn are likely to play a greater role in the future in the manufacture of paints, adhesives, rubber, and detergents. In Europe, especially England and Ireland, wheat straw is used in the making of thatch roofs.

Some of the important field food crops are discussed in alphabetical order.

Figure 17–5

Barley. Source: USDA Agricultural Research Service Image Gallery, *http://www.ars.usda.gov/is/graphics/photos/*

Barley (*Hordeum vulgare* L.) POACEAE

Barley, a widely adapted small-grain cereal, is used for human food, livestock feed, and malting (Fig. 17–5). Barley and wheat are the most ancient of all cereal grains, both dating back 9,000 years. Barley apparently was grown in Mesopotamia in prehistoric times. It is believed to be native to southwestern Asia, but many wild species grow in Ethiopia and southern Tibet. There are two main species: *H. vulgare,* the common six-rowed barley (six rows of kernels per spike) grown mainly for food; and the two-rowed barley *H. distichon,* whose ancestor is probably the wild barley *H. spontaneum,* used in many parts of the world for malting. Malting is the process of germinating and then drying of the grain for use in converting starch to alcohol during fermentation. The crop is classified according to its growth habit as spring (earlier maturing) or winter barley.

Barley is widely adapted because of its drought resistance, tolerance to alkaline and saline soils, and early maturity. Some cultivars grow in subarctic climates, some only in the temperate zones, and others in the subtropics, although barley is not well adapted to hot, humid conditions. The crop may ripen in as few as sixty to seventy days, but normally ninety to 120 days are required. The principal production areas are the former Soviet Union, United States, Europe (Germany, France, United Kingdom, and Spain), and Canada. In the United States, barley is grown mostly in North Dakota, Idaho, Washington, Minnesota, and Montana.

Beans (*Phaseolus* spp. L., *Pisum, Vigna* spp. *Vicia* spp. L., etc.) FABACEAE

The seeds of several FABACEAE species of field beans and related field peas or cowpeas are grown for food and are a very important caloric and protein source (Fig. 17–6). These include members of several genera. Various types of beans of the genus *Phaseolus* include the lima, scarlet runner, snap, shell, white, black, pea, blackeye, kidney, tepary, black gram, and mung. Other important beans are the broad bean *Vicia faba*; chickpea or garbanzo *Cicer arietinum*; the cowpea, also known as the blackeye pea *Vigna unguiculata* or *V. sinenis*; the dried pea *Pisum sativum*; lentil *Lens esculenta*; pigeon pea *Cajanus indicus* and peanut *Arachis hypogaea.* Many other legumes having local importance are grown for human and animal food. Pulse crops are legumes whose seeds and/or pods are consumed.

The blackeye pea *Vigna unguiculata* is one of the oldest cultivated crops. It is native to Central Africa, and it spread to Asia and the Mediterranean basin; it is now grown in many areas. The red kidney bean *Phaseolus vulgaris* originated in Central and South America and was unknown in Europe before Columbus. The broad bean *Vicia faba,* with its origin probably located in North Africa or the Near East, is an old world species grown by the ancient Egyptians and Greeks. The pea *Pisum sativum* was probably domesticated in central and western Asia.

Beans are warm-season crops and are best grown for dry seeds in areas where temperatures are warm but not excessively high, and where the growing season is long enough to avoid frosts. Such areas are found in many temperate regions of North and South America, Europe, Asia, Africa, and Australia. Major production areas are China and India, where dried peas are important. India is also a large producer of lentils. Latin America, the United States, Japan, Italy, and Turkey are also major bean producers. Italy is a leader in the

Figure 17–6

Several different types of beans. Source: USDA Agricultural Research Service Image Gallery, *http://www.ars.usda.gov/is/graphics/photos/*

A

B

Figure 17–7

(A) The tassels at the top of the corn plant are the male flowers. (B). The cob develops from the female flowers. Source: USDA Agricultural Research Service Image Gallery, *http://www.ars.usda.gov/is/graphics/photos/*

production of broad beans. The major production areas of the United States are Michigan, Nebraska, California, Colorado, Idaho, and North Dakota.

Corn/maize (*Zea mays* L.) POACEAE

The botanical origin of corn, known in much of the world as maize, is vague and is mostly based on evidence from specimens discovered in caves in the Tehuacan Valley of Mexico. The earliest samples appear to date from 5000 BCE, although maize pollen fossils have been found that are about 80,000 years old. The modern history of corn begins with the first voyage of Columbus, who discovered not only the Americas but also corn. Later explorers found Native Americans growing corn in all parts of the Americas from Canada to Chile. The first Indians to plant and cultivate the crop lived in Mexico and South America.

Corn is a tall annual plant with strong erect stalks, a fibrous root system, long narrow leaves spaced alternately on opposite sides of the stem, and separate male and female flowers on the same plant (monoecious) (Fig. 17–7). Now grown worldwide in many environments, corn originated in the Americas as a short-day, long-season crop. The increase in diversity of production area environments is the result of the breeding and selecting of cultivars that mature rapidly and are nearly day neutral. Several botanical varieties of *Zea mays* having broad usage and economic importance are grown. The various types of corn can be differentiated on the basis of endosperm composition—usually the relative proportions of soft versus hard endosperm or starch (Fig. 17–8). Field corn is grown to full maturity and the starch-filled dried seeds are harvested or the stalks are cut green for silage. Sweet corn is grown for its immature seed which contain a high amount of sugar.

- Dent corn (*Z. mays* var. *indentata*) is the principal commercial feed type grown in the United States. The grain is normally yellow, hard, and horny. The kernel is comprised of hard starch with an overlay of softer starch in the crown. The soft endosperm in the crown shrinks on drying more than the hard innermost endosperm, resulting in a dent forming in the top of the kernel. White kernel corn use is increasing in the United States because of increasing Latino populations and the popularity of their foods, which use this kind of corn.
- Flint corn (*Z. mays* var. *indurata*) kernels are not indented when dry and are comprised mostly of hard starch.
- Flour corn (*Z. mays* var. *amylacea*) is also known as soft or squaw corn. It has little hard starch. It was preferred by early Native Americans because it is easily hand-ground into flour.

Figure 17–8

Different types of corn. Source: USDA Agricultural Research Service Image Gallery, *http://www.ars.usda.gov/is/graphics/photos/*

- Popcorn (*Z. mays* var. *praecox*) is a type of flint corn with a very high proportion of hard starch endosperm. The kernels and ears are smaller than those of dent corn.
- Pod corn (*Z. mays* var. *tunicata*) kernels are enclosed in husks or pods, and the ears are covered with husks. This curiosity is not grown commercially.
- Sweet corn (*Z. mays* var. *saccharata*) kernels have sugary kernels when immature; when dried the seeds are wrinkled and somewhat translucent. Both the immature (corn on the cob) and dried seeds are used as food. This type is a popular and important vegetable crop in some nations. The production of sweet corn is discussed in Chapter 19.

Corn has a multitude of food and industry uses. Corn finds use in the production of flour, starch, sugar and syrups, corn oil, alcohol as a beverage and a fuel, adhesives, coatings for paper and fabrics, and so on. Considerable attention is given to improvement of its nutritional value because of its extensive use as food for humans as well as animals. Much of this is aimed at improving its protein value by increasing the content of the amino acids, lysine, and tryptophan, which are normally low in corn grain.

Corn oil content is typically 4 percent. However, high-oil corn contains up to 7 to 8 percent oil. High-oil corn is attractive as livestock feed because it has greater energy than conventional dent corn and can replace more expensive dietary sources of fat. High-oil corn hybrids have not been widely used because their grain yield potential is usually lower than conventional corn hybrids however production practices are being developed to increase yield.

Optimal production of corn requires an ample and continuous supply of available soil moisture. The annual precipitation in the US corn belt varies from 60 to 115 cm (24 to 45 in.), with about one-fourth occurring during the summer months; thus, irrigation is usually not provided. Corn is a warm-season, C4 crop, requiring relatively high day and night temperatures. The best daytime high temperatures range from 20°C to 27°C (68°F to 81°F), and with night-time lows not less than 14°C (57°F). Seeds do not germinate well at temperatures lower than 10°C (50°F), and temperatures above 40°C (104°F) harm pollination.

In the United States, corn is well adapted to the Midwestern states where the term *corn belt* appropriately includes Iowa, Illinois, Nebraska, Minnesota, Indiana, Ohio, Missouri, and the Dakotas. Considerable quantities are also produced in other parts of the United States. In the corn belt, a large proportion of the crop is

Figure 17–9
Open air corn crib. The corn continues to dry naturally after harvest because the crib is ventilated, however rodents and insects can get into the crib causing loss. Source: USDA Agricultural Research Service Image Gallery, *http://www.ars.usda.gov/is/graphics/photos/*

fed to animals, notably hogs and cattle, for meat production. Most corn is used as livestock feed, but an increasing percentage is used for industrial purposes, including ethanol (as a gasoline substitute) and high fructose corn syrup (sugar for soft drinks). On-farm use accounts for about 35 percent of that produced. About 10 percent of the corn grown is converted into silage and/or feed as fresh fodder to animals. Other major corn producing countries include China, Brazil, the European Union, and Mexico.

As with many grains, corn must be dried before storage. A corn-drying method on the decrease, because it requires extra handling, is the placement of ear corn in covered cribs with slotted sides that permit ventilation (Fig. 17–9). The kernels are shelled at a later time. A moisture level of 20 to 22 percent will permit safe storage in cribs because there is air-flow through the slots. The problem of rodents entering the crib, however, now has the bulk of the corn production stored in enclosed bins or buildings that require the grain be dried to a moisture content below 14 percent for long-term storage (see Fig. 16–20 on page 354). There is a grade and price penalty for corn with higher moisture. High-moisture corn is subject to heat damage and spoilage. For a general discussion of storage and drying in general see Chapter 16.

Oats (*Avena sativa* L.) POACEAE

The origin of oats is unclear, but they are thought to have originated in Asia Minor or southeastern Europe.

Oats grew wild in western Europe during the Bronze and Iron Ages. Following domestication, their culture spread to other temperate-zone regions. Oats are principally grown for animal feed, especially horses, because the crop is easily produced, and the grain and foliage are nutritious. Substantial amounts are also processed for human foods. World production has declined because tractors have replaced horse-drawn implements and thereby the need for feed for those animals.

Oats grow to 60 to 120 cm (2 to 4 ft) tall, and have a fibrous root system. The flowers are borne on panicles, either nearly symmetrical or one-sided. The panicle is made up of numerous (20 to 120) small branches and spikelets composed of two glumes and usually two florets (except hull-less cultivars, which contain more).

Oats, like barley, are well adapted and are produced throughout the temperate zones. Important oat-producing countries are the former Soviet Union, the United States, Germany, Canada, and Poland. Major US producers are South Dakota, Minnesota, North Dakota, Iowa, and Wisconsin. Northwestern Europe, some northern areas of the United States, and southern Canada produce some late-maturing common white oats.

Rice (*Oryza sativa* L.) POACEAE

Knowledge of the ancient origin of rice is unclear, but most likely it was southeastern Asia. Some very early Chinese and Indian writings mention rice cultivation for more than 5,000 years in that part of the world. The relative importance of rice, wheat, and corn as human foods cannot be debated. Rice is most important in the tropic and semitropic zones and has significant production in some temperate regions.

The cultivated plant, customarily grown as an annual, is a semiaquatic annual grass that grows erect. It has narrow parallel-veined leaves. Spikelets are borne on a loose panicle (Fig. 17–10).

Rice is classified several ways. Based on the chemical characteristics of the starch and grain aroma, rice can be classified into three groups: (1) waxy or glutinous types (starchy endosperm contains no amylose[1]); (2) common types, more or less translucent nonglutinous (endosperm contains one-fourth amylose and three-fourths amylopectin[2]); and (3) aromatic or scented types, grown in India and southeast Asia. The glutinous

Figure 17–10
Rice seed heads beginning to ripen. Source: USDA Agricultural Research Service Image Gallery, *http://www.ars.usda.gov/is/graphics/photos/*

types, consisting entirely of amylopectin, are grown primarily for a specialty market in the United States but represent about 10 percent of the total production in China and 8.4 percent in Japan. Worldwide, the first two types comprise over 90 percent of the total rice grown.

Rice cultivars are classified as lowland rice (continuously or pond flooded) or upland rice (nonirrigated, or irrigated but not continuously pond flooded) (Fig. 17–11). Upland and lowland do not refer to elevation. More than 80 percent of the world's rice grown is the lowland type. Rice cultivars are also classified on the basis of kernel characteristics into short-, medium-, or long-grain types. The average length of unhulled kernels is 7.2, 8.4, and 9.9 mm respectively. Most northern Asian people who eat rice prefer the short-grain types (also known as pearl rice) because the kernels are sticky when cooked. Most Americans and Europeans prefer the nonsticking, long-grain types.

Another rice classification is based on maturity. The rate of maturity is genetically controlled and is measured by the number of days required for the plants to reach 50 percent heading.

Figure 17–11
Young lowland rice plants growing through the water held by earth levees in a flooded rice paddy.
Source: U.S. Soil Conservation Service.

---

[1] Amylose is a component of starch characterized by the tendency of its aqueous solutions not to gel.
[2] Amylopectin is a component of starch characterized by the tendency of its aqueous solution to set to a stiff gel at room temperature.

Rice is also classified on the basis of its cultured adaptation as japonica or indica types. The japonica cultivars are usually short-grain and adapted to a temperate climate, while the indica types are long-grain and tropical. Intercrossing of japonica and indica types has led to a breakdown of this traditional classification, and some long-grain varieties are well adapted to temperate areas.

Successful rice production depends on ample water availability, and thus important production areas lie along the great rivers and deltas of Asia and other regions. Asia produces over 90 percent of the world's supply. Major producers are China, India, Indonesia, Bangladesh, Thailand, Japan, Burma, and Brazil. Five states in the United States—Arkansas, California, Louisiana, Texas, and Mississippi—produce 95 percent of the US crop, this being only a little more than 1 percent of the world's total.

Rice grows well on moist soils. For best yields, the fields are flooded for most of the growing season; thus clay, clay loam, or silty clay loam soils are most frequently used to conserve water by minimizing seepage losses. Organic and light-textured soils are used provided they have an underlying hardpan or claypan that prevents water seepage losses. The soil pH ranges between 5.0 and 7.5 for satisfactory plant nutrient availability. Upon flooding, acid or alkaline soils shift slightly toward neutrality. Rice cultivars vary in their tolerance to salinity, but all are adversely affected.

In most Asian countries, the centuries of simple, nonmechanical methods of flooding, terracing, leveling, and other wetland rice cultural operations have produced soils with common characteristics (Fig. 17–12).

Figure 17–12
Simple equipment drawn by draft animals is still used in many areas of the world for rice production. Source: USDA Agricultural Research Service Image Gallery, http://www.ars.usda.gov/is/graphics/photos/

The immediate soil surface is oxidative and the subsoil strongly reducing (oxygen deficient). These soil conditions create what is called a rice paddy soil. The fields are harrowed crosswise and lengthwise until the soil is well puddled (a soft muddy mass). This helps create a hardpan that limits water percolation and facilitates the hand transplanting operation. Hand transplanting is done in tropical areas and much of Asia to establish the crop before seasonal rains (monsoons) occur, or to meet short growing season conditions, and/or to permit double cropping of the soil.

In most developed countries, seedbed preparation is completely mechanized. Direct seeding is practiced either by conventional drilling on dry soil or by seeding, often using aircraft, into flooded fields. Using this method, seedbeds are finished with large equipment that leaves the surface grooved, or the surface is left rough to prevent seed drift during flooding.

More and more rice farmers worldwide are recognizing and using the improved, high-yielding rice cultivars developed by the International Rice Research Institute (IRRI) in the Philippines or cultivars developed for local adaptation from IRRI germplasm. The choice of seed is an important consideration, and many rice-growing areas have seed certification programs designed to provide high-quality seed. Such seed is varietally pure, produces at least 80 percent germination, and is free of weed seeds and other impurities. In recent years a nutritionally enhanced rice was produced that contains more than 20 times more beta-carotene (a precursor of vitamin A) than most currently grown varieties. This transgenic type, referred to as Golden Rice, is being developed for use in rice- growing areas where local diets often result in vitamin A deficiencies that can lead to blindness, increased susceptibility to diseases, and other medical disorders.

Rice culture has always been and remains a labor-intensive endeavor in many Asian and African countries. Production requires more than 740 labor-hours per hectare (300 hr/ac) in most of Asia and Africa. In the United States and Australia, only 18.5 hours/ha (7.5 hr/ac) are needed because of mechanization (Fig. 17–13). This reduction in back-breaking hand labor is possible because of the substitution of machines for human labor. However, it may not be feasible or even desirable for all countries to mechanize rice production to the extent it is in the United States because of their social and economic situations. Many rice-growing countries are overpopulated, the people need jobs, and the farms are too small for efficient use of large machines.

Figure 17–13
In developed countries or regions, machinery such as this harvester are used in rice production. Source: USDA National Resource Conservation Service, *http://photogallery.nrcs.usda.gov/*

Rye (*Secale cereale* L.) POACEAE

Southwestern Asia is assumed to be the area of origin for rye. There is evidence of its early cultivation in western Asia, but domestication was relatively recent because rye is not mentioned in early Egyptian records. Because of its extreme hardiness it was, and is, widely grown in much of Europe, Asia, and North America, though usually not in large quantities. Rye does best in a cool dry climate and tolerates cold temperatures very well. It also performs well on poor and marginal soils. Therefore, it is often grown where environmental conditions are unfavorable for other cereal crops.

Rye is used for human consumption mainly as flour for bread, but most is grown for hay, pasture, and stock feed (Fig. 17–14). It also finds wide use as a cover crop (Fig. 17–15) and for the manufacture of beverage alcohol.

There are both spring and winter cultivars, but because rye is very winter hardy, most of the rye grown is fall-seeded. Major world production areas are the former Soviet Union, Poland, Germany, the United States, and Canada.

Ergot is a fungal disease often associated with rye. It produces small black bodies called *sclerotia* in the grain. The sclerotia, if present in quantity, are poisonous to animals and humans but can be used in pharmaceutical products.

Sorghum (*Sorghum bicolor* Moench), *S. vulgare* Pers, POACEAE

Sorghum is believed to have originated in Africa. Grain sorghums are known by several names, such as durra, Egyptian corn, great millet, or Indian millet. In India, sorghum is known as jowar, cholum, or jonna. In the United States, different types of grain sorghum are known as milo, kafir, hegari, feterita, shallu, and kaoliang.

In both Africa and Asia, sorghum is one of the major cereal grains. This important crop was introduced to the United States in the middle 1800s. Major world producers are the United States, India, China, Argentina, Mexico, and Nigeria. Resistant to heat and drought, sorghums having effective and extensive root systems are best adapted to warm and semiarid regions. The principal US production areas are Kansas, Texas, Nebraska, Missouri, and Arkansas, but significant acreages are grown in other states.

Figure 17–14
A rye crop ready for harvest. Source: USDA Agricultural Research Service Image Gallery, *http://www.ars.usda.gov/is/graphics/photos/*

Figure 17–15
Rye being grown as a cover crop. Source: USDA National Resource Conservation Service, *http://photogallery.nrcs.usda.gov/*

Agronomically, sorghum is a common catch-all name applied to all plants of the genus *Sorghum,* which includes numerous grain and foliage types (Fig. 17–16). However, the plants do have markedly different characteristics and uses, and have often been grouped into four types:

1. The grain sorghums (Caffrorum group) are the nonsaccharine plants, including milo, kafir, feterita, hegari, and hybrid derivatives, among others. These plants are grown for grain used principally for poultry and livestock feed. The grain is similar in composition to corn, but somewhat higher in protein and lower in fat. Grain is ground into meal and made into bread or porridge. Whole grains are sometimes popped or puffed for cereal. Grain sorghums are also used to make dextrose, starch, paste, and alcoholic beverages. The stalks have a dry pith and are not very juicy, except for milo and kafir, which are semijuicy, dual-purpose types for grain and forage.

2. Sweet or forage sorghums or sorgos (Saccharatum group) are used mainly for forage and silage and to make molasses. The stalks are juicy and sweet. This type is principally grown in the United States and South Africa.

3. Broom corn (Technicum group) is a sorghum and a panicle woody plant. It has dry pith, little foliage, and fibrous seed branches 30 to 90 cm (1 to 3 ft) long that are often made into brooms.

4. Grass sorghum (*S. Sudanense*) or Sudangrass is grown for pasture, green chop, silage, or hay. Sudangrass is usually ready to pasture in five to six weeks, but to avoid prussic (hydrocyanic) acid poisoning, it should be at least 45 to 60 cm (18 to 24 in.) tall when grazed. The acid in the form of glucosides occurs in young plants and in new shoots. It is highly toxic because it inhibits cellular oxidative processes.

Plant breeders have developed numerous new grain sorghum hybrids and cultivars with specific adaptation to local regions or conditions. Many cultivars are designated by their rates of maturity. The development of hybrid seed and short stalk cultivars has greatly improved yields and harvesting efficiencies. The most important production factor in grain sorghum is selection of the correct cultivar for a given area.

Sugar Beets (*Beta vulgaris* L., J. Helm)
CHENOPODIACEAE

Beets are believed to have their center of origin in the Mediterranean area. Cultivated beets were probably domesticated from the wild beet (*Beta maritima*). Within the one species *B. vulgaris,* there are several useful types. These include the coarse mangel-wurzel (also known as fodder beet—grown for animal feeding), the sugar beet, the table beet, chard, and other foliage-type beets. The last three types are commonly used vegetables.

The sugar beet is a biennial plant grown as an annual plant. The root enlarges the first year as a storage organ containing high amounts of sucrose to provide growth of the plant the second year. The root is harvested for its sugar content at the end of its development the first year. Sugar beets are by far the most important beet grown commercially. Their average sugar concentration is about 15 percent, with many crops exceeding 20 percent. The roots are typically sharply tapered, white-skinned, and white-fleshed (Fig. 17–17). Major

Figure 17–16
Different types of sorghum grain. Source: USDA Agricultural Research Service Image Gallery, http://www.ars.usda.gov/is/graphics/photos/

Figure 17–17
Sugar beet. Source: USDA Agricultural Research Service Image Gallery, http://www.ars.usda.gov/is/graphics/photos/

growing areas are the former Soviet Union, France, Germany, the United States, Poland, and Turkey. Leading US producers are California, Minnesota, Idaho, North Dakota, and Michigan.

Hybrid cultivars have been developed that are resistant to bolting (development and growth of a seed stalk) and to certain viral diseases. They were also developed for improved adaptation to specific areas and conditions. The recent introduction of sugar beet cultivars with Roundup® herbicide resistance offers another method of weed control.

Beets are best harvested when the sugar content is highest, but often the optimum harvest date does not coincide with when the processing mill can or will accept the crop. Because sugar beets are usually grown on contract with a sugar-processing plant, the crop, instead of having the harvest delayed, may have to be harvested before optimum maturity to meet contract terms.

Sugarcane (*Saccharum officinarum* L.) Poaceae

Sugarcane supplies much of the world's sugar as sucrose. In many areas where production and labor costs are low, sugarcane provides sugar at a lower cost than the sugar provided from sugar beets. Sugarcane is said to be one of the most efficient converters of solar energy and carbon dioxide into chemical energy, possibly producing more calories per unit of land area than any other crop. Sugarcane is the basis of Brazil's ethanol for biofuel industry.

Sugarcane probably originated in India, where it has been grown since ancient times. Its earliest mention is found in Indian writings from about 1400 to 1000 BCE. From India, sugarcane spread to China, Java, and other tropical Pacific islands. It also spread westward from India to Iran and Egypt. Columbus introduced the crop to the West Indies. Sugarcane was first planted in the United States in 1751 near New Orleans.

Sugarcane is a tropical plant that matures in twelve to eighteen months. It is a tall perennial grass, which often attains a height of 2 to 4 m (Fig. 17–18). Sugarcane has essentially the same structure as other members of the grass family.

Sugarcane is grown in tropical regions around the world. The main production areas are the warm, humid, tropical lowland regions of North, Central, and South America, the southern United States, the Caribbean, Africa, Asia, and Oceania, including Australia. The leading countries producing sugarcane are India, Brazil, Cuba, China, Mexico, Thailand, Australia, the United States, and South Africa.

Figure 17–18
Field of sugar cane. Source: USDA Agricultural Research Service Image Gallery, *http://www.ars.usda.gov/is/graphics/photos/*

Sugarcane is propagated vegetatively by planting sections of the stems containing three or four nodes into furrows. The stem sections are sometimes planted by hand but are usually planted mechanically. Plantings are made when soil temperature and moisture are suitable. In many areas, a perennial-like cropping is practiced, with the new crop re-sprouting from previously cut stems left in the soil. However, new plantings of selected and treated stem cuttings are more productive and reduce disease problems.

Most of the sugar in sugarcane is found in the stems which have to be crushed to extract the sugar. Before harvest, randomly selected canes are tested for maturity and the juice is assayed for sugar content with a hand refractometer. Testing generally starts four to six weeks before the proposed harvest date. As with sugar beets, the mill needs a steady supply of canes, which farmers provide by coordinating their harvest. Thus, the rate of harvest is governed by the crushing capacity of the mill. Sugarcane is harvested either by hand or mechanically. In the West Indies, the canes are generally sent to the mills remarkably free from undesirable leaves and stems because they are removed when canes are harvested by hand. In other areas, especially where labor costs have hastened the adoption of mechanization, delivery of trashy canes to the mill has been a serious problem, although the trash and waste (bagasse) is used as a fuel by the mill. To reduce trashiness and facilitate mechanical harvesting, firing the cane before cutting is a normal practice. Burning mature cane does not harm it unless the fire is exceptionally hot. The work of cutting and loading is greatly reduced by preharvest burning. The top of the plant and the attached young leaves then need to be removed because they contain two invert sugars (fructose and glucose), nitrogen compounds, and starch—all of which interfere with the extraction of sucrose sugar.

Wheat (*Triticum aestivum, Triticum* spp. L.) POACEAE

Wheat and barley are recognized as being among the most ancient crops. The Egyptians and Mesopotamians grew wheat as well as barley and oats. Two wild wheat species are still found growing in Syria and Turkey, where wheat probably originated and was domesticated. It is known that present-day species originated from the hybridization of several different species (Fig. 17–19).

Wheat leads all of the cereal grains in total volume. Countries that lead in production are China, the countries of the former Soviet Union, the United States, India, France, Canada, and Australia. Wheat is grown in almost every state in the United States, with the leading states being Kansas, North Dakota, Oklahoma, Washington, Texas, South Dakota, Minnesota, and Colorado.

Wheat is classified into market classes by the color and composition of the grain and the plant's growing habits; the latter also determines the production area of each class. The four main classes are (1) hard red spring, (2) durum, (3) hard red winter, and (4) soft red winter. A fifth class, white, is also grown in some regions of the world including parts of the United States.

Wheat has a relatively broad adaptation, is very well adapted to harsh climates, and will grow well where rice and corn cannot. Early growth is favored by cool and moist conditions, with warmer and drier weather toward crop maturity.

Figure 17–19

Field of wheat ready for harvest. Source: USDA Agricultural Research Service Image Gallery, *http://www.ars.usda.gov/is/graphics/photos/*

Generally the winter climate of a particular area determines whether winter or spring types are grown. If winters are severe, spring-type cultivars are planted in the spring to be harvested in the late summer and the fall. If winters are not extreme, winter cultivars are planted in the fall for spring harvest. If winter temperatures are mild, spring-type cultivars are planted in the fall for spring harvest. Another generalization is the different composition of the grain of the different cultivars. The hard wheats contain more protein (13 to 16 percent) and are usually grown in the drier climates. The soft wheats are more starchy, have a lower protein content (8 to 11 percent), and are usually grown in more humid climates. Each of these wheats has different characteristics and different uses.

In the United States, the hard red spring cultivars are grown in the northern Great Plains, the Dakotas, Montana, and Minnesota, and produce a high-grade wheat used principally for bread flour. Durum cultivars, used to make semolina flour for macaroni and other pasta products, are also grown in these areas and some other areas. Hard red winter wheat, which leads in the total volume of production, is used for bread flour and is grown across a broad area ranging from Utah to Illinois. Soft red wheat production is concentrated in Ohio, Indiana, and southern Illinois. This wheat is milled into flour for cakes and pastries. White wheat also has hard and soft types. In the United States, the soft type grown in the Pacific Northwest and parts of Michigan and New York. White wheat is produced in several regions of the world, including the United States though in much less quantity than the other types. It is used in whole wheat products, some breads, and oriental noodles. It is gaining popularity for use in whole grain products because the bran does not contain the tannins that give traditional whole wheat products a bitter taste that is unpleasant to some consumers. Grain fed to livestock is most often that of the soft wheat types.

Wheat grain is processed by milling, a procedure that removes the outer bran layer. In doing so, much of the grain's protein is lost because the bran layer contains the highest concentration of protein. This is the concession made to produce white flour with better baking characteristics. Whole wheat refers to wheat that is partially milled, and it has become a much more popular product because of its greater nutritional contribution.

Because of its importance as a basic food staple to so much of the world's population, extensive breeding programs have been undertaken in practically every wheat-growing country in the world. Breeders continue

to introduce new cultivars that are more resistant to diseases, insects, drought, lodging, and shattering. Plant breeders have improved quality (size, texture, weight) and increased yields and winter hardiness.

Wheat is grown on a wide range of soils in temperate climates where annual rainfall ranges between 30 and 90 cm (12 to 36 in.). Such areas constitute most of the grasslands of the world's temperate regions. Many of these soils are deep, well-drained, dark-colored, fertile, and high in organic matter, and they represent some of the world's best soils. The prairie soils of the United States and Canada and the steppes of the former Soviet Union are examples of such soils.

Oil Crops

In addition to the use of animal oils, early humans also used various oils from crop plants as a food source and for a primitive fuel. Some of the earliest written records indicated the use of plant oils for illumination, heat for cooking and warmth, and anointing the skin. Olive oil was widely used for these purposes by early Mediterranean civilizations. Linseed oil from flax was recognized for its usefulness in paints. Castor bean oil was used for lubrication and also in paint and varnish products, cosmetics, and as a cathartic. With technological development, the use of various oil crops expanded. These oils function in many applications that can be very specialized and often not very obvious. After the discovery of petroleum and the technology for its use in the internal combustion engine, the primary source of oil for energy purposes became the earth's underground supply. Concern about the future supply of petroleum compels a strong interest in the oil crop plants as a supplemental source. These crops represent renewable sources, whose importance will increase even more in the future. The better-known uses of plant oils are for cooking, flavoring, margarine, and salad dressings. A multitude of other uses include the manufacture of plastics, paint, varnishes, lacquers, soaps, detergents, inks, cosmetics, lubricants, medicines, fabric, and paper.

Oil is found in all living plants, even in bacteria and fungi. Certain plant tissues, usually the seed, are the most abundant in oil content. Although some seeds have very little oil, others have an oil content greater than 50 percent. Relatively few plant species provide the majority of the vegetable oil production.

The oil is usually removed by crushing and pressing; other extraction methods, often combined with pressing, include extraction with steam and various solvents.

Vegetable oils, readily seen as small droplets when expressed from tissues, are mostly a mixture of triglycerides with small amounts of mono- and diglycerides that are characterized by the content of their various fatty acids. They become degraded (rancid), some more rapidly than others, because of the breakdown of glycerol into other compounds.

A characteristic of oil is its degree of saturation of the fatty acid molecules. Unsaturated fatty acids have one or more double bonds in their structure and therefore bond less hydrogen. These generally are liquid, or have a melting point about 20°C. The melting point depends on the degree of saturation and also on the molecule's carbon chain length. An increase in saturation and in the number of carbon atoms increases the melting point. From a dietary standpoint, unsaturated is preferred because it is less likely to cause high levels of cholesterol in humans than saturated fat. Hydrogenation is a chemical process that is used to increase the degree of saturation. This is usually done to solidify an oil to a fat. This practice is decreasing because it also increases the amount of trans fat in a product. Trans fat is linked to poor heart health and has been banned from food sources in many areas.

The various crop plant oils are often roughly grouped by their degree of saturation and also by their ability to absorb oxygen (drying characteristics).

- Oils of highly saturated fatty acids (nondrying) are largely composed of glycerides of mostly saturated fatty acids such as palmitic and oleic that are found in plants of mostly tropical and subtropical origin, for example: palm, coconut, peanut, olive, and castor bean.
- Oils of highly unsaturated fatty acids (drying) are largely composed of higher amounts of unsaturated fatty acids such as linoleic and linolenic which are found in plants of temperate origin such as canola, corn, sunflower, linseed, safflower, and soybean.

Canola (*Brassica* spp.) BRASSICACEAE

Canola (also called rapeseed or rape) contains better than 40 percent oil, and the pressed meal cake makes an excellent feedstuff, containing about 40 percent protein. This crop is an important and significant oil crop in many areas because of its low temperature and broad soil adaptation (Fig. 17–20). Breeding research has greatly improved the quality of canola crop oil by reducing its glucosinolate and erucic acid content. Canola is grown in China, Canada, India, Europe, Australia, as well as the United States.

Figure 17–20
Canola field in flower. Source: USDA Agricultural Research Service Image Gallery, *http://www.ars.usda.gov/is/graphics/photos/*

Castor Bean (*Ricinus communis* L.) Euphorbiaceae

The castor bean is grown for the oil in its seed. The petalless flowers are produced on racemes, from which brown spine-covered capsular fruits normally containing three large seeds develop (Fig. 17–21). Seed oil content is about 50 percent. Although commercial hybrid seed is available, seed shatter and highly indeterminate maturity limit harvest mechanization and expansion of production. The seeds contain the alkaloid ricin, which is poisonous and makes the expressed cake unsuitable for animal feeding. The oil has many commercial uses in paints, resins, plastics, inks, cosmetics, greases, and hydraulic fluids.

Coconut (*Cocos nucifera* L.) Arecaceae

The coconut, in addition to being an important nut crop, is an important oil crop (Fig. 17–22).

Olive (*Olea europaea* L.) Oleaceae

Olives have produced fine oil for centuries, but they are also grown for their fruits.

Palm Oil (*Elaeis guineensis* Jacq.) Arecaceae

This tree was called the prince of the plant kingdom by Linnaeus because of its majestic appearance. The oil palm is botanically related to the coconut palm. When mature, the oil palm tree may attain a height of nearly 30 m (100 ft) but generally it grows no taller than 10 m (32 ft). The flowers grow on a short spadix that develops into a cluster of more than 1,000 drupes (fruits). The fruit matures about six months after pollination.

Figure 17–21
Castor bean pod. Source: USDA Agricultural Research Service Image Gallery, *http://www.ars.usda.gov/is/graphics/photos/*

The oil palm, native to tropical West Africa, is an important source of vegetable oil. It is grown most abundantly along the west coast of Africa. Oil palms are also cultivated to some extent in the rain forest regions of the Congo, Kenya, Indonesia, and Malaysia. There are plantings in Central and South American countries.

The oil palm grows well on a wide variety of soil types. The trees do best on deep, well-drained soils that are neutral to slightly alkaline in reaction, but they are grown successfully on acid soils. In many southeastern Asia areas, the fruit is harvested throughout the year, heaviest in late summer to early fall, lightest in late winter to early spring. Both the shell (pericarp) and the kernel contain oil.

Kernel oil is light yellow in color and is used principally in the manufacture of edible products such as margarine, chocolate candies, and pharmaceuticals. Palm oil from the pericarp tissues is deep yellow to

Figure 17–22
Coconuts on a coconut palm. Source: USDA Agricultural Research Service Image Gallery, *http://www.ars.usda.gov/is/graphics/photos/*

red-brown in color, and thick in consistency. It is used for making soap, candles, and lubricating greases. It is also used in processing tin plates and as a coating for iron plates. In recent years, palm oil's increasing use as a biofuel has resulted in the major clearing and destruction of pristine rainforests in Southeast Asia (especially Indonesia, and Malysia) to make way for palm oil plantations.

Peanut (Arachis hypogaea L.) FABACEAE

The peanut is native to South America. It was introduced into Africa, where it contributed a large part to the diet of the peoples of East Central Africa. From Africa, the peanut was taken to India, China, and the United States, presently the production leaders. Peanut (groundnut) is a widely grown crop because of its adaptation to tropical, subtropical, and warmer temperate regions.

The peanut is an annual plant with sturdy, hairy branches with growth habits from nearly prostrate to upright. After pollination, the flower withers and drops off, and in a few hours, the base of the flower stalk elongates into a unique structure called the peg. The tip of the peg contains the fertilized ovules, which the growing peg carries down and pushes into the soil. After the peg penetrates into the darkness of the soil, it turns horizontally and the ovary begins to swell and grow into the mature underground fruit (Fig. 17–23).

Figure 17–24
Safflower field. Source: USDA Agricultural Research Service Image Gallery, *http://www.ars.usda.gov/is/graphics/photos/*

In the United States, peanuts are produced mainly for grinding into peanut butter, for roasted and salted nuts, and for candy and bakery goods. Some limited amounts are used for livestock feed. In other parts of the world, peanuts are grown mainly for their edible oil and as a valuable protein source. Peanut seed has a content of 40 to 50 percent oil, and from 25 to 30 percent protein. After the nuts are harvested, the stems and leaves are often used for hay.

Safflower (Carthamus tinctorius L.) ASTERACEAEI

Safflower is another of the world's oldest crops. The plant is thought to be native to the Middle East and southwest Asia, where it has been known for centuries. The flowers were first used as a source of red dye for cloth (Fig. 17–24). Safflower, a spiny annual plant, is now grown principally for its seed oil, which yields two types of oil: a polyunsaturated (linoleic) type and a monounsaturated (oleic) type, each having varied uses, such as for margarine, salad oil, and mayonnaise. Some cultivars yield a monounsaturated oil similar to olive oil that is used for cooking.

Safflower is grown commercially in India, Egypt, Spain, Australia, Israel, Turkey, Mexico, Canada, and the United States.

Sesame (Sesamum indicum) PEDALIACEAE

The seed, containing about 50 percent or more oil of excellent quality and stability, is most often used as a salad or cooking oil, and for flavoring. The seed is also used as a garnish for bakery products. The pressed cake is an excellent protein source for livestock feeding.

Figure 17–23
Peanuts after being pulled from the ground. Source: USDA Agricultural Research Service Image Gallery, *http://www.ars.usda.gov/is/graphics/photos/*

Soybean (*Glycine max* [L.] Merrill) FABACEAE

The soybean, also known as the soja or soya bean, is native to eastern Asia. It was cultivated in China and Japan long before written history. Because of its great importance as a high-protein food source, the soybean became one of the five sacred crops of China, joining rice, wheat, common millet, and glutinous millet. Europeans first learned of the soybean about 1700, but not until 1875 was there any great interest in the plant. The soybean in the United States was first mentioned about 1800 in Pennsylvania, where it was reported to grow well.

The soybean is an annual plant of the legume family. It is erect and bushy with many branches, fewer at normal field populations. It varies in height from about 30 to 150 cm (1 to 5 ft) and its root system extends to 150 cm (5 ft) if the soil is permeable. The leaves are alternate and trifoliate except for a pair of opposite simple leaves at the first node above the cotyledons. Most cultivars are pubescent (hairy). Flowers are self-pollinated and are borne on racemes (clusters) of three to fifteen flowers that are either white or purple, or a blend of these colors). The seeds develop in pods. The pods are mature and ready to harvest when the leaves have dried and fallen off (Fig. 17–25).

The annual production of soybeans worldwide has more than tripled since World War II. The United States is the leading producer, growing about 55 percent of the world total of more than 130 million metric tons. Because the different cultivars of soybean mature their fruits over a wide range of photoperiods, the soybean is adaptable throughout many temperate areas.

Soybean cultivars are grouped according to their response to photoperiod, and it is important to select cultivars adapted to local conditions to take full advantage of the available season. Cultivars with maturity times corresponding with the latitude have been developed. Those requiring long photoperiods are in the higher latitudes; those requiring short photoperiods are in the lower latitudes. Cultivars are also classified as determinate (terminating vegetative development before flowering) or indeterminate (vegetative development continuing for several weeks after the beginning of flowering).

Though few soybeans are now grown for hay, there are forage-type cultivars with fine stems used for this purpose. Soybean hay is cut when the bean pods are about half filled. At this time, the leaves are about maximum size and the stems are not excessively woody. Good-quality soybean hay is about equal in feed value to other legume hays, but it is difficult to cure without loss of leaves, reducing quality. Special food grade soybean cultivars have been developed for use in preparation of various Asian cuisine foods that are increasing in popularity, including tofu (soybean curd) as well as natto and miso which are made from fermented soybeans. Soybeans are also used for soy milk and in infant formulas. The milk is made by grinding the beans and adding water.

Most soybeans contain high levels of linolenic acid, which reduces the shelf life and stability of products made from soy oil. To overcome this problem, soy oil is often partially hydrogenated to reduce linolenic acid levels. With the growing need to remove trans fats from food formulations, plant breeders have responded by developing varieties with less linolenic acid to eliminate the need for hydrogenation. Consumers benefit from healthier options because oil that does not require hydrogenation contains nearly no *trans* fats. Farmers benefit from planting low linolenic soybeans because they are increasing in demand and farmers also receive premiums for growing the low linolenic soybeans.

Sunflower (*Helianthus annuus* L.) ASTERACEAE

The sunflower is a native of North America and may have been domesticated before corn. It is a tall annual plant with rough hirsute stems. The flower heads are a compound inflorescence composed of many individual flowers in a large disc, subtended by large ray flowers (Fig. 17–26). In wild specimens, the flower heads are 8 to 10 cm wide, but larger in commercial cultivars. The flowers are cross-pollinated by insects. Wide use of sunflower as an edible oil crop began about 1830 in Russia,

Figure 17–25
Soybean pods nearly ready to harvest. Source: USDA Agricultural Research Service Image Gallery, *http://www.ars.usda.gov/is/graphics/photos/*

Figure 17–26

Sunflowers. Source: USDA Agricultural Research Service Image Gallery, http://www.ars.usda.gov/is/graphics/photos/

and it has become the main source of edible oil in the former Soviet Union, other eastern European countries, and Argentina. Reintroduction to the United States was for its use initially as silage. Its use as an oil crop has greatly increased its production, with over 1 million hectares now grown. This increase was assisted by the use of Russian germplasm providing oil content above 40 percent and the ability to produce hybrid seed. The introduction and wide usage of hybrid seed improved crop uniformity, yields, and earlier maturity, as well as disease resistance. Breeding for shorter stems (120 to 150 cm) reduced lodging and mechanical harvesting problems.

The cultivated plant has at least three uses: oil, confectionary products, and fodder. The oil form produces a valuable and desirable oil having polyunsaturated and monounsaturated characteristics similar to corn oil and olive oil, respectively. The confection-type cultivars are usually large-seeded and are roasted like nuts, with or without the hull. The smaller whole seed is used as a feed component for pet and wild birds and for small animals. The fodder-type plant is very tall with large stems and leaves, and the entire plant is chopped green for livestock feed. The cake or meal from the oil-pressing process is used for animal feed.

Fiber Crops

Many kinds of plants are cultivated for their fibers, which are used to make yarn, fabrics, rope, paper, insulation, raw cellulose, and hundreds of other products. Fiber-producing plants can be categorized as:

1. Surface fibers—those in which the fibers are produced on the surface of the plant parts in association with floral structures. The principal examples are cotton and kapok, where the fibers develop as outgrowths of the seed coat epidermal cells.
2. Soft or bast fibers—those in which the fibers are located in the stems or, more precisely, in the outer phloem tissues of the bark. Botanically, these are phloem fibers, which are groups of very long, thick-walled cells just external to the conducting phloem sieve tubes. Plants producing such soft-stem fibers include flax, hemp, jute, kenaf, and ramie.
3. Hard fibers—those that are rougher, more lignified strands of veinlike bundles of phloem and xylem supporting cells primarily from leaves of monocotyledons. These are longitudinal rigid fibers. The agave plant is an important example of this group.

For centuries people obtained many needed materials from these plants for clothing, ropes and sails for ships, and cord for fishing nets. In more recent times, numerous other products have been made from such fibrous materials.

Various types of cotton plants were naturally dispersed throughout many of the warm parts of the world for thousands of years, and their lint was spun and woven into cloth. The Egyptians made linen cloth from flax fibers, and ancient writings show that the hemp plant was being grown for its fiber in China north of the Himalaya mountains as long ago as 2800 BCE. Ramie was cultivated by early civilizations; the cloth made from it was used to wrap mummies in ancient Egypt. Early Chinese literature also mentions the cultivation of ramie for its fibers.

Plants Producing Surface-Fibers in Association with Floral Parts

Cotton (*Gossypium* spp.) MALVACEAE

Cotton has been one of the world's most important crops since the beginning of civilization. It continues to make immense contributions to people's comfort. In spite of competition from synthetic fibers, cotton is still basic in the world's textile industry. Cotton is grown principally for its lint, although the seed produces a valuable and widely used food oil. Cottonseed has an oil content of 30 to 40 percent. The seed residue, called cottonseed meal, is used as a livestock feed and is processed into a high-protein flour. Some seeds are roasted and enjoyed as a snack or as an ingredient in cookies and candy.

The various cotton species originated in several warm regions of the world. Tropical by nature, this heat-loving crop has been extended by humans into the warmer temperature regions. There is evidence that cotton was grown and processed into cloth as long ago as 5000 BCE in Mexico, 3000 BCE in Pakistan, 2500 BCE in Peru, and 2000 BCE in India. Today, the leading cotton producing countries are China, India, and the United States. The leading US cotton-producing states are Texas, California, Mississippi, Arizona, and Louisiana.

Only four of the twenty or so cotton species are cultivated for their spinnable fibers. Two are Old World species—*G. arboreum* L. and *G. herbaceum* L.—which are believed to have originated in southern Africa. These produce a short—1 to 2 cm (0.4 to 0.8 in.)—coarse lint. The other two are New World species—*G. hirsutum* L. and *G. barbadense* L. About two-thirds of the cotton grown in the world and almost all that is grown in the United States is known as American Upland cotton and belongs to the species *G. hirsutum*. This species originated in Central America and southern Mexico as a perennial shrub, although for production it is grown as an annual. American Upland cotton produces fibers varying from short to long—2.2 to 3.1 cm (0.9 to 1.2 in.)—according to the cultivar. *G. barbadense* L., originating in the Andean region of Peru, Ecuador, and Colombia, accounts for no more than 1 percent of the cotton grown in the United States, where it is known as American Pima. It is an extra-long staple cotton with fibers up to 3.3 cm (1.3 in.) long that are very good for fabrics.

Planting is done after all danger of frost has passed and soil temperatures have reached at least 16°C (61°F), preferably 20°C (68°F). In the United States, this time frame ranges from early March in southern Texas to early May in the northerly areas of the cotton belt. Maximum productivity is favored with high temperatures during the growing season, high light intensity, ample soil moisture, and good soil fertility. Cotton cannot tolerate frost and generally requires at least 200 frost-free days from planting to crop maturity. Optimum day temperatures during the flowering period range from 30°C to 38°C (86°F to 100°F). Because of breeding and selection programs in the temperate regions, cotton is no longer subjected to major photoperiodic control, but high light intensity does promote maximum production of fiber and seeds.

Depending on the cultivar and environment, flowering branches in cotton originate at about the seventh node on the main stem. Floral buds, called squares, are visible on those branches about three to

Figure 17–27

Immature cotton bolls. Source: USDA Agricultural Research Service Image Gallery, *http://www.ars.usda.gov/is/graphics/photos/*

four weeks before flower opening, which occurs about two months after the seed is planted. The cotton plant initiates far more flower buds than will develop into flowers, and far more flowers than will mature into profitable bolls—the ovaries or fruits that contain the seeds (ovules) bearing the fibers (Fig. 17–27) that develop from protuberances on epidermal cells of the ovule seed coat. The cotton flower is perfect, containing both male and female organs, and is either self- or cross-pollinated. The majority are self-pollinated. In most commercial cotton, flower petal color is cream to pale yellow on the day the flowers open. The notable exception is Pima cotton, whose flowers have deep purple spots at the base of the corolla, and bright yellow petals.

All commercial cultivars of Upland cotton produce two types of fibers: lint and fuzz (the latter called linters). At maturity, the lint dries, and the fibers become twisted and thereby are more suitable for spinning. As bolls dry, they split and open, exposing the lint and fuzz (Fig. 17–28). At maturity, lint fibers have narrow and delicate shanks at their bases. Fuzz fibers thicken at their bases and are more strongly retained by the seed. The lint is removed by ginning (Fig. 17–29). After lint removal, seeds for certification and future planting are treated with acid to remove the fuzz fibers, or these fibers are removed by a different set of gin saws. The fuzz is used for mattress stuffing or for cellulose. Removal of fuzz fibers from the seed is essential for the operation of precision planting equipment. The seed-free cotton lint may be further cleaned and is then baled and moved to the textile mills for processing into yarns and fabrics. The seeds left behind are collected and moved to seed-processing houses for fuzz removal. After cleaning, some seeds are saved for next season's planting and others are used for oil.

Figure 17–28
Mature cotton bolls that have split open to expose the fiber.
Source: USDA Agricultural Research Service Image Gallery, *http://www.ars.usda .gov/is/graphics/photos/*

Figure 17–29
Unginned and ginned cotton. Source: USDA Agricultural Research Service Image Gallery, *http://www.ars.usda.gov/is/graphics/photos/*

Hand harvesting of cotton is laborious but is still practiced in some areas. However, for most large production areas, cotton is mechanically harvested (Fig. 17–30). Before machine harvesting starts, chemical defoliants (for the spindle-type harvesters) or chemical desiccants (for the stripper-type harvesters) are sometimes applied. The defoliants accelerate leaf drop so that the spindles can remove the bolls easier and lessen contamination. Desiccants rapidly kill the leaves but they stay attached to the plant. Both types of chemicals are applied seven to fourteen days before harvest. After the cotton has been picked from the plants by the harvesters, it is either taken directly to cotton gins or pressed into large bales or modules (Fig. 17–31) then taken to the gin.

Kapok (*Ceiba pentandra* [L.] Gaertn.) BOMBACACEAE

Moisture-resistant fibers called kapok or silk cotton are derived from the seed hairs in the pods of the kapok tree. These trees grow wild in tropical American forests and have been introduced to other tropical regions throughout the world. Almost all commercial kapok production—several thousand tons annually—now comes from Asia, particularly Thailand, Indonesia, and Kampuchea. Most kapok is sold to the United States, where it is used as insulation in sleeping bags and as filling in life preservers. Synthetic fibers, foam rubber, and plastics are rapidly replacing kapok for such purposes. The hollow kapok fibers are similar to those produced by the cotton plant. The fibers are not twisted and are

Figure 17–30
Cotton harvester. Source: USDA Agricultural Research Service Image Gallery, *http://www.ars.usda.gov/is/graphics/photos/*

Figure 17–31
Bales of cotton waiting to be taken to the gin. Source: USDA Agricultural Research Service Image Gallery, *http://www.ars.usda.gov/is/graphics/photos/*

too smooth and brittle to be spun into cloth but have a remarkable buoyancy and resiliency and are impervious to water. The large kapok trees bear a great many of the football-shaped pods, up to 15 cm (6 in.) long, which are filled with the fiber-producing seeds. The pods either fall to the ground or are cut from the tree. After drying, the fibers are extracted from the seeds by hand. An edible oil, similar to cottonseed oil, can be pressed from the seeds.

Plants Producing Soft Stem-Fibers

Flax (*Linum usitatissimum* L.) LINACEAE

The flax plant is a slender, herbaceous annual grown in many subtropical and temperate zone countries for its oily seed and for its stem fibers. The fibers are used to make linen cloth, book and cigarette papers, and paper currency.

In the United States, culture of flax for its fibers became unprofitable and ceased in the mid-1950s. Currently, all US flax is produced for seed. Different flax cultivars are used for linseed oil rather than fiber production. Linseed oil is pressed from the seed for paint manufacture, and the residue, called linseed meal, is used as a livestock feed.

The former Soviet Union is by far the leading producer of flax fibers. Poland, France, Czechoslovakia, Rumania, and Belgium are also major producers.

Flax is a cool-season plant. In areas with mild winters, seeds are sown in the fall and the crop harvested early the following summer. In cold-winter regions, seeds are sown in early spring and, if used for fiber production, the stems are harvested eighty to 100 days later, when about half the seeds are mature and the leaves have dropped from the lower two-thirds of the stem. At harvest, weather should be dry to cure the plants properly.

The fibers are obtained from the stems of the flax plant by a process called retting. Commonly, the retting involves simply leaving the stems in the field, where they are exposed to weathering from the dew and to the action of soil-borne bacteria on the straw. This partial rotting for one to three weeks dissolves gums that hold the fibers to the woody xylem tissues and destroys soft tissues around the fibers. A more intensive rapid method is cold water retting by complete immersion of flax stem bundles in cold water for several days. The stems then pass through machines that break up the woody parts but retain the flexible fibers largely intact. Other machines separate the short, woody sections from the fibers, which are then baled and shipped to spinning mills.

Hemp (*Cannabis sativa* L.) CANNABACEAE

The hemp plant is a seed-propagated herbaceous annual adapted to mild temperate zone climates. The unusually stout stem fibers of certain hemp cultivars are used to prepare tough threads, twines, ropes, and textile products. Cloth from hemp can be used in clothing and other products. Most hemp produced for fiber is grown in temperate countries such as the former Soviet Union, Yugoslavia, China, India, Korea, Poland, and Turkey, but various hemp cultivars can be found cultivated or wild in almost all countries in the temperate and tropical zones.

For fiber production, the plants are seeded thickly and grown to a height of 1.5 to 3 m (5 to 10 ft). Hemp seed is planted early in the growing season, and the plants are uprooted or cut off for fiber harvest during the period from start of bloom until full maturity. The stems are hollow except near the base. The fibers are located in the phloem and pericycle tissues of the bark.

The straw is retted like flax to aid in separating the fibers from the other tissues in the stem. Machines are used to separate the fibers further. Manila hemp (*Musa textilis*), also known as abaca, is used in paper manufacturing.

Because certain low-growing types of *Cannabis sativa* are the source of the drugs marijuana and hashish, cultivation of all hemp is prohibited in many countries, including the United States. However, the demand for hemp fiber has changed the laws in some countries such as Canada where growers can obtain permits to produce the fiber type of *Cannabis*.

Jute (*Corchorus capsularis* L. and *C. olitorius* L.) TILIACEAE

Plants of these two jute species are herbaceous, seed-propagated annuals cultivated in hot, moist climates with at least 7.5 to 10 cm (3 to 4 in.) of rain per month. The plants grow from 1.8 to 4.5 m (6 to 15 ft) high. The fiber strands are located in the stem just under the bark, some running the full length of the stem, and are embedded in nonfibrous tissue. The fibers are held together by natural plant gums.

As a fiber-producing crop, jute probably ranks next to cotton in importance. It is grown primarily on small farms in Bangladesh and in several districts in India. Jute is chiefly used in the manufacture of bulky, strong, nonstretching twines, ropes, and burlap (Hessian) fabrics, bags, and sacks for packaging many industrial and agricultural commodities. It has a great many other industrial uses—such as backings for carpets and linoleum

coverings, webbings for upholstered furniture, packing in electric cables, and interlinings in tailored clothes.

Jute seeds are broadcast over the soil in the spring and the plants are later thinned to about $33/m^2$ ($3/ft^2$). The crop is harvested by cutting off the stems about the time the flowers start to fade. The bundles of stems are left in the fields for a time to shed their leaves.

To release the fibers from other tissues, the stems must be retted. To do this, bundles of the stems are submerged under water in pools or streams for ten to thirty days—long enough for bacterial action to break down the tissues surrounding the fibers, but without damaging the fibers themselves. At the proper time, the fibers are loosened from the wet stems by beating the small bundles with paddles, then breaking the stems to expose the fibers, which are pulled from the remainder of the stems. The fibers are then washed in water and hung on poles or lines to dry. The dried fibers are taken to baling centers, where they are sorted and graded according to strength, cleanliness, color, softness, luster, and uniformity, then pressed into bales. The bales are shipped to local spinning mills or they are exported.

Kenaf (*Hibiscus cannabinus* L.) MALVACEAE

Kenaf has been cultivated for centuries in many places throughout the world between about 45° north and 30° south latitudes. Thailand, India, Brazil, China, and the former Soviet Union are the major kenaf producers. Kenaf competes with jute as a stem fiber crop but has less exacting soil and climatic requirements. Kenaf produces an excellent fiber, tougher and stronger than jute but somewhat coarser and less supple. Kenaf fibers are used for making twines, ropes, and fishing nets and are also suitable for paper manufacture.

The kenaf plant is a herbaceous annual with a strong taproot and a long, unbranched stem reaching to a height of 1.5 to 4.5 m (5 to 15 ft) (Fig. 17–32). The plants require a growing season of 100 to 140 days with considerable moisture from either rainfall or irrigation. Harvesting, retting, and drying are done during drier weather. The stems are harvested just as flowering starts by uprooting the plants and tying them into bundles that are placed horizontally in water for retting, as described for jute. Retting may take ten to twenty days. The fibers are then stripped off the stalks and dried.

Ramie (*Boehmeria nivea* [L.] Gaud.-Beaup.) URTICACEAE

Ramie fibers are long strands in the inner bark of the ramie or China grass plant, a many-stemmed perennial

Figure 17–32
Kenaf plants. Source: USDA Agricultural Research Service Image Gallery, *http://www.ars.usda.gov/is/graphics/photos/*

shrub with slender shoots about 2.5 cm (1 in.) thick and up to 2.4 m (8 ft) long and heart-shaped leaves along the upper third of the stem. New stems arise from the crown after the older ones are harvested. Three harvests can be obtained annually, and the plant can live for several years before it has to be replaced.

Ramie has been cultivated for thousands of years, and is mentioned in Chinese writings as early as 2200 BCE. It apparently is native to the Chinese area of eastern Asia. It was also cultivated by the ancient Egyptian civilizations. Present-day ramie production centers mostly in warm, humid regions of China, the Philippines, Japan, Indonesia, and Malaysia with fertile soils.

The cells making up ramie fibers are among the longest known—up to 0.3 m (1 ft). The fibers are eight times stronger than cotton and have a fine durable texture and a good color. Ramie fibers are superior in many ways to flax, hemp, and jute. However, the chief problem with ramie has been the difficulty in freeing the fiber bundles from the gummy tissues surrounding them, along with problems in mechanizing the processing of the fibers—the extraordinarily smooth surface of the fibers makes spinning difficult with machinery developed for other fibers.

As with other stem-fiber crops, retting is required to free the desirable fibers from the other stem tissues. This is done by bacterial action, which decomposes the thin-walled surrounding cells and leaves the

thick-walled fibers intact. Ramie is more difficult to rett than flax and hemp. Just wetting the stems is insufficient to break down the cementing gums. Pounding and scraping is also required, followed by chemical treatments with acid or lye to remove the tenacious gums and resins to produce completely smooth fibers. After degumming, the fibers are washed, then softened with glycerine, soaps, or waxes. Ramie fibers are generally spun on machinery designed for silk. Ramie fabrics are usually blends with other materials, such as cotton or wool.

Plants Producing Hard Leaf-Fibers

Agave (Agave sisalana Perr.) AGAVACEAE

The long hard fibers used in making twines, cords, and ropes are obtained from the 0.6 to 1.2 m (2 to 4 ft) leaves of agave plants. Sisal is produced mainly in East Africa, Brazil, Mexico, and Haiti. A related plant known as henequen (*A. fourcroydes* Lem.) is also grown for its fibers, which are much weaker than sisal. Plants of these two species are similar in appearance to the common century plant (*A. americana* L.).

The agave plant consists of a rosette of stiff, heavy, dark green leaves 10 to 20 cm (4 to 8 in.) wide, 60 to 120 cm (2 to 4 ft) long, and 2.5 to 10 cm (1 to 4 in.) thick, arising from a short trunk. The plant grows very slowly, but after about four years, it reaches full size and harvesting of the lower leaves begins. About 200 leaves can be harvested for fiber extraction before leaf growth ceases and a flower stalk grows upward rapidly to a height of 4.5 to 7.5 m (15 to 25 ft) and bears light yellow flowers. During the next six months, flowering and fruiting take place, followed by the death of the entire plant. Agaves are vegetatively propagated by suckers arising around the trunk of the plant or by bulbils (small bulbs) that develop in the flower clusters after flowering, thus allowing growers to maintain improved forms.

In harvesting for fiber, the lower leaves are cut off and bundled after the spines are removed. Cleaning machines scrape the leaves, removing the pulp and waste material and leaving the fibers exposed. The fibers are then dried in the sun, or artificially, and are brushed, graded, and baled. The final product is yellow to yellow-white in color, flexible, and strong. Sisal is used mostly in making cords, such as binder twine; it has only limited use in fabric manufacture.

SUMMARY AND REVIEW

Although many different plant species are grown as field crops and each has specific requirements for optimum production, some key concepts are common to all. The first is understanding the types of interactions that occur among all the production factors, the second is to develop management systems that maximize the beneficial aspect of each interaction. Tillage, crop sequencing, variety selection, crop establishment, water and nutritional management, and weed and pest control are the most common factors that have to be considered.

Cropping sequences can be monoculture (the same crop over and over) or rotational cropping (rotation of crops within a season or annually). Each has advantages and disadvantages. Variety selection is an area over which a crop producer has considerable control. It is critical to select the variety of the species that is best suited to the area in which the crop will be grown. Varieties are often distinguished by the factor that is most influential in their development. Often this factor is temperature, but it can be sensitivity to daylength, or resistance to pests or other stresses.

Successful crop establishment requires that a grower know when conditions are right for planting and then follow proper planting procedures. These procedures include using the right planting equipment, sowing on the right date, seeding at the proper rate and spacing, and planting at the proper depth.

After establishment, water (irrigation) and nutritional management is critical. In much of the United States east of the Mississippi, irrigation of field crops is not necessary, however in the western United States, erratic rainfall usually requires irrigation of most crops. If nutritional elements are not naturally available in sufficient amounts, fertilization is required. A grower has to understand what nutrient elements a crop uses and what is removed at harvest from what would be normal nutrient cycling in an undisturbed ecosystem.

Weeds are the major pest in most field cropping systems. No single control program handles all weed problems effectively. Most weed-control problems follow three strategies: prevention, eradication, and control. Insects are usually not considered to be a serious problem in most field crop commodities because the damage the insects do is below a threshold level of significance. When it does become necessary to control insects, management of pest populations requires a

combination of preventive and timely responsive actions based on the risks associated with various cultural practices and pest activity observations collected through periodic field inspections. Diseases and weed and insect pests usually require multiple strategies of control. Be aware that a change in cultural practice can have an effect, either good or bad, on disease and other pest problems.

Proper harvesting is as important as the actual growing of the crop. Harvesting must be done when the crop is at the proper stage of development and when weather conditions are appropriate. For most crops there is a preferred moisture content of the part being harvested. Some harvest residue such as straw can be harvested itself for use as animal bedding and thatch roofs.

Important field crops used for human food are: barley, various beans (dried), corn, oats, rice and wild rice, rye, sorghum, sugar beets, sugarcane, and wheat. Many of these are cereal grains (members of the POACEAE) or are grain-like in how their seeds are used. Oil crops include castor bean, coconut, olive, canola (rapeseed), palm, peanut, safflower, sesame, soybean, sunflower. Fiber crops are categorized as producing (1) surface fibers, such as cotton; (2) soft/bast fibers, which are phloem fibers in the stem, such as flax and hemp; and (3) hard fibers that are lignified vascular fibers, such as agave. Fiber crops include cotton, kapok, flax, hemp, jute, kenaf, ramie, and agave. The use of some field crops, especially corn and soybeans, for industrial use such as fuels and polymers has been increasing dramatically and is expected to continue to increase.

KNOWLEDGE CHECK

1. Cropping sequences are based on what two considerations?
2. What is the difference between a monoculture and crop rotation?
3. What are three advantages of growing two different types of crops, for example corn with soybeans, in rotation compared to just growing one or the other every year?
4. What are three reasons for using a monoculture system?
5. What are: ratoon cropping, double cropping, and intercropping?
6. What are three factors that are important in variety selection?
7. What is the advantage of using growing degree days over the days-to-maturity system of determining corn maturity?
8. What is lodging?
9. What are the major transgenic crops?
10. What are the two major traits placed into transgenic crops?
11. What is a risk of continually growing transgenic crops? Give an example of how good management can reduce the risk.
12. What are some problems when planting soil that is too wet? When soil is too dry?
13. What are the two main considerations when determining planting date in the temperate regions? In tropical and subtropical monsoon regions?
14. When are winter types of crops planted? When are spring types planted?
15. Why would a newly germinated soybean field more likely have to be replanted after a late spring frost than a newly germinated corn field beside it?
16. Why are winter and spring types of grains usually not interchangeable in regard to when they are planted?
17. What determines seeding rate of a crop?
18. What are the main determining factors in determining optimal plant population?
19. How does plant population affect: lodging, tillering, ear (corn), or seed head (grain) numbers?
20. Why is some increase in plant height an advantage for soybeans and cotton?
21. Why is seeding rate for late plantings usually increased for small grains and soybeans but decreased for corn?
22. What is the "rule of thumb" for seed planting depth?
23. How do the following affect planting depth: seed size, soil texture, soil temperature?
24. Why is a seed that has hypogeal emergence planted deeper than a seed of similar size that has epigeal emergence?
25. What is the maximum level of grain moisture allowed for most crops?
26. In what family do true cereal grains belong? Why are seeds from other families sometimes called cereal grains?
27. Why are the cereal and "pseudo" cereal grains the principal food source of humans?
28. List five genera of FABACEAE beans (do not include soybean) that are important to human nutrition.

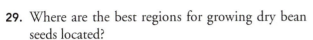

29. Where are the best regions for growing dry bean seeds located?
30. What are the differences among dent, flint, and flour corn?
31. What are three uses of sorghum?
32. What part of the sugar beet is harvested for its sugar content? What part of sugar cane is harvested?
33. What are the four main classifications of wheat? Give a production area and culinary use for each type.
34. What are the native areas of plants high in saturated oil? Unsaturated oil?
35. Why is unsaturated oil preferred for human diets over saturated oil?
36. List three uses of plant oils not including plant and animal food.
37. Name the plant part that is the source for fiber in each of the following: cotton, flax, hemp, and agave.
38. Why is growing hemp such a controversial topic in many countries, including the United States?

FOOD FOR THOUGHT

1. What are some ways that a homeowner could practice rotation in her or his landscape or garden to get the benefits that come from rotating crops?
2. If a soybean was developed using transgenic technology that grew as well as other soybeans but because of the the genetic engineering contained no linolenic acid thus eliminating the need for hydrogenation, as a farmer would you plant it or as a consumer would you buy products that contained it? Explain your decision.
3. If you were a hemp grower in Canada and you wanted long stems to get as much fiber as possible, would you try spacing the plants a little closer to encourage longer stem development? Explain why you would or would not. Could there be negative effects of going to a closer spacing?

FURTHER EXPLORATION

Martin, J. H., W. H. Leonard, D. L. Stamps, and R. P. Waldren. 2005. *Principles of field crop production,* 4th ed. Upper Saddle River, NJ: Pearson Education.

Sheaffer, C. C., and K. M. Moncada. 2008. *Introduction to agronomy: Food, crops, and environment.* Clifton Park, NY: Delmar Cengage Learning.

*18

Forage Crops and Rangelands

Dᴀᴠɪᴅ Bᴀʀᴋᴇʀ ᴀɴᴅ Mᴀʀᴋ Sᴜʟᴄ

Forage, browse, or herbage are the edible parts of plants, primarily the leaves and digestible stems that can be fed directly or following storage to livestock. Forage crops, **pasture**, meadows, prairie, **grazing land**, and **range** are a continuum of forage-producing grasslands (swards) that vary in characteristics such as their geographical location and intensity of production. Generically, all are termed grasslands and although they are usually dominated by grass (**monocotyledonous**) species, they also include **dicotyledonous** (**legume** and **forb**) species. Alfalfa is sometimes termed the queen of the forages. Forage crops and rangelands have unique and distinct characteristics compared to other crop systems:

1. Worldwide, more land is devoted to grasslands than to all other crops combined (8.4 vs. 3.7 billion acres) and totals 26.1 percent of the world's land area (FAO statistics). Much of this land is too rocky, hilly, dry, or wet for growing other crops but is suitable for forage production (Fig. 18–1). Grasslands can also have environmental benefits such as stabilizing land against erosion (Fig. 18–2), carbon sequestration, and providing wildlife habitat.

2. Most forage and rangeland species are perennial plants, and part of year-round production systems. Usually these production systems are required to support livestock for successive years. To be sustainable, management to ensure production in future years is a priority over maximizing production in any single year.

3. Forage crops and rangelands encompass tremendous diversity. Rangeland systems typically comprise five to fifty plant species, and worldwide more than 500 plant species are used for forage. These grasslands support a wide diversity of livestock. Forage crops and rangelands occur in every country of the world and on almost all soil types in the world. They encounter the range of climate within and between years and are employed by almost all the world's cultures.

4. Forage crops and rangelands have no immediate value to humans in terms of direct food products. It is only through consumption of the plant by livestock that biological and financial production can be captured from grasslands (Fig. 18–3). In the

key learning concepts

After reading this chapter, you should be able to:

- Describe the different types of forage and rangeland crops.
- Explain the principles of hay and silage growing, harvesting, and storage.
- Discuss rangeland ecology and the principles of rangeland management.
- Describe the diverse uses of rangelands.

Figure 18–1

This rangeland in Idaho is too rugged for farming most crops.
Source: USDA Natural Resource Conservation Service, *http://photogallery.nrcs .usda.gov/*

Figure 18–2

The grass and other plants in this pasture are protecting this hilly land from erosion. Source: USDA Natural Resource Conservation Service, *http://photogallery.nrcs.usda.gov/*

Figure 18–3

Worldwide, more land is devoted to grasslands than to all other crops combined. Grazing of perennial grasslands is productive and economic, and has many environmental and social benefits. Source: D. J. Barker, Ohio State University.

UTILIZATION OF FORAGE CROPS

Forages in Traditional Confinement Systems

In the United States, livestock are usually housed in feedlots (for beef) (Fig. 18–4) and milking barns (for dairy) and have all their rations brought to them. These rations usually contain forages as well as other high-energy concentrates and protein, such as by-products from other industries (e.g., distillers grains after ethanol fermentation, soybean meal after oil extraction) or cereal grains (e.g., corn). A specialized industry has developed around

Figure 18–4

The animals in this feedlot are feeding on the rations delivered by truck. Source: USDA Natural Resource Conservation Service, *http://photogallery.nrcs.usda.gov/*

United States, the value of animal products is more than 25 percent of all farm cash receipts, and the value of the forage crops fed to animals is greater than any other single crop. It is often suggested that it would be more efficient for people to consume plants (e.g., grain) directly rather than having the plants fed to livestock and then consuming the meat and dairy products. However, most types of forage consumed by livestock are high-cellulose vegetative material that can't be digested by humans. Most areas of grasslands can only be converted to human food products by pasturing animals.

5. Animal access to pasture is a requirement for most organic livestock production systems.

the dedicated production of hay and silage by some farmers, and the blending of these crops into totally mixed rations (TMRs) by feed companies. Although these animals gain weight more rapidly and produce more milk, they could survive totally on forage crops alone. Confinement systems are rare in New Zealand, Australia, South America, and South Africa, where livestock exclusively graze forages. Grazing systems usually have lower animal production than confinement livestock systems, but they are justified because the costs of production are proportionally lower, and proponents argue that they are more profitable. Regardless of the system, forage production is the basis of livestock production.

Grazing

Throughout the world, most forages are utilized by grazing. **Ruminants** (e.g., sheep, goats, and cattle) have four digestive chambers, the first of which (the rumen) is filled with microorganisms that digest and break down forage fiber (cellulose, hemicellulose, and pectin). Other groups of livestock, such as camelids (camels and llamas) and monogastrics (horses and pigs), can also thrive totally on forages but employ different digestive mechanisms.

The most important objective in grazing is to match the supply of forage production to the nutritional requirements of livestock. Many management strategies achieve this balance; however, the most important issue is to match stocking rate to forage production. Short-term deficits in forage production (e.g., during drought or over the winter) are usually met by feeding forage that has been conserved and stored (see later in this chapter) during a period of surplus (e.g., spring). Long-term deficits in forage production (i.e., overstocking or overgrazing) result in deterioration of the forage and soil resource.

There are many strategies for forage utilization by grazing; however, these can usually be simplified to variations of either rotational stocking or of continuous stocking. Rotational stocking is the movement of livestock between pastures according to a prescribed strategy. Most rotational stocking employs frequent (daily to weekly) movement of livestock; however, an infinite number of more and less frequent rotations exist. Once all pastures have been grazed, livestock return to the first pasture grazed, which has had sufficient time to regrow to a harvestable mass (Figs. 18–5, 18–6, and 18–7), thus completing the rotation. Rotational stocking is usually achieved with temporary electric fences that can be easily moved and replaced to allow livestock access to the correct area of pasture. Continuous stocking involves the allocation of

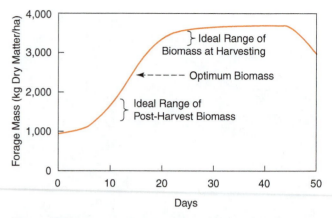

Figure 18–5

Generalized growth curve for forage production. Forages show slow initial growth, a period of high-growth rate, and a period of reducing growth rate. After long regrowth periods, forage mass can decrease as leaves low in the canopy wither and fall off. Source: David Barker, The Ohio State University.

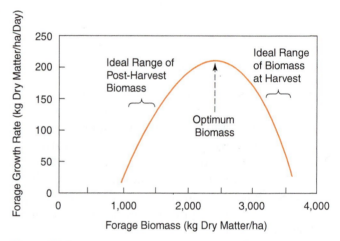

Figure 18–6

The data in Figure 18–5 is redrawn to show the relationship between forage growth rate and forage mass.
Source: David Barker, The Ohio State University.

livestock to a fixed area for a prolonged period. In a carefully managed situation, sufficient mass and growth rate of the forage crop support the livestock for the required period. As a generalization, intensive (high population of animals on relatively small amount of land) grazing systems use rotational stocking, while extensive grazing systems (few animals on a large area of land) use continuous stocking. Many grazers claim higher forage production from rotational stocking; however, it is difficult to identify the precise mechanisms for the response. For example, in addition to different patterns of forage removal compared to continuous stocking, rotational stocking also includes factors such as more regular livestock inspection, ease of herding livestock, more uniform manure dispersal, and less selective grazing.

Figure 18–7

The cattle on the left are grazing on a rotational field and will soon be moved to the adjacent field or another field in the rotation where the forage has had time to re-grow.

Source: USDA Natural Resource Conservation Service, *http://photogallery.nrcs .usda.gov/*

CONSERVED FORAGE

Forage crops are often conserved or preserved and fed later to livestock. This is the case for confinement systems, where livestock are not on pasture. Conserved forages also provide a useful function by providing forage during periods when grasslands might be dormant or not growing rapidly enough to support livestock needs (e.g., during drought and winter). Forages are conserved in two primary forms: as dry hay or as moist silage.

Hay   Hay is defined as the shoots and leaves—and in some cases, the flowers, fruits, and seeds—of forage plants that are preserved by field drying, harvested, and stored for future feeding to livestock. Hay usually consists of grasses, legumes, or a combination of the two. The harvested material is dried to a moisture level of 15 to 20 percent. Proper dehydration is important to preserve quality and to prevent barn fires that result from spontaneous combustion of wet hay as it decomposes. Hay is the most important type of stored forage and provides considerable flexibility in animal feeding operations. Properly dried and stored, it can be kept for several years with minimal loss of nutritive qualities. In severe-winter regions, it permits feeding livestock a nutritious bulk of material when outside pastures are not growing and are covered with snow or ice. Hay is a cash commodity that can be bought and sold, whereas the value of pasture can be converted only by animal feeding.

The best quality hay in the United States is usually made from alfalfa, however the greatest quantity of hay is made from grasses (meadow hay). One technology—as of 2009—still awaiting approval for use in the United States is Roundup Ready®, genetically modified alfalfa. Roundup® is a broad-spectrum herbicide, and Roundup® Ready varieties have been artificially created that have a resistance gene to Roundup®. Thus, Roundup® Ready alfalfa has the potential for easier weed control when treated with Roundup® herbicide.

In general, the best time to harvest hay for maximum yield and still maintain acceptable quality is just prior to the heading (seed development) stage. Hay fields can often be harvested several times during a growing season. Some forage crops, such as timothy, give only one or two harvests during the growing season, while others, such as alfalfa, give three to ten harvests depending on the region.

Hay quality is determined by plant maturity, leafiness, color, odor, and amount of foreign material. During harvest and transportation, every effort should be made to retain as much leaf as possible because most of the protein is in the leaves. High-quality hay has a fresh green color, a good aroma, and a pliable texture. It is nutritious, digestible, and palatable. Detrimental foreign material includes weeds, poisonous or thorny plants, spiny seeds, and objects such as wire, nails, rocks, and soil. Effort should be made to eliminate these from the hay.

Hay may be stored loose, chopped, baled, or pressed into cubes, pellets, or wafers. Harvesting is now highly mechanized. Many specialized machines are available for handling hay. The hay is cut, and laid down into **windrows** (Fig. 18–8), dried over two to

Figure 18–8

The hay is this field has been cut and raked into windrows to dry to the proper moisture level to preserve quality and reduce the risk for spontaneous combustion in storage.

Source: USDA Natural Resource Conservation Service, *http://photogallery .nrcs.usda.gov/*

Figure 18–9

In spring, forage growth exceeds the requirements of livestock. Surplus forage production can be made into hay and sold or used later during periods of deficit. Round bales suit mechanical manipulation better than the traditional small, rectangular bales. Source: R. M. Sulc, The Ohio State University.

Figure 18–10

Forages can be conserved by several methods. Here, alfalfa has been wilted to reduce its water content. It is being blown into a wagon and will be transported to a silo to be made into silage. Source: R. M. Sulc, The Ohio State University.

five days, and baled by mobile machinery (Fig. 18–9). This saves considerable labor over earlier methods of hauling hay to stationary machines. Most hay in the United States is now stored as small rectangular bales (14 × 18 × 38 in., 40 to 60 lbs), large rectangular bales (3 to 4 × 3 to 4 × 7 to 8 ft, 900 to 1,800 lbs), or round bales (4 to 6 ft diameter, 850 to 1,900 lbs). These bales are a commodity that can be bought, sold, or stored for later distribution.

Hay can also be chopped, green or dry, in the field into particles small enough to be blown in an air stream into trucks. The hay is then transported to feeding areas to be fed green to livestock or to dehydration equipment for processing.

Hay can be cubed, pelleted, or wafered and these practices are more common in regions of the United States where summers have low humidity, such as California, Arizona, New Mexico, and eastern Washington. In these areas, hay can be field-cured down to 12 percent moisture.

Silage The feeding of silage to livestock is an ancient practice from Europe dating back many hundreds of years, although the first **silo** was not constructed in the United States until 1876. **Silage** is moist forage, preserved by bacterial fermentation under anaerobic conditions. The corn plant is used most often for making silage, but almost any other forage species can be used, alone or in mixtures. Some commonly used species are oats, cereal rye, triticale, sorghum-sudangrass, smooth bromegrass, Italian ryegrass, orchardgrass, timothy, alfalfa, and red clover.

In silage fermentation by the direct cut method, the green chopped forage material—at 60 to 70 percent moisture—is blown into the silo. This is the procedure used for corn silage. A variation of this is haylage or hay crop silage: when the green forage is cut and laid in the field to wilt until it is about 45 to 50 percent moisture. It is then picked up, chopped, and blown into the silo. The practice of field wilting before chopping is required for perennial grasses and legumes, or any standing forage with moisture content above 70 percent (Fig. 18–10).

Silos for silage must be airtight structures and are either vertical or horizontal. They may be made of concrete or glass-coated steel plates; they may be large, plastic, baglike containers, or they may be mere trenches in the ground. After the oxygen in the mass of tightly packed chopped material is used up by plant respiration (or aerobic bacteria if spoilage occurs), anaerobic bacteria multiply and act on the carbohydrates in the plant tissue to form lactic acid, which essentially ferments the plant material. The pH drops to 4.2 or below, inhibiting spoilage bacteria and enzyme action, thus preventing deterioration. The fermentation process is completed in two to three weeks. Silage can be kept in good condition for several years if air is kept out, moisture stays high, and the pH remains below 4.2.

Modern silage preparation is highly mechanized. It starts with cutting the corn plants (or other forage material) by special harvesters that chop and blow the particulate material into trailers for hauling to the silo. At the silo, it is mechanically unloaded and blown into

vertical silos, or bags are spread and packed down in bunker silos.

Baleage   With the advent of bale-wrapping technology in the mid-1990s, farmers have the option of replacing silos with individual plastic-wrapped bales. A variation of this technique is wrapping many round bales within a continuous plastic tube in a single row. Wrapped bales have the advantage of being weather resistant (as long as the wrap is not broken), thus allowing a shorter drying period in the field compared to hay, and they can be transported. Typically, the fermentation process is similar, although less efficient than in silos because more air (O_2) is usually present in bales than in a well-packed silo. Baleage has become popular on small to medium-size farms in the humid areas of the United States such as the East Coast and southern states, where drying times in excess of three days increase the risk of rainfall damaging traditional hay crops. The additional cost of plastic, disposal issues with the plastic, and the costs of the wrapping machinery make this method less popular in western states of the United States, where field-dried hay is easier to make.

Integrated Systems (Crop Rotations with Cereals, Hay, or Livestock)

The diversity of forage production systems also includes many options for integrating forages into other types of production systems. Forage crops are often included as short-term components within a more complex crop rotation. This technique can work in many situations:

1. Various options for double cropping can be used. When a cereal crop is harvested early (e.g., winter wheat, or silage corn), a short-term annual forage crop might be planted to provide forage for livestock in the period until the main crop is replanted the subsequent year. Forage species that can be used in this situation include warm-season annuals such as sudangrass or cool-season annuals such as cereal rye, oats, annual ryegrass, forage triticale, and brassica species (e.g., turnips, kale, and rape).

2. Forages can be used as part of a rotation with cereal crops. Legume species within grasslands build organic soil nitrogen that can be used by subsequent cereal crops. A possible rotation might include several years in forage crops, followed by several years of grain crops.

3. Hay and grazing systems are often integrated into a single production system. Although examples of grazing-only and hay-only systems are common, these are more generally integrated. Forage growth usually exceeds livestock requirements in

Figure 18–11
Livestock production systems on forages are very flexible. Here, cattle are grazing corn stubble immediately following harvesting. Cattle consume both the crop residue and any grain that escaped harvest. Source: R. M. Sulc, The Ohio State University.

the spring, so farmers usually conserve this excess for other periods of the year when livestock demand exceeds supply. These periods include the short (one- to two-month) period of summer drought, and a longer (up to four-month) period of cold during winter when forage growth is slow or zero.

4. There are also situations when livestock can graze crop residues remaining after grain harvest (Fig. 18–11).

FORAGE QUALITY

Although the single largest influence on the profitability of forage systems is total production (or yield), the next most important factor is forage quality. The quality of forage is affected primarily by the amount of **fiber** it contains and its digestibility. Fiber in forage can be present as **pectin** and **hemicellulose** (moderately digestible by livestock), **cellulose** (poorly digestible by nonruminant livestock), and **lignin** (indigestible). Total fiber is measured in the laboratory using neutral detergent fiber (NDF) digestion. Digestibility can be measured directly in animals as the proportion of digested forage compared to total forage consumed (the difference being forage excreted); however, this procedure is time-consuming and expensive. Digestibility can also be predicted following incubation of a ground forage sample in **rumen** fluid in a laboratory and is currently considered the best available laboratory method for estimating forage digestibility.

Digestibility of a forage sample is commonly predicted by laboratories from empirical equations based on chemical composition of the forage (such as NDF and lignin).

Protein content, nutrients, and vitamins also affect overall forage quality. In most cases, these are present in sufficient quantities that they have less effect on forage value than does fiber content and digestibility. The main plant factors affecting forage fiber and other quality factors are listed below:

- *Forage species.* Forage species vary in their amount of fiber and the chemical composition of that fiber. Generally, grasses have more fiber than legumes, and although grass fiber is generally more digestible (more hemicellulose) than are legumes, it can be insufficient to compensate for the higher amount of fiber. The result is that legumes generally produce better quality forage than grasses. Variation exists among grasses as well, and species such as perennial ryegrass and timothy are higher quality than orchardgrass and tall fescue.
- *Leaf-to-stem ratio.* Almost without exception, stems have higher fiber (lower quality) than leaves. Vegetative grasses do not have true stems; their pseudostem is actually leaf sheaths wrapped together. Nonetheless, grass pseudostem, the true stems of legumes such as alfalfa, and the reproductive stems of grass all have higher fiber and lower quality than leaves. Good forage management aims to reduce these components in forage and to maximize leaf content.
- *Maturity.* As forages mature, yield increases but quality declines. Stem yield increases dramatically with advancing maturity in most forage species. Flowering initiates developmental changes in all forage species that include higher amounts of fiber. Furthermore, the fiber produced (i.e., lignin) is less digestible. Although one component of this loss in quality is the prevalence of grass seed heads and flowering structures, leaves also have more fiber in flowering forages. Good harvesting strategy is to harvest forages prior to or right as flowering occurs. One field recommendation for obtaining the best balance of yield and quality in a forage crop is to harvest at 10 percent bloom.
- *Antiquality components.* Almost every forage species can contain antiquality components with varying toxicity to livestock. The list of potential toxic components is lengthy but some important compounds follow:

  a. *Leaf saponins.* Most legumes contain saponins that can produce a foam complex that causes **bloat**.

Bloat is distortion of the rumen by excess gas production. If left untreated, it will result in death. Livestock should not be allowed pure diets of legume species.

  b. *Ergot* **alkaloids** (e.g., ergovaline). These are produced by fungal endophytes of some grass species. Levels vary with heat and drought. In most years, they cause subclinical growth retardation in young livestock, and in extreme cases, can result in animal death.

  c. *Nitrates.* Many C4 annual species have the potential to accumulate nitrates that can be toxic to livestock. The greatest risk occurs with the flush of forage regrowth that can follow a temporary drought.

  d. *High protein.* Livestock cannot digest protein in excess of 20 percent and forages with levels higher than this can result in impaired livestock growth and even abortion. High-protein forages should be diluted with low protein sources.

  e. *Natural estrogens.* Some legume species (e.g., red clover and subterranean clover) can produce natural estrogens that can impair reproductive performance of sheep and cattle. Modern varieties have been selected to avoid this problem.

ESTABLISHMENT

Forage establishment is a necessary component of a forage management and forage improvement program. The many forage establishment methods can be reduced to three main options:

1. *Full cultivation.* The most expensive and most reliable establishment option of forage crops involves destruction of the existing vegetation by mechanical (plowing) or chemical means, a mechanical operation to bury this vegetation (disking or plowing), mechanical treatment of the soil to establish a fine and firm seedbed (by disking and rolling).
2. *No-till planting.* As implied in its name, **no-till planting** involves pasture establishment without tillage of the soil. In most cases, existing pasture land already has desirable physical characteristics (e.g., high porosity, infiltration, and organic matter), and only the pasture species need to be changed.
3. *Frost seeding.* Also called **broadcast seeding** and **oversowing**, **frost seeding** applies seed directly to existing vegetation (or, if applicable, to the soil surface). This method has the lowest success rate but is very inexpensive.

PLANT DIVERSITY IN GRASSLANDS

Forages and rangelands typically comprise a mixture of five to fifty species. One important aspect is the benefit that this diversity (**biodiversity**) provides to the pattern of forage production. The benefits include better tolerance of environmental stresses, a more uniform forage growth pattern, increased stand persistence, and fewer losses of nutrients to streams and groundwater. Biodiversity can be considered at various levels, and one theory is that biodiversity is maximized by adding species with different and unique functions rather than adding forage species that are essentially similar to those already present (Fig. 18–12). There are many methods or systems for grouping species; however, some common functional groups for grasslands are listed below:

- *C3—cool season grasses.* This is the largest and most important group of forages and includes species such as perennial ryegrass, orchardgrass, timothy, Kentucky bluegrass, tall fescue, reed canary grass, crested wheatgrass, and smooth bromegrass (Fig. 18–13). These species can be characterized by typically higher quality for livestock, moderate to poor drought tolerance, good spring and autumn production, and moderate to low tolerance of low fertility. The name for this functional group results from the photosynthetic pathway these species use for fixing CO_2; that is, the first step for fixing CO_2 involves 3-carbon molecules and the enzyme **ribulose bisphosphate carboxylase (RUBISCO)**.

- *Legumes.* Legumes are species of the plant family FABEACEAE and are vital components of most forages and rangelands. Examples of legume species include alfalfa, red clover, white clover, birdsfoot trefoil, and lespedeza. Legumes provide many useful characteristics such as nitrogen fixation by *Rhizobium* bacteria, high-digestibility forage, high-protein concentration, and often a growth pattern complementary to their companion grass species (Fig. 18–14 on page 408).

- *C4—warm-season grasses.* This group of forages includes species such as Bermuda grass, big bluestem, switchgrass (Fig. 18–15 on page 409), indiangrass, limpograss, and paspalum. These species can be characterized by excellent tolerance of drought and heat, and high water use efficiency. In addition, many of these species are native prairie species and can have both conservation and productive value. Disadvantages of these species include their difficult establishment and lower forage quality. The name for this functional group results from the photosynthetic pathway these species use for fixing CO_2; that is, the first step for fixing CO_2 involves 4-carbon molecules and the enzyme phospho-enol pyruvate (PEP) carboxylase (see Chapter 11).

- *Forbs.* This is a largely maligned group of species often regarded as weeds in pastures. We now realize that these species might provide useful functions including filling vacant spaces, protecting bare soil, and providing forage (often with high nutrient value) to livestock (Fig. 18–16 on page 409). Two species relevant to mention are chicory and plantain (also called buckthorn) (Fig. 18–17 on page 409), which have been selected for forage production and are being used in forage mixtures. In many extensive pastures, other forb species can provide unique characteristics to livestock products such as unique (niche) flavors to meat and dairy products.

- *Annual species.* Annual forages are used at two extremes of forage production:
  a. In intensive systems, annual forages such as cereal rye, wheat, oats, and sorghum-sudangrass can be

A

B

Figure 18–12

(A) This healthy pasture is a mix of grasses and legumes. (B) This prairie grassland is being restored. The addition of forbs to the ecosystem is intended to stabilize and strengthen the system. Source: USDA Natural Resource Conservation Service, *http://photogallery.nrcs .usda.gov/*

Figure 18–13
Some grass species forage crops. Above (left to right): smooth bromegrass, orchardgrass, tall fescue.
Below (left to right): timothy, bluegrass, western wheatgrass.

Figure 18–14

Examples of legume species forage crops. Above (left to right): alfalfa, sweet clover, red clover. Below (left to right): birdsfoot trefoil, lespedeza, bur clover.

A

B

Figure 18–15
(A) Switchgrass with seed heads.
(B) Big bluestem seed heads.
Source: (A) USDA Agricultural Research Service
Image Gallery, *http://www.ars.usda.gov/is/
graphics/photos/* (B) USDA Natural Resource
Conservation Service, *http://photogallery.nrcs
.usda.gov/*

Figure 18–16
Mixture of grasses and forbs in a prairie. Source: USDA Natural
Resource Conservation Service, *http://photogallery.nrcs.usda.gov/.*

used for short-term forage production. In these
cases, use of these species is often in rotation with
traditional forage crops.

b. In extensive systems, annual forages such as subterranean clover and annual bromegrass can also provide significant short-term forage production. This
land often includes stressed environments that
encounter low fertility, drought, and cold winters.
Annual species can also occur as opportunists
within perennial pastures. Perhaps the most common species worldwide is *Poa annua.* Although
often characterized as a pasture weed, it does provide forage and useful ground cover when it is
growing; however, its propensity to produce seed
heads (of low quality) and its failure to produce
forage during summer make this species undesirable in most cases.

• *Shrubs.* Shrubs can occur in some grasslands such
as in tropical grazing systems and rangelands, but
they are not common in intensive grasslands in the
United States.

A Chicory

B Plantain

Figure 18–17
(A) Chicory and (B) plantain are two
forbs that though usually considered
to be weeds are now being used in
some forage mixes to give stability
to the forage ecosystem. Source:
Margaret McMahon, The Ohio State University.

SYMBIOSIS WITH MICROORGANISMS

Many forage species form symbiotic relationships with other microorganisms that are important and vital for the success of these plants.

- Rhizobia are bacteria that form a symbiosis with legumes, in the form of nodules on the roots. The species of bacteria are unique for each species of legume. The bacteria benefit from a protective host which provides a source of energy and nutrition, while the plant benefits from a source of nitrogen. The nitrogen is fixed from the air and converted by the bacteria into a useful form in the plant. This symbiosis is described in greater detail in Chapter 13.
- Mycorrhiza comprise a broad class of fungi that form various relationships (mutualistic, symbiotic) with a range of grass, legume, and forb species. In general, the mychorrhiza are more efficient at the uptake of essential nutrients, especially phosphorus, and can aid plant survival in low-fertility soil. In the case of most prairie grasses such as big bluestem and indiangrass, the presence of the appropriate native mychorrhizal species is essential to the persistence of the grass.
- **Endophytes** are a classic seed-borne fungi that form a symbiosis in the intercellular spaces of grass leaves, stems, and seeds. Although there are about fifty fungi species infecting 200 grass species, most interest focuses on the genus *Neotyphodium,* and the grass species tall fescue and perennial ryegrass. When growing in the grass host, fungi produce various alkaloids that are feeding deterrents to both insects and livestock. The symbiosis has the ecological benefit of providing to the host increased production, persistence, insect resistance, and drought/heat tolerance, but it has the practical disadvantage of being toxic to livestock. To avoid the toxic alkaloid effects, the forage seed industry has developed mechanisms for the supply of endophyte-free seed. Since 2001, a new endophyte option has become commercially available. Nontoxic endophyte is a fungal isolate identical in all respects to the toxic endophyte except that it does not produce the toxic alkaloids. Pastures established with nontoxic endophyte have the benefits resulting from the fungus being present, but without the toxic effect of the alkaloid to animals.

GRASSLAND ECOLOGY

Plant Dispersal and Propagation

Although grasslands include many types of vegetation such as turf, hayfields, pasture, rangelands, and prairie, they are all perennial and are expected to last from several to many years. In contrast to this expectation, the individual growth units (**tillers** or **phytomers**) are relatively short-lived and any grassland that seems stable and uniform is actually undergoing a continual cycling of individual plants. Added to this complexity is that most grasslands comprise many different species, each with different patterns of growth. This complex cycling is termed population dynamics.

Grassland species can propagate sexually (via seed dispersal) or asexually (via clonal growth and dispersal from tillers, stolons, and **rhizomes**). The functional properties of any grassland for example, the color uniformity of a golf course fairway, stand thickness of a meadow, or survival of a prairie from drought, all depend on an understanding of population dynamics.

Sexual Propagation   Propagation by seed is one way that grassland populations are maintained. However, depending on the species, propagation by seed can create management problems. For example, one common species in pasture is *Poa annua*—an annual grass species. The life cycle of this plant includes the production of seed in early summer, which germinates during fall, and seedlings grow during spring and produce seed in early summer. Bare areas during summer can result from *Poa annua* dying after seed set and there is insufficient water for germination. Many other annual weed species have a similar life cycle and control of these weeds requires that their seed production be prevented. Other perennial grasses produce seed; however, this seed does not make a major contribution to the overall plant population.

Tillering   The basic component of a grass is the tiller—and a grass clump is a collection of tillers. These are identical clones of the original parent seedling. Tillering is stimulated by light at the base of the plant, and keeping a grass short (e.g., close grazing) promotes a high tiller density. Allowing it to become tall (e.g., hayfield) results in fewer, larger tillers.

Stolons   Stolons (or runners) are specialized stems of grasses, legumes, and forbs that creep along the ground surface and have the ability to produce roots and initiate new plants. This important plant dispersal mechanism allows plants to explore and exploit areas favorable for growth (e.g., gaps in a grassland). Important species with stolons are white clover and Bermuda grass.

Rhizomes   Rhizomes are similar to stolons, but they exist underground. Rhizomes have the same function as stolons in spread of a species to new areas, but have

the advantage of being protected from heat and freezing. Rhizomatous plants are very hard to kill, and the agricultural examples are restricted to a few forage species. The most common rhizomatous plants are the weeds quackgrass and Canada thistle. Forage species that produce rhizomes include Reed canarygrass, tall fescue, and Kentucky bluegrass.

Grassland Growth Dynamics

With few exceptions, forage crops are plants whose stems and leaves are grown to feed livestock. Strategies for the removal of the herbage have been the subject of extensive research.

The forage mass (or leaf area) present when the crop growth cycle begins determines the initial rate of growth (see Figs. 18–5 and 6 on page 401). Overgrazing maintains a grassland with a low mass (<1,000 kg dry matter/ha), slows the growth rate, and reduces productivity. Residual leaf area is required for the plant to continue **photosynthesis**. Some grasslands can tolerate removal of most of the forage (e.g., alfalfa), but the harvests must be spaced far enough apart to allow accumulation of excess energy reserves in the root systems that are used to generate regrowth. Undergrazing has a high pasture mass (>3,500 kg dry matter/ha) and does not result in additional crop growth. In extreme cases, it might even result in yield loss if leaves in the base of the canopy senesce. The ideal range of forage mass ensures continued high growth rate without compromising quality. This ideal range of forage mass allows maximum harvested yield but does not result in slowed growth rate from shading of leaves lower in the crop canopy.

Expert managers can monitor their crop as near as possible to the optimum forage mass that keeps the crop in a state of rapid vegetative growth. In reality, this balance is more complex than only the pre- and postharvest forage mass. Other factors such as availability of machinery, labor, and livestock requirements also affect the farmer's decision about when to harvest forage; it is a managerial art to balance the multiple objectives and factors successfully!

Balance between Species

Grasslands appear to provide uniform green vegetation, but in many cases, they comprise a dynamic balance with different ratios of species in each season. The precise ratios between species depend on which species are present. One of the most important ratios is the balance between grasses and legumes. Legumes have the useful functions of fixing nitrogen from the atmosphere and providing high-quality nutrition to livestock, but they cause bloat. Most recommendations are to maintain 30 percent legume in a pasture, and pasture managers work continually to manage the balance when it deviates from this ideal. Other important ratios include the proportion of forbs (typically weeds), the balance between desirable and undesirable grasses, and the invasion of woody weeds.

NUTRIENT BALANCES

Grasslands generally have lower fertility than most other cropland; thus, they generally have lower losses of nutrients to the environment than other cropping systems. The low nutrient status occurs because the land is usually of lower quality (shallow soils, sloping hillsides), and grasslands typically have fewer agricultural inputs such as fertilizer. Soil fertility is the most important management tool for control of forage production. Soil fertility also affects the quality of forage for livestock and the balance between desirable and undesirable species. The most significant nutrients in forages are nitrogen (1 to 4 percent by weight), potassium (2 to 4 percent by weight), and phosphorus (0.2 to 0.5 percent by weight).

One of the most important differences between forages and other crops is the contribution of livestock manure (nutrient cycling) to soil fertility. In most cropping systems, nutrients are provided by synthetic fertilizers or organic sources. In grazing systems, the inclusion of livestock has a dominating effect on nutrient cycling. While some nutrients are incorporated into the animals' body mass, 98 percent of nutrients pass through the animals and are excreted in urine and dung. The effect that livestock has in removing forage (and the nutrients it contains) from a largely uniform distribution around the pasture and depositing it into piles has a significant effect on the grassland system. Typically, the return of manure by livestock is concentrated in areas where they ruminate, for example, near water sources, and seek shelter, such as near trees and buildings. This redistribution is called nutrient transfer and it can have positive and negative benefits, depending on the manager's objectives.

Important factors affecting hay and silage systems are the nutrients harvested and removed from the field with the crop. Hay and silage crops can yield as much as 10 tons per acre, although yields of 3 to 4 tons per acre are more common. These yields are greater than for any other crop and can represent a huge loss of nutrients from these systems. Sustainable hay and silage production from an area is usually possible only where these nutrients are balanced by inputs of fertilizer. Collecting animal waste and applying it to pastures or hay fields can reduce the need to apply commercial fertilizer (Fig. 18–18).

A   B

Figure 18–18

(A) Manure and urine from these cattle is being collected in the pond to be (B) spread on the hay fields, as shown here, that will feed the cattle, thus completing the nutrient cycle.
USDA Natural Resource Conservation Service, http://photogallery.nrcs.usda.gov/

OTHER USES OF GRASSLANDS

We have focused primarily on the productive capacity of grasslands. As the world's population grows, we are facing increased complexity in our land use, demanding multiple uses from the same unit of land. Our objective as land managers is not only to ensure food production, but also to maximize alternative and often competing land-use options. The dedicated use of grassland for food production only is becoming less and less important as society appreciates the diversity of land-use options that grasslands provide.

• *Lifestyle.* Regardless of our cultural background and overriding our detailed biological understanding of plant production systems, the image of contented livestock on pasture is one of the enduring images we all recognize (Fig. 18–19). Stereotypical images of New England, New Zealand, England, and Switzerland are characterized by pastoral landscapes that provide value to society (e.g., economic value through tourism). Even margarine products are marketed with images of cows on pasture! Many small grazing farms list the lifestyle as one of the primary benefits of their business. Values such as working with family members (including children), a belief in enhancing the environment, and making a good profit are based on forage and rangeland plants. Increasingly, products produced in these systems (e.g., free-range product) are marketed as being differentiated from bulk commodity production and thus can gain a market premium.

• *Carbon sequestration.* Grasslands have an organic matter content of 2 to 5 percent greater than other cropping systems on identical soil. As we are becoming aware of the environmental impact on the world's rising carbon dioxide (CO_2) levels, there is interest in the use of grasslands to sequester carbon in the soil. Currently in the United States, there is little economic incentive to farmers to use this capability; however, if a system of carbon credits were introduced, one option might be to make greater use of forage crops.

• *Soil preservation.* Most forages are perennial crops with large root systems. These features provide grasslands with extremely desirable characteristics for year-round soil stability, slowing the surface movement of water and improving water infiltration. Grasses are invaluable in riparian management and are frequently planted in the drainage channels of cropping systems to reduce soil loss during extreme rainfall events.

• *Water harvesting.* As a result of their dense structure and short vegetation, grasslands provide an ideal combination for water harvesting. The yield of water from land and its quality (freedom from soil particulates) are greater than from land producing any other vegetation.

Figure 18–19

A pastoral setting like this provides many values to society from nutritional to aesthetically pleasing. In other words, food for the body and the soul. Source: USDA Natural Resource Conservation Service, http://photogallery.nrcs.usda.gov/

Figure 18–20

Grasslands provide habitat for wildlife as well as livestock feed. Source: USDA Natural Resource Conservation Service, *http:// photogallery.nrcs.usda.gov/*

Figure 18–21

The preservation of prairies as is shown here is an important aspect of grassland management. Source: USDA Natural Resource Conservation Service, *http://photogallery.nrcs.usda.gov/*

- *Wildlife.* Grasslands have the potential to provide nutritious forage for large livestock such as deer and wild horses, nesting habitat for birds (although well-fertilized introduced species can have excess stand density for some prairie birds), and clean water to streams. Grasslands are an important component of a wildlife management program (Fig. 18–20).
- *Conservation.* Native prairie is threatened vegetation throughout the US plains. Traditionally, prairies supported intermittent grazing by bison and deer; however, the potential for both livestock production and prairie conservation remains underutilized (Fig. 18–21).

- *Food quality.* We are becoming more and more aware of the unique contributions that forages can make to the nutritional value for humans. Quality characteristics such as lean meat, organic produce, conjugated linolenic acid (CLA) levels (in the class of omega-3 fatty acids that have cancer fighting properties in laboratory animals), and vitamin A levels can all be influenced by the amount of forage in an animal's diet.
- The raising of forage crops for seed production is a specialized industry that is concentrated in the western states in the United States. Seed crops comprise less than 0.1 percent of total grassland area but are important in providing the seed for newly established grasslands.

SUMMARY AND REVIEW

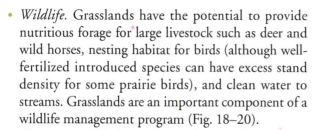

Forage is the edible parts of plants that can be fed either immediately or after harvest and storage to livestock. Forage crop and rangelands have unique characteristics compared to other crops. They are predominately perennial production systems, encompass tremendous diversity, and have no immediate value to humans in terms of direct food products. Worldwide, more land is devoted to grasslands than all other crops combined.

Animals can consume forages directly when grazing or as preserved food (hay and silage) provided by farmers. Grazing systems generally require little in the way of land preparation and there is no harvesting, but these fields still require considerable attention and management to keep them productive. The production of harvested forage (hay and silage) is much like other field crops and requires land preparation, planting, growing,

harvesting, and storing the crop. Hay is the preservation by drying of the edible plant parts. Silage is the moist storage of forage crops in silos or plastic-wrapped bales (baleage) under anaerobic conditions and fermentation. The production of a quality forage crop depends on the grower's understanding of the interactions of the crop and the environment, along with production and grazing or harvest and storage practices, and maximizing the positive benefits of those interactions.

Forage quality is determined by the amount of fiber it contains and the digestibility of that fiber. Fiber can be present as pectin, hemicellulose, cellulose, and lignin. Pectin and hemicellulose are moderately digestible, cellulose is poorly digested, and lignin is indigestible. Forage crop species, plant maturity, and the parts consumed by the animals influence fiber type.

Biodiversity is an important concept in forage and rangeland crop management. C3 and C4 grasses, legumes, and forbs are the primary broad groups of plants that contribute to forage/rangeland biodiversity. The benefits associated with a production system that is biodiverse include better tolerance to environmental stresses, more uniform growth patterns, increased stand persistence, and fewer nutrient losses to runoff or leaching.

Unlike most field crops, forage systems require that plants regenerate from year to year (even hay fields are often perennial in nature). Perennial grassland crops usually propagate naturally from seeds or the spread of tillers, stolons, and rhizomes. Whether animal feeding or harvesting removes the edible parts, forage growers must manage the fields so that the plants have the chance to reproduce adequately for the next season's or year's growth. Limiting the amount of grazing, maintaining the best array of forage species (biodiversity) for the area, and maintaining proper nutrient balance are all essential for forage productivity.

Grasslands have many uses in addition to providing livestock feed. They promote a desirable lifestyle for humans. Their ability to sequester carbon is likely to have an impact in the effort to reduce atmospheric CO_2. They aid in soil and water conservation efforts. Providing wildlife habitat and enabling the conservation of threatened species are additional beneficial uses for grasslands. Improvement in human nutrition can be provided by the use of forages over other types of animal feed.

KNOWLEDGE CHECK

1. What is forage?
2. In comparison to other crops, how much land worldwide is devoted to forages?
3. What types of species dominate forage fields and rangeland?
4. What is the typical range in the number of species in a forage crop or rangeland system?
5. How do forages and rangelands have value for humans?
6. What is the difference between a grazing and a confinement system for raising livestock?
7. What is the most important objective of grazing?
8. What is stock rotation in a grazing system and what are its benefits?
9. What kind of rotation system is used with intensive grazing? With extensive grazing?
10. What is hay and how is it used as forage?
11. In the United States, what species usually makes the best quality hay? What makes up the greatest quantity of hay?
12. In general, when is the best time to harvest hay?
13. What determines hay quality?
14. Give two reasons why proper dehydration is necessary when preserving hay.
15. What is silage?
16. Describe the process that occurs in silos to make silage.
17. Describe two ways that forage production can be incorporated into a crop rotation system.
18. What is the primary factor affecting forage quality?
19. What are the different types of fiber found in forage and what is the digestibility of each?
20. How do each of the following affect forage quality: species, leaf-to-stem ratio, and maturity?
21. List three things that are considered to be anti-quality factors in forage.
22. What are the three main ways that forage systems can be established?
23. Give three reasons that biodiversity is important to forages and rangelands.
24. Give two characteristics of each of the following types of forage species: C3 grasses, C4 grasses, and legumes, forbs.
25. What symbiotic relationship do endophytes have with plants? How does this affect forage quality? Since 2001, how has this problem been reduced?
26. What is population dynamics in a grassland?
27. Why can the reproductive cycle of *Poa annua* be a management problem in grasslands subject to summer drought?
28. What are tillers, stolons, and rhizomes?
29. Why does overgrazing reduce pasture growth rate? Why can undergrazing sometimes be detrimental to pasture growth and quality?
30. What is the main difference in nutrient cycling between most crops and grazing systems?
31. Where are most of the nutrients deposited by livestock in a pasture? How does this affect nutrient recycling and the use of fertilizer inputs?
32. List four other land-use options for grasslands other than as pastures for livestock.

FOOD FOR THOUGHT

1. You are raising livestock in an area of the country that is characterized by cool wet springs and autumns but hot, dry summers. You are getting ready to renovate your pastures and have decided to use C4 grasses. Your neighbor is also renovating her fields but has chosen to go with a mix of C3 and C4 grasses. Whose pasture will most likely allow grazing for a longer period of time during the year? Why? Whose may have had better growth during the summer? Why?

2. You are growing livestock in a confined system and a neighbor who just moved to the country from the city (to enjoy the fresh air) is complaining about the smell when you spread the manure and urine waste from the animals on your hay field. You know this neighbor is very much into nature and understands ecological principles to some degree. What would you tell him to make him "appreciate" the odor from the manure?

3. Your alfalfa hay field has been looking less and less vigorous after every cutting this summer. There has been enough rain and a soil analysis shows the nutrient level of the soil to be OK. There is no sign of serious disease or insect infestations. What do you think is the reason for the loss of vigor and what can you do to correct the problem?

FURTHER EXPLORATION

BALL, D. M., C. S. HOVELAND, AND G. D. LACEFIELD. 2002. *Southern forages,* 3rd ed. Atlanta: Potash and Phosphate Institute and Foundation for Agronomic Research.

BARNES, R. F., C. J. NELSON, M. COLLINS, AND K. J. MOORE. 2003. *Forages: An introduction to grassland agriculture,* 6th ed. vol. 2. Ames, IA: Iowa State University Press/Blackwell.

HODGSON, J. 1990. *Grazing management—Science into practice.* New York: Longman Scientific and Technical/John Wiley & Sons.

HOPKINS, A. (ed.). 2000. *Grass—Its production and utilization,* 3rd ed. Malden, MA: Blackwell Science/Oxford England.

MURPHY, B. 1998. *Greener pastures on your side of the fence,* 4th ed. Colchester, VT: Arriba.

VALENTINE, J. F. 2000. *Grazing management,* 2nd ed. Burlington, MA: Elsevier Science and Technology.

*19
Vegetable Production

Matt Kleinhenz

key learning concepts

After reading this chapter, you should be able to:

- Explain how vegetable production benefits society.
- Discuss the differences among field, tunnel, and greenhouse vegetable production.
- List the basic steps to successful vegetable production.
- Describe the basic characteristics of the major vegetable crops.

PRINCIPLES OF VEGETABLE PRODUCTION

The Importance of Vegetable Production

Vegetable production is important for many reasons. For example, vegetables are important to peoples' diets. Hippocrates is credited with founding the practice of medicine and offering advice such as "let your medicine be your food and your food be your medicine." This long-ago recognition of the diet–health relationship is supported by a large amount of modern scientific evidence. Some of this evidence points to the specific dietary value of vegetables as derived not only from what they contain—vegetables are a prime, convenient, and natural source of minerals, vitamins, fiber, and energy—but also from the apparent ability of a diet rich in vegetables to reduce the risk of the onset of diseases and disorders. Therefore, it is clear why many different health professionals consider vegetable consumption central to nearly all types of healthy eating, including the personalized eating plans outlined in the USDA's MyPyramid (Fig. 19–1). (For more information about the pyramid, go to http://www.mypyramid.gov) Making vegetables available through production, distribution, and marketing contributes strongly to their volume of consumption.

Vegetable production can contribute to our psychological or emotional well-being as well as the health of our bodies. Edward Wilson has called our necessary connection with nature "biophilia." Many people derive pleasure, satisfaction, pride, relaxation, camaraderie, and other benefits from vegetable gardening, whether done individually or in community garden settings (Fig. 19–2). Indeed, in addition to gardening activities done for pleasure or to produce food, gardening activities can add to therapy and rehabilitation programs designed to help maintain or improve emotional stability, self-image, or cognitive function. Vegetable gardening can be an ideal form of individual or group exercise for people of all ages and abilities (Fig. 19–3).

Regardless of location, vegetable gardens and farms can also have positive effects on our environment that, in turn, benefit people. For example, by composting organic waste, natural resources can be recycled and maintained on vegetable gardens and farms. Vegetable gardens and farms help to maintain open, rural, or green spaces—used in various ways, including relaxation and revenue generation—on our landscapes.

Figure 19–1

The USDA's MyPyramid is an assessment of personal dietary needs. From left to right, the sections of the pyramid represent grains, vegetables, fruits, milk, and meat and beans. The USDA website at www.mypyramid.gov/ allows you to determine your dietary needs.
Source: USDA, *http://www.mypyramid.gov/.*

| Grains | Vegetables | Fruits | Milk | Meat & Beans |

Along with food for humans, these enterprises can provide habitat for some wildlife.

The different scales of vegetable farming and the history of vegetable production can give us a better understanding of societies and the natural world. Over time, successes or failures in the production of vegetables and other crops have had major impacts on societies, especially in terms of the health and movement of people. For example, failures of potato crops in Ireland in the middle nineteenth century from a disease called blight, along with other socioeconomic factors, led to the death of many (estimated to be from 500,000 to 1.5 million) people and emigration of others from Ireland. Ireland lost nearly 25 percent of its population to emigration or death. Many people in the United States and other countries of Irish ancestry can trace the emigration of those ancestors to the potato famine. The large number of poor Irish immigrants to the United States led to many discriminatory practices including the infamous "Irish Need Not Apply" signs that were placed in the windows of business establishments.

Today, the likelihood of similar, biologically based crop failures has been reduced through the development of improved production methods. However, some of these methods involve the use of crop protectants (pesticides) that have the potential to harm people

Figure 19–2
Vegetable garden. Source: Jane Martin, The Ohio State University Extension.

Figure 19–3
Vegetable gardens provide exercise and education for people of all ages. Source: Margaret McMahon, The Ohio State University.

or the environment. To protect people and the environment, even newer tactics designed to help reduce the use of pesticides were developed, but some of these tactics involve controversial manipulations of plant genomes. As you can see, advancements in production methods and crop variety development can offer tantalizing benefits but may have other, potentially undesirable consequences. Nevertheless, efficient production of an abundant supply of high-quality food in environmentally sound ways—whether in small gardens or on large farms—is important to all societies. Vegetable plants respond strongly to their growing environment in ways that can be easily seen and measured, making them an ideal commodity for educating citizens from all walks of life in the laws of nature and how to grow crops safely within those natural laws.

Although home gardening is practiced by many, most of the vegetables we eat are grown on commercial farms and perhaps are processed in factories. Therefore, the fact that vegetable production and processing are large industries is another reason why vegetable farming is important. In fact, economies (family, corporation, community, state, region) throughout the United States and world are based on income generated from the production and sale of vegetables and vegetable products. The heavy financial impact of commercial vegetable industries is derived from the sale of products of vegetable farms and processing facilities and the large amount of supplies, equipment, labor, and services they purchase. For example, the total farm-gate value for fresh and processed vegetables grown in the United States in 2008 was estimated at $10.4 and $1.9 billion, respectively. However, these estimates of crop value do not include the total economic impact of the US vegetable industries; for example, additional, significant private and public revenue is generated through vegetable processing.

Overview of Vegetable Farms and Industries

Vegetable farms and industries are (1) widespread but are more prevalent in certain areas; (2) variable in size, crop diversity, and production method; and (3) managed ecosystems. Databases compiled and managed by the USDA at http://www.usda.gov/nass/, the United Nations at http://faostat.fao.org/, and others underscore that vegetables are produced commercially in all US states and throughout the world. The size of the area used in commercial farming, including for vegetables, is typically measured in acres or hectares. A hectare contains 2.47 acres. Acreage devoted to commercial vegetable farming tends to be greatest and most concentrated

where the environment (physical, chemical, and biological components) consistently favors crop growth, and natural and other resources required for production are readily available. Physical environments permit crops to be grown in many regions; however, commercial production of many vegetable crops on the largest scale is concentrated in areas where crop marketable yields are consistently high and/or production costs are relatively low.

US farms vary widely in their characteristics, and vegetable farms are no exception. A USDA typology of US farms includes five types of small family farms, two types of larger family farms, and nonfamily farms. Farms, including vegetable, within and among these groups differ in their contribution to overall agricultural production, product specialization, farm program participation, and dependence on farm income. Vegetable farms ranging in size from less than one acre to hundreds of acres produce hundreds of pounds to thousands of tons of one or more crops, and may or may not be the sole source of income for their owner. Some of these farms are in rural areas, but many are located close to urban areas.

Demographic characteristics of the farming population in the United States seem to be changing. Statistics suggest that a rising number of farms are owned and/or managed by women and minorities, that the number of part-time farmers is increasing, and that larger farms tend to have more than one owner–operator.

Characteristics of Vegetable Farming and Gardening Systems

Managed Ecosystems   A defining characteristic of vegetable farms and gardens is that they, like other types of farms, operate as managed ecosystems (agroecosystems). Agroecosystems combine elements of business (e.g., accounting, finance, marketing) and the natural world (e.g., biology, chemistry, physics). In a business system, relationships among different industries may become mutually beneficial and self-sustaining. Such relationships involving abiotic and biotic participants are also found in agroecosystems. For example, vegetable crop plants interact with nonliving components of soil and air and numerous members of hundreds of micro- to macroscopic species from different kingdoms—insects, weeds, pathogenic and nonpathogenic microbes, birds, rodents, and so on.

The genetic potential of vegetable plants for crop yield and quality is often high. However, this genetic potential is rarely realized in commercial production systems or gardens. Plant diseases and pests interrupt or retard crop growth or reduce the marketability of

A

B

Figure 19–4

(A) Overhead irrigation systems such as the one pictured here can be used in large-scale vegetable production. (B) Spray-type overhead system. Flood and drip systems are also used in both large- and small-scale production facilities. Sources: (A) USDA Agricultural Research Service Image Gallery, www.ars.usda.gov/is/graphics/photos/; (B) Margaret McMahon, The Ohio State University.

harvested units. Weeds compete with crop plants for light, water, nutrients, and other growth factors. And the levels of light, temperature, nutrients, and/or water may be too low or too high relative to crop need. To capture as much of the genetic potential of vegetable plants as possible, farmers and gardeners supply them with growth factors (e.g., nutrients, water) and protect them from adverse temperatures and light intensities, diseases, insects, and weeds. Approaches used by growers to achieve these goals take into account production, market, and personal factors.

Open Field

Vegetable farms and gardens are diverse in overall production scheme. Farming and gardening in open fields is by far the most common approach in the United States and elsewhere. Using this approach, the ability of farmers and gardeners to control the natural environment is limited, and their crops are fully exposed to weather, soil, and pest and disease conditions common to their area. However, these systems usually include an irrigation system (Fig. 19–4), a fertilization program, and utilization of pest- and disease-control strategies to manage the system somewhat.

Climate Controlled

Other farmers and gardeners grow their crops in semienclosed and somewhat climate-controlled structures (e.g., high tunnels) (Fig. 19–5). Using this approach, vegetable growers limit the exposure of their crops to undesirable natural conditions while, in some cases, creating conditions more supportive of plant growth. For example, placing growing crops within a high tunnel covered by a single layer of clear plastic shields them

Figure 19–5

High tunnels are used to extend the vegetable growing season in some areas of the country.
Source: Dr. Matt Kleinhenz, The Ohio State University.

from wind, rain, and some pests. And it exposes crops to temperatures higher than those outside the high tunnel. In fact, temperatures in a high tunnel in northern Ohio (USDA Plant Hardiness Zone 5b) in February may closely resemble normal outside temperatures typical of parts of Tennessee (USDA Plant Hardiness Zone 6a). Therefore, vegetables can be grown when conditions outside the tunnel would otherwise not allow it (e.g., early spring, late fall). Semiprotected vegetable production is increasingly common.

Other vegetable growers enclose their crops in fully climate-controlled structures, such as greenhouses (Fig. 19–6). The production of vegetables in greenhouses differs from that in fields or semicontrolled environments in several important aspects. For example, plants grown in commercial greenhouses are often

Figure 19–6
Greenhouse tomato production.
Source: Margaret McMahon, The Ohio State University.

grown hydroponically. Hydroponic systems can either have plants growing in a water and fertilizer solution only or in manufactured, soilless, inert growing medium that requires a nearly constant supply of water and fertilizer. The fertilizer is carefully regulated so that plants are constantly receiving the mineral nutrients in optimum concentrations. Vegetables grown in fields or semi-enclosed structures are typically grown in naturally occurring soil that needs much less fertilizer monitoring and adjusting. Field-grown crops generally have unrestricted root growth too. However, plants in greenhouses are often grown in containers that provide firm, physical boundaries to root growth which can also affect nutrient and water uptake. Another type of hydroponic system is to grow the crop (typically lettuce) on floating trays that allow the roots to grow into the nutrient solution that fills the troughs on which the trays float. Greenhouse vegetable production also often relies on artificial lighting to supplement sunlight or shading material to reduce sunlight when it is very intense and temperatures are too high. Commercial greenhouse vegetable production tends to be highly mechanized and involves computerized heating, cooling, lighting, irrigation, spraying, and planting and harvesting systems.

Establishing and managing all vegetable production systems typically involve numerous interventions by growers. Different strategies and practices are used to meet goals of yield, profitability (in the short- and long-term), and environmental stewardship. Collectively, these strategies and practices comprise an overall system or approach to farming and gardening. Current major approaches to farming and gardening, including vegetables, are often described as conventional, organic,

and sustainable. (See Chapter 16 for a detailed discussion of the three approaches.)

Conventional

Overall, in conventional approaches, strategies for protecting crops from diseases, pests, and weeds include the application of synthetic pesticides, which may or may not be based on naturally occurring compounds, as a central component. Pesticides are applied to soils, seeds, plants, and harvested units. Strategies for protecting crops in conventional systems may also include the use of crops genetically modified to resist or tolerate diseases, pests, or herbicides. Some genetic modifications may combine portions of genomes that may not otherwise combine naturally. Crop nutrient management strategies in conventional systems typically involve the application of synthetic fertilizers, often made from the mining and chemical combination of minerals, as a main component. The content of synthetic fertilizers is tailored to meet the needs of specific combinations of vegetable crop, soil, climate, water source, and other factors. Synthetic fertilizers tend to be clearly described and consistent in terms of makeup, are relatively inexpensive, and are widely available.

Organic

Organic agriculture has been defined as "an ecological production management system that promotes and enhances biodiversity, biological cycles, and soil biological activity" by the National Organic Standards Board (NOSB). Approval of an application for certified organic status requires payment and extensive documentation of production practices.

Organic vegetable farming and gardening involves a strong reliance on cultural practices and biological principles for weed, disease, pest, and nutrient management. For example, weed growth can be limited by preventing light from reaching the weeds by using a mulch (Fig. 19–7) and by timely mechanical or hand cultivation. The demand for organic vegetables has risen dramatically in the past two decades. Evidence of this is no better seen than in the produce department of large supermarket chains that now have a sizable area devoted to organic fruits and vegetables. Many vegetable farms, large and small, are certified organic (Fig. 19–8).

Sustainable

Sustainable approaches to vegetable production integrate components of conventional and organic management. The USDA Sustainable Agriculture Research and Education (SARE) program describes sustainable agriculture as encompassing diverse methods of farming and

Figure 19–7
Straw and plastic mulch in a tomato field. Mulches help to suppress weeds, moderate soil temperatures, and reduce water loss from the soil. With time, organic mulches such as straw add nutrients and organic material to the soil.
Source: Dr. Matt Kleinhenz, The Ohio State University.

ranching that are more profitable, environmentally sound, and good for communities (see http://www .sare.org/coreinfo/consumers.htm). Sustainable production may involve the selective, targeted use of pesticides (synthetic or organic-approved) to augment biologically and culturally based disease, insect, and weed management, a key component of integrated pest management. Vegetable growers may also release, apply, or attract beneficial organisms that act as competitors, predators, or parasites for organisms that would otherwise damage crops.

Figure 19–8
Certified organic vegetable farm.
Source: Margaret McMahon, The Ohio State University.

Likewise, farmers and gardeners employing sustainable approaches often employ a combination of rotation, cover crops, and synthetic fertilizers to meet crop fertility requirements. Sustainable approaches are increasingly common among vegetable growers. Still, it is important to note that sustainability has multiple dimensions (e.g., financial, social, environmental) and may best be considered as a guiding principle for the evaluation of cropping practices within a context larger than production.

Markets for Commercially Grown Vegetables

Once grown and harvested, farm-grown vegetables enter one of two leading markets. In the *fresh* market, vegetables are packaged, at facilities on or off the farm, and distributed to buyers or consumers with relatively minimal pretreatment or preparation. Pretreatment or preparation includes sorting, trimming, washing, cooling, and packaging. Fresh market vegetables may be sold by the farmer directly to consumers at markets and produce stands, via the Internet, or as shares in a community-sponsored agriculture (CSA) farm. Fresh market vegetables are also sold to grocers, market managers, restaurateurs, or wholesalers–distributors for resale, possibly on a contractual basis.

Processing markets add value and diversity of use to freshly harvested vegetables. Processing vegetables are packaged and distributed to consumers after being, for example, frozen, precooked, or dehydrated. During processing, vegetables may be used to create another product (e.g., potato chips, french fries). Processed vegetables may be packaged alone or in combination with other vegetables (sauces, soups). Processing plants are highly engineered facilities owned by individuals, groups, or multinational corporations employing few or many people. Transactions culminating in the transfer of vegetables from farms to processors often involve contracts set up before harvest.

The distances and time differences between sites of production and consumption of vegetables, particularly fresh market, may be minimal or large. In most cases, consumption occurs many miles from and days after production and harvest, requiring sophisticated harvest and postharvest procedures to maintain product quality as viewed by consumers. However, a growing percentage of vegetables are being grown and consumed within smaller geographic areas as interest in local food systems increases (see, for example, http://www.cias .wisc.edu/; http://www.sustainable.org/index.html/).

Within a county, state, region, or country, whether a vegetable commodity is grown commercially and at what scale depends on consumer- and production-related

factors. Perhaps the most important among consumer-related factors may be the overall demand for the product: greater demand creating a potential need for more growers. Demand for vegetables appears to be driven by their availability, convenience, and diversity, as well as by consumer perceptions of their eating quality, potential health or nutritional value, and appeal to a particular ethnic group. In general, farmers' interests in the production of various vegetables are based on their potential to generate profit and the farm's ability to meet market demands in terms of quantity, quality, and consistency.

STEPS IN VEGETABLE PRODUCTION

Challenges in vegetable production, particularly outdoors, arise from the fact that production is strongly influenced by the natural environment (e.g., soils, climate, weeds, plant diseases, and pests) and that the natural environment is variable in time and space. Gardeners and farmers deal with these challenges through careful planning and execution in key areas. Successful vegetable farmers must also follow principles integral to the success of other types of businesses, including (1) studying the market and being prepared to meet its requirements; (2) producing a high-quality product efficiently; (3) packaging the product appropriately and delivering on time; and (4) innovating. The key steps in vegetable production are discussed in the following sections.

Site Selection

Factors important in selecting the site for successful vegetable production include climate; soil type; the size and total cost (land, utilities, taxes, etc.) of the site; and the cost and ease of access to water for irrigation, pesticide application, post-harvest washing, and packing. The proximity of the site to suppliers, support industries and personnel (labor), and the market is also important to farmers. Vegetables can be classified as warm- or cool-season types. Warm-season crops require warm growing temperatures and a relatively long growing season. Cool-season crops do not tolerate high temperatures. These are often grown as spring or fall crops in areas with hot summers or year-round in areas with mild or cool summers. It is important to keep in mind the types of vegetables you want to grow when selecting a site.

Site Preparation

Once selected, the site's drainage must be checked and improved, if necessary. Installing tile or raised beds or leveling land can improve drainage. Thereafter, in field production, plowing, disking, and other tillage is often needed to prepare the soil for planting operations. Preparing soil usually also involves fertilizer application and steps to minimize weeds and soil-borne diseases and pests through, for example, the application of pesticides and fumigants or the use of cultivation, solarization, plastic mulches, or trap crops. Solarization is the installation of a clear plastic cover over the soil prior to planting. Sunlight causes temperatures under the plastic to reach levels that destroy many weeds and pathogens. Trap crops are crops that are more attractive to pests than the main crop. The pest goes to the trap crop first where it can be managed on a small scale rather than trying to manage the pest in the main crop which may be many acres. Usually the trap crop is eventually destroyed to prevent it from becoming a breeding ground for the pest.

Variety Selection

In summary, the outcome (e.g., marketable yield) of vegetable production at any scale results from the interaction of the crop's genotype with its growing environment (physical, chemical, biological). In farming, fresh and processing markets maintain strict standards for the delivery date, appearance, shape, dimensions, weight, and sensory and chemical properties of crops and products. Crops not meeting standards set by the market typically are unsalable, creating financial and other burdens for farmers. Often, market-based crop standards and desirable outcomes for gardeners can be met only through specific combinations of genotypes and growing environments. Therefore, variety selection is key for all vegetable growers in part because varieties differ in their ability to withstand abiotic (e.g., temperature, moisture, daylength extremes) and biotic (e.g., disease, insect) stress.

Planting

Vegetable propagules include seed and transplants or slips. Crops are established through direct seeding or transplanting, with transplants and slips set by hand or machine. Fertilizers and pesticides may be applied simultaneously. In farming, the population of crop plants per unit area and their arrangement is manipulated to achieve production and marketing goals. Most crops are planted in rows, with the distance between adjacent plants within the same row and adjacent rows specific to the crop, genotype, growing environment, and market.

Management during Crop Growth and Development

Vegetable crops are irrigated; fertilized; mechanically or chemically pruned or trained; and protected from damage due to diseases, pests, or weeds. In greenhouses,

light and temperature levels are also optimized. The use of remote or on-site scouting or monitoring (soil, crop, weather, pest, disease, weed) tools and services is an established and increasingly common practice in vegetable farming. Information gained through scouting and managing can have significant, variable benefits to individual farmers and regional industries.

Harvest

Crops are visually assessed for harvest readiness or **harvest maturity**. Harvest maturity may not be the same as plant maturity which is the ability to reproduce. Harvest maturity means that the part to be harvested is ready for harvest. The harvestable stage of maturity depends on the crop. We want the part we are harvesting to be mature enough to have the qualities we desire (sugar or starch content, flavor, color, etc.). If harvest occurs too early, these qualities may not develop. If harvest occurs too late, undesirable traits such as starch formation in sweet corn or over-softening of a tomato may develop. Vegetables such as tomato that continue to mature after harvest can be harvested slightly immature, when it is firmer and less easily bruised during harvest and transport. If harvesting stops the maturation process, then that vegetable must be harvested when it is at harvest maturity. Farmers may employ instruments (e.g., refractometer to measure color, scales to measure weight) to assist in evaluating harvest maturity. Careful handling and timeliness of activities are components of harvest procedures that proceed at crop maturity so that crop yield and quality losses are minimized. Keeping crops free of abiotic and biotic contaminants during harvest and postharvest handling is also important. Crops are harvested by hand and/or machine. In fact, the significant amount of hand labor required to harvest and prepare many vegetable crops for market is an important distinction between vegetables and other crops. The need for production and harvesting equipment specifically designed for use on a single crop is also a characteristic of vegetable production. In contrast, equipment such as planters and combines can be used with only a little modification on several agronomic crops.

Packaging and Postharvest Management

Regardless of the market for which they are destined, commercial vegetable crops are placed in containers or trucks before delivery or shipment. They may also be sorted, trimmed, washed, cooled, and packaged in units of market-specified weight or volume. For a more detailed discussion of the above steps, see Chapter 16.

Figure 19–9

Professional and amateur vegetable growers can learn about the latest research discoveries and recommended growing practices at field days sponsored by educational institutions.
Source: Dr. Matt Kleinhenz, State Vegetable Specialist, The Ohio State University.

Record Keeping and Maintenance and Repair of Facilities and Equipment

These activities are needed throughout all aspects of vegetable farming. Vegetable farming requires the tracking, documentation, and analysis of a wide range of production, market, financial, sociopolitical, regulatory, and other information. Farmers use increasingly powerful digital and paper-based tools and resources to assist them in these activities. Vegetable farming also involves the use of specialized buildings and equipment that require ongoing maintenance and repair.

Continuing Education

Vegetable farming is a professional occupation requiring technical and practical understanding of an increasingly complex array of areas (e.g., biology and ecology, chemistry, engineering, marketing and economics, public relations, organizational or association development, and management). Many gardeners also have a strong interest in improving their production systems. Therefore, many growers commit to receiving ongoing education in formal and informal settings ranging from conversations with successful peers to credit courses at universities and other institutions (Fig. 19–9).

VEGETABLE CROPS GROWN FOR FRUITS AND SEEDS

Many of the most popular vegetables we eat, for example, green and dried beans, corn, cucumbers, tomatoes, and many others, are botanically classified as fruits or seeds.

Convention, though, has us overlooking the botanical classification in favor of a dietary concept. We generally think of fruits as being eaten for their sweetness and dessert-like quality, while vegetables usually are not very sweet. Some of these types of vegetables, for example, tomatoes, can be eaten raw or cooked, while others such as sweet corn, green beans, and dried beans are usually cooked.

Beans, Snap or Green (*Phaseolus vulgaris* L.) and Lima Beans (*Phaseolus limensis* Macf.) FABACEAE

The genus *Phaseolus* includes a number of "vegetable" bean species whose immature fleshy pods and immature seeds, and/or dried seeds are a highly valued food. Dry beans in particular are a nutritionally important protein and carbohydrate food source. The common bean and other *Phaseolus* members are native to Central and South America, and were domesticated and widely used long before the arrival of European explorers.

The better-known common bean types are the snap bean and lima bean, and various types of field beans (shell beans, kidney beans, mung, etc.) grown for their mature dry seeds. Snap beans and limas are grown as annuals for their immature pods. They can have two different growth types: the pole form, which is usually indeterminate and grown on supports (poles or trellises); and the bush form, usually determinate and grown unsupported. Because the immature pods and seed of snap beans are the edible portions, selection was directed toward developing cultivars having thicker, fleshy pods of low fiber and slowly developing seed (Fig. 19–10). Pod fiber development occurs concurrently with seed maturation and is undesirable. For the dry-seed

bean types, rapid seed development is the more important criterion, and pod fiber is not of interest. Lima beans, whether grown for fresh market or for processing, are harvested when their seeds have enlarged but before they are fully mature. However, fully mature dried lima bean seeds are also an important product. Beans are self-pollinated and propagated from seed.

Both snap and lima beans are warm-season crops requiring frost-free growing periods, preferably without large temperature fluctuations. Lima beans are very sensitive to both cool and high temperature fluctuations, which interfere with flower fertilization and may cause abortion of blossoms or young pods.

Cucumber (*Cucumis sativus* L.) CUCURBITACEAE

The cucumber is a prostrate, branching vine native to Asia, most probably northern India, and introduced to Africa and Europe before written history. The plant is an annual with hairy leaves and tendrils, commonly with a **monoecious** flowering habit. The first flowers, generally staminate, are followed by pistillate flowers that, when fertilized, develop into fruit. In field production, insect pollinators are necessary. Many commercial cultivars have predominantly pistillate flowers, producing many fruit per plant, and therefore are especially desirable for mechanical harvesting of processing-type cucumbers (Fig. 19–11). Processing cucumbers are mostly used for making pickles and relish. Cultivars used for glasshouse production are gynoecious (all female) as well as vegetatively parthenocarpic meaning

Figure 19–10
The fleshy pods of green beans. Source: USDA Agricultural Research Service Image Gallery, *http://www.ars.usda.gov/is/graphics/photos/*

Figure 19–11
Processing cucumbers. Source: USDA Agricultural Research Service Image Gallery, *http://www.ars.usda.gov/is/graphics/photos/*

the fruit develops without pollination. Cucumbers grow best in warm temperatures and do not tolerate frost. The plants can become quite large as the vines grow. Bush types are available that are suitable for small gardens and containers.

Eggplant (*Solanum melongena* L.) SOLANACEAE

The eggplant, also known as aubergine or brinjal, is thought to be native to southeast Asia and has been grown in China for many centuries. Eggplant is grown commercially in many US states, however, with most sold to nearby markets. Although traditionally eggplant fruits are purple, a great variation of types, varying in skin color and shape, are produced (Fig. 19–12). Eggplant is a

warm-season crop. Some eggplant production is done in greenhouses, but it is mainly an outdoor crop.

Okra (*Abelmoschus esculentus* L. Moench)

Okra, also called gumbo and lady's finger, is African in origin, with early introduction into Mediterranean areas and India, where it continues to be very popular and appreciated. Like its relative, cotton, okra is a warm-season crop that requires a long growing season with hot days and warm nights. A significant quantity of okra is grown because of its gummy or thickening characteristics in the preparation of soups and stews (Fig. 19–13).

Peas (*Pisum sativum* L.) FABACEAE

The garden pea, sometimes called the English, green, or common pea, is an annual cool-season vine-type plant grown for its edible immature seed, although some cultivars are also grown for their immature edible pods. Neither the wild progenitor nor the early history of the pea is well known. Probable centers of origin are considered to be Ethiopia, the Mediterranean area, and central Asia, with a proposed secondary center of diversity in the Near East. Greek and Roman writers mentioned peas, but not until the seventeenth century were varieties described.

Dried peas, later reconstituted and used in soups and other products, are an important crop worldwide but less so in the United States. In contrast to green peas, which are harvested when immature, dried peas are allowed to mature and dry, which allows for easier storage and transport.

Peppers (*Capsicum annuum* L.) SOLANACEAE

A

B

Figure 19–12

(A) White eggplant. (B) Purple eggplant. Source: USDA Agricultural Research Service Image Gallery, *http://www.ars.usda.gov/is/graphics/photos/.*

Figure 19–13

Okra pods. Source: Margaret McMahon, The Ohio State University.

Peppers are an important vegetable commodity highly prized for the flavor, color, vitamin C, and pungency (heat) that they provide to the human diet. Peppers are a valued spice throughout the world and an important contribution from the Americas to the world's spices. The plants are shrubby perennials, although usually grown as herbaceous annuals in tropic, subtropic, and temperate regions. Peppers are native to the Central and South American tropics. Columbus brought the pepper to Europe. From there, it was introduced to other areas and into China in the late 1700s.

Capsicum annuum is the most extensively grown species. *C. frutescens* is represented by the tabasco type; *C. chinense*, *C. pubescens*, and *C. baccatum* are also cultivated. Common black pepper (*Piper nigrum*), also used for seasoning foods, belongs to a different botanical family and should not be confused with the Solanaceous pepper.

The plant is frost-sensitive and grows best at a warmer temperature, preferably within a range of 25°C to 30°C (77°F to 86°F). Excessively high temperatures, above 38°C (100°F), especially at flowering, can result in poor fruit set.

Peppers are interesting in the number of types grown, each having various preferred uses (Fig. 19–14). These types are distinguishable mostly by their characteristic shape and include types commonly known as bell, pimiento, squash or cheese, ancho, anaheim, cayenne, cuban, jalapeno, cherry, wax, and tabasco. Fruit color is also used as a distinguishing characteristic.

Figure 19–14

Peppers come in many different shapes and colors as well as pungency levels. Source: USDA Agricultural Research Service Image Gallery, http://www.ars.usda.gov/is/graphics/photos/

Peppers are green, yellow, various shades of brown, purple, and red in color. Often, maturity influences color as does breeding to achieve certain cultivars. For example, most bell peppers are green when immature, yellow as they near maturity, and red at maturity; but there are cultivars that exhibit other colors during the maturation process. Peppers also differ in pungency; some are extremely pungent, others very mild or sweet. Pungency is contributed by the capsaicinoid content, the principal compound being capsaicin. Pungency is measured in Scoville units, a test based on taste that was developed in the early twentieth century. A more accurate measure of capsaicinoid concentration is now done using high-performance liquid chromatography (HPLC). Pungency can range from zero Scoville units for bell peppers, 200,000 for habaneros (*C. chinense*), to over 500,000 for the Naga Jolokia (*C. frutenscens*).

Pumpkins and Squashes (*Cucurbita* spp. L.)
CUCURBITACEAE

The term *pumpkin* or *squash* does not have a precise botanical meaning. These vegetables are *Cucurbita* species of the *Cucurbitaceae* family, and the terms *squash* and *pumpkin* are used interchangeably. While there is less confusion regarding summer squash (cultivars used for their immature fruit and having soft rinds), there remains confusion between types of winter squash and pumpkins (mature fruit having hard rinds) (Fig. 19–15). These commodities are important in the diets of much of the world's population, and they are grown from the tropics and into much of the temperate zones. These commodities, in addition to their multiple uses as a vegetable, are also used for their seed, as ornamentals, and for livestock feeding.

The center of origin of the genus is primarily Central America, ranging into Mexico and the northern parts of South America. The major volume produced is contributed by four species, with *Cucurbita pepo* as the largest contributor, other species being *C. moshata*, *C. maxima*, and *C. mixta*. Species are distinguishable by their stem, peduncle, seed, and leaf variations. All are monoecious with generally stable sex expression characteristics and are dependent on insect pollinators for fertilization of the flowers. Having a tropical origin, the genus is frost sensitive. Warm temperatures above 25°C (77°F) are more favorable for growth and development.

Winter squashes and pumpkins are generally larger-size fruit. A major distinction is that these are harvested after they have developed hard rinds and well-developed seed and store well. Their holding or

A

B

Figure 19–15
(A) Yellow, straight-neck summer squash. (B) An array of winter squash and pumpkins. Source: (A) Margaret McMahon, The Ohio State University; (B) USDA Agricultural Research Service Image Gallery, *http://www.ars.usda.gov/is/graphics/photos/*

storage capabilities are highly important. This characteristic is assisted by the development of a hard rind that resists disease and deterioration. Better-known types of winter squash and pumpkins are acorn, butternut, Hubbard, banana, marrow, spaghetti, and, of course, the popular pie and Halloween pumpkins. Fruit shape and color vary greatly. These variations along with their irregular surfaces (warts) and their durability make some cultivars popular for use as ornamentals. Some cultivars achieve enormous size, some intentionally grown for exhibition purposes. Single-fruit weights as much as 1,500 lbs have been obtained.

Summer squashes are harvested at an immature stage when the rind is still soft and the seed are underdeveloped. Delayed harvest results in further development of the fruit with over-enlargement, toughness of the rind, and hardening of the seed—all contributing to loss of quality and value. Because they are softer than winter squash and pumpkins, they do not store well after harvest. Common summer squash include zucchini, yellow crook neck, and yellow straight neck.

Sweet Corn (*Zea mays* L. var. *rugosa*, Bonaf.)
POACEAE

Corn, which originated in the Americas, was domesticated at least 5,000 years ago. It was a staple grain crop for the Native Americans and remains a major world grain crop. Only since the middle of the nineteenth century have sweet corn types of corn achieved status as a vegetable commodity. The popularity of sweet corn is at a maximum in the United States, where it is a frequently grown vegetable in many home gardens and where a substantial level of commercial production occurs in many parts of the country. The crop has steadily increased in popularity in other parts of the world, notably in Europe.

The crop is a warm-season grass that thrives in humid environments having hot 30°C (86°F) days and warm nights. Through plant breeding efforts, the temperature range has been broadened so that cultivars are available that perform well at lower temperatures without the penalty of lower crop performance. Extensive use of hybrid cultivars provides producers with greater adaptation, yield, and kernel quality attributes.

The range of characteristics, especially those related to eating quality, available in sweet corn cultivars has steadily increased since the early 1980s. Originally, sweet corn was immature field corn (roasting ears) that was harvested before the sugars in the kernel had been converted to starch. Following the identification of several mutations controlling various kernel traits, plant breeders and others have used an array of techniques to create sweet corn kernels with different combinations of sweetness, aroma, texture, and pericarp thickness. Some types are so sweet and succulent that they can be eaten raw, though this is not commonly done (Fig. 19–16). The time that the kernels remain sweet after harvest has also been increased, facilitating the development of long-distance shipping to fresh markets.

Tomato (*Lycopersicon esculentum* Mill.)
SOLANACEAE

Tomatoes are one of the most popular, versatile, and widely grown vegetables—known and grown throughout the world and in nearly every home garden. Although not a new crop to the native populations of tropical and subtropical America, the tomato is relatively new to the rest of the world. Columbus returned with tomatoes to Europe, where they were first grown for ornamental purposes. Cultivation for a food crop was soon established along with its dispersion throughout

Figure 19–16
Modern sweet corn cultivars can be eaten raw, though this is not commonly done. Source: Margaret McMahon, The Ohio State University.

Europe and other areas. The crop began to be cultivated in North America in the early 1700s.

Tomatoes are warm-season perennials grown as annuals. They are frost-sensitive, and many cultivars require about three to four months from seeding to produce mature fruit. Production is generally optimized under conditions that provide warm, above 27°C (86°F), temperatures with clear dry days.

Tomatoes are grown for processing and fresh market use. Processing tomatoes are harvested mechanically. Determinate types are usually used as well as an application of ethylene to ripen the fruit uniformly so all the fruit can be harvested at one time. Fresh market tomatoes are usually picked by hand and the vines can be determinate or indeterminate, depending on other factors such as market window.

The production of fresh market tomatoes in structures such as glass or plastic houses is a significant and very important activity. This operation is expensive and highly specialized and requires a high level of technology and management, especially when hydroponic systems are used. Indeterminate types are usually grown. The vines are trained around string that can be continually lengthened to drop the vines and keep the fruit at a level easily picked (Fig. 19–17).

VEGETABLE CROPS GROWN FOR FLOWERS, LEAVES, OR STEMS

A diverse and large number of crops are included in the group of vegetables grown for the consumption of their floral, leaf, or stem parts. One similarity is that most crops in this grouping are cool-season vegetables.

A

B

Figure 19–17
(A) Greenhouse tomatoes are attached to strings that allow the vines to be lowered to keep the fruit at an easy-to-reach height. (B) These vines originate many feet behind where they rise from the ground here. The strings are lengthened to allow the stems to be dropped.
Source: Margaret McMahon, The Ohio State University.

Some, such as cabbage and lettuce, are produced in significant amounts. However, many others—for example, artichokes, asparagus, or rhubarb—are rather specialized, and their production is relatively limited. These vegetative types of vegetables are not considered to be major caloric providers to human diets. Nevertheless, they do have considerable nutritive value, notably for their contribution of vitamin A, vitamin C, other vitamins and minerals, as well as dietary fiber. Their other attributes are the texture, flavor, color, and variety they provide to meals.

Artichoke (*Cynara scolymus* L.) ASTERACEAE

The globe artichoke, cultivated for over 2000 years, is a thistle-like herbaceous perennial plant native to North Africa and other Mediterranean areas. In early times, the foliage was eaten. However, the preferred edible portion is the enlarged but immature flower bud, comprised of numerous overlaid bracts, and its fleshy receptacle (Fig. 19–18). The globe artichoke is sometimes confused with the Jerusalem artichoke, *Helianthus tuberosus.* Although both are members of the ASTERACEAE family, they are different, the edible portion of the Jerusalem artichoke being the underground fleshy tubers. When eaten, these have a taste that resembles that of the globe artichoke, perhaps a reason for the confusion.

A frost-free climate with cool, foggy days is a requirement for achieving quality production of artichokes. Hot, dry weather causes rapid growth accompanied by rapid development of the internal floral tissues and fiber, both detrimental to tenderness. High temperatures also cause the bracts to separate and spread apart, which is detrimental to market appearance. Freezing temperatures blister the epidermal tissues of the bracts and are a detraction to market appearance. Because of the climate restrictions, artichoke production in the United States is limited almost exclusively to the Monterey Bay area of California. Careful and rapid post-harvest handling permits marketing during periods when the crop is in short supply.

Artichokes are propagated vegetatively from stem pieces or from rooted offshoots from the base of older plants. Plantings last for five or more years before reestablishment is necessary. Farmers generally cut back foliage down to the soil surface following final harvest in late spring or early summer. Regrowth during the summer is delayed by restricting moisture for several months. This method allows for scheduling the resumption of production.

Asparagus (*Asparagus officinalis* L.) LILIACEAE

The edible portions of this perennial plant are the stems (spears) that arise from fleshy underground rhizomes called crowns. The plants are dioecious, having either staminate or pistillate flowers. Asparagus is thought to be native to the eastern Mediterranean region.

Better quality and higher yields occur when the crop matures during periods when temperatures are about 16°C to 25°C (60°F to 77°F). In colder climates, the growing season is shortened because of the limited period for fern (foliage) production. A reasonable period of active fern growth is necessary to supply the crown with the products of photosynthesis that the fern produces following the harvest period. The photosynthate produced is translocated to the crowns to be utilized for the following year's production of spears. In warmer areas, sufficient fern growth occurs and the crowns can sustain a longer harvest period without exhaustion of their food reserves. During warmer periods, spear growth is very rapid, with greater fiber development within the spears and with a tendency for the spear tips to lose some of their compactness. Therefore, harvest is usually limited to spring and early summer in most areas. Plantings are maintained for five, ten or more years. White asparagus production is a popular choice in some markets (Fig. 19–19). Such production

Figure 19–18
Whole artichoke and cross section. The edible parts are the bases of the flower bracts and the fleshy heart or receptacle to which the bracts are attached.
Source: Margaret McMahon, The Ohio State University.

Figure 19–19
Green and white asparagus spears. The white spears were covered with soil while growing to prevent them from developing chlorophyll. Source: Margaret McMahon, The Ohio State University.

depends on blanching the elongating spears, usually with a mounding of soil cover that prevents the development of chlorophyll in the spears. Harvest is more difficult because it must be done just before the spear tips emerge from the mound. The high cost for such production is the main reason that the production of white asparagus is considerably less than the green variety.

Cole Crops

The flowers of the cole crops—broccoli, Brussels sprouts, cabbage, cauliflower, collards, kale, and kohlrabi—have four petals arranged in the form of a cross. This feature gave the family the name CRUCIFERAE (now BRASSICACEAE), which in Latin means "cross bearer." The plants are often referred to as cruciferous or crucifers. They are all *Brassica oleracea* but belong to different groups. Most cole crops are cool-season crops and are planted in the early spring for harvest in late spring and early summer. In many areas they are replanted in late summer for a fall harvest.

Broccoli (*Brassica oleracea* L., Italica Group) BRASSICACEAE

The word *broccoli* comes from the Latin word *bracchium*, meaning arm or branch. The tightly grouped fleshy terminal stems, young leaves, and young flower buds and bracts are the plant parts of broccoli that are consumed (Fig. 19–20). The plant is native to eastern Asia and the Mediterranean areas and was introduced a relatively short time ago into the United States, in about the 1920s. Previously it was not widely cultivated in Europe. However, because of its high nutritive value and ease of preparation, it has become very popular,

Figure 19–20

Broccoli. Source: USDA Agricultural Research Service Image Gallery, http://www.ars.usda.govlis/graphics/photos/

and its production has expanded to many other parts of the world. The developing inflorescence (head) is cut from the stem by hand while it is still immature, compact, and absent of open flowers, resulting in a head about 15 to 25 cm (6 to 10 in.) in length, and about 10 to 20 cm (4 to 8 in.) in diameter.

Brussels Sprouts (*Brassica oleracea* L., Gemmifera Group) BRASSICACEAE

This vegetable, which tastes like cabbage, probably evolved from a primitive nonheading Mediterranean cabbage, after its progression to northwestern Europe. The plant received its name from the city of Brussels, where it was first reported to be grown; it continues to be a popular vegetable throughout that area. The plant develops leafy buds resembling miniature cabbage-like heads about 2.5 to 5 cm (1 to 2 in.) in diameter in the leaf axils along the single tall unbranched stem (Fig. 19–21). The plant grows 60 to 90 cm (2 to 3 ft) tall and requires a cool climate to develop compact, quality buds. The crop is relatively tolerant to cold temperatures and can withstand slight freezing.

Cabbage (*Brassica oleracea* L., Capitata Group) BRASSICACEAE

Cabbage is well adapted for growth in cool climates. It is a biennial plant and tolerant of slight freezing. On the other hand, cabbage is somewhat more tolerant of warmer temperatures than many other cool-season crops. The edible portion is essentially a large vegetative terminal bud, formed by overlapping of numerous leaves developing over the growing point of its shortened stem, and is called the head. The head is harvested before stem elongation begins. Cabbage is believed to have developed from a wild type of a nonheading plant on the east coast of England and on the western European coast. Evidence indicates that the early Egyptians used cabbage for food and medicine, and that the Mediterranean area was its origin.

Cauliflower (*Brassica oleracea* L., Botrytis Group) BRASSICACEAE

The origin of cauliflower is believed to be the eastern Mediterranean area, with subsequent development perhaps from sprouting broccoli in northwestern Europe, where it has been known for centuries. Both annual and biennial forms are cultivated. Among the Brassica crops, cauliflower has the most exacting climatic and cultural requirements. Moderate uniformly cool climates

However, cultivars known as tropical types having better warm-temperature adaptation are grown in large volume in many warmer temperate regions, namely, India and China.

The edible product, called the head or curd, is mistakenly considered to be floral tissue but is instead prefloral, fleshy, apical meristem tissues (Fig. 19–22). The curd consists of the repeatedly branched terminal portion of the main axis of the plant, comprised of a shoot system with short internodes, branch apices, and bracts. High temperatures result in poor head formation and quality, with accompanying market defects such as bract development within the curd, loss of compactness, and ricy curds.

Direct exposure of the curd to sunlight tends to discolor the otherwise white curd surfaces to a creamy yellow, which is a market defect but does not otherwise affect the edible quality. Leaves cover and protect the curd as they form, but as the curd enlarges, the inner leaves are forced apart and the curd is exposed to sunlight. To protect and blanch the enlarging curd, the long outer (wrapper) leaves are gathered together over the top of the curd and tied together to keep out light. Sometimes self-blanching cultivars are used.

Collards and Kale (*Brassica oleracea* L., Acephala Group) BRASSICACEAE

Collards and kale are plants that resemble cabbage but are nonheading. They are closely related to wild cabbage and have been cultivated for many centuries; they

A

B

Figure 19–21
A Brussels sprouts form from buds in the leaf axils along the stem. (B) Harvested Brussels sprouts. Source: (B) Margaret McMahon, The Ohio State University.

are best for cauliflower, and production sites are frequently near large bodies of water to benefit from their climate-moderating influence. Preferred mean temperatures during growth and crop maturity are 15°C to 20°C (59°F to 68°F). Unlike cabbage, cauliflower is much more intolerant of frost or high temperatures.

Figure 19–22
Freshly harvested cauliflower with some of the outer leaves still attached. Source: USDA Agricultural Research Service Image Gallery, http://www.ars.usda.gov/is/graphics/photos/

are considered to be the least-developed members of the *Oleracea* species. Collards and kales are winter-hardy biennials and tolerate both somewhat lower and somewhat higher temperatures than cabbage. These plants are grown for their abundant foliage, which is usually consumed as a cooked vegetable or as an ingredient in other foods. Some kale cultivars, in addition to having attractive green coloration, also have foliage with color variations ranging from purple, red, and pink to white (Fig. 19–23). They also have very crinkled leaves and find use as ornamental plants.

The distinction between collard and kale is not clear. Collards have a relatively smooth leaf blade, whereas kales, with some exceptions (forage kales), generally have crinkled leaves. Some cultivars have pronounced curly foliage. Kale cultivars also vary in height, with dwarf, medium-tall, and tall types. Some kale cultivars, which include the thicken stem or marrow-stem types, are used for animal forage.

Kohlrabi (*Brassica oleracea* L., Gonglylodes Group) BRASSICACEAE

Kohlrabi is a cool-season biennial plant grown for its turnip-like, above-ground enlarged stem. Its delicate flavor compares with that of the turnip, although it is milder. Kohlrabi is harvested when the enlarged stem is between 2 to 3 in. in diameter. Stems much larger than 3 in. develop woody fibers that are unpalatable.

Celery (*Apium graveolens* L. var. *dulce* Mill.) APIACEAE

Figure 19–23
Kale is both an edible and ornamental plant. Source: USDA Agricultural Research Service Image Gallery, *http://www.ars.usda.gov/is/graphics/photos/*

The former family name, *Umbelliferae,* came from the characteristic umbel form of the inflorescence, and the family includes celery, carrots, parsnips, and parsley. Writings indicate that the first use of celery was as a medicine in the fifth century CE and as a cultivated food crop in the early 1600s. The plant is thought to be native to the Mediterranean area, although wild types are found widely distributed in other areas. The crop's world importance is relatively small, both as a vegetable and for the use of its seeds and foliage in seasoning foods. Celery is a biennial with rather exacting climatic requirements. Optimum daytime growing temperatures range between 16°C to 21°C (60°F to 70°F), with night temperatures slightly lower. Exposure of temperatures less than 10°C (50°F) initiates vernalization and seed stalk development and should be avoided. Celeriac (*A. Graveolens* var. *rapaceum),* also known as knob or root celery, is closely related and finds similar use, particularly among northern Europeans.

Chive (*Allium schoenoprasum* L.) AMARYLLIDACEAE

Chives are small, bushy, onion-like perennials grown for their delicate tube-like leaves, which are used in salads, soups, stews, omelets, and various cheese products, and to flavor numerous other food products. It appears that chives are native to the northern hemisphere, and wild forms are found in North America and Eurasia. The crop is of minor economic importance, although it has been cultivated for centuries in Europe and the Orient. Chives are cool-season, cold-tolerant plants with climatic requirements similar to onion and garlic.

Endive (*Cichorium endivia* L.) and Chicory (*Cichorium intybus* L.) ASTERACEAE

Early records indicate that forms of endive and chicory were grown for many centuries and were well known to the early Egyptians. The probable origin of endive was eastern India, whereas chicory appears to be a native of the Mediterranean region. Both are cool-season crops, somewhat similar in appearance to lettuce (Fig. 19–24). Endive and chicory initially produce a rosette of leaves that develop from a short stem. When mature, some types form a head, others remain as an open rosette, and some are intermediate between these forms. Additional variation exists with regard to leaf size and shape: some are broad-leaved; others have narrow, divided, and rather crinkled leaves; some have smooth, broad-leaved foliage. Those endives with broad crinkled leaves are commonly known as escarole. Some forms of endive and chicory also develop anthocyanin

Figure 19–24

Chicory (pictured here) and endive resemble lettuce. Source: USDA Agricultural Research Service Image Gallery, *http://www.ars.usda.gov/is/graphics/photos/*

and have reddish-tinged leaves, such as radicchio. Many endive and chicory cultivars are annuals or are grown as annuals. They are used in salads. The chicory root can be roasted and ground to be used as a coffee substitute or additive. The distinctive blue flower of wild chicory can be seen on the side of the road and in fields in summer in many parts of the United States (Fig. 19–25).

> **Herbs: Basil (*Ocimum basilicum*, LAMIACEAE),**
> **Oregano (*Origanum vulgare*, LAMIACEAE),**
> **Rosemary (*Rosmarinus officinalis*, LAMIACEAE),**
> **Thyme (*Thymus vulgaris*, LAMIACEAE),**
> **Spearmint (*Mentha spicata*, LAMIACEAE),**
> **Peppermint (*Mentha* × *piperata*, LAMIACEAE)**

Figure 19–25

The flowers of wild chicory are a common sight in open fields and roadsides during the summer in many parts of the United States. Source: Margaret McMahon, The Ohio State University.

The above herbs are perennials; but basil and rosemary are usually grown as annuals in areas with cold winters, which they do not tolerate. All grow best in warm, sunny locations. The leaves can be used fresh or dried.

> **Leek (*Allium ampeloprasum* L., Porrum Group)**
> **AMARYLLIDACEAE**

Leek, a biennial plant closely related to the onion, is grown for its sheath-like arrangement of elongated, closely overlapped, blanched leaves (Fig. 19–26). The leaves are fleshy and solid, but generally do not form bulbs as onions do. Under long-day conditions, however, a slight thickening near the base of the leaves resembling bulbing may occur. Exposure to low temperatures, if of sufficient duration, may result in bolting. Leek is believed to be a native of the eastern Mediterranean region, where its food use was known as much as 3,000 to 4,000 years ago.

> **Lettuce (*Lactuca sativa* L.) ASTERACEAE**

Lettuce is a plant with an ancient history. Its origin appears to have been in Asia Minor. There is evidence that forms of lettuce were used in Egypt during the period about 4500 BCE. The Romans grew types of lettuce resembling the present romaine cultivars as early as the beginning of the Christian era, and the crop was well known in China about the time of the fifth century CE.

Cultivated lettuce is an annual plant closely related to the common wild or prickly lettuce (*L. serriola*) weed.

Figure 19–26

Whole leek and cross-section.

Source: Margaret McMahon, The Ohio State University.

The many forms of lettuce are grouped into several types. These types include the usually large and dense heading crisphead cultivars that have brittle-textured foliage that is tightly folded. The outer leaves are dark green and the inner foliage pale and mostly absent of chlorophyll. The butterhead cultivars, whose heads are smaller, are relatively soft and less dense, and are comprised of broad, soft, smooth, or slightly crumpled oily textured leaves. The romaine or cos cultivars have elongated coarse-textured leaves that form a loaf-shaped, semicompact, low-density head. Outer leaves are darker green, coarse textured, with heavy ribs. Inner leaves are smaller and lighter in color, with interior leaves nearly absent of chlorophyll. The leaf lettuce cultivars, which are leafy but loose, are nonheading cultivars of varied color, leaf types, and textures. Figure 19–27 shows some of the many types of lettuce. Lettuce is a cool-season crop and grows best within a temperature range of 12°C to 20°C (55°F to 68°F). The cool Salinas Valley and other central coastal regions in California are major production areas for lettuce during the summer. Leaf lettuce is also grown in large quantities in float-type hydroponic systems in greenhouses in the United States and Canada.

Parsley (*Petroselinum crispum* Mill.) APIACEAE

Parsley, a biennial plant native to the Mediterranean area, has been known as a cultivated crop for more than

Figure 19–28
Parsley plant. Source: Margaret McMahon, The Ohio State University.

2,000 years. The crop, a close relative of celery, is widely used as an herb or garnish and in flavoring various foods. The plant produces a rosette of divided leaves on a short stem (Fig. 19–28). Distinctive foliage types are produced, such as the plain or single-leaf type used mostly for food flavoring, and the double-leaf type and the moss or triple-curled leaf type preferred for garnishes. Parsley is a highly nutritious crop and an excellent source of vitamins A and C.

Rhubarb (*Rheum rhabarbarum* L.) POLYGONACEAE

Rhubarb is a large-leafed plant grown for its thick leaf petioles, which are used as a dessert vegetable in pie fillings or sauces. The attractive deep red coloration of the rhubarb petiole is an important market feature (Fig. 19–29). Rhubarb is forced in the dark to avoid

Figure 19–27
Many different types of lettuce from the USDA lettuce germplasm preservation center in Salinas, CA.
Source: USDA Agricultural Research Service Image Gallery, http://www.ars.usda.gov/is/graphics/photos/

Figure 19–29
Rhubarb petioles. Source: Margaret McMahon, The Ohio State University.

chlorophyll development, which would interfere with the expression of the red pigment. *The leaf blades are poisonous and must not be eaten.*

The plant is a cool-season perennial, and it is one of the first vegetables ready for consumption in the spring. By using plant forcing procedures such as a row cover or tunnel to stimulate early rapid growth, a crop can be produced during the late winter–early spring. In mild climates, warm early spring temperatures encourage early production in field conditions. The plant is not harvested the first year to allow the root system to develop. The plant does not grow well at temperatures above 25°C (77°F), and summer-produced rhubarb is usually of poor quality.

Spinach (*Spinacia oleracea* L.) CHENOPODIACEAE

Spinach is a popular and nutritious vegetable frequently used raw in salads or cooked. Spinach is an annual plant, dependent upon wind for pollination of its flowers. The plant is unusual because it exhibits dioecious characteristics, male or female flowers on separate plants, and also can be monoecious, with varying proportions of male and female flowers on the same plant in between. The greater the proportion of male flowers, the greater the tendency is to flower earlier. The male plants have relatively little vegetative growth and die soon after flowering; therefore, they contribute very little to crop production. The preferred types are the female and vegetative male plants (male but with well-developed leaves), which are much slower to flower. These plant types can produce an abundant amount of foliage, which can usually be harvested before flowering occurs.

Spinach is native to central Asia, most likely in the area of Iran. It is a cool-season, long-day plant that produces its best vegetative growth under cool 15°C to 18°C (59°F to 64°F) temperatures and a short daylength. Long days, especially if coupled with higher temperatures above 25°C (77°F), cause the plants to bolt and flower, which is detrimental to production. Spinach plants have some frost tolerance and are grown in many areas as an overwintering crop.

Types of spinach are distinguished by their foliage characteristics. Some cultivars are smooth-leaved, others are savoy-(curly) leaved, and some have intermediate leaf textures. Other important cultivar characteristics are leaf color, leaf blade size, leaf petiole length, growth habit—prostrate or upright, seasonal adaptation, bolting, and disease resistance.

Swiss Chard (*Beta vulgaris* L., Cicla Group) CHENOPODIACEAE

Swiss chard or chard is a biennial plant grown for its leaves, which can be used as a substitute for spinach. It is very similar to the table beet in developing large, crisp, fleshy leaf stalks and large leaves, but it differs in not producing the enlarged tap root. Swiss chard is a widely adapted, cool-season crop that tolerates hot weather better than spinach. It is a well-known crop and one of the easiest to grow.

VEGETABLE CROPS GROWN FOR UNDERGROUND PARTS

Underground plant parts used for food consist of storage roots, bulbs, rhizomes, corms, and tubers. Many plants store part of the excess photosynthate (food) produced during their growth in these plant parts. Rhizomes, corms, and tubers are specialized stems that have buds and the ability to develop roots, leaves, and other stems, and can regenerate fully into new plants. Tubers are thickened stems that exhibit compressed or small internodal elongation. The eyes of the potato tuber are buds that develop into stems, which can also develop into fully functional plants. Bulbs, such as the onion, are comprised of specialized storage leaves attached to a short compressed stem. The base of such leaves enlarge because of the accumulation of food synthesized by the plant and stored in these leaves. This enlargement produces the structure known as a bulb. The sweet potato, carrot, beet, and turnip are examples of vegetables where the enlarged storage root serves as the accumulation point of plant food reserves.

Underground plant parts have worldwide importance as a food source. These crops are staples, extensively used in the food supply of all nations, whether developed or developing. Characteristically, these plants are efficient converters of solar energy into chemical energy, principally as carbohydrates, but also as proteins. For example, the Irish potato plant, using an equivalent land area, produces carbohydrates second to sugarcane and protein second to soybeans. Other plants grown for their edible underground parts are also efficient in their synthesis and storage of food products. Most root crops develop the best product when grown in soil that is deep, well-drained, and loose enough that the underground part can expand.

Beets (*Beta vulgaris* L., Crassa Group J. Helm) CHENOPODIACEAE

Table beets, native to Europe and North Africa, are cool-season biennials grown mainly for their fleshy roots that are harvested the first year. However, the

tender young tops are sometimes used as greens. Beets were first used for food about the third century CE.

Beets are rather hardy and tolerate some freezing, but they also grow well in warm weather. Excessively hot weather, however, causes zoning—the appearance of alternating light and dark red concentric circles in the root. Beet seeds germinate within a broad soil temperature range from 5°C to 25°C (41°F to 77°F). Beets are generally ready to harvest 2 to 3 months after planting.

Carrots (*Daucus carota* L.) APIACEAE

The carrot, a cool-season crop, is grown for its fleshy storage root. It is a native of Europe and parts of Asia. The carrot was introduced in North and South America in the early 1600s. The wild type (Queen Anne's Lace) is an annual, but cultivated types are biennials (Fig. 19–30). The root is slender and varies in length from 5 to 25 cm (2 to 10 in.) when harvested at the end of the first growing season. Carrots are often divided into types according to shape and length of root, with each type represented by several cultivars.

Carrots grow best at temperatures ranging from 15°C to 20°C (59°F to 68°F). In the seedling stage, the plants are sensitive to both high and low temperatures, and mild freezes at harvest cause some leaf damage. Many young plants are injured or killed on hot, sunny days. Long periods of hot weather can cause objectionable strong flavor and coarse root texture. Temperatures below 10°C (50°F) tend to cause longer, more slender, and paler roots. Color develops best at 15°C (59°F).

Celeriac (*Apium graveolens* L., var. *rapaceum* Gaud-Beaup.) APIACEAE

Celeriac and celery are closely related, both belonging to the same species. Celeriac is commonly known as turnip root. The above-ground parts look similar to celery except that the foliage is less compact, more dwarfed, and darker green. In contrast to celery, the leaf petioles are less fleshy and more fibrous. The edible part is the enlarged stem and root axis. Celeriac is mainly used to flavor other foods, such as soups and stews, but it is also used raw or cooked in salads. The crop is not grown extensively in the United States. In Europe, the crop is better known and more widely used. The climatic requirements for celeriac are similar to those for celery.

Garlic (*Allium sativum* L.) AMARYLLIDACEAE

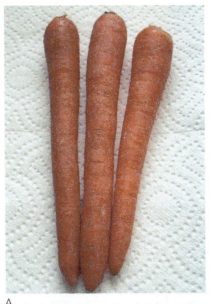

A

B

Figure 19–30

(A) Cultivated carrots are biennials, but the roots are harvested the first year. (B) Queen Anne's Lace is the wild type of Daucus carota. *Its white flower is often seen growing with wild chicory in many parts of the United States (see Fig. 19–25).*

Source: Margaret McMahon, The Ohio State University.

Garlic is grown especially for its strong and popular flavoring characteristics, which make it a universally important spice. The edible part is a compound bulb consisting of segments called cloves, each surrounded by a thin white or pink sheath. The cloves are formed in the axil of the inner foliage leaves; the outer leaves form the sheath. The leaves are solid and flattened, rather than hollow and round as in onions. Garlic has been grown for centuries. Its origin is believed to be central Asia.

Garlic is a cool-season crop usually planted in the fall or early spring. This time of planting allows the plant to develop sufficient size and to be exposed to low temperatures, both important, prior to experiencing

the progressively longer summer photoperiod needed to initiate bulbing. Harvest occurs in the fall or late fall depending on planting dates.

Horseradish (*Armoracia rusticana* P. Gaertn., B. Mey. & Scherb.) BRASSICACEAE

Horseradish is a hardy perennial, native to southeastern Europe. It is a perennial but is commercially grown as an annual to prevent it from becoming a weed pest. The leaves, rhizomes, and roots have been used as food or condiment since the Middle Ages and before that for medicinal purposes. The plant is grown for its pungent compound, allyl isothiocyanate (C_3H_5CNS).

There are two types of horseradish. The common type has broad, crinkled leaves and produces high-quality roots. The Bohemian type has narrow, smooth leaves, is more disease-resistant, but produces lower quality roots.

Onions (*Allium cepa* L.) AMARYLLIDACEAE

Onions are one of the oldest vegetables and are known to have been used before 3000 BCE. Onions are probably a native of the area from southwestern Asia eastward to Pakistan.

Onions are cool-season biennial plants that grow well at temperatures ranging from 13°C to 25°C (55°F to 77°F); they have some frost tolerance. Seeds can germinate from 7°C to 30°C (45°F to 86°F) but do best at about 18°C (64°F). During the early stages of growth (before bulbing), onions grow better at relatively cool temperatures, but during bulbing, harvesting, and curing, higher temperatures and low relative humidity are desirable. Plant growth declines at temperatures greater than 27°C (81°F). Bulbing is initiated primarily by daylength and not by the age of the plant. There are short day types that require less than twelve hours of daylight to start to bulb as well as long-day types which need at least fourteen hours. Once initiated, warmer temperatures hasten bulb growth and enlargement. Most commercial onion bulbs are harvested using mechanical aids.

Green onions are young immature plants that have not been exposed to the daylength required for bulbing. They are grown at very high densities for fresh market use in salads. These plants are generally hand harvested, several being bunched together, the tops uniformly trimmed, and marketed directly (Fig. 19–31).

Parsnips (*Pastinaca sativa* L.) APIACEAE

A   B

Figure 19–31
Green onions (A) are the tops of immature bulbing onions (B) that have not been exposed to bulb-inducing photoperiods. Source: Margaret McMahon, The Ohio State University.

Parsnips are considered a native of Europe and Asia and have been used for food and medicinal purposes since the Greek and Roman eras. They were introduced into the United States in Virginia and Massachusetts in the early 1600s. The plants are cool-season biennials, but the roots are harvested at the end of the first growing season. Parsnips are slow growing and require a long season (100 to 120 days) to form their white, fleshy edible roots. Parsnips grow best when they are planted and grown during a relatively cool summer and harvested in a cool fall.

Potatoes (*Solanum tuberosum* L.) SOLANACEAE

The potato plant is a bushy, herbaceous annual 60 to 90 cm (2 to 3 ft) in height. The edible tuber is a swollen underground stem.

As a source for human food, the potato is a leading vegetable crop. Potatoes are South American in origin and were cultivated in Chile and Peru by the Indians long before the arrival of European explorers. Early Spanish explorers transported them to Europe about 1575, and from there they returned to North America with the colonists settling along the Atlantic coast. In those early days, however, potatoes were not a significant food source in North America until the influx of Irish immigration, which began in the early 1700s and reached a maximum later, after the potato famine in Ireland in 1846. Potatoes then became America's favorite carbohydrate food because their cost was low and most people liked them. Today, potatoes are grown on every continent and in every state

of the United States. Like carrots, potatoes are harvested somewhere in the United States every month of the year.

Potatoes are a cool-season crop, slightly tolerant of frost, but easily damaged by freezing weather near maturity. Tubers form on the end of stems and botanically are rhizomes (underground stems). The tuber is a fleshy stem with buds in the axil of leaf scars (Fig. 19–32). These scars are often referred to as eyes. Maximum yields of high-quality tubers are produced when the mean temperature is between 15°C and 18°C (59°F and 64°F) during the growing season. Tuberization (tuber formation) is also favored by long, cool days of high light intensity. An optimum temperature for tuber development is about 18°C (68°F). Tuberization is progressively reduced when night temperatures rise above 20°C (68°F) and is totally inhibited at 30°C (86°F).

Flowers are common but fruits rarely set unless precise climate and a long daylength period are met. The fruits are green berries, somewhat similar to small tomatoes, within which seed are produced. Most potatoes are started from "seed" potatoes, which are pieces of tuber. Each piece must have an eye with a bud. True seeds are used in breeding programs but have not, until recently, been planted for commercial production. Recently there has been much interest in using seeds for propagation in developing countries. This interest is with regard to the need and cost of importing disease-free seed potatoes for propagation, where disease-free seed stock cannot be produced.

Figure 19–32

The buds in the axils of the leaf scars (eyes) develop into sprouts which can be a problem in storage but are the source of stems when the tuber is cut up as a seed potato.

Source: Margaret McMahon, The Ohio State University.

Radish (*Raphanus sativus* L.) BRASSICACEAE

Of all vegetable crops, the garden radish is one of the easiest and quickest vegetables to grow. A crop can be produced within four to six weeks of seeding. The radish, probably native to the eastern Mediterranean region, was cultivated for centuries. Its use can be documented beyond 2000 BCE.

Radishes are a cool-season crop tolerant of some slight frost. Best growth occurs when the monthly temperature averages about 15°C to 18°C (59°F to 64°F). They are somewhat intolerant of temperatures exceeding 25°C to 27°C (77°F to 81°F). When grown during high temperatures, the interior root tissues tend to be pithy, and the plants may bolt. Overmaturity causes a softening or pithiness of the normally crisp tissues.

Rutabaga (*Brassica napus* L., Napobrassica Group) BRASSICACEAE

Rutabagas are a cool-season biennial root crop grown mainly in northern Europe, England, and Canada. Limited amounts are grown in the United States for human consumption and some for stock feeding. They are sometimes called Swedes, Swedish turnips, Russian turnips, Canadian turnips, or yellow turnips.

Salsify (*Tragopogon porrifolius* L.) ASTERACEAE

Salsify is a hardy biennial, sometimes called oyster plant or vegetable oyster. It has long, cylindrical roots, used mainly in soups for its delicate oysterlike flavor. The plant grows wild around the Mediterranean Sea, in southern England, and along roadsides in the United States and Canada. During its second season, it bears a purplish flower that looks like an enlarged dandelion head.

Salsify is a cool-season crop tolerant of some freezing weather. Seeds are planted about the time of the average date for the last frost in the spring. This slow-growing plant requires the entire season to develop, and like parsnips, its quality is improved after a heavy frost or cold storage for at least two weeks. The tapered roots are 20 to 23 cm (8 to 9 in.) long; are 2.5 to 4 cm (1 to 1.5 in.) thick; and feature creamy, white flesh.

Sweet Potato (*Ipomoea batatas* Lam.) CONVOLVULACEAE

A   B

Figure 19–33

(A) Sweet potatoes are grown for their edible fleshy root. (B) Some cultivars have attractive foliage and are grown as ornamentals. If they are grown in an area that has a long, warm growing season, they will produce edible roots.

Source: Margaret McMahon, The Ohio State University.

Sweet potatoes are grown for their sweet, tuberous roots. The moist, soft-textured sweet potato cultivars are often erroneously called yams to distinguish them from the dry-textured sweet potato cultivars. This misnomer is unfortunate because true yams are different plants entirely, not even slightly related to sweet potatoes. True yams belong to the DIOSCOREACEAE family.

Sweet potatoes are native to tropical America, and were transported to the Pacific islands and to Asia early in history. They were cultivated as food in the southern parts of the United States by the Native Americans and were later taken to Europe. Sweet potatoes are perennial vines that grow prostrate on the ground. They belong to the morning-glory family. They are unique in producing adventitious shoots (slips) used to propagate the crop. When grown in temperate and some subtropic regions, the crop is handled as an annual. In the tropics and much of the subtropical regions, the crop is cultured as a perennial.

Sweet potatoes are of particular importance as a food crop throughout subtropical and tropical regions. It is one of the most important carbohydrate sources for many millions of people, particularly those in developing nations. Ornamental sweet potatoes are grown for their attractive foliage, but they will produce edible roots if the growing season is warm and long enough (Fig. 19–33).

Sweet potatoes are warm-season plants that do not tolerate frost nor grow well in cool weather. Temperatures below 10°C (50°F) cause chilling injury. They require a long, warm growing season and grow best when mean monthly temperatures are above 20°C (68°F) for at least three months.

Turnip (*Brassica rapa* L., Rapifera Group)
BRASSICACEAE

Turnips are one of the easiest vegetables to grow and one of the more widely adapted of the root crops. They are native to northern Asia and are extensively grown in Europe, Asia, and almost everywhere in the United States. They are used as food for both animals and humans. The roots may be eaten raw but are usually cooked while the leaves make delicious greens. The plant is a cool-season biennial. Seeds are sown either early in the spring (for a fall crop) or in the fall (for a spring crop) for the roots to mature during cool weather. Continued temperatures below 10°C (50°F) are likely to cause bolting. The plants resist frost and mild freezing. In most areas spring and fall crops are grown because only sixty to eighty days are required for the roots to reach maturity.

SUMMARY AND REVIEW

Vegetable production is important for many reasons. Vegetables are major components of a healthy diet. Growing vegetables can contribute to a sense of well-being; gardening and farming vegetables can have positive effects on the environment. The greatest concentration of area devoted to commercial vegetable production tends to be where the environment consistently favors good crop growth and where the resources needed for production are readily available. Although many large commercial farms are concentrated in specific areas, vegetables are grown in almost all regions of the world, making the growing, processing, and selling of vegetables a major economic factor in local to global economies.

Vegetable farms are managed agroecosystems that are widespread and diverse in many aspects, including size, crops grown, and methods used. Vegetable growers establish and manage biological communities within larger natural systems so that high quantities of desirable vegetables are produced efficiently and profitably, and with as much integration among other ecosystems as possible. Like other agroecosystems, vegetable-farm ecosystems combine elements of business and the natural world (e.g., biology, chemistry, physics).

Vegetable production can be done outdoors with minimal environment management or in covered structures ranging from simple row covers to sophisticated greenhouses with highly managed environmental systems. Production methods can be traditional, organic, or sustainable. The designations among these systems is not distinct, although to be certified as organic, a farm must meet stringent requirements.

Vegetable farmers must also follow principles integral to the success of other types of businesses, including (1) studying the market and being prepared to meet its requirements; (2) producing a high-quality product efficiently; (3) packaging the product appropriately and delivering on time; and (4) being innovative. Key steps in successful vegetable production are site selection, site preparation, variety selection, planting, management during growth and development, harvest, packaging and postharvest management, record keeping, and continuing education.

Vegetables can be classified by whether they are warm- or cool-season types. They can also be classified by the part of the plant that is harvested. Plant part classifications include: fruit and seed; stem, leaf, and flower; and root. Vegetables that are grown for fruit and seed include snap/green and lima beans, cucumbers, eggplant, okra, garden peas, peppers, pumpkins and squashes, sweet corn, and tomatoes. Vegetables grown for their flowers, stems, or leaves include artichoke, asparagus, the cole crops (broccoli, Brussels sprouts, cabbage, cauliflower, collards, kale, and kohlrabi), celery, chives, endive and chicory, kohlrabi, leek, lettuce, parsley, rhubarb, spinach, and Swiss chard. Many herbs are also grown for their leaves; these include basil, oregano, rosemary, thyme, and mint (spearmint, peppermint). Plants grown for their edible underground parts are beets, carrots, celeriac, garlic, horseradish, onions, parsnips, potatoes, radish, rutabaga, salsify, sweet potatoes, and turnips.

KNOWLEDGE CHECK

1. List four things that vegetables contribute to human nutrition and physical health.
2. How do vegetables contribute to our emotional well-being?
3. How did the potato famine impact Ireland and the United States?
4. Why is using vegetable growing a good way to help educate people about growing plants safely and within natural laws?
5. Give two characteristics of vegetable farms and industries.
6. What are the two factors that determine where large-scale vegetable production is concentrated?
7. What is the most common approach to vegetable farming?
8. What is the advantage of using high tunnels for vegetable production?
9. Why is much more precise fertilizer control necessary in hydroponic systems compared to field and semi-enclosed growing systems?
10. Has the demand for organically grown vegetables increased or decreased over the past two decades?
11. What are three factors that drive consumer demand for vegetables?
12. What are the factors that determine a grower's interest in producing vegetables?
13. What is the reason for solarizing soil?
14. What is a trap crop?
15. What is harvest maturity and why is it not always the same as plant maturity?

16. What is a major difference between the equipment that is used on vegetable crops and that used on agronomic crops?
17. Why are many vegetables that are botanically fruits or seeds still considered to be vegetables?
18. What is the difference between pole- and bush-type beans?
19. Why are cucumbers with predominately pistillate flowers desirable?
20. How does maturity influence the color of many pepper fruit?
21. What are Scoville units?
22. What is the difference in maturity of summer squash compared to winter squash and pumpkins?
23. What is the maturity difference between sweet corn and field corn?
24. How does harvesting differ between processing and fresh market tomatoes?
25. What type of climate is needed for globe artichoke production?
26. Why must enough time be allowed for the ferns to grow in asparagus production?
27. How is white asparagus produced?
28. What are the main cole crops? What time or times of the year are they harvested?
29. Why is celery difficult to grow?
30. What are the growing conditions for many herbs?
31. What part of the leek is harvested for eating?
32. Describe crisphead, butterhead, romaine, and leaf types of lettuce.
33. What is the edible part of rhubarb? Why are the leaf blades not eaten?
34. Why is a short daylength important in spinach production?
35. What plant can be used as a substitute for spinach?
36. What is zoning in beets?
37. What is the difference between wild carrots (Queen Anne's Lace) and domestic carrots?
38. Where do garlic cloves form?
39. What environmental factor is the primary signal for bulbing to start in onions?
40. How are most potato plants started?
41. How are sweet potato plants started?
42. When are seeds sown for a spring beet crop? For a fall crop?

FOOD FOR THOUGHT

1. If you were growing fresh market tomatoes on a small (five acre) farm in northern Illinois not far from a large city where you take your produce to a farmer's market, do you think you would take one-half acre and construct several high tunnels on it? Each high tunnel would have water and a small heater. What is your reasoning for building or not building?

2. As a backyard vegetable gardener in an area that has moderately cold winters (first hard frost late October, last frost early May), devise a sequence of crops that you could plant to be harvesting something from midspring to mid- to late fall.

3. You are a vegetable grower who wants to develop a niche market for a large immigrant population in a city nearby. Choose an ethnic group and research the kinds of vegetables that the group eats and choose three that you might choose to grow.

FURTHER EXPLORATION

DECOTEAU, D. R. 2000. *Vegetable crops.* Upper Saddle River, NJ: Prentice Hall.

MAYNARD, D. N., AND G. J. HOCKMUTH. 2006. *Knott's hand book for vegetable growers,* 5th ed. New York: John Wiley and Sons.

✱20

Temperate Fruit and Nut Crops

Joseph Scheerens

key learning concepts

After reading this chapter, you should be able to:

- List the steps needed to establish a successful fruit or nut production business.
- Discuss the importance of cultivar and rootstock selection.
- Describe the basic principles of planting and maintaining fruit and nut crops.

The first essential step in the successful establishment of a fruit or nut planting is to be certain that the crop to be planted is adapted to the climate and soil characteristics of the region and has an economically viable market. After the suitability of the crop has been ascertained, several horticultural and economic factors should be carefully considered before planting is initiated. Tasks to be accomplished include:

1. Selecting a suitable site.
2. Evaluating market characteristics and developing a market strategy.
3. Projecting, in detail, capital investments, operating costs, and income over time.
4. Thoroughly reviewing and understanding the flowering and fruiting process.
5. Designing the most appropriate cultural management strategy.
6. Choosing cultivars and rootstocks for optimum performance.
7. Preparing the site for planting.

Because the production of most fruit crops is a long-term undertaking, poor initial decisions can be costly or impossible to correct later. Therefore, all available pertinent information should be sought out before final commitments are made. Many of the following concepts you learned in a general way in Chapter 16, but the special considerations needed for fruit crops are given in greater detail here.

SITE SELECTION

A thorough study should be made of the following aspects of the proposed site. Any one of them could affect the success or failure of the enterprise.

Climate

It is extremely important to learn as much as possible about the weather patterns of the proposed planting site. Fruit crops currently or historically grown in an area may indicate whether or not the climate is suitable for a particular crop. Several weather conditions are of particular importance in fruit growing.

Temperature and the Effect of Site Location The first determination needed in evaluating a site is the

minimum temperature that frequently occurs and if the fruit crop will survive that temperature. For example, if a site frequently experiences a midwinter temperature of −15°F to −20°F, peaches or vinifera grapes should not be grown; however, some cultivars of apples would survive these temperatures. In areas subject to severe winter cold, only the most hardy kinds of fruit crops should be attempted.

Second, the length and timing of the growing and dormant seasons are important climatic considerations for two reasons: (1) they may influence the accumulation of heat or chilling units, which affect important physiological processes such as growth, hardening off, dormancy, bud break, and fruit maturation; and (2) they determine the likelihood that maturing fruit or flowers will experience damaging temperatures during the fall and spring transitional periods, respectively.

Climatic extremes experienced by the crop can be greatly influenced by the features of a particular site. Avoid sites in low-lying areas, such as river bottoms or low spots in rolling hills, where cold air settles during frosty nights. This can be particularly dangerous when frosts occur during blooming periods in the spring or for cultivars with late-maturing fruits, which could be damaged by early fall freezes. It is much safer to select an orchard site on the upper portions or slopes of rolling terrain. Orchard heating, wind machines, or irrigation systems can often moderate or prevent low-temperature injury problems in frosty sites. While controlling frost is an additional expense, a more important factor is the cost of crop loss or injury.

In regions with hot summers, site location can influence the temperature; a northern or eastern slope may be a few degrees cooler than southern and western slopes (in the northern hemisphere). For growing crops that cannot tolerate high summer heat, such as sweet cherries and apples, the site location and orientation of tree rows is of considerable importance. (The south side of east–west rows can be excessively hot.)

The location of the proposed planting site in relation to large bodies of water should be considered. A planting site on the leeward side of a large body of water is likely to have a microclimate modified considerably both in summer and winter, with the temperature lower in summer and higher in winter than similar sites at a distance from such bodies of water. For example, fruit such as peaches and some types of grapes can be grown along the south shores of Lake Erie but cannot be grown a few miles inland.

Wind   Avoid sites that have a history of strong winds. Wind can be detrimental from many aspects. Reduced bee activity during windy days in the pollination season can seriously reduce fruit set and yields. Wind can damage young, tender shoots in the spring and can scar and bruise young fruits. It can limit application of pesticides with speeds over 10 miles per hour. Wind can interfere with the uniform application of sprinkler water. Young trees planted in windy areas must be staked and tied. Windbreaks can help reduce this problem, but they can be a costly added expense and can complicate air drainage problems.

Precipitation and Available Water   Fruit plants require adequate soil moisture throughout the growing season to maintain vegetative growth and to produce a full crop of quality fruits. If no supplemental irrigation is possible, attention must be paid to the rainfall history of the proposed site to determine whether the total and summer rainfall is likely to be adequate and consistent. A better situation exists where water supplies for irrigation are available during times of drought. Some of the best fruit-growing areas in the world are located in areas where little or no rainfall normally occurs during the growing season. These regions depend entirely on irrigation (Fig. 20–1).

A pattern of continual rains during the pollination period in the spring could result in poor crops by interfering with bee activity. Continual rains during the fruit harvesting period lead to problems, not only in getting the fruit picked but also in promoting various physical defects (e.g., cracking, epidermal blemishes) or fungal diseases on the maturing fruits. In cold winter areas, accumulated snowfall during the dormant season

Figure 20–1

An irrigated almond orchard in California. Source: USDA Natural Resource Conservation Service, *http://photogallery.nrcs.usda.gov/*

provides protection against low soil temperatures and potential root injury. Snow can also delay bloom in the spring and reduce the possibility of frost damage to the buds and flowers.

The proposed site for a fruit planting should not be subject to periodic flooding from nearby rivers or streams. Fruit plants will not tolerate water around their roots for any length of time because the water stops air penetration to the roots. Areas with a high water table are usually unsuitable for fruit growing because only the soil mass above the water table is available for root development, and this is usually quite limited.

Hail   Hail can cause serious damage to orchards and other fruit crops from both mechanical injury and disease infections entering open wounds. Small hailstones early in the growing season can reduce crop value, whereas large hailstones can destroy fruit, shred foliage, break spurs and new shoots, and remove bark from branches. Thus, it is very important that the frequency and severity of hailstorms at the proposed site be determined in advance.

Land and Soil Characteristics

Slope of the land is important. Ideally it should be over 2 to 3 percent for good air and water drainage but no greater than 10 to 15 percent for successful operation of equipment without contouring.

Topography should be relatively even, without nondraining depressions and no more than 10 percent grades. A north slope reduces potential for frost losses and sunburn but delays bloom and opportunities for early harvest. West slopes should be avoided except in cold climates or at high elevations where heat units are limiting.

The ideal orchard soil should be a deep (1.8 m or 6 ft) well-drained, nonsaline, fertile silty loam to a fine sandy loam (see Chapter 5). However, a rooting depth of between 2 and 3 feet is usually quite adequate for most fruit crops. There should be no impervious hardpans or claypans under the surface. Fine-textured clay soils generally make poor orchard soils and should be avoided.

Planting on unsuitable soil handicaps productivity for the life of the planting and can make the enterprise marginal or unprofitable. If a less than ideal soil type is used, such as a clay loam, it is best to consider planting a fruit species that does relatively well on fine-textured soils, such as plums, pears, or apples, and avoid planting peaches or almonds, which will not tolerate such soils. If the available soil is a loamy sand or sandy loam, plant peaches or almonds (if the climate is suitable). Some species, however, such as grapes and strawberries, do well on a wide range of soil types.

Because soils differ with respect to parent material, ion exchange capacity, organic matter content, and physical properties, the native fertility of the proposed cropping site should be determined as early as possible. It is also important to consider soil pH (acidity or basicity) because it greatly influences mineral availability and soil solution composition (see Chapter 5). Most fruit crops grow best in mineral soils that are slightly acidic (pH of 5.8–6.5), which provides a balanced window of nutrient availability. In soils that are more acid, the availability of nitrogen (N), phosphorus (P), potassium (K), sulfur (S), calcium (Ca), magnesium (Mg), and molybdenum (Mo) drops off. Conversely, in basic soils, N, P, iron (Fe), manganese (Mn), boron (B), copper (Cu), and zinc (Zn) are less available. Extremes in pH can also cause specific minerals to accumulate in the soil solution at toxic levels. For instance, low soil pH (4.5 or less) results in availability of Mn at superoptimal levels and its uptake at toxic concentrations, one of the interacting factors responsible for internal bark necrosis disorder in apple. On the other hand, certain fruits of the ERICACEAE (heath) family are acid-loving and grow best at a soil pH between 4.0 and 5.5. The most notable temperate zone ericaceous fruit crops are blueberry and cranberry. Blueberries also prefer a soil organic matter content of 3 percent or greater for adequate root development. Most mineral soils typically contain 2 percent or less decomposed organic matter.

It is essential that all available information be obtained about the soil in the proposed site before planting. A soil map of the area should be consulted at the library of the nearest agricultural college or the office of the cooperative extension or the Natural Resources Conservation Service. The Internet is also a valuable resource for obtaining site-specific information about soil and soil quality. The soil of the area to be planted should be systematically sampled with a backhoe and enough pits opened to inspect the soil profile to a depth of about 1.8 m (6 ft). The sample will divulge any hardpan or rock layers or sandy pockets and will show whether the soil is deep enough to support fruit trees. A shallow soil may support smaller fruit plants such as strawberries or bushberries.

A study of the history of the site, including information of other crops previously grown, can be useful in analyzing soil problems. If cotton, potatoes, or tomatoes have been grown there, expect trouble from verticillium wilt. If an old orchard has been pulled out,

the soil could be compacted, infected with *Phytophthora* fungus, or high in pesticide residues. Often it is necessary to fumigate the soil to eliminate fungus and nematode problems before planting a new orchard to obtain good, vigorous tree growth. However, because of the negative effects of some soil fumigants on the ozone layer, very few are now available for use. Good alternatives have yet to be developed, although considerable research is being conducted.

In tree fruits and nuts, resistant rootstocks can often reduce problems with soil pests. For example, many sandy loam soils, well suited for growing peaches, are infested with root-knot nematodes, and large-scale fumigation may not be practical. But in some instances, by planting peach trees grafted on a nematode-resistant rootstock, such as Nemaguard, good tree growth and productivity can be obtained in spite of the nematodes in the soil. Often, however, such resistant rootstocks are not winter hardy in cold climates.

Irrigation Water: Quality and Availability

In areas with low rainfall during the growing season, assurance should be obtained that there is a potential source of ample high-quality irrigation water. The water should not contain high soluble salts, including sodium, chlorides, boron, calcium, and magnesium bicarbonates. Water samples can be analyzed by state or commercial laboratories and unsuitable water sources detected before the planting is made. It is difficult, if not impossible, to correct poor-quality water without high costs.

The quantity of irrigation water to be applied to a fruit planting will vary according to the climatic zone, the crop, and the irrigation system. The frequency of irrigation will depend upon the moisture-holding capacity of the soil. Young trees, coarse-textured soils, and shallow soils require more frequent irrigations but not greater amounts. Most critical are the last thirty days before fruit harvest, when water shortages can significantly reduce fruit size and quality, while increasing various fruit disorders.

For some fruit crops, irrigation systems are also necessary as a means to avoid spring frosts during flowering. As water changes state (from liquid to solid), it releases energy in the form of heat (i.e., the heat of fusion). When the blooming crop is sprinkle-irrigated, this heat energy transfers to open or nearly open flower buds as ice forms on the flower surface. This heat of fusion, along with dissolved sugars and salts in the cell sap, protects the flower tissues from ice damage. For sprinkler irrigation to be effective against frost damage,

it must be initiated when temperatures reach $1.1°$ to $0.6°C$ ($34°$ to $31°F$) and continued throughout the period of freezing temperatures. On windy nights, the effectiveness of irrigation for frost control is greatly reduced.

MARKET AVAILABILITY AND MARKETING STRATEGIES

Utilization of the crop is generally assured for plantings in the home garden—by the family, friends, and neighbors, plus canning, drying, or freezing and, perhaps, sales of surplus fruits and nuts. In a commercial planting, however, it is essential to know that a market will be available for the crop. The potential marketability of a fruit crop is simply a matter of supply and demand. It may be influenced by global, regional, and local production capacity and competition for customers at the time of harvest. In most large supermarkets, fruit is available from countries in the Far East, South America, Europe, and Africa and from all regions of the United States. Locally produced fruit may be in direct competition (e.g., local blueberries versus those from another state) or indirect competition (peaches from the United States versus papayas from Brazil) with other fruits if marketed at the same time. Also, trends in fruit crop consumption and use have changed over time. Therefore, it is extremely important to appreciate the worldwide production and marketing trends for any crop being considered for planting. Table 20–1 and 20–2 and Figures 20–2, 20–3, and 20–4 on 450 to 451 show how acreage can change over time in the United States and globally for several fruit and nut crops.

The marketing situation for most fruit crops is often quite fluid and should be studied thoroughly before heavy planting commitments are made. New plantings of a particular fruit crop in a certain area would be questionable when experienced growers in that area are either not planting or are pulling out existing plantings. Perhaps the influx of an insect or disease problem has added control costs that eliminated the profit margin for that crop. Heavy plantings of a crop in a given area because of enticingly high returns at the moment may lead to market gluts when the plants come into production. Prospective fruit growers should realize that market demands for their produce can shift over the years, so they should study and determine what are the present, and possibly the future, demands for fruit products—by commodity and by cultivar. Figure 20–4 shows that in the United States from 1979 to 2004, fruits marketed as fresh and

(*text continued on page 451*)

Table 20–1
CHARACTERISTICS OF COMMON TEMPERATE TREE FRUIT AND NUT CROPS

| Species | Origin | Production Centers | | Plant Habit | Winter Hardiness | Edaphic Requirements |
		International	US			
Almond (*Prunus dulcis*) ROSACEAE: edible portion is a seed	S.E. Asia cultivated BCE in Mediterranean region	United States Spain Turkey Greece Italy	California	Vase-shaped tree (4–6 m)	Requires mild winters	Light well-drained soil preferred
Apple (*Malus X domestica* Borkh.) ROSACEAE: edible portion is a "pome fruit"	W. Asia cultivated in Mediterranean region 1200 BCE	China United States France Italy Turkey	Washington New York California Michigan Pennsylvania	Central leader tree (height determined by rootstock)	–25° to –29°C	Deep well-drained loam
Apricot (*Prunus armeniaca* L.) ROSACEAE: edible portion is a "stone fruit" or drupe	W. China, cultivated in W. China 2000 BCE	Turkey Iran Italy France Pakistan	California Washington	Vase-shaped tree (3–8 m)	–29° to –34°C	Deep well-drained, fertile soils
Cherry (sweet) (*Prunus avium* L.) ROSACEAE: edible portion is a "stone fruit" or drupe	S. Central Europe, Asia Minor, cultivated prehistorically in Europe	Iran Turkey United States Italy Spain	Washington Oregon Michigan	Pyramid-shaped tree (4–5 m), moderate control	To –29°C	Deep well-drained loam
Cherry (tart) (*Prunus cerasus* L.) ROSACEAE: edible portion is a "stone fruit" or drupe	S. Central Europe, Asia Minor, cultivated prehistorically in Europe	Russia Poland United States	Michigan New York	Tree more spreading than sweet cherry (<4.5 m)	Greater than sweet cherry	Deep well-drained loam
Hazelnut (*Corylus avellana* L.) BETULACEAE: edible portion is a nut	N. America, Europe, Asia, cultivated in ancient Greece	Turkey Italy United States Spain	Oregon	Multistemmed bush or spreading tree (to 9 m)	–9° to –12°C	Tolerates various soils, well-drained preferred
Peach and Nectarine (*Prunus persica* L.) ROSACEAE: edible portion is a "stone fruit" or drupe	China, cultivated in China 2000 BCE	China Italy United States France Japan	California S. Carolina Georgia Pennsylvania New Jersey	Vase-shaped tree (to 6 m)	To –23°C	Deep well-drained, fertile soils
Pear (*Pyrus communis* L.) ROSACEAE: edible portion is a "pome fruit" or drupe	W. Asia, cultivated by early Egyptians and Greeks	China Italy United States Japan Spain	Washington California Oregon New York Michigan	Columnar-shaped tree (4–6 m), moderate central leader	To –32°C	Silt loam, somewhat tolerant of heavy soils
Pecan (*Carya illinoensis* Wang) JUGLANDACEAE: edible portion is a nut	S. central United States, cultivated in 1800s	United States	Georgia Texas New Mexico Arizona	Central leader tree (18–24 m)	To –25°C for mature tree, to –5°C for young tree	Deep well-drained loam
Plum (*Prunus* spp.) ROSACEAE: edible portion is a "stone fruit" or drupe	N. Hemisphere, cultivated in Greece 600 BCE	China United States Romania Germany France	California Washington Oregon Michigan	Vase-shaped tree (6–8 m)	To –30°C highly dependent upon species	Deep well-drained fertile soils
Walnut (*Juglans* spp.) JUGLANDACEAE: edible portion is a nut	Europe, Asia, N. America, S. America, throughout the Roman Empire	China United States Iran Turkey Ukraine	California	Up to 15 m under cultivation, leader	To –42°C highly dependent upon species	Deep well-drained, fertile soils

Table 20–1 (Continued)

Chilling Requirements	Reproductive Behavior				Major Diseases in United States	Major Insect Pests in United States
	Fruit Producing Wood	Flower Types	Compatibility	Harvest		
200–500 hours	Lateral-bearing on spurs	Complete	Mostly self-incompatible, some cross-incompatible	Mechanized	Blossom and twig blight	Navel orange worm, peach tree borer, various mites
800–1,700 hours	Lateral-bearing on spurs, some terminal-bearing on shoots	Complete	Cultivar-specific compatibility, cross pollination is beneficial	Hand	Scab, fire blight	Coddling moth, plum cucurlio, wooly aphids
400–1,000 hours	Lateral-bearing on spurs	Complete	Self-fruitful except for specific cultivars	Hand	Blossom blast, brown rot	Species of boring insects, leaf rollers, spider mites
1,000–1,500 hours	Lateral-bearing on shoots	Complete	Self-incompatible except for specific cultivars	Hand	Brown rot, *Verticillium* wilt, canker	Cherry fruit fly, black cherry aphid, plum curculio, spider mites
1,000–1,500 hours	Lateral-bearing on spurs, some terminal-bearing on year-old shoots	Complete	Self-fruitful	Mechanized	Brown rot, *Verticillium* wilt, canker	Cherry fruit fly, black cherry aphid, plum curculio, spider mites
1,500 hours	Terminal-bearing on year-old shoots	Separate male (catkins) and female flowers on same plant	Self-incompatible, some cross-incompatible	Mechanized	Eastern filbert blight	Filbert aphids, filbertworm
200–1,000 hours	Lateral-bearing on shoots	Complete	Self-fruitful except for specific cultivars	Hand	Peach leaf curl, *Verticillium* wilt, *Armillaria* root rot	Plum curculio, borers, scale, aphids, Oriental fruit fly
1,200–1,500 hours	Terminal-bearing on spurs, some lateral-bearing on shoots	Complete	Predominantly self-incompatible	Hand	Fire blight, pear decline	Codling moth, mites, pear psylla
500 hours or less	Terminal-bearing on current season shoots	Separate male (catkins) and female flowers on same plant	Requires pollinizers due to protandry or protogyny	Mechanized	Scab, leaf diseases	Pecan weevil, hickory shuck-worm, aphids
700–1,100 hours	Terminal-bearing on short spurs, lateral-bearing on year-old shoots	Complete	Variable and highly dependent upon species and cultivar	Hand (fresh market), mechanized (processing)	Bacterial canker, crown gall, oak root rot, crown rot, plum pox	Scale, peach twig borer, codling moth, mites
400–1,600 hours	Terminal-bearing on current season shoots	Separate male (catkins) and female flowers on same plant	Requires pollinizers due to protandry or protogyny	Mechanized	Crown rot, crown gall, oak root fungus, walnut blight	Walnut husk fly, navel orange worm, codling moth, aphids, scale

Source: USDA National Agricultural Statistics Service.

Table 20–2
CHARACTERISTICS OF COMMON TEMPERATE SMALL FRUIT AND VINE CROPS

| Species | Origin | Production Centers | | Plant Habit | Winter Hardiness | Edaphic Requirements |
		International	US			
Blackberry (*Rubus* spp.) ROSACEAE: edible portion is an aggregate of drupelets	Europe and N.America, primarily domesticated in N. America in 1800s	United States Canada Chile New Zealand Guatemala	Oregon California	Perennial plant with biennial canes, erect, semi-erect, and trailing types	–18°to –23°C	Deep well-drained, fertile soils
Blueberry (highbush) (*Vaccinium corymbosum* L.) ERICACEAE: edible portion is an epigynous or "false" berry	Eastern N. America, domesticated early in the twentieth century	United States Canada Poland	Michigan New Jersey Oregon Washington	Perennial, deciduous shrub (1–2 m in cultivation)	–21°to –32°C	Sandy, acidic soils high in organic matter
Blueberry (lowbush) (*Vaccinium angustifolium* Alt.) ERICACEAE: edible portion is an epigynous or "false" berry	Eastern N. America, under domestication, harvested from wild stands	United States Canada	Maine	Creeping, rhizomatous shrubs (12–20 cm)	Very hardy	Sandy, acidic soils
Cranberry (*Vaccinium macrocarpon* Ait.) ERICACEAE: edible portion is an epigynous or "false" berry	Eastern N. America, domesticated early in the nineteenth century	United States Canada	Wisconsin Massachusetts New Jersey Oregon Washington	Creeping, rhizomatous evergreen shrubs (12–20 cm)	–20°to –40°C, bogs flooded for winter protection	Humanmade bogs of acidic sand and organic matter
Currant and gooseberry (*Ribes* spp.) SAXIFRAGACEAE: edible portion is an epigynous or "false" berry	Europe, Asia, and N. America, cultivated in Europe in the seventeenth century	Germany Poland Russia United Kingdom Czech Republic	Commercial acreage < 100	Perennial, deciduous shrub (1–2 m in cultivation)	–30°to –35°C	Deep well-drained, fertile soils
Elderberry (American) (*Sambucus nigra* s. canadensis L.) CAPRIFOLIACEAE: edible portion is a small "berry-like" drupe	Eastern N. America, domesticated early in the twentieth century	Not applicable	Limited commercial production	Perennial stoloniferous shrub (to 4 m)	Winter injury of roots or year-old canes not reported	Deep well-drained loam
Grape (European) (*Vitis vinifera* L.) VITACEAE: edible portion is a berry	Asia Minor, cultivated prehistorically and spread throughout Europe in BCE	Italy France United States Spain China	California	Perennial vine, size varies with cultivar and growing system	To –23°C	Variable, dependent on cultivar and use
Grape (Am. Fox and muscadine) (*Vitis labrusca*, *V. rotundfolia*, etc.) VITACEAE: edible portion is a berry	N. America, domesticated by early European settlers	United States	Washington New York Michigan Southeastern states (*V. rotund.*)	Perennial vine, size varies with cultivar and growing system	To –31°C for *V. Labrus*. not less than –12°C for *V. rotund.*	Variable, dependent on cultivar and use
Kiwifruit (*Actinida deliciosa* A. Chev) ACTINIDIACEAE: edible portion is a berry	Mountainous regions of Central China, cultivated in New Zealand in 1930	New Zealand Italy Japan United States France	California	Vigorous, perennial, deciduous, woody vine (18–24 m)	To –15°C	Deep sandy loam, high in organic matter
Raspberry (*Rubus* spp.) ROSACEAE: edible portion is an aggregate of drupelets	N. Hemisphere, First domesticated by Romans in fourth century, gathered from wild BCE	Russia Serbia/ Montenegro United States Poland	California Washington Oregon	Perennial plant with biennial canes, height varies with cultivar and region	–34°to –40°C	Deep well-drained, loamy, slightly acidic soils
Strawberry (*Fragaria X ananassa* Duch.) ROSACEAE: edible portion is a modified stem or receptacle	Species hybrid created in Europe 1750 from two species native to the Americas	United States Spain Japan S. Korea Poland	California Florida Washington Oregon	Stoloniferous, herblike perennial with short stem (crown), 3–5 cm	–12°to –40°C, highly dependent upon cultivar	Well-drained, sandy or loamy soils

Table 20–2 (Continued)

| Chilling Requirements | Reproductive Behavior | | | | Major Diseases in United States | Major Insect Pests in United States |
	Fruit Producing Wood	Flower Types	Compatibility	Harvest		
50–800 hours	Primarily terminal-bearing on current season's lateral growth borne on biennial canes	Complete	Self-fruitful except for specific cultivars, cross pollination beneficial	Hand (fresh market) and mechanized (processed)	Verticillium wilt, anthracnose, cane blight, orange rust, viruses	Aphids, borers, weevils, Japanese beetle, mites, raspberry fruit worm
150–1,100 hours	Terminal- and lateral-bearing on year-old shoots	Complete	Self-fruitful except for specific cultivars, cross pollination is beneficial	Hand (fresh market) and mechanized (processed)	Mummy berry, stem cankers, Botrytis rot, anthracnose, viruses	Blueberry maggot, cranberry fruit worm, plum curculio, mites
1,000 hours	Terminal-bearing on year-old shoots	Complete	Self-incompatible	Harvested with hand held rake	Twig and blossom blight, Botrytis rot, red leaf disease	Blueberry maggot, beetles, red-striped fireworms, Tussock moths
600–700 hours	Terminal-bearing on year-old shoots	Complete	Self-fruitful	Mechanized (fresh and processed)	Early, end, and black fruit rots, leaf spots, twig blight and shoot dieback	Root grubs and weevils, black-headed fireworm, fruit worms
800–1,600 hours	Terminal- and lateral-bearing on year-old shoots, flowers borne on strigs	Complete	Self-incompatible (black currants) or self-fruitful (gooseberries and red currants), cross pollination is beneficial	Hand (fresh market) and mechanized (processed)	Viruses, leaf diseases, oak root rot	Gall mite, clear-winged borer
Not determined	Terminal-bearing on current year's shoots, lateral-bearing on perennial shoots	Complete	Predominantly self-fruitful, cross pollination is beneficial	Hand	Tomato ringspot virus, stem and twig cankers	Elder shoot borer, sap beetles, mites
100–500 hours	Lateral-bearing on current season's growth borne on year-old canes	Complete	Self-fruitful	Hand or mechanized harvest, dependent upon use	Powdery mildew, viruses, Pierce's disease	Borers, beetles, leafhoppers, leaftollers, mites, Phylloxera
1,000–1,400 hours—V. Labrus.	Lateral-bearing on current season's growth borne on year-old canes	Predominantly complete, some gynoecy	Predominantly self-fruitful	Hand or mechanized harvest, dependent upon use	Anthracnose, black root rot, Phomopsis blight, viruses	Borers, beetles, galling insects, moths, grape curculio, leafrollers
500 hours or less	Lateral-bearing on current season shoots borne on year-old flowering laterals	Dioecious, separate male and female plants	Cross-pollinated	Hand	Blossom blight, root rots, fruit rots	Leafrollers, scale, root-knot nematodes, passion vine hoppers
800–1,600 hours	Terminal- and lateral-bearing on current season's lateral growth borne on biennial canes	Complete	Self-fruitful, cross pollination is beneficial, requires pollinating insects	Hand (fresh market) and mechanized (processed)	Phytophthora root rot, cane blight, viruses, anthracnose, orange rust	Aphids, borers, Lygus bug, mites, raspberry fruit worm
200–500 hours	Terminal-bearing from crown or branched crown. Floral buds initiated in previous or current season	Complete	Self-fruitful, pollinating insects are beneficial	Hand	Red stele, anthracnose, Photophthora rots, Verticillium wilt, viruses	Aphids, mites, thrips, Lygus bugs, beetles, strawberry clipper

Source: USDA National Agricultural Statistics Service.

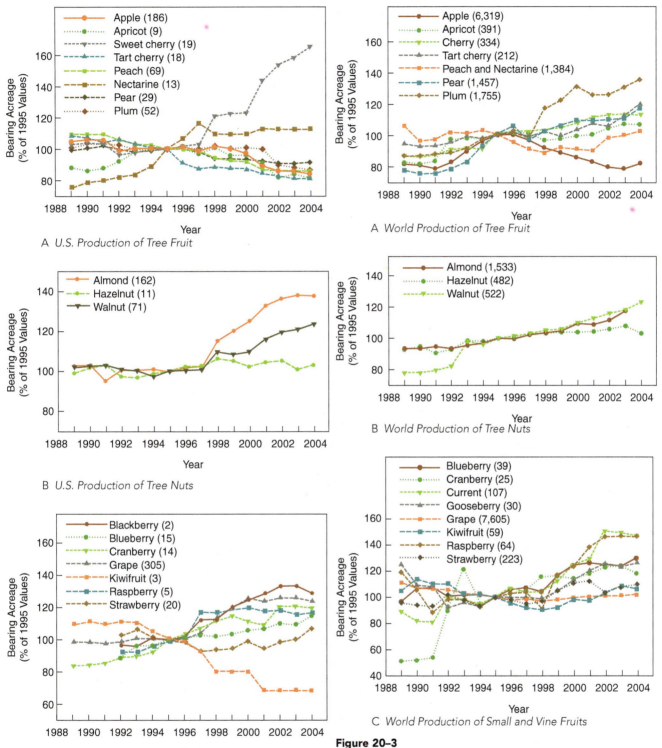

Figure 20–2

Trends in temperate zone (A) tree fruit, (B) tree nut, and (C) small fruit and vine crop production (1989–2004) in the United States. Graphed values represent the percentage increase or decrease in acreage with respect to that reported for 1995 (the reference year). Values in parentheses following legend designations represent the bearing area in hectares × 1,000 reported for 1995. Source: USDA National Agricultural Statistics Service.

Figure 20–3

Global trends in temperate zone (A) tree fruit, (B) tree nut, and (C) small fruit and vine crop production (1989–2004). Graphed values represent the percentage increase or decrease in acreage with respect to that reported for 1995 (the reference year). Values in parentheses following legend designations represent the bearing area in hectares × 1,000 reported for 1995. Source: United Nations FAOSTAT 2005.

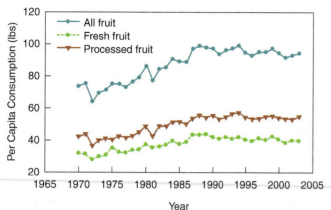

A *Consumption of Fresh Versus Processed Fruit*

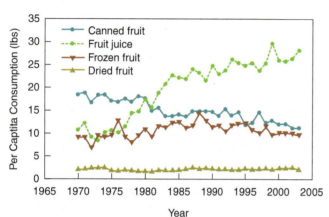

B *Consumption of Processed Fruit*

Figure 20–4

US trends in per capita consumption (lbs) of temperate zone fruit and fruit products, 1970–2003. Crops included in the data set are apple, apricot, blackberry, blueberry, cherry, cranberry, grape, nectarine, peach, pear, plum, raspberry, and strawberry. Source: USDA Economic Research Service.

as juice increased slightly, while canned products declined. Frozen and dried fruit consumption increased in the early 1980s and then declined in the 1990s. Recent consolidation of large wholesale grocers has dramatically reduced the number of buyers and thus competition. Overall, this has depressed the prices received by fruit producers. Also in the late 1980s, produce sections of supermarkets in the United States and Europe started displaying fresh fruit and vegetables that were unknown a few years earlier, such as kiwifruit, lychees, Asian pears, carambolas, and mameys. This diversity of products coupled with increasing world production and concentration of buyers has created pressure on fruit growers to change cultivars to achieve higher prices. Standard cultivars such as 'Delicious' apples or the 'Hayward' kiwifruit are treated as commodities and receive much lower prices compared with newer cultivars such as 'Gala' or 'Braeburn' apples or yellow-fleshed kiwifruit.

In light of all these factors, marketing strategies for the crop should be considered in advance. Are there marketing cooperatives or private packers and shippers available who will take the crop, or will the grower need to grade, pack, store, and then transport the fruit to city fruit markets? Or, perhaps, in locations with considerable highway traffic, particularly close to large cities, on-the-farm roadside or pick-your-own sales may utilize most or part of the crop. Agrotourism is gaining popularity in many regions of the United States. A new fruit producer considering a pick-your-own marketing strategy might benefit from offering amenities, inducements, or entertainment (e.g., corn mazes, hayrides, etc.), which can greatly increase the pleasure of an on-farm experience for local customers. The type and extent of amenities offered is limited only by cost, liability concerns, and the imagination or ingenuity of the producer. Sometimes mail-order or Internet enterprises can be developed with proper advertising.

COSTS IN ESTABLISHING AND MAINTAINING A FRUIT OR NUT PLANTING

There are certain costs to be considered in establishing and maintaining an orchard, vineyard, or berry planting. Unfortunately, many of these costs occur during the planting's establishment. The period from planting to first production can be as short as one season for annually produced strawberries to greater than four or five years for apples, pears, cherries, and many nut crops. Whatever the length of the preproduction phase, fruit producers must develop a business plan that allows for the survival of the enterprise during this period while little or no income is being generated. Once fruit is being produced, yearly expenditures are still necessary to maintain the productivity of the planting and the viability of the business. Costs associated with both pre- and postproduction phases of the planting vary considerably among production regions, localities, crops, and planting systems, and from year to year. These costs are largely the result of:

1. **Capital investment**
   - Land.
   - Roads, buildings, irrigation and drainage, fencing, etc.
   - Equipment, including, tractors, trucks, tillage, spray and harvesting equipment, and tools.
   - Growing system materials, such as staking, trellising, and mulches.

2. **Overhead costs**
   - Interest, taxes, depreciation, and insurance.
   - Utilities.
   - Management.
3. **Production or direct operating costs**
   - Supplies.
   - Land preparation.
   - Chemicals and fertilizer.
   - Harvest supplies, including crates or bins, boxes, cartons, and so on.
   - Marketing costs, such as advertising, packaging, and transportation to markets.
4. **Labor**
   - Permanent employees for management and production operations.
   - Seasonal employees for pruning, harvesting, and packing-shed operations.

In a fruit-growing enterprise, harvest labor is usually the most costly production expense and may determine the profit (or loss) margin. Harvest labor costs and even the availability of harvest labor can fluctuate widely and may remain unknown until harvest is actually under way.

The harvest of some fruit and nut crops such as sour cherries, prunes, almonds, walnuts, pecans, canning peaches, cane fruit and blueberries for processing, and cranberries has been wholly or partly mechanized (Fig. 20–5), and efforts are being made to mechanize the harvest of other fruit and nut crops. Mechanization greatly stabilizes fruit-growing production costs.

Figure 20–5
Mechanically harvesting cranberries. Source: USDA Agricultural Resource Service Image Gallery, *http://www.ars.usda.gov/is/graphics/photos/*

Capital investment and production costs can also be influenced by the production system, and in some cases, the cultivars chosen for the new planting. However, before important decisions concerning production strategies or cultivars are made, the producer should ensure that he or she fully comprehends the integrated relationships among plant genotype, horticultural manipulation, and the physiological processes controlling the flowering and fruiting for the new planting's intended crops. A thorough understanding of these relationships is central to the production of commercial quantities of high-quality fruit and ultimately to the success of the enterprise.

UNDERSTANDING THE FLOWERING AND FRUITING PROCESS

There are four initial steps that are critical to the production of large quantities of high-quality fruits and nuts. Briefly, these are:

1. The initiation of flower buds in the summer, followed (in most deciduous fruits) by the development of a physiological "resting" condition. This is overcome by chilling winter temperatures, and the flower buds continue development early the following spring.
2. Flower opening and pollination in the spring.
3. Fertilization of the egg in the flower, fruit setting, and the beginning of fruit development.
4. Fruit maturation and seed maturation (nuts).

Flower Initiation

Initiation (also called **differentiation**) involves the change of a vegetative growing point deep inside a bud in the axil of a leaf on a shoot (as in peaches) or on a fruiting spur (as in apples) into miniature flower parts. While some small fruit species produce flower buds in the planting year, initiation does not occur in fruit trees until they have reached a certain size or age (three years or so for some peaches, and up to ten or twelve years for some apples). The transition from the vegetative to the reproductive state and early flower bud development involve a highly integrated series of morphological and physiological events that are highly conditioned by levels of stored carbohydrates (energy), mineral nutrient status, and hormonal balances within the plant during the initiation period. In some fruit plants, this change from a vegetative to a reproductive state is triggered by certain environmental cues. Strawberries, raspberries, and many other fruit crops require short days in the fall

(i.e., increasing night length as fall approaches). With most deciduous fruit species, however, no definite environmental factor is known that triggers the change in a bud from a vegetative to a reproductive (flowering) growing point.

Studies many years ago showed that initiation of flower parts in the buds of apples, pears, peaches, cherries, plums, raspberries, blueberries, and so forth, begins between late spring and late summer during the year preceding bloom. Flower initiation starts after the new shoots have attained a certain diameter and length and a portion of their leaves have matured. Figure 20–6 shows various stages in flower bud initiation and development in the sour cherry.

Horticultural practices that optimize the physiological condition of the fruit plant during the initiation period ensure adequate fruit production for the next

fruiting season. In other words, the size of the subsequent fruit crop depends on the number of buds changing from a vegetative to a reproductive state. This, in turn, depends on the general health, hormonal balance, and nutritive condition of the fruit plant.

Trees that were pruned and fertilized heavily the winter before with nitrogen, then copiously irrigated, are likely to become strongly vegetative, producing long, succulent shoots. Buds on such shoots are not likely to form flower buds for the next year. Weakly growing, slender shoots, especially on older trees that are low in vigor, also form few flower buds, particularly if the leaf area has been damaged by insect or disease attacks, thereby reducing the tree's photosynthetic capacity.

Severe and prolonged drought during the critical period of flower bud initiation in early summer can

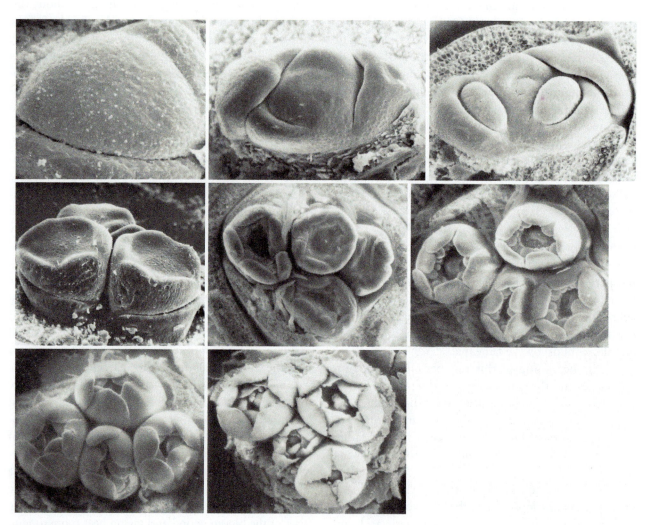

Figure 20–6

Early stages in the development of flower parts in the buds of the sour cherry (Prunus cerasus), sampled in Hart, Michigan, as shown by scanning electron micrographs. Sampling dates and magnification are: Top row (left to right): June 10 (× 266); June 16 (× 293); July (× 233). Center row (left to right): July 30 (× 213); August 15 (× 120); September 29 (× 110). Bottom row (left to right): November 4 (× 55); March 27 (× 40). This shows that the cherry flowers that open in the spring have developed slowly in the buds during the preceding summer and fall. Source: D. H. Diaz.

STRAWBERRIES

The strawberry is an example of a fruit species whose flowering is triggered by a definite, easily defined, environmental factor. Most of the important strawberry cultivars are short-day plants. That is, with the onset of short days (and long nights) in the fall, vegetative growing points in the crown of the plant begin changing to reproductive growing points—or flowers. Such plants then bear a single crop the following spring. When the daylength increases during the summer, flowering stops and the plants become vegetative and start producing runners. However, such cultivars also respond to temperature.

These short-day plants grown under long days still produce flowers if the temperature is reduced from 21°C (70°F) to 15°C (60°F). This is the situation in the cool, coastal strawberry districts of central California, where very high yields are obtained because plants of short-day cultivars grown there produce fruit all summer long, even with long days. Certain strawberry cultivars, however, are not responsive to changing daylength. They do not produce runners but continuously form flowers through the long days of summer. These cultivars are termed everbearers.

create water deficits in the trees that interfere with flower-bud formation. Slight water deficits are not likely to be harmful and can even stimulate flower initiation by reducing shoot elongation and causing carbohydrate accumulation from photosynthesis. Pruning or similar activities, such as the renovation of matted, row-cultured strawberries, should be accomplished in a timely manner to maintain an adequate leaf canopy and to avoid drastic disruptions in carbohydrate and hormonal balances at the time of flower initiation. Biotic stresses (insects, disease, weeds, etc.) should also be controlled to maximize photosynthetic capacity. It is evident that the care and management that fruit plants receive strongly influence their productivity.

Rest and Chilling Requirements

Once the flower parts are fully formed in the flower buds of deciduous fruit plants, the buds enter a physiological **rest period** in which they will not open even if the plants are subjected to favorable temperature, moisture, and light conditions. It is important to note that the same is true for vegetative buds (those that did not differentiate into flower buds) on the same tree.

The beginning of this rest period depends on the species and the general vigor of the plant. The rest develops slowly, reaches a peak, then diminishes; the onset of the rest period can range from midsummer to late fall. In shorter, slow-growing shoots, it starts earlier than in the buds of longer, more vigorously growing shoots.

There is some evidence that the physiological resting condition develops because the buds accumulate certain natural growth inhibitors, such as abscisic acid, and lose native growth promoters, such as the gibberellins.

The physiological buildup of such growth blockages in buds of deciduous fruit trees—which also occurs in temperate-zone ornamental deciduous trees

and shrubs, as well as in certain bulb species—is an evolutionary development that increases the plants' chances of survival through the winter. Without this self-blocking mechanism for bud growth in the autumn, tender, succulent shoots and tender, newly formed flowers developing from the buds would start to grow during the winter and would be killed by sub-freezing temperatures. This physiological internal inhibition of growth (rest) occurs only in the buds—not in the roots.

Chilling of the buds (both vegetative and flower) through the winter is needed to reverse the rest influence. If the chilling is long enough, the influences blocking bud growth disappear. Then, with the beginning of warm spring temperatures, plus adequate soil moisture, both vegetative and flower buds grow rapidly and vigorously. It is believed that the chilling temperatures may act on the buds to lower the levels of growth inhibitors—such as abscisic acid—and increase the amounts of growth promoters, such as the gibberellins.

If the amount of chilling through the winter is marginal, as it can be in regions with normally mild winters, such as the fruit-growing areas of South Africa and the Central Valley of California, bud growth in the spring is erratic. Some flower buds may drop before opening, thus reducing the crop potential, or bud opening in the blooming period may be late and prolonged. Delayed foliation refers to slow development of vegetative buds due to the lack of sufficient winter chilling. A marginal winter chilling may not always be a disadvantage because it can lengthen the blooming period, increasing the chances of good weather during part of the bloom period and giving bees more time to pollinate the flowers. It can also cause the flower buds to open later and avoid being killed by late spring frosts. In areas with long, cold winters, such as the East, Northeast, and Midwest parts of the United States, the

buds always receive sufficient chilling and this problem does not exist.

Much study has gone into determining the amount of chilling required by buds of the different fruit species to overcome the rest influence. Tables 20–1 and 20–2 (pages 446–449) summarize much of this information, which is expressed as the number of hours below 7°C (45°F) but above 0°C (32°F), required by the various species to overcome the rest influence. Subfreezing temperatures are not necessary. However, high winter daytime temperatures greater than 16°C (61°F) can be detrimental for deciduous fruits because they tend to counteract chilling temperatures. Fog or overcast weather during the winter days can be beneficial in obtaining good bud chilling by keeping direct sun rays from the buds, which would raise bud temperature above the air temperature.

To summarize, the usual sequence of events for temperate-zone plants is for the buds to enter the rest phase in late summer or fall; then, after sufficient winter-chilling, the rest influence is terminated. The buds are then said to be quiescent but will resume growth with the onset of favorable growing conditions in the spring.

There are exceptions to the normal pattern of floral initiation, differentiation, rest, and resumption of bud growth in the following spring. Elderberry flowers, for instance, are borne terminally on new canes or arise within a growing season from axillary positions on second-year and third-year canes. Each flower bud produces an inflorescence (cyme) in early summer that measures from 8 to 25 cm in diameter. Each cyme holds hundreds of self-fruitful flowers that develop and mature into very small, dark purple fruit in late summer or early fall.

Moderate to severe biotic or abiotic stresses can trigger precocious (early) flower development. Trees of some species, particularly almonds and plums, will often bloom in late summer or fall if they are partly defoliated by mites or drought and then wetted by rain or irrigation. Some cane and berry fruit cultivars (e.g., strawberries) have been reported to develop a small portion of their flowers (2 to 3 percent) in the fall, presumably from flower buds that have somehow circumvented the dormancy requirement.

In a more dramatic way, strawberry and cane fruit breeders have developed cultivars with the rather unique characteristic of day neutrality. Day-neutral strawberries are sometimes called everbearers, although the terms may not be synonymous. Unlike their June-bearing counterparts, day-neutral strawberries initiate flower buds in the long, warm days of summer. The buds do not require chilling for continued development. In cooler climates, such as the US Midwest, growers have been able to use day-neutral cultivars to obtain a fruit during the same year as planting and to extend the harvest season from five weeks to five months. In the establishment year, peak fruit production occurs in late August or September. In seasons thereafter, two fruiting peaks occur in a day-neutral strawberry planting, one in June and the other in August. In warmer production areas such as California, day-neutral varieties have been used to produce fruit earlier in the winter (south coast) and to enhance production in the spring and summer months (central coast).

Cane fruit (e.g. blackberry and raspberry) crops are most often grown as perennials with biennially bearing shoots; as such, most cultivars follow the typical temperate perennial floral development pattern. A newly emerged cane (primocane) grows vegetatively during its first season of growth until compound buds containing floral primordia are initiated in the fall in response to daylength. The flower buds cease to develop when the cane enters dormancy and then resume

TERMS ASSOCIATED WITH DORMANCY

Dormancy A general term denoting an inactive state of growth.

Rest period This occurs when buds fail to grow, even under optimal environmental conditions, because of internal physiological blocks. Exposure of the buds to a sufficient amount of above-freezing chilling temperatures can terminate rest.

Quiescence This term describes a condition when nonresting buds fail to grow due to unfavorable environmental conditions such as low temperatures, unavailable water, or an unfavorable photoperiod.

Correlative inhibition In this situation, buds do not grow because of an inhibitory influence of another plant part, for example, failure of lateral buds to grow due to inhibition by the terminal part of the shoot.

Chilling requirement The number of hours a plant (bud) must experience temperatures between 0°–7°C (32°–45°F) to achieve bud-break in the spring. During the accumulation of chilling hours, the physiological (hormonal) conditions affecting dormancy are gradually reversed.

growth and flower in the following spring. These two-year-old fruiting canes are called floricanes. However, a small but important percentage of red raspberry and, more recently, blackberry cane fruit cultivars also carry genes for day neutrality. These primocane fruiting (or fall bearing) cultivars precociously initiate their flower buds in midsummer, and most of them continue to develop without a period of rest or dormancy. Flowering usually proceeds from the apical nodes to the base of the cane. In most primocane fruiting cultivars, a portion of the flower buds initiated do undergo a chilling treatment, then bloom in the following spring.

Location of Fruit Buds and Fruiting Structures

Although most temperate fruit and nut crops are produced from flowers initiated the previous season, the location of flower buds on the plant and the plant structures on which they develop vary widely among fruit-producing species. They can even differ significantly among cultivars within species. Some fruit and nut trees bear their fruit on fruiting spurs rather than laterally or terminally on predominantly vegetative shoots (Fig. 20–7). Spurs are essentially short shoots composed of floral, vegetative, or mixed buds separated by a few shortened internodes. Fruiting spurs can be located throughout the fruiting surface of the tree and can live for several years. Apple spurs terminate in mixed buds that contain leaf, shoot, and floral primordia. As they develop in the spring, these buds produce leaves, at least one new shoot (a bourse shoot), and approximately six to eight flowers, only one or two of which will develop into a mature fruit. In addition to specific apple cultivars, pears and mature pecans and

walnuts bear fruit developed from terminal spur buds. Cultivars of almonds, apricots, sweet cherries, and plums also fruit heavily on spurs, but the floral primordia of these species are positioned in lateral buds on the spurs rather than in the terminal bud.

Some fruit and nut trees predominantly set floral buds at lateral or, more rarely, terminal positions along the growing shoot (Fig. 20–8). Predominately shoot-bearing deciduous tree crops include the non-spur-forming apple cultivars, peaches and nectarines, sour cherries, and young pecans and walnuts. Bush fruits commonly set floral buds along new shoots as well.

In contrast, the extensive upright stems of cranberry initiate floral development in mixed terminal buds in early to midsummer; floral tissues continue to develop in these buds throughout the summer and during the following spring. In spring, the bud breaks and continues to develop vegetatively. Flower stalks bearing solitary flowers on stiff pedicels develop at the base of the new shoot growth.

Currants and gooseberries also initiate floral buds in midsummer, which develop in the axils of main shoots or at the terminus of secondary shoots. In the following spring, these buds produce short, branchlike fruiting structures called strigs that hold one to three or eight to thirty flowers for gooseberries and currants, respectively.

Grapes produce two types of buds in the leaf axils of the current season's growth: vegetative buds that may give rise to lateral shoots within the same season, and compound buds (eyes) that contain both vegetative and floral primordia for next season. These compound buds are fully developed by late fall and the onset of dormancy. In the spring, after chilling requirements have been satisfied, buds that remain after dormant pruning

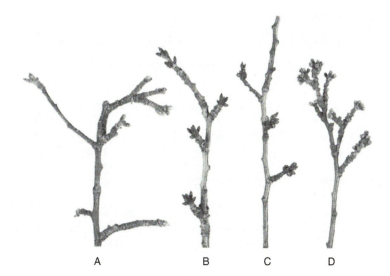

Figure 20–7
The development of fruiting buds terminally on (A) fruiting apple spurs and laterally on the fruiting spurs of (B) sweet cherry, (C) apricot, and (D) plum.

A   B   C   D

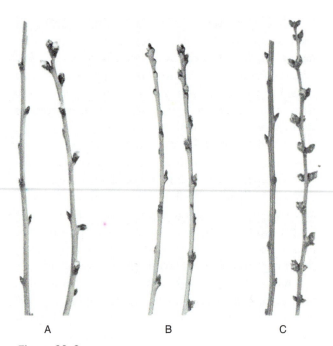

Figure 20–8

The development of lateral fruit buds on one-year-old shoots of (A) almond, (B) peach, and (C) apricot. Each pair consists of a shoot with vegetative buds (left) and fruit buds (right). Vegetative buds are small and pointed, whereas fruit buds are large and plump.

develop into the new shoots that will bear the grape clusters. The number and position of these clusters on the new shoot is genetically controlled but environmentally and culturally influenced. European grapes and many other grape species often produce only two clusters per shoot, whereas American grapes typically produce two to four clusters borne at nodes two to five of the shoot. French–American hybrid wine grapes are prolific, producing four or more clusters per shoot and additional clusters from buds at the base of the shoot and from latent buds on the cordon (permanent scaffold) of the vine.

As stated earlier, most cane fruit develop compound buds on first-year canes (primocanes) during late summer to fall, or in the case of blackberries, even as late as the following spring. After dormancy is satisfied, these buds develop into fruiting laterals, which can be several nodes long. Flowers are borne on branched inflorescences formed terminally or in leaf axils of the fruiting lateral. Each inflorescence bears from one to many flowers, depending on the species, cultivar, and position on the lateral. In red raspberry, fruiting laterals arise primarily from the leaf axils of the unbranched cane (now a second-year cane, or floricane),

---

TERMS ASSOCIATED WITH FRUITING AND FRUITING STRUCTURES

Apical floral bud A bud forming at the shoot apex that has undergone a transition to produce floral rather than vegetative tissues.

Mixed bud A bud that contains both vegetative and floral tissues.

Compound bud A bud that contains more than one growing point; usually, one growing point is dominant. The bud may contain floral tissues, vegetative tissues, or both.

Shoot-bearing A term describing a plant that predominantly bears fruit terminally or axially on shoots developed during the current or previous season.

Spur-bearing A term describing a plant that predominantly bears fruit on fruiting spurs.

Fruiting spur A short, thick, fruit-bearing stem terminated by an apical floral, vegetative, or mixed bud at the end of the growing season, depending on the species. Fruiting spurs usually develop flowers, leaves, and new shoot growth.

Bourse shoot Shoot arising from the mixed bud on a fruiting spur of apple. Bourse shoots terminate growth by forming floral or vegetative buds.

Upright A decumbent shoot arising terminally or axially from a cranberry runner that will bear fruit from terminal buds.

Strig A current or gooseberry inflorescence.

Primocane A vegetative cane of raspberries, blackberries, or their hybrids that arises from crown or rhizome buds. Primocanes develop floral buds in response to physiological or environmental cues but typically do not fruit until their chilling requirements have been met.

Floricane A second-year cane (previously a primocane) of raspberries, blackberries, or their hybrids, on which floral buds initiated in the previous season develop flower and fruit.

Crown A strawberry stem with severely shortened internodes that develops a terminal floral bud in response to environmental cues.

Branched crown A branch arising axially from the strawberry stem. Branched crowns also develop terminal floral buds.

Scape A strawberry inflorescence.

with about 70 percent of the fruit-bearing potential concentrated in the central one-half of the cane. In contrast, black raspberries and blackberries and other cane fruits are often cultured to produce vegetative laterals on primocanes because many of the fruiting laterals produced in the following season will arise from compound buds in the leaf axils of these vegetative laterals.

The strawberry crown is essentially a stem with severely shortened internodes resulting in leaves that are arranged in a rosette pattern. Each axillary bud of the strawberry crown may give rise to a branch crown or a runner. However, in response to environmental cues, the terminal growing points of the main crown or branch crowns are transformed into terminal floral buds. In the US Midwest, the floral bud is well-developed by late fall, before entering a dormant period. The strawberry inflorescence or scape (botanically a cyme) develops from this bud in the following spring. Crown growth is reinitiated from a lateral bud, close to the terminal. In continental climates, each crown or branch crown can, at best, produce one scape per year, whereas in warmer regions, several flowering flushes can occur.

Pollination and Pollenizers

When the previously formed flower parts start to develop and the flowers open, a new critical stage begins in the production of the crop. For most fruit species, the flowers must be adequately pollinated before fruit can set and develop. The pollination requirements of the various major fruit species were given in Tables 20–1 and 20–2 on pages 446–449.

Some fruit species have definite pollination requirements. Successful fruit growers know the needs of their fruit crops and provide for adequate pollination. For example, in both almonds and sweet cherries, the flowers of any given cultivar must be pollinated by those of a particular different cultivar. This means that pollen must originate in the pollenizer cultivar and be carried by bees (or other insects or wind) the flower of the cultivar that is to produce the main crop.

Some fruit species, such as most apple cultivars, must be cross-pollinated, but the pollen can originate in flowers from any other cultivar of that species. But cultivars selected for cross pollination must have overlapping blooming periods. In contrast, other fruit species, such as sour cherries and apricots, are self-fruitful. The pollen can originate in any flower on the same cultivar, either on another tree, another flower on the same tree, or even from the anther of the same flower. These species are normally self-pollinated, but pollen coming from flowers of different cultivars in the same species may also cause fruit to set.

Before the trees are purchased for a planting of the fruit species, the grower must determine the pollination requirements of his or her proposed cultivars. Where cross pollination is required, the planting should consist of the main fruiting cultivar, interspersed with trees of the pollenizer. Quite often, the latter trees are set in some arrangement, such as every third tree in every third row, or if the main cultivar is one that tends to set fruit heavily, every fourth tree in every fourth row.

Temperature is an important factor during all the stages of pollination, pollen tube growth, fertilization, and fruit setting. A temperature range of 15.5°C to 26.5°C (60°F to 80°F) is considered optimum for deciduous fruits. Temperatures much above or below this range impair good fruit setting. Temperatures much above 26.5°C (80°F) inhibit pollen germination. Pollen grains themselves are quite stable at low temperatures—when dry, they can be kept viable for years at −18°C (0°F)—but temperatures dropping to −3°C or −2°C (27°F or 28°F) can kill the ovules in the open flowers of most fruit species.

Insects and Pollination

Fruit species with large, showy flowers generally depend on insects to transfer pollen. Bees, particularly honeybees, are the most important type of insect involved in the pollination of commercial crops. Some fruit plants, especially those with nonshowy flowers, such as the walnut, pecan, and filbert, are wind-pollinated. They generally produce large amounts of very light pollen that is carried considerable distances in the wind onto the stigmas of other flowers.

In fruit orchards that require cross pollination, even when the proper mixture of pollenizing cultivars is present, bees must be in the orchard during the blooming period to carry pollen from the flowers of one cultivar to the flowers of the other (Fig. 20–9). For one or two trees in a home garden, enough bees are generally present for successful pollination, but an orchardist with a large number of trees that absolutely require cross pollination (self-unfruitful) will need to have bees brought in. For a rental fee, beekeepers will provide the necessary bees during the blooming season.

Honeybees work the flowers to collect pollen and nectar, which they use as food. Since the bees generally stay within about a 100-yard radius of their hives, hives are placed in an orchard no more than 150 to 180 m (500 to 600 ft) apart. Weather conditions affect bee activity. Below about 13°C (55°F) they are inactive; the optimum temperatures are about 18° to 27°C (65°F to 80°F). Winds much above 15 mi/hr (24 km/hr) keep bees from flying, and continuous rains interfere with

Figure 20–9
To obtain good crops of many fruit species (e.g., sweet cherries, almonds, plums, and apples) cross pollination between cultivars is necessary. Fruit growers often set hives of bees in their orchards during bloom, as shown here. The bees, working the flowers for nectar and pollen, which they use as food, also transport pollen from tree to tree. Source: USDA.

A

B

Figure 20–10
Grape trellis systems for (A) raisin grape and (B) wine grapes. Source: USDA Natural Resource Conservation Service, *http://photogallery.nrcs.usda.gov/.*

their activity, but intermittent showers do not. If bees or other insects are important pollinators, pesticides must be used with extreme caution to protect the pollinators.

CHOOSING A GROWING SYSTEM

Many temperate fruit and nut crops can be cultured in a variety of ways, so producers have the opportunity to tailor a growing system to fit their climate, capabilities, situation, and needs. Growing systems are often chosen to maximize yield, facilitate horticultural practices, and/or alter fruit harvest to capture market opportunities. Successful use of a growing system is crop- and even cultivar-dependent. For example, both grapes and cane fruits can be grown using a wide variety of trellising techniques that affect the plant's physiology, yield potential, or hardiness, as well as the horticultural practices necessary to effect sustained productivity and planting longevity (Fig. 20–10). Vigorous American or French–American hybrid wine and table grape varieties, such as "Concord," are perhaps best cultured using a Geneva double-curtain bilateral cordon or training system (see Training and Pruning later in this chapter), whereas grapes of moderate vigor can be cultured successfully using the popular high bilateral cordon (single-curtain) training system. Both of these systems are adaptable to both hand and machine harvest. In contrast, French vinifera wine grapes are often trained most successfully using the vertical shoot

position system because they have a strong tendency toward an upright growth habit.

Strawberry growing systems vary widely among production regions in response to climatic differences (see Fig. 20–11). Additional information is available under Flower Initiation described previously. In much of the Midwest, strawberries are cultured perennially in matted rows established by the setting of runner plants. In southern California and Florida, individual plants are grown as annuals at close spacing on plastic-mulched, raised beds during the winter months, when short daylengths permit multiple flower flushes (cropping cycles) per season. In northern California coastal regions, cool night temperatures also allow for multiple cropping within a single season. The annual strawberry production system is gaining popularity in the southeastern states and some midwestern states as well. In these areas, fall-planted strawberries are overwintered with natural or synthetic mulches, which results in an early spring harvest when market demands are high.

A

B

C

Figure 20–11
Strawberry production systems: (A) perennial, matted row culture in the Midwest, (B) annual two-row winter production on black plastic in Florida, and (C) annual four-row winter production on clear plastic in southern California.
Source: Joseph Scheerens, The Ohio State University.

Cane fruit crops are most often grown as perennials with biennially bearing shoots, requiring growers to manage primocanes and floricanes simultaneously. To reduce primocane interference with the fruiting canes, however, a few commercial producers have adopted

a two-year cropping system where new primocanes in the second or fruiting year are partially or totally suppressed. All canes are removed after harvest to begin another cycle in the following season. Cycle timing is often staggered among rows or areas within the production field to provide a harvestable crop each year. Pruning activities are simplified and spray costs are reduced using the two-year system, but overall yield is reduced because only half the planting produces fruit in a given year. Reductions in berry size and quality have also been reported using this system.

In recent years, the so-called high-density orchard plantings, particularly with apples and to a lesser extent with pears and stone fruits, have become popular. Trees are planted close together, as hedgerows, or tree walls. Dwarfing rootstocks are used to keep the trees small. The land can be utilized to the maximum by high-density plantings, especially when the trees are young. High-density plantings of the stone fruits—peaches, plums, apricots—have not been as successful mainly because no completely satisfactory dwarfing rootstock is available for these species.

The different categories of planting densities and management systems in use today, particularly for apples, are:

1. *Low density.* Trees are widely spaced (fewer than 250 trees/hectare; 100 trees/acre) so that after maturity, each tree has ample space and light contact around it. Pruning is kept to a minimum to allow rapid development of maximum tree size. Dwarfing rootstocks are not used. Maintenance labor is minimal, but the yields and gross returns per unit area are also likely to be minimal, particularly for the first fifteen to twenty years of orchard life, compared to higher-density plantings (of apples and pears). Fifteen to twenty years may be needed to reach full production. For the stone fruits, tree nuts, and citrus, low-density plantings may be the most profitable, although there has been considerable interest and experimentation in developing high-density management systems with these crops.

2. *Medium density.* Tree spacing (250 to 500 trees/hectare; 100 to 200 trees/acre) is at least 1.2 m (4 ft) closer than for low-density plantings, and pruning and training are more intensive. Dwarfing rootstocks, such as M.M. 106 or M. 7A, may be used with apple, or the trees may be spur-type cultivars budded on seedling rootstocks. More labor is required in pruning and training the trees, particularly during the early developing years. More care and supervision are required, and the investment

per hectare—in nursery stock and perhaps irrigation equipment—is greater than for low-density plantings. However, commercial yields are achieved sooner after planting, which reduces the cost of establishing the planting.

3. *High density.* Trees are planted very close together (500 to 1,235 trees/hectare; 200 to 500 trees/acre) and specific training systems are used, such as slender trees, multiple rows, or trellis-trained plantings. Training and pruning are very important. Reduced tree size can be aided by the use of the more dwarfing rootstocks, such as Angers quince with pear and M. 9 or M. 26 with apple.

Trees grown at high densities or those destined to be harvested mechanically are often intensively trained using a variety of high input (materials and labor) training systems that promote early fruiting and/or the exacting placement of fruiting surfaces. Often these growing systems employ stakes or trellises to control shoot growth carefully. Some examples of high-intensity training systems include the slender spindle, the French axe, the hybrid tree cone, the Heinicke method, the palmette leader, the Lincoln canopy, and the Tatura trellis. Although pruning and training are discussed below, a thorough discussion of these methods, their techniques and applications, and their effects on tree physiology and growth are well beyond the scope of this text. However, the interested reader can explore this topic further using the references listed at the end of this chapter and/or obtaining other appropriate resources.

The yield efficiency of a planting may be described mathematically as marketable yield per unit of production input (cost, time, etc.). Obviously, the choice of a specific growing system will have a marked impact on the yield efficiency. High-density or high-input systems may raise production efficiency, but they are often harder to manage. Because high-input systems usually involve increased establishment and production costs, poor management decisions are more likely to have drastic consequences for the success of the enterprise. High-input system profitability often requires that the crop exhibit an optimum response to horticultural practices to maximize yield. Therefore, an experienced and skilled operator with an in-depth understanding about how a fruit plant's physiology is affected by horticultural manipulation is more likely to succeed with high-input systems than are producers attempting to raise fruits or nuts for the first time.

Organic production of fruits and nuts has increased steadily for many years. As you learned in Chapter 16, there are stringent regulations that must be followed to have a product labeled as organic. Some considerations for organic fruit and nut production are discussed here. Seed or transplants used to establish production fields should themselves be organically produced. Sources of organically produced nursery stock may not be available for many fruit crops; if conventional nursery stock is planted, the fruit cannot be sold as organic for a period of twelve months following transplanting. The organic integrity of the produce must be stringently maintained to prevent contamination with prohibited materials. For instance, organic fields must be isolated from conventionally managed fields at a distance that would prevent accidental spray drift or other sources of contamination. Also, machinery, equipment, and processing and storage facilities must either be designated specifically for organic produce or cleaned extensively after use in conventional systems.

Obviously, suitability for organic production differs among fruit crops, although most can be organically cultured successfully under favorable conditions. In general, the nitrogen requirements of fruit crops are modest; the availability of excess nitrogen actually overstimulates vegetative growth at the expense of fruit production. Therefore, fruit crops are well-suited for culture using natural fertilizers or composts that tend to release nitrogen incrementally over time as they are decomposed by soil organisms. The main challenge for organic fruit growers is perhaps the control of crop pests. Under an organic production system, pests are managed culturally through the use of genetically resistant cultivars, crop rotation, ground covers and cover crops, soil and crop nutrient practices that restrict pest establishment, additional growing practices that promote crop health and limit pest infection or infestation rates (e.g., proper light penetration and air movement), sanitation and physical removal of crop pests and their habitats, the use of biological control organisms and trap crops, and (as a last resort) the application of natural pesticides.

Site characteristics are extremely important indicators of potential economic success for all fruit plantings; this is especially true if a producer is considering an organic growing system for his or her new planting. In general, it is somewhat easier to produce organically grown fruit in arid or semiarid regions because disease pressures are commonly not as severe in these areas. This may be especially true for grapes, for which consumer preference tends toward wine and dessert cultivars that are not highly resistant to insects or diseases that are prevalent in wetter climates. In response to consumer demand, several well-known western U.S. vintners have developed successful, large-scale organic vineyards. Because of the prevalence of fruit-rotting

fungal diseases, organic production of bunch grapes is not generally recommended for the US South, except for the indigenous muscadine grape that bears its fruit in small open clusters.

The demand for organically grown strawberries and the premium price they command offset lower yields commonly attained with the organic production system. Regardless of production region or growing system, one of the most serious challenges faced by strawberry growers is the control of weeds. To control weeds, large-scale strawberry production systems often involve the preplant application of dangerous and ecologically unsound soil fumigants, such as methyl bromide, as well as the use of plastic mulches throughout the life of the planting. Although not specifically prohibited in organic production, many in organic agriculture do not consider the use of plastic mulches as weed barriers or the use of other petroleum-based materials such as row, tunnel, hoop, or greenhouse covers used in colder climates to be ecologically friendly or sustainable practices. As with any growing system, weed problems can be diminished by selecting a site where effective weed management has been employed in the past and by using cultural practices that promote plant and soil health. Hand or mechanical removal of weeds is effective for organic production yet often costly and labor-intensive. There are several other approaches to weed control for organically produced strawberries and similarly grown fruit. Steam pasteurization or solarization of the soil prior to planting can be done. To steam pasteurize soil, the soil is covered and steam is injected into the soil. To solarize soil, it is covered with clear plastic before the crop is planted. As weed seedlings germinate under the plastic, the high temperatures resulting from heat trapped under the plastic on sunny days kill the seedlings and, if the temperatures get high enough other soil pests are also killed. Other novel—and sometimes expensive—alternative methods include geese, which can be brought in to eat the weeds when ripe fruit are not present; organic mulches such as straw and shredded newsprint; flame weeding; and natural herbicides composed primarily of vinegar with lemon juice or citrus oil additives.

Preliminary studies suggest planting strawberries and other herbaceous fruit into living mulches as a component of a weed-management strategy. The mulch crop is cut prior to planting the fruit crop (Fig. 20–12). Preliminary evidence also suggests that corn gluten, a by-product of corn milling containing high levels of nitrogen and a natural herbicide, is effective against selected weed pests, especially grasses (Fig. 20–13). Under conditions where the natural herbicide is likely

Figure 20–12
This crop of rye and vetch is being cut to serve as a mulch weed barrier for a fruit crop that will be planted in the field. Source: USDA Agricultural Resource Service Image Gallery, *http://www.ars.usda .gov/is/graphics/photos/*

Figure 20–13
Corn gluten being tested for use as an herbicide. Source: USDA Agricultural Resource Service Image Gallery, *http://www.ars.usda.gov/is/graphics/ photos/*

to be leached, however, the increased soil nitrogen supplied by this material may actually promote weed growth. Finally, many organic strawberry growers are opting for shorter life spans for their planting to control the buildup of perennial weeds.

Highbush and rabbiteye blueberries are perhaps the best suited among fruit crops for organic culture. In most production fields, the row middles are sodded and the blueberries are deeply mulched, which discourages weed seed germination and weed growth. Blueberries are more resistant to insects and diseases than most fruit crops, which makes them easier to culture organically. They have a relatively low nitrogen requirement and thrive on organically based fertilizers. Because blueberries are naturally adapted to soils with low pH and high organic matter, the use of composting materials that release nitrogen as NH_4^+ are ideally suited to meet blueberry nutritional and soil requirements. The increased soil organic matter associated with organic-production systems also favors the proliferation of ericoid mycorrhizae, the symbiotic fungi necessary for adequate blueberry nutrient uptake.

In summary, the choice of a growing system is an important one. Drastic changes in production systems are often difficult or impossible to effect after a planting is established. Therefore, a grower establishing a fruit or nut planting must choose the growing system wisely, remain committed to his or her chosen system throughout the life of the planting and be determined to make it work.

SELECTING FRUITING CULTIVARS AND ROOTSTOCKS

Perhaps one of the more important preplant decisions to be made involves cultivar and/or rootstock choice. This decision should not be taken lightly. It should not be left to the local nursery owner, and it should not be based on what surplus nursery stock happens to be on hand for a bargain price at the time. Sometimes this decision must be made and an order placed a year or two in advance so that the propagating wholesale nursery will have time to propagate the desired cultivar on the desired rootstock and to permit obtaining disease-free plant material.

Because cultivar performance affects success at every step of the enterprise, cultivars should be chosen that function well during many facets of the fruit production cycle. Specific characteristics of cultivars to be considered before the choice is made include:

1. *Adaptation response.* Within fruit and nut species, each cultivar has unique physical and physiological traits preconditioned by its genetic profile. The expression of these traits is influenced by the planting environment to produce a phenotype. The adaptation response of a cultivar can be defined as how well the genetic and environmental components of phenotype complement one another to form a highly productive plant with desirable characteristics. Many cultivars perform well in only specific climatic regions or in specific types of soils, whereas others are more widely adapted.

2. *Response to growing system.* A cultivar's adaptation response is also highly dependent on the growing system under which it is cultured. For instance, highly efficient cultivars that partition more of their energy to reproduction than to vegetative growth may be the most suitable for high-density plantings. Some fruit cultivars require extensive support or trellising to perform well, whereas others can be cultured as free standing plants. In some climatic areas, small fruits, such as strawberries or raspberries, can be grown as annuals. However, successful annual production schemes require the use of cultivars that are uniquely suited to, and sometimes specifically bred for, production under this system.

3. *Growth form (habit), vegetative vigor, and size at maturity.* These three interrelated cultivar characteristics can influence light penetration and air movement within the planting and often direct horticultural factors such as plant spacing, training, trellising and pruning systems, fertilization and irrigation practices, equipment use, and harvest techniques. The importance of considering these characteristics when selecting cultivars is obvious, especially for tree fruit genotypes, which often differ widely in size and vegetative vigor. Apples, cherries, grapes, pears, walnuts, and many other woody fruit and nut species are often propagated by budding or grafting cultivars with desired growth and fruiting characteristics to related rootstocks that exhibit superior climatic and edaphic (soil) adaptation responses (Fig. 20–14). Rootstocks are frequently from a different species within the genus, and they often affect the size, precocity, and reproductive capacity of the cultivar under production. The ability of the East Malling and Malling Merton series of apple rootstocks to produce ultimate tree sizes ranging from dwarf to full stature are perhaps the most well-known examples of this phenomenon (Figs. 20–15 and 20–16). For some fruit species, cultivar choice is conditioned partly by selecting for the most appropriate growth habit. For instance, cane fruit cultivars can be classed as trailing, semitrailing, or erect, and they differ widely with respect to their ability to produce fruit without trellising. Half-high blueberry cultivars (interspecific hybrids of highbush and lowbush species) are short enough in stature that fruit buds are often protected from winter temperature extremes by snow

Figure 20–14

Grafting a young fruit tree. Source: USDA Agricultural Resource Service Image Gallery, *http://www.ars.usda.gov/is/graphics/photos/*

Figure 20–16

These two trees are the same on top but have different rootstocks. The rootstock on the tree in foreground has induced a dwarf characteristic. Source: USDA Agricultural Resource Service Image Gallery, *http://www.ars.usda.gov/is/graphics/photos/*

cover. For growers in many northern regions, choosing a half-high cultivar is imperative because cultivars that are standard size are unprofitable to grow.

4. *Flowering and fruiting characteristics.* Beyond a doubt, aspects of flowering and fruiting are extremely important determinants of cultivar performance. A grower has to have a solid understanding of flowering and fruiting characteristics of the cultivars he or she is growing. For example, many well-known standard apple cultivars such as 'McIntosh,' 'Delicious,' and 'Golden Delicious' also have strains that are spur-bearing, allowing for relatively direct comparisons between the two flowering types. In general spur-type strains are generally smaller than their standard counterparts; as such, their within-row spacing is often reduced by comparison. Spur-type strains also exhibit greater fruit-setting ability and higher yields per unit trunk cross-sectional area. However, because spur-types often produce a greater yield (i.e., higher number of fruit per tree), their fruit, on an individual basis, are generally smaller than those of the standard strains.

Flower initiation, as conditioned by levels of stored carbohydrates (energy) and hormones, is often balanced in the plant with the needs of the developing fruit. The alternate bearing phenomenon often seen in fruit and nut tree crops results from a physiological imbalance usually brought about by the loss of a season's crop to frost or a similar physiological calamity. The lack of a crop during this initial off year shifts the tree's physiology to an extent that an overabundance of flower buds are initiated for the following season (the on year). If not severely thinned, these flowers will develop excess fruit of poor size and quality. Large numbers of developing fruit during floral initiation will severely limit the number of flower buds produced for the next season, thus triggering another off-year. This cyclic phenomenon is difficult to correct once it has been initiated. Susceptibility varies by species and cultivar.

Growers must also consider differences in the dormancy or chilling requirements of species and cultivars when planning a fruit planting. Plants with relatively low chilling requirements may not be suitable for areas with dramatic climate fluctuations in early spring. These plants often resume

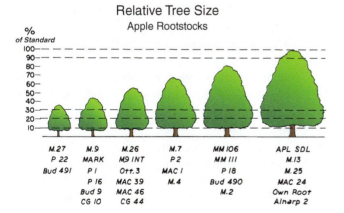

Figure 20–15

Relative apple tree size produced by selected new and proven rootstocks. Source: *HortScience* 1988, 23(3). Alexandria, VA, American Society for Horticultural Science.

growth too early during warm weather spells and then are easily damaged by subsequent frosts. Cultivars with relatively high chilling requirements and that flower later than others are better choices for these areas. Often a flowering delay of even seven to ten days can result in a significantly better yield.

Conversely, successful fruit growing in southern locations often requires plants with low chilling requirements. Until a few decades ago, blueberry production in southern states of the United States was based entirely on cultivars of rabbiteye blueberry. However, federal and state fruit breeders have developed specific low-chill highbush blueberry cultivars (e.g., 'Sharpblue,' 'Gulf Coast,' and 'Cape Fear') with improved fruit quality for these regions. These cultivars are steadily gaining popularity among growers.

Self-fruitfulness is also an important cultivar consideration. Most apricots, blackberries, currants and gooseberries, highbush blueberries, grapes, peaches, raspberries, sour cherries, strawberries, and walnuts are predominantly self-fruitful, but some cultivars within this group require cross-pollination for adequate fruit set (see Tables 20–1 and 20–2 starting on page 446). Even though cultivars may be self-fruitful, fruit production and/or fruit size may be increased when cross pollination occurs. Cultivars of apple, almond, hazelnut, muscadine grape, rabbiteye blueberries, sweet cherries, and other species are considered to be predominantly self-unfruitful, but the degree of unfruitfulness may also vary among cultivars. For instance, the sweet cherry "Stella" can be produced without additional pollinator cultivars. In contrast, some cultivars of almond must be self-pollinated. Cultivars of European and Japanese plum, pear, and chestnuts are highly variable with respect to their cross pollination requirements. Regardless of self- or cross compatibility, cultivars may vary in the timing or spatial arrangement of pollen shed and stigma receptivity, thus requiring the movement of pollen from one flower to another for adequate fruit set. Species and cultivars may also differ in their requirement for insect vectors to accomplish this task.

5. *Disease and insect tolerance or resistance.* The incorporation of increased insect and disease tolerance or resistance into desirable cultivars and rootstocks has perhaps received more emphasis and effort by both public and private hybridizers than any other fruit crop improvement goal. The diversity and extent of biotic stresses that may affect fruit culture is too large a topic for an adequate discussion of specific examples of tolerant or resistant cultivars or

rootstocks in this textbook. However, the major biotic pests of tree and small fruit crops are listed in Tables 20–1 (page 446) and 20–2 (page 448), respectively. Climatic and soil conditions affect the incidence and severity of biotic pests. For instance, *Phytophthora* root rots may be common on heavy soils, whereas nematodes are more frequently a problem in sandy soils. Choosing tolerant or resistant cultivars or rootstocks is always wise, especially if they are available for specific disease or insect problems in the proposed production area. The culture of resistant or tolerant genotypes increases the probability of harvesting a marketable crop of high quality and lowers production inputs for chemical or biological control measures. Environmental concerns surrounding the use of agrichemicals and public awareness of and concern about health and food safety issues emphasize the desirability of tolerant or resistant cultivars and rootstocks for the production of fruit.

6. *Winter hardiness and/or tolerance to abiotic stresses.* **Winter hardiness**—the ability to withstand severe midwinter temperatures—is often the product of a plant's unique physical and physiological characteristics and is conditioned by the success of the hardening-off process as the plant enters dormancy in late fall. Within most temperate fruit species, a broad range in winter hardiness can be found among cultivars and among rootstocks. For example, the crowns of many strawberry cultivars are severely injured or killed if they are exposed directly to temperatures lower than −9.5°C (15°F), but cultivars such as 'Fort Laramie' and 'Ogallala' developed in the United States, in Wyoming, tolerate winter extremes approaching −40°C (−40°F). The degree of winter hardiness displayed by a given cultivar or rootstock often limits its range, with only the most hardy cultivars being suitable for planting in colder climates of either hemisphere. For instance, most European wine grape cultivars can be cultured successfully in California and the Pacific Northwest, but only the hardiest among them, such as 'White Riesling,' 'Lemberger,' 'Cabernet Franc,' and perhaps 'Chardonnay,' are recommended for production in northern and central US states. Fruit cultivars and rootstocks also differ in their ability to withstand heat; wind; and adverse soil conditions such as drought, waterlogging, low or high pH, and mineral imbalances.

7. *Ripening period.* Fruit harvest, handling, marketing, and storage are often labor-intensive, time-consuming events, so it is fortunate that cultivars differ substantially with respect to when they ripen

their fruit. In most production regions, a selection of cultivars and/or cultivar–rootstock combinations can be planted to extend the harvest season through one to two months, or perhaps longer. For instance, in the US Midwest, early apple varieties (e.g., 'Pristine') are harvested as early as mid-August, whereas late varieties (e.g., 'Fuji,' 'GoldRush') are not harvested until very late October or early November. In the same region, highbush blueberry cultivars such as 'Earliblue' or 'Bluetta' can be harvested mid- to late June, whereas 'Elliott' berries ripen in mid-September. The harvest of primocane fruiting raspberry or blackberry cultivars begins in late summer and can continue until frost. Aside from horticultural advantages (e.g., greatly simplified pruning procedures), primocane fruiting cane fruit cultivars increase the market presence and consumer awareness of these crops. With the range of standard (summer-bearing) and primocane fruiting cultivars available, it is now possible to produce fresh red raspberries in the Pacific Northwest from mid-June to frost.

8. *Quality and desirability*. Fruit quality can be defined in many ways, and the term can be interpreted differently by producers and consumers of fresh or processed fruit products. Producers often consider fruit quality characteristics that affect the harvestability, ease of shelling (in nut crops), storage, shipping, and shelf life of the product, whereas ample evidence suggests that consumers value fruits and nuts for nutritional and health benefits as well as for their appearance, flavor, and other sensory characteristics. Fruit cultivars often differ considerably in firmness at harvest and susceptibility to bruising during handling, storage, and shipment. Furthermore, differences in post-harvest respiration rates and other physiological characteristics make some cultivars suitable for long-term storage, whereas others cannot be stored for more than a few months, or even days in the case of soft fruits. Rapid cooling and subsequent refrigeration after harvest as well as the employment of controlled atmosphere storage and modified atmosphere packaging can prolong the useful life of most cultivars. Nevertheless, cultivars respond differently to these treatments. For instance, some apple cultivars can be stored successfully for only a few months; others, such as 'Golden Delicious,' can be marketed eight or nine months after harvest if they are stored properly.

Cultivar name recognition has been a long-standing marketing instrument for crops such as apples, grapes, and pears. Consumers often purchase specific cultivars of these crops based on their intended end use (eating, cooking and baking, canning and sauce or preserve making, etc.). Newer cultivars of apples such as 'Gala' and 'Honeycrisp,' with improved flavor and texture qualities, are demanding a premium price and greater share of the market than older cultivars with good but less desirable characteristics. The use of name recognition as a marketing tool is likely to increase with time. Breeders are releasing many of their new fruit cultivars with trademarked names specifically for this purpose. Many major food outlets have begun identifying the fruit cultivars they market with specific numeric designations in cases where specific names are not available or used. Consideration of a cultivar's quality characteristics, therefore, is extremely important when producers plan new plantings. This may be especially important if the product is to be marketed directly to consumers via farm markets or pick-your-own operations.

Admittedly, the perfect fruit cultivar does not exist. Therefore, when planning a new production field, a grower must necessarily weigh the importance of various cultivar characteristics to the ultimate success of the enterprise. Often, it is a good idea to plant several cultivars with a balance of characteristics to offset the vulnerability of establishing only one.

Information concerning the various fruiting cultivars can be obtained from several reliable sources. Most extension bulletins produced by state and federal agencies contain cultivar recommendations for the areas they serve. Likewise, reliable cultivar information is often available from organized producer groups and/or individual growers, buyers, and packers in the area. Reputable nurseries also publish accurate catalog descriptions of the fruit crops they offer. Opinions may differ among sources, however, so it is often wise to consult several sources concerning the suitability of various possibilities before committing to a specific set of cultivars. It is risky to accept new, untried cultivars or to base the decision solely on nursery advertisements.

ACTIVITIES PRIOR TO PLANTING

In addition to choosing an appropriate site, conducting an in-depth market analysis, developing a sound business plan, understanding aspects of flowering, and researching potential cultivar and rootstock characteristics, several important tasks must be completed prior to planting the new production site. Many of these tasks should be completed, or at least begun, one to two years before the anticipated planting date. Because they often require much capital and time to install, the construction of roads and infrastructure (fencing, packing sheds, etc.) necessary for production and handling

should be initiated as soon as possible. The site itself should be thoroughly examined for potential problems concerning slope, exposure, drainage and low-lying areas, and other defects in order to correct them in ample time. The soil should be tested by a reputable state or commercially operated soil-testing facility to determine accurately its physical and chemical characteristics and recommendations implemented. If the site is infested with weeds, especially perennial species, control measures should begin at least one year prior to planting.

Closer to the scheduled planting date, the land should be prepared. For some fruit crops, these operations may include the addition of fertilizers to correct nutrient imbalances, tilling, and/or raised bed construction. Materials needed for plant establishment (e.g., trellis, mulch, etc.) should be purchased and/or constructed.

It is highly advisable to purchase certified virus- and disease-free nursery stock from a reputable nursery rather than obtaining propagules from existing plantings or accepting plants from other sources that may have been previously infected with pathogens. A reputable nursery producer uses many techniques to help insure the plants are healthy and will grow well after transplanting. Nursery stock can be purchased within a year of planting, but the delivery date of the stock should be negotiated to coincide as closely as possible with the scheduled planting date. On arrival, the shipment should be checked for clear cultivar and rootstock designations on plants or plant lots and then stored at the recommended temperature and well away from any fruit or other ethylene-generating plant materials in storage. The roots of most, if not all, nursery stock should be continually protected against desiccation by using moist packing material or wood shavings until they are planted.

High-quality, deciduous nursery stock possesses an abundance of well-distributed roots. Tree nursery stock should display a strong, vigorous trunk free of damage from careless handling. A small shallow slice into the trunk should show bright green tissue below the bark with no evidence of brown areas from winter damage or sunburn. A slice into the roots should reveal a moist whitish color. No root tissue should be shriveled or look brown, gray, or black below the bark. In budded or grafted plants the union should be well-healed and strong, with no more than a slight bend at the union. The graft union should be at least 10 cm (4 in.) above the previous soil level so that scion rooting (roots developing above the graft union) is unlikely after planting. In Figure 20–14, the graft union is visible as the swollen area just above ground level. Scion rooting can lead to a loss of any desired effects of specific rootstocks such as dwarfing, resistance to nematodes, diseases, and so on.

In deciduous nursery stock, the dormant buds should be plump and well-developed and should look bright green when cut into. Dead buds may indicate low temperature or herbicide injury or lack of water during the growing season.

Cane fruit, bush fruit, and strawberry nursery stock can be purchased as dormant plants or crowns, but they are typically available in actively growing forms such as bare-rooted plants, tissue-cultured transplants, transplant plugs, and containerized plants. Often, this nondormant nursery stock is produced and purchased for use with specific growing systems. These materials should arrive with healthy looking leaves that are free from insects or diseases, or evidence of herbicide damage or nutrient deficiency symptoms. Optimal, preplant care and storage conditions are specific to each nondormant nursery product available.

This discussion of preplant considerations is by no means exhaustive. Specific crops and areas will require additional steps to be accomplished prior to planting. It is best to obtain copies of relevant extension guides or to consult with local producers and producer groups for pertinent preplant information concerning the crops to be planted. New producers should also familiarize themselves with all federal, state, and local ordinances governing the production, harvest, and marketing of the prospective crop. Be as ready as you can be!

PLANTING AND CULTURE

Major decisions must be made when the time comes to plant a new orchard, vineyard, or berry planting. Planting distances and patterns must be determined. The distance between plants depends primarily on the choice of growing system and cultivars. However, several factors can modify recommended distances. Greater spacing should be used with conditions of high soil fertility; long growing seasons; vigorous, large-size cultivars: invigorating rootstocks; ample rainfall or irrigation; and heavy use of fertilizers. Spacing should be closer in the opposite situations.

Spacing between rows should also be considered. Between-row spacing should allow for ample sunlight penetration and air movement through the planting and be wide enough to easily accommodate tractors, spray rigs, and harvesting equipment. Some recommended within- and between-row spacings for various fruit and nut crops are listed in Table 20–3. Row orientation should also be considered to maximize sunlight interception and air movement, but protect against potentially damaging winds and or erosion. In planting fruit trees of a species requiring cross pollination to set good commercial crops, it is of the utmost importance that

Table 20–3
SUGGESTED PLANTING DISTANCES FOR COMMON TEMPERATE FRUIT AND NUT CROPS[*]

Species	Planting Distances
Almond	7.5 × 7.5 to 9 × 9 m (25 × 25 to 30 × 30 ft).
Apple	Highly variable—depends on production system, cultivar, rootstock, soil vitality, and cultural practices.
Apricot	6.6 × 6.6 m (22 × 22 ft) (on plum roots); 7.5 × 7.5 m (25 × 25 ft) (on apricot roots).
Blueberry	1.2 m (4 ft) apart in rows 3 m (10 ft) apart.
Cherry, sour	6 × 6 m (20 × 20 ft).
Cherry, sweet	7.5 × 7.5 to 9 × 9 m (25 × 25 to 30 × 30 ft).
Grape	1.2 to 2.4 m (4 to 8 ft) apart in rows 2.4 to 3 m (8 to 10 ft) apart.
Hazelnut	4.5 × 4.5 m (15 × 15 ft).
Kiwifruit (vines on trellis)	5.4 to 6 m (18 to 20 ft) apart in rows 4.5 m (15 ft) apart.
Peach	3 × 6 to 5.4 × 7.2 m (10 × 20 to 18 × 24 ft).
Pear	6.6 × 6.6 m (22 × 22 ft).
Pecan	9 × 9 to 15 × 15 m (30 × 30 to 50 × 50 ft).
Prune	6 × 6 m (20 × 20 ft).
Raspberry, black	0.6 to 1.2 m (2 to 4 ft) apart in rows 2.1 to 3 m (7 to 10 ft) apart.
Raspberry, red	0.75 m (2.5 ft) apart in rows 1.8 m (6 ft) apart.
Strawberry (matted-row system)	61 to 71 cm (24 to 28 in) apart, permitting a matted row, 38 to 61 cm (15 to 24 in) wide to develop from runners.
Strawberry (double-row bed system)	Beds 96 to 112 cm (38 to 44 in) apart, center to center; two rows in each bed 20 to 30 cm (8 to 12 in) apart. Plants in each row 23 to 36 cm (9 to 14 in) apart.
Strawberry (single-row bed system)	Beds 100 to 107 cm (39 to 42 in) apart, center to center; plants 20 to 25 cm (8 to 10 in) apart in rows.
Walnut (Persian)	6 × 6 m (20 × 20 ft) to 10.5 × 10.5 m (35 × 35 ft) (Paynet type); 10.5 × 10.5 m (35 × 35 ft) to 12 × 12 m (40 × 40 ft) (Hartley type).

[*]For information on apple spacing, consult regional extension bulletins (e.g., Pennsylvania Tree Fruit Production Guide at http://tfpg.cas.psu.edu/).

trees of the pollinizing cultivar be appropriately spaced among trees of the principal fruiting cultivar.

Laying Out the Planting

Once the desired plant spacing has been determined and the planting infrastructure has been installed (e.g., raised beds, irrigation systems, etc.), the site can be prepared for planting. Rows and plant positions within the row should be marked just prior to when plants are due to arrive from the nursery. It is important that the rows are lined up properly. This will facilitate many future operations. A plant out of line can be a target for cultivating disks and other equipment moving down the rows.

Planting the Crop

In temperate climates, the best time to plant the nursery stock is any time during the dormant season, when the ground is not frozen and air temperatures are above freezing. It is important to plant as early as possible so the roots will be well established by the time hot weather arrives and the plants will have a full growing season before they face cold weather.

Alternately, some cultural systems, especially those employed in warmer climates, require planting to be initiated in other seasons. For example, annual production of strawberries in California and Florida are planted from mid- to late fall. In areas with relatively mild winters, such as North Carolina, fall-planted systems are successful when used in conjunction with spun-bonded row covers and other frost-protecting cultural practices. Late summer or fall-planted systems are also used in cooler climates where plants are produced in high tunnels covered with polyethylene or similar materials (see Chapter 19 for a description of high tunnels).

In planting tree, vine, or bare-rooted strawberry or cane crops, it is important that the planting hole be dug to the proper depth. The base of the main supporting roots, which usually have been trimmed back, should rest on solid, undisturbed soil. If the hole has been dug too deep, necessitating some backfilling before planting, then the plant is apt to sink after watering and settling, putting the graft union or plant crown below the soil level and leading to attacks of crown rot fungi, principally of the *Phytophthora* species.

The planting hole should be wide enough to easily accommodate the roots without bending and twisting. Tractor-operated soil augers are often used for digging tree holes. These have the advantage of working fast and saving labor, but if they are not operated properly, the holes can be dug too deep, leading to settling. Also, if the soil is too wet when the auger is used, the sides of the hole become severely glazed, making air and water permeability and root penetration difficult. A shovel should be used to break up the compacted sides of the holes when the trees are planted.

Filling loose soil around the roots once the plant has been set in the hole is an important operation. The soil should be worked around the roots as the hole is filled to ensure good soil-root contact, with the soil pressed firmly about the roots. If there are persistent summer winds from one direction, the plant can be leaned slightly into the wind at planting. Alternately, young plants can be staked to ensure proper growth and development. In systems that employ intensive training regimes, the control of plant form begins at planting or soon thereafter.

Many of the planting techniques for bare-rooted plants are also applicable to the planting of soil-bearing nursery stock such as container plants, tissue-cultured plants, or plant plugs. Typically, the depth of planting should be at or near the depth of the original soil ball. Shallow planting, leaving a significant portion of the soil ball exposed, promotes its dessication; planting at a depth greatly in excess of the original soil ball can expose young tender canes or stems to disease organisms. If the nursery stock was produced or maintained in a controlled environment such as a greenhouse, plants should be hardened off by gradual exposure to full sunlight over a period of a few days or weeks prior to planting in the field. Moisture levels should be monitored carefully during the hardening-off process.

For large-scale plantings, mechanical tree planters are used. The tractor driver must exercise care to ensure that the rows are as straight as possible. In-row spacing is never as accurate with mechanical planting, and it is critical that personnel follow the transplanter to make slight adjustments in graft union or crown height and to firm the soil around the plant.

Some cultural systems require additional operations to be performed at or near the time of planting. For instance, for some raised-bed planting systems, the mechanical planter covers the bed with a synthetic (plastic) mulch just prior to setting the plants. Synthetic mulches are employed in these systems in part to conserve and regulate soil moisture, moderate soil temperatures, discourage weed growth, and simplify crop harvest. When raised beds are hand planted, mulch can be laid just prior to planting and then slit by hand or machine to provide openings in which to set the plants. If the crop has specific soil requirements, soil amendments may be added at the time of planting. As mentioned above, blueberries thrive best in acidified soil with high levels of organic matter. Therefore, fruit production guides typically recommend adding up to 1 lb of soaking wet peat moss to the planting hole when planting blueberries on mineral soils (soils that typically contain less that 2 percent organic matter). The moss must be thoroughly wet because dry peat moss is difficult to hydrate after planting and could potentially draw water away from the roots during the critical hours after transplanting. Finally, the fibrous root systems of dormant strawberry crowns are often soaked in water for a short period just prior to being set, again in an effort to avoid moisture stress to the root system during the planting process.

Almost without exception, newly set fruit plants require immediate or nearly immediate irrigation (Fig. 20–17). For tree crops, a shallow basin or furrow should be left at the top for filling with water.

The soil basins should be filled with water within a few hours of planting to prevent dehydration of the roots and also settle the soil around the roots. Watering by filling the basins should continue until vigorous shoot growth is well underway, when the regular irrigation system or rainfall can be used. In heavy soils with drainage problems, care must be taken to avoid overwatering, which can impede good root aeration and lead to attacks by *Phytophthora* and other soil pathogens. For those plants with a soil ball around their roots, the top of the soil ball should be slightly exposed at the surface of the basin so that added water can enter directly

Figure 20–17
A newly planted vineyard being watered by drip irrigation. Notice that the vines are tied to stakes for support. Source: USDA Natural Resource Conservation Service, *http://photogallery.nrcs.usda.gov/*

into and through the root ball. This is done because the nursery soil often found in such root balls is light and porous and if, after planting, they are covered by the heavier clays of native soils, irrigation water will remain in the clays because of their smaller pores and high capillarity, leaving the lighter soil with the roots completely dry. A few days of hot desiccating winds can increase transpiration and severely injure trees planted too deeply, even though the basins have been filled with water.

Bush fruit, cane fruit, and strawberries also require irrigation immediately after planting. However, basins or furrows are not often employed as a means to direct water to the newly planted root systems. Rather, these small fruit plantings are often irrigated with portable or solid-set sprinklers or with drip-irrigation systems. These systems should be operated soon after planting and until the soil is wetted to the depth of the transplant root system.

Newly planted fruit plants often must be shielded from harsh environmental conditions during their period of establishment. The trunks of young trees and grapevines must be protected from sunburn by wrapping them with paper, cardboard, or styrofoam coverings during the summer or painting with whitewash or water-base paint (one part interior white latex paint and one part water). Inspect for bark damage at intervals during the winter by rabbits, mice, or gophers at the soil level and take the necessary preventive measures. Special care should also be given to control weed growth during the establishment of a young fruit planting because competition for moisture; nutrients; and, in extremely weedy conditions, light can prevent adequate crop growth.

It is best not to fertilize or apply herbicides to the trees at planting time. The developing roots could be injured by excessive salts or by herbicides. Generally, the young trees obtain enough mineral nutrients from the soil for their first year's growth. However, recommendations for agrichemical use during the establishment season vary significantly among crops and growing systems. Growers are advised to consult one or more respected state, regional, or federal crop production guides to determine the best management practices specific to their planting as it undergoes establishment.

AN INTRODUCTION TO THE CARE AND MAINTENANCE OF A FRUIT PLANTING

By definition, horticultural crops are intensively cultivated. The care and maintenance of a fruit planting after establishment requires a host of inputs (materials and labor) and horticultural procedures that must be implemented in an integrated fashion by the producer. The ultimate goal of care and maintenance procedures is to optimize the physiological status of the plants, and thus to achieve a sustained production of high-quality fruit throughout the life of the planting.

Common care and maintenance procedures include training and pruning; crop-load management; irrigation; weed-, insect-, and disease-control measures; and fertilization. This list is by no means complete, and most fruit crops require additional inputs that are crop-specific (e.g., the routine organic mulching of blueberries grown on mineral soils). The timing of application for each input and/or procedure is often critical, as is the integration of maintenance strategies because choices made concerning one often dictate the proper recourse for others. For example, the optimum (balanced) pruning procedure for individual grapevines is determined by last year's cropping history, and the weight of pruned material is often used to determine crop-load adjustments for the following season. Beyond all doubt, the growing system employed by the producer profoundly affects care and maintenance strategies. High-intensity versus moderate-intensity systems, annual versus perennial production systems (strawberries and other small fruit crops), organic versus conventional production systems—all require very specific sets of management practices. In a perennial fruit planting, management practices are often altered as the plants progress from juvenility, through the peak bearing years, to their eventual decline, to accommodate changes in the crop's physiology during each period.

A complete discussion of all aspects for all crops in all growing systems is well beyond the scope of this introductory text. Inexperienced producers should seek technical information specific to the crops and growing systems they plan to use from state and governmental extension agencies, fruit growers' societies, or qualified consultants before planting. However, even the most well-written and detailed production guides will not substitute for knowledge gained through personal experience. The best advice for a new producer might be to begin with a small planting, grow the crop to the best of his or her capability using recommended practices, be as attentive to detail and as observant of crop behavior in response to care and maintenance procedures as possible, and learn from outcomes that are less than optimal. As the producer experiences the cropping cycle, intuition concerning crop behavior is likely to be gained; once a grower feels a degree of confidence about producing the crop using his or her management strategy, the scale of the planting can be increased.

As mentioned previously, crop physiology and management practices are inextricably intertwined. To illustrate this relationship (one of the key learning concepts listed at the beginning of this chapter), two care and maintenance topics will be discussed in moderate detail in subsequent chapter sections. They are the crop-load management strategies of training and pruning.

TRAINING AND PRUNING

Training and pruning of fruit plants is an integral part of the procedures used for sustained and profitable production of high-quality fruits. **Training** can be thought of as a procedure that results in the positioning of limbs, branches, or canes, whereas **pruning** can be considered the act of cutting limbs, branches, or canes. However, because the early training of tree and bush crops almost always involves pruning, these two definitions are somewhat simplistic. Because they are inextricably linked, training and pruning procedures are discussed together herein.

Reasons for Training and Pruning

Fruit and Nut Trees

1. To develop a strong trunk and a balanced scaffold system of branches, well distributed around the tree, which can support heavy loads of fruit without limb breakage. The tendency for limb breakage is partially a function of crotch (branch) angle. In general, crotch angles of 60° to 90° result in strong branches that are vigorous, fruitful, and capable of carrying mature, heavy fruit loads, whereas branches with wider or narrower crotch angles are inherently weak and often unfruitful. Improper pruning can actually increase the frequency of narrow crotch angles. Crotch angles wider than 90° result in branches that are weak and unfruitful and should be removed.

2. To regulate fruit production. Proper pruning encourages development of the type of shoot system that produces the fruit. In older trees with little vegetative growth rejuvenation, pruning can force the development of productive fruiting shoots. Pruning can also be used to limit excess numbers of fruit by removing some fruit-bearing branches, giving a thinning effect that can improve fruit size and quality. Pruning and training also provide important means of balancing vegetative growth to fruiting, which aids in the development of high-quality, high annual production, and increased resistance to disease, heat, and cold. With some fruit tree cultivar/rootstock combinations in high-density

training systems, bending or weighting young branches to increased crotch angles alters the hormonal balance, resulting in a significant increase in fruit bud development.

3. To remove dead, broken, or interfering branches.

4. To improve light penetration to the inside of the canopy and lower parts of the tree so as to improve yield and quality for many species.

5. To facilitate insect and disease control by opening the tree, thus increasing penetration of spray materials to the interior branches.

6. To limit tree size and shape to the space allocated to it and to limit tree height to that from which fruit can be conveniently harvested. As extreme examples, intensive tree-training systems in which trees are essentially trellised were developed in part to facilitate the mechanical harvest of fruit. In addition, for some species (e.g., apples) cultured at high densities, tree size can be controlled by yearly mechanical root-pruning treatments, which reduce overall vegetative growth and trunk girth. Some studies have shown root pruning to improve fruit firmness, color, and eating quality; however, reduced fruit size and photosynthetic capacity, and increased susceptibility to drought conditions have also been observed. Root-pruning treatments may be most applicable after the loss of a crop to curtail vigor and reduce alternate bearing tendencies.

Grapevines

1. To establish strong trunk and cordon systems that are maintainable in a form that facilitates vineyard cultural operations.

2. To distribute the fruiting wood to obtain maximum production of high-quality fruit.

3. To maintain vigor and production of fruiting canes and to balance vegetative growth with fruiting for high-quality fruit and to impart greater resistance to heat and winter cold.

4. To aid in control of crop size and increase berry size by reducing the number of fruiting clusters.

5. To increase light penetration throughout the canopy and to increase sunlight exposure for leaves, clusters, and basal buds that will provide next year's crop.

6. To remove old, nonproductive canes.

Bushberries (Raspberries, Blackberries, Blueberries, Currants, and Gooseberries)

1. To remove dead, weak, or diseased canes and old shoots that die following fruiting.

2. To thin out weak canes to give adequate light and space to the remaining canes, resulting in larger, better-quality fruit.

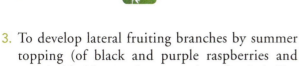

3. To develop lateral fruiting branches by summer topping (of black and purple raspberries and upright blackberries, but not red raspberries). (See the discussion of fruiting habits earlier in this chapter.)

4. To control crop load and increase fruit size.

5. To improve light penetration and photosynthetic capacity. Both training systems and pruning techniques can increase sunlight exposure to actively growing shoots, thus ensuring adequate carbohydrate availability and a physiological balance optimal for the initiation of fruit buds. For example, many trellising systems for cane fruit effect a separation of floricanes from newly emerging primocanes that will produce next year's crop.

6. To control plant shape and the position of fruiting surfaces to ease harvest, especially for crops that are mechanically harvested.

Physiological Responses to Pruning

All plants, if not pruned, tend to develop a balance between growth of the shoot and the root systems. Cutting away part of the top, including the plant's photosynthetic apparatus and food storage tissues, together with reducing the number of vegetative growing points and flower buds while leaving the root system intact leads to some interesting physiological reactions. The fruit grower should be aware of these reactions to understand how to prune the plants properly.

1. *Removing shoot terminals causes lateral branching.* Actively growing meristems produce an abundance of the natural growth-promoting substance, auxin, which promotes cell division and elongation and is primarily responsible for the phenomena of phototropism (the bending of a shoot toward light) and apical dominance. In the latter, the downward movement of auxin transported in the phloem inhibits the development and breaking of axillary (lateral) buds and the elongation of side branches. Natural apical dominance, therefore, results in a plant form with fewer branches (and perhaps fewer fruiting surfaces). Heading back a shoot by removing its terminal portion is the most effective pruning cut for promoting the formation of new branches. Thinning cuts which remove entire shoots or limbs are generally used to open the canopy to air and sunlight or to remove old, unfruitful, interfering, or diseased wood.

2. *There is a reduction in total vegetative growth.* Removing a portion of the top reduces the total amount of subsequent growth, compared to an unpruned plant. The total number of growing points is reduced, resulting in fewer developing shoots, fewer leaves, reduced photosynthesis, reduced amounts of carbohydrates translocated to the roots, reduced root growth, followed by a reduction in mineral and water absorption, which, in turn, further decreases shoot growth. These effects dwarf the entire plant. Generally, the more severe the pruning, the greater the dwarfing.

3. *Continual heavy pruning each year of young fruit trees can delay the onset of bearing.* For many fruit species or for specific cultivars, precocity and pruning severity are negatively related, perhaps due to the decreased availability of carbohydrates for flower and fruit development or from reduced development of fruiting structures (e.g., spur development). Although some pruning is required for proper plant training, the extent of tissue removal should be limited to what is necessary.

4. *Pruning effects from removal of smaller shoots and branches tend to be localized.* Heading back of shoots during the dormant season, reducing the length of a shoot by 50 percent, for example, results in the development of one or more new shoots from buds just below the pruning cut. The total growth by the end of the summer will not be nearly equal to the total growth of an adjacent shoot that was not dormant pruned. This response is often used in pruning to retard the growth of one of two equal-growing shoots and thus prevent the development of a weak crotch.

5. *Severe pruning in early summer is more likely to weaken the tree and reduce total growth than dormant (winter) pruning, especially with young trees.* Much of the energy for new spring shoot growth comes from stored foods in the roots, trunk, and branches of the tree. These stored foods are not replaced until an appreciable number of new leaves have formed and photosynthesized enough to reverse carbohydrate movement into the larger branches, trunk, and roots. If summer pruning is done, photosynthesis and the replenishment of carbohydrates are reduced intensely, debilitating the tree. Tree response to pruning in the late summer (August) is similar to the response from dormant pruning.

6. *Dormant pruning of bearing deciduous fruit trees can be invigorating.* Although vegetative growing points are removed, flower buds are also removed, thus reducing the crop, consequently reducing the demand on the plant's stored foods. The surplus is then utilized for new vegetative growth.

7. *Moderate dormant pruning of bearing deciduous fruit plants can increase production.* Flowers are

produced either laterally or terminally on one-year-old shoots and/or laterally or terminally on fruiting spurs (see Table 20–1 and 20–2, starting on page 446 and Fig. 20–7 on page 456 and 20–8 on page 457). Such fruit-producing growth (shoots and spurs) must be stimulated to obtain continual crops. Moderate pruning plus fertilizers and ample soil moisture provides such growth stimulation.

Training and Pruning Deciduous Fruit Trees

Over the years, many training techniques have been studied and ultimately employed to raise tree fruit crops. Each method is suitable only for specific crops, cultivar/rootstock, and growing system combinations, and each has unique advantages and disadvantages. Therefore, the compatibility of possible training

TERMINOLOGY FOR PRUNING AND TRAINING

Pruning and training has its own vocabulary. Here are some important terms and their meanings.

Heading back Shoots or limbs are not removed entirely, but the terminal portions are cut off at varying distances from the end. This procedure forces out new shoots from buds below the cut and retards terminal growth of the branch.

Thinning Shoots or limbs are completely removed at the point where they attach to the next larger (and older) limb. Thinning out corrects an overly dense area or removes interfering or unneeded branches.

Trunk That portion of the tree up to the first main scaffold branch.

Primary scaffold branches The main branches arising from the trunk of the tree.

Secondary scaffold branches Supporting branches arising from the primary scaffold branches.

Central leader training system A training system for fruit trees where one main vertical trunk remains dominant, with all scaffold branches arising from it in an almost horizontal direction.

Modified leader training system A central leader continues upward from the trunk, but its identity is lost as it becomes one of the primary or secondary scaffold branches.

Open center or vase training system A trunk is developed, then at the top of the trunk (0.6 to 0.9 m; 2 to 3 ft from ground level), several primary scaffolds develop outward and upward at a 30° to 45° angle to form a vase configuration.

Weak crotch A situation that can develop when two equal-size branches with a narrow angle between them grow for some years at the same rate. Eventually a considerable amount of bark inclusion develops between the two branches at their junction, with no connecting wood. With heavy loads of fruit—and strong winds—the limbs are almost certain to split apart.

Stub Pruning cuts of the thinning-out type must be properly done. Cutting off the branch leaves a wound that must heal over by callus growth proliferating from the surrounding live tissue to prevent decay organisms from becoming established in the wound. If the cut is not made close to the adjacent limb and a stub is left, the wound cannot heal properly.

Watersprouts Vigorous shoots usually arising from latent buds in the trunk or older limbs in the lower parts of the tree, growing upright through the center of the tree. They require immediate removal to maintain desired pruning characteristics.

Suckers Shoots arising from the underground parts of the plant, usually coming from adventitious buds on roots. In grafted trees, they generally arise from the rootstock below the graft union. They should be removed as soon as they are noticed.

Head The aerial terminus of a grapevine trunk; in head-pruning systems, it is the position where canes are selected for fruiting.

Cordon A permanent lateral extension of the grapevine trunk.

Grape cane A one-year-old woody shoot bearing buds for next year's vegetative and reproductive growth.

Grape spur A cane pruned to a length of a few nodes bearing buds for next year's vegetative and reproductive growth.

Prepruning The initial removal of most one-year-old canes before selection of specific canes or spurs for next year's vegetative and reproductive growth. In the balanced pruning system, the mass of prepruned shoots is used to determine the optimum crop load for the upcoming season.

Balanced pruning A pruning system that balances vegetative growth and fruiting to maintain optimum vine health and sustained productivity.

Hedgerow A cane fruit management system wherein primocanes arise from rhizome buds anywhere within a confined area. The identity of individual plants is not evident.

Stool A cane fruit plant managed as an individual, with most primocanes arising at or near the plant crown.

Primocane suppression The management system in which early-emerging primocanes are sacrificed to maximize the level of carbohydrate available for developing fruit.

techniques with other production factors should be carefully considered before a training method is chosen. For instance, as mentioned previously, orchards planted at high densities or those to be mechanically harvested often require high-input, high-maintenance training systems that ensure precocity and proper placement of fruiting surfaces. Growers are also advised to familiarize themselves with all procedures associated with specific methods and to understand fully the physiological ramifications of employing them before training begins. Because training involves the removal of wood, it is difficult, if not impossible, to change methods after training is initiated.

Logically, the choice of training system should consider the natural growth habits of the tree to be trained. Because of their strong apical dominance and upright growth habits, apples and pears grown in moderate density plantings are most often trained using the central leader or modified central leader training systems (Fig. 20–18 and 20–19). Peaches and other stone fruits are more bushlike in growth habit than are apples and pears, and therefore they are often trained using an open-center or vase-training system (Fig. 20–20). If not done so by the nursery, newly planted peach trees should be headed back to force lateral branching that will eventually become the primary

scaffolds. A typical tree will possess an open center and about eight scaffold branches distributed evenly to form a vase shape.

Regardless of the system used, bearing deciduous fruit trees require moderate annual pruning to:

1. Stimulate production of new fruiting wood.
2. Allow light penetration into the tree and prevent the development of dense pockets of vegetative growth. This is especially important for spur-bearing fruit trees because the primary source of carbohydrates for developing fruit are supplied by spur leaves. Fruit set, growth, and ultimately fruit quality are negatively affected if these leaves are in constant shade.
3. Reduce excessive amounts of fruiting wood and lessen the need for expensive hand thinning of fruits.
4. Remove broken, diseased, dead, or interfering branches.
5. Keep the trees from growing so tall that harvesting becomes difficult.
6. Confine the trees to the space available to them.

A light annual pruning of bearing trees is better than a heavy pruning every three or four years because it results in more regular cropping.

Figure 20–18

Central-leader apple tree after several years' growth. Left: Before dormant pruning. Right: After pruning. Some lateral branches were removed. In those retained, forked tips were cut to one outward growing branch. Spreader boards have been used with some laterals to develop a wider growth habit. Source: M. Gerdts, A. Hewitt, J. Beutel, J. Clark, and F. Cress, *Pruning Home Deciduous Fruit and Nut Trees*, University of California, Division of Agricultural Science Leaflet 21003 (1977).

Figure 20–19

Some fruit trees, such as pears, tend to develop an upright, narrow shape, as shown in the unpruned tree at left. Often, this cannot be corrected by pruning alone. With notched, spreader boards, as shown in the pruned tree at right, the primary scaffold branches can be trained to give a wider tree shape. Inner limbs should be removed and those retained should be pruned to outward facing branches or buds. Source: M. Gerdts, A. Hewitt, J. Beutel, J. Clark, and F. Cress, *Pruning Home Deciduous Fruit and Nut Trees,* University of California, Division of Agricultural Science Leaflet 21003 (1977).

Figure 20–20

Young, vigorous peach tree before (left) and after (right) pruning. Many scaffold limbs have been removed, particularly those that grow low and horizontal. The stronger, moderately upright limbs were selected for the permanent primary scaffolds and were headed back to force out additional branching. Source: M. Gerdts, A. Hewitt, J. Beutel, J. Clark, and F. Cress, *Pruning Home Deciduous Fruit and Nut Trees,* University of California, Division of Agricultural Science Leaflet 21003 (1977).

Care must be taken not to prune or break off fruiting spurs, although in trees that consistently over-bear to the detriment of fruit size and quality, there may be some benefit in partial spur removal. Spurs do not live indefinitely, and some renewal is constantly taking place. Annual pruning tends to stimulate new spur formation.

Fruit species producing fruiting buds laterally or terminally on shoots that developed the previous growing season should be pruned to encourage the growth of such shoots. This entails a moderate annual dormant pruning, together with moderate nitrogen fertilization and ample soil moisture. Excessive pruning plus heavy applications of nitrogen can cause excessive rank vegetative growth, which will not form flowers.

The severity of pruning needed by bearing trees to keep them in a state of optimum fruitfulness can be judged to some extent by the length of annual shoot growth. Table 20–4 gives the amount of growth associated with good fruiting. If the annual growth of the average shoot exceeds these amounts, the severity of annual pruning should be reduced and nitrogen fertilizer withheld. If shoot growth is much below these amounts, severity of pruning should be increased as well as, perhaps, the amount of fertilizer applied annually.

Training and Pruning Grapevines

In general, grapes are more intensively trained and heavily pruned than fruit and nut trees. Grapes are almost always trellised using a variety of training systems to enhance light penetration and air movement through the canopy and to support the relatively heavy fruit load. On a yearly basis, effective pruning of mature vines requires the removal of most of the previous year's growth to balance this year's vegetative growth against energy devoted to setting and maturing a crop. Grape training and pruning are relatively labor-intensive,

exacting processes, but if accomplished properly, they can ensure the productivity and profitability of the vineyard over an extended period of time. For instance, operating 'Concord' vineyards in the US East and Midwest have been cropped continuously for over a century.

Proper cultural management (weed and disease management, fertility, etc.) and training are critical during vineyard establishment to achieve optimum vegetative (cane and root) growth necessary to produce strong, straight shoots that will form the trunk and scaffold on which future crops will be borne. Newly planted vines are left intact or are sometimes pruned to six or eight buds, depending on cultivar, vineyard conditions, and the expected growth. If conditions are very favorable, unpruned plants will generally develop more leaf area, thus producing longer, thicker shoots. If less-than-optimal growth is anticipated, plants should be pruned to a few buds, and only the two strongest shoots should be kept. At the beginning of the second year, only the healthiest canes are retained as developing trunks. If the vines are growing from their own roots, there is benefit to choosing canes that arose from below-ground buds; however, if the cultivar is grafted to a disease or insect-resistant rootstock (e.g., all European or vinifera grapes are grafted to phylloxera-resistant stock), only canes originating from buds above the graft union should be chosen. The number of canes kept depends in part on the training (trellising) system to be used. Often these canes are trained to be straight and perpendicular to the trellis wire using twine or bamboo stakes as temporary plant supports (see Fig. 20–17 on page 469). For all cordon-based training systems, rapidly growing shoots are trained (wound) around the trellis wire to form the horizontal scaffolds or cordons during the second year of growth (Fig. 20–21). Some producers remove the apical buds of these shoots when they

Table 20–4
DESIRABLE AMOUNTS OF AVERAGE SHOOT GROWTH FOR BEARING TREES TO GIVE MAXIMUM FRUIT PRODUCTION

Species	Young Trees—Under 10 Years of Age		Older Trees—Over 10 Years of Age	
	Centimeters	Inches	Centimeters	Inches
Freestone peaches and nectarines	50 to 100	20 to 40	30 to 76	12 to 30
Clingstone peaches	76 to 100	30 to 40	30 to 76	12 to 30
Apricots	30 to 76	12 to 30	25 to 60	10 to 24
Plums (except prunes) and quinces	25 to 60	10 to 24	23 to 46	9 to 18
Almonds, prunes, and cherries	23 to 46	9 to 18	15 to 25	6 to 10

Figure 20–21

These young grapes are being trained on a trellis. Source: USDA Natural Resource Conservation Service, *http://photogallery.nrcs.usda.gov/*

cordons can be headed back to half their final length (approximately ten nodes) to promote uniform bud break along the cane. New growth is then wrapped around the trellis wire as it develops.

Common grape-growing systems vary from region to region, but grape growers predominantly choose trellis and training regimes that either capitalize on or compensate for inherent characteristics of specific grape types or cultivars. Vegetative vigor and fruitfulness of the grape to be grown are perhaps the characteristics most often considered when choosing a growing system. Producers must also consider the grapevines' growing habit, cold hardiness, and disease resistance.

Although the positioning of trunks, scaffolds, and fruiting wood differ significantly among training regimes, all systems are based on either cane pruning or spur pruning strategies to affix the correct number of fruitful buds and thus achieve a desired crop load (Figs. 20–22 and 20–23). For spur-pruned systems, one-year-old wood is trimmed back to two to three bud spurs positioned evenly along the length of the cordon (see Fig. 20–23). Cane-pruned systems achieve the same end, but longer

reach the desired length for the cordon to promote lateral growth. If lateral growth is sufficient and not winter-damaged, it can be pruned to one- or two-node spurs in the third season. The shoots chosen to be

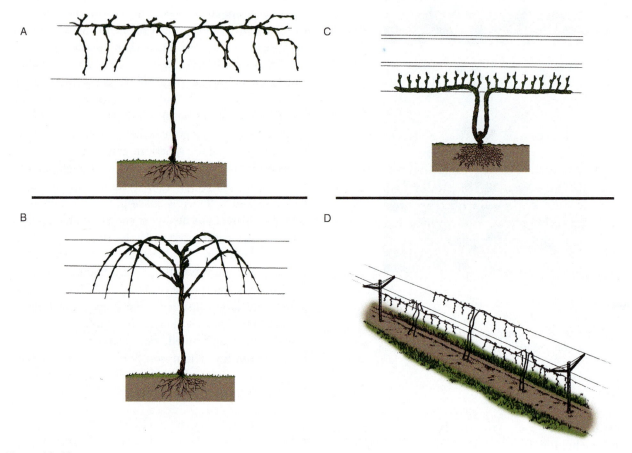

Figure 20–22

Common grape-training systems in the eastern United States: (A) high bilateral cordon or single curtain (HBC—spur-pruned); (B) umbrella Kniffen (UK—cane pruned); (C) vertical shoot position (VSP—spur pruned); and (D) Geneva double curtain (GDC—spur pruned). Source: *Midwest Grape Production Guide,* Bulletin 919. Ohio State University Extension.

Figure 20–23
Balanced pruning of French–American hybrid grapevines: (A) collection of prepruned, year-old wood to be weighed, (B) adjusting spurs to the correct number of buds, and (C) the vine after the completion of the balanced-pruning operation.
Source: Joseph Scheerens, The Ohio State University.

thus photosynthetic capacity. The weight of prepruned (initially removed) one-year-old canes is highly indicative of this vigor. Formulas for balanced pruning procedures based on prepruning weights have been developed for many grape cultivars and specific growing regions. Balanced pruning ensures that vines will not be cropped beyond their ability to mature the developing grapes adequately and that sufficient vegetative growth will be added to maintain vine health and sustain productivity for the life of the vineyard.

Among spur-pruned systems, the high bilateral cordon (HBC), or single curtain, system, is most suitable for grape varieties with moderate vegetative vigor. It is commonly used in eastern and midwestern US production regions, especially for French–American hybrid wine grapes. The advantages of this system are many: it has low establishment and production costs; it facilitates mechanized prepruning and fruit harvest; it places fruiting and renewal zones high in the canopy, thus optimizing sunlight penetration and fruit yield and quality; and it may be less susceptible to frost injury and deer damage. The vertical shoot position (VSP) system is the most commonly used trellis system in the world for European grape cultivars because it is complementary to their natural upright growth habit. In this system, the distance between rows is reduced and overall productivity is improved. It is also easy to manage and mechanize. The Geneva double curtain (GDC) is essentially a double trellis system with four fruiting planes (outer and inner on each trellis wire). It supports approximately 30 percent more vine growth than other systems, and it is commonly used for American grape cultivars such as 'Concord' with high vegetative vigor and high potential fruit yields per vine. This system requires wider between- and within-row spacing, with curtains typically 4 ft apart. Fruit quality is enhanced and susceptibility to disease is reduced using this system. In addition to these systems, two head-training systems are used for certain grape cultivars in the western United States. The head-trained, spur-pruned system is used for some table grapes and large-clustered wine grapes. No trellis is required for this system after the trunk is well established. The head trained, cane pruned system is frequently used for 'Thompson Seedless' grapes because buds located near the base of the canes are typically not very fruitful for this cultivar.

Training and Pruning Cane Fruits

With few exceptions (e.g., erect blackberries, some eastern-grown raspberries), all cane fruit crops are trained to one of various trellising systems. Foremost, the trellis provides support for fruit-laden floricanes

canes arising from near the trunk head are retained, each cane holding from ten to twenty potentially fruitful buds. The number of buds that should be retained for the current fruiting season depends directly on the vigor of last year's vegetative growth, leaf cover, and

and protects all canes from physical damage due to wind, snowloads, and other mechanical stresses. From a physiological standpoint, trellising allows for greater light and air penetration into the canopy, which increases net photosynthesis and flowering and limits disease infection periods and insect population growth. As stated earlier, several trellis designs allow for the separation of primocanes from floricanes, resulting in better primocane growth and ease of harvest. Other horticultural considerations surrounding cane management via trellising include ease of cropping operations, such as spraying and mowing; pruning; and control of row width. Most cane fruit cultivars will tip layer (asexually reproduce) if the cane apex is allowed to be in contact with moist soil.

Trellising systems range from the simple to the complex in design. Perhaps the simplest construction is the I-trellis, composed of periodically placed posts that hold a single horizontal monofilament or metal wire placed approximately 1 to 1.5 m above the soil surface (Fig. 20–24). The I-trellis provides basic support and protects against mechanical damage, but it does little to improve the plant's physiological status or to ease horticultural manipulation of the crop. Trellis system designs such as the V-trellis include two supports that are horizontally separated by approximately 1 m. These trellises allow more light to penetrate through the canopy, allow for the separation in space of primocanes and floricanes, simplify horticultural management and harvesting, and often improve yields. The V-trellis system is versatile because it provides several options for the positioning of floricanes and primocanes. Both the I- and V-trellises can be constructed using T-posts if desired. More complex trellis designs have also been examined for cane fruit production under specific circumstances, with some degree of success. For example, scientists at the US Department of Agriculture have engineered a mechanical harvester specifically for eastern-grown thornless blackberry cultivars and concurrently developed a trellis system that is specifically suited to optimize the harvester's performance. The size of the trellis and the strength and extent of trellising materials used depends on the vigor of the crop. Typically, blackberries and blackberry–raspberry hybrids (e.g., boysenberry, loganberry, etc.) require a more substantial trellis than do raspberries. The trellis design and construction should consider other facets of the growing system. For instance, the fruit-laden canes of primocane-fruiting raspberries are often supported by a temporary trellis constructed of T-posts and twine, which can be removed after harvest to facilitate pruning.

Proper training and pruning procedures for cane fruits depend on the growing region, growing system, crop, and cultivars used. Training and pruning are perhaps least complicated for primocane-fruiting cultivars. Most growers of primocane-fruiting berries simply mow off the previous season's growth during midwinter, when plants are deeply dormant. Using this approach, they sacrifice whatever percentage of the crop would have been borne on the floricanes (usually of inferior quality anyway) for ease of management, substantially reduced production costs (i.e., no hand-pruning labor costs), winter damage avoidance, stronger primocane growth in the following season, and sustained productivity for the life of the planting. Old canes must be pruned as closely to the ground as possible.

Floricane-fruiting red raspberry primocanes are not generally trained per se, but they can be protected from wind and mechanical damage by the training of floricanes to various two-wire trellises (e.g., the V-trellis described above). Summer pruning of red raspberry primocanes is generally not necessary, except when a cultivar's adaptation response results in excessive, unwieldy vegetative vigor. However, growers may mechanically or chemically remove many of the early-emerging primocanes to limit competition for carbohydrates between vegetative meristems and developing fruit. This practice is called primocane suppression, and it is usually practiced in the Pacific Northwest region, where the growing seasons are long, the climatic and soil conditions are generally ideal for cane fruit growth, and midseason primocanes can develop fully prior to dormancy.

Spent floricanes are typically removed sometime after harvest. The timing of their removal is situation-dependent. Under conditions that favor heavy fungal disease or insect pressure, or if the crop is being grown

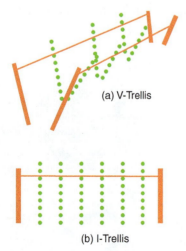

Figure 20–24
The I- and V-trellis systems.
Source: Joseph Scheerens, The Ohio State University.

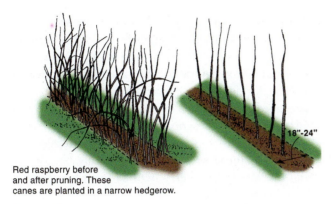

Red raspberry before and after pruning. These canes are planted in a narrow hedgerow.

Figure 20–25

Dormant pruning the red raspberry hedgerow.

Source: *Brambles—Production Management, and Marketing,* Bulletin 782, Ohio State University Extension.

organically, the spent floricanes should be removed as soon as possible after harvest. If left in place until dormant pruning practices are initiated, however, harvested floricanes may contribute to the carbohydrate reserves of the plant and continue to protect primocanes from damage.

Most red raspberry pruning is accomplished in midwinter, while plants are dormant. If still present, spent floricanes are removed first. Then winter-damaged canes or cane portions are removed. Next, the canes that will bear next year's crop are chosen. Ideally, only the most vigorous (thick) and healthy canes are kept, whereas weak and spindly canes are removed. Thinning the number of next year's floricanes is essential; if too many canes are left, the plant may become stressed, resulting in the production of small, poor-quality fruit. If red raspberries are grown in a hedgerow, as is typical in the eastern United States, about three to five canes per foot should be kept (Fig. 20–25). In the western United States, where plants are cultured in hills or stools, about five to eight canes per crown are kept. The chosen canes are then headed back and trained

(attached) to a trellis wire simply in a linear fashion or using more complex cane arrangements (see Figs. 20–24 and 20–26).

Steps in the training and pruning of black raspberries, blackberries, and blackberry-raspberry hybrids are similar to those employed for red raspberries, with one important exception. In these crops, primocanes are summer tipped, or headed back via the removal of 7–10 cm of growth. Tipping releases the axillary buds from apical dominance and thus stimulates lateral branching. Most fruit buds for the following season will develop on these lateral branches. Black raspberry primocanes are usually tipped when they reach 0.5–0.6 m in length, whereas blackberry primocanes are taller when tipped, approximately 0.9–1.2 m. During the dormant season, lateral branches are trimmed back, and canes are chosen for next year's production (Fig. 20–27).

Training and Pruning Bush Fruits

Most bush fruits are not trained per se but are grown as freestanding bushes. However, their size and shape are often controlled by judicious pruning. As with other fruit crops, bush-fruit pruning is practiced to maintain vigor and productivity, improve growth habit, distribute fruiting wood throughout the plant, increase sunlight to the plant's interior for improved fruit size and quality, and increase air circulation for decreased incidence of disease. Plants are pruned in winter or early spring while dormant. The amount of wood removed depends on the species, the cultivar, plant vigor, and age. Pruning of mature bushes involves the removal of weak, low-growing, or twisted canes or branches; winter-injured tissue; canes that rub together; older canes that are less productive; and branches that receive little light. Pruning cuts should be made as close to the crown or to a main branch as possible to prevent the entrance of disease organisms.

Figure 20–26

(A) Weaving and (B) tepee arrangements of Washington-grown red raspberry floricanes as they are attached to the trellis. Source: Joseph Scheerens, The Ohio State University.

A

B

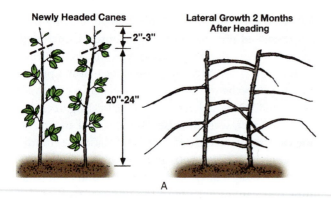

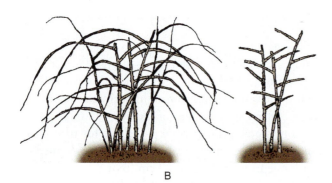

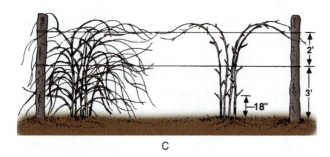

Figure 20–27

Black raspberry and blackberry pruning: (A) summer tipping of black raspberries, (B) dormant pruning of black raspberries, and (C) dormant pruning of thornless semi-erect blackberries.

Source: *Brambles—Production Management, and Marketing*, Bulletin 782, Ohio State University Extension.

FRUIT THINNING

Many temperate tree fruit crops, especially pome and stone fruits, produce an overabundance of flowers each year, perhaps as an evolutionary consequence of the potential for flower and flower bud losses during spring frosts. If pollinators are active, most of these flowers are pollinated and many are fertilized. After fertilization, tissues surrounding the embryos (which will eventually become the fruit) begin to enlarge in response to hormonal signals from the developing seed. Obviously, the growth and development of both seed and fruit tissues relies on an adequate supply of energy. However, carbohydrates are typically in limited supply, and this supply is not sufficient to grow and mature all the fruit that was initially set. In response to this stress, trees typically experience a period commencing a few weeks after bloom wherein many of the newly developing fruits abscise. This abscission period is called June drop and it can last as long as one month. June drop can be considered a natural thinning process whereby the tree adjusts its crop load to enhance seed and fruit development and thus ensure reproductive success. In some years, when environmental conditions are unfavorable, too many of the fruits drop, and yields are reduced. In some cases, however, the trees retain more fruit than is acceptable from a horticultural standpoint. In the latter case, fruit-thinning procedures are advocated.

Chemical or mechanical fruit-thinning procedures for apples, apricots, grapes, nectarines, peaches, pears, and plums are routinely practiced. The horticultural goal of all fruit-thinning procedures is to adjust crop load so that it is balanced between overall yield and fruit size and quality (Fig. 20–28). Fruit size is an extremely important commercial issue, with large fruit usually being preferred by consumers. In almost all fruit crops, the size of the fruit harvested is negatively correlated to the number of fruits developed on the plant.

Final fruit size more or less depends on the leaf–fruit ratio on a branch. The more leaves per fruit, the larger the final fruit size. Fruit size, nevertheless, is increased at the expense of total yield—but thinning can be highly profitable where a premium is paid for larger-size fruits.

Fruit thinning is the best way to increase fruit size, although flower and bud thinning (by pruning out fruiting shoots) is also effective. Overirrigation or excessive nitrogen fertilization will not increase fruit size; in fact, these approaches can stimulate new shoot growth, which competes with fruits for carbohydrates.

Ripening fruit undergoes a series of chemical and physical changes (e.g., increased pigmentation, increased sugar levels, the development of volatile flavor components, etc.) that determine the eating or processing quality of the product. Typically, if the level of carbohydrate supplied by vegetative plant tissues is inadequate for the number of ripening fruits present, the quality of those fruits will be diminished.

As with training and pruning, maintaining a balance between vegetative and reproductive growth

Figure 20–28

Left: *Plum fruits on a branch that was properly thinned. Proper spacing has allowed them to develop properly for fruit size, quality, and early maturity.* Right: *Unthinned fruits are small, late maturing, and of poor quality. In addition, such limbs often break from the weight of the fruit.*

through fruit thinning is an important physiological goal. In addition to its effects on fruit size and quality, removal of excess fruits by thinning results in a greater allocation of carbohydrates for the root and shoot growth that is important for sustaining the productive life of the orchard. Also, if fruit loads are not thinned effectively, floral bud initiation for next year's crop will

likely be decreased, inducing the alternate bearing pattern discussed previously in this chapter. Controlled fruit thinning to prevent excessive fruit numbers during any one year is one of the most effective cultural procedures for an orchardist or a home gardener in overcoming alternate bearing. Even if alternate bearing does not occur, however, several consecutive years of overcropping can greatly weaken trees or vines. Last, the inordinate weight associated with an overabundance of fruits may structurally damage or break shoots and even large limbs that cannot withstand the physical stresses the fruits impose.

Thinning Requirements

It may be obvious that a fruit tree is overloaded with young fruits and needs thinning—but how much? It is difficult to judge because several factors are involved, such as the cultivar, time of fruit maturity, availability of water, general vigor and age of the tree, as well as growing conditions. Some general guidelines for fruit thinning can be given, however. Peaches are often thinned to about one fruit every 15 to 20 cm (6 or 8 in.) of shoot. Another rule is twenty-five to forty leaves per fruit for apples and about fifty leaves per fruit for peaches.

Thinning to control yields and fruit size is so important in crops such as peaches and apricots that thinning tables have been developed to determine the optimum number of fruits to leave on the tree. These tables consider the distances between trees, fruit size desired (number per pound), and number of tons of fruit desired per acre. For example, if a 22.4 MT/hectare (10 t/acre) harvest is desired for apricots and the trees are set on a 7.2 × 7.2 m (24 × 24 ft) planting distance, no more than about 3,200 fruits should be left per tree to attain a fruit size of at least 200 g (0.4 lb). Fruit counts are made on sample trees throughout the orchard to determine if the fruit set is uniform.

REFERENCE DATE

In the stone fruits, fruits that are small early in the season are also small at harvest time. Commercial growers of clingstone peaches and apricots have made use of this relationship to predict much earlier in the season average fruit size at harvest. The young, developing fruits are measured at a specific time, called a reference date. For apricots, this time is determined by cutting through the ends of the young fruits at intervals. The reference date is seven days after the pit begins to harden, when representative fruit samples are measured for size. The reference date for clingstone peaches is the same except that ten days are added to the date that the tip of the pit begins to darken. For clingstone peaches an average suture diameter of 34 to 35 mm (1.36 to 1.40 in) at the reference date is needed to produce an average diameter at harvest of 67 to 68 mm (2.68 to 2.72 in). This is sufficient to make 90 percent of the fruits larger than the minimum harvest size. If the young fruits at reference date are too small, the fruits are thinned to bring the remaining fruits to a satisfactory size by harvest time.

Chemical Thinning

The idea of using chemical sprays applied to trees to remove some of the fruits is definitely appealing, and much research has been given to develop thinning sprays to replace hand thinning. For apples, several natural or synthetic, hormone-based thinning sprays are available and are used commercially. However, the two most effective and widely used thinning chemicals are naphthaleneacetic acid (NAA) and carbaryl (1-naphthyl N-methyl carbamate.)

It is often difficult to obtain consistent results with chemical-thinning sprays, and general recommendations applicable to different situations cannot be made. First, there is often a cultivar-specific response to chemical-thinning agents. For instance, NAA can be a successful thinning agent for late-maturing apples, but its use on summer cultivars results in foliage injury, premature ripening, and fruit cracking. The use of NAA on 'Fuji' and spur-type 'Red Delicious' often results in the formation of pygmy (small, seedless) apples. Carbaryl is a preferred thinning agent for the latter cultivar, but its use with 'Gallia Beauty' and 'Rome' can result in overthinning. 'Honeycrisp' may be overthinned if thinning agents are applied in combination. Environmental conditions also affect the efficacy (potency) of spray compounds. In general, chemical-thinning agents remove more developing fruit if (1) the trees are not vigorous or have been only lightly pruned; (2) the bloom is particularly heavy or the flowers have not been well-pollinated; (3) the humidity is high and the spray's drying time is slow; and (4) the weather is cloudy and cool before or after bloom. There are exceptions to these generalities, however; for example, NAA is a more efficient thinner if it dries quickly. Adding a surfactant (wetting agent) to the tank mix increases foliar absorption rates and minimizes the effects of adverse climatic conditions.

With any chemical-thinning applications, proper timing is critical. Typically chemical-thinning treatments are applied at petal-fall or postbloom (ten to thirty days after full bloom). The critical periods for thinning are both cultivar- and product-specific. In general, it is recommended that late-maturing cultivars be sprayed when the predominant fruit in each cluster is approximately 11 to 13 mm in diameter and that early maturing cultivars be treated at petal-fall With any chemical-thinning applications, proper timing is critical.

Mechanical Thinning

Many other fruit crops are mechanically thinned. Hand thinning is the most precise method of thinning because the decision to remove a fruit is made on a fruit-by-fruit basis. However, it is laborious, time-consuming, and expensive and perhaps is a suitable method for the homeowner with only a few trees to thin.

As an improvement over slow and expensive hand thinning, long poles with sections of rubber hose at the end can be used to hit fruiting branches and knock off some of the fruit. Hand thinning to remove missed fruiting clusters may follow. Mechanical tree shakers, used in harvesting operations, are sometimes used to shake the trees to remove a portion of the small, immature fruits. Shaking can be satisfactory, but it is not very precise and the desirable larger fruits tend to be removed. A light machine shaking, followed by hand thinning or pole knocking, works well in some cases. Hand thinning gives the best results, however, and should be used whenever practicable.

In addition to tree fruits, some vine crops are also thinned. To obtain better yields of high-quality grape berries, one extra bud on half or more of the fruiting spurs (spur pruning) and an extra cane or two per vine (cane pruning) can be retained. Thinning of grapes can be done as flower thinning—removal of part of the fruiting clusters before bloom—or cluster thinning—removal of clusters after the berries have set. Clusters or fruits should be thinned to reduce the crop to the amount it would have been had the extra fruiting canes not been left. The increased leaf area produced by retaining the extra spurs or canes to support the same number of fruits per vine raises the yield and the quality of the berries.

Because of the production of fruit-bearing shoots from latent or noncount buds, some cultivars of the French–American hybrids (e.g., Seyval blanc, Vidal blanc, and Chambourcin) need annual cluster thinning. This cluster thinning is necessary to maintain sufficient vine growth for subsequent production and to maintain grape quality.

Mature sweet and sour cherries, prunes, almonds, walnuts, pistachio and hazelnuts, cane fruits, bush fruits, and strawberries ordinarily do not require fruit thinning. During plant establishment and juvenility, however, some species benefit greatly from hand removal of flower buds or flowers to conserve carbohydrates for root and shoot development. For instance, it is recommended that flower buds and flowers be hand stripped from newly planted blueberries for the first two years of the planting. Strawberries grown in a matted-row system also benefit from flower removal in the planting year to proliferate runners (stolons) that will be used to establish the solid bed.

FRUIT DEVELOPMENT TERMINOLOGY

Maturation The final stage of development while fruits are still attached to the plant. It includes cell enlargement plus the accumulation of carbohydrates and other flavor constituents and a decrease in acids. The flesh may or may not soften.

Ripening The changes taking place after full maturation, involving a softening of the flesh, the development of characteristic flavors, and an increase in the juice content. Ripening may occur either before or after the fruit has been picked.

Senescence The period following ripening, during which growth ceases and aging processes replace ripening processes. Senescence may occur before or after harvest.

FRUIT MATURATION, RIPENING, AND SENESCENCE

As fruits near their maximum size, significant changes lead to the end product—a fruit with an attractive color, soft enough to be palatable, sweet and juicy, with an accumulation of the other components that give it its own distinctive flavor and aroma. The increase in sugars—or soluble solids—can be measured by a refractometer or hydrometer. Decreasing hardness can be measured by various kinds of pressure testers. Changing color, as the masking chlorophyll disappears, can be measured by color charts or color meters. Changing acids can be determined by chemical means or pH meters. All these procedures are used to determine quantitatively when the fruit is ready for harvest. In some states in the United States, legal maturity standards demand such measurements to determine when fruit can be harvested.

Optimum harvest maturity depends on the intended use of the fruit. For example, strawberries harvested for processing or local markets are picked at or near peak ripeness to maximize their sugar–acid ratio, volatile flavor constituents, and overall eating quality. However, ripe or nearly ripe berries are far too soft for long-distance shipping and prolonged shelf life. Commercial strawberries grown for the larger marketplace are typically picked when they are full size but still very firm and only partially colored.

The home gardener has an advantage over the commercial grower because he or she can allow the fruit to reach optimum maturity right on the tree for eating at the optimum degree of ripeness. For commercial production, however, perhaps involving shipments in containers to great distances, the fruits generally must be picked firm enough to endure shipping stress and arrive at their destination in an acceptable condition.

All fruits on a tree do not mature at the same time. Fruits on the top and outside are usually ready to pick before those on the inside. For the home gardener, this is an advantage because it prolongs the harvest period, but for commercial once-over harvest, some of the fruits are harvested at a less-than-ideal time.

The various kinds of fruits have different ripening patterns and require different storage procedures. An understanding of these is essential to know how to harvest and handle the fruits properly. Most fruits reach maturation and ripen (become softer and sweeter to be more edible) on the tree, vine, or bush, at which time they should be picked. If not picked, they senesce and deteriorate. However, other fruits, such as pears, are best harvested when they reach maturity on the tree then are ripened to eating condition off the tree.

It must be emphasized that maturing and ripening fruits are living organisms. Their cells are respiring, and many other complex chemical and physical changes are occurring. Some types of fruits are said to be climacteric and others are nonclimacteric (Fig. 20–29). In both types, the respiration rate of the cells slowly decreases as maturation proceeds. However, in climacteric fruits, the respiration rate rises abruptly as the fruit ripens, reaches a peak (the climacteric point), and then declines. Senescence ensues, leading eventually to the death of all the fruit cells. In nonclimacteric fruits, the respiration rate gradually declines during ripening, with no particular peak.

Apples, pears, peaches, apricots, plums, and blackberries are examples of climacteric fruits. Most climacteric fruits show the same respiration pattern whether ripening on or off the tree. Cherries, strawberries, grapes, and raspberries are examples of nonclimacteric fruits.

Refrigeration prolongs fruit life chiefly because the lowered temperatures reduce respiration rates.

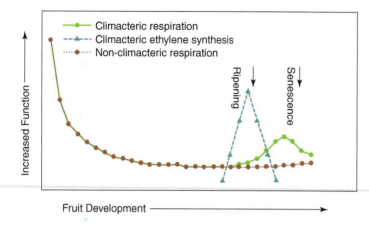

Figure 20–29
Schematic representation of the physiological differences between climacteric and nonclimacteric fruit. Source: Joseph Scheerens, The Ohio State University.

Controlled atmosphere (CA) storage can further lower respiration rates and greatly extend the storage life of fruits. The levels of O_2 and CO_2 that can be tolerated vary with species and cultivar. CA storage procedures have come into widespread use in recent years, especially for apples and to a lesser extent for pears. Although harvested mainly in autumn, apples are now available in markets year-round because of CA storage facilities.

Ethylene (C_2H_4), which is produced by all plant tissues, including fruits, can stimulate fruit ripening. An ancient Chinese custom was to ripen fruit in rooms where incense was burned and, in the early twentieth century, kerosene stoves were widely used in California to stimulate color development in lemons. Later it was discovered that ethylene is released by such combustion and was the cause of the ripening. Ripe fruits, such as apples, give off ethylene and stimulate the climacteric ripening of other fruits confined in the same containers with them. Ethylene is much involved in fruit ripening and is considered a ripening hormone. Now, ethylene is used to stimulate the ripening of most climacteric and nonclimacteric fruits. Strawberries, however, are not affected by ethylene. Gibberellin, a natural plant hormone, is known to counteract ethylene effects and delay fruit ripening.

The way fruits ripen varies greatly among species. The following examples will show the variability among fruit species and will point out the necessity for understanding these patterns for proper harvesting and fruit ripening.

Table grapes mature and ripen on the vine; the peak of fruit quality is reached and, ideally, the clusters are picked and consumed at that time. Once the clusters are removed from the vine, deterioration sets in. This can be slowed by placing the clusters in cold storage— 0°C (32°F) at 95 percent relative humidity. Such stored grapes can remain edible from one to six months, depending on the cultivar. In addition, the European (but not American) grapes require sulfur dioxide fumigation treatments to adequately prevent decay.

Apples also develop optimum maturity and ripeness while on the tree and senescence begins with harvest. This can be delayed by cooling the fruits to 0°C (32°F) within two to three days after harvest and holding them in storage at a relative humidity of 90 to 95 percent. By retarding the respiration rate, CA storage enables fruits of some apple cultivars to remain in good eating condition for as long as nine months.

For ideal eating characteristics, pears should be harvested before they ripen and then ripened in a cool place (15.5°C to 22°C; 60°F to 72°F) at 80 to 85 percent relative humidity. Or they can be placed directly into cold storage (−0.5°C; 31°F) at 90 percent relative humidity, where they will keep for several months. They can be taken from cold storage to higher temperatures at any time for ripening. Some pear cultivars, like 'd'Anjou,' do not ripen properly unless they are held in cold storage for a time or they are treated with ethylene gas.

When harvested for commercial shipments, plums are too firm for eating. They continue softening and ripening after harvest to an optimum point for eating. Soon after this point is reached, deterioration commences. Plums are often stored at 0°C (32°F) just after picking to prolong the life of the fruits for two to four weeks.

Fruits of many species, if stored below certain temperatures but above the freezing point, will develop typical damage symptoms. Such as internal browning, pitting, scald, dull skin color, a soggy breakdown, susceptibility to decay, and failure to ripen properly. Such fruits become worthless. Although influenced by many factors, optimum storage conditions are given in Table 20–5 for most fruit species.

Table 20–5

RECOMMENDED STORAGE TEMPERATURES AND RELATIVE HUMIDITY, APPROXIMATE STORAGE LIFE, HIGHEST FREEZING POINT, AND WATER CONTENT OF FRESH FRUITS IN COMMERCIAL STORAGE

Commodity	Temperature		Relative Humidity (Percent)	Approximate Storage Life	Highest Freezing Point		Water Content (Percent)
	°C	°F			°C	°F	
Apples	−1–4	30–40	90–95	1–12 months	−1.5	29.3	84.1
Apricots	−0.5–0	31–32	90–95	1–3 weeks	−1.0	30.1	85.4
Berries;							
Blackberries	−0.5–0	31–32	90–95	2–3 days	−0.7	30.5	84.8
Blueberries	−0.5–0	31–32	90–95	2 weeks	−1.2	29.7	83.2
Cranberries	2–4	36–40	90–95	2–4 months	−0.8	30.4	87.4
Currants	−0.5–0	31–32	90–95	1–4 weeks	−1.0	30.2	84.7
Dewberries	−0.5–0	31–32	90–95	2–3 days	−1.2	29.7	84.5
Elderberries	−0.5–0	31–32	90–95	1–2 weeks	—	—	79.8
Gooseberries	−0.5–0	31–32	90–95	3–4 weeks	−1.0	30.0	88.9
Loganberries	−0.5–0	31–32	90–95	2–3 days	−1.2	29.7	83.0
Raspberries	−0.5–0	31–32	90–95	2–3 days	−1.0	30.0	82.5
Strawberries	0	32	90–95	5–7 days	−0.7	30.6	89.9
Cherries, sour	0	32	90–95	3–7 days	−1.7	29.0	83.7
Cherries, sweet	−1 to −0.5	30–31	90–95	2–3 weeks	−1.8	28.8	80.4
Grapes, vinifera	−1 to −0.5	30–31	90–95	1–6 months	−2.1	28.1	81.6
Grapes, American	−0.5–0	31–32	85	2–8 weeks	−1.2	29.7	81.9
Nectarines	−0.5–0	31–32	90–95	2–4 weeks	−0.9	30.4	81.8
Peaches	−0.5–0	31–32	90–95	2–4 weeks	−0.9	30.3	89.1
Pears	−1.5 to −0.5	29–31	90–95	2–7 months	−1.5	29.2	83.2
Plums and prunes	−0.5–0	31–32	90–95	2–5 weeks	−0.8	30.5	86.6

Source: R. E. Hardenburg, A. E. Watada, and C.Y. Wang, *The Commercial Storage of Fruits, Vegetables, and Florist and Nursery Stock,* USDA Agricultural Handbook No. 66 (1986).

SUMMARY AND REVIEW

The successful production of temperate fruits and nuts requires several considerations, including site selection, marketing and cost analysis, and developing an effective cultural management strategy.

Climate and soil characteristics are factors to consider when selecting a site. Climate determines if temperatures, wind, and precipitation are appropriate for proper growth of the crop. If a crop has a specific temperature requirement, such as chilling, the site must provide that without exposing the plant to deleterious temperatures. Wind can affect pollination, pesticide applications, and irrigation, and can cause damage to young trees and fruit. Most fruit and nut crops require a deep, fertile soil that drains excess water quickly yet retains enough water to supply the crop. Disease organisms that may be present in the soil from previous vegetation is an important consideration. Precipitation should provide adequate water throughout the growing season if irrigation is not feasible. Precipitation that

regularly interferes with proper growth, for example, hail or continual rainfall during pollination, presents a serious problem for fruit and nut production.

The marketability of a crop is influenced by global, regional, and local production capacity and the competition for customers. These factors have to be studied and analyzed before producing a crop at the site under consideration. A crop production cost analysis includes estimating the costs of labor, capital investment, overhead costs, and production or direct operating costs. Many of these costs are incurred before the crop is ready to harvest. In some crops, especially herbaceous or semiwoody plants, there is a relatively short time between investment and harvest. For the tree fruits and nuts, however, several years can elapse between investment and harvest.

Effective cultural management begins with the selection of proper cultivars and rootstocks. They must thrive in the climate conditions and soil of the area.

Understanding the growth, flowering, and fruiting physiology of the plants is critical to implementing further cultural management practices. The life cycle (annual, biennial, perennial) of the crop determines many of the cultural practices that are implemented. Pruning practices rely on the pruner understanding when and where the plant develops fruit. A grower has to know if the plants are self-fruitful. If pollenizers are required, they must be planted appropriately.

Collectively, the cultural practices implemented for a fruit or nut crop are called the growing systems. There are major categories of growing systems that a grower can choose from. These are often developed for growing crops in specific regions, or for specific crops or specific purposes. For example, Midwestern strawberries are produced as perennials and in Florida, as annuals. A Geneva double-curtain bilateral cordon may be best for American and French–American hybrid grapes, while French vinifera wine grapes are often grown most successfully with the vertical shoot position system. Organic growing systems are gaining popularity in many fruit and nut crops.

The appropriate time for harvest varies greatly on the species and even cultivar being grown and depends on the stage of development of the fruits and whether or not the fruit will proceed with ripening after it is harvested. Carbohydrate content, firmness, respiration patterns, and use of the fruit all influence the stage at which a fruit is ready to be harvested.

KNOWLEDGE CHECK

1. Why is the length and timing of the growing season important in fruit and nut production?
2. Why is it not recommended to plant orchards in low spots?
3. For heat intolerant fruits such as cherry, would you plant the orchard on the western or eastern slope of a hillside?
4. Why is accumulated snow often an advantage for dormant plants growing in very cold areas?
5. What soil pH is preferred by ericaceous (from the family ERICACEAE) fruits such as blueberry and cranberry?
6. When is the availability of water generally most critical for fruit production?
7. What are two different types of competition a fruit grower could encounter in the marketplace?
8. Why do the preproduction costs extend over a much longer time for crops like apples, cherries, and nuts compared to crops like strawberries?
9. Give two examples of capital investment factors.
10. Give two examples of production/direct operating costs.
11. What is usually the greatest expense in fruit growing operations?
12. What are three factors that influence the transition from vegetative to reproductive status in a plant?
13. What will likely happen to fruit production the following spring if trees are fertilized heavily with nitrogen the before winter?
14. What is meant by "rest" period in regard to deciduous fruit plants? What is the advantage to plants to have a rest period?
15. What happens if a deciduous fruit does not receive enough chilling during the dormant period?
16. What is the difference in fruiting times for an everbearing strawberry during its establishment year and subsequent years?
17. What is a fruiting spur, bourse shoot, primocane, floricane, and crown?
18. Why are at least two cultivars needed in almond and sweet cherry orchards?
19. What is the main cultural difference between annual and perennial strawberry production?
20. What are the drawbacks to producing cane fruits in a two-year cropping cycle?
21. Give an advantage and disadvantage each for: high-, medium-, and low-density orchard plantings
22. Which one of the above three planting systems requires the greatest understanding of fruit physiology?
23. Why are fruit crops better suited for organic fertilizing methods than some other crops?
24. What makes highbush and rabbiteye blueberries particularly well suited for organic production?
25. Give three characteristics that grafted rootstocks can influence in fruit trees.
26. What is alternate bearing in fruit trees? What happens to fruit quality and numbers if this pattern gets established?
27. Why can highbush blueberries now be grown in the southern United States?
28. What root problem can be common on clay soils? On sandy soils?
29. What is winter hardiness?

30. What factors does a producer use to judge fruit quality? What does a consumer value in fruit quality?
31. Why is it recommended to purchase certified virus- or disease-free nursery stock rather than from existing plantings?
32. Give two factors to consider when deciding what row spacing to use.
33. What can happen if a fruit tree is planted too deep?
34. Why is wet peat moss often placed into the planting holes of blueberries?
35. How does the difference between the soil of a fruit tree rootball and the soil around the hole where it is being planted affect irrigation?
36. When is the best time to prune most dormant fruit trees and canes?
37. Give two reasons for pruning fruit trees.
38. Why is annual light pruning preferable to a heavy pruning every few years?
39. Give two reasons why grapevines are trellised. Give two reasons cane fruits are trellised.
40. What benefit is there to fruit thinning?
41. Why is it said that optimal harvest maturity depends on the use of the fruit?
42. How is ethylene used in fruit ripening? What is a fruit that is not affected by ethylene?

FOOD FOR THOUGHT

1. Peaches and apples both have a rest requirement but peaches come out of rest much sooner than apples. Which crop would be more suitable for an area where there are often frosts following late winter warm-ups?
2. Precocious flowering and fruiting can occur in some fruits and nuts when the plants are stressed. How do you think precocious fruiting in late summer would affect the winter hardiness of plants that normally flower and fruit in the spring and early summer?
3. Why would it be wrong to think that since apples and pears are both climacteric pome fruit (the edible part is a fused hypanthium, which is part of the flower), they should be harvested at the same stage?

FURTHER EXPLORATION

GALLETA. G. J., AND D. G. HIMEBRCK. 1994. *Small fruit crop management.* Englewood Cliffs, NJ: Prentice Hall.

JACKSON, DAVID I., AND N. E. LOONEY (eds.). 2009. *Temperate and subtropical fruit production,* 3rd ed. Cambridge, MA: CABI Publishing.

LYLE, S. 2006. *Fruit and nuts: A comprehensive guide to the cultivation, uses and health benefits of over 300 food-producing plants.* Portland, OR: Timber Press.

*21

Tropical and Subtropical Crops and Crop Production Systems

Richard Pratt

We often imagine the tropics as lush and full of exotic plants and animals—the tropical paradise evoked by images in film and advertising. In fact, the tropics and subtropics (the region of the earth lying within 30 degrees of the equator—between the Tropic of Cancer and the Tropic of Capricorn) comprise a broad expanse of climatic regions characterized by rich cultural and biological diversity. The majority of the world's population lives in the tropics, and some form of plant agriculture is a primary occupation.

The primary factors determining the feasibility of crop production are land forms, soil conditions, and climate. Climate exerts profound control over the distribution of crops and the manner in which they are produced. Some tropical environments are well suited to crop production, but others present harsh challenges to the producer. In many tropical countries, resource-poor farmers must also cope with poor soils. Crop production is feasible when adapted crops have been selected and/or the limiting factors in the environment can be modified. Many strategies are employed; for example, organic amendments or fertilizers are used to correct soil nutrient deficiencies, multiple crops are grown in the same field, and planting dates and harvest seasons are adjusted to the length of the rainy season.

Tropical plant agriculture assumes countless forms—from complex subsistence-agro ecosytems to large monocultures geared toward global markets. It would be impossible to cover the rich spectrum of crops (Figs. 21–1 and 21–2) and crop systems found in the tropics in one chapter. This chapter will introduce you to the principal tropical crops and to several complex polycropping systems that are of particular importance in the tropical and subtropical regions. This chapter should expand your understanding of how human ingenuity and crop diversity come together in tropical cropping systems.

CLIMATES OF THE TROPICS AND SUBTROPICS

Temperature and rainfall are the most important determinants of suitable climate for tropical crop production. The duration of arid seasons is also an important factor. The humid tropics are typically classified as

key learning concepts

After reading this chapter, you should be able to:

- Discuss the principal components of the tropical environment that differ from those of temperate environments, and how they influence the manner in which crop production is undertaken in the tropics and subtropics.

- Explain the rationale for the adoption of diverse techniques and the manner in which these techniques are employed by small-landholder farmers to optimize crop production systems in diverse tropical environments.

- Identify leading tropical fruit and nut species produced on both a large scale for export as well as in small settings near the home.

Figure 21–1
Fruit basket showing an array of tropical and subtropical crops. Source: Richard Pratt.

Figure 21–2
Market day in Bayaguana, Dominican Republic, with yams (front), sweet potatoes (rear left), and plantains (rear right). Source: Richard Pratt.

those with dry periods less than 2.5 months per year. Wet equatorial climates characterized by deep, moist equatorial air masses, frequent rainfall, constant heat, no marked seasons, and rainforests are found 5° north to 5° south of the equator. Examples include parts of Brazil, Surinam, Guyana, and Venezuela; the Congo River basin; parts of Malaysia, Sumatra, Java, Borneo, the Celebes, New Guinea, and many Pacific islands. In the intermediate tropics, wet–dry seasons (dry 2.5 to 5 months) and tropical savannah (dry 5 to 7.5 months) are characteristic. In the dry tropics, such as the Sahelian region and semideserts, the dry seasons may extend 7.5 to 10 months. True tropical deserts are described by dry periods of ten to twelve months. Tropical desert areas

may be found between 15° and 30° latitude north and south. Examples include North Africa, the Arabian Peninsula, Iran, northwest India, Australia, and South America. Alternately wet and dry **monsoon** climates occur between 5° to 15° north and south latitude. Monsoon climates are wet and dry about six months of each year, depending on prevailing winds.

Areas can have good rainfall (100 to 200 cm/year) distributed in two rainy seasons with only short dry seasons between them (these regions tend to be nearer to the equator, with no large seasonal fluctuations in temperature) or with two shorter rainy seasons and pronounced intervening dry seasons. There may also be considerable variation in annual rainfall (within 60 cm to 125 cm/year) distributed as one fairly long rainy season (70 cm to 125 cm/year) and one long dry season, or one short rainy season and a long dry season.

Rainfall and Humidity

One of the most difficult things tropical farmers have to contend with is the variation in rainfall. The rainfall regime frequently has the largest effect on which crops may be grown in particular areas, the farming systems to be employed, and the sequence and timing of farming operations. Most crop systems in the tropics are closely matched to the local rainy seasons. Shifts in rainfall patterns in the tropics could be devastating.

Rainfall generally decreases as distance from the equator increases. This variation is primarily an effect of temperature. The amount of water in the air (water vapor pressure) is highest at the equator. It decreases as the latitude increases and also falls with increasing elevation. High humidity reduces the evapotranspiration (water loss) demand on crops. High humidity over long periods, combined with high temperatures, also have a detrimental effect in that these conditions also favor rapid development and spread of fungal pathogens on crops and in stored grain and produce.

Average annual rainfall received in tropical environments can vary widely. Land forms have a distinct effect on rainfall through the effect of altitude and the effect of mountains serving as obstacles to movement of air masses whether moving to (with moisture) or from (without moisture) land masses. In regions of equatorial or wetter monsoon climates—insufficient rainfall is not often a limiting factor to crop production. In parts of the tropics, a large proportion of the rainfall falls in heavy showers and storms of high intensity—much more so than in temperate zones. These heavy and frequent rains become a disadvantage when they increase the leaching of nutrients, carry away soil and applied fertilizers in surface runoff, or interfere with tillage or

other operations. Rainfall is also important, especially in areas with wet and dry seasons—how well distributed and how reliable it is will be of paramount importance to crop production.

Light and Temperature

The input of solar energy sets the ultimate limit to dry-matter production by crops. Solar radiation is highest in the tropics, so the potential productivity is highest. Unfortunately, higher plant populations frequently do not achieve higher yields because water or nutrients are often limiting.

The surface temperatures of some soils may increase considerably when they are exposed to intensive solar heating called **insolation** (a value expressing the amount of light energy striking one square centimeter of the earth's surface per hour). Soil temperatures may even become hot enough to inhibit or retard the germination and early growth of some crops. The amount of insolation depends primarily on latitude, altitude, season of the year, and a location's proximity to large bodies of land or water. Insolation at the equator varies only slightly from month to month because the angle of the sun's rays does not deviate greatly from vertical with the change in season, as it does at higher latitudes.

In the tropics, the reflection coefficients (loss of light available for photosynthesis due to reflection back to the atmosphere) of crops tend to be lower because the higher elevation of the sun results in light penetrating deeper into the canopy and more being absorbed after reflecting between leaves.

Insolation is less at increased latitudes because of the greater angle at which the radiation waves strike the earth's surface. (There can be some compensation due to longer days at higher latitude.) Because temperature varies with elevation, changes in altitude tend to substitute for changes in latitude. The rate of crop development decreases as altitude increases. For example, in lowland East Java, the time needed for rice to mature (from transplanting to harvest) can range from 90 to 120 days. The same cultivar at 1,600 m (5,250 ft) of altitude could take as long as 220 days because of cooler temperatures at the higher altitude.

Low temperature is rarely a factor in tropical crop production except where it is markedly reduced by altitude. Some tropical crops are susceptible to high temperatures, but this susceptibility is usually associated with drought susceptibility. High temperatures combined with long periods of bright sunshine can also cause scorching of certain fruits, notably pineapples.

At the equator, the day length is almost constant at twelve hours throughout the year. Flowering is especially affected by very short differences in the length of day versus night (photoperiod). (See Chapter 7 for an in depth description of photoperiodic effects on plants.) Night temperature interacts with the photoperiod in influencing the flowering of many plants. Pineapple, for example, is a quantitative short-day plant in which low temperatures, especially cool nights, may enhance the effects of short days.

Temperature regimes in the tropic include equatorial, tropical, tierra templada (frostless highlands), tierra fria (nonfrostless highlands), andine, semitropical, marine, temperate, pampean-patagonian, continental, polar, and alpine. Most of these are the result of changes in altitude rather than latitude. An example of an equatorial climate is Chennai (Madras), India, which is over 100 miles north of the equator, where the temperature during nine or more months exceeds 33.5°C (92°F). At the other extreme, Quito, Ecuador, just 12 miles from the equator at a height of 2,851 m (9,350 ft) and surrounded by snow-capped peaks, has an alpine climate.

Important Agro-Climatic Regions

The term savanna is used to refer to a variety of ecosystems. Tropical savannas include both open grassland with widely spaced shrubs or trees, and more closed woodlands. Rainfall (both amount and distribution), and to some degree the soil water-holding capacity, are the primary determinants of the density of the woody vegetation. Savannas cover many large areas around the world: much of central and southern Africa, western India, northern Australia, large areas of northern Brazil (known as *cerrados*), Columbia and Venezuela (known as *llanos*). Savanna is limited in Malaysia. Sometimes savannas are the result of human degradation of original forest land (e.g. central India).

Savannas typically occur on land with little relief. Soils found in these regions tend to be low in nutrients. This characteristic is partly due to the low value of parent material and also to the process of long weathering. Savanna regions are associated with a warm continental type of climate and are subject to large variation in soil water content. They may become extremely dry, making them prone to fires. The dominant vegetation tends to be fire adapted, and the life cycles of the woody species are usually short (typically not more than several decades). A notable exception is the baobab tree (*Adansonia digitata*).

Tropical rainforests are among the most productive ecosystems in the world and are known for their

diversity of plant and animal life. They are restricted mostly to the equatorial climatic zone between latitudes 10° north and 10° south. The rainforests occupy regions of the world where it is warm and rainfall can be measured by the meter! Annual variation in temperature and rainfall is minimal, but daily variations in heat, rainfall, and humidity may be large. Most tropical rainforests grow below 1,000 m altitude (3,280 ft). There are three main groups of rainforest: Amazonian basin (the largest), Indo-Malaysian area (west coast of India and southeast China through Malaysia and Java to New Guinea), and West Africa (extending to the Congo basin). There are many subtypes of rainforests. The most luxuriant is the multilayered lowland rainforest. At higher elevations, the lowland forest grades into mountain forests. Fingers of rainforest, called gallery forests, may follow river courses into savannas. The tropical rainforest can be divided into five general layers, but stratification is usually poorly defined. The large standing biomass is a store of nutrients protected from leaching as well as repositories for carbon fixed from the atmosphere.

Large areas of the rainforests are now being converted for forestry, ranching, and farmland uses. The relationship between high rainfall and soil erosion is a very important issue when tropical areas are cleared for crop production. The cleared land is exposed and vulnerable to erosion by heavy rains until vegetation can be reestablished. Tropical rainfall is intense due to the larger size of raindrops and the increased number of drops falling per unit time than in other biomes. Therefore, tropical rains have more erosive power than temperate rains. Poor soil infiltration capacity can exacerbate the problem.

Considerable areas of tropical land are being increasingly exposed to erosion and nutrient loss as demands for food production increase. As clearing and pressure on the land escalate in response to growing human and livestock populations, more and more land is denuded of its natural vegetation. Cultivation, overgrazing, and the removal of trees all contribute to the dilemma. Eventually, as soils become increasingly less productive, cultivators are forced onto hill slopes and other marginal lands. The increasing proportion of the land that is left bare and exposed to high-intensity rainfall is prone to erosion unless countermeasures are taken.

Crops may influence soil erosion potential as well—not only roots and surface organic matter but the canopy too. The greatest potential for agriculture in the tropics is in the 1.5 billion hectares (3,705,000,000 acres) where there is nearly constant temperature and a dry period not longer than four months, and a tropical rainforest (or seasonal tropical forest) as native vegetation. The major constraint to realization of the potential productivity in these regions is attributable to the poor-quality soils found there.

TROPICAL SOILS

Most tropical soils differ from those in temperate regions because they are highly weathered—low in organic matter, high in aluminum and manganese, low in cation exchange capacity (CEC), and low in base saturation (the percentage of the CEC that is occupied by basic cations such as calcium, magnesium, potassium, and sodium), and low in nutrients (major and minor, especially phosphorus). In the humid tropics, where there is more intensive weathering, there is a more extreme loss of silica and bases.

Also typical of tropical soils is the presence of very little accumulation of organic matter on the surface because the organic matter tends to be rapidly attacked by primary decomposers. Termite mounds are common, especially in Africa. They cycle the organic detritus but may also become pests of crops that are planted following clearing. Trees tend to add some organic matter and nutrients to the soil beneath them.

Many of the parent materials of tropical soils are of granitic origin or of sedimentary rock. They generate soils of poor nutrient status. These parent materials also give soil a low iron content and poor structure. For example, most of Africa is underlain by a vast shield of ancient pre-Cambrian granitic rock. The shield has been lifted up and eroded down repeatedly over time. Kaolinitic clays form in most tropical soils that undergo leaching.

Agriculturally Important Soil Orders of the Tropics

Oxisols (Latosols in Brazilian Classification)   These soils are widespread in the tropics. **Oxisols** are highly weathered and leached, and they lack distinct horizons. They are found most extensively in Latin America and Africa. Brownish-red and reddish soils are common in East Africa. In Malaysia, the soils may be yellow (depends on the degree of hydration of ferric oxide). Most are high in clay content. Oxisols vary from coarse sands to fine clays and have a low CEC, low base saturation, and low pH. Because of leaching, they are usually low in most nutrients. The low nutrient retention capacity of oxic horizons is one of the major weaknesses of oxisols. Some cultural practices can alleviate this problem. CEC can be developed by protecting the soil surface from loss of organic matter through erosion,

and by addition of organic matter (high CEC). Oxisols are high phosphate fixers, which means that they do not release phosphorus (P) very easily. In capital-intensive systems, large quantities of phosphorus (P) fertilizers may be added, with continued applications in subsequent crops. Organic matter breaks down quickly in oxisols and mineralization is rapid. Adding lime is also beneficial as long as it does not unduly accelerate decomposition of the organic matter. Thus, these soils require careful lime and fertilizer management practices. They are free draining and tend not to have good water-holding capacities. Because of the low water-holding capacity, annual crops that grow in deep oxisols are more exposed to drought than in other soils of comparable clay content. Perennial crops (with deeper roots and better access to more stable subsoil moisture) are less exposed to drought stress in oxisols than are annuals.

Ultisols and Alfisols   These soils usually occupy younger geomorphic surfaces than do oxisols. The **alfisols** have considerably higher natural fertility than do the ultisols. The alfisols cover a large area in West Africa and much of East Africa. **Ultisols** are found in relatively more humid areas of Africa such as Liberia, Guinea, Uganda, and the eastern portion of the Democratic Republic of Congo. These are typically red and yellow soils.

Usually within 1 m (3.3 ft) of the soil surface is a zone with a significant increase in clay. The higher clay content of the **argillic** horizon (the most common type of B horizon in soils) can slow down water percolation. The coarse-textured surface may develop a crust during the dry season that may be difficult to till. In general, these soils have more weatherable minerals than do oxisols. Phosphorus fixation can be severe in ultisols because of high amounts of exchangeable aluminum (Al). Potassium is seldom deficient except after several years of intensive cropping. Alfisols can sustain more cropping than ultisols because of higher weatherable minerals, CEC, exchangeable calcium (Ca), and magnesium (Mg).

Vertisols   This soil order is characterized by high clay content and poor physical structure. **Vertisols** typically contain over 30 percent clay. They are known for their propensity to shrink and swell with changes in moisture content. This process creates typical soil heaving and cracking. When they are wet, they become very sticky; and when they are extremely dry, they may crack and break into blocks. These dark gray and black soils have common names including cracking clay, grumusols, black cotton soils, and regurs.

The largest expanses of vertisols are located in Australia, India, Sudan, and parts of Ethiopia—usually on flat or level topography in savanna regions. They occupy approximately 4 percent of the land area of the tropics. Marked dry and rainy seasons are needed for their formation. They tend to be high in Ca, Mg, and potassium (K) but are low in nitrogen (N) and P. Vertisols tend to have high water retention capacity, but because of their high clay content they hold the water so tightly that it is not readily available to plants. Their cracking and swelling characteristics make them hard to till and those forces may prune and damage crop roots.

Andisols   **Andisols** are glassy, highly permeable soils derived from volcanic ash. They may occur at any elevation where volcanoes are active. Usually they are very young soils, of low bulk density and variable texture, and with highly stratified layers. Internal drainage is excellent and they possess a high nutrient reserve. These soils are generally the most productive upland soils in the tropics, although N and P deficiencies can be problems. They represent a small fraction of the land, but they tend to be highly farmed because of their suitability for crop production.

Alluvial Soils   Alluvial soils occur in river plains, estuaries, deltas, and low coastal regions. They are the result of relatively recent sedimentary deposits and have little or no horizonation (differentiation of soil profiles). **Alluvial soils** may show considerable stratification and occur in complex patterns because they are deposited by flowing water. Their quality is highly dependent on the source material from which they were derived. They tend to be less desirable than the alluvial soils found in temperate climates. Floodplains may harbor diseases, and flooding is generally uncontrolled, except where rice cultivation is practiced. Major areas of occurrence are along river systems such as the Ganges, Mekong, Congo, Niger, and Amazon.

Inceptisols and Entisols   These are pedologically young soils. The horizons of **entisols** are only slightly developed or underdeveloped. Entisols are widespread in southwest Africa. Some are productive soils, whereas others may require artificial drainage and flood control if they are cropped. **Inceptisols** are present in the Central African Republic, southern Sudan, and parts of India. They are slightly more developed than entisols and not as strongly weathered.

Spodosols (Red and Yellow Podzols in Other Classification)   These soils are found in two unique tropical areas. One is at high altitudes where the areas have often been influenced by volcanic ash. Cool tropical

spodosols are typical for mountain slopes that are hit by rain-producing clouds. These soils may also be found in the low tropics (e.g., quartz sand beaches in southeast Asia) and on the lower footslopes of recent valleys in lowland rainforests. **Spodosols** are highly leached, are low in nutrients, and display poor water-retention characteristics. They are not well suited to agricultural endeavors. Erosion is greater than on oxisols because of poor permeability. High elevation spodosols are usually best protected by forest vegetation.

Aridisols (Sirrozems, Desert, and Red Desert Soils)

Aridisols soils are formed in areas where there is low leaching as a result of low rainfall. They are fertile soils, although they lack nitrogen. Calcium carbonate accumulates near the surface.

Tropical soils can present unique challenges to cropping systems. A major problem is their inherent infertility. Nutrient deficiencies are manifested in many regions of the tropics. For example, of the soils in tropical America, 96 percent are phosphorus deficient, and 89 percent are nitrogen deficient.

The oxisols, ultisols, alfisols, and entisols comprise about two-thirds of tropical soils and they are basically infertile, acidic, and with low base saturation. Also, the poor organic matter and poor infiltration characteristics of many tropical soils, and the heightened magnitude of the intensity of tropical storms, makes them especially vulnerable to erosion.

The increasing demands of population pressure and industrial development have resulted in the loss of protective vegetative cover in many regions. The rate of deforestation in the moist tropical regions during the early 1990s was about 8.5 million hectares (20,995,000 acres) per annum. Potentially important consequences of large-scale deforestation, aside from direct erosion losses and loss of biodiversity, are global warming and the resultant potential disruption and increased variability of the wet and dry seasons in tropical areas. If these consequences were to arise, they could have dramatic effects on crop production in the tropics and subtropics. Experts have advised the adoption of sustainable cropping systems and reductions in deforestation to avoid increased economic and environmental failures in the tropics.

CROPPING SYSTEMS

Shifting Cultivation

Shifting cultivation is also referred to as slash-and-burn, or swidden, agriculture. This system is highly variable and may be found in the forest, bush, or savanna regions. Traditionally, individuals do not own the land. Instead, the community crops the land together. *Shifting* refers to the fact that people may change the location of cultivated fields around a permanent home. Many populations do not rely exclusively on shifting cultivation and may apply it only on locations more distant from the permanent dwelling.

One essential feature of shifting cultivation is that nutrients held in the vegetative biomass are released by cutting, drying, and burning. Litter fall, comprised primarily of the old leaves of shrubs, woody vines, and trees, is considered the most important because it is nutrient-rich. Burning, however, causes a loss of nitrogen to the air. During the fallow period (when fields are left to return to natural vegetation), nutrients are taken up by the vegetation and then returned to the soil surface as organic litter when the plants die.

Under this system, in tropical forest regions, woody vegetation is felled and burned during the dry season. The ash is left in place and the crops are interplanted (not in rows) with digging sticks and/or hoes. Livestock are usually not part of the system. Yields usually decline by the third year, and fields are abandoned after three to four years to allow the natural vegetation to regenerate, and the cycle is started again elsewhere. In many areas, the fallow period is no longer of sufficient duration to recharge the nutrient levels to satisfactory levels before they are again brought into cultivation.

In traditional West African forest region systems, lianes (woody vines) and small trees are chopped down, but larger trees are left to provide light shade. Maize is planted into the forest topsoil using a digging stick. Weeds are removed. The maize is intercropped or followed with cassava (*Manihot esculenta*), cocoa (*Theobroma cacao*), yams (*Dioscorea* spp.), tannia (*Xanthosoma* spp.), and bananas (*Musa* spp.). Small areas are planted with a wide range of vegetables.

In a bush system, stumps and large roots are allowed to resume growth and will attain about 6 m (15 ft) height after five years, 10 m (25 ft) after ten years. The resting period may vary from five to twenty years; an eight-year resting period is considered necessary to restore fertility after three years of cropping. The longer the regrowth is allowed, the better is the subsequent level of nutrients available to the crop.

In a tallgrass savanna, the soil needs more vigorous cultivation to deal with the grass roots. Soil is traditionally scraped into mounds using hoes. The mounds are then planted with climbing yams and interplanted with maize (*Zea mays*), beans (*Phaseolus* sp.), and vegetables. The yam mounds are broken down the second year, and sorghum and maize are planted on narrow

ridges. Peanuts (*Arachis hypogaea*) are interplanted with millet (*Pennesitum americanum*) in the third year.

The rate of decline of crop yield is related to the natural fertility of the soil, the amount of nutrition contributed by the ash, and how demanding the crops are of nutrients. Usually the cash crops or preferred subsistence crops are grown during the first phase of the cropping cycle, followed by legumes in the later stages as fertility declines, and finally by nutrient scavengers such as cassava toward the last stages of the cropping cycle.

Weed pressure tends to increase during the cycle so taller, robust crops that compete well with weeds may also be chosen toward the latter part of the cycle. Some biennials or perennials may also be grown with the hope that they may persist even after the field has been abandoned. Cassava fits the bill in every regard.

Typically, the most diverse and irregular cropping patterns in shifting cultivation systems are to be found in the more humid, lowland forest areas. Where the soil water regime is less favorable, the crop diversity usually decreases and the planting patterns become somewhat more regimented.

The exploitation of microclimatic conditions in these diverse systems has been described as "ingenious." For example, shade-tolerant species are planted on the perimeter of a plot, and moisture-demanding species on the bottom of sloping sites, and fertility-demanding species where the most ash has accumulated. In addition, plants may be planted at various positions on a mound of hoed-up topsoil, vines may be planted on upright tree trunks, and so on. These systems, now considered to be ingenious, were once described as "primitive" in a pejorative way by so-called experts. See the book by Norman et al. listed at the end of this chapter for more information.

Today, there appears to have arisen a greater appreciation for the human-generated precision farming techniques that have been refined for centuries. As William Allan (see book listed at end of the chapter) says: "The shifting cultivator has always required a detailed knowledge of the environment to survive." The shifting cultivator typically has a wealth of ecological information, such as which types of vegetation indicate physical or nutritional characteristics of the soil, how long a soil may be cropped, and how long it should remain fallow before it can be cropped again.

Shifting cultivation is practiced on about 30 percent of the world's arable land and about 45 percent of the tropical land area. In the past, when population pressure was not high and land tenure was not an issue, these systems were considered sustainable. Shifting cultivation was at one time able to provide sufficient food and products to satisfy the cultivators' needs while maintaining soil fertility for the long term.

The critical elements of the system were that desirable locations were not limited and the regrowth (or fallow) period was long enough to maintain soil fertility. With increasing population pressure came a progression toward increasing cropping intensity, and declining periods for regrowth, reduced species diversity, and increasing soil degradation. Typically, when farmers are confronted with low productivity due to declining soil fertility, they switch to less demanding crops, for example, cassava. Faced with ever worsening productivity however, farmers then begin expanding the practice to newer and usually more marginal lands (e.g., to steeper slopes and drier areas). If that does not work, they attempt to increase the nutrient inputs or resort to conservation techniques that slowly build up the fertility of the soil once again. The hesitation to adopt conservation techniques is probably related to the increased labor demands and constraints those practices may place on planting patterns, and so on.

Typically, the vast majority of African farmers cultivate 0.2 to 0.4 hectare (0.5 to 1.0 acre) of subsistence crops per person. A family of five to six people would cultivate approximately 1 to 5 hectare (2.5 acre to 12.5 acre) to provide some surplus as well. Problems arise, however, when novices begin the process of slash-and-burn agriculture. Often, in these instances, or when indigenous systems are perturbed (e.g., forced to marginal areas because of increased population or development schemes, etc.), the systems become unsustainable or further degraded.

Making Shifting Systems More Sustainable   Some practices can make these systems more sustainable, for example, low impact land clearing, the use of mulches, plant cover crops, understory crops, and reduced tillage planting techniques. The proportion of legumes as food crops, cover crops, and fields in fallow can also be increased. Improved fallow management techniques may be useful, and breeding crops adapted to cropping system, for example, aluminum tolerance, less sensitivity to low pH, and so on, can have a positive impact. In some areas, farmers practice contour cropping or use integrated plant health management (IPHM) techniques or even agroforestry. It is also important not to abandon mixed cropping.

The speed and effectiveness of the regeneration of fertility can be improved in response to increasing population pressure. Farmers may sow seeds of, or plant,

leguminous trees. The cultivation cycle may be ended with a shade-producing crop such as plantain (*Musa* spp. AAB group) or cassava (*Manihot esculenta*) to aid tree establishment. Also, ending the cultivation cycle with pigeon pea (*Cajanus cajan)* provides nitrogen, and it is a good nutrient accumulator. Farmers can plant selected grass species for the fallow period, and grazing during the resting period can be beneficial.

Fallow System

Where land tenure is held by individuals, shifting cultivation tends to be replaced by a fallow system. The fallow system may be found in the forest, bush, or savanna and is more important in Africa than is shifting cultivation. Grass fallows are common in East African savanna regions. Bush fallows are common in West African and South American savanna regions. Boundaries of fields (2 to 4 hectare or 5 to 10 acre) are usually present near a permanent house compound. Changes in cropping patterns are associated with a fallow system. Intercropping systems are often arranged in a more ordered or geometric pattern than in shifting cultivation. Relay cropping is found in the fallow system (delay planting, overlapping crops) and differs from sequential cropping (one crop following another during the same year).

Brush is burned during the dry season. Rains are then necessary to soften the soil, allowing hoe cultivation. Planting may be delayed, resulting in a missed nitrogen release (often termed a flush) into the system. Cash crops may be grown but the high labor demand for weeding and harvest periods may delay planting of cash crops. Typically cropping systems in the wetter tropical climate areas are based on noncereal energy corps, which changes to systems based more on cereals in the wet and dry climate areas. The cycle time is usually four years cropped followed by four years fallow. Animals are usually supplementary, grazing communally on undesirable land.

Permanent Upland Cultivation

Permanent upland cultivation is often practiced in semiarid regions with distinct dry seasons. Planting is delayed until the onset of the rainy season appears to be established. Varieties are chosen as much for reliability as for their yield. In areas with longer rainy seasons, intercropping may be practiced and cattle may be integrated into the overall farm system. Farmers often attempt to irrigate a portion of their fields, but there is typically little investment in agrochemical inputs due to the high-risk factor. Legumes are emphasized in the rotation, as is the use of green manure crops (typically legume crops with vigorous growth that can suppress weeds and provide valuable nutrients for the subsequent crop). Farmers also allow some legume trees or shrubs (e.g., *Acacia*) to remain for fodder and nitrogen fixation. In more humid regions, it is hardly possible to maintain fertility without fertilizers or animal manure. Weeds also tend to grow out of control in permanent systems.

Arable Irrigation Farming—Upland Cropping Systems

This system is typically irrigated by gravity-flow systems with water from a reservoir or stream (sprinkler systems are less common in the tropics), which bring water to fields by a system of canals and ditches. The water is distributed through ridge and furrow systems in fields, and surplus water is removed through natural drainage systems (e.g., a river). Irrigation allows the production of multiple crops per year and is used for high-intensity production of high-value crops.

EXAMPLES OF TROPICAL CROP PRODUCTION SYSTEMS

Flooded Systems—Tropical Wet Rice

Wet paddy rice systems have supported dense populations in southeast Asia for many centuries. Rice is an ideal crop in many ways—it stores well, needs little fuel for cooking, and is convenient and nutritious. It is also uniquely and highly adapted to flooded conditions. Wet rice systems can be subdivided into various categories: shallow-water rice only; shallow-water rice with upland crops; and long-standing floodwater rice. The degree of water control varies widely. The farmer may grow one or more crops per year.

Rice is the most important food crop in the world. Several species of *Oryza* are cultivated, but *O. sativa* is the most widely grown. The species consists of three ecogeographic races: indica, japonica, and javanica. The indica types have been bred extensively and are now highly productive. They are grown widely in wet-rice systems. The predominant systems are irrigated (wet season) and the rain-fed (shallow) system found in Asia.

The average rice farm is rather small and for many years existed as a subsistence system, with little rice sold off the farm. Notable exceptions are the vast wet rice systems of Thailand and Myanmar, where much of the crop is exported. High-yielding varieties were introduced in the 1960s and many of the traditional systems changed. The Green Revolution cultivars were highly responsive

to fertilizers and farmers began purchasing off-farm inputs for their systems. The farms' sizes have not changed dramatically, but intensification of the systems has increased dramatically in many areas. The changes have brought with them myriad benefits and costs.

In addition to higher-yielding cultivars, the availability of earlier maturing cultivars has allowed changes to the wet rice cropping systems. Some of the more important one-year systems that have been adopted include (1) one rice crop followed by a season of fallow, (2) one rice crop and one cash crop during the dry season, and (3) two rice crops (where irrigation is available or rainfall is sufficient). A two-year cycle may be planted with up to five continuous crops of rice—although this is a clear invitation to outbreaks of pests or diseases. The main cropping pattern in the rice bowl of southeast Asia and the rain-fed lowland areas of Indonesia and Bangladesh is still one rice crop during the rainy season or two rice crops; one during the rainy season followed by another using irrigation during the dry season.

Land preparation for rice is unique because manipulation of the field to achieve appropriate water management during the cropping cycle is extremely critical. In contrast with farmers in temperate regions who strive to preserve good soil **tilth**, lowland rice farmers prepare fields for planting by puddling the soil (purposely working the soil while it is wet to break down its structure) to *decrease* water infiltration. Maintenance of optimal water levels in the field is critical to successful rice cultivation, and excess loss through infiltration is to be avoided.

When water has been drained off the puddled soil, rice is then transplanted by hand from nurseries. Direct seeding of dry seedbeds, or casting of pregerminated kernels into shallow water are techniques also practiced. The initial planting operations must be timed to coincide with the commencement of the monsoon season in tropical areas. Puddling of the soil and transplanting are considered to be good practices from a weed-management standpoint. If weeding is not performed during the first forty days after planting, yields can be adversely affected.

Yields during the wet season are actually lower than those during the dry season (assuming irrigation has taken place). The lower yields are ascribed to less than optimal partitioning of energy into the canopy versus the grain during the wet season.

Rice blast is considered to be the most devastating disease of rice. It can kill seedlings and infect the panicles, resulting in appreciable yield losses. Many diseases of rice are incited by viral pathogens, perhaps the most

serious of which is rice tungro. Two viruses, vectored by several insects, are causal agents of the disease. Small landholders are still the primary producers, so host resistance is the most appropriate strategy for management of these pathogens.

Harvesting of rice in lowland tropical systems is still primarily by hand. Mechanical threshing and winnowing is gaining in practice, although in many villages, hand or animal threshing is still performed. Sun drying is used to reduce the moisture content to around 14 percent so that the rice may be stored safely.

The limits of expansion are being reached for rice production in Asia, but in Latin America and Africa, additional land could still be brought into production. In Asia, additional population growth will require further intensification of production systems in a manner that could become unsustainable.

Cereal-Based Systems of Semiarid West Africa

The cereal cropping systems of West Africa are based primarily on pearl millet (*Pennisetum americanum*) and sorghum (*Sorghum bicolor*). Maize may be added in the higher rainfall or more fertile environments. Cowpea (*Vigna sinensis*) and peanut (or groundnut; *Arachis hypogaea*) are important grain-legumes traditionally interplanted in the cropping system. These cereal agroecosystems comprise the critical food base for Senegal, Mali, Burkina Faso, Niger, Nigeria, Chad, and Cameroon.

Rainfall is a primary determinant of the cereal of choice, and its distribution shows a steep north–south gradient from 40 to 120 cm annual precipitation. Pearl millet is grown mainly in the most arid regions receiving only 20 to 80 cm precipitation, sorghum in those regions receiving between 70 and 140 cm, and maize in those with excess of 100 cm. Rainfall events in this region are typically highly variable and unpredictable due to the randomness of convective storms. Rain frequently occurs in short, intense storms. Droughts of various degrees of magnitude and duration are to be expected.

The high radiation load in the semiarid tropics of West Africa also results in elevated temperatures. The temperatures also increase from north to south in an inverse relationship with annual rainfall. In the Sudano-Sahelian region, mean monthly maximum temperatures can exceed 40°C at both the time of sowing and harvesting. Extremely high temperatures capable of causing seedling death, poor stands, and the need to resow crops sometimes occur. These harsh and uncertain climatic conditions pose a constant threat to stable food production.

Cereals are typically planted in a variety of soils. Aridisols, alfisols, and entisols are found in millet-based systems, and alfisols and small areas of vertisols, entisols, and inceptisols are utilized where sorghum and maize-based systems predominate. Millet is the crop of choice on the sandier soils. A wide variety of soil textures is employed in sorghum production, and maize is the preferred crop on the heavier clay soils, where rainfall is not limiting. Upland soils are generally preferred for millet production, and lowland soils with better organic matter content and water-holding capacity are more desirable for sorghum and maize production. Systematic utilization of the soil variation is a key element of the local farming strategy and contributes to the stability of crop production.

Cereal-cropping systems in semiarid tropical West Africa experience a seasonal availability of nitrate-nitrogen. The rapid release or flush occurs at the onset of the wet season due to the burst in soil microbial activity. Farmers must work hard to ensure their crops are planted in a timely manner to receive the full benefit of the nitrogen while it is available. Marked deficiencies of phosphorus are common in the alfisols of West Africa. Farming strategies should take into account the low inherent fertility of sandy alfisol soils where millet is grown. In these very poor soils, modest amounts of fertilizer can result in substantially increased yields.

Cultural and economic constraints also limit the degree to which the cereal-based agroecosytems may be altered. The basic social unit in many traditional societies is the compound, or assemblage of living areas organized along familial lines. Compound members may work on both community fields and their own fields. The farms are generally from 2 to 25 hectares (5 to 62 acres) in size. Traditionally, decisions concerning crop production are formed in the context of this complex unit of related families. The size often depends on whether traditional complex family units or simple nuclear family units are operating them. Renting, pledging, leasing, and purchasing of land are becoming more common, but farming remains largely at the subsistence level, with hand labor the primary input. Sufficient labor can be scarce during the rainy season, for weeding, land preparation, and planting. Little capital is available in most enterprises for either chemical fertilizers or hired labor.

Traditional fallow-rotational systems permit the buildup of organic matter and make nitrogen and phosphorus available. Without a fallow period, reliance on the planting of legumes such as cowpea and peanuts to enhance the system becomes even more critical. Practices such as accumulating removed weeds in mounds or ridges between cereal rows are useful for adding organic matter. Another traditional method is the burning of crop residues and cut brush. Animal wastes are also added to the fields, either by transporting them or through relationships with herdspeople who graze their cattle on the farmers' fields for payment in-kind or for cash.

Environmental conditions are primary determinants of whether or not disease occurs on millet and sorghum. The timing of flowering is critical. The start of the rainy season, and hence the planting date, is always a bit uncertain. Photoperiodicity of local varieties has ensured that flowering commences around the expected end of the rainy season so that grain maturation occurs during dry weather. Disease incidence tends to be higher with increasing rainfall. The primary disease of sorghum and millet is downy mildew (*Sclerospora graminicola*).

Mixed Annual/Perennial Systems

In the majority of regions where tropical root crops are cultivated, these crops are typically produced in mixed-cropping systems. Among the more important systems are the African and Caribbean mixed-root crop systems. These systems are low-input, extensive systems and are focused less on economic returns than on maximizing benefit from limited resources. Asia systems tend to have cereals intercropped with legumes and are highly intensive. Planting is usually on fertile alluvial soils, and purchased inputs such as fertilizers and pesticides are frequently utilized.

African Mixed Root Crop Systems

Tropical root crops are diverse and ancient crops. The major tropical root crop species are cassava, true yams (*Dioscorea* spp.), sweet potato (*Ipomea batata*), the edible aroids dasheen and eddoe (ARACEAE family), tannia (*Xanthosoma* spp.), and white potato (*Solamum tuberosum*). Cassava (sometimes called manioc) is a starchy tuber crop that has significant amounts of some nutrients but also has high concentrations of a cyanide compound in the tuber, which has to be removed (detoxification) by soaking, fermenting, or other methods to make cassava edible. Flour from the cassava tuber is known as tapioca. Root crops are traditionally cultivated in mixed systems on small family farms. They are important as an energy source for people, as survival crops, and as industrial crops.

Cassava and both yellow guinea yam (*Dioscorea cayensis*) and white guinea yam (*Dioscorea rotundata*)

are of considerable importance in the yam zone (wetter regions) of sub-Saharan West Africa. Globally, cassava production far outstrips that of yams. Nigeria is the biggest producer of both cassava and yams. The next highest world producers of cassava are Brazil, Indonesia, and Thailand. Other world-leading yam producers are Ghana and Côte d'Ivoire (Ivory Coast). Traditional mixed-root cropping systems planted on newly cleared land in the forest zone are generally planted with a mixture of three or more species. Yam, cassava, maize, or upland (does not require flooding) rice usually predominate in these mixed cropping systems. The choice of yam or cassava is often predicated on the amount of time the field has spent in fallow prior to clearing, and the inherent quality of the soil. Cassava is selected for poorer soils or following shorter fallow periods, whereas yam is planted in the better soils following a longer fallow period. Yam requires more laborious and costly crop husbandry than does cassava. Cassava is easily propagated and more drought-tolerant than yam and this too can affect selection. Yam is popular in many African societies, but cassava cultivation is expanding.

Wetland yam mound systems are centered around the planting of white yams (*Dioscorea rotundata*) on the top of large mounds averaging 1 meter in height. The associated annual crops and vegetables are then planted at different positions on the sides of the mounds depending on their ability to tolerate waterlogging. Wetland (flooded) rice can be planted between the mounds where flooding is apt to occur during part of the growing season.

Cassava-based mixed systems are planted mainly in the fields most distant to the family dwelling. The system tends to be less diverse than yam-based mixed cropping systems, and frequently only one additional crop is planted to take advantage of the open area prior to ground covering by the cassava crop. Common crops planted with cassava are maize (e.g., in Nigeria and Ghana), upland rice (e.g., Liberia and Sierra Leone), and groundnut (e.g., Democratic Republic of Congo and the Central African Republic.)

Maize and rice are interplanted at a reduced density, and secondary crops such as cocoyams, leafy vegetables, melons, okra, or legumes are used to fill in remaining open areas. The species diversity in this system may sometimes be impressive—with as many as fifty species identified in farmer plots of southeastern Nigeria.

Rotation with bush fallow is the most common practice, but true shifting cultivation is becoming less frequent because of increasing pressure on the fields to produce more food for growing populations. Thus, pressure to produce more food decreases shifting cultivation with regenerative fallow periods, resulting in reduced field productivity over time. The precise times of cropping and fallow thus vary considerably depending on soil type and the amount of pressure on the land to produce adequate food supplies.

Caribbean Mixed Root Crop Systems

Major root-crop production in the Caribbean is found in the larger islands of the Greater Antilles (Cuba, Jamaica, Haiti, and the Dominican Republic). The most important center of yam cultivation in the Caribbean is Jamaica.

Jamaican hill-land culture of yams typically includes multiple varieties representing several species of yam, often intercropped in the same field. Typical planting is on hills or mounds approximately 20 to 90 cm (8 to 35 in.) in height and spaced from 1.5×1.5 m to 3.0 to 3.0 m (approximately 5×5 ft to 10×10 ft) apart. Intercropping is done both on the hills and in the space between hills. Yams are set at the top of the hills, and shorter-season crops such as tomatoes or *Phaseolus* beans are planted at the base of the hills. A variety of crops, including white potatoes, tannia, sweet potatoes, and *Amaranthus* spp., may be planted in irregular patterns between the hills. The land is typically cultivated for three to five years and then left fallow for one or two years.

The system is designed to provide yams year-round by concentrating the planting of rotundata yams from November to March, and greater Asian (*D. alata*), cush-cush (*D. trifica*), and lesser Asian (*D. esculenta*) yams from April to June. Throughout the year cayensis yams may be planted. Yams are typically staked with poles 1 to 6 m (about 3 to 20 ft) long. Weed control is achieved by mulching and employing hand-weeding as needed. Yams are usually available for harvest approximately six to twelve months after planting. Some of the first yams are used for propagation and then are immediately replanted. A major strength of the system is the inherently different rhythms of the different yams and their adaptation to the bimodal rainfall pattern. Differences in the dormancy and sprouting of tubers, combined with differences in growing seasons, lend themselves to the harmonious rhythm of the system. The major costs include capital for yam planting and staking materials. In this system, the extensive amount of labor required for preparation of the hills limits the extent to which this system can be utilized.

Comparisons between Systems

The methods of clearing the land are much the same in Jamaica and in Africa, with hand-labor predominating. Seedbed preparation techniques involve the formation of small mounds or low ridges in shallow or infertile soils so that mineral nutrients and organic matter can be aggregated shortly after clearing and burning. In more fertile alluvial soils, large mounds 0.5 to 1 m (3.3 ft) in height are constructed to avoid flooding.

Key management considerations of mixed cropping systems hinge on the ability of the farmer to minimize competition for nutrients and light. Where regions are characterized by strongly weathered and infertile soils, the mixed root crop systems offer several advantages over monoculture systems. They include higher food and nutritional production on small areas of land, lower risk of crop failure due to pests and diseases, higher utilization of soil nutrients on land previously left fallow and cleared using traditional methods, and the stability of crop yield.

The major advantage of cassava-based systems is the inherent capacity of cassava to withstand poor environmental conditions and still produce good crops. The major disadvantage is the need to detoxify the tubers. A unique feature of mixed cassava systems is the opportunity to mix several crops of varying life-cycle duration. This technique results in a gain in total yield in time and space. Traditional cassava cultivars are upright and ground coverage is insufficient for adequate weed control. This system does permit the opportunity to interplant other crops of shorter maturity. Crops such as maize, melon, cowpea, and beans provide rapid coverage that can suppress weeds, but they complete their life cycle before serious competition with cassava is realized.

Pests and Diseases of Root and Tuber Crops

Nematodes are important pests of yams in Africa and the Caribbean for the damage they do and the diseases they transmit. Termites are also a problem in African. Common diseases and pests of cassava in Africa are cassava mosaic virus (CMV), bacterial blight (*Xanthomonas campestris*), variegated grasshoppers (*Zonocerus* spp.), and silverleaf whitefly (*Bemesia tabaci*). Some cultivars of these species have been developed to be resistant to pests. Programs to provide more inexpensive disease- and nematode-free or resistant planting material can have a tremendous impact on the efficiency of production.

PERENNIAL CROPS

Avocado (*Persea americana Mill.*) LAURACEAE

Evergreen avocado trees are native to Mexico and Central America. Avocado trees are fast-growing, reaching heights between approximately 9 and 20 m (30 to 66 ft). Tree limbs are easily broken by strong winds or heavy crop loads.

Avocados, also known as *aguacate* (Spanish) or alligator pears, are assigned to three ecological races: Mexican (*Persea americana* var. *drymifolia*), Guatemalan (*P. americana* var. *guatemalensis*), and West Indian (*P. americana* var. *americana*). The races differ in several traits including cold tolerance, salinity tolerance, iron chlorosis tolerance, fruit size, skin thickness, oil content, and flavor. Mexico is the leading producer of avocados, followed by Indonesia, the United States, Colombia, Brazil, and Chile. Chile is the principal exporter of avocados to the US market. The chief factors limiting avocado consumption worldwide are lack of familiarity and high cost.

Avocado production is limited by its sensitivity to climatic extremes and its vulnerability to root rot. Trees of the West Indian race produce well in tropical climates, but they freeze at or near 0°C (32°F). The other two races generally fail to flower or set fruit in the tropics. On the other hand, the West Indian race sets little or no fruit in subtropical climates, such as that of southern California. In regions where minimum winter temperatures of −5.5°C to −3.5°C or below occur, only trees of the Mexican race can be expected to survive. Guatemalan types are native to cool, high-altitude tropics and are hardy to −1°C to −3.5°C. If the proper race and cultivar are chosen, avocados thrive and produce well in climatic conditions from truly tropical to the warmer parts of the temperate zone.

Avocado cultivars must be propagated vegetatively as grafts. A major need in California and some other regions is for a rootstock highly resistant to Phytophthora root rot.

Avocado trees grow well in a wide range of soil types provided drainage is good. Loose sandy loams are ideal. They will not survive in locations with poor drainage, nor will they tolerate drought stress, so supplemental irrigation is important in many drier regions, for example, southern California and Israel. The trees grow well on hillsides and are tolerant of acid or alkaline soil. Avocado tree growth and production is best suited to acidic soils with a pH of 4.5 to 5.5. However, some West Indian race rootstocks allow production on

marginal calcareous soils with an alkaline pH, such as those found in Israel and Florida. Avocado trees of all ages are pruned sparingly, mainly by clipping back upright shoot tips to prevent excessive height. Heavy pruning encourages strong vegetative growth, thus reducing yields.

An incredible number of flowers per tree is produced, but only one to three fruits per panicle will mature. Avocado flowers are perfect (having both male and female parts), but they exhibit an unusual behavior. Avocado flowers are either receptive to pollen in the morning and shed pollen the following afternoon (type A), or are receptive to pollen in the afternoon, and shed pollen the following morning (type B). Flowers of a given cultivar tend to behave uniformly as type A or type B. Cross pollination may occur when female and male flowers from type A and type B varieties open simultaneously. Production is best with cross pollination between types A and B. Self-pollination appears to be caused primarily by wind, whereas cross pollination is secured by large flying insects such as bees, wasps, and hoverflies. Some cultivars bloom and set fruit in alternate years.

The avocado fruit is a berry, consisting of a leathery skin (exocarp), the fleshy mesocarp (which is eaten), and a large seed consisting mostly of two cotyledons. West Indian type avocados produce enormous, smooth, round, glossy green fruits that are low in oil and weigh up to 0.9 kg (2 lb). Guatemalan types produce medium ovoid or pear-shaped, pebbled green fruits that turn blackish-green when ripe. Fruit of Mexican varieties are small, weighing 170 to 280 g (6 to 10 oz), with paper-thin skins that turn glossy green or black when ripe. Varieties vary in the degree of self- or cross pollination necessary for fruit set.

Cultivars of the three horticultural races differ in oil content: the Mexican race has the highest, the Guatemalan race an intermediate amount, and West Indian the lowest. Mature fruits can be "stored" on the tree for several months. The fruits remain hard as long as they stay on the tree, softening only after harvest. Mature avocado fruits, even though firm, bruise easily and must be handled carefully. Avocados can be held for about a month in cold storage at −7°C (45°F) to (40°F), depending on cultivar. The fruit softens at room temperature. California harvests and ships avocados throughout the year, although the peak season is usually from March to August. The Florida season is from August through December and into January. Production in Chile (southern hemisphere) compliments the US season. Production in the Mexican state of Michoacán also compliments the US season

somewhat. The combination of opportunities for both storage and importation of fruit result in availability of avocados to US consumers year-round.

Avocados are a mainstay in the diets of Mexican and Central American people. Consumption of avocado in salads and as guacamole (avocado sauce or dip) has grown in the United States and other parts of the world as people discover its taste and nutritional benefits. Avocados contain higher quantities of fiber and protein than most other fleshy fruits and are an excellent source of potassium and vitamin A. Avocado fruit do not contain cholesterol. They are high in monosaturated fatty acids (heart-healthy oils), and the oil content of avocados (3 to 30 percent) is, on average, second only to that of olives among fruits.

Pests and Diseases Root rots caused by *Phytophthora, Armillaria,* and *Verticillium* are the major causes of poor tree health and death. These pathogens are easily transported from infected soils. Once a tree is infected (signs include yellowing and dropping leaves), little can be done other than to cut back on water. Some *Phytophthora*-resistant rootstocks are now available, but considerable research is still needed on this problem. It is important to use virus-free propagating wood. Cercospora spot on fruits and leaves is the most important avocado disease in Florida. Avocado brown mite can be controlled by dusting with sulfur. The six-spotted mite is very harmful; even a small population can cause massive leaf shedding. Natural predators can help manage the mite population if they are present.

Banana (*Musa* spp.) MUSACEAE

Only a fraction of banana production enters the world market, yet it is the leading fruit in world trade, with sales totaling more than $2.5 billion annually. To those who have enjoyed only sweet dessert bananas, it may come as a surprise to know that about half the world's production is starchy bananas for cooking. Starchy bananas such as plantains (*platanos* in Spanish) are very popular in South America, the West Indies, and East and West Africa. East African highland bananas are a preferred staple. Uganda is the leading producer of starchy bananas. Per capita consumption of highland bananas (*matooke*), a staple food, reaches an amazing figure of 200 kg/year (441 lb/year) in Uganda. Bananas are now cultivated throughout the tropics and in selected areas of the subtropics, by the smallest landholders to the largest corporate exporters. The largest producer nations of

sweet bananas are India, Brazil, China, Ecuador, and the Philippines.

The origin of banana species can be traced back to different ancestral species.. The cultivated species are polyploid (having more than two [diploid] sets of chromosomes), whereas the wild species are all diploids. Sweet bananas are low in starch and high in sugar when ripe; unripe fruit are sometimes used for cooking. True plantains are starchy even when ripe and are usually cooked.

The origin of bananas is in southeast Asia and the southwest Pacific. It is likely bananas were taken to the East Coast of Africa about 450 CE; and the first introduction in the new world appears to have been in Haiti in 1516. The banana industry grew in the West Indies and Central America in the nineteenth century.

Growth and Development   Bananas are tall, rhizomatous perennials. Their monocot leaves makes them readily discernible. Short underground rhizomes grow horizontally and rather slowly. Aerial shoots (also known as suckers or followers) arise from lateral buds on the rhizome. Aerial shoots are *pseudostems* built out of overlapping leaf bases rolled tightly around each other. When the pseudostem is about six to ten months old (about forty leaves), the meristem then becomes reproductive. Each shoot is determinate—once a flowering shoot is produced, no further leaves are initiated. However, secondary shoot growth occurs and a second crop, or ratoon, develops.

Fruit develop parthenocarpically (without fertilization) from the ovaries. A banana plant produces a bunch every six months in the lowland wet tropics, although fruit development may take as long as 210 days in cooler climates or under overcast conditions.

Plants are inherently shallow rooted and are susceptible to drought stress and lodging due to high winds. Rainfall at or above 125 cm (49 in.) per year is needed. The optimum temperature for growth is about 27°C (81°F). Frost causes rapid death of the trees and damages the fruit. Temperatures above 38°C (100°F) stops plant growth and causes leaf burn.

Production   Bananas are grown mostly in lowland wet tropics, but they are also grown in wet and dry, and cooler tropical areas. They may be grown successfully in a wide variety of soils, but good soil drainage characteristics are critical. They tolerate a wide range in pH and may even be grown in soils with pH as low as 3.4. (They are not sensitive to aluminum and manganese toxicity, which is common at such low pH.) Bananas also grow in slightly alkaline soils. High organic matter and fertility produce a high yield of bananas, and mulching of bananas is also considered important to maintain constant soil moisture conditions.

Bananas are planted as rhizome pieces or aerial shoots at a density usually at or below 2,000 plants/ hectare (8,100 plants/acre). Good crops remove a lot of nutrients from the soil. Balanced fertilizer (NPK) is usually recommended where available. Drought restricts the root system and diminishes yield; it also predisposes the plant to lodging due to high winds. Waterlogged soils also predispose plants to root lodging.

In subsistence systems, bananas are popular because land clearance and cultivation inputs are low. Planting preparation usually entails digging a large planting hole and placing the planting material (propagules) in the hole then adding manure over time. Propagules typically consist of a sucker with a piece of corm that contains a growing point. Peeling and cutting away of old or diseased portions is generally practiced. Dipping in very hot water (52°C for 20 minutes) (126°F) or treatment with pesticides may be practiced in an attempt to eliminate nematodes and borers. In some areas, farmers are fortunate enough to have disease-free clones available to them for planting.

Banana trees are usually intercropped with annuals when first planted. They are also common in adjacent rows or clumps and in mixed culture gardens. Bananas growing around compounds and villages usually continue to receive organic wastes and mulch on an ongoing basis. Fruit are produced within one year and they are typically harvestable for many months. Bananas are harvested when the fruit are fully mature but still hard and solid green. They do not ripen well if left on the plants. The large bunches or, hands, of bananas may be shipped whole to local markets, or they may be cut into clusters of four to sixteen fingers for export in cartons. Upon arrival at distributing centers, the cartons first go to ripening rooms. The fruit is exposed to ethylene for twenty-four hours to induce ripening. The temperature is varied during ripening to achieve the desired combination of color, pulp texture, and flavor, according to market demands. Desert bananas are ready to use when a full yellow color is achieved; however, the best flavor and nutritive value is indicated when flecks of brown appear on the surface of the peels. The banana is a highly nutritious food, low in sodium and fat but high in potassium and easily absorbed carbohydrates.

In East Africa, a common cash crop is the banana–coffee intercrop. Banana trees provide shade for the coffee plants. Banana is also associated with a variety of subsidiary crops. For example, near East African compounds, it is frequently associated with vegetables, and both crops are manured. Farther from the compound, it will likely be planted in association with, for example, beans, maize, groundnuts, or cassava. The banana-coffee balance usually shifts toward coffee at the periphery. Subsidiary crops are highly dependent on regional differences in climate, food preference, and so on.

Utilization   Sweet bananas are enjoyed around the world as desserts, snacks, and breakfast cereal toppings. They are especially important in the diets of infants and young children because of their nutritional value, sweetness, and soft texture. Cooking typically involves boiling and mashing or frying. Leaves are commonly used as food wrappers in many regions, and they may also be used for brewing beer, fodder, or roof construction.

Pests and Diseases   Diseases have frequently dictated what the commercial industry can and cannot grow. The most serious diseases have been Sigatooka leaf spot (infectious agent *Mycosphaerella fijiensis*), which causes small lesions that coalesce under favorable conditions, resulting in browning or scorching of the leaves. In large plantations, black Sigatooka is managed using frequent applications of fungicides, an unrealistic option for subsistence farmers. Panama disease (elicited by *Fusarium oxysporum*) causes plants to wilt. The Gros Michel group is susceptible to Panama disease. Yellow and black Sigatooka leaf spot can result in almost complete defoliation in susceptible cultivars of banana and plantains. Panama disease and Sigatooka forced prioritization of resistance breeding programs on these pathogens. Cercospora leaf spot and bunch top virus are also problems in some banana-growing regions. Banana bacterial wilt, or Xanthomonas wilt (causal agent *Xanthomonas campestris* pv. musacearum), has become a serious problem in Uganda. Nematodes (e.g., *Radopholus similes* are also a serious problem in some areas, particularly those growing plantain. New cultivars with resistance to pests and diseases are being developed and tested in both the private and public sectors.

Cacao Theobroma (*Cacao L.*) STERCULIACEAE

The center of the genetic diversity of cacao is the Amazon Basin region of South America. The species also seems to have its origin there. Cacao evolved as a perennial evergreen understory tree in the tropical forest. It was naturally dispersed by foraging animals and people, who ate the sweet pulp from the pods and dropped the seeds. It was first cultivated by the Mayan and Aztec civilizations of tropical Central America. Both groups used the ground beans, frequently mixed with chiles, to make a stimulating beverage. Archaeological information suggests that the Mayans cultivated cacao 2,000 to 4,000 years before the arrival of the Spanish explorers. Columbus, on his fourth voyage in 1502, saw cacao beans, but it was Cortés who, a few years later, took the beans back to Spain. From the European perspective, the beverage was improved considerably by the omission of chile and the addition of sugar. It gradually spread throughout Europe and became very popular. The Dutch invented chocolate candy about 1830. Cacao has mild stimulating properties from its theobromine and caffeine content. It is now also appreciated for its antioxidant (health-promoting) properties.

Cacao production has developed into a billion dollar industry of worldwide importance in the manufacture of cocoa and chocolate. Until about 1900, most cacao plantations were in Central and South America and the Caribbean Islands. In 1879, however, cacao was introduced into West Africa, where it started on small farms of 0.8 to 2.0 ha (2 to 5 acres) each. Production in this region increased phenomenally. The Côte d'Ivoire (Ivory Coast) is now the leading producer, followed by Ghana, Indonesia, Nigeria, and Brazil.

Cacao trees are broad-leaved evergreens that grow to a height of 5 to 8 m (16 to 26 ft) under cultivation, but may attain a height of more than 15 m (49 ft) in neotropical forests. They require a truly tropical climate: all production areas are located within 20° of the equator and at altitudes below 300 m (1,000 ft). Cacao trees grow best at a mean annual temperature varying between 21° to 27°C. Growth slows at average temperatures of 15°C and is minimal at average temperatures of 10°C. Production is highest in areas with high humidity and rainfall well distributed throughout the year (about 127 to 152 mm [5 to 6 in.] per month) and with little or no dry season. Winds harm cacao trees by increasing water loss from the leaves and causing defoliation; thus, windbreaks are often used. Cacao trees, like some other tropical crops (coffee and tea), are usually grown in the shade of taller trees, planted especially for this purpose or retained when forests are cleared for new plantings. The optimum degree of shade depends on the fertility of the soil. On fertile or fertilized soils, no shade is needed and yields increase.

Cacao trees grow best in deep, slightly acidic soils that are high in organic matter and are well drained but with a high water-holding capacity. Heavy clay soils are

unsuitable. Cacao trees respond to nitrogen, phosphorus, and potassium fertilizers when shade is reduced or removed.

Cacao plantings consist largely of seedling trees that develop a wide, branching growth habit and start bearing at two to six years of age. Some high-yielding clones have been developed that are propagated vegetatively by root cuttings. The flowers and fruit are borne directly on the older, leafless parts of the branches and on the trunk. Flowering and fruiting may occur throughout the year, and both are determined largely by the rainfall pattern, but the majority of the fruit harvest occurs from September to March.

Cacao flowers are perfect and may be self- or cross-pollinated for fruit setting. Pollination of flowers is by small-bodied flying insects in the Ceratopogonidae family. Lack of adequate cross pollination may sometimes limit crop production. A mature cacao tree may produce 50,000 flowers, but less than 5 percent set pods, and many of those do not reach maturity. The pod-like oval fruits are 10 to 50 cm long containing thirty to fifty seeds, each surrounded by sweet, white mucilage. Pods usually mature in five to six months on cultivated clones and are harvested weekly when they change to a red or yellow color. They are cut open when fully ripe to remove the seeds. The mucilaginous pulp around the seeds is removed by a fermentation process, which takes three to eight days and causes temperatures to rise to about 51°C. This process kills the embryo in the seeds and develops certain chemical precursors that give the chocolate flavors when the beans are later sun-dried and finally roasted.

Cacao trees are attacked by several fungal pathogens. The pods are susceptible to infection by *Phytophthora palmivora,* which results in a disease called black pod. Another pathogen, *Marasmius* spp., causes witches' broom. Recommended management to reduce the deleterious effects of witches' broom on cacao production includes the use of phytosanitation (removal of diseased plant parts), application of chemical fungicides, and the use of pathogen-resistant varieties.

Citrus (*Citrus* spp.) RUTACEAE

Citrus fruits, especially oranges, are among the world's highest ranking fruit crops, along with grapes, bananas, and apples. Commercial citrus species are native to southeast Asia and eastern India. Commercial citrus is primarily produced at low elevations in the world's subtropical climatic zones between 20° and 30° north and south of the equator. Production can occur at higher latitudes if the orchard is near a large, relatively warm body of water such as the ocean. Citrus has comparatively little cold resistance and is not grown commercially where minimum temperatures are likely to fall below −6.5°C.

During the past 100 years, world citrus production has increased from less than 1 million to over 100 million metric tons (MT). Brazil is the largest producer of oranges—followed by the United States, Mexico, India, and Italy. The United States is the largest producer of grapefruit and pommelos; Mexico is the largest producer of lemons and limes and China is the top producer of mandarins and tangerines. About 80 percent of all citrus grown is consumed in the producing countries themselves. Oranges account for the bulk of the citrus produced, followed by mandarins, including tangerines. Lemons, limes, grapefruit, and pommelo account for most of the rest. Florida is the leading citrus-producing state in the United States, especially in orange and grapefruit production. The vast majority of the Florida orange crop is used for juice production. California is second in production, and the majority of its orange crop is produced for direct consumption. Citrus production in the United States has declined somewhat during the last several years.

Most present-day orange cultivars have been grown for many years. The famous seedless 'Washington Navel' orange was found in Brazil in the early 1800s, and propagating material was introduced to Washington, D.C., by the USDA in 1870. It is believed to have originated as a limb sport of the seedy 'Seleta' sweet orange in Brazil. 'Seleta' was introduced to Brazil by the Portuguese via the Iberian Peninsula from settlements in the Orient. Citrus species interbreed easily, and several interspecific and intergeneric hybrids, for example, tangelos, have been developed from a cross between mandarin orange and the grapefruit or pommelo. New cultivars targeted at specialty markets, for example, "blood" oranges (oranges containing red and/or purple pigments), have recently been introduced. The common and scientific names and the important citrus cultivars grown in the United States are provided in Table 21–1.

A deep, well-drained, fertile soil is optimal for growing citrus, but many kinds of soils are used. Soils with low fertility can be made more productive by adding fertilizers. In alkaline or high saline soils, iron deficiency appears and the trees require iron chelate sprays. Some areas where citrus grows have long, dry periods that make irrigation mandatory. All types of irrigation are used: furrow, basin, sprinkler, and drip. Because much of the citrus crop is on the trees during the winter months, frost damage is a real hazard. In

Table 21–1
SOME COMMERCIALLY IMPORTANT CITRUS SPECIES IN THE UNITED STATES

Common Name	Scientific Name	Important US Cultivars
Grapefruit	*Citrus paradisi* Macf.	'Marsh,' 'Duncan,' 'Redblush,' 'Thompson', 'Henderson and Ray,' 'Rio Ruby,' 'Star Ruby'
Lemon	*C. limon* (L.) Burm. f.	'Eureka,' 'Lisbon,' 'Bearss'
Lime	*C. aura otifolia* (Christm.)	'Tahiti'
Orange (sweet)	*C. sinensis* (L.) Osbeck	'Valencia,' 'Washington Navel,' 'Hamlin,' 'Pineapple,' 'Parson Brown,' 'Marrs'
Mandarin	*C. reticulata* (Blanco)	'Satsuma,' 'Dancy,' 'Temple,' 'Minneola,' 'Orlando,'* 'Nova,'* 'Owari'

*Known hybrids, not pure mandarin.

selecting the site for the grove, low-lying frost pockets should be avoided. Oil-burning heaters or wind machines are often installed in bearing orchards to protect against local radiation frost damage to blossoms, fruits, and trees. Wind machines are operated at 0°C and below, except when there is no temperature inversion. Where water is available for flood, furrow, or sprinkler irrigation, it may be applied for heat release in cold weather.

Most commercial kinds of citrus set adequate crops without cross pollination. A few of the hybrids do require cross pollination. Bees work among citrus flowers, and hives are placed in citrus groves for the collection of honey.

In subtropical regions with cool winters, most citrus species bloom once a year in the spring. In the tropics and warm-winter areas, the flowering period may be prolonged or it may occur several times during the year. Some seed-propagated citrus trees produce acceptable fruit, but most superior cultivars must be propagated by vegetative methods. Most citrus is grafted onto vigorous rootstock. Much of that rootstock comes from embryos that originate in nucellar tissue. Nucellar embryos are produced asexually in the seed (in addition to the sexual embryo) by apomixis. The nucellar embryos are the same genotype as the female parent and thus maintain the maternal clone. The type of rootstock used in citrus can strongly influence fruit quality, yield, and tree size of the fruiting cultivar, as well as other horticultural characteristics such as disease resistance and soil adaptation.

Rootstocks generally used are seedlings of rough lemon, sour orange, 'Troyer' citrange, trifoliate orange, 'Cleopatra' mandarin, and 'Rangpur' lime. Which rootstock is used depends on location and species to be grafted.

Pruning young citrus trees delays bearing and should be done only enough to develop a trunk and strong scaffold system. Bearing orange and grapefruit trees require only light pruning. On the other hand, bearing lemon trees need heavier pruning to facilitate orchard operations and limit their height.

Most citrus fruits store best on the tree, but they can be held for a time under refrigeration. Most California oranges are marketed fresh, but in Florida about 90 percent are processed, mostly as frozen concentrate juice. Florida and Brazil together produce more than 85 percent of the orange juice consumed in the world. About two-thirds of Florida's grapefruit are marketed in a processed form. Many insect pests attack citrus. Some important ones are citrus red mite, citrus thrips, citrus mealy bug, red scale, yellow scale, snow scale, purple scale, citrus whitefly, and aphids. In addition, certain species of nematodes attack citrus roots.

Citrus disease problems are more devastating than insect pests, and their management is critical for production of high-quality fruit. Virus diseases include tristeza, exocortis, xyloporosis, vein enation, and yellow vein. All of the viruses are transmitted by budding and grafting with infected wood, and some are distributed by insect vectors. The best control is to use only propagating material or nursery trees produced under conditions free of such viruses.

Coconut (*Cocos nucifera* L.) ARECACEAE

The coconut is by far the most important nut crop in the world. The coconut is a tall, unbranched monocotyledonous tree (palm) grown throughout all the tropical regions. The native home of the coconut is difficult to determine because over the centuries, the nuts

(seeds) were easily dispersed by ocean voyagers and by ocean currents from island to island and continent to continent. Some believe it originated in the islands of the Malaysian Archipelago, but others believe it had a Central American origin.

Although the coconut is a tropical plant with commercial production within 15° north and south of the equator, it grows and fruits some distance from the equator, as shown by plantings in southern Florida at latitude of 26° north. The world's leading producers are Indonesia, the Philippines, India, Brazil, and Sri Lanka.

Coconuts withstand some frost. Growth and production are best under high humidity with mean annual temperatures around 26.5°C and daily fluctuations of no more than 5°C. The coconut palm is a light-requiring tree and does not grow well under shade or in very cloudy conditions. Coconuts need ample, well-distributed soil moisture and suffer from long, rainless periods. Annual rainfall of at least 152 to 177 cm is required, although irrigation can be used to supplement low rainfall. The trees grow in a wide range of soil types—beach sand, coral rock, or rich muck—provided they are at least 1.2 m deep. Coconuts tolerate high salt levels in the soil and salt sprays found along the seashore. Shallow soils with a high water table, which prevents good root development, are unsuitable because the tall palms are likely to blow over in tropical typhoons and cyclones.

The coconut tree has a tall, flexible trunk marked by leaf scars and topped by a crown of leaves. A growing point in the central crown produces new leaves and flowers. This growing point is the only bud on the palm; hence, if it is killed, death of the entire palm follows. These growing points, called hearts of palm, are a very nutritious and tasty food product, reminiscent of artichoke hearts. Trees of some palm species, including the coconut palm, can be grown specifically for this product, even though extracting it kills the tree.

Nearly all trees are propagated by seed because no part of the tree can be used for vegetative propagation. The coconut palm does not produce offsets as does the date palm. Tree characteristics of the coconut are reproduced fairly well by seed so that the best strategy is to select seeds for new plantings from trees that produce large crops of high-quality nuts. Fully matured nuts (at least twelve months old and still enclosed in the husk) are usually germinated in a seedbed by planting them flat on their sides and covering them with soil. They are kept moist during germination; after eight to ten leaves have formed, the new plant plus the husk is transplanted to its permanent location. Trees are usually set about 7.5 m (25 ft) apart.

Several heterogeneous cultivars are recognized. Tall palms tend to be slow maturing, flowering six to ten years after planting, with a life span of eighty to 100 years. Dwarf coconut palms start flowering in their third year and have a productive life of thirty to forty years. Coconut palms respond to nitrogen and potassium fertilizers and, in some soils, to phosphorus, although in many naturally fertile soils no benefits have been obtained from added fertilizers. Little is known of their need for trace elements.

Both male and female flowers are borne on the same many-branched inflorescence arising from a leaf axil in the top of the palm. Each inflorescence can have up to 8,000 male flowers along the terminal part of the inflorescence and one to thirty female flowers near the base. Transfer of pollen from the male to the female flowers by wind, birds, or various insects is required for nuts to develop.

The coconut fruit is classified as a drupe—just as is the fruit of the peach or apricot. The hard shell commonly seen when one buys a coconut in the store is the inner layer of the matured ovary wall of the fruit (the endocarp). Outside this is the husk (the mesocarp plus exocarp), which is usually removed when the nuts are harvested. The husk is now being processed and sold as coir fiber, used in the manufacture of soilless potting mixes. Inside the shell (endocarp) is the true seed with a thin brown seed coat. The white solid part (meat) of the coconut is part of the endosperm, a food storage tissue. Coconut milk is also endosperm in a liquid form. The tiny embryo is buried under one of the three eyes in the endosperm at the end where the fruit was attached to the plant. The coconut is mature and ready to eat about a year after pollination of the flower.

An important trade item is copra, the dried endosperm. Copra is used principally as a source of coconut oil and of shredded and sweetened products used in confections. It contains 60 to 68 percent oil, of which approximately 64 percent is extractable. The copra cake left after oil extraction is ground to a meal that is a high-protein cattle feed. Fresh coconuts can be stored for up to two months at 0° to 1.5°C at a relative humidity of 75 percent or less. They will keep for two weeks at room temperature.

The low content of unsaturated fatty acids makes coconut oil resistant to oxidative rancidity. It is used in the preparation of margarines, shortenings for cooking, frying oils, and imitation dairy products. Coconut stearin is valued as a confectionery fat and as a substitute for cocoa butter. Coconut oil is also used in the manufacture of liquid and solid soaps and detergents, cosmetics, hair oil, and various lubricants.

Pests and Diseases   Coconut trees are highly susceptible to various diseases. Viruses have eliminated entire plantations. Lethal yellowing, possibly caused by a mycoplasma-like organism, is particularly severe in the Caribbean area and has killed thousands of acres of coconut palms. A similar disease has occurred in West Africa. In Florida and some other regions, the common coconut is being replaced with 'Malayan Dwarf' (*Cocos* sp.), which is resistant to lethal yellowing. In the Philippines, a disease called cadang-cadang has wiped out many large coconut plantations. *Phytophthora palmivora* fungus can kill the terminal growing point, leading to the death of the entire palm. Bronze leaf wilt causes death of the older leaves, eventually working its way to the younger leaves. It is a nonpathogenic problem due to unfavorable weather conditions, often appearing after a long drought. Some insect pests attack coconut palms, but only a few are serious and can generally be controlled by sanitary measures.

Coffee (*Coffea* sp.) RUBIACEAE

The genus *Coffea* consists of more than seventy species, but only four species (*Coffea arabica, C. canephora, C. liberica,* and *C. dewevrei*) are involved in world coffee production. *Coffea arabica* (Arabica coffee) and *C. canephora* (Robusta coffee) account for more than 70 percent and 25 percent of the world market, respectively. *C. arabica* is native to the highlands of southwestern Ethiopia and southeastern Sudan. *C. canephora* originated in the African equatorial lowland forests from the Republic of Guinea and Liberia eastward to Uganda, Kenya, and southern Sudan, extending into central Africa. The Arabs were using coffee as a beverage as long ago as 600 CE. During periods of expansion between the eleventh and sixteenth centuries, coffee appeared in Turkey, the Balkan states, Spain, and North Africa. It found its way into western Europe about 1615 (tea was first used there about 1610 and cocoa about 1528).

Coffee plants were brought to Brazil in 1727. After forty years, it became, and still is, the world's leading coffee grower and exporter, now producing one-quarter of the world's coffee supply. Vietnam ranks second in coffee production, followed by Indonesia, Columbia, and Mexico. Latin America produces mostly *C. arabica,* whereas the African continent and southeast Asian coffee production are dominated by *C. canephora.* Although arabica beans represent 70 to 75 percent of the world's production, only about 10 percent qualify as specialty coffees sold by gourmet retailers. Coffee grown in highland areas, where the temperatures are somewhat cooler, produce a milder-type bean from which a better drink can be prepared than that from trees grown at lower and hotter elevations. Robusta beans produce a stronger, less flavorful coffee with a higher caffeine content and are now in great demand for use in blending with *C. arabica* coffees and in the manufacture of instant coffees. *C. liberica* is an important species in some areas but produces an inferior coffee. Coffee is the mainstay of the economy of the Central American countries, and it is grown to a limited extent in the West Indies and in Hawaii, where the Hawaiian Kona coffee is well known. The typical *C. arabica* plant is a large bush to small tree with dark green leaves. It is an allotetraploid and differs from other coffee species that are diploid. Optimum altitudes for *C. arabica* are 1,000 to 2,000 m and a temperature range of 15° to 24°C. *C. arabica* trees withstand temperatures near 0°C (32°F) for only a short period. Frost is one of the principal hazards in growing coffee. Cold-damaged trees that have been exposed to temperatures below freezing have considerable difficulty recovering. Temperatures above 27°C tend to reduce flowering and fruiting, while temperatures below 13°C (55°F) cause cessation of growth and tree stunting. *C. arabica* is self-fertile. Fruit shape is oval and generally reaches maturity in seven to nine months. It is grown throughout Latin America, in Central and East Africa, in India, and to some extent in Indonesia.

C. canephora prefers altitudes of 0 to 700 m, and optimum temperatures are 24° to 30°C. It ranges in size from a small shrub to a small tree reaching heights up to 10 m. The rounded fruits mature in ten to eleven months and contain oval-shaped seeds, which are smaller than the seeds of arabica coffee. *C. canephora* is diploid ($2n = 22$) and self-sterile. It must be cross-pollinated and is highly heterozygous. Robusta coffee is grown in West and Central America, throughout southeast Asia, and to some extent in Brazil, where it is known as "Conilon." Because robusta grows at lower elevations, it matures more rapidly than arabica and can thus be cultivated more easily and brought to market more economically.

Coffee trees need a continual soil moisture supply. Annual rainfall of about 165 cm (65 in.) is required, although supplemental irrigation can be given in drought periods. Optimum precipitation for *C. arabica* is 150 to 200 cm, while *C. canephora* requires 200 to 300 cm annually.

In many countries other than Brazil and Hawaii, coffee trees are planted in the shade of other tall-growing trees. Coffee, apparently, has a low light saturation level for photosynthesis so the trees grow adequately in the

shade. However, sun-grown coffee trees give the highest yields provided all other cultural factors, particularly soil fertility, are optimal.

Growth and Development   Although the coffee plant is technically an evergreen shrub, it is referred to as a tree. One reason is that unless pruned, a coffee plant can grow more than 6 m (20 ft) high. The root system will penetrate down to 4 or 5 m in depth. Coffee trees and their foliage are tender and are quickly injured by strong winds. Wherever such winds blow, rows of windbreak trees must be used.

Most coffee trees now are from seedlings grown in a nursery for one year before planting in the field. Coffee trees are planted about 2.4 m apart and are kept low—about 1.8 to 2 m—by pruning to facilitate harvesting.

The trees grow best in a slightly acid, loamy soil at least 0.9 m deep with good aeration and drainage. To grow and yield, coffee trees require a high level of soil fertility. They readily show symptoms of a lack of any of the essential elements. Plantations managed without added fertilizers show a steady decline in production. As pH moves above 5.1, iron deficiency becomes a problem. At pH below 4.2, calcium becomes deficient. Deficiency symptoms of all the essential mineral elements have appeared in coffee plantations.

Coffee plants start bearing full crops on the lateral branches at about five years and reach full production at fifteen years. Clusters of two to twenty perfect flowers are produced in the leaf axils. The white flowers are very attractive and fragrant. The blossoms last only about three days, pollination is accomplished by wind or insects, and little berries appear six to nine months later. A coffee tree can produce for up to twenty five years, so fertilization is required.

C. arabica is self-fertile and tends to reproduce fairly true from seed. *C. canephora* trees are self-sterile and depend on cross pollination; therefore, the seedlings vary considerably.

Botanically, the coffee fruit is a drupe. The two greenish seeds (the coffee beans) in each fruit are surrounded by a mucilaginous pulp and consist mostly of a hard, thick, and folded endosperm food-storage tissue covered with a thin silvery seed coat. The fruits (called cherries by the growers) first are green and then become dark red when ripe.

Harvesting of coffee begins in Brazil in May at the end of the rainy season. In harvesting the coffee fruits, which are mostly picked by hand, the pickers must remove only the mature red fruits, leaving the green ones for further development. A coffee tree may have blossoms, green fruit, and ripe fruit all at the same time on the same branch. Following harvesting, the pulp must be removed from the coffee cherries to obtain the seeds (beans) inside. There is both a dry and a wet process for this stage. In the dry process, the fruits are washed and then spread out on concrete slabs in the sun to dry. They are turned several times a day. Dry fruits are then repeatedly run through fanning and hulling machines to free the coffee beans inside. In the wet process, the cherries are run through a depulping machine that breaks them open and squeezes the beans out of the pulpy skin. Then they go into large tanks for twenty-four to forty hours and the jelly like substance surrounding the beans ferments slightly. Then the beans are thoroughly washed, spread out in the sun to dry, and continually turned and mixed. In wet weather, drying machines are used, tumbling the beans in perforated drums through which warm air blows.

Finally, the green coffee beans are polished and graded according to origin, size, quality of preparation, and taste or cup quality. Once quality is determined, and the green beans are inspected to remove any that are defective, they are packed into 60 kg bags for shipment to coffee-importing countries.

Beans on the same tree but picked six months apart may have a perceptibly different flavor. Different growing climates during the year affect the taste of the beans. Bean quality at a given location also varies from one year to the next. Coffee companies must sample beans continuously and buy from different areas to keep the taste of their products the same. Roasting the green coffee beans further develops the characteristic tastes and color and must be skillfully done. Coffee purchases are often based on quality, as determined by highly trained coffee tasters who test-sample lots of roasted and ground coffee.

Millions of bags of unroasted coffee beans are shipped to the consuming countries each year for blending, roasting, grinding, and packing. Marketing of coffee produced by thousands of small farmers must necessarily be done by cooperative organizations. *Organic, sustainable,* and *fair trade* coffees have also entered in the market. This is in response to consumer interests in ensuring that production takes plance on farms that conserve resources and protect the environment while striving to enhance the quality of life for the producers.

Many disease and insect pests affect coffee trees. Probably the worst disease is the leaf-rust fungus (*Hemileja vastatrix*). This rust first appeared in Sri Lanka

in 1869, destroyed all coffee plantations throughout the island, and spread to other coffee-producing countries in that part of the world. Coffee planters in Sri Lanka dared not risk replanting coffee so they turned to tea, developing Sri Lanka into the world's second-largest tea producer (India is first). Vigorous trees well fertilized with nitrogen seem to resist coffee leaf rust, and sprays with copper compounds aid in controlling the fungus. *Coffea arabica* is susceptible, while *C. canephora* is considered to be resistant.

Another fungus, *Mycena flavida,* is the causal agent of American leaf spot. This disease causes defoliation and actually kills the plant faster than leaf rust. The disease is most prevalent in Mexico, Guatemala, Costa Rica, Colombia, and Brazil, and it can be severe. The spores live indefinitely in the ground. The coffee bean borer (*Stephanoclores*) is a problem in Brazil, where it is referred to as the coffee plague. The coffee leaf miner and the Mediterranean fruit fly commonly attack coffee trees in some countries.

Mango [*Mangifera indica (L.)*] ANACARDIACEAE

The cultivated mango is a member of the ANACARDIACEAE (cashew family). Mangoes have been cultivated in India for at least 4,000 years. In northeastern India, the small-fruited ancestral species, *Mangifera indica,* is indigenous.

The mango has been associated with the economic, cultural, religious, and aesthetic life of the Indian people since prehistoric times. The Gautama Buddha is said to have been presented with a mango grove by Amradarika (500 BCE) as a quiet place for meditation. The fruit appear to have spread eastward during 400 to 500 BCE to East Asia and they reached the Philippines in the fifteenth century. Westward spread occurred during the sixteenth century via Portuguese explorers, who carried them to Africa and Brazil.

Mango trees are long-lived, erect, and symmetrical perennial trees. They typically reach about 10 to 30 m (33 to 98 ft) in height, but sometimes grow to 40 m (131 ft). Mature leaves are leathery, glossy, and deep green in color. The tree has a deep tap root (to 6 m [20 ft] in good soil) and a wide-spreading feeder root system and several anchor roots. The efficient root system enables the tree to tolerate drought. Mangoes are adapted to a range from 25°N to 25°S, and up to 1,000 m (3,289 ft) elevation. Tropical climates with a wet–dry season (monsoon climate) are preferable because a dry season is best for fruit set.

It may be cultivated in small or large commercial plantings throughout the humid and semiarid lowlands of the tropics. The tree is nearly evergreen and is a favorite in dooryards and gardens of rich and poor alike. Leading producers are India, China, Thailand, Pakistan, and Mexico.

Although the mango is a tropical plant, mature trees have withstood temperatures as low as –4°C for a few hours. Young trees and actively growing shoots are likely to be killed at –1°C. Flowers and small fruits are damaged if temperatures drop below 4.5°C for a few hours. Temperatures of 24° to 27°C are considered optimal for mangoes during the growing season, along with high humidity. They tolerate temperatures as high as 48°C.

Propagation Mangoes can be propagated in many ways. Monoembryonic types must be propagated vegetatively if the cultivar characteristics are to be retained. Polyembryonic types can be propagated by seed or by one of several vegetative methods. Mangoes are large trees and should be planted 10.5 to 12 m (35 to 49 ft) apart. Mango is not particular about soil as long as it is well drained though it may become chlorotic in alkaline soil. It prefers sun and will not flower in the shade. During tree establishment, phosphorus is important for root development, and nitrogen is applied to sustain the first years of growth. Nitrogen and potassium are needed by bearing trees for good yields. Applications of nitrogen to mature trees should be performed with caution because excess application may prevent blooming. The amount of rainfall is not as important as when it occurs. The optimal pattern is four months with 750 to 2,500 mm rainfall followed by a dry season. Irrigation of young trees may be required until roots are well established.

The mango inflorescence is an erect, pyramidal, branched terminal panicle up to 0.6 m (2 ft) long, having several hundred to several thousand flowers. A tree may have from 200 to 3,000 panicles so that tremendous numbers of flowers are produced. The small flowers mostly function as males by providing pollen, but a fraction are able to set fruit. The flowers are pollinated by fruit bats several insects. Rain or heavy dew during bloom are deleterious. High temperature, wind, and low humidity prevent flowering.

Until ten years of age, trees tend to bear fruit each year. After this point, they tend to alternate bear. This tendency may be on a whole tree or branch basis. Some cultivars may flower and fruit irregularly throughout the year. The best approach to getting a consistent harvest is to plant cultivars with regular bearing habits like 'Tommy Atkins' (Fig. 21–3), 'Keitt,' 'Kent,' and 'Pope.'

Nombre Común   : Mango
Nombre Cientifico : Mangifera indica
Variedad   : Tommy Atkins

Figure 21–3
'Tommy Atkins' mangoes. Source: Richard Pratt.

The fruit are variable in color, size, and shape. Mangoes ripen and may be picked when the flesh inside has turned yellow, regardless of exterior color. Fruit may be round, oval, ovoid-oblong, or somewhat kidney shaped. Fruit flesh is very juicy and usually light yellow to deep orange in color. In some cultivars, masses of fibers extend from the stony pit into the flesh, making the fruit difficult to cut.

It has been estimated that there are between 500 and 1,000 named varieties in India. Most of them arose through seedling selection; some have arisen through hybridization and selection. Most leading Indian cultivars are yellow skinned. The Indian race, grown in India, Pakistan, and Bangladesh, has highly flavored, brilliantly colored fruits well suited to commercial production. The Indochina, or Saigon, race produces smaller, less attractive fruits with a yellowish-green color and nonfibrous flesh with a delicate flavor.

Utilization   The mango is a highly esteemed tropical fruit. It is generally peeled to be eaten fresh, but it can also be frozen, dried, and canned; used in pies, jams, and jellies; or processed for juices and nectars. Eastern and Asian cultures use unripe mangoes for pickles, chutney and relishes. In India, unripe mangoes are sliced, dried, and made into powder for amchoor, a traditional preparation used for cooking. Seeds can be milled for flour and may also be eaten during periods of food shortages. The fruit are good sources of Vitamins A and C. Vitamin A content is the highest in cultivated crops, even exceeding that of red pepper. The timber is used for boats, flooring, furniture, and other applications.

Pests and Diseases   Anthracnose (*Colletotrichum gloeosporioides*) affects fruits, inflorescences, and foliage.

Anthracnose may be a problem, especially at fruit-setting time. Powdery mildew (*Oidium mungiferae*) on inflorescences, common in India, is a problem in Brazil. Trees may also be attacked by mango scab (*Elsinoe* sp.).

Scale is considered the worst pest. Mango hopper is a serious pest in India. Mediterranean (*Ceratitis capitata*) and Mexican fruit flies are frequently a problem. To manage fruit flies, many commercial producers have implemented more stringent field sanitation, the use of lures and bait, and release of sterile flies to mate with wild flies helps to suppress populations. Internal breakdown of the fruit is an important problem. Its cause has not yet been determined.

Papaya (*Cariaca papaya* L.) CARICACEAE

Common names of cultivated papaya are paw-paw (Australia), mamao (Brazil), and tree melon. *C. papaya* plants have never been found in the wild, but a close relative, *C. peltata*, grows wild in southern Mexico and Central America, leading to the supposition that *C. papaya* may have originated in this region. The early Spanish and Portuguese explorers carried the papaya to most tropical countries throughout the world.

Papaya is cultivated primarily for its ripe fruits. It is favored by tropical people as a breakfast fruit, as an ingredient in jellies and preserves, or cooked in various ways. The enzyme papain, which is used in meat tenderizers, is extracted from green fruit. The world's leading producers are Brazil, Mexico, Nigeria, India, and Indonesia. Most papaya fruits are consumed locally because of the difficulties in long-distance transportation of the tender fruits. The practice of air shipments to the US mainland and to Japan, however, has given impetus to the Hawaiian industry. Most US supplies of papaya come from Hawaii, with smaller amounts from Mexico and the Dominican Republic.

Papaya is reported to tolerate annual precipitation of 650 to 4,300 mm (mean = 1,920 mm, or 76 in.) and annual mean temperature of 24.5°C. It grows best below 1,500 m in well-drained, rich soil of pH 6 to 6.5. Cultivation is known in soils ranging in pH from 4.3 to 8.0. It is reported to tolerate drought and high pH. It is extensively cultivated—as far north and south as 32° latitude. At extreme latitudes, however, the weather is often too cool in most years for papayas to ripen properly. Papaya becomes almost weedy in some areas of the tropics.

Growth and Development   The papaya is an erect, short-lived, fast-growing, woody tree or shrub. It usually grows to a height of several meters but may reach 8 m to

9 m (26 ft to 30 ft) tall. The hollow green or deep purple trunk is straight and cylindrical with prominent leaf scars. The leaves emerge directly from the upper part of the stem in a spiral. The leaves remain clustered near the top of the plant. Leaves are large, typically 40 to 60 cm (16 to 24 in.) wide, soft, palmate, and alternate, and are borne on long petioles. Trees bear clusters of delicious, mild-flavored large or small fruits varying in shape from spherical to pear-like.

Flowers and fruits appear nearly continuously all year. Papayas are usually dioecious in nature; thus, trees differ from each other because some are either male (staminate) or female (pistillate). Some are perfect, having both male and female parts in the same flower. At certain seasons, some plants produce short-stalked male flowers, at other times, they produce perfect flowers. This change of sex may occur temporarily during high temperatures in midsummer. Certain varieties have a propensity for producing certain types of flowers. For example, the 'Solo' cultivar has flowers of both sexes 66 percent of the time, so two out of three plants produce fruit, even if planted singly. How pollination takes place in papayas is not known with certainty. Wind is probably the main agent because the pollen is light and abundant, but thrips and moths may assist. Hand pollination is sometimes necessary to get a proper fruit set. For heavy production of high-quality fruits, papayas must be grown in warm, 21° to 26.5°C, frost-free climates in full sun. Lower temperatures give poor results.

Fruits of papaya are sweet, juicy, and of orange color. The seeds, numerous in the central cavity, are rounded, blackish, and about 0.6 cm (0.2 in.) in diameter. Each is enclosed in a gelatinous membrane (aril).

Propagation   Papayas are normally propagated by seed. Soil used for seedbeds should be absolutely free from root knot nematodes (*Meloidogyne*). Early planting is much to be desired to make a vigorous plant before the beginning of the following winter.

The average production life of a papaya tree is considered to be two or three years, but trees may live in the wild twenty-five years or more. Yield declines after the first few years. All inferior and wild male trees close to a production area should be destroyed so that their pollen cannot fertilize blossoms of trees from which seed is to be selected. Seed for new plantings should be saved from perfect-flowered plants whenever possible so that one can be reasonably assured of the source of pollen. Propagation by tissue culture is also possible.

Production   The usual planting distance is about 3 to 4 m (10 to 13 ft) apart each way, giving about 1,750 trees/hectare (700 trees/acre). The more vigorous growing plants are usually the males, and some may be discarded when selecting plants for field planting. It is necessary to leave about one male plant for each twenty-five females to ensure pollination. Transplants must be watered and shaded. Mulch helps to reduce weeds, preserve moisture, and shade the soil from the hot summer sun.

Papayas grow on many soil types, but they do not tolerate salinity in either irrigation water or soil. Good aeration and drainage are necessary. A light, well-drained soil with a pH of 6.5 to 7.0 is optimal. Papaya trees need ample soil moisture at all times but they also can be killed by excess moisture. If rainfall is lacking during parts of the year, irrigation is required to supplement the rain. For good plant growth and production, the various mineral nutrients must be readily available in the soil. The kinds and amount of fertilizers required depend on the soil type. Ideally, the soil should be moist in hot weather and dry in cold weather. A plant that has been injured by frost is particularly susceptible to root rot.

The fast-growing papaya responds to regular applications of nitrogen fertilizers, but exact rates generally are not well established. Phosphorus deficiency causes dark green foliage with a reddish-purple discoloration of leaf veins and stalks. Generous applications of manure or commercial fertilizers are considered beneficial.

Papayas do not need to be pruned, but some growers pinch the seedlings or cut back established plants to encourage multiple trunks. Papaya trees are somewhat delicate and should be protected from strong winds. Recovery of partially lodged trees may require one or two months.

Mature fruits are produced about ten months after planting and then year-round. Individual fruits are harvested when they are at a mature-green stage and show a tinge or more of yellow at the apical end.

Utilization   In the tropics the papaya is used for many purposes. Fruits are eaten fresh and the juice makes a popular beverage; young leaves, shoots, and fruits are also cooked as vegetables. Green papayas should not be eaten raw because they contain latex, although they are frequently boiled and eaten as a vegetable. Many manufactured products are made, especially papaya juice concentrate and dried slices. They are good sources of vitamins A and C and of potassium.

Immature fruits contain papain, a protein-digesting enzyme, which is extracted from the latex of the skin. It is obtained by scoring or injuring the fruit,

followed by collecting and drying the latex. Presently Uganda and Tanzania are the principal producers of papain. The United States is the principal importer of papain. It is utilized for meat tenderizers, beer treatment, chewing gum, and in the textile and tanning industries.

Pests and Diseases The papaya fruit fly (*Toxotrypana*) is a serious pest in some areas, so that bagging fruit while still quite small is the only means of protecting it. Both fungal and viral diseases attack papayas. Anthracnose (*Colletotrichum*) is a common fungal disease in Hawaii that attacks the fruits. Some local packinghouses use a hot water bath to help minimize anthracnose as a post-harvest problem. A *Phytophthora* blight is the most important disease. It attacks above-ground portions of the plant as well as the roots. Powdery mildew (*Oidium*) can sometimes present a problem. Two viral diseases, papaya mosaic and papaya ring spot, caused by papaya mosaic virus (PMV) and papaya ring spot virus (PRSV), respectively occur in Hawaii and have been an ongoing threat to the industry. Initially, papayas were moved from Oahu to Hawaii to escape PRSV infection during the 1960s. By the 1990s, infections were so widespread that growers were replanting trees annually because of the severity of PRSV, which is transmitted by the green peach aphid (*Myrus*). Production had fallen by nearly 40 percent by 1997. However, two cultivars 'Rainbow' (yellow-fleshed, transgenic) and 'Sun Up' (red-fleshed, transgenic cross), have been made available and the industry has bounced back. The industry in Florida has also been hit hard by viruses and the papaya wasp. Nematodes species isolated from papaya are numerous.

Mites are a serious pest on papayas but can be controlled by sulfur sprays. The melon, Oriental, and Mediterranean fruit flies attack papaya fruits in Hawaii and are the cause of the strict quarantine and fumigation requirements. Before Hawaiian papayas can be shipped to the US mainland, they must be treated to destroy fruit flies.

Pineapple [*Ananas comosus* L. Merrill]
BROMELIACEAE

The pineapple is today one of the best known of all tropical fruits. It is a low-growing, herbaceous perennial bromeliad. The pineapple is the only bromeliad producing tasty fruits, although many bromeliads are used as ornamental plants in outdoor and indoor settings. The pineapple is thought to have originated in southeastern Brazil, Paraguay, and northern Argentina in central South America, although other researchers argue for a more northern center of origin. Improved cultivars selected and propagated by the indigenous people in this region had become widely distributed throughout tropical America when Columbus encountered the fruit on the island of Guadeloupe in 1493 on his second voyage. The Portuguese and Spanish explorers apparently disseminated the pineapple throughout the world's tropical regions in the 1500s by carrying the fruit crowns, which remained viable throughout the long voyages. Pineapple was grown in Hawaii as early as 1813, but it was not until the middle 1880s that a significant industry began to develop. At the turn of the nineteenth century, Florida was the leading producer of pineapples, but the industry was decimated by a presumed disease, which later was found to be mealybugs. Hawaii became the leading producer. Today, the leading producers of pineapple are Thailand, the Philippines, China, Brazil, and India.

Adaptation Pineapple grows best under uniformly warm temperatures year-round. While plants might survive −2°C, significant leaf damage would severely weaken the plant. Temperatures higher than 32°C can lower fruit quality, and higher daytime temperatures can sunburn exposed fruits. Because pineapple is a xerophytic plant, it uses water very efficiently, so the crop is grown in areas having relatively low rainfall, from 50 to 200 cm annually (20 to 80 in.). The optimum for commercial production is in the range of 100 to 150 cm annually. In the past, irrigation was used only during very dry periods or in very dry areas. It is now recommended whenever less than 5 to 10 cm of rainfall occurs during a one-month period, and particularly as an aid for the establishment of young plantings. Commercial pineapples are generally grown at somewhat higher elevations close to the equator.

Propagation Pineapple plants are propagated vegetatively using suckers that arise below ground (ratoon suckers) or in the leaf axils, slips growing from the fruit itself or along the stalk below the fruit and crowns. The crown is the vegetative shoot on top of the fruit (Fig. 21–4). New plants from crowns require about twenty-four months to fruit; from slips, about twenty months; and from suckers, seventeen months. Plantings are generally made to coincide with periods of heavy rainfall. Slips and suckers are preferred in commerce. Plants may be dispersed here and there in a compound garden, or they may be planted in long

Figure 21–4
Pineapple plant with immature fruit. The uppermost fruit shows the crown, which can be used for vegetative propagation. Source: Ken Chamberlain, OSU-OARDC.

rows on beds under plastic mulch in commercial plantings.

Each original plant set out produces one fruit. The pineapple fruit forms in a complex fashion. At the terminal vegetative growing point of the plant, a central axis core develops, bearing an inflorescence of 100 or more closely attached lavender flowers arranged spirally, each subtended by a bract. This central axis core terminates in a crown or rosette of small leaves. All parts of each flower, together with the bract, develop into the fleshy fruitlet tissue. As it develops, the entire inflorescence converts to the single pineapple fruit (botanically, a sorosis) by the coalescence of the inner stem tissue and fruitlets (developed ovaries and other parts of the flowers).

Production   Pineapple plants absolutely require well drained soils. They grow and fruit best in slightly to somewhat acidic (pH range 4.5 to 5.0) sandy loams and volcanic soils. Field preparation of large commercial fields is usually done with heavy machinery. Fields are often covered with black plastic and fumigated to manage nematodes. Restrictions and concerns regarding the use of fumigants have encouraged the development of alternative management techniques. Nonvolatile nematocides can be introduced through drip irrigation systems with success. Pineapples generally require added fertilizers, particularly nitrogen and potassium. Low levels of phosphorus are needed, and excess levels may reduce yields and lower quality. Spiny leaved varieties can make hand weeding difficult. Chemical induction

of flowering by treating the plants with ethylene is frequently practiced in plantations to allow synchronous harvest of fruit.

Cultivars   'Smooth Cayenne' types are grown extensively in the tropics and are the major processing varieties. They are also found fresh in many markets but are the most sensitive to internal browning caused by low temperatures. Internal browning is also known as blackheart or chilling injury. The 'Queen' group has superior fresh-market quality, and it is more frequently cultivated in subtropical areas because it is less sensitive to chilling injury. The group is, however, more prone to sun-scalding of the fruit. The 'Natal Queen' cultivar produces fruit for the fresh market. 'Red Spanish' is a major (white-fleshed) fresh-market pineapple in the Caribbean region, and it is a little hardier than other cultivars There are varieties of pineapple with much better eating quality than 'Smooth Cayenne' or 'Red Spanish.' However, those with better eating quality do not ship very well, so they are not likely to be encountered in export markets. Among the better pineapples for fresh consumption are 'Pernambuco' (pale yellow-flesh) and 'Eleuthera', and 'Abakka.' 'Cabezona' is a triploid variety from the Spanish group producing very large that is grown in the state of Tabasco, Mexico, and a small part of Puerto Rico. The 'Maipure' group provides fresh-market fruit in Central and South America.

Pests and Diseases   The major insect problem is the mealybug (*Dysmicoccus* sp.), which is tended and protected on pineapple plants by big-headed ants (*Pheidole megacephaola*). An effective ant control program makes it possible for the mealybug to be kept under control using baits or by biological means, mostly by predation. Ant-control costs usually are less than those incurred in controlling mealybugs directly. Diseases of pineapple include root rot and heart rot caused by *Phytophthora parasitica* Dastur and *P. cinnamomi* Rands. Root and heart rot are exacerbated by poor soil drainage and high rainfall. Black rot, caused by *Thelaviopsis paradoxa,* is characterized by a soft, watery rot. The rot is also called butt rot, blister rot, and soft rot. Interfruitlet corking caused by species of *Fusarium* and *Penicillium* fungi, and pink disease caused by a Cetamonas bacteria may also occur periodically. Black heart, and root rots are controlled by a pre-plant dip containing fungicides and by postplant sprays where problems are expected to be severe. Root rot is also controlled to some degree by keeping soil pH low, in the range of 4.5 to 5.5. Pineapples are also sensitive to attack by nematodes.

SUMMARY AND REVIEW

The tropics and subtropics are characterized by rich cultural and biological diversity. A large portion of the world's population resides in these regions, and agriculture is a primary occupation. Land forms, soil conditions, and climate determine the feasibility of crop production. Temperature and rainfall are the most important determinants for crop production. Humid tropics, intermediate tropics, tropical savanna, dry tropics, and desert describe the range of dry period durations from shortest (less than 2.5 months) to longest (ten to twelve months). The pattern of wet/dry periods is also an important factor in crop productivity. Solar radiation is highest in the tropics and subtropics so potential productivity is high, but water or nutrient limitations reduce productivity. The intense solar radiation also causes insolation (intense heating) of the soil, again limiting productivity. Insolation decreases with an increase in latitude as the angle at which solar radiation strikes the earth increases. Temperature regimes in the tropic include equatorial, tropical, tierra templada (frostless highlands), tierra fria (nonfrostless highlands), andine, semitropical, marine, temperate, pampean-patagonian, continental, polar, and alpine.

Savannas are mostly flat and include open grassland with widely spaced shrubs or trees that are mostly short-lived, and more closely spaced trees. Rainfall and the ability of the soil to hold water determines the density of the trees. There are several types of rainforests including multilayered lowland, lowland, and mountain forests. Many of the rainforests are being converted to other uses such as farming and ranching. When the soil is not quickly covered with new vegetation, erosion occurs. The intense rainfall of the region and poor infiltration characteristics of the soil create heavy erosion. The warm temperatures and plentiful rainfall of the tropical and subtropical rainforests create great potential for agriculture, but the poor quality soils there jeopardize the productivity.

Tropical soils differ from temperate soils in several ways. Tropical soils are low in organic material, high in aluminum and manganese, low in CEC and base saturation, and low in major and minor nutrients. Important soil orders found in the tropics are the oxisols, utisols and alfisols, vertisols, andisols, alluvial soils, inceptisols and entisols, spodozols, and aridisols. Oxisols are highly weathered and leached and lack distinct horizons. Utisols and alfisols are typically red and yellow. Alfisols are considerably higher in fertilityand can sustain more cropping than utisols. Vertisols have a high clay content and poor physical structure. Andisols are derived from volcanic ash and are highly permeable. They tend to be some of the most productive soils in the tropics. Alluvial soils occur in low-lying areas such as river plains and low coastal regions and are the result of recent sedimentation. Their quality depends on their source material. Inceptisols and entisols are "young" soils. Spodosols are highly leached, low in nutrients, and have poor water-retention characteristics. They are vulnerable to erosion. Aridisols occur in areas where is little rainfall and consequently little leaching. They tend to be fertile, though they lack nitrogen.

Several different cropping systems are found in the tropics and subtropics. Shifting cultivation is practiced at the community level. The location of the production fields shifts around a permanent home. During the fallow period, natural vegetation regrows. When put back into production, nutrients held in the vegetative biomass are released by cutting, drying, and burning. The sustainability of the system depends on how long the regrowth period occurs. As long as it is long enough to maintain soil fertility, it can continue conceivably forever. Making shifting systems more sustainable can be done by using low-impact land clearing and tilling methods, increasing the number of legumes grown, and using mulches as well as cover and understory crops. Fallow system farming is similar to shifting but is practiced by individuals rather than communities.

Permanent upland cultivation is practiced in semi-arid regions with distinct dry seasons. Planting is delayed until the rainy seasons appear to be established. Arable irrigation farming, as its name implies, relies on irrigation. In this case, water is gravity fed and allows production of multiple high-value crops per year. Flooded systems produce crops of tropical wet rice. The fields may be flooded by rainfall or irrigation. Rice may be the only crop or it may be grown with other crops. One or more crops may be grown per year. Cereal-based systems rely mainly on sorghum and millet, but maize may also be grown where water and nutrients are more plentiful. Cowpeas, beans, and peanuts are legumes that are traditionally interplanted with cereal crops. Mixed annual/perennial systems involve some type of root or tuber crop and another root crop or a cereal or legume.

Some important tropical and subtropical crops that are harvested as annuals include the following: yams, sweet potatoes, cassava, dasheen and eddoe, tannia, white potatoes, legumes, cereals (millet, rice, sorghum, maize, and several vegetables. Perennials that are important include: avocado, banana, cacao, citrus (oranges, grapefruit, lemons, limes, etc.), coconut, coffee, mango, papaya, pineapple.

KNOWLEDGE CHECK

1. What are the two dominate climate factors for tropical crop production?
2. What is the length of the dry season in humid tropics? Tropical savanna? Tropical desert?
3. What distinguishes monsoon climates?
4. How does the intensity of rainfall in the wetter areas of the tropics compare to the intensity of rainfall in temperate zones? How does this affect the soils in the wet tropical areas?
5. What is insolation and how does it affect crop productivity?
6. List six of the temperature regimes in the tropics.
7. What geographic factor accounts for most of the differences among the regimes in question 6?
8. What are savannas?
9. Where are the three main groups of tropical rainforests located?
10. What makes tropical rainfall more intense than typical rainfall in temperate zones?
11. How does rainfall in the tropics affect soil erosion?
12. List the eight major categories of tropical soils and give a characteristic of each.
13. Why is the term "shifting" used to describe that type of cultivation?
14. In a shifting system, who traditionally owns or crops the land?
15. Describe the essential feature of a shifting cultivation system.
16. Why is the length of the fallow period critical to the sustainability of a shifting system?
17. Describe the mound-type shifting system used for growing crops in the tallgrass savanna.
18. Why are taller, more robust plants grown in the latter part of a shifting cultivation cycle?
19. What is meant when it is said that exploitation of microclimates occurs in diverse shifting cultivation systems?
20. What are two negative effects on shifting cultivation that are coming from increased population pressures?
21. When is planting done in a permanent upland cultivation system?
22. What form of energy and what source of water is used for irrigation systems in arable irrigation farming?
23. What effect does flooding have on weed management in rice fields?
24. What are the three main cereal grains grown in the cereal cropping systems of West Africa? What legumes are traditionally interplanted with them?
25. How does soil type influence which grains and legumes are grown in the cereal cropping systems of West Africa?
26. Why is time of planting very important in the cereal cropping systems of West Africa?
27. What is the difference in input with the Asian mixed annual/perennial system compared to the mixed-root crop systems and wetland yam mound systems?
28. What crops predominate in African mixed-root crop systems?
29. How does length of fallow time influence whether cassava or yams are grown in a mixed-root crop system?
30. Where are the different intercropped species planted in a wetland mound cropping system?
31. Why is using so many different species of yams critical to the year-round productivity of the hillland growing system? What other crops are also traditionally grown in this system?
32. What are three advantages to mixed crop systems compared to monocultures in areas with poor soils?

FOOD FOR THOUGHT

1. What do you think plant scientists who are interested in the ecology of tropical and subtropical cropping systems could learn from the indigenous people who practice that type of farming?
2. You have recently joined the Peace Corp (or some other similar organization) and have been assigned to help farmers in a remote tropical rainforest of West Africa to develop a plan to harvest some of the forest for timber and create land for farming but retaining the characteristics of the forest that benefit the environment and keep the soil from degrading. What are some things you would consider as you begin the plan?
3. Your neighbor in Southern California has two avocado trees of the same cultivar in her yard because she heard it takes two to make fruit and there didn't seem to be any other avocadoes nearby. The trees have matured but are not producing any fruit. What would you tell her to do to correct the problem?

FURTHER EXPLORATION

ALLAN, W. 2005. *The African husbandman*, 2nd ed. London: Lit Verlag.

COMMITTEE ON SUSTAINABLE AGRICULTURE AND THE ENVIRONMENT IN THE HUMID TROPICS, BOARD ON AGRICULTURE AND BOARD ON SCIENCE AND TECHNOLOGY FOR INTERNATIONAL DEVELOPMENT, AND NATIONAL RESEARCH COUNCIL. 1993. *Sustainable agriculture and the environment in the humid tropics*. Washington, DC: National Academy Press.

JUO, A. S. R., AND K. FRANZLEUBBERS. 2003. *Tropical soils: Properties and management for sustainable agriculture*. New York: Oxford University Press.

NORMAN, M. J. T., C. J. PEARSON, AND P. G. E. SEARLE. 1995. *The ecology of tropical food crops*, 2nd ed. Cambridge: Cambridge University Press.

UNITED NATIONS AND FAOSTAT (Food and Agriculture Organization Statistics). 2007. *FAOSTAT*, http://faostat.fao.org/faostat/ FAOSTAT provides time-series and cross sectional data relating to food and agriculture for some 200 countries.

WRIGLEY, G. 1982. *Tropical agriculture*, 4th ed. London: Longman.

*22

Nursery Production

Gary Bachman

The nursery industry is an exciting business and provides many opportunities for students interested in plant production. The production of nursery plants for profit has the potential of providing many personal and financial rewards to those who are willing to work hard. On the surface, operating a nursery appears to be fairly simple, but the nursery business is very complex. Operating a nursery requires not only knowledge of and skills in plant basics, but experience in labor management and marketing of nursery crops.

The nursery industry is diverse. People often think that nurseries produce shrubs and trees, but increasing numbers of nurseries are producing flowering perennials, annual plants, ground covers, and hardscaping materials. **Hardscape** consists of the nonplant components of a landscape, for example, patios, decks, and so on. Always remember that a nursery is a business. Like any other business, success depends on imagination, determination, planning, and good management of resources available to the nursery.

key learning concepts

After reading this chapter, you should be able to:

- List the factors that go into site and product selection for a nursery.
- Explain the principles of field and container (including pot-in-pot) nursery crop production.
- Discuss the importance of and the methods for testing media fertility for container production.

NURSERY ESTABLISHMENT

The evaluation of a proposed nursery site is vital to the success of the nursery. Four categories must be considered before establishing a nursery. These categories are ecological, economic, sociological, and biological in nature.

The first ecological consideration that must be examined is water. An adequate supply must be available through the growing season either from municipal systems or from surface or subsurface sources of good quality for irrigation purposes. What to do with runoff from the nursery is as important as the water supply. Many states and countries now require that the runoff from nurseries be closely monitored and often contained. As a result, retention ponds are often built to collect the runoff. Often these are used as a source of irrigation water but care must be taken to make sure that the water is of suitable quality and does not contain unacceptable amounts of fertilizers and other chemicals. Achieving high quality is often done by collecting all runoff including rain water. The rain water can dilute any chemicals in the pond to a level that is acceptable for use in irrigation.

The other ecological consideration: the physical features of the property and the prevailing weather patterns of the area are discussed in Chapter 1.

The site analysis includes economic factors, with probably the biggest expense being land cost, depending on the size of the nursery. From the start-up situation, a nursery often has to wait three to five years before seeing a return on investment dollars. Labor is another very important consideration when looking at the economics of a nursery, with the availability of labor being the biggest issue. The nursery owner must realize that he or she cannot do everything.

Marketing is vital to any nursery business. Deciding how the plants will be sold is as important as how they will be grown. Wholesale nurseries sell plants only to other businesses, not to the general public. Retail sells to the public. Some nurseries will do both. Both retail and wholesale nurseries require sufficient space for both growing and sales areas, along with parking for both staff and customers. The landscape nursery provides landscape and landscaping services as well as retail and/or wholesale plant sales. A third marketing decision is the sales area to be serviced: local, regional, and/or national. **Bare root** plants (plants that are dug from the ground and have most of the root soil removed) can be shipped economically nationwide via postal service or commercial carrier due to reduced weight. **Ball and burlap** plants (plants that are dug from the ground and the root and soil ball is wrapped with burlap or other material to keep the soil in place) have a limited shipping range because of the weight of the soil ball containing the root system. This range is usually within 300 to 500 miles of the nursery, and, like bare root plants, shipping is limited to the plants' dormant season. Container-produced plants (plants that are grown and usually shipped in pots) have the advantage because they can be marketed all four seasons although weight of the soil in the container can add to shipping costs. Potential niches within the nursery industry must also be explored.

A question that must be answered before any production is begun at a nursery is, which plants should I grow? Or phrased another way, which plants can I sell and to whom do I sell them? There are literally thousands of different species and/or cultivars from which a nursery owner can choose. Not only must the nursery grower decide what species to grow, but he or she must be able to predict the needs of the market anywhere from three to ten years in the future. This timeline should allow ample time for propagation and subsequent growth (in other words, production time) of adequate numbers of the desired plants. A nursery with a good marketing plan can try to create market demand for a particular species of plant, but change in the nursery industry requires time. The rule of thumb for nursery

inventory considerations is that 70 percent of production should be industry standard species. These are species for which there is a known market. The remaining 30 percent of the inventory can include plants that are currently being promoted as plants having a bright future in the landscape industry. Plants not to grow include:

- Easily produced plants that every other nursery is already producing.
- Plants that the nursery grower personally likes (unless they fit one of the above 70/30 criteria).
- Slow-growing plants, at least in the initial stages of a nursery.

All plants should pay for themselves regardless of marketing strategy.

Sociological factors must also be taken into consideration. The demographics of an area must be familiar to the nursery. Population numbers and income levels are important if the nursery has targeted local sales. Zoning and local regulations must be adhered to by the nursery. Examples of such regulations include proximity of neighbors, and any restrictions regarding local road weight limitations for streets and bridges, as well as any water runoff laws.

Biological factors that must be taken into consideration have a more direct production-oriented effect for the nursery. In the initial stages of nursery establishment, it is important to know if a site has a predisposition for certain pests and if there are beneficial organisms in the ecosystem. If the area has a large deer population, you can rest assured that the deer will cause browse damage on your most expensive items. Pest management is discussed in greater detail in Chapter 15.

Once you have decided which crops to grow and have performed a proper site analysis, the next decision is which production system will best suit the potential business: nursery field production; nursery container production; or pot-in-pot, which is a hybrid production system between field and container production strategies. These options will be discussed in the following sections.

NURSERY FIELD PRODUCTION

A field nursery grows nursery crops to marketable size in fields laid out similar to other field crops. Roadways, field borders, fence rows, wooded areas, and areas unsuitable for cultivation can easily account for 30 to 40 percent of a field nursery's land area, leaving only 60 to 70 percent of the space available for actual production. A field is generally divided into blocks, separated

Figure 22–1
Field-grown nursery trees in rows with overhead irrigation.

by 10 to 12 ft wide grass or gravel-covered roadways. Primary roadways in a field nursery need to be at least 15 to 20 ft wide to accommodate trucks, tractors, and various harvesting machinery used to perform maintenance, planting, and harvesting tasks (Fig. 22–1).

The design of the production blocks is vital for the success of producing plants through field production. Spacing plants in the field should make efficient use of land, facilitate traffic and maintenance needs, and provide easy access to the field. The most common and most expensive mistake new growers make is to plant **liners** (young plants) too close together in very narrow rows in the field. This practice is done with the good intention of digging every other plant early in the production cycle to allow additional growing space for the remaining plants. It is unlikely, however, that every other plant will grow equally with straight, well-branched trunks and canopies. Close spacing also makes harvesting extremely difficult, and digging one tree often damages two or three, making them unsalable.

The distance between the rows and between each plant depends on desired market size, width of maintenance equipment, and the type of harvest (either hand or mechanical). Wider spacing between rows ensures that plants will not be shaded and will have longer and stronger lower side branches, thus achieving a more normal shape. Proper plant spacing allows plants to remain in the field longer, if needed, if sales do not meet projections. Spacing within rows depends on final market size and if intermediate harvesting is to be practiced. A practiced rule of thumb is to plant the trees 3 ft apart for each inch of anticipated trunk diameter. For example, a grower plants Red Maple liners with 3 ft spacing within the row. When the plants reach an inch

in diameter, every other tree is harvested and the remaining trees are grown to a 2 in. diameter because they are now spaced 6 ft apart.

All field stock is harvested during the dormant season. Trees that are transplanted when dormant will establish a new root system capable of supporting the new foliage that will emerge in the spring. Harvesting field-grown nursery stock requires a mechanized tree spade or laborer experienced in hand digging. Tree spades can be purchased in a wide array of sizes to dig balls from 15 to 96 in. and bigger. The size of the root ball should be in proportion to the diameter of the trunk caliper. Nursery trees are measured 6 in. above the soil. For example, a 1¼ to 1½ in. (3.2 to 3.8cm) caliper tree should have an 18 in. ball, a 2½ to 3 in. (6.4 to 7.6 cm) caliper tree should have a 28 in. (71 cm) ball, and so on.

Using a mechanical tree spade to harvest a tree is a straightforward process. The tree spade is positioned around the trunk of the tree and large blades are hydraulically forced into the soil, cutting all lateral and tap roots. In many cases, 80 to 90 percent of the root system is left behind during this process. If it is to be balled and burlapped, the roots and attached soil (rootball) is lifted and placed into a wire basket lined with burlap. Once the tree spade is removed, the burlap is wrapped tightly around the rootball to hold it in place (Figs. 22–2 and 22–3). If the plant is to be shipped bare-root, the soil is washed or blown off the roots after digging. The air spade is special type of mechanical spade that removes soil from around the roots during the digging process which decreases the damage to the roots compared to conventional digging

Figure 22–2
Using a mechanical tree spade to harvest broadleaf evergreen trees.

Figure 22–3
Securing a mechanically harvested tree in a wire basket to hold the root ball together for shipping.

Figure 22–4
Digging plants using an air spade.
Source: Photo courtesy of Barbara Fair and Ahlum and Arbor Landscaping.

(Fig. 22–4). After digging, the bare-root mass is wrapped in a moist material such as sphagnum moss for storage and transport.

As with other crops, it is important to rotate species when replanting to help manage pests.

NURSERY CONTAINER PRODUCTION

Almost every species and type of herbaceous and woody ornamental plant produced in a nursery setting can be grown in containers that sit on top of the ground (Fig. 22–5). Growing plants in containers is vastly different from growing plants in the field and provides production advantages to the nursery grower. The grower can increase the number of plants per unit area when compared to field production: container-grown plants may be placed closer to each other because their root systems are controlled. Also crop rotation issues are eliminated because the plants are not growing in the soil itself. Injury caused by cultivating machinery used in field production is reduced. The grower can produce more uniform plants in containers when compared to plants produced in the field because optimum growing conditions may be maintained more effectively. Container production provides the grower complete control over the media, nutrition, irrigation, light, and spacing.

The root systems of container-grown plants suffer less damage compared to plants that are harvested from the field. A tree that's dug in the field has its root system reduced between 80 and 90 percent. Because of this, field harvesting can be accomplished only when the

trees are in a dormant state, typically in late fall through very early spring. Container-grown plants, on the other hand, can be harvested and marketed throughout the year because there is little or no loss of roots so the roots can continue to take up water and nutrients.

Container production does have disadvantages for the nursery grower to consider. The biggest disadvantage is the plant's total dependence on irrigation because the root system is limited to the container size and is not in actual contact with any source of groundwater. Another disadvantage is the limited development of a root system within the container. Normally, the root system grown in the field is approximately two to three times the diameter of the canopy of the plant. Because of this growth potential, plants that remain in containers for too long become root bound and may

Figure 22–5
View of a large container nursery.

not grow well when transplanted unless the root mass is loosened and the roots pulled out from the form of the container.

Exposure of the root system to cold temperatures becomes an issue during winter months in cold climates. Various techniques have been used by nursery growers for overwintering containerized nursery stock: clear or milky plastic covered houses or tents (polyhouses) over the plants; consolidating the containers with or without covering; mass consolidation with mulch around the containers, mass consolidation a covering of a special type of foam; and consolidation with a top covering of poly-coated plant foam, white plastic, clear plastic/straw/clear plastic ("the sandwich"), clear plastic/grass, geotextiles, white plastic/microfilm, or microfoam (Figs. 22–6 and 22–7).

Inexpensive greenhouses covered with a single layer of polyethylene film (poly houses) are also used routinely to provide winter protection for containerized plants. Temperatures within poly houses depend primarily on the amount of sunlight entering the houses. More sunlight means higher temperatures within the structures. Unfortunately poly does a very poor job of holding heat within the houses and tents. On a clear, cold winter day, heat can radiate out of a poly house in late afternoon or early evening as fast as it increases in the morning with the rising sun. Even so, plants in a poly structure generally remain several degrees warmer than the ambient outside temperature. In nearly all situations, except in the most extreme cold temperatures, this winter protection is enough. Usually the plastic film is removed in the early spring to allow the plants inside to acclimate to ambient conditions. The discarded plastic can create a serious disposal problem,

Figure 22–7
Applying opaque white polyethylene across container-grown plants for winter protection.

however, there are now companies that will take the plastic away and recycle it.

POT-IN-POT PRODUCTION

Pot-in-pot production is a recent development in container-grown plant production. In this production system, a planted container is placed into a slightly larger holder pot that has been permanently placed in the ground. Pot-in-pot production was first started in the southern states to help provide root protection from extreme summer temperatures, but it has really caught on in northern states because of the advantages of root protection during winter (Fig. 22–8).

The pot-in-pot production system can eliminate many of the other difficulties associated with

Figure 22–6
Nursery structure used for winter protection of container-grown herbaceous perennials.

Figure 22–8
Pot-in-pot production field showing socket pot.

Figure 22–9
Blowover of container-grown plants. Blowover can be pre-vented by using pot-in-pot production.

conventional container plant production. Wind tipping or container blowover is another time-consuming, laborious drawback of conventional container production (Fig. 22–9). It is detrimental to plant quality because top-dressed fertilizers and media are knocked out of the pot. Irrigation applications can be missed or delayed, resulting in drought-stressed plants. Pot-in-pot production can prevent blowover. In addition, the sunken pots require less irrigation than do conventional containers; with micro-irrigation, each pot has an individual emitter or spray stake, with almost no runoff. Pot-in-pot production requires no overwintering structures because the pots are insulated from freezing temperatures by the ground around the outer pot. Pot-in-pot production requires more management than does field production, but less than conventional container production. Pot-in-pot production in a way mimics field production.

MEDIA TESTING FOR CONTAINER-GROWN PLANTS

Certain physical and chemical properties of soilless media are necessary for growing high-quality nursery plants. The basic physical and chemical characteristics of soils as well as nutrition that are important to a field nursery are covered in Chapters 5 and 14; however, there are some aspects of soil and nutrition that are unique to nursery production especially for container systems.

Each nursery grower should monitor four media properties as a course of normal production management. The four properties are the physical property of the media aeration and the three chemical properties of electrical conductivity (EC), pH, and nitrate nitrogen (NO_3^-) content. Each media property has simple techniques available for on-site nursery testing. The impact of all four on media management, and ultimately on plant quality, is important.

The composition of the growing medium is an extremely important consideration for growing plants in containers. Not only is the plant root system restricted to the limited volume of the container, but the perched water table created in the bottom of all containers further restricts the total root growing space. The perched water table is an area where all the pore spaces in the media are filled with water. This saturated area occurs no matter how many drainage holes are in the bottom of the container. It results from equilibrium between gravity and the capillary attraction of the media particles and the water molecules. The deeper the container, the lower the impact of the saturated area at the bottom of the container and the greater the overall media aeration. It is recommended that media aeration porosity be measured before any new media formulation is used. It is also recommended that aeration porosity be checked periodically during the production cycle, with frequency depending on the components of the media mix and how fast they decompose.

Aeration porosity is a measure of the amount of air space in the media mix at container capacity after a complete irrigation. The optimum aeration porosity for most woody plants is between 20 and 30 percent. Analysis of media porosity is very simple and can be completed by the nursery grower using the following procedure:

1. To measure the container volume, seal the drainage holes in the bottom of the container and fill it with water to the level it would normally be filled with container mix. Record this volume as "container volume."

2. Empty and dry the container and fill it with growing mix. Slowly add water with a graduated cylinder until the mix is completely saturated. A very thin slick of water will appear on the surface once saturation is reached. Record the total volume of water added as "total pore volume."

3. Place the container over a watertight pan and remove the seals from the drain holes. Allow the free water to drain out of the container. This may take several hours. Measure the amount of drained water and record this amount as "aeration pore volume."

4. The preceding three steps will allow you to calculate total porosity, water-holding porosity, and aeration porosity using the following three equations:

$$\text{Aeration porosity (\%)} = \frac{\text{Aeration pore volume}}{\text{Container volume}} \times 100$$

$$\text{Total porosity (\%)} = \frac{\text{Total pore volume}}{\text{Container volume}} \times 100$$

$$\text{Water-holding porosity (\%)} = \text{Total porosity} - \text{Aeration porosity}$$

Electrical Conductivity (EC) and pH

Fertilizers and other dissolved salts change the ability of the solution to conduct electricity. This change in electrical conductivity allows growers to measure roughly the fertility of their media. Portable electrical conductivity and/or pH meters can be used by growers to measure the electrical conductivity and pH of the soil solution within their container mixes. These portable meters for the nursery industry range in price from $60 to over $1,000, with the more expensive meters generally being more durable and lasting many years. Growing media should be tested at least every two weeks, preferably every week, for electrical conductivity and pH. Keep a chart of the pH and electrical conductivity values collected. With this information, you can establish trends, for example, whether the electrical conductivity and/or pH is increasing, decreasing, or remaining constant. Nursery growers need to pay close attention to electrical conductivity levels. Growers need to realize, however, that not all electrical conductivity is related to the fertilizers used. It is also important to establish the nonnutritional background content of the irrigation water. Another thing to be aware of is that the relative amount of each nutritional element in the media can change, but the EC reading may not indicate that. For example if Ca is decreasing but another element is increasing, the EC reading may not change very much.

Many nursery growers use controlled-release fertilizers. Frequent monitoring of the media electrical conductivity is the easiest way for a grower to track the nutrient release from controlled-release fertilizers. When the electrical conductivity begins to approach that of the irrigation water, the controlled-release fertilizer is becoming exhausted and supplemental fertilization is required.

As you learned in earlier chapters, nutrient availability to plants is affected more by pH than by any other factor. Figure 5–12 shows how pH influences nutrient availability in natural (mineral-based) soil. However, in artificial media pH impacts nutrient availability differently

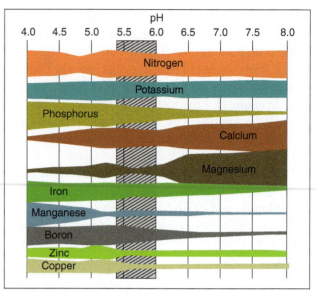

Figure 22–10

pH range of macro- and micronutrient availability in artificial media. The thicker the line, the more available the nutrient.

than in mineral soils (Fig. 22–10). Plants in a high-pH soilless media may express micronutrient deficiencies of iron (Fe), boron (B), zinc (Zn), manganese (Mn), and molybdenum (Mo).

Media Testing Methods

Ornamental nursery growers have been slow to conduct regular EC and pH monitoring of container-grown plants. One of the major reasons has been a lack of a uniform and simple media-testing procedure during the crop production cycle. Two media testing methods are now in use: the saturated media extract method and Virginia Tech extraction method.

The saturated media extract (SME) procedure has been used for many years by commercial testing labs and some growers. This procedure is relatively time-consuming, and growers are hesitant to use it. The procedure requires a sample of the media to be removed from the container and placed in a collection vessel of known volume. Water is slowly added to the collection vessel until the media is saturated and there's a slight sheen on top. The water is allowed to equilibrate for about an hour. pH is measured by inserting the pH probe directly into the saturated media. To measure electrical conductivity, the saturated media is placed in a Buchner funnel, and the water is vacuum-extracted, filtered, and measured with a conductivity meter. If you are using controlled-release fertilizers, false readings may occur from damage to the fertilizer prills.

The Virginia Tech extraction method (VTEM), or pour-through method, is gaining acceptance with

nursery growers because it does not require any special handling of the container media. Within two hours after irrigation, a minimum of three containers from each block of plants should be selected for testing. The containers are placed on a PVC ring in a collection vessel, and a volume of water with which the crop is normally irrigated is applied. The volume applied depends on the volume of the container; typically about 150 ml of water is added for each gallon size of container, that is, 150 ml/1-gallon container, 300 ml/2-gallon container, and so on. Leachate from the collection vessel is analyzed for pH and electrical conductivity.

For the nursery grower, the Virginia Tech extraction method has five main advantages over other monitoring procedures:

1. The sample extraction time is relatively short, as little as five minutes.
2. pH and electrical conductivity measurements may be made in the field.
3. No media handling is required.
4. Sample extraction requires no special equipment.
5. Controlled-release fertilizer is not in contact with the probe; thus, it does not create false readings.

The Virginia Tech extraction method should be performed every one to two weeks. It is important to keep complete records and to chart the pH and electrical conductivity values measured. Graphically charting the pH and electrical conductivity provides the grower with information about whether the pH and electrical conductivity are changing. The results of a single sample should never be used to make a change in culture and/or management. The strength of the Virginia Tech extraction method is its help in establishing a picture for the crop, thus helping the grower to make informed decisions about watering frequencies, fertilizer needs, and leaching requirements.

Plant growth and ultimately sales have already been lost if the nursery grower analyzes media only when nutritional deficiencies appear. Whichever media testing technique a nursery uses, a media analysis program must be established that is consistent in sample frequency. Table 22–1 serves as a guideline for comparison of different media testing procedures.

Nitrate Nitrogen Testing

Nitrogen is one of the most widely distributed elements in nature, but in container production, nitrogen is the most limiting element related to plant growth. Many plants take up nitrogen most easily when it is in the form of nitrate (NO_3^-), and many fertilizers supply this form of nitrogen to plants. However, nitrate ions, like other anions, are easily leached out of the container media by rain and/or irrigation. The nitrates can then get into ground or surface water where it is considered to be a pollutant. Because of the importance of nitrate nitrogen to plant growth as well as its potential to be a water pollutant, the nursery grower should manage

Table 22–1
INTERPRETATION OF SOLUBLE SALT AND pH MEASUREMENTS BY EXTRACTION METHOD[*]

Method	Soluble Salt	pH	Electrical Conductivity (dS/M or mMhos/cm)
VTEM	Sensitive crops (liquid feed)		0.50–0.75
	Nursery crops (liquid feed)	5.2–6.2	0.75–1.50
	Nursery crops (controlled-release)		0.20–1.00
Saturated extract method (nursery crops)	Low		0.00–0.74
	Acceptable		0.75–1.49
	Optimum	5.8–6.8	1.50–2.24
	High		2.25–3.49
	Very high		3.50+
Saturated extract method (greenhouse crops)	Low		0.00–0.75
	Acceptable		0.75–2.0
	Optimum	5.6–5.8	2.0–3.5
	High		3.5–5.0
	Very high		5.0+

[*]The ranges of pH and soluble salt levels should be used as guidelines only. Irrigation water should be <0.75 dS/M. The soluble salt level of water used in the VTEM procedure should be subtracted from the final leachate value.
Source: The University of Georgia College of Agricultural and Environmental Sciences Cooperative Extension Service.

nitrate nitrogen closely. The nursery grower has two choices for determining nitrate nitrogen levels in the media or run-off. The first is to send media samples to a commercial lab and wait for the results. The second choice is to use an ion-specific probe, which is readily available from most horticulture supply companies. Ion specific probes analyze a specific nutritional element.

NUTRITION MANAGEMENT

Good nutrition management is required for the production of quality nursery plants, whether plants are produced in the field, containers, or pot-in-pot production systems. This section on fertilization will concentrate on container and pot-in-pot production because the total supply of federal elements available for plant growth is limited by the size of the container. Good container media and nutrition management are basic requirements in the production of quality container grown plants. The decisions involved in providing good nutrition to container stock are complicated by several factors: the multitude of fertilizer products available, the variations in container media, the great number of species involved, and the various cultural practices used.

The impact of fertilizers on the environment and groundwater is an important concern. To minimize environmental impact from nutrient run-off, nursery growers are using controlled-release fertilizers (CRFs). The two most common methods of using controlled-release fertilizers are incorporation into the growing media and top-dressing. When incorporating controlled-release fertilizers into the media, a previously determined amount is uniformly mixed into the media prior to potting of liner plants. Top-dressing of controlled-release fertilizers is usually performed on plants later in the production cycle, after the original incorporated fertilizer has been exhausted of its nutrients. A previously determined amount is spread evenly across the top of the

Figure 22–11
Controlled-release fertilizer top-dressed on a container-grown plant.

media in the container. Most controlled-release fertilizer manufacturers supplied growers with scoops or spoons that hold a measured amount of the fertilizer to help facilitate application (Fig. 22–11).

Growers using CRFs gain several advantages over traditional granular and/or liquid fertilization including: initial fertilizer levels that are lower in the growing media, availability of nutrients during the entire crop cycle, the great reduction in the amount of nutrients lost from runoff and leaching, and a localized and efficient supply of nutrients to the nursery crops. Although nitrogen is the primary element of concern in nursery production, growers should be aware that the other required elements need monitoring to make sure no imbalances occur. Commercial laboratories are generally used to test for these elements. If an imbalance is found, one or more applications of a dry or liquid fertilizer of the proper formulation will most likely be needed to correct the problem, even there is still CRF remaining in the media.

SUMMARY AND REVIEW

Site selection for a nursery includes evaluation of four categories: ecological, economic, sociological, and biological. Ecological factors include water availability and quality, physical features of the property, and weather patterns. Economic factors are land cost, labor, marketing, and choice of products to grow. Sociological factors include demographics of the neighboring populations, the targeted market population, zoning and other development regulations, and restrictions that may limit access to the property. The indigenous population

of insect and disease organisms and the deer population are biological factors to be considered.

Field production is the growing of nursery stock directly in the ground. Spacing between and within rows is determined by the type of plant being grown and the state of maturity it will reach at harvest time. Access roads into all parts of the field are critical for the movement of trucks, tractors, and maintenance and harvesting machinery. Field stock is harvested during the dormant state. The plants are dug using a mechanical

digger, which separates up to 80 to 90 percent of the roots from the plant. The remaining roots and attached soil are wrapped tightly with burlap to hold them in place.

Traditional container production of nursery stock has the containers placed on top of the ground. Spacing is generally denser than with field production, thus increasing the number of plants that can be produced. Crop rotation is not necessary because the plants are not grown directly in the ground. Another advantage of container production is the reduction in the number of roots lost during harvest. Almost no roots are lost when the plant is removed from its container for transplanting, which means that the plants can be harvested and marketed throughout the year, not just during the dormant season. Disadvantages of container production include the possibility that the plants may become pot-bound and irrigation systems are an absolute requirement. The likelihood of freeze-damage to the roots occurring during overwintering is an issue in colder climates, necessitating the use of overwintering structures. Pot-in-pot production was developed to reduce the heating of the container during the summer, but it has become adopted to reduce freeze-damage during the winter. Pot-in-pot production has other advantages, including a reduction in the number of plants blown over by the wind and reduced irrigation frequency.

Management of soilless growing media and its fertility is one of the key elements in successful container production, whether traditional or pot-in-pot. The choice of media is crucial to producing high-quality plants. Media and aeration porosity are important factors to be considered, as are the EC and pH of the media, both at planting and throughout the production cycle. Growers can monitor these factors as well as nitrate/nitrogen levels themselves. Other nutrient elements are usually tested by outside laboratories. Fertilizers can be applied in many different ways, but controlled-release formulations are preferred because they have a minimal impact on the environment.

KNOWLEDGE CHECK

1. What are three other categories of plants besides trees that are now often produced in nurseries?
2. What are hardscape materials?
3. What are the four categories to consider when evaluating a potential nursery site?
4. Why are retention ponds being built as part of nurseries?
5. How long does it often take for a new nursery to start having a return on investment dollars?
6. The grower has complete control over what aspects of container-grown plants?
7. What are the disadvantages of growing plants in containers?
8. How does pot-in-pot overcome many of these disadvantages?
9. Why is it important to monitor container media for EC and pH?
10. Why does a grower use use a different pH versus nutrient availability charts for field-grown and container-grown plants?
11. What is a media-testing method that is easiest for growers to do on-site?
12. What are the advantages of using controlled-release fertilizers?
13. Why is nitrate-nitrogen considered so important in nursery production?
14. What is an ion-specific probe?

FOOD FOR THOUGHT

1. If you were getting ready to start a new nursery, what would you tell your loan officer to convince him or her that a water retention pond is an essential part of the start-up cost of the nursery even though there is plenty of underground water available?
2. You are growing nursery plants in above ground containers in an area with low temperatures that are usually in the teens. Which of the following winter protection methods would you choose and why would you choose it? No protection; consolidated with mulch, or consolidated in unheated polyhouses?
3. Your plants are showing definite signs of nitrogen deficiency. You have regularly monitored EC and it has not changed. What could be happening?

FURTHER EXPLORATION

AVENT, T. 2003. *So you want to start a nursery.* Portland, OR: Timber Press.

DAVIDSON, H., R. MECKLENBURG, AND C. PETERSON. 2000. *Nursery management administration and culture.* 4th ed. Upper Saddle River, NJ: Prentice Hall.

YEAGER, T., C. GILLIAM, T. BILDERBACK, D. FARE, A. NIEMIERA, AND K. TILT. 1997. *Best management practices guide for producing container-grown plants.* Marietta, GA: Southern Nursery Association.

*23
Floriculture

Jeff Kuehny

Jeff Kuehny

key learning concepts

After reading this chapter, you should be able to:

- Describe the basic greenhouse structure and components.
- Explain how the greenhouse environment is manipulated to regulate plant growth and development.
- Discuss the principles of growing several greenhouse crops.

THE GREENHOUSE INDUSTRY

Greenhouses (or glasshouses) date from Roman times when wealthy people grew plants in small enclosures, which allowed for spring flowers in the winter and fruit out of season. This desire spread across Europe, helping to create the largest greenhouse industry in the world in the Netherlands. The initial structures were built so that during the day, when the sun shone brightly, the glass could be lifted slightly to allow the trapped hot air to escape. Low, unheated glass-covered structures that could be opened were called cold frames by the English. To add heat and make it a hot frame, decaying manure was buried about 0.5 m (18 in.) under the structure (Fig. 23–1). Both the cold and hot frames were constructed to allow a person to reach in and tend to the plants (Fig. 23–2). These units developed into pit houses, which were similar to today's greenhouses and were constructed with a peaked roof, doors at each end, and a center walkway cut about 1 m deep into the ground to accommodate a standing person.

The true greenhouse with its high sides and glass roof, was developed later (Fig. 23–3). Greenhouses were constructed to allow the rays of the sun to enter the glass roof perpendicularly in the winter but partially reflect them when the sun was high during the summer. Heat was supplied with circulating hot water and, at a much later date, with steam. Forced air and radiant heating came even later. Early greenhouses had sash bars usually made of thick wood, especially in areas with high snowfalls, to provide structural support (Fig. 23–4). Lumber resistant to rot such as cypress and redwood was found to be ideal, but because these woods are not particularly strong, the girth of the sash bars had to be large and the panes of glass small, creating many shadows and thereby reducing the light intensity.

The even-span greenhouse (has eaves or gutters on both sides) was the most popular glass greenhouse design for many years (Fig. 23–5). Even-span greenhouses connected along the eave are termed ridge-and-furrow greenhouses (Fig. 23–6). For decades, greenhouses did not vary much in design except that the width of the glass increased with improved glass-making techniques and strong, narrow extruded aluminum bars replaced the wide wooden sash bars. Narrow sash bars and wider glass allowed more light to enter during the winter, when it was needed most.

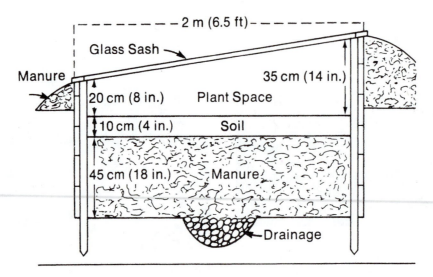

Figure 23–1
A hotbed warmed by decaying manure. Modern hotbeds are warmed by an electrical cable that is thermostatically controlled. The glass sash is lifted during the day to allow excess warm air to escape.

Figure 23–2
Cold frame against the side wall of a greenhouse used to overwinter plants. Source: Jeff Kuehny, Louisiana State University.

Figure 23–4
Interior structure of a glass greenhouse consisting of painted cypress sash bars with iron columns providing structural support. Source: Jeff Kuehny, Louisiana State University.

Figure 23–3
Detached even-span glass greenhouses constructed in an open area to maximize light intensity and expansion capability. Source: Jeff Kuehny, Louisiana State University.

Figure 23–5
Even-span glass greenhouse attached to a head house. The head house is used for storage of planting materials and provides a space for planting and shipping.
Source: Jeff Kuehny, Louisiana State University.

Figure 23–6

A ridge-and-furrow greenhouse range with glass covering and polycarbonate end walls, which provide maximum light intensity and good insulation.

Source: Jeff Kuehny, Louisiana State University.

Figure 23–7

A squirrel-cage fan pulls outside air through the attached flexible tube and into the air space of the two layers of poly-ethylene to maintain a constant pressure.

Source: Jeff Kuehny, Louisiana State University.

As plastics increased in popularity, so did their use as greenhouse glazings. Polyvinyl chloride was used briefly, but it had an unacceptable life expectancy. Fiberglass, with a life expectancy of twenty years, became more popular and is still used if ultraviolet (UV) light-resistant protection is added. Although polyethylene film has a short life expectancy (five years or less), it has become the most popular glazing because of its weight, low heat loss (double layers), and less-expensive price. The quality of polyethylene has increased with added UV inhibitors, infrared inhibitors and absorbers, anticondensate additives, and photoselective inhibitors. Polyethylene film can also be purchased in varying thicknesses: 3, 4, 5, or 6 mil. The greater the mil (0.001 in.) or thickness, the stronger and more durable the film. Single-layer polyethylene is used for overwintering perennials in temperate climates and for greenhouses (usually for propagation) in subtropical or tropical climates. Double-layer polyethylene is used in colder climates; it should have a dead-air space between the layers of approximately 4 in. to provide insulating properties. The outer layer of polyethylene is usually 6 mil, while the inner layer is 4 mil. A small fan is used to maintain the air space (Fig. 23–7).

Acrylic and polycarbonate single- and double-layer panels have gained in acceptance. The double layers have good insulating properties. The two materials have many of the same properties as polyethylene. However, these panels are more rigid, with a life expectancy of up to twenty years. Acrylic has very high

Figure 23–8

Quonset-style greenhouse with a single-layer polyethylene covering, exhaust fan for cooling, and unit heater for heating.

Source: Jeff Kuehny, Louisiana State University.

light transmission, similar to glass. Because of their light weight, both acrylic and polyethylene panels can be very large, which, in conjunction with narrow aluminum sashes, allows for good light entry into the greenhouse.

The use of plastic film as greenhouse glazings led to the development of quonset-style (Fig. 23–8) and gutter-connected greenhouses (two or more quonset-style greenhouses connected, Fig. 23–9), which are less expensive to build than even-span greenhouses. They remain the most popular greenhouses today, with slight design changes that continually improve the growing of plants.

Figure 23–9

Gutter-connected greenhouse range with a double-layer polyethylene covering, polycarbonate end walls, and automatic vents providing natural air circulation and cooling. Source: Jeff Kuehny, Louisiana State University.

Figure 23–11

Greenhouse range with exhaust fans in the side walls used to pull air through the greenhouse for greater cooling and air circulation. Source: Jeff Kuehny, Louisiana State University.

CONTROLLING THE GREENHOUSE ENVIRONMENT

Temperature

Greenhouses are heat traps that are cooled by applying white shading compound to the glass, by applying shade cloth on the outside of the greenhouse, by using shade curtains on the interior (Fig. 23–10), by convection ventilation [allowing the hot air to rise through the overhead ventilator and cool air through the side vent (Fig. 23–11)], by forcibly removing air with exhaust fans (Fig. 23–11), or by evaporative cooling using a fan-and-pad system. The internal shade curtains can be automated to facilitate their use.

Figure 23–12

Cross-fluted cellulose cooling pads with water passing over them. As outside air is pulled through the pads, heat is absorbed, thereby cooling the air. Source: Jeff Kuehny, Louisiana State University.

The fan-and-pad cooling system operates by using a pad, usually made of cross-fluted cellulose, with water passing through it along one wall and exhaust fans evenly spaced along the opposite wall (Fig. 23–12). Outside air is pulled through the pads into the greenhouse. The water passing through the pad absorbs the heat from the air, thereby cooling it. The exhaust fans help pull this cool air through the greenhouse and remove the warmer air.

The greatest heat load in the greenhouse occurs when the air is heated by the sun's rays and then trapped by the enclosure of glass or plastic. Short-wave radiation passing through the greenhouse covering strikes the objects in the greenhouse. The reflected radiation in the

Figure 23–10

An automated shade system in a greenhouse. Source: Margaret McMahon, The Ohio State University.

Figure 23–13
Centralized boilers used to heat water that is piped through-out the greenhouse range for heating. Because centralized heating provides a single source of heat, most greenhouse ranges have at least two boiler units so that if one fails, the other can be used. Source: Jeff Kuehny, Louisiana State University.

Figure 23–14
Single-unit, forced-air furnace used for heating a single greenhouse. Most greenhouses have more than one unit heater to provide even heat distribution.
Source: Jeff Kuehny, Louisiana State University.

form of long-wave radiation (heat) does not pass through the covering and is trapped in the greenhouse, causing an increase in the temperature during the day. However, some form of heat must be added at night during the winter season because too little daytime heat is retained for that purpose. In extremely cold climates, supplementary heat is necessary even in daylight hours. A central heating system derives its heat from a boiler circulating hot water or steam (Fig. 23–13). This type of system is primarily used in large greenhouse ranges. Localized heating can be accomplished through the use of unit heaters, forced-air furnaces (Fig. 23–14), infrared radiant heaters (Fig. 23–15), and root-zone bottom heating.

Convection tubes and horizontal airflow fans are also important for heating and air movement in the greenhouse. Convection tubes are made of transparent polyethylene and are attached to a blower on one end and sealed at the other end. These tubes have 5 to 7.6 cm (2 to 3 in.) round holes along the bottom of the tube. The air that is pulled from the blower is forced through the holes at a high velocity, creating a mixing or stirring with the surrounding air (Fig. 23–16). The blowers may pull cool air through a gable end vent to cool the air in the greenhouse or pull hot air from unit heaters to warm the air in the greenhouse. This system may also be used simply to circulate the air to prevent stagnation. **Horizontal airflow fans (HAF)** (Fig. 23–17) may also be used to help circulate the air in the greenhouse or help mix warm air from unit heaters with the cooler air. These fans gently move the air around the greenhouse in a horizontal direction. For individual greenhouses,

Figure 23–15
These heaters emit infrared radiation that, on striking an object, is converted to heat. Thus, the plants and containers are heated without having to heat the air—saving energy and money. Source: Jeff Kuehny, Louisiana State University.

Figure 23–16
A convection air tube made of transparent polyethylene is attached to a blower on one end and sealed at the other end. Air pulled from the blower is forced through the small holes on the bottom of the tube at a high velocity, thus mixing the surrounding air. Source: Jeff Kuehny, Louisiana State University.

fans should be installed in two rows, one side directing air opposite the other, to help stir the air around the greenhouse. Unit heaters are installed 3 to 4.6 m (10 to 15 ft) behind the first fan at each end of the greenhouse. This positioning helps mix the warm air from the heater with the cooler air of the greenhouse. For a gutter-connected greenhouse, the fans can be installed down the center of each bay and facing the opposite direction for each bay. In addition to mixing warm and cool air, HAFs also reduce the relative humidity in the plant canopy and as a result, decrease the incidence of disease.

Environmental controls for operating the cooling and heating of greenhouses range from very simple, such as a basic thermostat, to a very complex computerized system that controls nearly every aspect of the greenhouse environment. The more complex environmental control systems can control cooling, heating, shade curtains, fertilization, irrigation, irradiance, relative humidity as well as other environmental factors in the greenhouse.

Day and night temperatures are both very important for flowering pot plant production and are usually managed independently. Green foliage plants and some orchids require night temperatures as high as 21°C (70°F) and day temperatures as high as 29°C (85°F). Temperatures not only affect the growth of plants but also flowering. For example, *Cattleya* orchids grown at a 29/21°C (85/70°F) day/night temperature will flower much more quickly than those grown at a 21/13°C (70/55°F) temperature. Roses, chrysanthemums, and

azaleas do best at night temperatures in the 17°C (63°F) range, whereas carnations, snapdragons, cinerarias, and calceolarias grow well at 10°C (50°F). If height control is not a problem, day temperatures are set 0 to 5 degrees above night temperature on cloudy days and 10 to 15 degrees above night temperature on sunny days to maximize photosynthesis and minimize respiration. Maintaining these temperature differences is easy on overcast days but difficult on bright days because of an extreme heat buildup from the sun's radiation. When bright conditions prevail, plants may have to be (1) shaded during the summer months (e.g., May to September), (2) watered more often to compensate for excessive transpiration, or (3) sprayed with an overhead mist system to increase the humidity. All three of these procedures compensate for the high light intensity conditions that may cause sunscald damage to leaves or flowers.

DIF (day temperature–night temperature) can be used to help control the plant height of many greenhouse crops. As the day temperature increases above the night temperature (a positive DIF), internode (stem) elongation increases. Inversely, maintaining a night temperature greater than the day temperatures (negative DIF) decreases internode elongation.

Light

Three aspects of light affect plant growth and development: light quality, light intensity, and light duration. These are discussed in detail in Chapter 7 but how they are controlled in a greenhouse will be discussed here. The **daily light integral** (DLI) is determined by intensity and duration. **Light quality**, or the light spectrum,

Figure 23–17
Horizontal airflow fans. Source: Margaret McMahon, The Ohio State University.

drives photosynthesis and can affect plant shape (photomorphogenesis). Plants use red and blue light most efficiently for photosynthesis, red and blue light inhibit stem elongation while far-red light promotes internode elongation. Plants under high levels of red light are short, branched, and have dark green green leaves. Plants grown under high levels of far-red light have longer internodes, are tall, unbranched, and have light green leaves. Properly spacing plants in a greenhouse reduces the effects of far-red light. This is a very important technique for growing high quality plants.

Light quantity or intensity refers to the amount of light that a plant receives and can be measured in photosynthetically active radiation (PAR $\mu mol\ s^{-1}$) or photosynthetic photon flux (PPF $\mu mole\ s^{-1}\ m^{-2}$). Quantum sensors are available commercially that allow a grower to inexpensively measure PPF. The greater the light intensity, the greater the rate of photosynthesis until the light saturation point occurs.

Solar light is the best light for growing plants in the greenhouse. However artificial lighting has been used in greenhouse production to supplement or replace sunlight at times. Fluorescent lights have low intensity but good quality for photosynthesis so they are used primarily for germination and growth chambers and in homes for interior lighting of houseplants. Incandescent lamps are used only for creating night breaks for photoperiod control because they are high in far-red light. High-intensity discharge (HID) lamps are the most common types of supplemental lighting in the greenhouse because they provide acceptable light intensity and quality for photosynthesis (Fig. 23–18). High pressure sodium or metal halide are the best types of HIDs for use in a greenhouse.

Figure 23–18

Overhead high-intensity discharge lamps used to supplement the lighting in the greenhouse for optimum crop growth.

Source: Jeff Kuehny, Louisiana State University.

Photoperiod control is very important for the timing of photoperiodic greenhouse crops to meet production demand and market windows. Short-day (SD) plants flower when the dark period is longer than a specified length. Plants such as azaleas, chrysanthemums, poinsettias, and some orchids require SD to initiate flowering. Long-day (LD) plants flower when the light period is longer than a specified length. China asters, Shasta daisies, *Gypsophila,* and *Liatris* require LD. Day-neutral (DN) plants, for example, *Pelargonium ×　hortorum,* require no specific daylength for flower initiation. Growers can time photoperiodic plants to flower for specific dates by manipulating the photoperiod. To do this, greenhouse growers should remember four dates that will help determine the necessity for photoperiod manipulation: June 2, the longest day of the year (give or take a day); December 21, the shortest day of the year (give or take a day); and September 21 and March 21 (both give or take a day), the autumnal equinox and vernal equinox, respectively, when the light period is equal to 12 hr. During a long light period (between March 21 and September 21), flower induction of SD plants or maintenance of vegetative growth of LD plants can be accomplished by covering the plants with black cloth (excluding all light) from 5 P.M. to 8 P.M. (cooler climates) or 7 P.M. to 7 A.M. (warmer climates). If the light period is short, flower induction of LD plants or vegetative growth of SD plants can be accomplished by night interruption lighting. Incandescent or fluorescent lights can be placed over plants and turned on from 10 P.M. to 2 A.M. (4 hr.) with a light intensity of approximately 2 μ mol. HIDs can be used if already in place for supplemental photosynthetic light.

The total quantity of light delivered over the course of an entire day, the daily light integral (DLI), is closely correlated to plant growth. This unit of measure (moles day^{-1}) is a combination of both light intensity and duration. The DLI inside a greenhouse is rarely higher than 25 to 30 moles day^{-1} and can vary widely depending upon the geographic location, time of year, and cloud cover. The minimum DLI for growing most greenhouse crops is 10 to 12 moles day^{-1} although the optimum DLI varies by the type of crop being grown. Branching, rooting, stem thickness, and flower number may increase as the average DLI increases. When natural DLI falls below acceptable levels, then the use of supplemental light may be necessary.

Control of Plant Growth and Flowering

Control of plant height and growth is important in greenhouse crop production. Plants should be of adequate size for shipping and handling as well as being

Figure 23–19

Poinsettia (Euphorbia pulcherrima) *grown for the Christmas holidays. The plant on the left was grown without a plant growth retardant (control) and the plant on the right was treated with Bonzi. Bonzi is a common plant growth retardant used to reduce plant size for ease of shipping as well as to provide a more aesthetically pleasing appearance.*

Source: Jeff Kuehny, Louisiana State University.

aesthetically pleasing to the consumer (approximately 1.5 to 2 times the height of the container) (Fig. 23–19). Plant height can be controlled by manipulating the greenhouse environment or production practices. Controlling light quality, quantity, and photoperiod provides a means of controlling plant height and growth. Temperature manipulation including using DIF, as discussed earlier, can help control plant height and growth. Applying or withholding a certain amount of water during irrigation along with changing fertilization formulations and concentrations can also be used to manipulate plant growth. However, many situations prevent the greenhouse grower from using the aforementioned methods. When controlling the environment is not possible to get the desired growth, chemical plant growth regulators and retardants (PGRs) can be used. However, if the crop is to be used for edible purposes (vegetables, herbs, edible flowers, etc.), most chemical plant growth regulators and retardants cannot be used.

Plant growth retardants can be applied to control plant growth. With the exception of ethephon, the chemicals following inhibit stem elongation by blocking the synthesis of gibberellic acid. Plant growth retardants can be applied by spraying the plant or drenching the growing medium (uptake by the root system). Following is a list of popular plant growth retardants:

- A-Rest® (ancymidol) is labeled for many flowering pot plants and bedding plants, and it has been a popular growth retardant of bulb crops for many years. A-Rest® can be applied as a spray or drench.
- B-Nine® SP (daminozide) is labeled for many flowering pot plants and bedding plants. It can be applied only as a spray to the surface of leaves.
- Bonzi or Piccolo (paclobutrazol) is labeled for many greenhouse crops and can be applied as a spray or a drench. Sprays should be applied to the shoots of the plant because the active ingredient is translocated through the stem and reduces internode elongation.
- Cycocel (chlormequat) is labeled for most greenhouse crops and can be applied as a drench or a spray. Combining B-Nine® and Cycocel in the same tank mix provides often better growth regulation than using either alone.
- Florel Brand Pistill (ethephon) is labeled for many floriculture crops and works differently from the previously mentioned PGRs by releasing ethylene to the plant. It has been used for many years to enhance flowering and ripening of fruit. More recently, it has been used extensively to reduce plant height, induce lateral branching, and delay flowering.
- Sumagic® (uniconazole) is labeled for some floriculture crops. It can be applied as a spray or drench and works in a similar manner to Bonzi.
- Topflor (flurpimidol) is labeled for most floriculture crops. It can be applied as a spray or drench and works in a similar manner to A-Rest.®

Many factors influence the efficacy of these plant growth retardants: timing of application, environmental conditions, dosage, target tissue, pH, uniformity of application, the concentration of the chemical, and the active ingredient.

Graphical tracking has become a popular method of monitoring plant height to help determine the timing of growth retardant application. A standard growth curve is established for a specific crop, and plant height is plotted against the standard curve on a weekly basis. When the plant height begins to exceed the height of the standard curve, a growth retardant application may be necessary to help the plant reach the specified height. If the plant is below the curve, strategies to increase growth are implemented.

Other plant growth regulators are not growth retardants but they affect other plant processes. ProGibb® (gibberellic acid) promotes cell elongation, cell division, and larger flowers. It can be applied as a spray and substitutes for cold in breaking the dormancy of azaleas and hydrangeas, for internode elongation and leaf expansion in plants that are too small, or as a bulb dip to promote flowering of calla lily. Fascination®

(gibberellin$_{4+7}$ and benzyl adenine) can be used as a spray to prevent leaf yellowing of Easter and hybrid lilies. EthylBloc (methylcyclopropene [MCP]) is an antiethylene compound applied as a gas to help prevent flower drop during shipping of flowering plants and cut flowers.

See Chapter 7 for more information on chemical plant growth retardants and regulators.

Growing Media

Choosing the proper medium to grow quality greenhouse crops is very important (Fig. 23–20). Although the root systems of plants are not visible, the health of the shoot depends on the health of the root. The media must have desirable physical characteristics in terms of bulk density or weight so that the container does not tip over easily. However, if the media weight is too great, the cost of shipping may be prohibitive. The total pore space, water retention, and air-filled porosity also contribute to the health of the root system by providing appropriate amounts of oxygen and moisture for optimum growth. The pH and electrical conductivity (EC) of media are important chemical properties. For most crops, the pH range of media is between 5.5 and 6.5, but some crops must have a more specific pH. The EC should be between 0.75 and 2.0 mmho/cm. If the EC of the media becomes too high (>2.0 mmho/cm), salt accumulation can cause water to be pulled out of the roots, resulting in root burn and death.

Most greenhouse crops are grown in soilless media, which can be mixed in an almost endless array of combinations to get the type best-suited for a specific crop. The benefits of using soilless media over media that contain soil are reduction in bulk density (weight),

increased uniformity of the media, as well as having other desired characteristics such as nutrient holding, water retention, and drainage. The most common components of soilless media are sphagnum peat moss, pine bark, perlite, coir (coconut) fiber, and vermiculite. Parboiled rice hulls (PBH) are increasing in use as a replacement for perlite. Sphagnum peat moss is low in salts, does not decompose readily, has high water- and nutrient-holding characteristics, is uniform, and is relatively disease-free as found in nature. Vermiculite is a sterile expanded mica that holds water and nutrients well and does not decompose. It is uniform and reasonably priced. Vermiculite does compress quickly and, therefore, is used most frequently in seedling mixes because of its water-holding capacity. Perlite is a heat-expanded aluminum silicate rock that is porous and lightweight. It does not hold water but provides excellent pore space. However, it has become an unsightly environmental nuisance in some areas because it does not decompose over time. Pine bark and hardwood bark make excellent media components. When composted or aged properly, both barks have high water- and nutrient-holding capacities, and can be graded to specific sizes (¼ in. and ⅜ in. across are the most popular). Parboiled rice hulls are a relatively inert organic byproduct of rice production and are biodegradable, which is why they are being used as a replacement for perlite. Most soilless media are amended with dolomitic limestone to increase the media pH. Phosphorus and trace elements may also be added so that the nutrient content is adequate for greenhouse crop production. Because artificial media characteristics can vary tremendously depending on the components and their percentages, water and fertility have to be closely managed.

Containers

The early floriculture industry built greenhouses to grow primarily cut flowers or vegetable transplants planted in the ground. As the industry began to mature and technology and transportation improved, more plants were grown in pots or other type of containers. Some commonly used types of containers are plug trays which are small cells in a larger tray. The tray is usually about 21 × 11 in. with from 98 to 512 cells per tray (Fig. 23–21). These are used for seed propagation. Cell packs are larger cells grouped together in what is called an insert. The insert is designed to fit a standard tray, or flat, for easier handling. Cell packs come in a wide array of sizes but they all are designed to fit in a flat of dimensions similar to a plug tray. Traditional pots are measured by their diameter at their top and range from as small as 2 to 16 in. (5 to 40 cm) or hanging baskets,

Figure 23–20

Soilless growing media moved by a conveyor for filling containers prior to transplanting. Source: Jeff Kuehny, Louisiana State University.

Figure 23–21
A ninety-eight-cell plug tray.
Source: Jeff Kuehny, Louisiana State University.

which are usually 6 to 12 in. (15 to 30 cm). Other types of containers include planters and urns. All containers must also be labeled for the volume they contain. This relatively new requirement makes it easier for growers to calculate the volume of media needed to fill the pots.

Originally, containers were primarily terracotta (kiln-fired clay). These containers provided growers and consumers with flowering plants that could be used indoor or outdoor. While these terracotta containers helped bring change and new markets to the floriculture industry, they had some drawbacks such as being heavy and fragile. Plastic containers helped to revolutionize the floriculture industry in that they were light weight, sturdy, and easily shippable. Plastic containers could be made into any shape, size, or color. This has been particularly true for the bedding plant industry, which utilizes large numbers of plastic containers in various sizes, shapes, and styles, as well as plug trays, flats, and inserts. However, as sustainable production practices have become increasingly important to the floriculture industry, there has been an increased interest by growers and consumers alike in the use of biocontainers. Biocontainers are generally made from a variety of organic components that readily decompose (although the time to decompose may vary widely depending upon composition) and are generally considered to be more environmentally friendly than traditional plastic containers. Biocontainers provide the floriculture industry with an opportunity to improve its level of adoption of sustainable products and practices. Depending upon their composition and intended use, biocontainers may reduce labor, provide new marketing strategies, and contribute to the mineral nutrition of the crop.

Irrigation and Water Quality

Water quality is extremely important to a greenhouse operation due to the use of soilless media, which have much less buffering capacity than that of soil. That means that any salts (nutritional or nonnutritional) in the water can quickly become a problem if not managed carefully. Irrigation water should be analyzed prior to beginning a greenhouse operation and at least once a year during regular operation. Samples can be submitted to the local extension agent for testing, and they can also be tested by the grower using a small handheld meter (Fig. 23–22). Salinity, the measure of sodium and chloride in irrigation water, can contribute to increased media EC and root damage if in high concentrations (>250 ppm Na). This has been an increasing problem, especially in areas where saltwater intrusion of aquifers occurs. Alkalinity, the concentrations of carbonates and bicarbonates in irrigation water (>3.25 me l^{-1}), can increase the pH of the growing media. There are methods of controlling salinity and/or

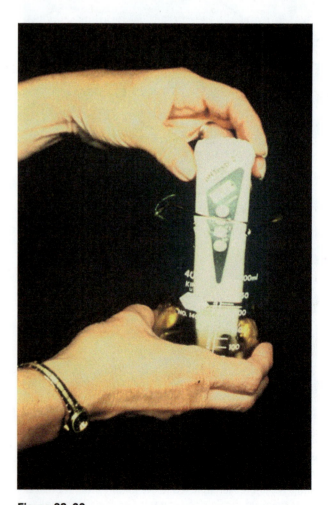

Figure 23–22
Handheld EC meter used to determine the electrical conductivity of irrigation water. Source: Jeff Kuehny, Louisiana State University.

alkalinity in irrigation water, but they can be costly if concentrations are high enough.

Irrigation is an art that is not learned quickly. The ability of a grower to look at the plants across an entire greenhouse range and determine the need for irrigation is of primary importance. Hand watering of greenhouse crops can be one of the best methods of irrigation, but it is not economical (Fig. 23–23). However, many growers still water all or some of their crops by hand. It is especially effective for "spot treating" small areas of plants that are drier than others around them. The use of overhead sprinklers for the production of some bedding plants is more economical, but the foliage of most crops must be kept dry to prevent disease. Automated overhead boom watering can be used effectively on the production of plugs and other crops (Fig. 14–38 on page 285). The use of microtubes has been popular for many years and is an efficient method of irrigating large numbers of plants without getting the foliage wet while minimizing water usage (Fig. 14–42 on page 286). Mat watering by the use of a capillary mat is another method of irrigating crops when the sizes of containers vary on the same bench. The use of an ebb-and-flood bench system and flood floors has become popular in recent years, but it can be cost-prohibitive due to the amount of materials used to build the system (Figs. 14–33 and 14–34 on page 283). The water can be recyled, but disease can also spread rapidly in this system; thus, it must be monitored carefully. As mentioned earlier, withholding water can help control plant growth. It is very important that whoever is employing this technique knows how dry to let the plant become before watering. Applying water too soon negates the control, applying water a little too late can cause damage to the plant.

Most greenhouse operations use some type of liquid fertilizer in the irrigation system, which is called **fertigation**. This practice adds another factor to irrigation and is discussed in the next section.

Mineral Nutrition

Providing greenhouse crops with all sixteen essential elements is vital for optimizing growth due in part to the use of soilless media and the need to produce a marketable crop as quickly as possible. Most greenhouse fertilization recommendations are based on the parts per million (ppm) of nitrogen (N). Therefore, commercial fertilizer mixes for greenhouses are formulated to have phosphorus, potassium and other nutrients in proportion to N, depending on the nutritional need of the crops to be fertilized. For example, some poinsettia cultivars are inefficient in the uptake of molybdenum (Mo); therefore, most poinsettia fertilizers have increased concentrations of Mo.

Commercial greenhouse fertilizers have been manufactured from primarily inorganic forms of various fertilizer salts and they have been applied in a soluble or liquid form. Liquid fertilizers are used on short-term crops (in the greenhouse for less than two to three months). Slow-release fertilizers, such as Osmocote® and Nutricote®, are commonly used to fertilize long-term crops, such as foliage plants, orchids, azaleas, and hydrangeas. Slow-release fertilizers may also be added in small amounts to the media as it is being mixed to provide a "starter charge" of fertilizer and supply nutrients during the period of time when a seed is germinating or a transplant is developing new roots and there is not much need for irrigation. Organic-based fertilizers, such as Daniels Plant Food®, are becoming more popular because they provide an alternative to inorganic-based fertilizers. Utilizing the beneficial

Figure 23–23

Hand watering of greenhouse crops using an extended wand with a water breaker attached to the end. The water breaker helps break up the flow of water to prevent washing the media and/or plants out of their containers.

Source: Jeff Kuehny, Louisiana State University.

aspects of the various types of fertilizers will help provide for a more sustainable fertilization program.

Plants have different nutrient needs during different stages of growth, and fertilization must be adjusted to meet those needs. For example, the proportion of nitrate nitrogen to ammonium can be changed during production to foster reproductive or vegetative growth. Because most greenhouse crops are flowering plants, the fertilizer concentrations are low at the planting stage of the crop, are increased during rapid crop growth, and are decreased at crop maturity or flowering or shortly before time of sale, regardless of maturity. Application of fertilizers in too high a concentration without leaching can increase the soluble salts in the media and be deleterious to root growth and thus crop quality. However, leaching can contribute to runoff of irrigation water and fertilizers, which has become an increasing problem for greenhouse operations. Therefore, growers are trying to match more closely the fertigation schedule to crop growth in order to reduce waste, runoff, and cost.

Media/soil and foliar testing are two methods used for monitoring the nutrient status of a plant. The first is testing the media/soil by either sending a sample to an extension agent for analysis or taking a pour-through leachate (PTL). When potting media is sent to a lab, it is important to indicate that the media is artificial and highly organic because that determines the type of test the lab will run. Highly organic media and inorganic soil are tested differently, with different interpretations of the results. A PTL can be conducted easily by pouring a small amount of water over the top of the media in the pot and collecting the leachate from the bottom of the pot. The EC and pH of the leachate can be measured and evaluated to determine if adjustments are needed to the fertigation program. (See Chapter 22 for a description of how to collect leachate and measure EC and pH.) The leachate can also be tested for specific nutritional ions, but for this analysis the samples are usually sent to a soil testing lab. A foliar analysis is another method for monitoring nutrient status. Recently matured leaves should be taken from the plant. If a plant is showing signs of what could be a nutrient deficiency, then leaves that show the symptoms should be taken and analyzed separately from those showing no symptoms. Foliar analysis must be interpreted according to plant species. Because of the enormous number of greenhouse crops, there often is little information available on correct foliar readings for the nutrients for a specific species of cultivar. Therefore, the comparison between apparently deficient leaves and those that appear healthy is critical for interpreting foliar analysis.

Diseases and Insects

Greenhouses are small microclimates that not only provide an optimal environment for producing quality plants, but also an optimal environment for the growth of diseases and insects. The most common greenhouse insects are aphids, mealy bugs, spider mites, white flies, thrips, fungus gnats, and scale.

Aphids can range in color from transparent to green to brown (see Fig. 15–2 on page 315). They can be wingless or winged, and they feed on almost all floriculture crops. Aphids feed on young flower buds and newly formed leaves This is where most of the soluble sugars are translocated and where the piercing–sucking mouthparts of this insect can easily penetrate for it to feed. Infestation may lead to distorted flowers and leaves.

Mealy bugs are fuzzy white insects that also have piercing–sucking mouth parts. They are primarily a problem on foliage plants. These insects can be difficult to control because of their outer waxy surface.

Spider mites are not insects but belong to the class Arachnida (see Fig. 15–2 on page 315). The two-spotted spider mite seems to be the most prevalent mite and feeds on the foliage of a plant, causing a yellow stippling of the leaves. If the infestation is great enough, they will spin webs around the leaves and flowers of plants. They seem to be most aggressive when environmental conditions are warm and dry. Spider mites are found first on the underside of lower foliage where they are difficult to see and control with most insecticides so closely monitoring a crop is very important in spider mite control.

White flies feed with piercing–sucking mouthparts and are primarily problems on poinsettia, chrysanthemums, lantana, and salvia. The adult form resembles a small white fly. They lay their eggs on the underside of leaves. The silverleaf whitefly is an especially important pest on poinsettia (see Fig. 15–2 on page 316).

Thrips are small winged insects that feed with piercing, sucking mouthparts, but they cause a streaking pattern on leaves and flower petals instead of stippling (Fig. 23–24). They tend to feed most heavily on flower buds and can cause desiccation of the buds. They are also the primary vector for impatiens necrotic spot virus (INSV) and tomato spotted wilt virus (TSWV). The use of a very fine-mesh screen over vents to keep thrips out of a greenhouse is often the only way to control them. The small size of the screen openings greatly reduces airflow so a large cage-like structure is constructed around the vents to provide more surface through which air can move (Fig. 23–25).

Fungus gnats are gray flies that lay eggs on moist soil surfaces. The adults do not harm the plant. However, the larvae that are hatched from the eggs feed on the roots of plants, causing injury to the roots and thus greater susceptibility to diseases. These insects have become a problem for all floriculture crops and are difficult to get rid of once an infestation has occurred. Shore flies are similar to fungus gnats in that the larva may feed on roots leading to spread of plant pathogens. Shore flies have little or no antenna while fungus gnats have a more visible antenna and are less fly-like in appearance. Keeping the greenhouse free of weeds (including algae), spilled media, and puddles can help control both pests.

Scale insects vary in form and size, but most of them resemble the scales on a fish. The scale covering is waxy and protects the adult from many insecticides and predators. Some forms of scale have wings. The immature stage is called a crawler and is the easiest stage to control because it has not yet formed the protective scale-like covering. These insects are most commonly problems on foliage plants and can be difficult to eliminate.

Nematodes can be a problem if media is not fumigated or pasteurized by composting, manufacturing processes, or steam.

The most common greenhouse diseases are caused by viruses, bacteria, and fungi. The most common viruses are INSV and TSWV, mentioned earlier. Because plants do not produce antibodies as do most animals, they cannot recover from a virus or develop immunity. Bacterial diseases are difficult to control, and fortunately there are fewer bacterial than fungal diseases. There are some bactericides; but the best method of control is prevention and elimination. Some

Figure 23–25

"Cage" made of thrip screen around a vent. The cage increases the surface area of the screen to allow more air to flow into the vent.
Source: Margaret McMahon, The Ohio State University.

of the more common bacterial diseases are bacterial blights and *Erwinia* (soft rot). There are many fungal diseases, but they can be controlled through the judicious use of preventive fungicides. Some common fungal diseases are powdery mildew, botrytis blight, pythium, rhizoctonia, and verticillium. The best methods for circumventing these pests are sanitation, integrated plant health management (IPHM), biological control, preventive fungicides, and spray programs.

Integrated plant health management (IPHM), as you learned in Chapter 15, is a method of controlling insects and diseases through surveillance, preventive measures, and corrective action. IPHM is especially important in greenhouses where pest management generally has to be much greater than outside. IPHM requires understanding the dynamics of the pest population, biology of the pest and its host, effective environmental control and cultural practices, and effective biological as well as chemical controls. Managing all these factors achieves healthy plants. The effectiveness of the program depends greatly on the scout, the person in charge of screening plant material for diseases and insects, and the grower working together to start control measures when a pest or other problem is detected.

For a thorough discussion of insects, diseases, and IPHM, see Chapter 15.

Propagation

Historically, seed germination, the growing of stock plants for cutting production, and the propagation of cuttings were commonly done at the same location where the plants were grown to finish for sale. This practice has

Figure 23–24

A greatly enlarged thrips. Normal size of a thrips is about 1.5 mm (0.05 in.) long. Source: USDA Agricultural Research Service Image Gallery, http://www.ars.usda.gov/is/graphics/photos/

changed over the years as our understanding of the physiology of germination and rooting has increased along with the technology that is used for those procedures. Seed germination and seedling growing, stock growing, and cutting harvest and propagation are now very specialized operations that are done by a few growers to sell to many other growers. Thus, the floriculture industry has added another facet to production.

Most greenhouse plants, including the wide array of annual bedding plants (plants grown for planting in gardens and landscapes each year), used to be propagated by hand-sown seed, which were then hand-transplanted into larger pots or containers, an extremely time-consuming and expensive process. The advent of mechanized plug production and transplanting revolutionized floriculture by allowing great numbers of potted and bedding plants to be propagated economically and efficiently. That mechanization includes seeders, transplanters, water booms, trolleys to move the plants, as well as many other types of equipment. Mechanization reduced labor needs, shortened production time, and eased the shipping of seedlings.

For many years, vegetatively propagated plants were mainly the potted plants: poinsettias, chrysanthemums, Easter lily, and several foliage plants. Now vegetative propagation has advanced beyond propagation of the primary potted plants mentioned earlier and is now also important to the bedding plant industry. The increasing number of new bedding plant types developed through hybridization has facilitated the need for clonal propagation of many of those new plant types.

A tremendous amount of breeding work has developed new plant forms, flowering colors and shapes, disease and insect resistance, and stress tolerance. This work continues to be important to the industry but is very costly. Plant patents can be obtained on asexually propagated plant material to help protect the breeders' work and recoup some of the costs put into the development of new plant material. Branding of plant material has also become an important tool in advertising and selling plant material from a specific company. Branding can be a form of trademark that can be combined with the plant patent.

GREENHOUSE CROPS

History

The greenhouse industry in the United States began in the Northeast close to large population centers. Cut flowers—primarily roses, carnations, and chrysanthemums—were the major greenhouse crops produced in the early 1900s. They were grown for their flowers, leaves, and stems mainly for use in floral arrangements. Production of fresh-cut flowers spread slowly across the United States to the Midwest, then to Colorado, and finally into California by the 1960s. During this shift of cut flower production from the East Coast to the West Coast, flowering potted plants began to replace cut flowers. Examples of popular flowering pot plants include poinsettia, orchids, chrysanthemums, African violets, azaleas, hydrangeas, Easter lilies, and other flowering bulb crops. Production of flowering pot plants became an even greater part of the United States greenhouse industry when foreign imports of cut flowers became a major competitor in the late 1960s. During the 1970s and 1980s, green plant or foliage plant production became a considerable part of the greenhouse industry in response to the so-called green revolution. Foliage plants are grown for their foliage rather than their flowers. Many of these plants are tropical and adapted to the warm, humid, and low light conditions found under the canopies of large trees in tropical rainforests. Foliage plant production is a large industry in south Florida, where much of the crop is grown either in greenhouses or under shade structures. As technology and transportation improved, so did the greenhouse industry. This was most evident in the production of bedding plants, with the introduction of plug technology. Bedding plants now make up a majority of total greenhouse crop production.

This brief history of greenhouse production reflects the dynamics of this field in horticulture. The production guidelines of the following crops are examples of some of the more important greenhouse crops grown now.

Cut Flowers

Central and South America, and South Africa are very important production areas. A majority of the cut flower production in the United States occurs in California, which produces primarily roses and carnations. *Rosa* L. hybrids flower profusely during the long, warm, sunny days of California. Greenhouse nighttime temperatures should be maintained at 15°C to 18°C (60°F to 65°F) for optimum growth and flowering. Flowering can be timed by pinching the terminal buds, which produce new flowering shoots in thirty-five to sixty days. The rapidity with which they flower depends on light intensity and temperature. A flower forms on every new shoot after the removal of a flower from the bush (Fig. 23–26). If properly cut and pruned annually, plants may not have to be replanted for five to eight years, depending on the health of the plants and the soil structure. Occasionally, disease hastens the need to replant.

Figure 23–26
Roses grown in a greenhouse for cut flower production.
The flower stems are a few days from the cutting stage.
Source: Jeff Kuehny, Louisiana State University.

Figure 23–27
Cut roses being graded for packing and shipping to a whole-
sale florist. Source: Jeff Kuehny, Louisiana State University.

Flowers are cut in the late-bud stage, usually when one or two sepals have turned downward or just as one or two petals have unfurled. They are sorted and graded into uniform stem lengths that also have similar flower head (bud) sizes and quality (Fig. 23–27). Flowers are packaged in bunches of twenty-five and kept in post-harvest coolers at 2°C to 4°C (35°F to 40°F).

More recently, specialty cut flowers (cut flowers that are not grown in large quantities by smaller growers) have increasingly become part of a local or regional floral industry. Specialty cut flowers include species such as *Achillea, Alstroemeria, Celosia, Dianthus, Helianthus, Ilex,* and *Solidago*. As the demand for these types of crops increases, so will the need for production of this type of flowering plant in the greenhouse industry.

Flowering Potted Plants

There are far too many flowering pot plants to give a description of growing each one. Therefore, poinsettias will serve as a general example of how to grow these crops. Poinsettias are one of the most important flowering pot plant species sold in the United States, even though they are only sold during the Christmas holiday season. *Euphorbia pulcherrima* Wild. Ex Klotzch is a tropical plant native to Mexico belonging to the family EUPHORBIACEAE (Fig. 23–28). The poinsettia was introduced into the United States in 1825 by J. R. Poinsett, from whom it gets its common name. Poinsettias have been grown as a Christmas holiday crop mainly due to their photoperiodic response for flower initiation. They are classified as short-day (long-night) plants that initiate and develop flowers as the nights become longer than the days. The inflorescence is characterized by a single female flower, without petals and usually without sepals, surrounded by individual

Figure 23–28
Various poinsettia cultivars in shades of red and white grown
for the Christmas holidays. Source: Jeff Kuehny, Louisiana State University.

male flowers all enclosed in a cup-shaped structure called a cyathium. The showy red, pink, or white portions of the plant, popularly referred to as the flower, are modified leaves called **bracts**.

Rooted cuttings are planted in well-drained media during August. Plants should be watered immediately after planting, and a fungicide should be used shortly thereafter to prevent disease infection. Poinsettia cultivars are classified according to the amount of time required from the start of short days to flowering, which are called response groups. Other photoperiodic plants such as chrysanthemum and soybean are also classified by their response groups. Poinsettia response groups range from seven to ten weeks. Therefore, timing of a finished crop can be manipulated by photoperiod control. Some plants, such as 'Freedom Red,' are very sensitive to changes in photoperiod and may need an extended daylength in the early fall to prevent premature flower initiation. Daytime temperature should be maintained below 25°C (77°F) to maintain leaf unfolding, with nighttime temperature maintained at approximately 20°C (68°F) for flower initiation and development. Reduced night temperatures near the end of the production cycle of 16°C to 17°C (60°F to 62°F) are necessary for bract coloration. Light intensities of up to 1,000 μmol/m2/s should be maintained for good bract coloration. After the roots have grown to the sides of the container (approximately two weeks), the terminal meristem should be pinched to six internodes to ensure lateral branching. Approximately two weeks after pinching (laterals of 2 to 3in. in length), a short daylength should be started for inducing flower initiation. One or two plant growth retardant applications are usually needed to produce a quality finished plant. Depending on response group, the plants should reach anthesis seven to ten weeks later.

Nearly all flowering potted plants sold in the United States are grown in the United States because of quarantine laws that prohibit the importing of plants that are in soil or media.

Foliage Plants

Foliage plants are grown primarily in south Florida in the United States (Fig. 23–29). They can be grouped by sensitivity to light intensity. The high-light group consists of light levels of 600 to 1200 μmol (i.e., *Ixora coccinea*, *Codiaeum variegatum pictum*, Cactaceae, Crassulaceae), the medium-light group of 150 to 450 μmol (i.e., *Begonia* spp., *Draceana* spp., *Ficus* spp.), and the low-light group of 10 to 100 μmol (i.e., *Aglaonema commutatum*, *Philodendron scandens oxycardium*, *Spathiphyllum*

Figure 23–29

Foliage production in Florida. Source: Margaret McMahon, The Ohio State University.

spp., *Dieffenbachia* spp., *Epipremnum aureum*). Foliage plants that are to be moved from an environment with a higher light intensity into one of lower light intensity must be able to adjust to this change. Some plants lose chlorophyll and then drop their leaves soon after being moved to a lower light intensity (i.e., *Ficus benjamina* and *Asparagus densiflorus*). These plants develop new leaves that are usually thinner and broader for more efficient photosynthesis. Many plants, including the aroid, palm, and lily families, do not abscise their leaves. Instead, the leaves become yellow-green and may wither or die but still cling to the stem. Therefore, plants must be acclimatized to the lower light intensities before removing them from environments with higher light intensities to maintain their appearance and quality. First, reduce the light by 50 percent for several weeks and, finally, reduce the light with shade to 20 percent of the original growing light intensity. The entire acclimatization process should take four to twelve weeks, depending on the species in question. Reducing fertilizer and watering during this period also aids the acclimatization process.

Flowering plants have relatively narrow and specific temperature requirements for flowering; however, because foliage plants are not grown for their flowers, they can be grown under temperatures that range from 16°C to 32°C. The lower the temperatures, the slower the growth, and vice versa. The optimum growing temperature for most houseplants is approximately 21°C to 26°C.

Most foliage plants prefer a relative humidity between 70 and 80 percent. This can be easily achieved in a greenhouse full of plants. However, when foliage plants are moved from the greenhouse to an interior environment, the low relative humidity of the new

environment can result in slower growth and insect problems. Misting these plants does not provide enough moisture in the air to compensate for this loss. Keeping plants out of the direct flow of air from heating and/or air-conditioning vents and doorways where large amounts of air exchanges occur, and providing adequate water to the container are the best means of compensating for the loss of relative humidity.

Rapidly growing foliage plants in a greenhouse utilize large amounts of water and fertilizer. However, when moved to an interior environment, growth slows, as does nutrient and water uptake. Thus, houseplants need much lower amounts of fertilizer and water. Overwatering house-plants is one of the most common causes of plant death. Salts may accumulate in the media of plants that are maintained in the same container for long periods of time (greater than one year). Leaching the media once a year or repotting can circumvent this problem. Houseplants can also benefit from rinsing the dust that accumulates on the leaves.

Bedding Plants

The term bedding plants was originally used by gardeners who wanted to use annual or perennial plants in a permanent border or in a specially designed flower bed. This term has been expanded to include flowering ornamental plants, vegetables, and herbs (Fig. 23–30). Bedding plants are used in large or small containers, on patios, in window boxes, in hanging baskets, in small gardens, and in beds and borders. The bedding plant industry has responded to the needs of a range of gardeners, from the sophisticated to the novice, by offering a variety of plants in large, medium, or small containers

Figure 23–31

A 288-plug tray with Impatiens wallerana *grown for transplanting in larger containers.* Source: Jeff Kuehny, Louisiana State University.

(six-packs are the most common) and a large number of species from which to choose.

Greenhouse production of bedding plants differs from the growing of other greenhouse crops because a larger number of many different species of plants are grown in the same space in a much shorter time. This means quick rotation of crops and rapid return on investment.

As you learned before in this chapter, early production of bedding plants was by hand-seeding seeding individual or small numbers of seed into pots or into rows in flats or by broadcasting seed into flats. The seedlings that grew in the flats had to be transplanted into containers. These traditional methods of bedding plant production were improved on and developed into the mechanized plug and transplant method (Fig. 23–31) which has meant the continued growth and expansion of bedding plant sales in the greenhouse industry.

Because of the large number of different species of bedding plants, scheduling, fertilization and irrigation, greenhouse temperatures, types and rates of plant growth regulators, and light levels vary significantly. Although bedding plant production can be highly automated and bedding plants are a quick crop to produce, the amount of knowledge that a grower must have to grow these crops is extraordinary.

Bulb Crops

Many species of plants have fleshy underground storage organs capable of carrying the plants through seasonal cold/warm or dry/wet periods. Such structures are popularly called bulbs, but they are defined more accurately as bulbs, tuberous roots, tubers, corms, or rhizomes (Fig. 23–32). They all produce one or more buds for flower production or renewed vegetative growth.

Figure 23–30

A gutter-connected greenhouse range full of bedding plants on benches in flats (containing six sets of six-packs) and hanging baskets overhead grown for spring sales.
Source: Jeff Kuehny, Louisiana State University.

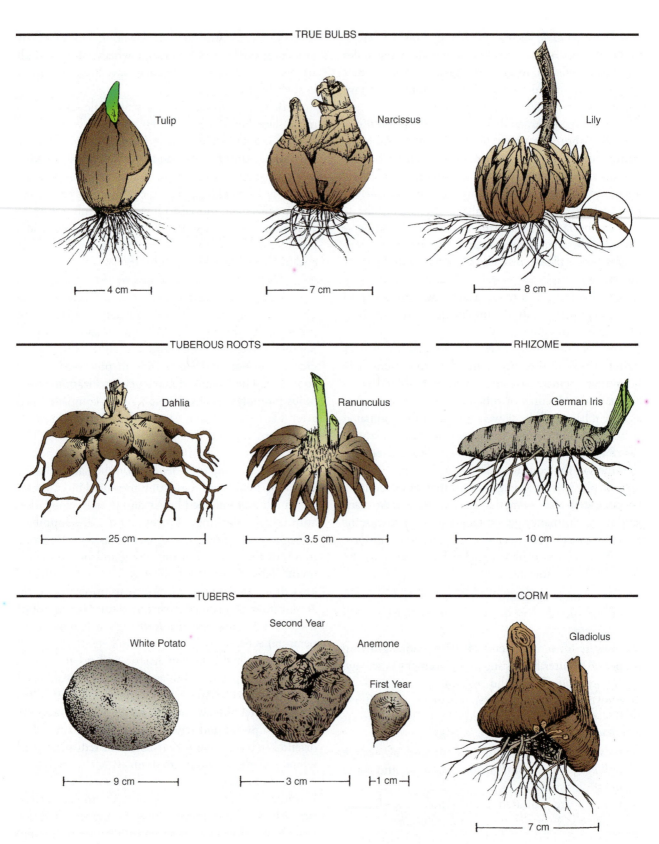

TRUE BULBS

Tulip

⊢— 4 cm —⊣

Narcissus

⊢— 7 cm —⊣

Lily

⊢— 8 cm —⊣

TUBEROUS ROOTS

Dahlia

⊢—— 25 cm ——⊣

Ranunculus

⊢—— 3.5 cm ——⊣

RHIZOME

German Iris

⊢—— 10 cm ——⊣

TUBERS

White Potato

⊢—— 9 cm ——⊣

Second Year

Anemone

First Year

⊢—— 3 cm ——⊣   ⊢1 cm⊣

CORM

Gladiolus

⊢—— 7 cm ——⊣

Figure 23–32

Examples of underground food-storage structures found in some herbaceous plants. Sizes given are approximate.

The *true bulb,* such as the lily, hyacinth, *Muscari, Narcissus,* tulip, and onion, has numerous fleshy scales or leaf bases attached to a distinct basal plate (stem) that gives rise to roots and shoots. They may or may not have one or more impervious covering layers (tunic).

Tubers are enlarged fleshy stems with adventitious buds (eyes) near the upper surface, as in the tuberous begonia, Eranthis or Caladium, or in a systematic pattern or arrangement, as in the Irish potato. Some organs are enlarged roots and are classified as *tuberous roots,* such as Agapanthus, Dahlia, and Hemerocallis.

Corms have solid shortened stems with buds systematically arranged under a paper-thin protective covering of the leaf base or scale. Crocus, Gladiolus, and Freesia are examples.

The *rhizome* is a fleshy, horizontal underground stem that grows laterally. Examples are the rhizomatous iris, calla lily, and ginger.

Roots originating from the basal plates or lowest portions on all of these structures are adventitious. The shoots that originate either inside the bulb or on the surface, as with the corm or tuber, give rise to stems that bear the foliage and the flowers. Flowers can be initiated within the bulb during the previous growing season (*Narcissus*) or just after the new shoots begin to emerge from the soil (bulbous iris, lilies, and gladioli). All these types of plants have storage tissues that can produce a flowering stem after going through the seasonal dormant period. The dormancy period ranges from a few months to almost a year under certain environmental conditions.

Bulbs can also be categorized according to the pattern of flower formation:

1. Flower buds form during the spring and summer of the previous year (*Narcissus, Galanthus,* and *Leucojum*).
2. Flowers form at the end of the previous growing period or after harvesting and placement in storage (*Hyacinthus, Tulipa,* and *Iris reticulata*).
3. Flowers are initiated after replanting and after a low-temperature treatment (bulbous iris, most lilies, *Triteleia,* and *Ornithagalum*).
4. Flowers are initiated toward the end of storage (*Allium cepa, A. escalonicum, Galtonia,* and some lilies).
5. Flowers are initiated after replanting (*Freesia, Gladiolus, Anemone, Ixia,* and *Ranunculus*).
6. Flowers are initiated more than a year before flowering (*Amaryllis belladonna, Nerine sarniensis,* and *N. bowdenii*).
7. Flower initiation occurs throughout the entire growing season but alternately with leaf formation (*Zephranthes*).

A method for determining the flowering bud stage is by dissecting the bulb to determine whether flower buds are present and whether they may have been injured by heat or disease.

Because bulbs vary in their stage of flower development, some bulbs such as daffodil already have immature flowers and already may have been precooled and are ready to be planted, rooted, and forced into bloom. Other bulbs must first be given the proper low-temperature treatment (precooling). Then the bulbs must be planted in containers and moved to low-temperature rooms to establish a good root system before they are forced in a greenhouse. Temperatures for precooling and forcing are critical for each species and cultivar.

The Easter lily, or *Lilium longiflorum* Thunb, is one of the most important flowering bulb crops in the United States. These true bulbs are native to Japan and grown in California and Oregon. They are dug in September and October and are either shipped to growers for potting then cooling or precooled ("case-cooling"). The cooling process or vernalization of the bulbs promotes floral induction and is completed just after Christmas. Bulbs should be planted in well-drained media and drenched with a fungicide. The bulbs are then forced into bloom in the greenhouse. The desired forcing time depends on the date of Easter in that particular year. Growth is measured by counting the number of unfolded leaves during stem elongation and then tracking flower bud development. Greenhouse temperatures can be increased or decreased to adjust the rate of leaf unfolding and bud development. This allows the grower to adjust the rate of growth so that bloom occurs just prior to Easter. Application of plant growth retardants or the use of negative DIF (night temperature higher than day temperature) is necessary to reduce stem height.

Lilium hybrids, both Asiatic and oriental lilies, are becoming popular both as flowering pot plants and as cut flowers. The Asiatic hybrids include flower colors of orange, red, yellow, tan, and white. Their flowers are oriented upward and have little or no fragrance. The oriental hybrids have flower colors of red, pink, and white, and their flowers are oriented horizontally and have a strong fragrance. Asiatic and oriental lilies easily hybridize so there are many cultivars with characteristics of both types. Asiatics, orientals, and their hybrids must be precooled for up to ten weeks prior to forcing. The storage duration and temperature is cultivar-dependent. The bulbs should be planted and forced in a greenhouse, like Easter lilies. Because these lilies are usually not grown for a specific holiday, however, leaf counting is not necessary for finishing. Many of these

lilies do require a growth retardant for producing a quality finished plant. These lilies are most often forced from March to June, and the time to finish depends on the growing conditions. The bulbs of some cultivars can be frozen for up to a year then forced, which extends the production season. Asiatic types take approximately thirty days to flower from visible bud, and oriental hybrids take approximately fifty days.

Amaryllis or *Hippeastrum* consists of leaf bases and no scales. The genus is indigenous to tropical and subtropical South America. Commercial bulbs 20 to 30 cm (7.8 to 11.8 in.) in circumference are produced by Israel, South Africa, the Netherlands, and Brazil. The bulbs are dug, cured, and shipped at a storage temperature of 9°C to 13°C and then are planted in a container for forcing in the greenhouse or placed in a bulb kit to be forced by the consumer. Depending on the cultivar and its handling, the flower stalk emerges and elongates simultaneously with four leaves. *Hippeastrum* is becoming a very popular Christmas flower in the United States.

SUMMARY AND REVIEW

Greenhouses were initially built to protect plants from cold temperatures while allowing sunlight to reach the plant. Through the years, however, greenhouse structures became much more sophisticated, enabling growers to manipulate the environment and thus control plant growth and development. Greenhouse structures can have many different shapes and coverings.

Greenhouse heating and cooling systems are used to regulate temperatures and consequently plant growth and development throughout the year. Negative DIF has become an important means to control stem elongation and development of plants. Light quality, intensity, and duration all influence plant growth and each to some degree can be controlled in the greenhouse. Shading systems allow the amount of light reaching a plant to be limited when light intensity is too great. High intensity discharge (HID) lamps can be used to provide supplemental light for photosynthesis during low light periods. Incandescent and fluorescent lights have limited, specific uses in the greenhouse. Many greenhouse crops are photoperiodic and their flowering is controlled by manipulating the daylength by use of opaque fabric covering or night interruption lighting. The daily light integral is the combination of the intensity of light and light duration in a day.

In addition to negative DIF, chemical plant growth retardants are often used in greenhouses to control stem elongation and maintain quality. These chemicals include ancymidol, chlormequat, daminozide, ethephon, flurpimidol, paclobutrazol, and uniconazole. All but ethephon inhibit stem elongation by blocking the synthesis of gibberellic acid in the plant. Ethephon releases ethylene in the plant. Gibberellic acid and methylcyclopropene are chemical plant growth regulators that increase stem elongation and inhibit ethyelene action, respectively.

Greenhouse growers most often use an artificial media that has been blended to provide the specific characteristics needed for each crop. Common ingredients include sphagnum peat moss, pine bark, perlite, vermiculite, coir fiber, and parboiled rice hulls. Many types of containers are used in greenhouse crop production. They range from plugs trays used for seed germination to very large and showy urns and planters used to display mature plants. Although plastic has been the main material from which containers have been made for many years, biodegradable containers are becoming more prevalent in the industry.

Water quality is extremely important in greenhouse production because the artificial media usually have less buffering capacity than field soils. Salinity and alkalinity has to be monitored constantly and corrected if they get out of the acceptable range. There are several methods of irrigating in the greenhouse. Hand watering with a hose or can is still practiced, especially in small businesses and for spot treatments. Overhead watering, including automatic booms, is used for many greenhouse crops as are drip and mat systems. Ebb and flood benches and floors are another type of irrigation found in greenhouse operations.

Fertigation is used in most greenhouses. Greenhouse fertility programs are usually based on parts per million of nitrogen. Therefore, commercial fertilizer mixes are formulated to have phosphorus, potassium, and other nutrients in proportion to nitrogen, depending on the crops to be fertilized. Liquid fertilizers are the most common type of fertilizer in the greenhouse, but slow release fertilizers can be used on long-term crops and as a starter charge in media. Traditionally, greenhouse fertilizers have been inorganic in origin, but the use of organic based fertilizers is increasing. For most flowering crops, fertilizer

concentrations are low at the beginning of the crop cycle, increase as the plant is actively growing, then decreased as the plant starts to mature or is ready for sale. Matching fertilizer concentration to plant need helps reduce fertilizer run-off and cost. Monitoring fertility by doing media/soil as well as foliar testing is key to proper fertility management.

Common greenhouse insect pests include aphids, mealy bugs, spider mites, white flies, thrips, fungus gnats, and scale. Common viral and bacterial diseases are impatiens necrotic spot virus, tomato spotted wilt virus, and bacterial blights and soft rot. Fungal diseases include powdery mildew, botrytis blight, pythium, rhizoctonia, and verticillium. Nematodes can be a problem if media is not fumigated or pasteurized by composting, manufacturing processes, or steam. Integrated plant health management is critical for controlling pests and diseases in greenhouses.

Propagation is a special aspect of greenhouse operation and is often done by specialists who then sell the young plants or unrooted cuttings to other growers. Mechanization revolutionized plug production. Vegetative propagation has expanded from the traditional potted plant crops to include many bedding plants.

Greenhouse crops can be grouped into four broad categories: cut flowers, potted or container crops, foliage plants, and bulb crops. Cut flowers are plants grown for their flowers, leaves, or stems, which are harvested and mainly used in flower arrangements. Central and South America, and South Africa are major important production areas. Most cut flower production in the United States is in California although specialty cut flowers may be grown locally or regionally. Potted or container crops are grown mainly for their flowers and are sold in their container. They include crops such as poinsettia, chrysanthemum, and bedding plants such as geraniums. Foliage plants are grown for their foliage rather than their flowers. Many of these plants are tropical and adapted to the warm, humid, and low light conditions found under the canopies of large trees in tropical rainforests. In the United States, foliage production is concentrated in south Florida. Bulb crops are grown from underground storage sites and can be categorized by their pattern of flower formation. They may be used as potted plants or for cut flowers.

Each category and the species within that category have specific cultural requirements that must be met in a greenhouse to produce plants of high quality.

KNOWLEDGE CHECK

1. What are cold frames?
2. What was the heat source for early hot frames?
3. What are three ways heat can be supplied in a more modern greenhouse?
4. Why did the use of aluminum and improved glass-making technology for greenhouse structures increase the amount of light entering a greenhouse?
5. Why is polyethylene film used in double thickness as a greenhouse covering in cold climates? Why is a small fan needed for double polyethylene coverings?
6. Describe two ways greenhouses can be cooled.
7. What is the difference between a central and a localized greenhouse heating system?
8. What are horizontal airflow fans and how do they affect temperature and disease control?
9. Generally, how much higher are day temperatures held above night temperatures on cloudy days? On sunny days?
10. What is negative DIF and does it relate to stem elongation?
11. What are the three aspects of light that affect plant growth?

12. What is the best light source for growing plants in a greenhouse? What is the best artificial light source for photosynthesis?
13. Why is it important to control the photoperiod for photoperiodic plants in greenhouse production?
14. What is daily light integral and what is the minimum for most plants to grow well?
15. For potted plant crops, what is generally the proportion of container to plant for an aesthetically pleasing size?
16. When are PGRs used in greenhouses?
17. On what types of crops can PGRs not be used?
18. How do most PGRs except ethephon work? How does ethephon work?
19. What are three components that can be used in artificial media? Give a characteristic of each.
20. Describe three different types of pots or containers used in greenhouse production and give a use for each.
21. Why is water quality management very important with greenhouse crops?
22. Describe three different types of irrigation systems used in a greenhouse.

23. On what type of crops are liquid fertilizers used in the greenhouse? On what type are low release fertilizers used?
24. What is the general pattern for fertilizing a flowering plant during its production cycle?
25. Why is it a good idea to include a sample of healthy leaves along with samples of leaves with an apparent nutritional disorder when having a lab analyze leaf nutrient content?
26. What are four of the most common insect pests found in greenhouses? What are three of the most common fungal diseases?
27. What lead to the specialization of seed germination along with stock plant growing and cutting propagation?
28. What are plant patents and why are they necessary?
29. Why are potted plants not imported into the United States?
30. At what stage are cut roses harvested?
31. Give two examples of specialty cut flower species.
32. Botanically, what is the colorful, showy part of the poinsettia?
33. What is a response group in terms of poinsettias (and other photoperiodic plants)?
34. Why are temperatures reduced somewhat at the end of the production cycle of poinsettias?
35. Why is acclimatization to low light often required in foliage plant production? How is this done?
36. Where did the term "bedding plants" come from?
37. What two methods of tracking growth are used to bring Easter lilies into flower for Easter?

FOOD FOR THOUGHT

1. Given the choice of building an even-span greenhouse made of aluminum and glass or a quonset type made of double-layer polyethylene film to grow Easter lilies in the winter and early spring, bedding plants after the lilies are sold, and poinsettias for Christmas, which structure would you choose and why would you choose it?
2. You are growing a crop of potted chrysanthemums and using graphical tracking to make sure that they are growing properly. The last time you measured the plant height, you saw that they were above the curve of the graph. What would you do to correct the problem? What would you do if you overcorrected and the height started to fall under the curve?
3. You are at a party and when the other guests find out that you are foliage plant grower, you are bombarded with questions from several people about why their foliage house plants are not growing well in their homes. What would you ask or tell them?

FURTHER EXPLORATION

DOLE, J. M., and H. F. WILKINS. 2005. *Floriculture: Principals and species,* 2nd ed. Upper Saddle River, NJ: Prentice Hall.

JARVIS, W. R. 1997. *Managing diseases in greenhouse crops* 3rd ed. St. Paul, MN: APS Press.

NELSON, P. V. 2002. *Greenhouse operation and management,* 6th ed. Upper Saddle River, NJ: Prentice Hall.

*24 Turfgrasses

David Gardner

key learning concepts

After reading this chapter, you should be able to:

- Define the terms commonly used in turfgrass science.
- Explain the principles for establishing and maintaining turfgrasses.
- Describe the different types of turfgrass and the environmental and cultural requirements of each type.

The word lawn originally referred to a natural area of grass without trees. Today it refers to any expanse of ground on which grass is growing. Turf originally meant a layer of matted earth formed by soil and thickly growing grass plants. This meaning has gradually changed over the years, so that the word turf is used by horticulturists to refer to grass that is mowed and cared for. Turfgrass refers to the grass used in growing a horticultural turf. Growing turf is a multibillion dollar industry. The replacement value of turf in the United States has been estimated to exceed $12 billion, and annual maintenance costs have been estimated at more than $10 billion. Much of the turf is grass growing in public areas such as schoolyards, parks, cemeteries, and golf courses; along highways; on military installations; and on other public lands (Fig. 24–1).

Turf is a unique crop because the product is not what is harvested but what remains. The crop is grown densely as an entire population instead of individual plants spaced apart so that each grows vigorously. An appreciation of these differences is basic to being a good turf horticulturist.

There are three principal reasons for growing turf: (1) as a carpet, to protect the home from mud and dust, and to soften glare and reduce heat; (2) for recreation; and (3) for beauty and pleasure. Studies have shown that lawns so please us that many of our earliest memories are of grass and trees.

Turf culture differs from other horticultural pursuits in one important way. In nature, plant growth is often limited by competition with many other plants for water, nutrients, and light. Thousands of seeds germinate and die for every one that lives and grows. Successful horticulture comes from spacing plants and eliminating weeds so that each plant has soil, water, and space allotted to it alone. Grass is the exception. In growing a lawn, even more plants are crowded into a given space than would grow there even naturally. We strive to grow dense lawns similar to a fine carpet. The more we succeed in growing a dense lawn, the more stress each individual plant gets through competition from its crowding neighbors.

On an infertile dry soil, lack of nutrients and water limits plant growth and results in a poor lawn. We can grow a better, denser lawn by fertilizing and irrigating. But as the number of plants (shoots) increase, there is a point where plants shade each

Figure 24–1
(A) Public park. (B) College campus. Source: Margaret McMahon, The Ohio State University.

other and compete for sunlight. Beyond that point, individual plants are smaller, have fewer roots, and use nutrients and water less effectively. Thus, cultural practices that result in a dense, tight lawn can also result in individual plants of reduced quality and vigor. In addition to suffering stress from competition, most lawn plants also suffer climatic stress at some season of the year. In managing a lawn, we can fertilize, mow, and use other practices to either increase or decrease stress. Because of the need to balance the level of stress turf grass receives, turf grass management is a very exacting and science-based profession.

The stress concept gives us a tool for evaluating the effects of cultural practices on grass. In a figurative way, there is a stress budget for turfgrass. If plant competition is high and midsummer weather imposes a heat stress, little more allowable stress is left in the stress budget. Cultural practices that reduce stress, such as watering, should then be followed. On the other hand, early fall temperatures are often ideal for the growth of turfgrass. Then certain stressful management practices—for example, power raking to thin out plants—that would have damaged a bluegrass lawn in summer's heat can be used. Throughout this chapter, different management practices are evaluated in terms of the stress they cause. The practices that increase stress are not necessarily bad, but care should be exercised to select and use these practices in appropriate seasons, and not to use several high-stress operations at inappropriate times.

Turf culture can be divided into two procedures: (1) establishment and (2) maintenance for decorative or family use with modest traffic, or for heavy traffic by the public, or for athletic competition.

ESTABLISHING A TURF

A first step in growing a turf is to learn something about the species of plants used and their requirements. The grasses are monocotyledonous flowering plants belonging to the family POACEAE. Within the POACEAE are twelve currently recognized subfamilies. Three of these subfamilies are important in turfgrass management. Subfamily POOIDEAE contains the cool-season turfgrasses that are adapted to cool climates such as the northen United States. The warm season turfgrasses that are adapted to the warm climates like the southern United States are classified into two separate, distinct subfamilies. Grasses of the CHLORIDOIDEAE (such as zoysia grass) are adapted to warm, dry (arid) climates, while grasses of the PANICOIDEAE (such as St. Augustine grass) are adapted to warm, wet (tropical) climates. Differences in the evolutionary adaptation of the grasses used as turf have numerous implications for proper turfgrass management.

Between the region where cool season grasses are best suited and the region where warm season grasses are best suited is a transition zone where neither cool-season nor subtropical grasses perform at their best. Cool-season grasses grow well during cooler parts of the year, but during hot summer weather, they become dormant or suffer heat injury and disease. Subtropical grasses in the transition zone do well in the hot summer but become dormant and turn brown in winter (Fig. 24–2).

In the western, arid portion of the US plains states, where water is not available for irrigation, native drought-tolerant grasses such as buffalo grass or grama grass are sometimes grown. While their season of

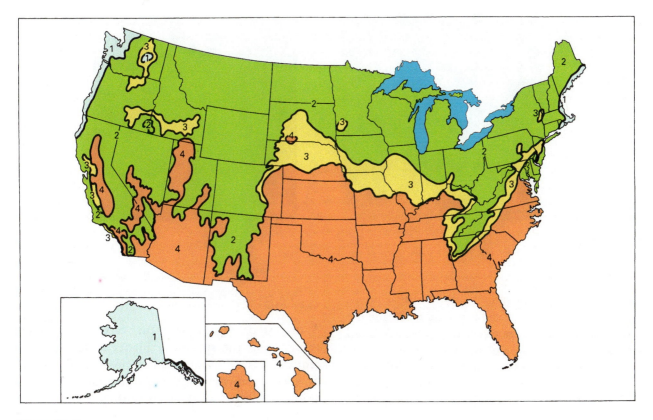

Figure 24–2

Areas of turfgrass adaptation. 1: Areas adapted to temperate grasses, particularly bent grasses (Agrostis). 2: Areas of general adaptation of temperate grasses. 3: Transition zone where subtropical grasses suffer winter cold stress and temperate grasses are stressed by summer's heat. 4: Area of adaptation for subtropical Bermuda and zoysia grasses. The more tender subtropical grasses are limited to hardiness zones 8, 9, and 10 (see Fig. 4–15 on page 54). This map is based on mean July temperatures, not the low winter temperatures of the hardiness zone map. The boundaries are not sharp and are greatly influenced by local microclimates, especially in the mountain states, where elevation and exposure affect adaptation. Alaska is too cold for all but one or two cultivars of turfgrasses, which survive in milder areas of the state. Hawaii is subtropical to tropical.

growth is limited, they do cover and protect the soil during the dry period.

Commonly used turfgrasses are described in Table 24–1.

TURFGRASS STRUCTURE AND IDENTIFICATION

Turfgrass anatomy is illustrated in Fig. 24–3 on page 556. Table 24–2 on page 557 gives a key for identifying common turfgrasses. The key is limited because it distinguishes only among the common cultivated turfgrasses.

Turfgrasses are well adapted to their role as a carpet. As with most monocots, growth of an individual shoot is **determinate** (ends in a flower), and the plants have fibrous roots without tap roots. Each shoot dies after a time but is replaced by new shoots growing from axillary buds. These new shoots produce adventitious roots at the nodes. In this way, turf continually rejuvenates itself. In many grass species, axillary buds produce horizontal creeping stems. If these grow over the surface of the ground, they are called **stolons**; below ground, they are termed **rhizomes**. Both structures enable grass plants to invade open areas and to spread. Other species of turfgrasses grow only new erect shoots or **tillers**, arising from axillary buds. Tillering grasses do spread but slowly.

Turfgrass flowers develop on elongated stems at certain seasons. The rest of the year, the tiller shoots are condensed with the vegetative growing point nestling among the leaves. The growing point produces **intercalary meristems**; that is, regions of cell initiation that lie across a stem or leaf and that, by dividing, interpose—or intercalate—new tissue between existing older tissues. The division and elongation of the intercalary meristems "extrudes" leaves and leaf sheaths. That is why grass blades continue to grow after mowing, the meristem continues to "push out" leaf tissue below the point where the leaf was

Table 24–1

CHARACTERISTICS, REGIONS, IMPROVED CULTIVARS, AND USES OF THE COMMON TURFGRASS SPECIES. THE FIRST EIGHT SPECIES ARE NORTHERN US GRASSES; THE LAST SIX ARE SOUTHERN

Turfgrass Species	Turf Characteristics	US Regions Where Used	Desirable Characteristics
Agrostis stolonifera L. (*Agrostis palustris* Huds.) Creeping bentgrass	Able to produce a dense turf of good color and fine texture under extremely close mowing (0.65 cm [0.25 in.] and less).	North, through the transition zone and, with care, into parts of the South.	Ability to stand low mowing. Withstands salinity.
Agrostis tenius Sibth. Colonial bentgrass	Variable grass adapted to northern coastal climates. Spreads. Color fair. Coarse-textured mowed high; fine-textured mowed low.	Well adapted to north coast, usable throughout the North.	Adaptable. Withstands acid soils of low fertility and wet soils.
Agrostis castellana Boiss, and Reuter Dryland bentgrass	In a mixture, it forms dense patches of blue-green grass of puffy character unless mowed under one inch. Tolerates more heat and less water than other bentgrasses.	Better adapted to transition zone heat and dryness than above.	Performs well in irrigated Mediterranean climate.
Festuca arundinacea L. Schreber. Tall fescue	A coarse-textured bunch grass suited only for growing in pure stands.	Adapted to transition zone. Not fully cold-hardy.	Deep rooted. Tolerates wear, drought, and neglect.
Festuca rubra L. var. *rubra* Creeping red fescue	Fine-leaved, drought-tolerant grass tolerant of some shade. Spreads by rhizomes. Favored by high mowing. Competitive under low fertility.	North. Does poorly in the transition zone.	Fine texture, spreading. Tolerates some drought, shade, infertility.
Festuca rubra L. var. *commutata* Chewing's red fescue	Fine-leaved, drought-tolerant grass. Tolerates some shade. Spreads slowly.	North.	More tolerant of summer stress than above.
Lolium perenne L. Improved perennial ryegrass	Turf cultivars have good color and fine texture like bluegrass. A patchy, hard-to-mow, clump grass unless well fertilized and watered.	Transition zone North. Not fully hardy. Winter grass in South.	Good color and texture. Tolerates traffic. Rapid establishment.
Poa pratensis L. Kentucky bluegrass	A dense, fine-textured, beautifully colored grass spread by rhizomes. Seeds produce uniform apomictic seedlings. The aristocrat of lawn grasses.	Best with cool nights and days under 84°F. Fair in transition zone.	Best color. Sod repairs itself from stolons. Heat and dry dormancy.
Cynodon dactylon (L) Pers. Bermuda grass	A dense vigorous grass creeping by stolons and rhizomes. Withstands neglect but a handsome turf when well cared for.	South and valleys of Southwest.	Vigorous. Withstands drought, wear, and salinity.
Eremochloa ophiouroides (Munro) Hack Centipede grass	A dense vigorous grass creeping by large stolons. Adapted to low maintenance lawns on acid or infertile soils.	Florida, Gulf states, coastal plains, parts of California.	Withstands low fertility and acidity. Low maintenance.

(Continued)

Table 24–1 (Continued)

Turfgrass Species	Turf Characteristics	US Regions Where Used	Desirable Characteristics
Lolium multiflorum Lam. Winter grass, annual ryegrass	A coarse short-lived grass used only for overseeding winter dormant grasses for winter color in the South and transition zone.	Winter grass in South.	Grows at low temperatures, near freezing.
Paspalum notatum Fliigge Bahia grass	Forms an open coarse turf.	Gulf states and coastal plains of the South.	Heat tolerant. Low water need.
Stenotaphrum secondatum (Walt.) O. Kuntze Saint Augustine grass	Strong, dense, coarse grass creeping by large stolons. Adapted to shade	Florida, Gulf states, and parts of California.	Vigorous. Withstands drought, some shade.
Zoysia japonica Steud. Zoysiagrass, *Z. matrella* (L.) Merrill	Extremely dense grower of good color. Spreads by rhizomes and stolons.	*Z. japonica* hardy through zones 2–10, *Z. matrella*, 8–10.	Very dense—crowds out weeds. Good color and texture.

Undesirable Characteristics	Improved Cultivars	Special Care Requirements	Uses
Subject to disease and insect pests. Invasive.	Penncross, Penneagle, A-4, G-2, Crenshaw, Providence, SR1020, L93	Regular day-to-day maintenance, mow short, control pests.	Fine turf for putting and bowling greens.
Coarse if neglected. Some disease and insect problems.	Astoria, Exeter, Bardot, Egmont, SW7100	Mow 2.5 cm (1 in.) or less; use low fertility for low maintenance.	Decorative lawns and utility turf.
Does not mix well with other grasses.	Highland	Mow 2.5 cm (1 in.) or less, dethatch in fall.	Decorative lawns and utility turf.
Coarse. Weedy in a mixture.	Arid, Bonanza, Falcon, K31, Mustang, Olympic, Rebel	Mow regularly 3.8 (1.5) in. or higher, use infrequent deep irrigation.	Decorative lawns and utility turf; sports turf.
Intolerant of heat, salinity, close mowing.	Boreal, Dawson, Ensylva, Flyer, Fortress, Merlin, Pennlawn, Ruby	Mow 5 cm (2 in.) or higher, light fertilizer. Unmowed for erosion control.	Decorative lawns and utility turf; overseeding subtropical grasses for winter color; unmowed for erosion control; sports turf.
Intolerant of heat, salinity.	Banner, Dover, Jamestown, Longfellow, Mary, Koket, Wilma	Mow 5 cm (2 in.) or less, fertilize lightly.	Decorative lawns and utility turf; overseeding subtropical grasses for winter color; unmowed for erosion control; sports turf.
Limited cold tolerance. Difficult to mow clean. Clumpy if unfertile.	Birdie II, Citation II, Manhattan II, Omega II, Palmer II, Prelude, Yorktown III, and others	Mow 3.8 cm (1.5 in.) plus with sharp mower. Fertilize and irrigate to keep dense.	Decorative lawns and utility turf; sports turf.
Disease susceptible.	Too many to name: Adelphi, Baron, Challenger, Eclipse, Flyking, Merion, Midnight, Pennstar, Touchdown, among many good ones	Mow 3.8 cm (1.5 in.) plus, keep up lime level, main fertilizing in fall.	Decorative lawns and utility turf; sports turf.
Invasive. Severe thatch builder. Disease susceptible in humid climate. Brown in winter where dormancy occurs.	Common, El Toro, Ormond, Santa Ana, Sunturf, Tifway, and others. Only common Bermuda is propagated from seed.	Close frequent mowing, dethatch, overseed for winter color.	Decorative lawns and utility turf; sports turf; fine turf for putting and bowling greens.
Coarse. A thatch builder.	Common, Oaklawn	Keep fertilizer low, no lime, dethatch if needed.	Decorative lawns and utility turf.

Table 24–1 (*Continued*)

Undesirable Characteristics	Improved Cultivars	Special care Requirements	Uses
Coarse. Disease susceptible.	Gulf, Tifton 1	Open turf so seed touches soil. Mow as needed.	Overseeding subtropical grasses for winter color.
Not cold hardy. Openness favors weed invasion.	Argentine, Pensacola, Tifhi 1, Wilmington	Mow 3.8 cm (1.5 in.) plus, low fertility. Reduced water need. Decorative lawns and utility turf.	Decorative lawns and utility turf; sports turf.
Coarse, invasive, disease- and insect-prone. Vegetative planting.	Bitter blue, Floratam, Floratine	Mow closely and dethatch. Control pests.	Decorative lawns and utility turf.
Vegetative planting. Slow recovery of injury. Too dense to overseed.	Meyer, Midwest (*Z. japonica*); Flawn (*Z. matrella*); Emerald (hybrid)	Low fertility and water need. Mow closely, or neglect for erosion control.	Decorative lawns and utility turf.

mowed off. Internodes of the stem of grass plants do not elongate except to produce stolons, rhizomes, or flower stalks.

CHOOSING TURFGRASSES

Proper selection of species and cultivars is critical to the long-term success of a turfgrass stand. However, selection of adapted species and cultivars does not guarantee long-term success. Proper cultural practices are still required. Local extension information should be consulted when selecting a turfgrass for establishment. Cool-season grasses are commonly established as a blend (two or more cultivars of the same species) or as a mix (two or more turfgrass species). Cultivars and species are mixed to take advantage of the good qualities of each while avoiding the extensive damage that could occur when a pure stand of a single cultivar develops a weakness. Warm-season grasses are rarely mixed, to avoid problems with segregation. Segregation occurs when the species start to appear in "clumps" rather than being evenly scattered throughout the area. When using a seed blend or mixture, three to five cultivars with differing genetic makeup should be used to provide adequate diversity. The components of a blend or mix should also be similar in appearance, reflect the local environment (heat, shade, etc.), and show resistance to the major diseases and insects (if possible) in a particular region.

For several of the turfgrass species, such as Kentucky bluegrass, perennial ryegrass, tall fescue, fine fescue, creeping bentgrass, and Bermuda grass, there are several cultivated varieties (cultivars). Each of these cultivars was developed because of some unique characteristic, such as darker green color or finer texture. Other grasses,

such as centipede grass and Saint Augustine grass, have fewer cultivars. Kentucky bluegrass has the largest number of cultivated varieties. Of the turfgrasses, Kentucky bluegrass is unique in producing a large percentage of seedlings by **apomixis**. In this type of reproduction, the embryo resulting from sexual fusion of an egg and pollen nucleus aborts, but an embryo is produced from somatic cells of the embryo sac (female tissue). In this way it is possible to get seedlings that are genetically identical to each other and to the female parent.

Superior plants are tested for turf characteristics such as low growth, slow growth rate, disease resistance, cold tolerance, drought tolerance, and so forth. If they are indeed superior, they can be released as a cultivar. As a result of selection and the ability of bluegrass to reproduce asexually, there are many clonal **cultivars** of bluegrass but few or none of the other turf species (Fig. 24–4).

Named clones of other grasses can be vegetatively propagated with sod or stolons. Named cultivars are sexually propagated as a genetic mixture of seed from parents carefully selected to a standard of excellence. Named cultivars are excellent, but they have the disadvantage of complete genetic uniformity so that a turf is totally subject to any weaknesses. For example, a pure stand of 'Merion' bluegrass appears orange at times because of spores of a species of stem rust disease to which it is particularly susceptible. That is why 'Merion' is blended with other bluegrass cultivars that are resistant to the rust.

With the exception of buffalo grass, none of the species commonly used as turfgrass in the United States are native to North America. Rather, they were introduced from other parts of the world. For example, Colonial bentgrass (*Agrostis tenuis*) is native to northern

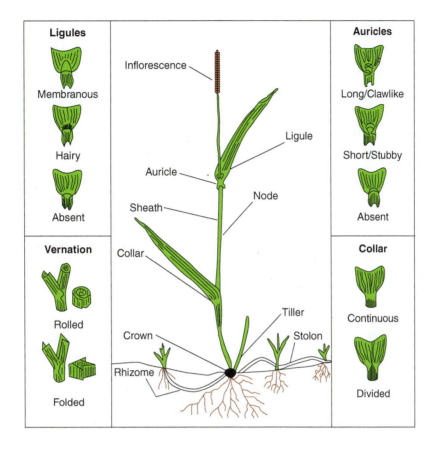

Ligules

Membranous

Hairy

Absent

Vernation

Rolled

Folded

Inflorescence

Ligule

Auricle

Node

Sheath

Collar

Tiller

Crown

Stolon

Rhizome

Auricles

Long/Clawlike

Short/Stubby

Absent

Collar

Continuous

Divided

Key structures, terms, and definitions used in the identification of turfgrasses

Phytomere: Basic unit of the grass plant (leaf blade, sheath, internode, axillary bud, adventitious roots, nodel plate).

Ligule: Appendage at abaxial junction of the blade and sheath.

Auricle: Appendages that extend from the collar and wrap around the leaf.

Collar: Distinct band of tissue at blade/sheath junction.

Vernation: Leaf arrangement in the bud shoot.

Sheath: Lower part of leaf attached to the node at the crown.

Internode: Stem segment between nodes.

Node: Joints of stems where leaves are attached.

Stem appex: Contains the apical meristem.

Inflorescence: Terminal reproductive shoot.

Shoots: Primary—arise from the embryo of the germinating seed.
Lateral—arise from buds on the crown.
Tiller—intravaginal lateral shoot (syn.: bunch type).
Rhizome—extravaginal subsurface lateral shoot.
Stolon—extravaginal above-ground lateral shoot.

Crown: Area of meristematic tissue between the shoot and the root.

Roots: Primary—arise from the embryo of germinating seeds.
Adventitious—arise from nodes on the crown, rhizomes, or stolons.

Figure 24–3

Turfgrasses are crown-type plants, and internode elongation on the stem occurs only when an inflorescence is generated, as shown in this diagram. Source: David Gardner, The Ohio State University.

Europe. When introduced by colonists to the New World, it quickly took over the coastal regions from Rhode Island to Nova Scotia and from Oregon to British Columbia. It also became widespread along the coasts of New Zealand. Kentucky bluegrass (*Poa pratensis*) is thought to have been introduced at Vincennes, Indiana, by the French from Europe in

about 1720. By the end of the Civil War, it was established throughout the Middle West, and the westward settlers thought it was a native grass. Similarly, Bermuda grass (*Cynodon dactylon*, thought to be native to Asia) was introduced into New Mexico in 1750 and within fifty years spread throughout the Southwest. Frank Meyer, a plant explorer with the USDA, brought

Table 24–2
IDENTIFICATION KEY TO TURFGRASS SPECIES

1. Ligule not a fringe of hairs
   2. Folded vernation
      3. Boat shaped leaf tip
         4. Ligule short or absent
            5. Rhizomatous, leaf blade with translucent midrib = Kentucky bluegrass (*Poa pratensis*)
         4. Ligule present
            5. Ligule prominent, sheaths slightly compressed, bunch type = Annual bluegrass (*Poa annua*)
            5. Ligule abrupt, sheath strongly compressed = Canada bluegrass (*Poa compressa*)
            5. Ligule 4–6 mm, stoloniferous, onion-skin sheath = Rough bluegrass (*Poa trivialis*)
      3. Pointed leaf tip
         4. Very fine leaves, bunch type = *Festuca spp.* fine fescues
         4. Very fine leaves, rhizomatous = Creeping Red Fescue (*Festuca rubra*)
         4. Leaves shiny on back, bunch type, red pigment at base of sheath = Perennial ryegrass (*Lolium perenne*)
   2. Rolled vernation
      3. Auricle present
         4. Long, clawlike auricle, bunch type = Annual Ryegrass (*Lolium multiflorum*)
         4. Narrow auricle, leaf blade 2–5 mm, pubescent upper leaf = Fairway wheatgrass (*Agropyron cristatum*)
      3. Auricle is difficult to detect or is absent
         4. Ligule short or absent
            5. Distinct collar, bunch type, prominent veination = tall fescue (*Festuca arundinacea*)
            5. Ligule evenly truncate, blades rough along edges, bunch type = colonial bentgrass (*Agrostis tenuis*)
         4. Ligule prominent
            5. Stoloniferous, light green color = creeping bentgrass (*Agrostis palustris*)
            5. Rhizomatous, prominent venation = Redtop Bentgrass (*Agrostis alba*)
            5. Swollen base (Haplocorm) = Turf Timothy (*Phleum pratense*)
1. Ligule a membrane with a fringe of hairs
   2. Margins ciliate toward base, thick stolons = Bahia grass (*Paspalum notatum*)
   2. Fine-textured, less aggressive stolons = Seashore Paspalum (*Paspalum vaginatum*)
1. Ligule a fringe of hairs
   2. Rolled vernation
      3. Light green leaf, surface pubescent on both sides = buffalo grass (*Buchloe dactyloides*)
      3. Distinct collar, evenly spaced internodes = zoysia grass (*Zoysia japonica*)
   2. Folded vernation
      3. Unevenly spaced internodes, vernation may be rolled, thin collar = Bermuda grass (*Cynodon dactylon*)
      3. Evenly spaced internodes
         4. Collar constricted with 90° twist = Saint Augustine grass (*Stenotaphrum secundatum*)
         4. Collar lacks 90° twist
            5. Hairs on lower 1 cm of leaf edge, cottony ligule = centipede grass (*Eremochloa ophiuroides*)
            5. Pubescent sheath, smooth leaf blade = Kikuyu grass (*Pennisetum clandestinum*)

Source: David Gardner, The Ohio State University.

Figure 24–4
Plots of different bluegrass cultivars being evaluated for their growth characteristics. Source: USDA Agricultural Research Service Image Gallery, http://www.ars.usda.gov/is/graphics/photos/

the first zoysiagrass seed to the United States from Korea in 1905 (Fig. 24–5). Meyer zoysiagrass is named for him.

In any location in the United States, several turfgrasses can be grown and, with good management, all can look good. In a few locations, one certain grass is particularly well adapted to the climate: it will gradually invade and dominate lawns in that area. It might be difficult to choose among several grasses when all grow well in an area or to see any reason why one cultivar or another should be preferred. Among well-adapted grasses, differences appear only under conditions of stress such as cold, heat, drought, wet soil, shade, low mowing, or disease; each of these stresses takes a toll at one time or another. The best species or cultivar for a given location is most likely to be determined, not by appearance or growth habit, but by its ability to survive the few days or weeks in

Figure 24–5
Frank Meyer, the USDA plant explorer who brought zoysia-grass to the United States from Korea. Source: USDA Agricultural Research Service Image Gallery, *http://www.ars.usda.gov/is/graphics/photos/*

the year when growing conditions are unfavorable. Such characteristics are given in Table 24–1 on page 553.

Heredity endows certain grasses with the ability to grow and survive stress. Culture determines how well a grass achieves its potential.

SOIL PREPARATION

Soil preparation is a first step in planting a turf, and it includes grading and tilling. Grading should result in a convex swell of the soil surface, free from dips, swales, or pockets where surface runoff water can puddle or pond. Tilling breaks up soil to form a seedbed with good porosity and aeration. In time, tilled soil settles back to its original density, but initial root growth of new turf is aided by the loosened soil.

During tillage, fertilizers and chemical or physical soil amendments should be incorporated in the soil. Chemical amendments include materials such as lime or gypsum used to improve the chemical and physical properties of the soil. Physical amendments are mineral or organic and are generally used to improve the physical properties of a heavy soil or to add organic matter to a biologically impoverished one. Because physical amendments must be used at very high rates to have beneficial effects (often 70 to 90 percent amendment), their use is questionable unless the particular soil problem has been thoroughly analyzed.

Organic amendments serve as food for soil organisms and usually improve soil structure. Peat or manure incorporated into the top 1 to 2 cm (0.4 to 0.8 in.) of the seedbed helps seedlings emerge from crusting soils. Incorporation of other organic wastes improves the root zone environment, but some materials can be toxic (e.g., fresh cedar sawdust). Many organic wastes result in severe temporary nitrogen deficiencies in the soil unless the carbon-to-nitrogen ratio in the waste is less than 20:1. Composting organic wastes for a time before application usually corrects such imbalances and toxicities.

When the seedbed is a sterile (no organisms present) subsoil, added organic matter and nitrogen fertilizer enhances biological activity. Where the soil is shallow, a few centimeters of additional soil are often added. A sterile soil may be a fill soil, or it may result from leveling or from spreading out soil from a basement excavation. Developers sometimes remove topsoil for sale to boost their earnings from a project.

The construction of a field for a sport such as football may require very sophisticated soil preparation. Synthetic materials are often added to the soil components to add stability and improve drainage. Competing weeds are a major problem in establishing a new turf. They shade and suppress the grass, and removing them requires time and effort. It is good practice to irrigate the seedbed to germinate weed seeds before sowing grass seed. Weed seedlings are then killed by cultivation or with herbicides. The seedbed should then not be disturbed to the extent that new weed seeds are brought to the surface, where they could germinate.

Grass seed is available either in mixtures or pure lots. Most mixtures are designed to be sold to home gardeners rather than to professional turf growers. These seed mixtures represent compromises by the seed companies, and the mixtures produce an acceptable lawn regardless of the management given. The label names the species or cultivar of each grass, the percentage of crop and weed seeds, and the percentage of inert matter (Fig. 24–6). The germination rate of the crop seed may also be given.

Germination varies from 75 percent for bluegrass seed produced in a poor crop year to over 95 percent for ryegrass seed produced in a good crop year. In general, seed germination should be over 85 percent. If germination rates are low, additional seed must be sown to compensate. Inert matter represents an inevitable amount of chaff or other material. Noxious weed seeds are of small concern in turf because most such weeds are subsequently destroyed by mowing. Turfgrass seed sometimes includes pasture grasses of no

Seed Mixture Analysis	
Fine Textured Grasses	**Germination**
26.24% Glade Kentucky Bluegrass	95%
24.00% Park Kentucky Bluegrass	90%
18.45% Midnight Kentucky Bluegrass	85%
Coarse Grasses	
27.81% Pennfine Perennial Ryegrass	92%
Other Ingredients	
0.45% Weed Seed	Tested 4/05
3.00% Inert Matter	50# net wt.
0.05% Other Crop	
No Noxious Weeds	Seed company address

Desirable crop species ⟶ (Fine Textured Grasses)

Undesirables not grown as a farm crop ⟶ (Weed Seed)
Anything not seed (soil, stems, etc.) ⟶ (Inert Matter)
Other plants normally grown for profit ⟶ (Other Crop)
• Often the most difficult weeds to control; weeds officially declared special problems ⟶ (No Noxious Weeds)

Figure 24–6

Example of a seed tag or label showing the information required by law in the United States. Note that seed quality is influenced by purity and germination. Seeding rate should be based on % pure live seed (PLS) = [(germination% × purity%)/100]. It is also important to note the date on which the seed was tested. Source: David Gardner, The Ohio State University.

concern in a meadow or pasture, but they form coarse, undesirable, persistent weeds in a fine lawn. To avoid contamination by such grasses, seed of sod quality, which is free of such contaminants, is often purchased at a premium price. When named cultivars of grass are used, certified seed carries a certification statement that the plants were inspected while growing in the seed field and found to be pure and true to type.

In the northern two-thirds of the United States, a typical packaged mixture of high-quality seed is likely to contain a large percentage of several cultivars of Kentucky bluegrass. This blend forms the basic grass. To this is added a blend of red fescue grasses. Red fescue mixes well with bluegrass and grows better in the dry, shady, and less fertile areas. Where winters are not severe, seed of fine-leaved cultivars of perennial ryegrass are also blended with the blue grass to help the mixture resist seasonal disease problems. A low percentage of a bentgrass is usually added. Bentgrass finally predominates if the resulting lawn is mowed too short for survival of the blue, rye, and fescue grasses. In addition, bentgrass often survives better than the others in areas with wet soils.

Such seed mixtures can be diluted, more or less, with seeds of a filler grass to adjust the price of the mixture. A good filler is meadow fescue (*Festuca pratensis*), which looks like bluegrass during its first few months. Later it becomes coarse but tends not to persist in a well-tended turf. Red top (*Agrostis alba*) is often used as a filler but is not desirable because it is coarse and persistent. Some mixtures are blended solely for a low price and often contain large amounts of pasture grasses or the less desirable turfgrasses.

A high-quality seed mixture selected by a knowledgeable horticulturist might consist of a blend of the first three grasses mentioned, but might not include bent or filler grasses. Bentgrass tends to be a weed in bluegrass (and vice versa). Along the coasts of New England, Oregon, Washington, and southern Canada, one might choose a colonial bentgrass alone. Bentgrasses are well adapted in those regions.

In the southern United States, a Bermuda grass, centipede grass, Bahia grass, or carpet grass lawn might be started from pure seed. Saint Augustine grass or hybrid Bermuda grass could be started from stolons or sod.

SEEDING

Once the kind of grass is chosen, the rate at which to sow the seed is considered (Table 24–3). Seedlings become crowded and cannot develop properly when seed is sown too heavily; it takes a long time to get a mature usable lawn. Sown too thinly, the plants are far apart, with space left for weeds to germinate. An initial stand of about twelve to fourteen plants per square inch will develop rapidly into a strong turf. If one sows off-season (e.g., late fall or early winter) when seed germination is slow, one should increase the rate.

Special machines are available for sowing seed, covering it, firming the soil, and even mulching it (Fig. 24–7). For hand operations, however, either box or cyclone fertilizer spreaders sow seed satisfactorily. The opening is adjusted to a size appropriate to the seed. Hand sowing tends to scatter seed unevenly because few persons are skilled seed sowers. When hand sowing is necessary, seed should be vigorously thrown

Table 24–3
CHARACTERISTICS OF SEEDS OF COMMON TURFGRASSES*

Species	Number of Seeds per Pound	Seeding Rate (pounds per 1,000 ft²)	Germination Time (days)
Cool-Season Species			
Creeping bentgrass	6,356,000	0.5–1.0	7–14
Colonial bentgrass	8,172,000	0.5–1.0	7–14
Velvet bentgrass	10,896,000	0.5–1.0	
Turf Timothy	1,135,000	1–2	
Tall fescue	227,000	7–9	5–15
Red fescue	546,000	3.5–4.5	5–12
Chewings fescue	546,000	3.5–4.5	5–12
Hard fescue	546,000	3.5–4.5	5–12
Perennial ryegrass	227,000	7–9	3–10
Annual ryegrass	227,000	7–9	3–8
Kentucky bluegrass	2,179,000	1–1.5	6–28
Rough bluegrass	2,542,000	1–1.5	6–21
Canada bluegrass	2,497,000	1–1.5	6–21
Fairway wheatgrass	318,000	3–5	
Warm-Season Species			
Buffalo grass (burs)	50,000	3–6	10–28
Bermuda grass (hulled)	1,787,000	1–1.5	8–15
Zoysia grass	1,369,000	2–3	10–14
Bahia grass	163,000	6–8	
Centipede grass	409,000	4–6	

*The number of seeds per pound, and therefore the recommended seeding rate, varies considerably among the species. The range of average germination times reflects the amount of time in which careful attention to surface moisture is required and varies depending on the time of year and conditions.

Source: David Gardner, The Ohio State University.

Figure 24–7
Seeding turf. Source: Pamela Sherratt, The Ohio State University.

forward in a sweeping arc. The falling seed is more apt to drift into a random pattern than when seeds are dribbled out close to the ground.

When both large and small seeds are used, they are best sown separately in two operations. On small areas, rather than calibrate the spreader, the operator may prefer to reduce the seeder opening to a low rate, then cover the area two or three times in different directions to ensure even coverage. Small amounts of seed are often diluted with sand to adapt the volume of seed to the area covered.

Seeding Depth

Another determination is the depth to cover the seed. Turfgrass seeds are small, varying from about 1 million per kilogram for rye- and fescue grasses to over 4 million for bluegrasses and over 10 to 17 million for bentgrasses. Relatively few seedlings emerge from depths over 1 cm (0.4 in), and seedling emergence for the smaller seeds is best when they are covered to less than 1 to 5 mm (0.04 to 0.2 in). Uniform coverage is not generally possible. A light raking or dragging of the soil surface after seeds are sown covers seeds at depths from 0 to 5 or 10 mm (0.2 to 0.4 in). Some seedlings on the surface die of desiccation, and some deeper ones fail

to grow to the surface, but seeding rates allow for such losses.

Time to Sow

At any given geographical location there is a best week in the year in which to sow grass seed. In the southern parts of the United States, subtropical (warm season) grasses are best sown in early summer after annual weeds are removed following their principal flush of seed germination. Seeds of temperate zone (cool season) grasses are best sown in late summer or early fall so that there is time for the seedlings to become well established and to cover the ground before freezing weather arrives. Fine, vigorous grass stands are easiest to obtain with fall-sown seed. Fall weed problems are reduced, and shorter days and cooler temperatures reduce evaporation and the need for frequent irrigation. The next best time to sow grass seed in the temperate zones is as early in the spring as the soil can be prepared. Grass seed germinates and the seedlings grow in cool soil; if a dense stand can be obtained before summer weed seeds germinate, the weed problem is reduced.

After the seed is sown, the seedbed should be rolled, unless you are relying on rainfall to germinate the seed. Rolling firms the soil around the seed and thus encourages capillary movement of available moisture toward the seed. If there is insufficient moisture for good seed germination, a mulch should be used. A light cover (2 to 5 mm [0.08 to 0.2 in.]) of clean, weed-free sand or organic waste slows moisture loss from the soil surface. However, light mulching materials such as peat moss tend to wash or blow or to gather in pockets. Such materials are best worked into the top inch (2.5 cm [1 in.]) of soil before seeding.

Seed Germination and Seedling Establishment

Water is the most critical factor during seed germination, whereas nitrogen fertilizer is more critical during seedling establishment. In warm weather, most turf seeds begin to germinate five to seven days after sowing and continue for another week. Bluegrass seed, however, is slower and continues to germinate for a month. Germination is slower in cold weather. During germination, soil moisture is necessary in the surface layer at all times. Differences in available water in the root zone soon appear as differences in color and stand of the grass seedlings. At the same time, excess surface water encourages damping-off pathogens, a complex of *Rhizoctonia, Phytophthora, Pythium, Fusarium,* and other species of fungi. Ideally the top soil layer should

remain moist, but if the soil surface is allowed to dry at least once a day, the mycelia of damping-off organisms shrivel and die.

Once germination has occurred, seedling roots begin to explore the soil. The period between irrigations is gradually extended so there is regular drying of the soil surface between irrigations. Irrigations should wet the soil deeper than the roots extend.

As grass seedlings grow, they may deplete the soil of nitrogen. Growth slows, and the seedlings' color becomes pale green to yellowish. Regular feedings of nitrogen fertilizer to provide about 0.25 kg of N per acre (0.5 lb/1,000 ft^2) keep the grass growing vigorously and help it suppress weeds. If seedling growth slows and the blade color is dark green, with some red anthocyanin pigment present, the seedbed probably contains insufficient phosphorus for initial seedling establishment.

As the turf grows and becomes thick and tall enough for the first mowing, the most suitable height for mowing must be decided. The mower should be sharp and set for the correct height. A dull mower pulls up seedling plants or tears leaves instead of shearing them. When the new grass is 2.5 to 5 cm (1 to 2 in.) higher than the desired mowing height, the soil should be allowed to dry for a day or two before the grass is mowed. If the clippings are scattered and the weather is dry, clippings shrivel and fall from sight. But if the clippings form heavy clumps and the weather is moist, they should be removed to prevent smothering and a good environment for diseases.

In the United States, it may take five months in southern states to twelve months in northern states after seed germination for the turf to form a vigorous mature lawn rugged enough for play.

SODDING

Points to Consider When Sodding

Sodding is the installation of a mat of turfgrass that has been cut from the ground where it was grown and moved to a new location (Fig. 24–8). It is an alternative to growing turf from seed. Sodding or seeding is a choice that turfgrass managers can make. The advantages and disadvantages of sodding compared to seeding are listed in Table 24–4. The ideal time of establishment from sod is the same as for seeding. However, spring sodding is usually much more successful than spring seeding. Sodding can be done in the summer in climates that do not have extended periods of hot weather (90°F+). Sodding can be also done much further into the fall season than can seeding.

Figure 24–8
Installing turf on a football field. Note the pieces of sod lying on the side of the field that were trimmed off to allow the strips to lie flat. Source: Margaret McMahon, The Ohio State University.

Quality sod should be purchased from a reputable dealer. The sod should be free of pest problems (weeds, insects, diseases). The grass should be in active growth and densely rooted, and the root system should be moist upon delivery. The sod strips should be uniform in width and thickness. The rhizomes should be severed at the bottom of the root mass because new roots form at the ends of severed rhizomes more quickly than fibrous roots can repair themselves.

Sod should be cared for properly after it is delivered. If possible, the sod should be laid the same day it is received to avoid what is called sod heating. Sod heating is an increase in temperature within the sod roll that results from plant and microbial respiration. Heating can result in high temperatures (38°C [100°F] or higher) that cause death of the grass. Sod may be stored for up to three days during cooler weather if exposed sides and ends are lightly dampened daily, or as needed, and if the sod is lightly covered with a light-colored material to shield it from direct sunlight.

STEPS TO FOLLOW WHEN SODDING

The soil should be prepared the same way as for seeded areas. The area should be graded to give a slope of more than 1 percent but less than 20 percent and away from buildings. If the soil is compacted, it should be loosened by tilling to at least a 15.2 cm (6 in.) depth. Debris such as sticks, stones, and weeds should be removed. Starter fertilizer and lime, as indicated by soil test, are then applied and incorporated to a 15.2 cm (6 in.) depth. The lawn area needs to be leveled to eliminate high and low spots. At the edge of the turf area, the soil should be sloped so that the sod is recessed to the thickness of the sod root system. This practice prevents the edge of the sod from drying out and dying. The final grade should be firm enough to avoid footprinting.

Sod should be planted on lightly moistened soil. The sod should be rolled out by staggering in alternate rows (like the running-bond pattern of a brick wall), with the edges placed tightly together. Overlapping sod pieces should be avoided; instead, the pieces should be cut to fit with a sod knife (Fig. 24–8). Also, one should not pull a strip of sod. Stretched sod will contract to its original size within hours and leave a space between pieces. Staking may be required on slopes with a grade of 8 percent of more. Three thin stakes (one at each end and one in the center) are placed about 3 in. from the upper edge of the sod strip. The stakes should be placed flush with the top of the sod root system. Laying long, narrow strips (5 to 7.6 cm [2 to 3 in.]) of sod (as may occur at the edges of flower beds) should be avoided. The edges of newly laid sod should not be allowed to dry out.

After the sod is planted, it should be rolled at a 45° angle with a 60 to 75 lb roller to remove air pockets and then thoroughly watered so that water penetrates at least 6 in. into the soil. Sod requires daily afternoon watering for the first seven to ten days. More frequent watering may be required during periods of warm weather. Sod begins to root after ten to fourteen days, and rooting can be tested by lightly pulling up on a sod

---

Table 24–4
ADVANTAGES AND DISADVANTAGES OF ESTABLISHING A TURFGRASS FROM SOD VS. SEED

Advantages of Sodding	Disadvantages of Sodding
• Sod provides an instant ground cover.	• Sodding is much more expensive.
• Watering frequency and duration is less at establishment.	• Sodding is much more labor intensive.
• Weed invasion is reduced.	• Sod is available with fewer cultivars and mixes.
• Damage from rainfall, animal footprinting, etc., is less likely.	• Improperly laid sod is much more difficult to repair than an improperly seeded area.
• Mulching is not required.	

Source: David Gardner, The Ohio State University.

Figure 24–9

Overseeding a warm season grass with a cool season grass.

Source: Pamela Sherratt, The Ohio State University.

strip. When the sod is well-rooted, begin deeper and less frequent irrigation to promote deeper rooting. Do not overwater rooting sod. The sod will not root properly if the soil is saturated. Mowing should begin after the grass begins to grow. Fertilization as for a mature turf should begin after the sod has rooted.

Overseeding

A common practice in the southern United States is to overseed dormant subtropical grasses, whose leaves become brown during the winter (Fig. 24–9). To keep the lawn green, seeds of a cool-season species are sown in the dormant turf in the fall. This practice provides winter color until the warming soil and vigorous spring growth of the subtropical grasses crowds and suppresses the winter grass. The traditional grass for such overseeding is annual ryegrass (*Lolium multiflorum*), but red fescue or perennial ryegrasses are also suitable and are finer textured. Other cool-season grasses are also used.

Timing is important in overseeding. There should be enough warm weather yet to come for germination and seedling growth, but if overseeding is done too early, hot weather is likely to encourage growth of **damping-off** pathogens.

Renovation

An old, thin, or weedy lawn can be renovated by introducing new seed with improved management practices. To germinate and become established, the new seed must contact soil. A seedbed for either overseeding or renovation is prepared in the existing turf by raking vigorously; by power raking; or by using a coring, thatching, or vertical mowing machine to tear out a

dense dead thatch and weak plants and to expose soil between the grass plants. Seed is sown, then kept moist during germination by sprinkling. Sown seed should also be lightly top-dressed with compost or sand. As soon as germination begins, a fertilizer should be applied to provide both nitrogen and a small amount of phosphorus. In renovating, a herbicide can be used in advance to kill undesirable grasses in portions of the lawn.

PRIMARY TURF MANAGEMENT PRACTICES

Maintenance begins once the grass is up. Maintenance consists primarily of mowing; fertilizing; irrigating; and controlling weeds, insects, and diseases. These procedures should be programmed to produce a beautiful dense turf and vigorous healthy plants. As noted earlier, these goals are not completely compatible. There must be some compromise area appropriate to the climate, to the equipment available, and to the level of maintenance one is prepared to pursue.

Not only must the chosen program be a compromise between beauty and vigor, but there is the added possibility that a goal can be achieved equally well with different management programs. If the desired result is achieved with reasonable economy of effort and resources, no program is more right or more wrong than another. The lack of positive answers makes turf management comparatively difficult or confusing for some and challenging to others. For this reason, turf culture is best considered in terms of principles rather than applications. There are three principal turf management practices and many secondary practices. The principal ones are mowing, fertilization, and irrigation. Each has a large effect on grass growth, and the manager manipulates them to change grass vigor and appearance.

Mowing

Mowing is a regular chore. There are several choices to make in mowing, for example, the kind of equipment to use, the mowing height, frequency of mowing, and whether to remove clippings or leave them.

Factors That Influence Mowing Frequency Several interacting factors influence the frequency of mowing required. Shorter cut turf requires more frequent mowing than taller cut turf. The time of year also affects mowing requirements. Cool-season turf requires more frequent mowing in fall and spring than in summer, when conditions are not as favorable for turfgrass growth. Also, when cultural practices such as fertilization and

irrigation are optimized, mowing frequency increases. Overapplication of fertilizer or the use of quick-release nitrogen will cause a growth flush. The species used can affect mowing frequency requirements. For example, zoysia grass grows more slowly than do other species.

Recommended Mowing Heights for Turfgrass Species   The recommended mowing height is determined by species and use. For example, tall fescue will appear open and stemmy if mowed shorter than 5 cm (2 in.). In contrast, creeping bentgrass mowed taller than 5 cm (2 in.) will result in puffy and somewhat course-textured appearance. Table 24–5 gives mowing height recommendations for commonly used turfgrasses. Slight variations to these recommendations may occur in different regions, based on climate and other factors. For example, shaded or otherwise stressed turf should be mowed taller. In general, management intensity increases as mowing height decreases.

Mowing Equipment   There are four kinds of modern mowing machines. Two kinds—flail and sickle bar—are used only for large-scale and heavy work.

The flail mower is used on roadsides and in rough park areas. It has a rapidly rotating horizontal axle that swings several vertical knives inside a housing. The mower requires high energy, mows tall grasses and weeds, and reduces them to a mulch.

The sickle bar mower resembles a hair clipper with a 2m (6 ft) horizontal blade. It uses low energy; mows grass, weeds, and woody seedlings; and lays them in a swath with the stems parallel, convenient for raking. This mower is the same type used to mow forage hay.

MOWING STRESS   ✱

Mowing places a stress on the grass and reduces its vigor. Root growth is slowed for a few hours to several days depending on the severity of mowing. The total leaf production for the season is reduced by each mowing. Mowing opens the grass canopy to allow more light to enter so more plants can grow in the same area, thus increasing competition. The greatest stress is added to the stress budget when grass is mowed very short. High populations of weak plants result. With exceptionally vigorous and invasive grasses, such as Bermuda grass or kikuyu grass, frequent short mowing can be used to deliberately weaken them.

Two types of lawn mowers are most commonly available: reel mowers and rotary mowers. Reel mowers are used when a high-quality cut or a low cutting height is required, such as on a putting green or fairway (Fig. 24–10). Reel-mower cutting action is similar to that of a pair of scissors. A series of blades forming a cylinder are mounted at a spiral angle to the axis. They spin over a parallel-mounted bedknife. Reel mowers require relatively less power than do rotary mowers.

Table 24–5
RECOMMENDED MOWING HEIGHTS FOR COMMONLY USED TURFGRASS SPECIES*

Turf Species	Mowing Height (inches)
Creeping bentgrass (*Agrostis palustris*)	0.125–0.5 (high intensity) 1.0–2.5 (low intensity)
Tall fescue (*Festuca arundinacea*)	1.5–3.0
Fine fescue (*Festuca rubra, F. longifolia*)	1.0–2.5
Perennial ryegrass (*Lolium perenne*)	1.0–2.5
Kentucky bluegrass (*Poa pratensis*)	
Common	1.0–2.5
Improved	0.75–2.5
Buffalo grass (*Buchloe dactyloides*)	1.0–2.0
Bermuda grass (*Cynodon dactylon*) (varies by cultivar)	0.3–2.5
Zoysia grass (*Zoysia* spp.)	0.5–1.0
Bahia grass (*Paspalum notatum*)	1.5–3.0
Saint Augustine grass (*Stenotaphrum secondatum*)	1.5–3.0
Centipede grass (*Eremochloa ophiuroides*)	1.0–2.0

*Mowing height varies depending on species and usage. Mowing height has a tremendous effect on management intensity required.

Source: David Gardner, The Ohio State University.

Figure 24–10
A reel mower. Source: Pamela Sherratt, The Ohio State University.

Rotary mowers are used when a moderate quality of cut at a higher height is acceptable, such as for home lawns, golf course rough, and utility turf. The cutting action is performed by the impact of a horizontal blade rotating at high speed. Rotary mowers can cut taller grass and seed stalks and are easier to maintain than reel mowers. They pose increased safety concerns, however, such as ejection of rocks and debris from the mower chase.

Factors That Influence Mowing Quality   Several factors, such as the species and cultivar utilized, influence the quality of cut when mowing. Certain species, such as perennial ryegrass and zoysia grass, have poor mowing quality because of their leaf structure characteristics. The result is shredded leaf blades and a dull, ragged appearance. Some improved cultivars have a demonstrated increase in mowing quality. Reel mowers provide a much higher quality cut than rotary mowers do in high maintenance situations.

It is also important to mow within the recommended height ranges for a species. Creeping bentgrass and annual bluegrass, when mowed above recommended ranges, produce undesirable thatchy and puffy turf. When all species are mowed below the recommended range, scalping and/or thinning of the turf results.

Properly sharpened blades are very important. Dull blades cause a ragged appearance and result in slower recovery of the turfgrass. When using a reel mower, the relationship between mowing height and the clip of the reel (the distance traveled between clips) is important. The highest quality cut results when the mowing height equals the clip of the reel. If the clip of the reel is greater than the mowing height, then marcelling occurs (the turf takes on a wavy appearance because of the uneven cut). If clip of the reel is less than the mowing height, the efficiency of a reel mower is reduced, resulting in a ragged, nonuniform cut.

Points to Consider When Mowing   To minimize stress to the turf, it should be mowed frequently enough so that no more than one-third of the leaf tissue is removed at one time. In other words, if the desired cutting height is 5 cm (2 in.), the grass should be mowed before the height exceeds 3 in. Mowing should be avoided when the soil is excessively moist because the weight of the mower will cause ruts to form in the turf. During excessively wet periods, mowing to remove no more than one-third of the leaf tissue is more important than avoiding ruts in the turf.

It is generally recommended that the mowing height be raised to the maximum for a particular species just before the onset of summer stress and winter. Do not scalp the grass prior to the onset of winter. The mowing height can be lowered to a more desirable height during the favorable growth periods of spring and fall. Lower mowing heights necessitate more frequent mowing. Gradually decrease mowing height (such as would occur in the fall after summer heat and moisture stress) by increasing mowing frequency, not by removing more than one-third of the leaf tissue at once. Alternate the direction or pattern of mowing each time to avoid ruts, uneven cutting of the grass, and clumps of tall grass in the wheel tracks. Select a mowing pattern that minimizes the number of sharp turns required. This practice improves mowing efficiency and decreases injury to the turf.

Collection of clippings is usually not necessary. Clippings contain 85 to 90 percent water and do not contribute to thatch development. Thatch is formed by stem tissue, for example, rhizomes, stolons, and also roots. However, one should collect clippings from diseased turf to help prevent spread of the disease or if uncontrolled weeds are setting seed. Similarly, collect excessively long clippings to prevent shading of the grass.

Turfgrass Growth Regulators

Several growth-regulating compounds are available for use in turfgrass management. These compounds fall into three general categories: Those that are hormones or produce hormones that affect growth, those that inhibit cell division (termed Type 1 materials), and those that inhibit cell elongation (Type 2, or gibberellin synthesis inhibitors).

One should read the label carefully before using growth regulators on turfgrass. Most growth regulators have specific turf species registrations. Certain growth regulators can be utilized on close-cut grass, while others cannot. Some have specialty uses. For example, certain growth regulators are registered for use in an overseeding program.

It is important to determine the economic feasibility of growth regulators before using any of them. Growth regulators are ideal management tools for use on golf course fairways. However, landscape maintenance firms that have a contractual obligation to mow a site on a predetermined (e.g., weekly) schedule may not realize any savings in labor or costs beyond that of reduced clipping hauling and disposal charges. In this situation, the economic benefit of use may be diminished. Growth regulators are also useful in areas that are difficult to mow or along edges of turf.

Growth regulators should not be applied to stressed turf, for example, if there is a disease or insect

infestation. In this case, the growth regulator interferes with the turf's ability to recover. Some undesirable yellowing of the turf may occur for a period of time after application. If in doubt, test a small area before making large-scale applications.

Undesirable growth characteristics (elongation of internodes resulting in puffy, thatchy turf) can result from using ethephon on turfgrass grown under shade. Mefluidide is a growth-inhibiting compound (not growth-suppressing, as are other Type I products for landscape use). Mefluidide causes good seedhead suppression.

On the other hand, Type II growth regulators (e.g., trinexepac-ethyl) are not generally effective at reducing seedhead development, though the height of the seedhead may be reduced. These materials suppress vertical growth for longer periods of time compared to Type I materials. In addition, they do not suppress lateral growth (tillering, rhizome) and root growth as much as Type I regulators.

Fertilization

The list of elements required for turfgrass growth varies (depending on the source consulted) from sixteen to eighteen (sodium and nickel are debated). The list can be subdivided into four categories. Carbon, hydrogen, and oxygen are found in highest abundance, yet they are not applied as fertilizer materials. Fertilizer elements are categorized as primary and secondary macronutrients and micronutrients. The primary macronutrients—nitrogen, phosphorus, and potassium—are the elements applied most often by the turf manager. The secondary macronutrients—calcium, magnesium, and sulfur—are generally available in the soil and are not routinely applied as fertilizers. The micronutrients include boron, chlorine, copper, iron, manganese, molybdenum, and zinc. With the exception of iron, these nutrients are generally not applied as fertilizers, because they are often present in sufficient quantities for turf growth in macronutrient fertilizers.

Factors that affect fertility requirements include cultural management (mowing, irrigation frequency), use of the turf, and environmental conditions. It is important to follow the recommended application rate for the particular species used and the location. These rates are usually based on pounds of element per 1,000 ft^2, and may include recommended frequency. Application when seeding or fertilizing an area for the first time should be based on soil-test results because changes in growth and color of the grass may not be due to a nutrient deficiency. Also, different species have

different fertility requirements and application schedules, which may further vary depending on location. Turf managers can consult their state extension literature for specifics. When applying fertilizer, it is important not to apply more than the maximum single application rate, or fertilizer burn will result. It is also a good idea to avoid fertilizing during unfavorable weather conditions such as high heat or drought and to avoid applications to wet foliage because some synthetic fertilizers will cause burn if applied to wet foliage.

Nitrogen (N) is the most important element in turfgrass fertility management. Average annual nitrogen requirements vary depending on numerous factors, such as species, use, soil type, and climate. Nitrogen requirements for several turfgrass species are given in Table 24–6.

Table 24–6
NITROGEN REQUIREMENTS OF COMMON TURFGRASS SPECIES*

Species	Lbs/1,000 ft^2 Management	
	Low	High
Cool-Season Species		
Creeping bentgrass (*Agrostis palustris*)	1–3	3–8
Tall Fescue (*Festuca arundinacea*)	1–2	3–5
Fine Fescue (*Festuca rubra, F. longifolia*)	0.5–2	2–4
Perennial ryegrass (*Lolium perenne*)	2–4	4–6
Kentucky bluegrass (*Poa pratensis*)		
Common	1–2	2–4
Improved	1.5–3	3–6
Warm-Season Species		
Buffalo grass (*Buchloe dactyloides*)	0–1	2–3
Bermuda grass (*Cynodon dactylon*)	1–4	3–8
Zoysia grass (*Zoysia* spp.)	2–4	5–7
Bahia grass (*Paspalum notatum*)	0–1	2–4
Saint Augustine grass (*Stenotaphrum secondatum*)	2–4	5–7
Centipede grass (*Eremochloa ophiuroides*)	0–1	2–4

*Higher management intensity increases annual nitrogen requirements.

Source: David Gardner, The Ohio State University.

Points to Consider When Conducting a Soil Test

When conducting a soil test in turf, nitrogen is usually not tested due to rapid fluctuations in the levels of this nutrient. While no accurate test for soil nitrogen exists, many fertility programs for other elements are based on the results of soil tests. Soil tests can determine whether mineral nutrients are available in sufficient quantities or if the addition of nutrients with a fertilizer is required. It is important to collect a representative sample for analysis. The usual recommendation is to collect several samples randomly from different parts of the area to be tested. Also, separate tests should be conducted on each soil type and in areas with different characteristics (e.g., separate wet and dry areas, high and low areas, etc.), and the collected cores from each sampling area should be uniformly mixed before being sent to the lab. In most areas, it is necessary only to conduct a complete soil analysis test once every five years. Basic soil tests to monitor conditions and access fertility needs should be conducted more often. Conduct complete tests yearly to monitor progress of pH modification or fertility buildup programs.

There are a couple of different soil test methods: sufficiency level of available nutrients (SLAN) and basic cation saturation ratio (BCSR). SLAN is an older method traditionally used in university labs. It relies on correlation of chemical extracts that approximate a plant's ability to extract nutrients from the soil and calibration of values generated by extraction data. Response curves are generated based on field-response studies. BCSR is used by many private labs. It is based on the concept that the ideal ratio of cations on cation exchange sites produces the best plant response (varies, but usually 65 percent Ca^{++}, 10 percent Mg^{++}, 5 percent K^{+}, 20 percent H^{+}). The test results are based on nutrient applications to restore this ratio. The best interpretation of actual fertility needs comes from combining the results from both types of soil test.

It is important for turfgrass managers to recognize that the test recommendations for phosphorus (P) and potassium (K) generated by some laboratories are based on field-crop requirements. The nutrient requirements of mature turfgrass are different than these values suggest. Phosphorus needs are generally overestimated and potassium needs are underestimated. Table 24–7 gives the nutrient status of soils at varying levels of phosphorus and potassium.

The secondary macronutrients—calcium, magnesium, and sulfur—are generally not applied as fertilizers because they are often present in sufficient quantities in the soil. An exception might occur on a golf green with

Table 24–7
RECOMMENDED PHOSPHORUS AND POTASSIUM LEVELS FOR TURFGRASS SOILS*

Nutrient Status of Turfgrass Soils	Phosphorus		Potassium	
	PPM	Pounds/Acre	PPM	Pounds/Acre
Very low	0–5	0–10	0–40	0–80
Low	6–10	10–20	40–175	80–350
Adequate	10–20	20–40	175–250	350–500
High	>20	>40	>250	>500

*These recommendations vary from many traditional agricultural row crops. It is important to have turfgrass soils tested by a lab familiar with turf management or to interpret the results for P and K using the values listed in this table.

Source: David Gardner, The Ohio State University.

a silica sand root zone. In this case, applications of calcium (in the form of gypsum) may be necessary.

Use of Sulfur to Decrease Soil pH

Sulfur is also sometimes used in areas with high soil pH to affect a temporary reduction in soil pH around the turfroot zone. Soils may also be too alkaline, causing nutrients such as iron and manganese to become unavailable to the plant. It is important to determine if high pH is due to a soil characteristic or application of lime. It is difficult, if not impossible, to lower the pH of naturally alkaline soils.

If the pH is high due to application of lime or other alkaline materials, then acid-forming materials can be used to decrease soil pH. For a given soil, the amount of sulfur required to decrease the pH one unit is roughly equal to one-third the amount of limestone required to increase the pH one unit. Sulfur may be applied at a maximum of 10 lb per 1,000 ft^2 every eight weeks. However, sulfur oxidizes and mixes with water which can then form a strong acid that can burn the grassroots. Therefore, it is important not to exceed the maximum application rate nor to apply more frequently than the minimum recommended interval.

Micronutrients

Iron is used by golf course superintendents and landscapers because it causes a rapid greening and darkening of the turf without increasing clipping production. However, too much iron can cause the turf to take on a temporary black appearance. Iron generally is applied as a chelated material to increase availability to the turf.

There is some debate about the accuracy of soil tests for other micronutrients, but deficiencies with them are generally uncommon in turf grown on soil with a pH between 5.6 and 7.0.

Points to Consider When Fertilizing

The most important step before applying fertilizer is to read and understand the fertilizer label (see Fig. 14–49 on page 289). A wide variety of fertilizer materials are available, and turfgrass managers can select fertilizer materials according to specific needs or maintenance goals. For example, water-soluble nitrogen sources (quick release) cause rapid green-up and growth, but precise application is more important to avoid fertilizer burn. In contrast, water-insoluble nitrogen results in slower response, long residual activity, lower burn potential, higher cost, low surface runoff and leaching potential, and low frequency of application relative to water-soluble nitrogen.

Certain fertilizers alter the soil pH. For example, if the soil pH is high, it may be beneficial to use potassium sulfate that temporarily acidifies the soil, rather than potassium chloride, which has no effect on pH.

Fertilizers are generally divided into natural and synthetic sources. Most organic fertilizers are slow-release and may yield higher visual quality, but the application rate and cost is higher than that of synthetic fertilizer materials. Also, less nitrogen from natural fertilizers is available in cool weather because of reduced microbial decomposition.

Determining Irrigation Requirements

All plants require water for growth. Several terms are used to define the overall irrigation requirements of turf. The total irrigation requirement for a turfgrass is the amount of water needed to meet the net irrigation requirement of the grass and also to compensate for losses from evaporation, percolation, and runoff. Irrigation requirements vary according to soil type. For example, sandy soils have a high infiltration rate, rapid drainage, and low water-holding capacity. The opposite is true of soils with a high clay content.

The turfgrass species used also influences irrigation requirements. Warm-season grasses, where they can be grown, require much less water than do cool-season grasses. Weather conditions (humidity, temperature, rainfall) and microclimate (amount of shade, slope, etc.) also influence irrigation requirements. Areas that are trafficked more intensely require more water. Cultural practices such as decreasing mowing height and/or increasing fertility also increase irrigation requirements. Finally, irrigation

Just rained   Getting dry   Time to water   Drought

Figure 24–11

Effect of water status on the appearance of a turfgrass leaf cross section. As the soil dries, the leaves of the grass plant fold and then curl over as a mechanism to conserve water. The difference in how a curled leaf reflects light compared to a flat leaf causes the turf to take on a bluish green cast when dry. Source: David Gardner, The Ohio State University.

frequency affects irrigation requirements because increased frequency causes increased water use by the turfgrass.

When determining irrigation requirements, adjustments must be made for rainfall received and the efficiency of the irrigation system. It is often better to wait to irrigate until it is required, rather than to irrigate daily with automated irrigation systems. Several signs indicate that irrigation is required. Footprinting is the easiest to use. Moisture-stressed turf is not as turgid and does not spring back to its original position as rapidly, causing footprints to remain for a longer period. Turfgrass should be irrigated just as the plants are beginning to wilt. A slight change in the color (to blue- or grayish green) or the quality of light reflected from the leaf precedes wilting. The appearance of the leaf in cross section can also be used to determine if the turf needs to be watered (Fig. 24–11). Equipment such as a soil probe (core sampler) can aid in determining irrigation requirements. One should irrigate if the soil is dry at a depth of 4 to 6 in. Tensiometers are devices that measure soil moisture potential.

Points to Consider When Irrigating

Turf should be irrigated in the early morning hours if possible (Fig. 24–12). Watering during the day is inefficient because of increased evaporation of the water spray. Watering in the evening increases the incidence of certain pathogens. On a home lawn, one should commit to regular irrigation or allow the lawn to go dormant. Sporadic watering reduces the carbohydrate reserves of the plant and decreases survival. On lower maintenance surfaces (home lawns), it is traditionally recommended that the turf be irrigated infrequently, that between 2.5 to 5 cm (1 and 2 in.) of water be applied per week. Finally, to avoid water loss due to runoff, the irrigation rate should not exceed the infiltration capacity of the soil.

Figure 24–12
Irrigating a golf course in Phoenix, Arizona, at 6:30 A.M. in July.
Source: Margaret McMahon, The Ohio State University.

Methods of Reducing Irrigation Requirements

Use drought-tolerant species or cultivars. Note that the water-use rate does not determine the drought tolerance of a species. For example, Kentucky bluegrass has a low water-use rate but is shallow rooted. It wilts more readily than tall fescue, which has a higher water-use rate but also a deep root system. In areas where warm-season grasses perform adequately (transition zone and south), the water-use rate is about half that of a cool-season grass.

Raise the mowing height to the maximum recommended for the species used. Taller grass has a higher water-use rate, but it develops a deeper root system and will also better shade the soil surface, thus reducing evaporation loss. Mowing frequency should be reduced because significant water loss occurs through mower wounds. Mow with a properly sharpened blade to hasten recovery of the plant.

Apply lower rates of nitrogen and higher rates of potassium during years when precipitation is below normal. Reducing nitrogen reduces growth and thus water requirements. Increasing potassium increases the osmotic potential within the leaves and in turn increases the plant's ability to extract water from the soil, thus improving drought tolerance.

Remove excess thatch to avoid shallow rooting and decreased water penetration to the soil. Cultivate compacted soils to increase water infiltration.

Syringing

Syringing is employed when the turf wilts despite adequate soil moisture. Wilting is caused when the evaporation rate exceeds the plant's ability to absorb moisture. A light amount of water is applied to the shoots of the plant to help reduce temperature of the plant and increase relative humidity around the plant. Not enough water is applied to contribute to soil moisture. Syringing is also employed to remove dew, frost, exudates, and foreign matter from turf leaves.

The use of syringing to alleviate moisture stress is misunderstood. The benefit of syringing is influenced by air temperature, canopy temperature, relative humidity, irridance, wind, the amount of water applied, timing of application, and water temperature. In areas where cool-season turf is grown, syringing, generally causes about a 1°F to 4°F decrease in canopy temperature for two hours. In southern regions, a return to presyringing temperature is observed within fifteen minutes.

Irrigation Water Quality

Secondary water sources (e.g., effluent, waste, reclaimed or recycled water) are becoming more common irrigation water sources for turfgrass in some parts of the United States. Secondary water is defined as the final liquid product from a sewage treatment plant. Pollutants are removed physically, chemically, or biologically before discharge. Several concerns and sources of increased cost are associated with secondary water use, including additional soil amendments to mitigate high salt and sodium levels, increased fertilizer and pesticide use due to overall poorer turf quality, and specialized irrigation components. As a result, addition, more frequent testing of water quality is necessary.

Several factors affect effluent water quality, such as the amount of salinity, sodium, toxic ions, and bicarbonate. Depending on the impurity found in the water, turf managers can mix gypsum, sulfur, lime, or acidifying agents or blend the effluent water with clean water to ameliorate the impurity. Other concerns to evaluate and monitor include the level of suspended solids, biodegradable organics, pathogens, nutrients, and stable organics.

Many golf courses are now using ponds to capture runoff from the property in addition to providing challenges for the players. When properly managed these ponds can reduce water and ground pollution from chemicals and fertilizers used on the course, provide a source of water for the course, as well as provide added aesthetic appeal and a habitat for wildlife (Fig. 24–13).

Figure 24–13

Ponds on a golf course in addition to providing challenges for the players, can collect runoff, be a source of irrigation water, and provide aesthetic appeal and wildlife habitat.

Source: USDA Agricultural Research Service Image Gallery, *http://www.ars.usda. gov/is/graphics/photos/.*

SECONDARY MANAGEMENT PRACTICES

Pest Control

Problems of pest control are universal. Weeds, diseases, insects, nematodes, and other pests all afflict turfgrasses. Because such organisms tend to occupy the environment to the full extent of its capacity to support them, turf always has insects feeding on grass leaves and roots, fungi and other microorganisms consuming dead grass clippings, and weeds filling in any bare spots of soil.

Insects   Insects attacking turf include various caterpillars, beetle grubs, bill bugs, wireworms, flea beetles, chinch bugs, fruit flies, leafhoppers, scales, and aphids. In the spider group, mites attack certain species of grass. Of all these, caterpillars and beetle grubs are likely to affect all lawns.

Caterpillars, the larvae of various species of moths, build silk-lined burrows among the grass crowns and feed nocturnally on the leaves. Caterpillars stay close to their burrows, and each thins out a small area of grass around its burrow, about 2 or 3 cm (0.8 to 1.2 in.) in diameter. Damage is inconsequential until populations build up to several dozens per square meter. At that point, turfgrass is rapidly thinned out as caterpillars mature and their appetites increase. To treat this problem, grass is mowed to reduce leaf area. Leaves are then sprayed with a stomach poison or contact insecticide. Irrigation water is withheld for a day or two so the insecticide is not washed from the leaves.

Grubs (beetle larvae) that live in the soil and feed on roots are a different problem because it is difficult to get insecticide into the soil at control levels. In their early growth stages, grubs are small and seldom a problem. But in the fall or the spring following their emergence, they are large and hungry and they eat so many roots that patches of grass die. The grubs can be found 2.5 to 5 cm (1 to 2 in.) deep in the soil at the edges of the brown dead patches.

Insecticide chemistry has changed considerably over the last twenty years. Pesticides such as chlordane, which were relatively broad spectrum and long lasting, have been replaced by newer, more selective materials with shorter residual life. Consequently, it is more important than ever to identify the insect pest correctly. In addition, one should have knowledge of the life cycle of the insect and its most vulnerable stage. Insecticides should be targeted for the stage when the insect pest is most vulnerable. For example, one should not try to control adult beetles; rather, the insecticide application should target the grubs. Cultural practices are less effective against insects because they prefer properly maintained turf. However, maintaining a healthy stand of turf by properly mowing, fertilizing, and so on, produces a turf that is more tolerant of insect damage. One should avoid planting tree and shrub species that harbor turfgrass insect pests.

Weeds   A weedy turf is symptomatic of poor management and often indicates that not enough attention has been given to practices that produce a vigorous turf. Weeds are frequent when turf is undernourished, overwatered, or mowed too short (Fig. 24–14). Soils

Figure 24–14

The open spots in this turf are the result of mowing the turf too short. Weeds can get established in the open areas.

Source: USDA Agricultural Research Service Image Gallery, *http://www.ars.usda .gov/is/graphics/photos/*

compacted by heavy traffic tend to grow poor turf, but they do support a large population of certain weed species.

Unwanted perennial grasses are the most difficult weeds to control in turf; broad-leaved annuals are the easiest to control. In some circumstances, a grass that is usually considered useful as a turf can be a serious weed problem. This outcome occurs most often when a turf species is grown out of place or in a situation for which it is not intended. Annual bluegrass is a serious weed in creeping bentgrass putting greens because its lack of tolerance for heat and drought stress causes it to die out during summer stress. Rough bluegrass is sometimes considered a weed because of its poor tolerance of traffic. Tall fescue is a serious weed in Kentucky bluegrass because of its course texture. Creeping bentgrass, when found in Kentucky bluegrass, forms unsightly clumps of thatchy, puffy turf because of its lack of tolerance to higher heights of cut. Annual ryegrass is commonly used in low-maintenance situations or for hillside stabilization during establishment. As an annual, it dies out after one season, resulting in bare patches in the turf. Annual ryegrass is commonly found in lower-cost seed mixes sold to homeowners. Warm-season grasses such as zoysiagrass grown north of their adaptive range or when present in a cool-season turf can produce unsightly clumps of straw brown turf during fall, winter, and spring, when temperatures are suboptimal for their growth.

Selective control of weedy turfgrasses is extremely difficult and in many cases impossible. Herbicide selectivity is based on physiological differences between species, and cool-season turfgrasses are all very similar physiologically. Certain herbicides can be used (e.g., Chlorsulfuron for tall fescue control), but success is often limited. In many situations, control with a nonselective herbicide followed by renovation as required.

The first step in weed control is to improve management to encourage the grass to grow vigorously. Then weeds can be dug, pulled, or treated with an herbicide. With the grass growing vigorously, space formerly occupied by a weed fills with grass. Control is then successful. If the space is recolonized by other weeds, management practices should be reexamined as the first step in additional control efforts.

Herbicides Herbicides used to control turf weeds should be used cautiously and with restraint. They are plant poisons that at recommended rates are more toxic to weeds than to the turfgrasses. Herbicides recommended for weed control on turf also injure and cause stress to the turfgrass plants, though the turf outgrows the injury in time. If herbicides are used repeatedly or at times when the grass is under stress or not growing well, grass growth might be so retarded that weed problems worsen.

Contact herbicides kill plants they touch and are sometimes used as spot sprays to kill a few individual difficult weeds in a turf. The results, however, are unsightly spots that the grass slowly recolonizes.

Preemergence herbicides control annual weeds by soil application before weed seeds germinate. The herbicide kills seedlings as they push through the treated surface soil layer.

Postemergence selective herbicides are used to remove broad-leaved weeds from grass. Such herbicides are most effective on weeds in the seedling stage. As weeds become older, they become more resistant. Many species of weeds are killed by routine mowing, which continually defoliates them. Weeds that persist tend to be low-growing species that spread out below the mower blades.

If a lawn has only scattered weeds, it is more prudent to dig them out than to mix chemicals and wash spray tanks and thus avoid the risk of spray drift onto garden flowers. The county agricultural agent or local garden center can recommend suitable herbicides to use on lawns.

Annual grasses such as crabgrass should be controlled using preemergence herbicides in late winter or early spring. It is important to apply products uniformly to prevent weed breakthrough. Annual grasses may also be controlled with a postemergence herbicide. However, the herbicides should be applied while weeds are young (before tillering). Caution must be exercised with postemergence products because application to stressed turfgrass may result in phytotoxicity. Perennial grasses such as quackgrass must be controlled with nonselective herbicides.

Perennial broad-leaved plants such as dandelion are best controlled with a postemergence herbicide applied in late fall. That weakens the plant making it more likely to be severely damaged or killed by cold and other winter conditions.

Combination herbicide products are used when more than one difficult-to-control weed species is present. With herbicides, fungicides, or insecticides, it is important always to read the label prior to use. One should be aware of reentry period, time until reseeding can occur, which species are controlled with the product, and site and use restrictions.

Diseases The most important step in disease management is to identify the pathogen correctly. Several characteristics should be used, including plant

symptoms and signs of damage, weather conditions, management practices, and so on, when attempting to make the determination in the field. Field identification can be difficult because many diseases have similar appearances. The most accurate diagnosis is achieved when sending a sample to a lab for analysis. However, conclusive results may take weeks to obtain, and sometimes you can formulate an educated guess about the pathogen while you are awaiting lab results.

Three important turf diseases are mildew, rust, and smut. Mildew forms a white dust on grass leaves in the shade and can be controlled only by opening up the area to provide the grass with better light and ventilation. Rust is sometimes prominent in the fall when the orange fruiting bodies discolor grass. Fertilizing with nitrogen usually controls rust. Smut occurs in the late spring or summer as a grass disease that produces a line of black greasy spores along the leaf blade.

Soils contain many saprophytic fungi, which live on dead leaves and other soil organic matter. Some of these are **facultative** parasites that cause turfgrass diseases if predisposing factors are present. These diseases are often most common on the best cared for lawns. For example, a *Pythium* water mold destroys turf when fertility, moisture, and temperature are all at high levels. Excess water soaks the soil, causing stress because of poor soil aeration. Add the stress of high temperatures favorable for the growth of *Pythium,* and the pathogen rapidly kills the nitrogen-rich grass.

Among diseases that commonly appear on the ordinary lawn are *Fusarium* and *Rhizoctonia* brown patch, both of which form rings of dying turf in the summer. These diseases are predisposed by heat stress on the grass. When snow covers turf for a long time, the melting snow reveals patches of dead grass killed by winter-active organisms, such as *Fusarium* and *Typhula* species. Cold and darkness encourage these organisms to cause turf disease. Long periods of wet overcast weather in the spring or fall favor growth of the *Helminthosporium* melting out pathogen. Growth of *Sclerotinia,* or dollar spot, (Fig. 24–15) is favored by nitrogen-deficient grass. Organisms causing all of these diseases are generally present but do not infect the grass until a particular factor or combination of factors develops, such as dark overcast days, winter snow cover, high summer temperatures, or excessively wet or compacted soils. The environment must be favorable for a certain disease to occur and affect turfgrass growth.

When fungal lawn diseases are a problem, observation shows that the grass areas most affected are those predisposed to fungal attack by factors such as compacted soil, reflected heat onto a lawn by buildings or

Figure 24–15
Dollar spot on turf grass. Source: USDA Agricultural Research Service Image Gallery, *http://www.ars.usda.gov/is/graphics/photos/*

fences, or excessively short mowing. In the southern states of the United States, hot spots are not a problem for the subtropical grasses; instead, shaded or wet spots lead to diseased turf.

Several cultural management strategies may reduce disease incidence. However, considerable variation exists in the effect of cultural practices on disease severity. For example, increased nitrogen may reduce the severity of dollar spot but increase the severity of another pathogen. Some cultural practices are almost universally beneficial. Attempt to reduce the amount of time that water is on the plant by watering in the early morning hours, not at night. Superintendents may syringe or drag turf to remove dew and guttation (liquid that exudes from leaf margins) fluids. If practical, the tree canopy in densely shaded areas should be pruned to open them up. Varieties of turfgrass with proven resistance to a particular pathogen can be utilized in areas with frequent occurrence of a particular disease. Finally, maintaining a healthy stand of turf by properly mowing, fertilizing, and so on, can make the turf more able to withstand disease.

Other cultural practices include cultivation, thatch control, and top-dressing. Cultivation and thatch control should not be conducted on a routine basis but only as needed to improve turfgrass stand performance by correcting soil compaction and excess thatch accumulation. The cause of decreased stand performance or stand failure should be determined. Many times coring and aerification are conducted when not warranted or necessary. Cultivation should be conducted if the reduction in turfgrass quality is caused by excess thatch or soil compaction, and if the turfgrass stand is otherwise in

reasonable condition, with minimal weed invasion, and no major soil problems are present.

Cultivation

Cultivation is performed to improve root zone conditions, primarily soil compaction, and to reduce excess thatch. Cultivation is also used to promote dense shoot growth. **Compaction** is the pressing together of soil particles into a more dense mass. Complications that arise when turfgrass is grown on compacted soils include reduced water flow, infiltration, and soil aeration. Root growth is restricted due to mechanical resistance. Compaction may be avoided by managing traffic patterns to avoid heavy wear patterns, alternating mowing patterns to avoid ruts, using equipment that is lightweight or with a larger tire width, and restricting traffic when the soil is wet.

Thatch is a layer of undecomposed organic matter located between the soil surface and the zone of green vegetation. A moderate layer of thatch may be advantageous (Table 24–8), but excessive thatch can cause problems. Accumulation of excess thatch occurs when organic matter production exceeds the rate of decomposition: when high amounts of nitrogen are applied, the soil pH is unfavorably low for microbial activity, or if aggressively growing species or cultivars are utilized.

Return of clippings does not contribute to thatch accumulation. Excessive thatch can be prevented by using an appropriate fertilization program, liming if necessary to maintain the soil pH neutral to favor soil microbes, or using species and varieties that are suitable and intended for site and use. Proper mowing practices and avoiding excessive use of pesticides also help control thatch accumulation.

Cultivation can improve growing conditions when turfgrass is grown on subsoil, which occurs when a residential contractor removes and sells the topsoil and uses subsoil excavated from the basement to replace the topsoil. Turf grown on subsoil grows poorly and is shallow rooted, resulting in frequent drought stress even if the turfgrass is well irrigated. Turf grown on subsoil tends to develop a thick thatch layer. Frequent core cultivation, with return of the cores, may improve the growing medium over time.

Alleviating both compaction and excess thatch is accomplished using core aerification. With core aerification, small soil cores (approximately 0.64 1.9 cm to [0.25 to 0.75 in.] diameter) are extracted. The name of the technique implies that aeration is directly improved. This assumption may or may not be accurate because the soil below and between the hole may actually become more compacted. The extracted cores are then broken up by weather or drag mat. The soil that is released incorporates into the thatch layer, causing an indirect improvement in aerification, higher cation exchange capacity, nutrient and water retention, and accelerated decomposition of organic matter (reduction in thatch). The vertical length of the cores varies by machine used and soil strength. Soil moisture is critical when conducting cultivation. Excessively dry soil reduces the depth of penetration of the coring tine. If the soil is excessively wet, soil tilth is destroyed by the cultivation equipment, resulting in sealed soil surfaces that are less permeable to gas exchange.

Several types of machines are available for core aerification. Vertical motion units are slower machines that cause relatively minimal disruption. They are used on putting greens. Circular motion units are faster machines that cause moderate to severe disruption of the surface. They produce shallower cores and are used

---

Table 24–8
MODERATE LAYER OF THATCH VS. AN EXCESSIVE LAYER

Advantages of Maintaining a Moderate Layer of Thatch	Complications from an Excessive Layer of Thatch
• More resiliency on sports turfs. • Improved wear tolerance. • Plants that are rooted in the soil may be insulated from temperature extremes. • Thatch may provide a barrier against weeds.	• Thatch does not retain water and nutrients as well as soil. • Thatch is not wetted by upward movement of water from the soil. • Plants rooted in thatch are more subject to drought and nutrient deficiencies as well as extremes of heat and cold. • Scalping from mowers. • Restricted water and pesticide movement into the soil. • Increased potential for disease and insect problems.

Source: David Gardner, The Ohio State University.

on athletic fields and home lawns. Drum types are fast and inexpensive, but they are not as effective as the other types. Finally, hydroject types inject water at high pressure. They are used on putting greens to disrupt layers caused by top-dressing. They cause minimal disruption but are actually a type of shatterhole cultivator.

Other methods of cultivation are available, but they do not reduce compaction or excess thatch. Shatterhole aeration involves driving a series of solid tines into the soil to create a shattering effect. Fracturing the soil improves water and air movement but may increase compaction along the sides and bottom of the hole. Slicing is less intense and causes less soil disruption. The turf is penetrated by a series of V-shaped knives mounted on discs. This method cuts stolons and improves tillering. Spiking is not used to aerify but to slice stolons and promote dense shoot growth. If hollow tines are used, the cores can either be returned or removed. Removal is common on golf course greens where returning cores results in substantial disruption of the playing surface or where sand top-dressing is practiced to keep the sand layer pure. Returning the cores is best, however, because they help to reduce thatch and soil layering.

Cultivation should be conducted during times of active turf growth. This approach allows the plant to respond to increased soil aeration and lack of soil resistance with increased root and shoot growth. Be aware, though, that desiccation damage can result from cultivation in summer because the roots are exposed to high evapotranspiration conditions around the coring holes. Desiccation damage can also result from late-fall cultivation because the coring holes may not close by winter. Desiccation also occurs under windy, low humidity conditions.

Mechanical Thatch Removal

Mechanical thatch removal may be used when a thick layer of thatch exists or if cultivation practices are inadequate. Mechanical thatch removal should not be considered a substitute for core aeration because it does not contribute to thatch decomposition. Current research suggests that mechanical thatch removal is not effective.

Machines used to remove thatch include the verticutter, which has vertically mounted blades that slice into the thatch layer. Solid blades are best, but they damage irrigation lines. However, spring verticutters are not very effective. Power rakes are used instead for areas that may have solid objects buried underground.

When removing thatch, it is best to avoid times favorable for weed seed germination. Dethatching

should also be avoided if turfgrass is stressed or not well-established. Excessive layers of thatch and/or shallow-rooted turf should be treated with multiple low-impact dethatching operations, thus allowing full recovery of the turf in between, rather than attempting a single, severe dethatching operation. Thatch removal causes injury to turfgrass; always follow dethatching with proper cultural practices that favor recovery of the grass.

Top-Dressing

Top-dressing refers to the application of a fine layer of soil over an existing turf surface. It is used to smooth and level playing surfaces such as putting greens. Top-dressing helps to control thatch problems by incorporating material into the thatch layer, thus improving the environment for soil microorganisms. Top-dressing can promote recovery of the turf from injury. Top-dressing can be used to alter the turfgrass growing medium (e.g., change a fine-textured, compacted soil to sand-based). Top-dressing can be used during establishment by sod or stolonizing to level uneven areas and make the surface more uniform. It may help protect soil from winter desiccation in regions with mild winters.

Top-dressing may contribute to problems caused when soil layers possess different physical characteristics. Core aerification or shatterhole aeration is usually performed in conjunction with top-dressing to alleviate this problem. Top-dressing with nonsoil materials such as crumb rubber can be effective in softening the soil surface and reducing traffic stress. Selection of the proper top-dressing material is critical to success. The material should be as close in characteristics to the underlying soil as possible. The top-dressing texture should be the same or coarser than the underlying soil.

The top-dressing rate must match the rate of turfgrass growth. If it is applied too slowly, alternating layers of thatch and soil result. If it is applied too quickly, the existing thatch can be buried and cause a barrier to root growth. There is debate about the merits of pure sand top-dressing. It is less expensive and much easier to use than a mix with exacting standards. Research suggests, however, that hydrophobic dry spots, nutrition problems, reduced water-holding capacity, and winter desiccation may result in the long-term use of sand.

Localized Dry Spot

Localized dry spots are small patches of turf where the thatch or soil turn hydrophobic. The most frequent occurrence is on sandy soils where bentgrass is grown,

but it can also occur on fine-textured soils with other turfgrasses. The cause of localized dry spot is not entirely understood. One theory is that microorganisms secrete complex polysaccharides that coat the sand particles, making them hydrophobic. Symptoms of dry spot include areas of localized drought stress that watering does not eliminate. Water tends to runoff rather than infiltrate the affected area. Strategies for dry-spot control include cultivation and the use of commercial wetting agents.

Miscellaneous Management Practices

When frost heaving has raised grass crowns out of the ground in cold-winter areas, they can be pressed back into the soil by rolling. Dyes are available for coloring brown or off-colored areas of turf. Such cosmetic practices are considered important when visible turf is part of television presentations, movies, or public spectacles. Plugs removed from sod can be planted into another turf to introduce a new grass species. Plugs are also used to repair small damaged areas. Devices are available to cut and remove plugs of various sizes and to cut similar holes in the soil to receive the plugs. Turf growth is often retarded by competition with tree roots. Judicious pruning of the tree's roots often greatly improves a turf.

Salt used to remove snow and ice damages turf bordering sidewalks and roads. The excess salt causes injury to turf and often death. The treatment for this problem is to apply lime on acid soils and gypsum on alkaline soils so that calcium replaces the sodium used in the snow-melting mixture.

Golf, tennis, and bowling greens are precise playing surfaces with grass under severe stress. Special management practices are used to maintain optimum playability of these critical surfaces. These practices are discussed in detail in some of the cited references.

Seasonal Growth

The seasonality of grass growth is illustrated in Figure 24–16. The dotted curve can be applied either to subtropical grasses growing in the southern United States or to temperate grasses growing in northern states or at high elevations. Low temperatures limit winter growth. As the season warms in the spring, growth increases, peaking in the summer months, if drought is not limiting. After late summer, growth declines until it is again stopped by low winter temperatures. Growth of temperate grasses over most of the urban United States is illustrated by the solid line. Depending on winter temperatures, growth can either

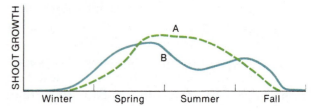

Figure 24–16

Seasonal growth of grasses. The solid line (B) represents a seasonal growth pattern for Kentucky bluegrass, a typical temperate grass. The start of spring growth occurs earlier or later depending on the local climate. Summer dormancy is more or less severe depending on the degree of heat and water stress. The dotted line (A) represents growth of the subtropical grasses such as Bermuda grass. Most of these make their greatest growth during the hottest weather, provided water is adequate. In the southernmost areas of the United States, growth continues through the winter at a low rate.

cease or slow down in the transition zone. The solid curve is similar to the dotted curve except that during the summer months, temperature optima are exceeded and growth slows. If high temperatures are accompanied by drought, grass can become semidormant, with little growth in summer. As temperatures cool in late summer, active growth resumes. Fall growth of Kentucky bluegrass differs from spring growth. Fall growth spreads and does not produce as many clippings. Growth comprises mostly new tillers and emerging rhizomes. The spreading form of growth decreases with dropping fall temperatures.

An awareness of the annual growth cycle in local regions can help in planning the best time for various operations. Coring, raking, and thatching are operations that should be done when growth is vigorous so recovery will be rapid. Herbicides are best applied during the rising growth curve in spring, when injury to the grass is low because of cool temperatures, and when rapid growth leads to rapid recovery.

Fertilizing Kentucky bluegrass during the rising growth curve in spring leads to excessive leaf growth. The same fertilizer applied during the rise of the growth curve in late summer results in more tillers and a more spreading growth. New leaves promoted by the fertilizer photosynthesize extra carbohydrate for winter storage.

The falling curve in summer represents a period of stress, and is an appropriate time to apply preventive fungicides—at the time predisposing factors develop. Reduced summer growth results from climatic stress, so the manager should be careful not to add extra stress at this time. One would not reduce mowing height nor increase mowing frequency at this time; instead, mow higher and less often.

It is best not to apply nitrogen fertilizer during summer stress. Heat stress results in part from the depletion of carbohydrates through higher respiration rates accompanying higher temperatures. Mowing also reduces carbohydrate reserves by reducing the leaf surface available for photosynthesis. Nitrogen fertilizer also depletes carbohydrate reserves, diverting them to protein synthesis. Nitrogen fertilizer applied during the summer puts an extra demand on the plant that can weaken its resistance to disease. Thus, during summer, it is best to withhold fertilizer or, if necessary, apply it only in small amounts of 0.125 kg/acre (0.25 lb/1,000 ft^2) or in forms that are slowly available to the plant.

When the growth curve rises in the fall, plants can take extra stress. The low spreading growth pattern and the favorable weather enable the turf to recover from all sorts of abuse.

SUMMARY AND REVIEW

Lawn, turf, and turfgrass in modern use refers to the ground where grass is growing, grass that is mowed and cared for, and grass growing in a horticultural turf, respectively. Turf is a unique crop in that it the product is what is left, not what is harvested. Turf is grown as a population of plants, not for individual plants. The population is subjected to many stresses. The stress concept is a tool for evaluating the effects of cultural practices on grass.

Turf culture is divided into two procedures: establishment and maintenance. Establishment requires first learning about the different turfgrass species and their requirement to be able to select the best plants for an area. Turfgrasses classified as warm season grow best in subtropical and tropical regions. Cool-season turfgrasses grow best in temperate regions. Warm-season turfgrasses can be further divided into those that are adapted to arid or humid conditions. Transition zones are the regions between areas where neither warm- nor cool-season grasses grow best.

Determinate, rhizome, stolon, and intercalary meristem refer to the type of growth associated with turfgrasses. Turfgrass has intercalary meristems in the leaves and leaf sheaths that cause leaf growth. Stem internodes do not elongate except to produce stolons, rhizomes, or flower stalks.

When planting turfgrass, several species or cultivars of cool season grasses are mixed in the planting to provide diversity to take advantage of the good qualities of each. Warm-season grasses are rarely blended because the species and cultivars tend to segregate. Kentucky bluegrass produces some seedlings by apomixes, which allows clones with desirable characteristics to be developed asexually.

Almost all of the turfgrass used in the United States was introduced from other countries. Many adapted very well to the region where they were introduced. As a result, in nearly every location in the United States, several turfgrasses can be grown.

Establishing turf first requires proper preparation of the seedbed or ground where sod is to be laid. Selection of the appropriate turfgrass species or cultivar(s) is critical. Good quality seed is important. The label gives the species or cultivar name, percentage of crop and weed seeds, and percentage of inert matter. Germination rates are sometimes included on the label. Correct seeding rates, depth of seeding, and time of seeding determine how well the turf will establish and cover the area. Water is the most critical factor in seed germination while nitrogen is most important for seedling establishment. Sodding is the installation of a mat of grass that has been cut from the ground. Over seeding is done in the southern United States to keep lawns green year-round by sowing cool-season grasses over warm-season grasses in the fall.

Maintenance is done to produce a beautiful, dense, healthy and vigorous turf. The three maintenance management principals are mowing, irrigating, and fertilizing. Secondary practices include pest control, cultivation, thatch control, top-dressing as well as others. Mowing must be done with the proper equipment at the appropriate intervals, at the proper height, and at the proper frequency. These are all determined by many factors including the characteristics of the species being grown and how much and what kind of stress the plants are under. No more than one-third of the turf should be removed in any mowing. Growth regulators can be used at times to help reduce the need for mowing.

Fertilization and irrigation programs are determined by the type of grass being grown. the environment, including soil type; and other cultural practices in use. Nitrogen is the most important element in turfgrass fertility management. No accurate soil test exists for nitrogen, but tests are available for the other nutritional elements. Sulfur is sometimes used to reduce soil pH, while lime or another alkaline material can be used to raise it. Irrigation should be done in the early morning

to reduce loss from evaporation and reduce incidence of disease. The quality of the irrigation water is important and should be monitored closely.

Grubs and caterpillars are the most difficult pests. Modern pesticides tend to be more selective and have shorter residual life than old pesticides, making it imperative that pests are correctly identified before applying a pesticide. Weeds become a problem when turf is poorly managed and unhealthy. Therefore, the first step in weed control is to manage the grass in a way that has it growing vigorously. Unwanted perennial grasses are the most difficult weeds to control in turf. The fungal diseases mildew, rust, and smut are the biggest disease problems for turf. Several cultural practices may reduce disease incidence, but sometimes a practice that controls one problem fosters another. Choosing species and cultivars resistant to particular pathogens, especially those prevalent in the area, is one practice that can help as well as maintaining the health of the turf in general.

KNOWLEDGE CHECK

1. What is turf and how does it technically differ from "lawn"?
2. What makes turf unique from most other crops?
3. Give two reasons that turf is grown.
4. Why is turf grass management such an exacting and science-based profession?
5. When is the best time of year to control perennial broad-leaf weeds? What type of herbicide should be used?
6. Turf culture is divided into what two procedures?
7. What is the transition zone in turfgrass culture?
8. Using Table 24–2 identify a turfgrass that has: a ligule that is a fringe of hairs, rolled vernation, and a distinct collar with evenly spaced internodes.
9. What is the advantage of mixing cultivars and species in a cool season turf planting? Why is this not done with warm season grass?
10. How many of the commonly used turfgrasses used in the United States are native to the US?
11. Describe how an intercalary meristem explains why a grass blade continues to grow after mowing.
12. What is unique about how bluegrass can reproduce? How does this affect the ability to produce many cultivars of bluegrass compared to other turf grasses? What most likely will determine the best species or cultivars to grow in a given location?
13. What are 3 reasons that a soil that is to be planted with turf can be "sterile" and need to have organic matter and nitrogen fertilizer added?
14. Why are turf seedbeds sometimes irrigated to allow weeds to grow before the turf seed is planted?
15. What 3 important pieces of information are included on a turfgrass seed label? Give a reason each would be important to know.
16. Why is red fescue often added to seed mixes of bluegrass cultivars when it is to be grown in a temperate region like the northern two-thirds of the United States?
17. What is the best time of year to sow warm season grass seed in the southern United States? When is the best time to sow cool season grass seed in temperate regions?
18. Why are freshly sown turf beds rolled?
19. What is the most critical environmental factor during seed germination? During seedling establishment?
20. From Table 24–4 give 2 advantages and 2 disadvantages of sod compared to seed.
21. What is overseeding and why is it done?
22. How is a seedbed for a lawn that is to be renovated prepared?
23. What time of year do cool season grasses need the most frequent mowing?
24. What determines mowing height?
25. What type of lawn mower gives the highest quality cut and is also the best one to use for low cutting heights?
26. What is the maximum amount of leaf that should be removed during mowing?
27. How does a Type I growth regulator affect growth? A Type II growth regulator?
28. What are turf fertilizer rates based on?
29. From Table 24–6, compared to tall fescue does fine fescue require more or less N? Compared to Saint Augustine grass does Bermuda grass need more or less N?
30. What are the two types of soil tests used to measure the fertility of turf soil? Why is N usually not included in them?
31. In terms of the soil tests in question 30, what gives the best interpretation of fertility needs?
32. Why is there a problem for turf growers if recommendations for P and K are based on field crop requirements?
33. Turf be irrigated when the leaves are beginning to do what?

34. What are two methods for determining if turf is just beginning to wilt?

35. What are 3 methods that can be used to reduce irrigation requirements?

36. What are the two insect pests that are likely to infect all lawns?

37. What are the most difficult type of weed to control in turf?

38. Give an examples where one turfgrass is considered to be a weed when it is growing in with another turfgrass.

39. When is the best time of year to control perennial broad-leaf weeds? What type of herbicide should be used?

40. What are the three most important turf diseases?

41. Heat stress fosters what two disease organisms?

42. How can soil compaction be avoided in lawns?

43. What is thatch? How can it be prevented?

44. What is the method used to correct both soil compaction and excess thatch after they have occurred?

45. What is top-dressing?

46. Why is awareness of the annual growth cycle of turf grasses important for turf grass management?

FOOD FOR THOUGHT

1. As a golf course superintendent, your job is to decide what species or cultivars of grass to grow. Describe what you would consider when you make that decision.

2. You have just been hired to be the head groundskeeper of a major league baseball field that is about to be constructed. Funding for the field was generated in part by advertising that it would be environmentally friendly because it will be built on native soil. What would you tell the owners when you find out that they want to sell the top-soil and build the field on native subsoil?

3. Your landscaping business prides itself on installing and maintaining lawns that stay green and nice looking all summer. What would you tell a home-owner customer who is insisting that the lawn be exclusively a single cultivar of bluegrass that was advertised recently in a magazine?

FURTHER EXPLORATION

CHRISTIANS, N. E. 2004. *Fundamentals of turfgrass management,* 2nd ed. Hoboken, NJ: John Wiley and Sons.

DANNEBERGER, T. K. 1993. *Turfgrass ecology and management.* Cleveland, OH: Franzak and Foster.

EMMONS, R. D. 2008. *Turfgrass science and management,* 4th ed. Albany, NY: Delmar.

TURGEON, A. J. 2007. *Turfgrass management,* 8th ed. Upper Saddle River, NJ: Prentice Hall.

*25

Landscape Plants

Trees, Shrubs, and Herbaceous Plants

Laura Deeter

Close your eyes and picture a tree, a shrub, a flower, and a vine. We all know what these words mean. Or do we? A botanical dictionary tells us that a woody plant has a corky outer surface of bark covering the older stems, which generally survive a dormant period and expand their girth every year. A herbaceous plant lacks wood and bark. The overwintering structures of herbaceous plants are typically underground. An evergreen is a plant that retains its foliage year-round. Evergreens may have thin needles (pines, spruces, junipers) or broad leaves (rhododendron, holly). Evergreens can be 30.5 m (100 ft) trees or 0.3 m (1 ft shrubs). Trees typically have one stem (trunk), whereas shrubs typically are multistemmed. We tend to think of trees as tall and shrubs as less than 3.7 to 4.7 m (12 to 15 ft) in height. Vines can be woody or herbaceous, depending on the species. Herbaceous plants are often referred to as flowers. Although they can be tall or small, many species are between 0.3 m and 1.2 m (1 and 4 ft) tall.

Woody plants, whether tree, shrub, or vine, are all perennials. Herbaceous plants have one of three life cycles: perennial, biennial, or annual. Perennials live for more than two years and typically flower each year. They do not die after flowering. Biennials complete their life cycle in two years. The first year they have foliage only. The second year, they flower, set seed, and die. Annuals complete their entire life cycle (germination, foliage, flower, and seed) in one growing season.

Although a botanical dictionary gives us the absolute meanings of these terms (perennial, biennial, and annual), in reality the boundaries between them can be vague and hard to discern. Some trees have multiple trunks, such as the paper birch (*Betula papyrifera*) (Fig. 25–1), or are very small; many cultivars of Japanese maple are under 6 ft tall (*Acer palmatum* var. *dissectum* 'Red Dragon') (Fig. 25–2). Carolina silverbells (*Halesia tetraptera*) is a shrub often used as a small tree. Some shrubs get very large. American hazelnut (*Corylus americana*) is a shrub that can reach heights of 20 ft. Some multistemmed shrubs can be pruned or trained to grow as small trees. Crape myrtle

key learning concepts

After reading this chapter, you should be able to:

- List the environmental characteristics that affect the growth of trees, shrubs, and herbaceous plants.
- Describe how to choose and care for trees, shrubs, and herbaceous plants.
- Discuss the proper planting of trees, shrubs, and herbaceous plants.
- Explain proper maintenance of trees, shrubs, and herbaceous plants.
- List the names and characteristics of some of the common landscape plants.

Figure 25–1
Betula papyrifera, *paper birch, is a tree that can have multiple trunks.* Source: Laura Deeter, The Ohio State University.

(*Lagerstroemia* spp.), Russian olive (*Elaeagnus angustifolia*), and dappled willow (*Salix integra* 'Hakiro Nishiki') are all plants that can grow either as a shrub or a tree. We use some plants as herbaceous plants, but in reality the plants are woody shrubs; examples include roses (*Rosa* spp.) (Fig. 25–3) and lavender bushes (*Lavandula* spp.). Many plants have a perennial life cycle but do not live more than one growing season for environmental reasons. *Pelargonium xhortorum* is not hardy in areas with distinct winters; tulips (*Tulipa* spp.) do not survive

the summers in the areas with hot summers such as the southern United States. Conversely, pansies are perennials that can overwinter in areas with mild winters but will die during hot summers. Because of this, pansies are now a popular plant for fall planting in the south to maintain landscape color from fall through spring.

Careful breeding produces many landscape plants with varying characteristics. Thus, a large shrub may have selections that are dwarf, such as the compact winged euonymus (*Euonymus alatus* 'Compactus'), or even have a vastly different habit from the original. Birdsnest spruce (*Picea abies* 'Nidiformis') is a low-mounding plant rather than a tall conical tree. These cultivated varieties further blur the lines among the categories of tree, shrub, perennial, and annual.

Whether tree, shrub, or herbaceous, all plants have characteristics that vary from season to season. Flowers, fall color, size, growth habit, fruit, ornamental bark, and persistent foliage are all characteristics that provide landscape interest and value. Some trees provide a filtered shade; others provide a dense shade. Some shrubs provide an excellent habitat for wildlife; others may bear fruit valued by humans. Shrubs can also be utilized as hedges or screens. Many woody plants can be sheared and pruned extensively, thus forcing the plant into topiary or bonsai, for example. Herbaceous plants often provide the flashy show for a garden (Figs. 25–4 to 25–7). The incredible variety of shapes, sizes, colors, forms, and flowering times of landscape plants provide us with an almost unlimited amount of variety for the yard and garden. Regardless of their size or shape, *all* plants should be selected with care and consideration to ensure their continued survival and success in any landscape.

Figure 25–2
Acer palmatum 'Red Dragon,' *the Red Dragon Japanese maple, is a low-growing cultivar of a small tree.*
Source: Laura Deeter, The Ohio State University.

Figure 25–3
Rosa 'Bonica' *is a woody shrub that is treated and utilized more often as a herbaceous plant.* Source: Laura Deeter, The Ohio State University.

Figure 25–4
Junipers (Juniperus scopulorum) can be pruned easily into decorative topiary. Source: Laura Deeter, The Ohio State University.

Figure 25–5
Perennials can provide the flashy show in any garden. Many also make great cut-flower arrangements.
Source: Laura Deeter, The Ohio State University.

Figure 25–6
This Freeman maple (Acer x freemanii) has spectacular red fall color. Source: Laura Deeter, The Ohio State University.

Figure 25–7
Manzanita (Arctostaphylos manzanita) has highly ornamental, smooth red bark. Source: Laura Deeter, The Ohio State University.

ENVIRONMENTAL FACTORS AFFECTING PLANT SELECTION

Temperature

Proper plant selection is vitally important to plant success. "Right plant, right place" is a phrase horticulturists use frequently when discussing landscape plants. One part of proper landscape plant selection is plant usage. How the plant will be utilized and what function it will have in the landscape play a vital role in choosing a plant for a site. Function is one of the main differences between landscape plants and plants grown for food or other specific purpose. Landscape plants are used for many different functions such as providing aesthetic appeal and shade, serving as windbreaks, and they may even be part of a structure. These functions

are discussed in more detail in Chapter 26. For this chapter, it is enough for you to know that trees, shrubs, and herbaceous plants all have different characteristics that can be used to advantage depending on the function the plant needs to fulfill.

Regardless of the type of plant selected or its ultimate use, many environmental factors must be considered before choosing any plant. You learned about these in earlier chapters, but as a review the following is presented. One of the environmental factors most often mentioned is the USDA hardiness zone rating. The USDA hardiness zone is simply a listing of the average minimum winter temperatures for a given region. The USDA hardiness zone (see Fig. 4–15 on page 54) is simply a guideline for selecting plants that are winter-hardy for a given region. However, each broad region contains areas protected by hills, lakes, or even cities that may allow the growing of plants otherwise not hardy to that zone. Conversely, exposed sites may allow winter winds to create cold pockets; plants may need to be more cold-tolerant than the USDA hardiness zone indicates. Extremely hot regions may be more concerned with summer temperatures and not as concerned with winter temperatures. We generally think of deserts in terms of the high daytime temperatures. However, the night temperature in the desert can drop to just above freezing. Desert plants not only must be very heat-tolerant, but cold-tolerant as well.

Some areas of the Midwest often have late frosts, during which many early flowering plants lose their flowers to the cold temperatures. Forsythia (*Forsythia* spp.), star magnolia (*Magnolia x soulangiana*), and dogwoods (*Cornus* spp.) may have their flowers injured or killed during a late frost. As distressing as it is to lose the flowers of a favorite plant, keep in mind that if the flowers are killed, there will be no fruit later in the season either. Apricots, peaches, and almonds are not typically produced in areas with late spring frosts, although the plants are more than winter-hardy. Milder winters in southern regions prevent plants from receiving the cold treatments they need for proper flowering and fruit set. Spring bulbs require a cold treatment to flower. If the winters are mild, they will not flower properly or at all the following spring.

The American Horticultural Society (AHS) heat zone map was developed in 1997 by Dr. H. Marc Cathey to help describe the impact that heat and drought play in plant survival. The map has 12 zones based on the severity of summer heat. This map is very useful when choosing landscape plants, especially in the southern regions of the United States. While cold causes damage that is immediately apparent, heat is often far more subtle; heat stress can appear in many forms and can be mistaken for other types of stress. Plants suffering from heat stress may not flower properly; flowers may be deformed; foliage may droop or form necrotic lesions, plants may appear stunted or wilted. It is also possible for combinations of these symptoms to occur simultaneously. The continual modification and development of the heat zone map as well as classifying more plants by their heat tolerance will enable gardeners in hot arid regions to predict more accurately which plants might survive the summers.

Light

All plants require light for photosynthesis. Each species is adapted to a certain light regime. Some plants will grow in just about any lighting situation a garden might have. Plants are often listed under one of the following types of light requirement: full sun, partial sun, partial shade, full shade. Most large trees require full sun for proper growth and development. As a general rule, needled evergreens also require full sun to perform properly. Broad-leaf evergreens, shrubs, and herbaceous perennials are highly variable, and you can find species to fit most garden situations.

By simply determining the direction the garden faces, you can learn much about the amount of light in the area. A south-facing garden is typically a full-sun garden; a north-facing garden is generally considered a partial- to full-shade garden. An eastern exposure receives the milder, less intense morning sun, while a western exposure receives hot, intense afternoon and evening sun. It is important to keep in mind that as the season progresses, the amount, intensity, and quality of light will vary greatly. In addition, some areas of the country have many more cloudy days, which also affect the amount of light available.

Trees, hedges, fences, and buildings all affect the amount of light a garden receives. A large shade tree can easily turn a south-facing garden into a shade garden. As you watch the patterns of the sun over the season, take note of the sunny and shady areas at various times of the year. Over time, you will develop a talent for "guesstimating" the amount of light that an area will receive and be able to select appropriate plants.

Moisture

Water is absolutely essential to plant life. The amount of rainfall in a region is critical in determining which plants will survive. Many areas of the country receive more than enough moisture over the year to support a large range of plant material; however, the amount that falls during a given month can vary greatly. In some regions of the

country, most, if not all, the precipitation comes during the winter months, which means that plants must be tolerant to very dry conditions during the summer months as well as tolerant to large amounts of precipitation during winter months. The Pacific Northwest is an area that receives rain throughout the year. The Midwest and New England generally get rain during spring and summer, with varying amounts of precipitation during the winter. The Southeast is highly variable: some areas get rain throughout the summer; others areas are far more arid. You can find the details of precipitation patterns for your region online using key words such as *precipitation data,* *climate data,* or *meteorological data,* and your city or region. Because aesthetic appeal is important in landscapes, it is critical to choose plants that will maintain their appearance during drought as well as under nearly constant precipitation.

Wind

Wind causes plants to utilize more water. In areas of high wind, transpiration increases, which increases the amount of water the plant needs to survive and grow. Windy, exposed sites increase the need for irrigation, especially during hot, dry months. A gentle breeze typically does not increase water usage. A gentle breeze benefits plants by replenishing carbon dioxide levels around the foliage. Strong or sustained winds dramatically increase water usage. In addition, strong winds can damage the structure and even the form of trees and shrubs. We have all seen pictures of wind-swept trees growing on mountain tops. The strength and direction of prevailing winds is an important consideration when choosing landscape plants as plants vary greatly in their tolerance to wind and its effects on transpiration. The high strength of wood can be damaged during a storm, causing otherwise healthy trees to break and fall. The natural strength of the wood of different species varies greatly, and this characteristic needs to be considered when choosing plants. Oaks (*Quercus* spp.) typically have strong wood; poplars (*Populus* spp.) generally have weaker wood. This trait can vary within a genus as well. Within the maples (*Acer* spp), silver maple and boxelder maple (*Acer saccharinum* and *Acer negundo,* respectively) are two species with weak wood. Norway maple and red maple (*Acer platanoides* and *Acer rubrum,* respectively) are two species with stronger wood. When working with large trees in an exposed site, it is imperative to consider the inherent strength of the wood (Fig. 25–8).

Though staking new trees is generally not a recommended practice anymore, newly planted trees in exposed, windy areas may require staking to help

Figure 25–8

It is important to consider the inherent strength of the wood when planting large trees in a windy area; otherwise the tree may break in a storm. Source: Laura Deeter, The Ohio State University.

establishment. High winds not only affect large and newly establishing trees, herbaceous plants are also affected. In windy areas, herbaceous plants may be shorter or go dormant earlier. Staking might be required to keep flowers on tall plants from falling over.

Soil

How many gardeners are aware of the type of soil they have? Successful gardeners are probably keenly aware of many factors in the soil, including texture, pH, and drainage. A simple soil test tells you a lot about the soil. The pH of a soil greatly affects nutrient availability. Some plants require an acid pH; others prefer the pH to be more alkaline. Most plants perform best at a pH of neutral to slightly acidic (7.0 or slightly below). At this pH level, the most nutrients in the soil are the most available. Despite the marketing of chemicals for altering soil pH, in reality it can be challenging to make long-term modifications. The best solution is to select plants that are adapted to the soil pH in your vicinity. Soil test kits are available at garden centers; soil samples can also be sent to various agencies for testing. Contact your local extension agent for agencies in your area that provide soil testing. It is inexpensive and it provides a wealth of valuable information about the soil.

Knowing the soil texture (sand, silt, and clay) provides information about how water moves through the soil and gives a basic indication of nutrient availability. Sandy soils are usually very well drained but with limited nutrient availability. Clay soils are often poorly drained but with high nutrient availability. Soils are typically a mixture of all three types in varying

Figure 25–9
Salix alba 'Tristis' (weeping willow) grows in standing water.
Source: Laura Deeter, The Ohio State University.

Figure 25–10
Agave spp. are extremely drought-tolerant plants for arid regions. Source: Laura Deeter, The Ohio State University.

percentages. Plants are highly adaptable organisms, and many are capable of tolerating a wide range of soil types. Plants such as alder (*Alnus* spp) and goat's beard (*Aruncus dioicus*) are tolerant of poorly drained soils. Weeping willow (*Salix alba* 'Tristis,' Fig. 25–9) and yellow flag iris (*Iris pseudacorus*) grow in standing water. Most plants require good drainage, however, and will die if the drainage is poor.

Extremely well-drained soils also pose challenges for landscape plants. Although these plants readily tolerate periods of drought, all plants require water. Newly transplanted plants are more likely to suffer from drought stress and require far more water than older, established plants. Plants with many surface roots, such as honey locust (*Gleditsia triacanthos* var. *inermis*) may suffer from drought more readily than a species that forms a deep tap root, such as black oak (*Quercus velutina*). Plants adapted to very dry situations are more apt to perform better in arid regions than those species adapted to a moister environment (Fig. 25–10).

It is important to remember that all these environmental conditions are highly interactive. The amount of moisture in the soil greatly affects the ability of plants to obtain nutrients. Shady areas often have increased water in the soil; plants not adapted to increased moisture suffer and perform poorly in the landscape. Areas in full sun are often dry; even drought-tolerant plants may require additional moisture to ensure long-term survival. No plant interacts with only one environmental condition at a time. Each condition interacts with and is highly dependent on the others. Keep this fact in mind as you work through the chapter.

TREES AND SHRUBS IN THE LANDSCAPE

Woody plants provide the structure and the backbone of any garden setting. They are generally permanent residents, providing depth, vertical lines, and character as they grow and change through the seasons and over the long-term life of the landscape. Each large tree has a unique character, and they affect our lives on an emotional level that is difficult to express. If you have a hard time believing that humans become emotional over trees, take note the next time a large tree is removed from the landscape, either via natural forces or through so-called progress. People will band together and fight a municipality that decides to remove a large tree from public property. Many will go to great lengths to save a tree with sentimental value, or one that was planted by a famous person (Fig. 25–11).

Trees not only have an emotional or sentimental value, they also have practical value. They provide lumber, shade, and shelter; purify the air; prevent erosion; and provide habitat for wildlife and food for people. Trees can even be utilized to remove harmful heavy metals from the soil. Because of their longevity, it is imperative to choose trees properly and consider their lifelong characteristics, that is, from sapling to senescence. This precaution helps the plant perform to its fullest and to the satisfaction of the property owner. Starting with a healthy plant is the first step.

Woody plants are generally sold in one of three ways: balled and burlapped (B&B), bare root, or container grown. Each has its own unique set of advantages and disadvantages. Balled and burlapped plants

Figure 25–11

This mulberry (Morus alba) on the grounds of Longwood Gardens, Pennsylvania, was the favorite tree of Pierre DuPont, the main founder of the gardens. The large anchors have allowed the plant to survive much longer than it might have done otherwise. The anchors were installed because of the sentimental value of the tree. Source: Laura Deeter, The Ohio State University.

Figure 25–13

Five- to six-foot white pine (Pinus strobus) B&B plants.
Source: Laura Deeter, The Ohio State University.

can be evergreen or deciduous, small or large. They are generally field grown until they reach salable size (Fig. 25–12). The plant is then dug and the root ball is wrapped tightly in burlap to keep the roots together and to keep the roots moist. Plants grown this way can generally be purchased in an array of sizes, ranging from very small evergreens and shrubs to extremely large trees (Figs. 25–13 and 25–14). They are easy for the nursery to care for and can be stored in the B&B form for many seasons provided the root ball is kept moist and winter protection is provided. However, the act of digging and wrapping in burlap can cause undue stress on the plant

if proper procedures are not followed. It is also very labor intensive. The root ball cannot be allowed to dry out. In addition, the root ball is unwieldy for landscapers and homeowners to maneuver; a large root ball is extremely heavy, especially when wet.

Bare-root plants are generally small and typically deciduous. Roses, several species of fruits, and some tree seedlings (three years old or less) are sold this way. Bare-root plants are typically field-grown. Once they reach the appropriate size, they are dug, and the roots and shoots are pruned if appropriate. The plants are then packaged in a plastic bag filled with moist sawdust because the root system cannot be allowed to dry out. This type of small plant is best suited for late winter or early spring planting.

Figure 25–12
Field production of woody trees.
Source: Laura Deeter, The Ohio State University.

Figure 25–14
An oak (Quercus spp.) with a trunk diameter of 8 ft and a root ball that is over 4 ft tall and 8 ft wide.
Source: Laura Deeter, The Ohio State University.

Figure 25–15
Purple-leaf birch (Betula 'Crimson Frost') in 5-gallon containers at a retail nursery. Source: Laura Deeter, The Ohio State University.

Figure 25–16
Small perennials in quart containers coming out of winter after overwintering in the container in a greenhouse.
Source: Laura Deeter, The Ohio State University.

They are generally inexpensive and easy to transport because of their small size. However, it can be difficult for homeowners to discern if the plant is still alive at the time of purchase. The roots should be moist and not moldy at the time of purchase.

Much of the nursery industry is moving away from B&B and bare-root plants to container-grown plants. Plants of all sizes and species are suitable for containers (Figs. 25–15 and 25–16). Plants can be grown directly in the container, eliminating the need for field digging. Container-grown plants are easy for the nursery to keep as stock plants for several years, and they are easy for the homeowner to move. There are many other advantages to container-grown plants: they can be purchased and planted any time temperatures are appropriate, and plants are generally of uniform size and shape. Container-grown plants have a few minor disadvantages compared to B&B. The top may become too large for the container; thus, water availability becomes an issue. The root system may girdle the plant if it is left in the container too long. To address any problems with girdling, container-grown plants need to have their roots disturbed (the root ball somewhat broken apart) prior to planting to ensure transplant success, and plants grown in soilless mixes may have a difficult time becoming established in poor soil. See Chapter 22 for an in-depth discussion on nursery production methods.

CHOOSING A QUALITY TREE OR SHRUB

When you are standing in the nursery looking at a large number of trees and shrubs, it may seem as if all the plants are equally healthy. It may feel intimidating to look each plant over carefully to select the best one for you. Remember, these are living creatures that are supposed to last a long time in your landscape. They will provide the backbone and structure in the yard and garden for many years to come. A few minutes spent selecting the best tree or shrub available can save time, money, and trouble down the road.

Look for vigorous plants that are free from obvious diseases, pests, or mechanical damage. Trunks should be clean with no nicks; foliage should be a good color throughout the entire plant. The plant should show signs of adequate growth for the season. Poor twig extension (the length the twig grows each year) could be a sign of an underlying problem with the plant. If you have problems assessing plant vigor, ask a nursery or landscape professional for assistance.

Avoid plants with obvious girdling roots. If they can be seen encircling the trunk at the top of the pot or root ball, do not buy the plant. The plant may appear in good condition and in perfect health, but a girdling root system will slowly but surely kill the plant. Once

the large roots begin to girdle the trunk, it is very difficult to correct their growth. Root pruning can be done with some species, but others do not tolerate this treatment. In addition, any plant that undergoes root pruning is far more susceptible to drought stress and requires additional water until a new root system has formed. Root pruning is also not a guaranteed method of treating girdling roots. Plants with girdling roots are more prone to fall over in the wind, can take longer to recover from transplanting, and eventually die.

For container-grown plants, be sure to compare the size of the top of the plant to the container. A small root system may not be capable of supporting the number of leaves on the plant. Immediately after planting, the roots are going through several changes, the greatest of which is simply growing more roots. A small number of roots may not be able to supply all the water necessary for a large number of leaves. Plants that are top heavy are also more inclined to blow over while they are still young. Plants that are top heavy but in otherwise excellent condition can have approximately one-third of their foliage and branches removed at planting to help overcome this condition.

Look for plants that are evenly full all around. Plants that are unevenly branched will remain so throughout their life. The overall shape of the plant should be uniform throughout. For trees, be sure the trunk is straight and has a good taper at the base. The trunk should be widest at the soil level and slowly taper toward the top. Plants with little to no taper are less able to withstand wind than those with a good taper to the trunk.

CORRECT PLANTING OF WOODY TREES AND SHRUBS

Trees are best planted in spring when the soil is warming, but the air is still cool and the transpiration rate is low. Under these conditions, the roots will grow into the soil rapidly and provide ample water to the tops. Conifers and many container-grown deciduous and evergreen broad-leaved trees can be successfully transplanted in cold areas (zones 5 to 6) in periods from mid-August to early October. In warm areas (zones 8 to 10), fall planting of many hardy trees is recommended to allow the root system to grow during the mild winters. Bare-root trees are usually planted in midwinter in warm areas and in spring in cold areas.

Care of the Newly Purchased Tree

After purchasing the tree, you should take good care of it before planting. Bare-root trees should be placed in the shade to keep the tops cool, and the roots should always be kept covered and moist. If the trees are not to be planted at once, the roots should be plunged (heeled-in) in moist sawdust, moist soil, or peat moss. B&B trees should not be allowed to dry out. A leafy tree continues to transpire in the balled condition, and the soil can dry out quickly, depending on the **evapotranspiration** rate. Keep the tree in a protected location and tie it to keep it from tipping. Extreme care should be taken to avoid breaking the ball when transporting it because the small roots can be broken and lose contact with the soil.

Container-grown plants are perhaps the easiest to maintain temporarily. Still, the plants should be kept in a shady, wind-protected area to reduce evapotranspiration. The containers should be watered well and should be protected from direct sun by shading them with aluminum foil or by plunging them in sawdust. The temperature of the exposed side of a black container can reach 49°C (120°F) in the direct sun. Either method of covering prevents excessive heat buildup on the periphery of the container—which is apt to kill the roots.

The Native Soil

A family usually purchases a home for its proximity to employment or because of the climate, and often because it is what they can afford. Soils can range from fertile farm lands to rocky hillsides with a view. Little consideration is usually given to the garden soil when a family purchases their dream home. The problems become apparent only when a garden is started.

One of the biggest problems with planting trees and shrubs after new construction is soil compaction. Soils are compacted by trucks and other construction equipment during the building of the house. Compacted soil should be tilled with a rototiller or a spade below the zone of compaction. To this loosened soil, incorporate 10 cm (4 in.) of organic matter into the top 20 to 30 cm (8 to 12 in.) of soil or beyond the limits of compaction. It is then best to irrigate the soil and allow it to settle and dry out to a good **tilth** before planting the tree.

The soil should be probed to 1 m (3 ft) deep to find any natural impervious layer (**hardpan**), which may prevent root penetration that in turn could result in a shallow root system. Similarly, the impervious layer may not provide adequate drainage, and the roots will lack adequate oxygen. Layers such as these are often caused by an accumulation of cemented minerals (calcium, iron, or aluminum compounds) and can be penetrated if they are thin. A power soil auger or a posthole digger penetrates layers up to 10 cm (4 in.) thick. These holes can be filled with amended soil or plant residues so that the roots can enter the open soil layers under the hardpan. Layers thicker than 10 cm (4 in.) are very difficult to

penetrate without heavy equipment such as bulldozers. It is possible, however, to haul in a large amount of soil and make a mound at least 1 m deep on which to plant a tree. Often such a scheme can be fit into a landscape plan. In such cases, a **tile drain** should be placed above the impervious layer to drain away the excess water.

Preparing the Hole and Planting

It is to be hoped that the local soil is one that does not pose problems like those discussed above. The hole for B&B (Fig. 25–17) or a container-grown tree should be dug just deep enough to allow the soil ball to set firmly on the native soil and should be about twice the diameter of the ball. When one is digging, it is usually difficult to estimate both dimensions accurately. If the hole is too deep, backfill with soil and tamp it firmly with both feet until the ball can be placed in the hole

slightly higher than the original ground line. In heavy clay soils, a glaze from the shovel may appear on the walls of the hole; this glaze should be roughened with a stick or shovel before backfilling to allow good root penetration from the soil ball.

Recent research has shown that it is not necessary to mix in any organic matter into the backfill soil before placing it around the soil ball. The native soil should be broken up with the shovel and gradually placed back and tamped in with a stick or gently with a foot. If the ball is sitting on firm soil, the ball will not settle with subsequent irrigations.

For bare-root trees, make a hole just large enough to accommodate the roots when spread out (Fig. 25–18). Prune broken or very long roots before planting. Gradually fill in the soil and tamp with a stick; finally, only after all the soil is well above the roots, apply a firm

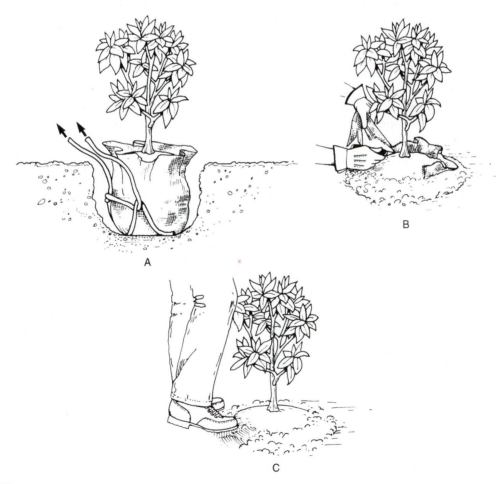

A

B

C

Figure 25–17

The ball of the tree should be placed on firm soil in the hole so that the ball will not settle lower than the intended planting depth. If the crown of the tree is below the final soil level, the trunk is subjected to decay organisms. To plant a balled and burlapped tree, (A) place the ball in the hole carefully, taking care that it is oriented properly. Remove ropes and cut the burlap in many places or remove it if it has been treated to prevent decay. (B) If the burlap is decomposable, as most are, cut away the top portion so that no burlap remains above the final soil level. (C) Fill in the hole and tamp the soil firmly until the desired soil level is attained. Water the soil ball and replaced soil.

Figure 25–18

Plant a bare-root tree in a hole large enough to accommodate the roots. Prune very long roots for ease in planting. Native soil taken from the hole should be broken up to make backfilling convenient. Firm the soil around the roots by tamping it with a stick as it is replaced. If the soil is properly firmed, the first irrigation will not settle the soil significantly.

foot to press the soil around the tree. Irrigate to settle the soil around the roots. Use the excess soil to make a berm (saucer) around the tree to contain water in a basin. Always plant the **crown** of the tree higher than the surrounding soil to ensure that water does not accumulate around the trunk and cause it to rot. Avoid overwatering newly planted trees before the roots become established in the undisturbed soil. The reduced oxygen in a saturated soil discourages growth of new roots.

When a bare-root tree is planted in a lawn area, enough **sod** should be removed to make room for a basin that should be constructed around the tree. The turfgrass should be kept away from the crown for at least two years to allow the young tree to become established. The bare soil area should be about 75 cm (30 in.) across. Three upright stakes about 30 cm (12 in.) high, spaced 30 cm (12 in.) from the tree trunk, reduces the risk of bark damage from lawn mowers. If the tree is staked, then these short stakes are not required.

Mulching the Soil Surface

All newly planted trees benefit from organic **mulches** of wood chips, fir bark, pine needles, or sphagnum peat moss. A mulch layer 5 to 10 cm (2 to 4 in.) thick may

be placed at the base of the tree, extending out to the edges of the longest branches. Bricks or redwood boards, which resist rotting, can be used to keep the mulch in place. The mulch eliminates most weeds, prevents the soil from caking in the hot sun, and moderates soil temperature and water loss near the surface. Water usually penetrates the soil more easily below a mulch.

Tree Staking and Trunk Protection

The newly planted tree may be staked, but it is not always necessary. The purpose of staking is threefold: to (1) protect the trunk from mechanical damage by mowers or other equipment; (2) anchor the root system; and (3) support the tree temporarily in an upright position.

Large leafy B&B trees are usually anchored with guy wires. The top of the tree offers so much wind resistance that the small root ball cannot prevent the tree from toppling. Excess movement keeps destroying any new roots but some movement is beneficial to establishing the tree. Three well-placed guy wires (Fig. 25–19) usually maintain stability and also keep machinery from running into the tree. When the roots have grown into

Figure 25–19
Three well-placed wires or plastic ropes prevent a B&B conifer with dense foliage from toppling in the wind. Wire should be covered to avoid abrasion of the trunk.

the native soil, the supports should be removed, usually after one year's growth.

Many newly planted trees require trunk support because they have been staked too tightly or have been grown too close together in the nursery. Such trees usually have tops that are too heavy for the trunk. The tree supports should be placed as low as possible but still maintain the top in an upright position. This point is found with the method shown in Figure 25–20. The tie is flexible enough to allow some trunk movement. The best material is an elastic webbing that allows movement without abrading the trunk. If the top is leafy and allows much wind resistance, it should be thinned to reduce the wind load.

If the tree lacks small shading branches on the lower trunk, the trunk should be painted with white interior latex paint (diluted with water) or wrapped with strips of burlap or paper wrap to prevent sun scald of the trunk on bright days. If small branches are still growing along the entire length of the trunk, those up to 15 cm (6 in.) above the soil should be removed, but the ones higher should be left. The small branches shade and supply food to the trunk; however, they should be pinched regularly so they do not develop into large limbs. These

lower laterals may be gradually thinned and removed over a two- to three-year period. A wire mesh encircling the tree to about 45 cm (18 in.) above the ground protects the bark against rabbits and mice if needed.

The support system should be checked occasionally. The ties can be removed after one full growing season, but the trunk should again be checked for strength, as shown in Figure 25–20. Some thinning of the top growth may be necessary during the first year to reduce the wind load on the canopy when it is leafed out.

Fertilizing Trees after Planting

There is a chance of applying too much fertilizer at transplanting time, which would burn the roots; therefore, trees are seldom fertilized when planted. However, a *small* handful of soluble nitrate-nitrogen fertilizer can be applied on the soil surface and watered in directly after planting. Fertilizer should never be placed near the crown of the trunk. Trees planted in turf must be fertilized more often than those growing in bare soil (without weeds). Turfgrass strongly competes for mineral nutrients added to the soil, leaving little for the tree roots.

Irrigation after Planting

It is important to irrigate recently planted trees deeply to encourage the roots to grow downward. In dry, hot climates careful irrigation is necessary for the survival of the tree during the first summer. Watering with a soaker hose, which provides low volume over many hours, allows the water to penetrate deeply in many soil types.

Container-grown or B&B trees should be irrigated frequently so that there is always a moisture contact between the soil ball and the backfill soil. This is especially true when the mix in the container is light and porous and the native soil tends to be clay. A bareroot tree, on the other hand, must be watered sparingly, especially during the first winter (warm zones) or spring (cold zones). The soil can become waterlogged with frequent irrigations, which reduces soil oxygen and discourages new root growth or may encourage root diseases, especially near the crown of the tree.

Tree Care during the First Year—A Summary

The first year is the most difficult period in the life of the tree. New roots must develop to anchor the tree and feeder roots must develop to supply the tree with water and mineral nutrients. Staking to keep the plant upright is necessary only for trees with weak trunks, trees planted in windy areas, or trees with large top-to-root ratios. Deep irrigation encourages deep rooting if the subsoil is similar to the top soil. Feeding the tree

Figure 25–20
Left: *To determine the point at which a tree should be staked, grip the trunk at various heights and bend the top to determine at which point the trunk springs back.* Right: *Tie the trunk a few cm above the point where the tree trunk returns to an upright position.*

during the first year is usually unnecessary because a small quantity of nutrients can be obtained from the native soil. Overfertilizing young trees the first growing season can damage and seriously hinder root growth. Some **thinning** of the top may be necessary to reduce the wind load on the top. The trunks of some trees need protection from sun scald or rodents, depending on the environment.

PRUNING METHODS

The two basic pruning methods of heading back and thinning out are discussed for fruit trees in Chapter 20, but the next few sections will discuss special considerations for landscape trees and shrubs. Before pruning any tree or shrub, one should ask three simple but basic questions:

1. *What do I want to accomplish by pruning?* Shall I head back the plant to keep it small? Shall I head it back to invigorate and strengthen it? Shall I reduce or increase the flowering or fruiting potential? Shall I thin out the canopy, etc.?
2. *How will it respond to pruning?* Will it greatly alter the growth habit? Will severe pruning markedly reduce or stimulate excessive growth? Will many latent or **adventitious buds** be stimulated to sprout?

Based on the normal habit of growth, is the method I am choosing for this particular tree suitable?

3. *How is it actually done?* Some basic pruning tools are necessary to make pruning easy. Bark splitting and ragged cuts should be avoided to allow for rapid healing.
   a. First remove any dead or dying branches or twigs. They usually have a gray lifeless (shriveled) appearance. Scratching through thin bark on small branches will show green tissue underneath on live branches.
   b. Thin out the cross-over branches or those that are growing toward the center of the crown.
   c. Remove narrow angle branches that weaken the tree (Fig. 25–21).
   d. Assess what you have done so far, then thin out the crown to give it the desired size and shape. At all times, keep in mind the natural characteristics and shape of the species.
   e. If not enough material is pruned this year, you can prune correctively next year after observing the resultant growth of this year's pruning operation.

Ornamental trees and shrubs should be chosen for their growth habits to fulfill the landscape function,

Figure 25–21

Left: *Branches with acute angles may split when the branch matures and (center) becomes heavy.* Right: *Narrow-angled branches should be removed and wide-angled branches should be encouraged.*

assuming that they are known to be hardy for the region. The trees and shrubs chosen to meet the needs of the landscape usually do not require much pruning after the first or second year when the shape is determined. Some pruning may be necessary annually in the case of some flowering shrubs, rose bushes, or conifers.

Pruning can be done almost any time of year except for trees such as the maples and the elms, which should be pruned in the late fall or early winter when the trees are not actively growing and the sap is not flowing. In areas with severe winters, pruning should be done after the cold weather has passed. Generally, late summer pruning should be avoided in cold areas because the new growth will not acclimatize properly and mature before an early frost. Typical methods of cutting twigs or large branches with shears and saw are illustrated in Figs. 25–22 and 25–23.

Container-grown plants often have roots that are circling in the container. Such plants should be rejected but, if used, the roots should be cut in several places to break the circling pattern and to stimulate new growth. The tops of newly-planted broad-leaved evergreens often need to be pruned to reduce the wind load and to reduce the transpiration surface. Remove about one-quarter of the leaf area by pruning immediately after planting. The tops of balled and burlapped shrubs should also be pruned for the same reasons, but the roots should not be disturbed other than loosening the burlap and cutting the ropes.

Pruning Narrow-Leaved Evergreens and Conifers

Narrow-leaved evergreens, or conifers, are shaped or kept within certain size limits by top pruning. Pine tops can be reduced in size by nipping the "candles" (the light-colored tips of the branches) of the current

Figure 25–22

When small branches are removed, they should be cut with pruning shears to obtain clean-cut surfaces. The cutting blade is held close to the main branch so that very little stub remains.

Figure 25–23

The method of cutting large branches to avoid splitting the trunk below the cut. Left: *Make a cut on the underside at (A).* Center: *Remove the branch at (B), leaving a stub.* Right: *Remove the stub at (C).*

year's growth to about half size with the thumb and forefinger. The remaining portion develops a group of short needles. The best time to prune the candles is in late spring when they have elongated but the needles are just beginning to grow. A new terminal bud will then form if the pruning is not done too late. Spruce and firs may also be pruned in a similar manner.

Junipers and other conifers without a strong radial symmetry or distant terminal buds may be sheared or irregularly pruned to form any shape or growth habit at most any time of year. If the top of a conifer is broken, generally a branch below begins to

bend upright. This can be encouraged by splinting the branch to hold it upright until it can remain erect without the aid of the splint. One full growing season is ample for this corrective procedure.

Pruning Deciduous and Broad-Leaved Evergreen Trees

Young deciduous or broad-leaved evergreens may have to be pruned after the first year. The purpose of tree pruning is to control the shape or the size, stimulate vigor, or, in the case of fruit trees, to affect the flowering and consequently the fruits that develop later.

The shape of the tree is directed early by selecting or choosing certain branches or buds to grow in a desired direction. The pruner can select one good leader branch and thin out badly placed branches so that a desirable, less-cluttered scaffold system remains (Figure 25–24). An irregular tree form is sometimes desired, and no one single leader is then chosen. At this time branches with narrow, weak angles (Figure 25–21) can be eliminated in favor of wide-angle branches. It is possible to prune certain trees so that there are multiple trunks and no single leader (Fig. 25–25). The crape myrtle is an excellent candidate for training in this fashion. A complicated form of pruning that requires much care is the espalier, in which fruit trees are trained against a fence or wall (Fig. 25–26).

Figure 25–25
Crape-myrtle (Lagerstroemia indica L.), *grown as a single- or multiple-stem tree. If pruned heavily early in its life, this tree can be grown as a shrub.*

A

B

Figure 25–26
Pruning and maintaining a fruit or an ornamental tree so that it grows in a single plane is called espalier. Shown here are pear trees trained in two different forms. Very often trees are trained on a wall having adequate light. The pruning is done mainly in the winter when the branches are bare, but vigorous troublesome shoots may be removed any time to maintain the shape of the plant. Photographed at Wisley Gardens, Great Britain.

Figure 25–24
Left: *A young established shade tree before pruning.* Right: *The same tree with those branches retained that will produce the scaffold branches of the tree. It is not a good idea to prune just the leader (topmost) branch to create low branching habits on an ornamental tree.*

Figure 25–27
Cutting a tree back severely each year during the dormant season (pollarding) is a way of keeping potentially large trees small. This London plane tree is one of the many on Nob Hill in San Francisco that are pollarded to keep their size down. The gnarled branches result from this severe pruning.

Figure 25–28
A type of topiary pruning in the Japanese section of Buchart Gardens, Vancouver Island, Canada. This Chamaecyparis pisifera 'Plumosa' often requires pruning during the growing season to remove the small tufts of growth from lateral buds.

The size and foliage density of shade trees may be controlled by pruning at least once a year, but more frequent pruning may be required for vigorous trees. Examples of such pruning control are creating a hedge from closely planted trees to produce a screen or pollarding trees (i.e., cutting them back severely each winter to let in more light in the winter but still to have shade in the summer). Sycamore and fruitless mulberry trees lend themselves to this severe pruning (Fig. 25–27). Because of their vigorous growth, the pruned trees still produce long leafy branches by midsummer. Controlled pruning may be done to shape the plant as animals, clumps, or odd forms, as in topiary pruning (Fig. 25–28). The ultimate in controlling tree size is the Japanese treatment of bonsai (Fig. 25–29). These trees are maintained and kept in small containers for many years, even centuries. Annual pruning of both roots and tops is necessary to maintain the bonsai form.

Even though pruning controls growth, it also invigorates individual shoots by reducing shoot number

and opening the center to more light. Thinning operations produce a better top-to-root ratio so that water and nutrients are distributed more efficiently. The remaining shoots have less competition for light, water, and nutrients. In hot dry climates, less leaf surface remains after pruning, thus reducing water loss by transpiration. The branch structure is strengthened by eliminating branches that produce narrow, weak angles (Fig. 25–21).

Figure 25–29
Normally large trees such as black pine and junipers can be grown in small containers and pruned to reduce their size. These containers average 30 cm (12 in.) long, and the trees are ten to twenty years old. They are known as bonsai trees.

PRUNING SHRUBS

Since shrubs grow differently from trees, they are pruned differently. Shrubs are best pruned just before planting. In the case of bare-root stock, broken roots should be removed and extra-long roots should be shortened to facilitate planting. If the top appears too large for the root system, the branches should be cut back to compensate for the root loss.

Once the shrubs are established in the landscape, maintenance pruning serves to control their size and shape by removing some stems from the multiple-stemmed crown (Fig. 25–30), improve the shrubs' health, and affect flowering. Some shrub species require more pruning than others because of differences in vigor and flowering habit.

Plant size is reduced by thinning out some of the branches. The tips of the branches may be headed back (only part of the branch is removed) to increase peripheral branching if a hedge or formal effect is desired. Thinning helps maintain the natural form and shape of the shrub. In thinning shoots, the wood should be cut above a bud that will grow in the desired direction (Fig. 25–31). Selecting the buds to remain allows partial control of growth even in an informal and natural setting of a rural home.

Shrubs may be thinned heavily by cutting some of the old branches back to the crown near ground level. The taller and older branches (i.e., those of greater diameter) should be cut out first, but not more than one-third of the branches should be cut out in any one year unless complete rejuvenation is desired. Old shrubs may be rejuvenated as shown in Fig. 25–32 by cutting back the entire plant. After many new shoots appear, weak ones are pruned out

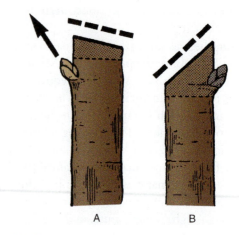

Figure 25–31

Twigs cut at too sharp an angle (B) may die back to the point that will also cause the bud to die. It is better to cut straight across (or nearly so), well above the bud desired for growth (A). That stub will only die back to a point where it will not interfere with bud growth (shown by the arrow).

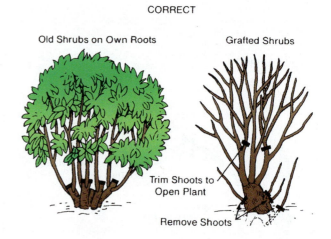

CORRECT

Old Shrubs on Own Roots

Grafted Shrubs

Trim Shoots to Open Plant

Remove Shoots

INCORRECT

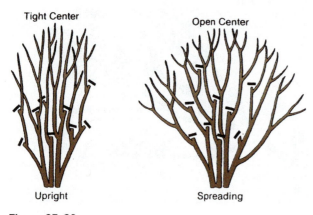

Tight Center — Open Center

Upright — Spreading

Figure 25–30

Two ways to prune a shrub to create different effects. The cuts are indicated by the black bars. Source: Adapted from USDA Home and Garden Bulletin 165.

Figure 25–32

Rejuvenating old shrubs. Upper left: In some cases, it is best to cut back severely (black bars) and develop a new top. Upper right: With grafted shrubs, do not cut back below the graft union (knobby growth at bottom). Remove suckers from the rootstock (lines) or thin shoots to open up the plant. Below: Cutting or trimming the tops of old shrubs, as indicated by the black lines (left), produces an unsightly leggy plant with a bushy top (right). Source: Adapted from USDA Home and Garden Bulletin 165.

to leave only the strongest and most desirable shoots to form the shrub.

Young hedges are pruned gradually by cutting about 6 to 8 in. above the previous cut. Mature hedges that have attained the desired size require frequent trimming to maintain their shape. The shape of the hedge should be slightly tapered, with the base wider than the top. This method of pruning allows light to reach the lower leaves and keep them alive. Keeping a hedge in a formal shape once it has attained the desired size may require as many as four clippings during the growing season. Pruning frequency of a hedge depends on plant vigor, light, temperature, and the water available to sustain vigorous growth.

Diseased wood is pruned out to keep the plant healthy. Diseased branches are cut back to nondiseased wood so that a healthy sprout will grow. This type of pruning should be done when the disease is first noticed, especially in the case of fireblight of pyracantha, cotoneaster, quince, and other members of the rose family.

One of the best reasons for pruning is to affect flowering. Some shrubs initiate their flower buds the summer before flowering (one-year-old wood) and thus should be pruned at a different season than those that flower on current year's wood.

Flowering on One-Year-Old Wood

Many fruit trees and shrubs initiate their flowers in summer, and the flower buds bloom the following spring when the weather becomes favorable. These species are called spring-flowering shrubs. Examples are *Cercis,* dogwood (*Cornus*), *Kalmia* lilac (*Syringa*), *Pyracantha, Rhododendron* (azalea), *Spiraea,* and *Viburnum.* Severely pruning the branches that have the flower buds in winter or early spring cuts away the potential flowers. One should prune such spring-flowering shrubs *soon after* they have flowered (Fig. 25–33). Removing the old flowers and shoots stimulates the plant to branch and produce additional flowering wood that summer for next year.

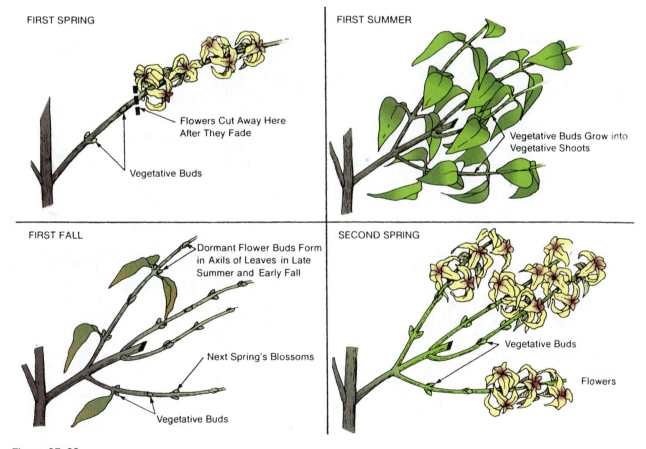

Figure 25–33

Spring-flowering shrubs and trees should be pruned after they have flowered and as vegetative growth begins. Unfortunately, many gardeners and homeowners prune these plants in the winter when the twigs are bare, and consequently they cut away the flower buds. Before pruning, study the flowering behavior of the shrub or tree.

This flowering behavior is usually identified by observing that the shrub flowers before the leaves appear, such as the flowering quince, flowering peach, forsythia, or *Chimonanthus praecox*). Some species like the lilac produce large flower clusters (inflorescences) instead of single blooms, but these too are borne on the previous year's shoots.

Flowering on Current Year's Wood

The branches of shrubs that flower on the current year's growth (sometimes called summer-flowering shrubs) should be pruned back in late winter or early spring before the buds have sprouted (Fig. 25–34). The resulting shoots have the capability of producing flower buds on the new growth. Some examples of summer-flowering species are *Abelia, Clethra,* crape myrtle, *Hibiscus, Nerium* (oleander), roses (floribunda and hybrid tea), and *Vitex.* This type of flowering is identified in some species by shoots that have light green leaves below the flower stalks. Flowers may also appear in several flushes or cycles during the growing season.

Pruning Nonflowering Shrubs

Shrubs not grown for flowers or fruits are pruned almost anytime of year except late summer. Late pruning often stimulates buds to sprout. Such shoots do not have adequate time to mature or "harden" before the onset of winter, leaving them susceptible to winter injury.

Shrubs, especially those grown as specimen plants, should be given ample space to develop properly. Most shrubs usually spread out as they grow taller. It is tempting to plant small shrubs close together to get the desired effect hurriedly. However, it usually becomes necessary to remove alternate plants when they begin to touch. Crowded plants do not develop the grace and beauty of those grown at recommended distances. Shrubs planted next to the house should be set far enough away to keep them from crowding into the building as they mature.

Figure 25–34

Summer-flowering shrubs and trees can be pruned either in the fall or in the winter when the twigs are bare. The flowers form and develop on current second growth that originates from vegetative buds on the previous season's branches.

DESCRIPTION OF COMMONLY UTILIZED SPECIES

It is vitally important to choose the right plant for the right place when selecting trees and shrubs. These long-lived organisms are generally thought of as the backbone of the garden or landscape. Thus, their survival is critical to the overall value of the landscape. Unless one enjoys replacing expensive plants on a regular basis, it is critical to select the proper plants for your region. The main decision to make is what function the tree or shrub must fulfill: foundation, specimen, backdrop, screen, and sound barrier are all potential landscape functions. Flowering, fall color, overall appearance, fruiting characteristics, evergreen foliage, and deciduous foliage also need consideration when choosing trees and shrubs. Chinese chestnut (*Castanea mollissima*, Fig. 25–35) is a medium-size tree with edible fruit that will attract squirrels and deer, and has beautiful flowers and a beautiful yellow-bronze fall color. However, the flowers have an extremely strong putrid scent, and the fruit is enclosed in a prickly involucre (covering). Despite the highly ornamental characteristics, this plant may not be the most appropriate for a small yard.

Black walnut (*Juglans nigra*) has edible fruit, is very strong wooded, and provides a nice filtered shade. However, the plant releases toxins into the surrounding soil. Thus, growing other plants under a walnut is a challenge. Stewartia (*Stewartia* spp., Fig. 25–36) are plants with beautiful white flowers, fantastic red fall color, and outstanding exfoliating winter bark. It is a

Figure 25–36
The bark of Stewartia (Stewartia spp.) is attractive all season, but especially so in winter.
Source: Laura Deeter, The Ohio State University.

Figure 25–35
The fruit of Castanea mollissima, or Chinese chestnut, is edible, but it is covered in a very sharp involucre (covering).
Source: Laura Deeter, The Ohio State University.

small tree appropriate for small yards. With its four-season interest and small size, it appears to be an ideal choice for just about any landscape situation. However, the plant has exacting cultural requirements: acid pH, extremely well drained soil with high organic content, and protection from winter winds and hot summer sun. If any of these requirements are not met, the plant will suffer and not survive long. It is vitally important, not only for our enjoyment of the landscape but also for the plant's ultimate survival, that we select plants appropriate for the environmental conditions of the given site.

North America is an extremely large continent with an incredible range of growing conditions. Nursery owners, landscapers, extension agents, and other reputable horticulturists can provide the most accurate information for your region and growing conditions; however, Table 25–1 provides a small sampling of deciduous and evergreen trees and shrubs common for many areas of the country. This table is meant only to illustrate the wide range of sizes, growth habits, and other ornamental characteristics available with woody plants. It is an extremely small sample of what is available in the nursery industry, and you should check with your local experts for a complete list of what grows in your area.

Table 25–1
A SMALL SAMPLING OF COMMON LANDSCAPE TREES, SHRUBS, AND EVERGREENS

Scientific Name	Common Name	Habit	Height (ft)	Ornamental Characteristics
Acer saccharum	Sugar maple	Tree	60+	Excellent fall color
Araucaria heterophylla	Norfolk Island pine	Tree	50–80+	Evergreen (houseplant in north)
Arctostaphylos manzanita	Manzanita	Shrub/tree	10+	Red bark, pink/white flowers
Betula nigra	River birch	Tree	50–70	Yellow fall color, exfoliating bark
Camellia japonica	Camellia	Shrub	10+	Many colors
Castanea mollissima	Chinese chestnut	Tree	40–60	Edible, yellow-bronze fall color, whitish flowers
Forsythia xintermedia	Forsythia	Shrub	6+	Yellow flowers
Fraxinus pennsylvanica	Green ash	Tree	60–80	Yellow fall color
Hibiscus rosa-sinensis	Rose of Sharon	Shrub	10+	Many colors available
Ilex opaca	American holly	Tree	20–40	Evergreen, red or yellow fruit
Juniperus virginiana	Eastern red cedar	Tree	20–50	Evergreen, can be pruned
Kolkwitzia amabilis	Beautybush	Shrub	10+	Pink flowers, dense plant
Lagerstroemia indica	Crape myrtle	Tree/shrub	10+	Pink or white flowers, exfoliating bark
Laurus nobilis	Laurel	Tree	20–40	Edible
Magnolia grandiflora	Southern magnolia	Tree	15–50+	Evergreen, white flowers
Mahonia aquifolium	Oregon grape-holly	Shrub	6	Yellow flowers, evergreen, blue fruit
Malus spp.	Crabapple	Tree	15–50	Fruit, pink or white flowers
Picea pungens	Colorado spruce	Tree	60+	Some with blue needles, evergreen
Pinus sylvestris	Scots pine	Tree	50+	Evergreen, orange bark
Potentilla fruticosa	Cinquefoil	Shrub	4	Yellow, white flowers
Pseudotsuga menziesii	Douglas fir	Tree	60+	Evergreen
Pyracantha coccinea	Firethorn	Shrub/vine	6+	White flowers, orange fruit, can be trained to a trellis
Quercus virginiana	Southern live oak	Tree	50–80+	Excellent foliage
Rhododendron catawbiense	Catawba rhododendron	Shrub	10+	Many colors available, evergreen
Sequoia sempervirens	Coast redwood	Tree	100+	Very large tree
Syringa vulgaris	Common lilac	Shrub	10+	Fragrant, many colors available
Taxus xintermedia	Anglojap yew	Shrub/tree	3–50	Evergreen, red fruit, can be pruned

HERBACEOUS PLANTS

Annuals

Annual plants have been an important part of gardens for a long time. These are often called bedding plants because they are usually planted in areas of carefully prepared soil called beds. They are grown for their visual appeal and to attract butterflies and hummingbirds, as well as for other characteristics. Because they are annuals, either botanically or because they are not hardy in an area, they must be replanted every year. This allows a great amount of flexibility in garden designs from year to year but it does require purchasing new plants, preparing the ground, and waiting for them to grow and cover the ground annually. Most annual species and cultivars have been bred or selected to grow and develop rapidly, which has reduced the waiting period significantly.

Although some annuals are grown directly from seed planted in the garden, most are grown from transplants which are plants grown in a greenhouse (to get an early start) then transplanted to the garden after the danger of frost has passed, or in the case of fall annuals, after the heat of summer is over. Transplants develop much faster than seedlings and are part of the reason that annual plantings can mature fairly rapidly. There are many species and cultivars of ornamental plants available as annuals. Although each one has its own cultural requirements, in general herbaceous annuals require the ground to be loose and well drained. The application of a slow release fertilizer worked into the soil at the time of planting will usually provide adequate nutrition during the growing season. Weeds can be a problem so some type of weed control is usually necessary. Mulching can help control weeds and reduce evaporation from the

soil, especially before the plants have filled in. Many annuals benefit from having old flower heads removed (**deadheading**). This not only improves appearance, but can help increase the number of flowers and prolong flowering.

Table 25–2 gives a list of some of the most common annuals (or plants grown as annuals) in the United States.

Herbaceous Perennials

Herbaceous perennial plants have become increasingly popular in recent years. Their popularity is probably due to the fact that they return from year to year; are available in an incredible array of sizes, shapes, flowering time, and color; and can be less expensive than annuals in the long run. There are *perennials* available for any garden situation: wet, dry, sandy, clay, sunny,

---

Table 25–2

SOME OF THE POPULAR ANNUAL PLANTS OR PLANTS GROWN AS ANNUALS IN THE UNITED STATES

Species	Description/Use
Ageratum (*Ageratum houstonianum*)	Two forms—low growing and tall; the low-growing form good for edging.
Allyssum (*Lobularia maritima*)	Short, good for edging.
Angelonia, summer snapdragon (*Angelonia angustifolia*)	Tall, flower spikes resemble snapdragons; several colors.
Bachelor button (Centaurea cyannus)	Moderately tall, blue flowers; good for cutting and drying.
Begonia (wax) (*Begonia × semperflorens-cultorum*)	Fairly low growing, dense habit, many colors.
Calibrachoa (*Calibrachoa × hybrida*)	Trailing, petunia-like plant; many colors.
Celosia (*Celosia* spp.)	Somewhat low growing; two forms—crested (cockscomb) (*C. argentea cristata*) and plume (*Celosia argentea plumosa*), many bright colors.
Coleus (*Solenostemon scutellarioides*)	Grown for its colorful foliage; many colors and patterns.
Cosmos (*Cosmos bipinnatus*)	Tall (3–6 in.), red, pink, lavender, and white; good for cutting.
Dahlia (*Dahlia* spp.)	Moderately tall, flowers can be cut, many colors; can be grown as a perennial if the tubers are dug in the fall, kept in a cool dry location then planted when the soil has started to warm in the spring.
Dusty Miller (*Senecio cineraria*)	Dusty colored fern-like foliage.
Gazania (*Gazania rigens*)	Low growing, heat tolerant; daisy-like flowers, many colors.
Geranium (*Pelargonium × hortorum*)	Large, showy flowers of red, pink, lavender, white; medium height.
Globe Amarantha, gomphrena (*Gomphrena globosa*)	Moderately tall, flowers are small globes; suitable for cutting and drying, purple, pink, red, white.
Impatiens (*Impatiens walleriana*)	Prefers shade; colorful flowers; fairly short.
Impatiens (New Guinea) (*Impatiens hawkeri*)	More tolerant of high light than regular impatiens, can have attractive variegated foliage.
Marigolds (*Tagates* spp.)	French type (*T. patula*) short; African type (*T. erecta*) tall; Both have colorful flowers of yellows and reds, some have multicolored flowers.
Nasturtium (*Tropaeolum* spp.)	Moderately tall, some forms vine and are suitable for trellises; colorful, edible flowers.
Pansy (*Viola × wittrociana*)	Prefers cool temperatures; short, colorful flowers, many of which have the characteristic "face" pattern.
Petunia (*Petnuia × hybrida*)	Very popular; many colors and patterns on flowers
Salvia (*Salvia splendens*)	Bright red flower on spikes; medium height.
Strawflower (*Helichrysum bracteatum*)	Flowers bright yellows and oranges, stiff petals with "dried" feeling to touch; good for drying.
Sunflower (*Helianthus annuus*)	Bright yellow flowers that follow the sun during the day; miniature cultivars available for small gardens.
Sweet pea (*Lathyrus odorata*)	Climbing, fragrant flowers; prefers cool temperatures.
Thunbergia, black-eye susan vine (*Thunbergia alata*)	Climbing, flowers orange with brown center though other colors available.
Verbena (*Verbena × hybrida*)	Compact, some forms may spread; flowers clustered, several colors available.
Zinnia (*Zinnia elegans*)	Composite flowers with many forms and colors; very popular garden plant; heat tolerant.

Figure 25–37
These desert plants are perfectly suited to arid conditions. They can also be utilized in a rock garden in more temperate climates. Source: Laura Deeter, The Ohio State University.

Figure 25–39
Perennials not only mix well with annuals, they can also be grown in decorative containers.
Source: Laura Deeter, The Ohio State University.

Figure 25–38
These perennials have been utilized to re-create a prairie garden. Source: Laura Deeter, The Ohio State University.

Figure 25–40
A gravel pathway bisects this calming, shady perennial garden.
Source: Laura Deeter, The Ohio State University.

shady, large, or small. Perennials allow the gardener to create specialty gardens: butterfly and hummingbird gardens, water gardens, cottage gardens, and rock gardens. An almost infinite array of other gardening styles can all be readily accommodated by utilizing perennials (Fig. 25–37 through 25–41). As with annuals, the possibilities are limitless!

This section should give you a brief overview of the possibilities available to the gardener who incorporates perennials into his or her garden. Although the term perennials will be utilized throughout this

section, annuals, bulbs, tender perennials, ornamental grasses, and ferns also follow many of the same concepts and readily grow with perennials. The term is not meant to be exclusionary; it is used here for convenience.

The concept of "right plant, right place" is still in evidence when utilizing perennials. Just as woody plants do not perform under stressful environmental conditions, neither do perennials. Because the maintenance requirements of these plants are more strict than those for woody plants, it is important to analyze your

Figure 25–41
This small container is actually a small water garden filled with specialty perennials. Source: Laura Deeter, The Ohio State University.

garden closely. A simple place to start is with honest answers to a few questions:

1. What function (beauty, cut flowers, color, hiding an unsightly feature) will the garden serve in the overall landscape?
2. How big a garden do I really need?
   a. How much work are you able/willing to do?
   b. How much can you afford?
3. What time(s) of the year do I want the garden to look its best?
   a. When am I outside most?
   b. Will the garden be visible from inside the house?
   c. Am I likely to take vacations during certain times?
4. What colors do I like best and which do I dislike: pastels, bright colors, or a smattering of everything?

Once you have determined your needs and desires, you need to look at your growing site. Take a careful inventory of all the environmental conditions. You might think your yard is quite sunny, but careful measurement of the amount of sun may reveal that your yard is shadier than you thought. Will existing trees or shrubs provide shade during certain times of the day or year? How fast will those plants grow and shade out sun-loving perennials? A basic site assessment will save money, time, and trouble in the future.

The same environmental conditions that allow trees and shrubs to grow also affect perennials. Perennials need light, water, soil, and nutrients just as woody plants do. You should be able to answer the following questions about your yard:

1. What is the average date of the first frost in fall or winter?

2. What is the average date of the last frost in spring?
3. Do you have reliable winter snow cover, or is it sporadic?
4. How hot and humid are the summers?
5. How much precipitation do you get and in what months?
6. How much sun does the garden get in the various seasons?
7. Will existing plants, buildings, or structures create shade or reflect light?
8. How will you water this garden?
   a. Is the natural rainfall sufficient and spread out enough, or will you need irrigation?
   b. Is there a hose nearby, or will you have to haul water long distances?
9. What are the traffic patterns in the area?
10. Do you have children and/or pets that might affect the garden or be affected by certain plants?

Preparing the Garden

More perennials die from lack of improper bed preparation than any other reason. Proper bed preparation seems to be the obvious way to get your plants to survive over the long run; however, it is not done often. Why? It is a lot of hard work, and it is expensive. It also isn't nearly as much fun as choosing plants! However, hard work in the beginning will pay off many times over in the long run. Bed preparation follows five basic steps:

1. Soil test
2. Elimination of perennial weeds
3. Proper drainage
4. Soil amendments
5. Additional fertilizers

Each step will be discussed in the following subsections.

Soil Test   Companies can perform simple soil tests for gardeners. Soil test kits are also available from local garden centers. Costs for professional testing range from $15.00 to $50.00, depending on location and types of testing done. Home kits can be purchased for as little as $5.00 or as much as $50.00, again depending on the complexity of the tests the kit performs. At the very minimum, you should know the soil texture, pH, and drainage of your garden soil. Without going to the expense of professional testing, you can perform some simple tests to determine roughly the texture and drainage. An inexpensive pH kit from the local garden center can provide you with a reasonably accurate measurement of the soil pH. Most perennials prefer a soil pH between 5.5 and 6.5, but in reality they are quite tolerant

of anything approximating neutral (7.0). Problems generally occur only when the pH is very alkaline or very acidic. There are exceptions, however. Japanese iris (*Iris ensata*) requires an acid pH to perform well. Baby's breath (*Gypsophila paniculata*) performs much better with a pH above 7.0. In fact, Gypsophila translates as "gypsum (or lime) loving" and refers to the fact that the plant prefers a more alkaline pH.

In the absence of a professional soil test, a homeowner may use two methods to determine soil texture: the soil ball test and the water jar test. For the soil ball test, water the garden thoroughly. Be sure to stop before the soil becomes soggy. Take a small handful of soil and squeeze it. Open your hand and carefully watch what happens to the soil ball. If it stays in a ball, it is predominantly clay. If it disintegrates and feels gritty, it's sandy. If it's somewhere in the middle, it's predominantly silt. The water jar test takes a bit longer, but it is more accurate in determining soil texture. Collect several soil samples (sample the soil, not the organic matter from the top) and remove the debris, pebbles, roots, and other large particles. Put one cup of the soil in a clear quart jar filled with water, seal it and shake vigorously; make sure it's mixed well. After twenty-four hours, examine the soil layers at the bottom of the jar. The heavy sand particles will sink to the bottom, followed by the silt, with the clay on top. Be sure you don't move the jar or the clay particles will float up again. Estimate the percentage of each compared to the whole to determine your soil texture. For example, a soil with one-quarter sand, one-half silt, and one-quarter clay would be 25 percent sand, 50 percent silt, and 25 percent clay.

Knowing the soil texture provides some information about how the garden drains. Again, a very simple test illustrates how well a garden drains. Dig a square hole 12 in. wide and 12 in. deep. Fill it with water and let it drain. Fill the hole a second time and record the length of time it takes to drain the second time. Soil that takes less than one hour to drain is very well drained and may require additional irrigation. If it takes over three hours to drain the second time, the area is extremely poorly drained and may require a drain tile to move excess water. Between one and three hours is good drainage and would support a wide array of plant species.

Proper Drainage   More perennials are killed from improper drainage than any other reason. Most perennials require well-drained soil to perform well. Thus, the soil drainage test mentioned above is an essential step when planting perennials. Several methods are available for improving the drainage of an area: install a drain tile, raise the bed, or add amendments to the soil.

Except in extreme cases of poor drainage, most gardeners do not have to worry about installing a drain tile. Although highly effective, they are expensive to install and should be done only by a trained professional to ensure that the new piping meets all the sewer and piping codes of your area.

Raising the bed is something that has been practiced since the times of the ancient Greeks and Romans. This technique involves simply adding from 0.6 to 0.9m (2 to 3 ft up to 12 ft) of height to the existing garden bed. A small wall or barrier may be required in cases where 0.9 m (12 ft) of height is added to a bed. Adding topsoil or peat moss to an existing bed can raise the level a couple of inches, which will probably not require a barrier of some sort to hold in; the roots of the plants will keep it in place. The last method for improving soil drainage is to add amendments, which is Step 4 in the list for bed preparation and is discussed in the next subsection.

Soil Amendments   Adding amendments to the soil is one of the simplest and easiest solutions to a mild drainage problem. Proper selection of soil amendments can help a soil that drains too quickly just as easily as a soil that drains poorly. They can also improve other characteristics of the soil such as pH and nutrient level. Soil amendments can be either organic or inorganic. Inorganic amendments include gypsum, lime, aggregate, or sand. For most garden situations, they are unnecessary additions because organic amendments can provide drainage along with many other benefits.

Organic amendments improve drainage by adding large particles to the soil, thus creating more air spaces. They also improve the water-holding capacity of a quickly draining soil. In dry areas of the country, this water-holding capacity means less irrigation. They create a favorable microhabitat for beneficial soil organisms. In addition, evidence suggests that compost can also dramatically decrease the incidence of several diseases in the landscape. Finally, organic amendments also add small quantities of nutrients to the soil. Regardless of the type of organic amendment utilized, be sure it is well composted and free of weed seed. Several types of organic amendments are available: homemade compost, commercial compost, grass clippings, various manures, peat moss, vermicompost (worm castings), mushroom compost, leaf mold, and so on. These products range in price from extremely inexpensive to highly expensive, depending on the product and the area of the country.

How do you mix the soil amendment into the soil? On an established perennial bed, the easiest method is simply to top-dress around all the plants with approximately 1 in. of amendment. This top-dressing will slowly work its way through the soil over the season. For a new bed, a rotary tiller is the most efficient method, especially if the area is large. Till the new garden area once prior to adding any amendment. Add half the amount of organic amendment, then till the area again. Add the other half and till a final time.

Additional Fertilizers   This section needs to be prefaced by the fact that perennials don't generally require large quantities of additional fertilizers. If a gardener is top-dressing every year with organic matter, no additional fertilizer is needed. A few exceptions are tulips, hyacinths, roses, and larkspur, which are heavy feeders. However, not all gardeners compost, nor is it always possible to top-dress every year for various reasons. In addition, nutrients may have become depleted and additional fertilizer is needed. Regardless of the reason, perennials don't generally require much fertilizer.

Elimination of Perennial Weeds   There is an old gardener's saying: "One year of seeds, seven years of weeds." For a new planting, it is far easier to get rid of the weeds prior to planting rather than dealing with them after the fact. Several methods are available to remove weeds: preemergent and postemergent herbicides, weed barriers, and hand pulling.

Although hand pulling can be effective for a limited number of weeds, it is generally not the most efficient method for large numbers, nor does it generally remove the entire weed. Certain weeds, like Canada thistle, dandelion, and crabgrass, to name a few, are renowned for their ability to return from the tiniest piece of root left in the ground. For an established perennial garden where herbicides may not be an option, hand pulling is often the only choice. A trowel or even a large knife can be very useful in removing as much of the roots as possible, especially for troublesome weeds.

Chemicals are very effective at removing weeds, especially prior to planting. Preemergent herbicides are designed to prevent seed germination and are effective at reducing the amount of seed endemic in the soil. In an established perennial bed, they are often highly useful in keeping weeds under control. Their main downfall is that many species of herbaceous plants self-seed and, indeed, the gardener may need the plants self-seeding to ensure continued presence of the plant in the garden. Johnny Jump-up and columbine (*Viola tricolor* and *Aquilegia* spp) are two perennials that

self-seed. A preemergent herbicide will kill those seeds as well. A postemergent herbicide kills the plant after it has emerged and is actively growing. Some of these herbicides kill any plant they are sprayed on (broad spectrum); therefore, it is imperative they aren't sprayed on desirable plants. Some postemergent herbicides are very specific, killing only grasses or only broad-leaf plants, and they can be useful for removing certain troublesome pests in the garden. Keep in mind that a grass killer also kills ornamental grasses and a broad-leaf weed killer also kills any of the perennials that it might be sprayed on accidentally. Caution must always be utilized when spraying any chemical; be sure to read and follow all label directions.

Barriers are a good way to kill weeds if you have enough time and you can plan well in advance. In a process called solarization, clear plastic is placed securely over the area and allowed to stand for several weeks during the hot summer months. The light and initial heat encourages plant and seed growth. The lack of water and the rising temperatures eventually kills not only the adult plants, but also the seeds. The only downside to this technique is the time and planning needed because it cannot be done once the perennials are in place. Newspapers piled on the site of the new garden can accomplish a similar effect. Spread 5 cm (2 in.) of newspapers over the new garden bed and pile the soil and amendments on top. The thickness of the papers prevents any weeds from penetrating; the new garden can simply be planted on top of the papers, which will not show once the garden is planted and will disintegrate, allowing the roots of the garden plants to extend deep into the bed. These and similar methods are excellent for eliminating weeds in situations where chemicals are not an option.

Perennial Maintenance

Once the garden is installed, regular maintenance chores must be performed. Mulching, weeding, checking for diseases and insects, staking, pinching, deadheading, and dividing are common garden chores for various times of the year.

Mulch   A layer of mulch in the garden provides several benefits:

- Conserves water.
- Regulates temperature (soils are cooler in summer and warmer in winter).
- Slowly adds nutrients to the soil as they break down.
- Prevents erosion.
- Is aesthetically pleasing.
- Controls weeds (but doesn't prevent them).

Several types of mulch, both organic and inorganic, are available. Gravel, marble chips, volcanic rock, and ground tires are all examples of inorganic mulches. These are good in permanent areas with limited plantings. They are generally *not* recommended for use around perennials. Organic mulches include various types of wood chips, pine straw, hay straw, and living mulches or ground covers. For perennials, organic mulches are the best.

Mulches have a reputation for preventing weeds in the garden. In reality, the actual number of weeds in a mulched and an unmulched garden is almost identical. However, the weeds in a mulched garden are easier to pull because the soil is generally moister. If weeds are blowing in after the mulch is down, the weeds root into the mulch, making them very easy to pull compared to weeds rooted in soil. Most mulches have one disadvantage that is generally not considered by homeowners: they are flammable. Although it's not common for mulches to catch fire, stray cigarettes or matches have been known to set the mulch on fire, occasionally with disastrous consequences. This issue is generally not serious enough to avoid mulching; the benefits greatly outweigh the disadvantages. In certain situations, however, it is a variable that must be considered.

Staking　Staking is nothing more than providing support for those plants that grow too large or whose flowers are too heavy for proper support. There is nothing pleasing about coming home from work in the evening to find that your larkspur or your peonies have flopped over and are flowering beautifully on the ground! Storms are another reason perennials fall over and might require support.

The trick to staking perennials is that the emphasis must always remain on the plant and the flowers, not on the stakes. Another trick with staking is to have the stakes in place before the plant requires support. Placement of the support after the fact invariably looks artificial. A properly staked perennial should not show the support (Fig. 25–42). Different methods of staking can be used, depending on the growth habit of the plant involved.

Tall flowers are best staked with a thin, single stake (Fig. 25–43). Foxglove and larkspur both have tall spikes of flowers. A tall branch, a thin bamboo stake, or a plastic or metal stake can all serve to hold these plants' flowers upright. Gently tie the flower stalks to the stake, making sure not to damage the succulent growth. As the stem grows, you can add more ties.

Bushy plants, such as aster or garden phlox, require a slightly different method. Circular, metal

Figure 25–42
This Apple Blossom yarrow (Achillea millefolium 'Apple Blossom') *is held up by a metal cage. Notice that the cage is hardly visible.* Source: Laura Deeter, The Ohio State University.

Figure 25–43
Large dahlia (Dahlia spp.) *are best held up by strong, single stakes.* Source: Laura Deeter, The Ohio State University.

frames are handy because they last a long time and are sturdy. Small sticks in the ground with twine between them can provide the same support, and you won't have to resort to spending money on metal cages. Airy plants, such as baby's breath, are best supported by small sticks set into the ground around the plant. Sticks with a network of string between them are also effective.

Deadheading and Pinching　As with annuals, perennials often need deadheading, but with additional reasons. Some perennials drop their old flowers willingly and neatly. Others are not so obliging. Faded flowers and brown or black petals can detract greatly from the

Figure 25–44

Tickseed (Coreopsis auriculata 'Nana') requires continual deadheading. The faded flowers detract from the overall appearance of the plant. Source: Laura Deeter, The Ohio State University.

overall beauty of the garden (Fig. 25–44). Thus, the prime reason for deadheading is aesthetics; however, the practice yields other benefits as well. Leaving the flowers on the plants only encourages them to make seeds. Unless you are hybridizing your own plants, there is no need for seed production in the garden. Some plants can become aggressive self-seeders, and deadheading prevents this tendency. Others spend energy attempting to reproduce at the expense of root production; thus, the plant may not survive the winter. Deadheading can also prolong the flowering time for many plants.

Pinching is removing new growth prior to flowering. This technique creates more compact, bushy growth (that might not require staking later in the season) and can increase flowering time. Simply pinch the top inch or two from each branch of the plant in mid- to late spring. You can pinch again in a couple of weeks if you want. Be sure not to pinch once you see flower buds or you'll pinch off the flowers.

Pinching can also be done after a plant has flowered to encourage another flush of flowers. Some species, such as tickseed (*Coreopsis lanceolata*), can be pinched up to three times during the growing season, for three almost full flushes of flowering. Pinching is also done to delay the flowering time. Hardy garden mums are pinched several times during the growing season to promote bushiness and to delay flowering until the fall, when the plants are sold for additional fall color.

Dividing   Dividing is the splitting of a single clump of a herbaceous perennial plant into many smaller segments. There are three reasons for dividing in the perennial garden. The first is to make more plants. The second is to rejuvenate an old plant. The final reason is to control an aggressive plant that is getting too large.

Plants with multiple stems, such as aster, hosta, garden phlox, and yarrow, are very simple to divide. Plants with a large tap root, such as butterfly weed (*Asclepias tuberosa*) and false indigo (*Baptisia australis*), are not easy plants to divide; it is best to leave them alone. In addition, some plants that we treat and use as herbaceous perennials are actually small woody plants. Lavender (*Lavandula* spp), Russian sage (*Perovskia atriplicifolia*), and zonal geranium (*Pelargonium xhortorum*) are all woody plants and should not be divided. These plants are the exceptions, however, rather than the rule. Most perennials are very easy plants to divide.

To divide a perennial (for any reason), dig the entire clump out of the ground. You might need someone's help: large, old perennials are very heavy. Utilizing a shovel, spade, sharp knife, or even a chainsaw (for large grasses), cut the plant into as many sections as you desire. Be sure to leave a good root system on all sections that are going back into the ground. The smaller you divide the plant, the longer it will take to reach flowering size again. Although some perennials require great care when dividing (sea pinks, or *Armeria maritima*, is one such perennial), most perennials are durable plants. Although a little care is needed, generally you need not worry about causing long-term damage to the plant.

Generally the best time to divide perennials is in the spring (early spring flowering plants are better done late spring or in the fall). This approach gives the plant the most amount of time to recover and put down a good root system again. Regardless of the timing, remember that all plants undergoing division need some extra tender loving care until they become established again. Make sure you water your plants regularly and often until they form a good root system. The following is a checklist to help ensure the future survival and growth of your new divisions:

1. Prepare the site for your new plants prior to dividing them.
2. Use sharp tools. Sharp tools make clean cuts, helping to prevent the spread of disease.
3. Discard any dead stems.
4. For large plants, cut the foliage back by half at least to help conserve water.
5. Replant divisions as soon as you can.
6. Water, water, water.

Perennial gardening is *not* a no-maintenance type of gardening. However, a small amount of effort and

maintenance ensures that your perennial garden will be beautiful and full of flowers for years to come.

THE WONDERFUL WORLD OF PERENNIALS

The vast number of perennials available in the nursery industry is staggering and can be intimidating for a new gardener. How do you choose a few plants to start your garden when hundreds are available at your local garden center? In addition, North America is extremely diverse in terms of the plant material that grows across the region. Southern regions have an amazing array of plant material available year-round, and these plants are treated as annuals in northern areas. Plants for full sun in the North may require shade in hot, arid regions. It is impossible to provide a list of perennials that will work in every garden situation across North America. There are perennials for every garden situation: wet, dry, arid, boggy, well drained, clay, sandy, sunny, and shady; small areas, large areas, and hillsides; erosion control; wildlife, butterflies, and hummingbirds; water gardens and herb gardens; to name but a few. A *very* brief listing of several specialty plants can provide you with the range of possibilities available when you include perennials in your landscape. If you are interested in the world of perennials, join the Perennial Plant Association (www.perennialplant.org).

The following list was compiled from nurseries across North America as their top ten best-selling plants, and it provides a general listing of plants that grow in as wide a range of growing conditions as possible:

1. *Hemerocallis* 'Stella D'oro'—Stella D'oro Daylily (Fig. 25–45):
   a. Golden-yellow flowers peak in early summer. They rebloom sporadically with deadheading.
   b. 15 to 18 in. tall and 24 in. in diameter.
   c. Grow in full sun or partial shade and well-drained soil.
   d. Foliage remains clean through the season.
2. *Hosta undulata* 'Medio-Variegata'—variegated hosta (Fig. 25–46):
   a. Purplish lilac flowers in midsummer.
   b. Green leaves with a pure white margin.
   c. 15 to 18 in. tall, 24 in. in diameter.
   d. Shade or part shade; moist, well-drained soil.
   e. Makes good cut flowers; hummingbirds are attracted to the flowers.
3. *Astilbe* spp.—false spirea, astilbe (Fig. 25–47):
   a. Foliage ranges from dark, shiny green through red and bronze.

Figure 25–45
Hemerocallis. Source: Laura Deeter, The Ohio State University.

Figure 25–46
Hosta undulata. Source: Laura Deeter, The Ohio State University.

Figure 25–47
Astilbe. Source: Laura Deeter, The Ohio State University.

b. Flowers late spring to midsummer depending on type. Flowers are open plume in white, red, pink, or purple.

c. Plants range in height from 8 in. to 3 ft, with a similar spread depending on type.

d. Partial shade; moist, well-drained soil.

e. Great cut flowers; excellent dried seed heads.

4. *Echinacea purpurea* 'Magnus'—purple coneflower (Fig. 25–48):

a. Pink to purple daisylike flowers in midsummer.

b. Dark green foliage.

c. Full sun; drought-tolerant.

d. 2.5 to 3 ft tall and wide.

e. Excellent cut flowers; attracts bees.

5. *Rudbeckia fulgida* var. *sullivantii* 'Goldsturm'—black-eye Susan (Fig. 25–49):

a. Yellow, daisylike flowers with black centers in late summer.

b. Dark green foliage.

c. Full sun; drought-tolerant.

d. 2 to 3 ft tall and 3 to 4 ft in diameter.

e. Excellent cut flowers.

6. *Coreopsis verticillata* 'Moonbeam'—moonbeam threadleaf tickseed (Fig. 25–50):

a. Light yellow flowers all summer (requires deadheading).

b. Full sun to part shade; drought-tolerant.

c. 18 to 24 in. tall and wide.

d. Airy foliage is attractive when the plant isn't flowering.

7. *Sedum* 'Autumn Joy'—Autumn Joy stonecrop (Fig. 25–51):

a. Succulent, fleshy foliage is bright green.

b. Flowers start out green in midsummer, change to light pink and finally to dark maroon in early fall.

Figure 25–48
Echinacea purpurea. Source: Laura Deeter, The Ohio State University.

Figure 25–50
Coreopsis verticillata. Source: Laura Deeter, The Ohio State University.

Figure 25–49
Rudbeckia fulgida. Source: Laura Deeter, The Ohio State University.

Figure 25–51
Sedum. Source: Laura Deeter, The Ohio State University.

c. 2 to 3 ft tall and wide.

d. Full sun; drought-tolerant.

e. Great for cut flowers; attracts bees and butter-flies; seed heads remain attractive through early winter.

8. *Heuchera* spp.—coral bells (Fig. 25–52):

a. Foliage colors range from green, purple, maroon, lime, golden, to silver, and to combinations of those colors.

b. Flowers well above the foliage in pinks, red, and white in late spring, and early summer.

c. 1 to 2 ft tall and wide.

d. Partial shade to full sun, depending on the variety.

e. Excellent for hummingbirds, cut flowers, and foliage color.

9. *Pennisetum alopecuroides* 'Hameln'—dwarf fountain grass (Fig. 25–53):

Figure 25–54
Calamagrostis xacutiflora. Source: Laura Deeter, The Ohio State University.

Figure 25–52
Heuchera. Source: Laura Deeter, The Ohio State University.

Figure 25–53
Pennisetum alopecuroides. Source: Laura Deeter, The Ohio State University.

a. Fine-textured, short, ornamental grass.

b. Seed heads are attractive in mid- to late summer.

c. Excellent winter foliage appearance.

d. 2 to 4 ft tall and wide (with seed heads); forms clumps.

e. Full sun; well-drained soil.

f. Great for winter interest.

10. *Calamagrostis xacutiflora* 'Karl Foerster'—Karl Foerster feather reed grass (Fig. 25–54):

a. Fine textured, upright ornamental grass.

b. Tan seeds are attractive from midsummer through midwinter.

c. 4 to 5 ft tall and 2 to 3 ft wide; forms clumps.

d. Full sun; well-drained soil; drought-tolerant.

e. Excellent for winter interest, foliage interest, cut flowers.

Specialty Perennials

In addition to the top-selling perennials, many gardens have specific themes and require plants with specialty characteristics. These garden themes include water, herb, and edible flower.

Water Garden Plants

1. *Nymphaea* spp.—water lily (Fig. 25–55):

a. Floating leaves; some varieties are mottled.

b. Flowers in a rainbow of colors; some varieties open only at night.

c. Some are hardy; others are tropical.

d. Full sun; water gardens.

Figure 25–55
Nymphaea. Source: Laura Deeter, The Ohio State University.

Figure 25–56
Eichornia crassipes. Source: Laura Deeter, The Ohio State University.

Figure 25–57
Thymus. Source: Laura Deeter, The Ohio State University.

Figure 25–58
Lavandula. Source: Laura Deeter, The Ohio State University.

2. *Eichornia crassipes*—water hyacinth (Fig. 25–56):
   a. Purple flowers in summer.
   b. Not hardy in the northern climates.
   c. Floating plant; does not need a container.
   d. Unique foliage and bladder for floating support.

Herb Gardens
1. *Thymus* spp.—thyme (Fig. 25–57):
   a. White to pink flowers in late spring to midsummer, depending on type.
   b. Upright or ground covers are available.
   c. Edible; highly fragrant with many different scents.
   d. Some are tolerant of minimal foot traffic.
   e. Full sun; very well-drained soil.
2. *Lavandula* spp.—lavender (Fig. 25–58):
   a. Many species available in various parts of the country.

   b. Purple, blue, red, and white flowers are fragrant and edible; attract bees.
   c. Attractive gray, edible foliage on a small woody shrub.
   d. Can be sheared heavily.
   e. Full sun; extremely well-drained soil.

Edible Flowers
1. *Tropaeolum majus*—nasturtium (Fig. 25–59):
   a. Round foliage can be green or with white variegation.
   b. Flowers all summer in yellow, orange, and red; flowers are edible and have a peppery taste.
   c. Partial shade; well-drained soil.
   d. Annual.
2. *Viola × wittrockiana*—pansy (Fig. 25–60):
   a. Flowers in cool weather (spring and fall in the north) in a rainbow of colors.

Figure 25–59
Tropaeolum majus. Source: Laura Deeter, The Ohio State University.

Figure 25–60
Viola x wittrockiana. Source: Laura Deeter, The Ohio State University.

b. Flowers are edible and taste like iceberg lettuce.
c. Great for adding color to salad greens and Jello–type salads and desserts, or using as a cake decoration; can be candied.

d. Partial shade; well-drained soil.
e. Prefers cool climates or protection from the sun in the south.

SUMMARY AND REVIEW

Botanical classifications of plants give fairly rigid definitions of categories, but in reality, many plants can exhibit characteristics from one or more category. For example, some trees can be grown as shrubs and vice versa. Some perennials behave as annuals if they are grown in areas that have seasons for which the plant is not adapted. The landscape interest of plants can come from its flowers, fall color, size, growth habit (natural or pruned), fruit, ornamental bark, persistent foliage, shade characteristics, ability to provide wildlife habitat, as well as other traits. Landscape plants are chosen based not only on their interesting characteristics but their ability to perform a function in the landscape.

As with all plants, there are environmental factors that have to be considered when growing landscape plants. These are temperature, light, moisture, wind, and soil. Most landscape plants have been designated for the USDA Cold Hardiness Zone for which they are suited. Some have also been classified by the American Horticulture Society Heat Zone for which they are suited. Knowing a plant's cold and heat limits helps to choose plants that more likely to survive in an area. Landscape plants are often classified by their light requirements. These classifications range from full sun to partial shade to partial sun to full shade. The type of shadow cast by trees as well the side of a building or other object on which plants are growing can influence shade

patterns. How shadows move during the day can affect the shade characteristics of a location. Understanding the precipitation patterns of location are critical to choosing the right plants. Landscape plants have wide range of tolerance for wind and the effect wind has on transpiration. Wind strength and prevailing direction need to be considered when choosing plants. The strength of the tree's wood is an especially important consideration for choosing trees for a landscape. Soil characteristics are as important for landscape plants as other plants. It is best to select plants that are adapted to the pH found in an area. Soil texture and its ability to hold and drain water is important because some landscape plants need well drained soil while some others prefer a constantly moist soil. Others can tolerate a wide range of soil conditions.

Trees and shrubs provide the "backbone" of a landscape. They are nearly permanent and have emotional and practical value. Choosing a tree for a landscape includes considering its lifelong characteristics. Woody plants are sold as bare root, balled and burlapped, or in containers. Each has advantages and disadvantages, so choosing which type to use depends on the situation and the plant being used.

When choosing a tree or shrub, look for healthy plants that have good color throughout their structure, good twig extension, no roots circling the top of the pot or rootball, the root system should be in proportion

to the rest of the plant. Root systems that are too small will not support the plant when it is planted.

Trees and shrubs should be planted in soil that has been prepared to have good characteristics. The plants should be planted to the correct depth, making sure to break up any hardpan or subsoil lying directly below the roots. Trees should not be planted any deeper than the original soil line (crown) on the trunk. Mulching can benefit newly planted trees. Staking may be staked but it is not always necessary. Staking can protect the tree from mowers and other machinery, anchor the root system, and support the tree temporarily in an upright position. Young trees that lack lower branches should have the lower trunk painted white with interior latex paint or wrapped to prevent sunscald. A small amount of nitrate fertilizer may be applied at planting but caution must be taken to not use too much. Newly planted trees need deep irrigation to encourage the roots to move downward.

Pruning is an important part of maintenance of shrubs and trees. Pruning can improve the strength and appearance of the plant. Choosing a tree or shrub with natural characteristics that can fulfill the plant's function reduces the need for pruning. Pruning techniques vary with the type of plant being pruned and its function in the landscape. If the plant is being grown for its flowers, particular care must be given to pruning at the right time of year so that flowers will develop during the next flowering period. There are very many species and cultivars of trees and shrubs available for use in landscapes.

Herbaceous annuals are also called bedding plants. They are usually grown for their colorful, visual appeal as well as other characteristics. Most annuals are started in greenhouses and transplanted after the danger of frost (or summer heat for cool-loving plants). Most annuals require loose, well-drained soil. The addition of a slow-release fertilizer at planting helps maintain plant appearance during the growing season. Many annuals benefit from deadheading.

Herbaceous perennials are very popular landscape plants, and their popularity is increasing annually. The same "right plant, right place" rules apply for herbaceous perennials that apply for woody plants. The environment must match the plants adaptability because the plants will be there for many years. Understanding the soil characteristics and then preparing soil for the needs of the herbaceous perennials to be planted there is an important first step. Soil texture, pH, and drainage are all important and can be determined either by professionals or the homeowner. To correct problems, the drainage can be improved by several methods including installing drain tiles, making raised beds, and adding soil amendments. Soil amendments can also improve pH and nutrient levels. Top dressing is the easiest way to amend established perennial beds. Tilling the amendments into the soil is the most efficient way to amend the soil in new beds.

Weeds can be very difficult to control in perennial gardens. Preemergent chemicals can only be used if the perennial comes back from an underground structure and is not self-seeding. Broad spectrum postemergent herbicides would most likely kill or damage the garden plants as well as the weeds. Herbicides selective for grasses or broadleaf plants would work as long as there were no desirable plants of the same type being sprayed as well. Weed barriers, mulches, and solarization can be effective if done at the right time.

Staking is used to support plants that get very tall or produce very heavy flowers. The stakes should not interfere with the visual qualities of the plants. Deadheading and pinching helps improve the appearance of the plant, reduces the spread of excessively seeding plants, and can help maintain the vigor of the plant. Pinching keeps the plants more bushy and can prolong and/or delay flowering time.

Many herbaceous perennials can be divided to make new plants, rejuvenate plants, and control aggressive plants. Plants that have multiple stems or a fibrous root system are the easiest to divide. Plants with taproots or are actually small woody plants should not be divided. The entire clump is dug from the ground, then divided. It may take only a knife or shovel to make the cuts but for some it requires chain saws. All pieces that are to be replanted need to have a good root system left intact.

There are perennials available for nearly every growing situation.

KNOWLEDGE CHECK

1. Why is it difficult to classify many plants by such terms as being a tree or a shrub, a perennial or an annual?
2. What are three plant characteristics that provide landscape value?
3. Why are cold hardiness and heat tolerant zones important when selecting landscape plants?
4. Besides heat, what other temperature do desert plants often experience?

5. Does an eastern exposure have a milder or harsher sun exposure than a western exposure?

6. Would a plant that needs sun and moderate moisture likely grow well in a Pacific Northwest garden?

7. Why is wood strength an important consideration when choosing trees for exposed sites?

8. What can happen to herbaceous plants in windy areas?

9. Name a garden plant that is tolerant of wet soil. Of standing water.

10. What makes a honey locust tree more susceptible to drought than a black oak?

11. What are the three ways trees are sold for planting in landscapes?

12. What are three things to look for to make sure a tree is healthy?

13. What are girdling roots?

14. What is the time of year for planting conifers in cold areas? Hardy trees in warm areas?

15. What should be done with bare-root trees if they cannot be planted at once?

16. How is soil compaction from construction equipment corrected for landscape planting?

17. Why is the tree crown planted above the soil line?

18. What effect does mulch around newly planted trees have on soil temperature?

19. What are two reasons newly planted trees are staked?

20. Why are the trunks of newly planted trees that do not have low branches painted white or covered with burlap or paper?

21. What is the danger from applying too much fertilizer to trees at the time of planting?

22. Why is deep watering newly planted trees important?

23. What are the two questions to ask when getting ready to prune trees and shrubs?

24. What is the first thing to remove when pruning?

25. Why are narrow angle branches removed during pruning?

26. What is espalier pruning and Pollard pruning?

27. What is the "ultimate" type of pruning?

28. In shrubs, which branches should be the first ones to be removed by pruning?

29. When are spring-flowering shrubs pruned? When are current year (summer flowering) shrubs pruned?

30. What is deadheading?

31. What are two probable reasons that herbaceous perennials have become popular?

32. How can a homeowner analyze soil texture in a garden without having it done by professionals?

33. What are 3 ways to improve soil drainage in a garden? Which is the easiest?

34. Why are preemergent herbicides not good for use in a perennial garden where the plants self-seed?

35. Besides removing old, unsightly flowers, what are two reasons for deadheading herbaceous perennials?

36. What are two reasons for dividing herbaceous perennials?

FOOD FOR THOUGHT

1. You are a homeowner who is moving from Seattle, Washington, to Atlanta, Georgia. You are thinking of taking divisions from several of the perennials in your old yard to plant them in your new yard. How well do you think these plants will grow in their new location? What problems do you think they might have?

2. When you bought your house in the spring several years ago, there were several overgrown forsythia in full bloom. You loved the color, but the plants were very unattractive. You had heard that it was best to prune in the late fall, so you've done that to the forsythia every year since the first year. However, you have not seen a forsythia flower since that first year either. Why do you think that could be happening?

3. From Table 25–2, choose three species of plants to put in a garden with one of the species being used as a low-growing border, one of the species being moderately tall and that you can cut to put in a vase in the house, and the third species that can be grown on a trellis at the back of the garden.

FURTHER EXPLORATION

DIRR. M. A. 1994. *Manual of ornamental woody landscape plants.* Champaigne, IL: Stipes Publishing.

DIRR, M. A. 1997. *Dirr's hardy trees and shrubs: An illustrated encyclopedia.* Portland, OR: Timber Press.

DISABATO-AUST, T. 2006. *The well tended perennial garden*, 2nd ed. Portland, OR: Timber Press.

PHILLIPS, E., and C. C. BURRELL. 2004. *Rodale's illustrated encyclopedia of perennials.* Emmaus, PA: Rodale Books.

STILL, S. M. 1994. *Manual of herbaceous ornamental plants.* Champagne, IL: Stipes Publishing.

✳26

Constructed Landscapes and Landscape Design Considerations

LAURA L. S. BURCHFIELD

INTRODUCTION AND HISTORY

Designed or constructed landscapes are influenced by many factors. Environmental factors such as climate, topography, geology, and hydrology, are a few of the limiting elements that influence the design of both public and private landscapes. Human factors, too, play important roles. Throughout recorded human history, society has shaped the landscape based on political, religious, and cultural beliefs and ideas as well as technological advances. People have grown plants for food, fiber, medicine, and forage since the earliest times of human settlement. With the rise of cites in China, Mesopotamia, and ancient Egypt, humans began to garden for pleasure, recreation, and ornamentation—much as we do now.

Function with beauty is a long standing ideal. The earliest gardens recorded contained plants for use as foods and medicines organized with a pleasing arrangement. Even in ancient times, people altered the landscape for more than just food and economic crop production. Landscapes for pleasure and recreation began very early in human history even if those landscapes were reserved for the wealthy and powerful of the time. The Egyptian tomb paintings depict the gardens of the wealthy containing both ornamental and agronomic plants. The Assyrians appreciated the natural environment and the rulers developed hunting parks for leisure, which included exotic plant species. In ancient Persia (modern Iran), gardens were walled oases centered around water which included flowering and fruiting plants often in a four-square configuration that was meant to represent the ideal world.

In Asia the tradition of pleasure gardens also started very early. The emperors of China built large hunting parks that were miniature versions of their empire and filled with constructed ponds, hills, exotic animals, and plants. Smaller gardens were also miniaturized versions of the natural landscape. Japanese landscapes were influenced by the Chinese traditions. Symbolism and philosophy were a driving force in Asian landscape. The later landscapes were meant for contemplation and meditation in keeping with the religious beliefs of the region (Fig. 26–1).

key learning concepts

After reading this chapter, you should be able to:

- Describe the history of the constructed landscape.
- Explain the terminology, elements, principles, and process of landscape design.
- Explain the functional uses of plants in the landscape.
- Discuss sustainable practices in landscape design.

Figure 26–1
Asian landscapes are meant for contemplation and mediation. Source: Laura Burchfield, The Ohio State University.

Figure 26–2
Monastic gardens preserved the practice of horticulture during the dark ages. Source: Laura Burchfield, The Ohio State University

Western European landscapes were influenced by the Persian paradise ideas and evolved into the courtyard gardens of the Greek and Roman cultures. The geometric designs enclosed by the walls of the house predominated. A typical villa garden might feature water fountains and artificial ponds as well as fruit trees, herbariums, kitchen gardens and the like. Statuary, trimmed plants, and topiary were often features of such gardens as well as potted—that is, containerized plants—which seem to have developed in this period.

After the fall of the Roman Empire, gardens reverted to more utilitarian purposes. Gardens were often smaller and within the walls required for protection and isolation by castles, convents, and monasteries. Within their walls, the monasteries of the medieval period preserved much of the horticultural knowledge of the Greeks and Romans during the Middle Ages (Fig. 26–2). Formal pleasure gardens reappeared during

the Renaissance in Italy, which influenced the design of European gardens for over a century. Italian gardens once again featured geometric designs and, because of the hilly terrain, they were often terraced. The emphasis was on the built structures—walls, fountains, and the like—as well as horticulture. The European nobility adapted the formal garden style of Italy throughout Europe. In the seventeenth century, the tastemaker was France. The gardens designed by André Le Nôtre (1613–1700) were extreme displays of wealth and power (Fig. 26–3). These landscapes, like other matters of taste, became the norm across Europe as the nobility of other countries often imitated the Frenchcourt. The English, however, became dissatisfied with the formal gardens in the early eighteenth century. Landscape architects designed more naturalistic, albeit inspired by idealized, landscapes much like those in classical paintings (Fig. 26–4).

A

B

Figure 26–3
Gardens such as Vaux-le-Vicomte on left and Versailles on right both designed by André Le Nôtre were created to display the wealth and power of their owners. Source: Laura Burchfield, The Ohio State University

Figure 26–4
The English Landscape Movement was an idealized natural landscape reminiscent of a landscape painting.
Source: Laura Burchfield, The Ohio State University

Figure 26–5
In the twentieth century the modernist landscape began as shown in this contemporary landscape.
Source: Laura Burchfield, The Ohio State University.

The gardens of the colonies in North America—part of which became the United States—reflected the styles of individual's homelands. Later, the wealthy took their gardening ideas from all parts of Europe, from publications and from travelling abroad. Andrew Jackson Downing influenced landscape design in the mid nineteenth century with his periodical *The Horticulturist* and his many books. Landscaping in the late nineteenth and early twentieth century was influenced by a number of people, among them Fredrick Law Olmsted who designed the landscapes for Central Park in New York City and the Biltmore Estate in Asheville, North Carolina.

Residential communities developed that had park-like settings, the precursor to the "lawn ideal" that we see today. The Modernists of the mid-twentieth century broke away from traditional symmetrical landscapes and developed asymmetrical landscapes that linked the indoor and outdoor spaces with a more casual style to match the needs of society (Fig. 26–5). With the post–World War era also came the predominance of suburban residences and what we still see in the landscape today.

RESIDENTIAL LANDSCAPES

Typical American residential landscape (Fig. 26–6) has three spaces; a front yard that is a public area viewed by everyone; the backyard that is a more private space used by the residents; and the side yards that are either unused or used for storage or as circulation between front and back.

The majority of modern residential landscapes are predominately lawn with plantings clustered up near the house to be viewed from the street (Fig. 26–7). These plantings generally include evergreen shrubs, possibly an ornamental tree, and, for the gardener types, annual and perennial flowers planted for color. The backyard often has a deck or a patio. Most lack privacy from the neighboring yards. There may be a few trees and maybe a vegetable garden, or a play structure and usually more lawn area. Because of all the lawn area, turf is the most prevalent ground cover in the constructed landscape if you do not include paved surfaces. There is much that can be done to diversify and

Figure 26–6
A typical residence has a front, back and side yards, but without good landscaping they are not an asset in terms of beauty or function. Source: Laura Burchfield, The Ohio State University

Figure 26–7
The typical residential landscape fails to create functional spaces. It is often a wide expanse of lawn with plantings tightly up against the house. These landscapes are indistinguishable from one region to another.
Source: Laura Burchfield, The Ohio State University.

improve the landscape aesthetically, functionally, and environmentally.

Several ideas about landscape design have developed over time to improve landscapes. One of them is the concept that the outdoor spaces should extend and connect the indoor spaces of the adjacent structures. This can apply to both public and residential landscapes. Another idea is that the landscape should fit with the surrounding natural and constructed landscapes so as to blend seamlessly. In addition, the style and architecture of the building or house should influence the style of the landscape. Form (aesthetic appeal or beauty) should follow function in landscape design

is, perhaps, the most important idea to have been developed. Landscape should first and foremost meet the needs for which it is intended. The concept of outdoor rooms can be useful in creating functional landscapes that have excellent form.

Outdoor Rooms

As stated in the previous section, it is important to think about the landscape as more than just arranging plants in an aesthetically pleasing manner for viewing. The plants and other elements of the landscape should create functional spaces for human use. The concept of space is the voids created by the area left after the masses are arranged; these spaces in the landscape can be thought of as outdoor rooms. Just as an interior room has three planes of spatial enclosure—floor, ceiling, and walls—so should the outside room (Fig. 26–8). The difference is the outdoor space is not as well defined. It varies in the amount of enclosure and light it receives. Outdoor spaces also change over time and throughout the day and year. In a public setting these spaces might be a large gathering space such as a plaza, or corridors for movement. In a residential landscape the outdoor rooms correlate to similar rooms found in the house. Typical outdoor rooms include entry, dining, entertaining, kitchen, and recreation rooms (Fig. 26–9). The connection between inside and out is an important feature of well designed landscapes. When possible there should be a logical relationship of spaces in and out. For example the inside kitchen should be in close proximity to the outside kitchen and dining areas. In creating outdoor rooms, the size and shape should fit the use and the number of users expected.

A

B

C

Figure 26–8
Creating outdoor rooms requires the use of three spatial planes to create ceiling, walls, and floor: (A) overhead plane, (B) vertical plane, and (C) base plane. Source: Laura Burchfield, The Ohio State University.

A   B   C   D

Figure 26–9

Typical "outdoor rooms" are similar to the rooms you would find inside the residence: (A) dining, (B) living, (C) entry, and (D) recreation. Source: Laura Burchfield, The Ohio State University.

FUNCTIONS OF PLANTS

Plants are used in the landscape for multiple functions. Plants serve aesthetic, architectural, and environmental protection and modification purposes in the landscape. Most often the aesthetic purpose of plants is what people think of when selecting plants. "Is it pretty?" "Will it make my landscape look good?" Plants offer so much more, and good landscape design can maximize the benefits of plants in the human environment.

Architecturally plants create space. Walls, ceilings and even floors of the outdoor rooms can be formed by plants. Choosing the right plants and planting them in the appropriate design, can create privacy and can encourage movement along circulation corridors. The structural framework of the landscape can be formed by plants. Plants can perform many other functions. Large spaces can be divided into smaller space to bring the great outdoors down to a comfortable human scale. Spaces can be enclosed, undesirable views screened, desirable views framed, and structures integrated into the landscape—all with plants (Fig. 26–10). Plants can provide a background for a focal point much as a wall provides a background for a painting. Architecturally, plants can serve many of the same functions in the outdoor space that built structures and furnishings provide in indoor spaces. Plants can also serve other architectural purposes, such as a screen or fence—a hedge, for example, to provide a backyard privacy fence—or to ensure safety as a barrier to keeping pedestrians away from traffic right-of-ways (Fig. 26–11).

Environmental protection functions of plants include air quality improvement along with erosion and flood control. Atmospheric purification is a well-known benefit of plants. In the process of photosynthesis, plants convert the carbon dioxide given off by

A   B   C   D

Figure 26–10

Plants can enclose space, frame a good view, screen an unsightly view and integrate structures into the landscape: (A) enclosing space, (B) framing a view, (C) screening a view, and (D) integrating structures. Source: Laura Burchfield, The Ohio State University.

Figure 26–11

Plants can be used to create a barrier for pedestrians from traffic. Source: Laura Burchfield, The Ohio State University.

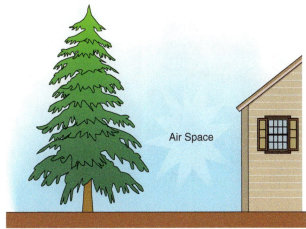

Figure 26–13

Airspace formed by plants insulates and moderates the temperatures inside buildings. Source: Laura Burchfield, The Ohio State University.

humans and animals into sugars and give off oxygen. Plants absorb odors and collect dust particles from the atmosphere (Fig. 26–12). Plants can be used to hold soil on slopes and reduce soil and water runoff. Roots penetrate the soil and create webbing that can hold soil particles. The vegetation creates friction that can slow the movement of water across the soil surface.

Environmental modification can be achieved with plants creating microclimates to make outdoor and indoor spaces more comfortable by warming or

cooling as desired. The use of deciduous trees can reduce heating and cooling costs by moderation of solar radiation. When large trees are planted on the south and west sides of a building, the foliage cover shades and reduces the temperature. In the winter, the leaves are absent, which allows solar radiation to warm the building. When dense evergreens are planted alongside a building, airspace is formed that moderates the temperature (Fig. 26–13). Wind modification is also possible—with proper plant placement, the direction

Glare Reduction

Acoustical Control

Atmospheric Purification

Traffic Control

Glare Reduction

Figure 26–12

Plants can be use in the landscape to reduce glare, purify the air, mask noise, and to control traffic.

Source: Adapted from, G. O. Robinette, *Plants, People, and Environmental Quality* (Washington, DC: U.S. Government Printing Office, 1972).

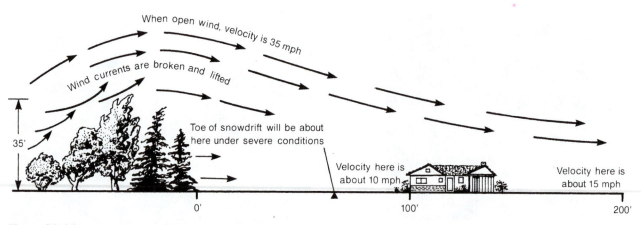

Figure 26–14

Planting trees and shrubs can reduce the wind velocity and alter the direction. Source: USDA.

and velocity of the wind can be directed or altered. By creating windscreens, the area to the leeward side can be made more comfortable. Trees and shrubs can channel winds either toward an area to create cooling breezes in the summer or away from areas where wind interferes with activities (Fig. 26–14).

Other environmental modifications include noise control and glare and reflection control. Noise can be reduced either by using plants to buffer the noise or to create "white noise" to mask the undesirable sounds (Fig. 26–15). Glare from streetlights and automobile headlights can be blocked by plants. Plants can also be used to minimize the reflection and absorption of light and heat on surfaces (Fig. 26–12).

Thoughtful selection and placement of plants can provide many enhancements to the human environment. When planning for a landscape, consideration should be given to the many possible functions that plants can fulfill.

DESIGN ELEMENTS AND PRINCIPLES

The effectively designed landscape is realized not only through the use plants to fulfill the function of the landscape but also with the elements and principles of design to arrange the landscape into a visually pleasing manner. These elements can be thought of as the tools for design. They are *line, form, size, pattern, texture, and color* (Fig. 26–16). The principles are order, unity, and rhythm. By arranging the elements with adherence to the principles, attractive landscapes are created.

Order, also referred to as balance, is the overall framework of a design. This balance can be achieved by symmetry and asymmetry. Symmetry is when objects are arranged equally around an axis with one side mirroring the other (Fig. 26–17). Symmetrical design

Figure 26–15

Masking noise can be accomplished by creating a barrier of dense plants or using plants such as grasses that create "white noise" when rustling in the wind.

Source: Laura Burchfield, The Ohio State University.

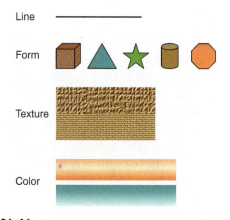

Figure 26–16

Design elements. Source: Laura Burchfield, The Ohio State University.

A

B

Figure 26–17
Symmetry is created by mirroring the design on either side of an axis which creates a formal feel to the landscape. (A) symmetry in a typical residence. (B) Symmetry on a grander scale. Source: Laura Burchfield, The Ohio State University.

Figure 26–18
Asymmetrical balance is a more informal landscape where the elements have equal weights without being identical.
Source: Laura Burchfield, The Ohio State University.

A

B

Figure 26–19
Focal points draw the eye. A focal point is a special element in the landscape and can be a grand structure as seen in (A) or a simple sculpture as seen in (B). It may also be a special plant, such as an ornamental tree.
Source: Laura Burchfield, The Ohio State University.

has a formal character. Asymmetry is when objects are arranged to have an equal weighted balance, but without an axial reference. The visual balance is achieved but the effect is informal or naturalistic (Fig. 26–18).

Unity is arranging of objects so the overall effect visually reads as one composition. Unity, or harmony, can be achieved by dominance, unity of three, interconnection, and repetition. Dominance is the use of one element that stands out from the rest. It is often referred to as the focal point (Fig. 26–19). Unity of three is the use of odd numbers of objects (Fig. 26–20). Interconnection is when elements are physically linked in the design (Fig. 26–21). Repetition is where an element is repeated throughout the design (Fig. 26–22).

Rhythm is the design principle that makes a design dynamic by creating movement. Rhythm is achieved by using patterns in the landscape. The patterns types are repetition, alternation, gradation, and inversion. Repetition is repeating elements in a set sequence. Alternation is when elements in the pattern

Figure 26–20
Unity of three. Source: Laura Burchfield, The Ohio State University.

Figure 26–21
Interconnection. Source: Laura Burchfield, The Ohio State University.

Figure 26–22
Repetition can be used to lead the viewer's eye as demonstrated by the use of purple flowers that repeat throughout this border.
Source: Laura Burchfield, The Ohio State University.

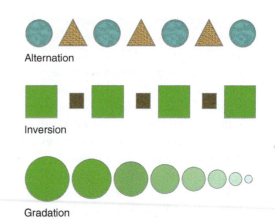

Figure 26–23
Types of rhythm used to create dynamic sequence of time and movement in the landscape.
Source: Laura Burchfield, The Ohio State University.

change in an altering sequence of size, texture, color, or shape. Gradation is when the pattern is a gradual change in size or color across a space. Inversion is a type of alternation where the pattern reverses in character for example light to dark or large to small (Fig. 26–23).

All of the design principles can be used in combination to create interesting landscapes that are attractive.

DESIGN PROCESS

The design process starts when the idea for a landscape project is first conceived. To carry the project to successful completion, an orderly process should be followed. The process can be broken down into five phases: research, design, installation, maintenance, and evaluation.

Research Phase

The research phase is the gathering of information about the users, the purposes intended for the site, and the physical aspects of the site itself. Things to find out from the users include, but are not limited to, the following. Do they have likes and dislikes regarding things like particular plants, colors, and the like? Are small children or pets going to use the space? What is the budget the users have for the project? Do they have a theme in mind for the landscape?

Site inventory is identifying the important conditions of the site as it pertains to the design. The purpose is to gather all of the relevant information about the site's physical conditions and characteristics to determine the opportunities and limitations. The site inventory includes creation of a base map. The base map shows the extent of the area to be designed, and locates the major existing features such as built structures, trees, and utilities.

The site inventory also involves making notes describing the site. The character of the region and

neighborhood, and the architectural style of the buildings should be noted. Local ordinances should be checked for regulations and restrictions. Climate, microclimate, soils, existing vegetation, topography, and drainage should all be observed and recorded. The views both in and out should be studied. Photographs can be a useful means of recording this information. Determine where the good and bad views are located and where views might be enhanced. When conducting the inventory consider all of the senses, sight, sound, smell, feel of the site.

Site analysis is taking all of the information from the site inventory and synthesizing it to determine the problems and potentials for the design. If there are low overhead utilities then the size of trees and shrubs will be limited. Areas with poor drainage will require plants that tolerate wet soils. Poor soil areas will require amendment of the soil or selection of plants that are adapted to poor soils. Unpleasant views will need to be blocked by dense evergreen trees and shrubs or perhaps tall grasses. Good views can be framed to direct the eye to the view. The regional and neighborhood character will need to be considered so that the landscape created fits in. A woodland garden would not be appropriate in the desert southwest, just as a cottage garden might appear odd if planted around an ultra modern house.

Another step of the research phase is to determine the needs of the people using the landscape. How many people will use the area? What are their ages and special interests? When and how often will they use the space? What do they want to achieve from the landscape design project? Budget and time constraints are another factor to be considered.

Design Phase

The second phase, the design phase, begins after all of the background information has been gathered. With the data collected in the research phase, the landscape designer now prepares a design that identifies where the various functional areas best fit on the site. Alternative sketches may be made to decide where things should go. These are referred to as functional or bubble diagrams (Fig. 26–24). Next form and space studies are done as well as selecting the plants to be used. From this information, preliminary design plans are developed. The preliminaries are then developed into a final master plan (Figure 26–25). This is the guide or map of what will be installed. Additional detailed plans may be required to support the master plant depending on the scope of the project. They would most certainly include a detailed planting plan that indicates the type, size, and quantity of plants and where they should be located. Other detail plans might include layout plan, grading plan, and construction drawings.

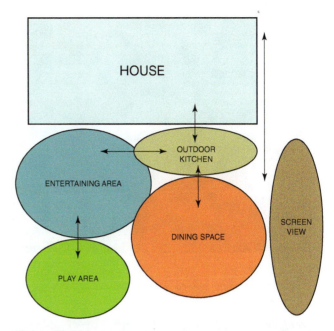

Figure 26–24
Bubble or functional diagrams allow the designer to quickly explore options to determine the best locations of elements and spaces of the landscape. Arrow represents circulation.
Source: Laura Burchfield, The Ohio State University.

Installation Phase

The installation phase is the planting of the plants and constructing the **hardscape**, that is, the nonplant features of the design such as a patio, pond, brick walls, and the like. Installation can begin immediately after the design phase is complete (Fig. 26–26). Installation may be done as a single operation or phased in over time depending on the site, weather conditions, and budget constraints. It is best to install the plants after all construction has been completed so as not to cause unnecessary stress or damage to the plants. Plants should be installed using best practice and procedures for the given region. Preparation of the planting areas to assure a good growing environment is required.

Maintenance Phase

The maintenance phase is critical in ensuring the landscape keeps its integrity over time (Fig. 26–27). Plants must be cared for with an understanding of their physiological needs. More care is required immediately after the plants are installed and during establishment than may be needed after establishment. Proper watering, fertilization, pruning, and monitoring and treating for pests and diseases will help ensure that the plants stay healthy and sustain the landscape's function and beauty. Proper plant selection can greatly reduce the amount of maintenance needed.

Figure 26–25
Final master plan includes all details for the design. Source: Virginia Oswald.

Figure 26–26
Installation phase. Source: Laura Burchfield, The Ohio State University.

Figure 26–27
Maintenance phase is critical to the longevity of the landscape. Source: Laura Burchfield, The Ohio State University.

Figure 26–28
Completed landscapes should be beautiful, functional, and compatible with the location.
Source: Laura Burchfield, The Ohio State University.

Figure 26–29
This rich tapestry of plants is an example of good planting design using a variety of textures, sizes and colors to a pleasing effect. Source: Laura Burchfield, The Ohio State University.

Evaluation Phase

Evaluation of the installed landscape design should be done periodically to determine what is working and what is not. This type of evaluation helps shape discussions for further change and improvements to the landscape as well as others that are designed in the future (Fig. 26–28).

PLANT SELECTION

There are many factors that must be considered when choosing plants for the successful landscape. The primary goal is for the plants to function for their intended purpose and thrive in their location. It is best if they can function and thrive with minimum input of labor, water, and chemicals. To determine what plant will be best suited for a particular landscape requires an understanding of plants' physiological needs, growth habits (size, shape, and growth rate), and ability to be adapted to the specific purpose.

Selection also requires an understanding of the site to ascertain whether or not the site will support the plant. Climate, soil type, moisture levels, and human and animal disturbances all determine what plants will grow in a given site. Cold and Heat hardiness maps can give guidance as to what plants will survive in a given region. Soil testing should be done to determine the type of soil (clay, sand, and silt), the pH, and the nutrient levels. The soil type can give information as to what the drainage might be like. Observation can also give clues to what the soil moisture may be in an area. Sun and shade patterns should be understood to assure the plants selected receive the proper amount of light. Availability will also affect plant selection. Landscape plants specified should be those found in local nurseries if possible.

A good starting point in plant selection is to determine which plants are native to a region or native to regions with similar conditions as the site. These plants are more likely to be adapted to the existing conditions and will require less time and materials. Fewer soil modifications and less irrigation will be required. This is not always the case, however, when a site has been altered significantly from the original existing conditions by human activity. Another good key to plant reliability is to select plants that have a history of being tried and successful in similar conditions. Information lists on plant selections are available that suggest plants for specific conditions such as deer resistance, urban tolerance, or specific uses such as hedging plants or ground covers.

Considering the mature sizes of plants at the design phase ensures proper spacing for a lasting landscape. Overcrowding and under planting are common mistakes that can reduce the health and beauty of the landscape. Also, the plant can change dramatically in size, shape, and other characteristics as it matures. It is very important to keep this in mind when designing a landscape.

By selecting the right plant for the right purpose and place, the landscape will realize its full potential for function, beauty, and longevity (Fig. 26–29).

SUSTAINABLE LANDSCAPE PRACTICES

When designing landscape it is important to consider the impact those landscapes have on the environment as a whole. The installation and maintenance of a landscape has not always been a benefit to the environment

as most people would believe it to be. Poor landscaping practices have resulted in a significant amount of environmental pollution, especially in suburban areas. Sustainable practices must be understood and utilized in future landscape design, installation, and maintenance. The evaluation phase must include evaluating sustainability of the practices needed to maintain the landscape.

Sustainability, as you learned in Chapter 16, is defined, "meeting the needs of the present without compromising the ability of future generations to meet their needs." The philosophy is that we must conserve the resources we have for the sake of the future, and that means not only using resources more wisely, but ensuring that we are not degrading the resources we have.

Landscapes are ecosystems and thrive under the same principles that create healthy natural ecosystems (see Chapter 2). Designing a landscape that requires the input of very few resources and does not harm existing resources can have a positive impact on the environment. If considered en masse, the positive impact of large numbers of individual landscapes using sustainable practices can be significant at the global level. There are many ways to incorporate sustainable practices into the landscape.

Understanding the site is the first step. Knowing the local conditions and limitations is necessary before making decisions. Conservation of water, energy, and soil is something that can be accomplished in any landscape.

Landscape plants can be used to filter and help purify water. This is especially true if "gray water" (water that comes from drains but does not contain human or animal solid and liquid waste) is used to irrigate the plants. If the landscape is large enough and the site appropriate, a small wetland can be created to provide wetland benefits. Using plants that are adapted to the precipitation patterns for the area where the site is found can reduce the need for irrigation during most, if not all, of the growing season. An example is selecting plants that need little water for arid areas. This is often referred to as **xeriscaping**. When irrigation is needed, the use of gray water or rain water collection systems such as rain barrels (Fig. 26–30) or cisterns reduces the water that must come from other sources. If another type of irrigation is required, using a system that wastes less water, such as drip irrigation, is recommended. Regardless of the system used, properly using the right amount at the right time will keep water consumption as low as possible. Rain gardens are designed to conserve water by absorbing and filtering runoff from roofs, parking lots, and driveways and other impervious surfaces and (Fig. 26–31).

Energy conservation can be as simple as using manual equipment when power equipment is not really needed. For example you should consider using a broom instead of a blower, or hand trimmers instead of

Figure 26–30
Rain barrels capture rain water that would ordinarily run off and allow it to be used for irrigating during dry periods.
Source: Sherrill Massey.

a string trimmer when the jobs are not large. Planting for temperature modification as described earlier in the chapter can reduce energy costs. Green roofs and green walls (Fig. 26–32) can create an insulating layer to

A

B

Figure 26–31
Rain gardens help reduce runoff. The runoff from the gutter (A) is directed to the garden (B) by the graveled area seen on the right side of the garden.
Source: Sherrill Massey.

Figure 26–32
Green roofs reduce energy costs, increase biodiversity, reduce the urban heat island effect, and reduce runoff.
Source: Laura Burchfield, The Ohio State University.

Figure 26–33
Wildlife can enhance the enjoyment of the landscape; and environmentally friendly landscapes can increase biodiversity.
Source: Laura Burchfield, The Ohio State University.

reduce energy used for heating and cooling. On warm days they indirectly lower the cooling costs of the building as they cool themselves through transpiration. Using materials from recycled or local sources reduces energy expended for manufacturing or transportation.

Soil conservation in landscaping includes using plants to control water erosion on hillsides and wind erosion on flat areas. Terraces can create very attractive hillside designs that control erosion. Masses of dense, low-growing, spreading shrubs can attractively control hillside erosion. Contour and strip-contour planting as is done in farming (see Chapter 14) to control erosion can be done with ornamental plants to create a functional and beautifully arranged landscape.

As with natural ecosystems, having an appropriate level of biodiversity creates a sustainable landscape ecosystem that has reduced input needs. Using a wide variety of plants in the landscape, instead of monocultures will increase biodiversity. Selecting plants that provide food and shelter for wildlife will also benefit the landscape by increasing biodiversity of the landscape ecosystem (Fig. 26–33). Integrated plant health management practices (see Chapter 15) will reduce the use of pesticides that pollute the air, soil, and water, and which can harm many necessary organisms in the landscape ecosystem.

Figure 26–34
Edible gardens can be functional as well as beautiful.
Source: Laura Burchfield, The Ohio State University.

Consider adding edible plants to the landscape to reduce the energy and material consumption that comes from the shipping and packaging of store-bought produce. The added benefits to edible landscapes are fresh, healthy foods that are produced using sustainable practices (Fig. 26–34).

SUMMARY AND REVIEW

Designed landscapes are shaped by environmental as well as human factors. Function with beauty goes back as far as the earliest human records of gardening, which show that the ancient cultures of the world developed designed gardens. Those designs often reflected religious beliefs and cultural values. Many components of modern landscape design can be traced back to those early designs and the types of designs that came after

them such as the French and English styles. However, in modern landscape designs, turfgrass often dominates the landscape, leaving opportunity to greatly diversify and improve the landscape aesthetically, functionally, and environmentally.

One of the concepts that can be used to improve landscapes is that landscapes should extend and connect with the indoor spaces of adjacent structures. Another concept is that the designed landscape should also blend seamlessly with the natural landscape. The concept of outdoor rooms can be useful in creating landscapes where form (beauty) follows function; where the landscape suits the use for which it is intended. Outdoor rooms can be in public or private settings. Plazas are an example of an outdoor space as a public room; while residential rooms can be kitchens, entryways, and entertaining or recreation areas.

Plants serve many functions in the landscape including aesthetic and architectural functions as well as environmental protection and modification. Aesthetics is what is generally thought of first in plant selection, but other factors are equally important. Architecturally, plants create the framework for the outdoor rooms and serve functions similar to those of nonplant structures such as screens, barriers, backdrops, and focal points. Environmentally, plants improve air quality and help control erosion and water runoff. Plants modify the environment by creating microclimates that are warmer or cooler than the surrounding area. They can reduce noise, glare, and reflection.

The aesthetic quality of a landscape is achieved through the use of the elements and principles of design. The elements are line, form (shape), size, pattern, texture, and color. The principles are order, unity, and rhythm. The elements and principles are used together to design attractive landscapes. Order is the overall framework and is sometimes called balance, which can be symmetrical or asymmetrical. Unity, or harmony, is the arrangement of objects to achieve a single composition. Rhythm makes the landscape dynamic by suggesting movement.

The design process starts with an idea for a landscape project followed by an orderly process that is broken down into five phases: research, design, installation, maintenance, and evaluation. In the research phase, information is gathered about the site, the users, and the purposes of the site. Design is the creation of the master plan and selecting the plants to be used. Installation is the planting of the plants using best practices and procedures for the area. It is best done after all other construction is finished and can be phased in over time if needed. Maintenance is done to keep the landscape's integrity over time. An understanding of plant physiological needs is critical for a successful maintenance program. Evaluation allows further changes and improvements to the landscape as well as future landscapes.

Plant selection is critical to design a successful landscape. Selection includes choosing plants that will perform the desired function and thrive where they are planted. It requires understanding the growth characteristics of the plant, its environmental requirements, and the environment where it will be planted. Plant selection also should keep in mind how the plant will change with time.

Sustainable landscape practices take into account the impact the landscape will have on the environment and reduce that impact. A sustainable landscape requires the input of very few resources and does not degrade resources. Conservation of water, energy, and soil is something that can be accomplished in any landscape. Xeriscaping, water collection systems, using hand rather than power equipment, and increasing biodiversity are some of the ways to create sustainable landscapes.

KNOWLEDGE CHECK

1. Name two environmental and two social factors that influence landscape design.
2. How long have people gardened for pleasure, recreation, and ornamentation?
3. What did the design of ancient Persian gardens represent?
4. How was horticultural and gardening knowledge preserved through the Middle Ages?
5. What type of garden did United States modernists develop in the mid-twentieth century?
6. What are the three spaces typically found in an American residential landscape?
7. What feature predominates in American residential landscapes?
8. What are two differences between indoor and outdoor rooms?
9. Architecturally, what are five functions that plants perform in a landscape?
10. Describe three ways that plants can modify the environment to make it more comfortable for humans.
11. What are the six elements of landscape design?
12. What are the three principles of landscape design?
13. What is the difference between symmetrical and asymmetrical design? Which is formal, which is informal?
14. What overall effect does unity have in a design?
15. What does rhythm do for a design?
16. What is the purpose of doing a site inventory?

17. What is a base map?
18. What is the site analysis?
19. Describe how the design phase begins.
20. What is the master plan in a landscape design?
21. Why is maintenance critical to landscapes?
22. Why is proper plant selection important for maintenance?
23. What is the purpose of periodically evaluating a landscape after installation?
24. Determining the plants best suited for a landscape requires understanding what three things about the plant?
25. In plant selection, why are plants native to the site—or an area with similar conditions to the site—important?

26. Why must the mature size and shape of plants be considered during plant selection?
27. Why do sustainable practices have to be understood and utilized in landscaping?
28. What are two sustainable practices that can be used to conserve water in a landscape?
29. What is xeriscaping?
30. What are two ways plants can be used to reduce energy costs?
31. What are two techniques for controlling soil erosion that can be used in landscapes?
32. How does selecting plants that provide wildlife habitat and food benefit the landscape ecosystem?
33. What are two benefits of adding edible plants to the landscape?

FOOD FOR THOUGHT

1. You are the creator and owner of a very successful landscape design company that also installs and maintains the landscapes that you design. You specialize in creating sustainable theme gardens based on historic styles of gardening throughout the ages for very wealthy and worldly clients. What would you tell a young college student interning for you about the importance of learning plant physiology, plant–environment interactions, and history if she or he wants to be hired as your well-paid employee upon graduation?

2. You are the proud owner of your first home and want it to look as nice outside as inside. However, you do not have a lot of money to spend on landscape plants and maintaining them. Nor do you have time to design it yourself. Assuming the cost of having the design drawn up and the number of plants used would be same, would you choose a landscaper who specializes in using inexpensive plants that require a lot of water and care, or one who specializes in plants that are more expensive but require less water and care? Explain why you made your choice.

3. During the site analysis of a landscape project, you discover that there is an area of very shallow soil over the septic tank—approximately 10 ft by 20 ft—which is in the middle of the site. Your client has been very clear that he wants only plants on the site, no patios or other paved areas. Nor is he willing to do any special watering of the plants, they will only get natural precipitation. What could you do to keep the area aesthetically pleasing using plants?

FURTHER EXPLORATION

ADAMS, D. W. 2004. *Restoring American gardens.* Portland, OR: Timber Press.

BOOTH, N. K., and J. E. HISS. 2007. *Residential landscape architecture.* 5th ed. Upper Saddle River, NJ: Prentice Hall.

BOURDON, D. 1995. *Designing the earth: The human impulse to shape nature.* New York: N. H. Abrams.

CHALKER-SCOTT, L. 2008. *The informed gardener.* Seattle: University of Washington Press.

SMITH, C. N. DUNNETT, and A. CLAYDEN. 2008. *Residential landscape sustainability: A checklist tool.* Hoboken, NJ: Wiley-Blackwell.

USEFUL WEBSITES

United Nations Food and Agriculture Organizations (FAO): Global information on all aspects of agriculture.

> http://www.fao.org/

United States Department of Agriculture (USDA), National Agricultural Statistics Service (NASS): Gives current and achieved information about agriculture in the United States.

> http://www.nass.usda.gov/

USDA Economics and Statistics System: Access to many national and international agricultural information sites.

> http://usda.mannlib.cornell.edu/

Locations of all land grant institutions in the United States.

> http://www.csrees.usda.gov/qlinks/partners/state_partners.html

Ohio State University's plant fact sheet, image, and video databases as well as a search engine for universities and departments that do research and/or teach plant science.

> http://plantfacts.osu.edu

NUTRITIVE VALUE OF THE EDIBLE PARTS OF FOODS

Foods, Approximate Measures, Units, and Weight (Edible Part Unless Indicated Otherwise)	Unit	Weight (gm)	Water (Percentage)	Food Energy (Calories)	Protein (gm)	Fat (gm)
FRUITS AND FRUIT PRODUCTS						
Apples, raw, unpeeled, without cores:						
3¼-in. diam. (about 2 per lb).	1 apple	212	84	125	Trace	1
Apricots:						
Raw, without pits (about 12 per lb with pits).	3 apricots	107	85	55	1	Trace
Dried, Uncooked (28 large or 37 medium halves per cup).	1 cup	130	25	340	7	1
Avocados, raw, whole, without skins and seeds:						
California, mid- and late-winter (with skin and seed, 3⅛-in. diam.; wt., 10 oz).	1 avocado	216	74	370	5	37
Florida, late summer and fall (with skin and seed, 3⅝-in. diam.; wt., 1 lb).	1 avocado	304	78	390	4	33
Banana without peel (about 2.6 per lb with peel).	1 banana	119	76	100	1	Trace
Blackberries, raw.	1 cup	144	85	85	2	1
Blueberries, raw.	1 cup	145	83	90	1	1
Cherries, sweet, raw, without pits and stems.	10 cherries	68	80	45	1	Trace
Cranberry sauce, sweetened, canned, strained.	1 cup	277	62	405	Trace	1
Dates:						
Whole, without pits.	10 dates	80	23	220	2	Trace
Chopped.	1 cup	178	23	490	4	1
Grapefruit:						
Raw, medium, 3¾-in. diam. (about 1 lb 1 oz):						
Pink or red.	½ grapefruit with peel	241	89	50	1	Trace
White.	½ grapefruit with peel	241	89	45	1	Trace
Grapes, European type (adherent skin), raw:						
Thompson Seedless.	10 grapes	50	81	35	Trace	Trace
Lemon, raw, size 165, without peel and seeds (about 4 per lb with peels and seeds).	1 lemon	74	90	20	1	Trace
Olives, pickled, canned:						
Green.	4 medium or 3 extra large or 2 giant	16	78	15	Trace	2
Ripe, Mission.	3 small or 2 large	10	73	15	Trace	2
Oranges, all commercial cultivars, raw:						
Whole, 2⅝-in. diam., without peel and seeds (about 2½ per lb with peel and seeds).	1 orange	131	86	65	1	Trace
Papayas, raw, ½-in. cubes.	1 cup	140	89	55	1	Trace
Peaches:						
Raw:						
Whole, 2½-in. diam., peeled, pitted (about 4 per lb with peels and pits).	1 peach	100	89	40	1	Trace
Dried:						
Uncooked.	1 cup	160	25	420	5	1
Pears:						
Raw, with skin, cored:						
Bartlett, 2½-in. diam. (about 2½ per lb with cores and stems).	1 pear	164	83	100	1	1

Carbohydrate (gm)	Calcium (mg)	Phosphorus (mg)	Iron (mg)	Potassium (mg)	Vitamin A (International Units)	Thiamin (mg)	Riboflavin (mg)	Niacin (mg)	Ascorbic Acid (mg)
31	15	21	0.6	233	190	0.06	0.04	0.2	8
14	18	25	0.5	301	2,890	0.03	0.04	0.6	11
86	87	140	7.2	1,273	14,170	0.01	0.21	4.3	16
13	22	91	1.3	1,303	630	0.24	0.43	3.5	30
27	30	128	1.8	1,836	880	0.33	0.61	4.9	43
26	10	31	0.8	440	230	0.06	0.07	0.8	12
19	46	27	1.3	245	290	0.04	0.06	0.6	30
22	22	19	1.6	117	160	0.04	0.09	0.7	20
12	15	13	0.3	129	70	0.03	0.04	0.3	7
104	17	11	0.6	83	60	0.03	0.03	0.1	6
58	47	50	2.4	518	40	0.07	0.08	1.8	0
130	105	112	5.3	1,153	90	0.16	0.18	3.9	0
13	20	20	0.5	166	540	0.05	0.02	0.2	44
12	19	19	0.5	159	10	0.05	0.02	0.2	44
9	6	10	0.2	87	50	0.03	0.02	0.2	2
6	19	12	0.4	102	10	0.03	0.01	0.1	39
Trace	8	2	0.2	7	40	—	—	—	—
Trace	9	1	0.1	2	10	Trace	Trace	—	—
16	54	26	0.5	263	260	0.13	0.05	0.5	66
14	28	22	0.4	328	2,450	0.06	0.06	0.4	78
10	9	19	0.5	202	1,330	0.02	0.05	1.0	7
109	77	187	9.6	1,520	6,240	0.02	0.30	8.5	29
25	13	18	0.5	213	30	0.03	0.07	0.2	7

(continued)

NUTRITIVE VALUE OF THE EDIBLE PARTS OF FOODS

Foods, Approximate Measures, Units, and Weight (Edible Part Unless Indicated Otherwise)	Unit	Weight (gm)	Water (Percentage)	Food Energy (Calories)	Protein (gm)	Fat (gm)
FRUITS AND FRUIT PRODUCTS						
Bosc, 2½-in. diam. (about 3 per lb with cores and stems).	1 pear	141	83	85	1	1
D'Anjou, 3-in. diam. (about 2 per lb with cores and stems).	1 pear	200	83	120	1	1
Pineapple: raw, diced.	1 cup	155	85	80	1	Trace
Plums:						
Raw, without pits:						
Japanese and hybrid (2⅛-in. diam., about 6½ per lb with pits).	1 plum	66	87	30	Trace	Trace
Prune-type (1½-in. diam., about 15 per lb with pits).	1 plum	28	79	20	Trace	Trace
Prunes, dried, "softenized," with pits: uncooked.	4 extra large or 5 large prunes	49	28	110	1	Trace
Raisins, seedless: cup, not pressed down.	1 cup	145	18	420	4	Trace
Raspberries, red: raw, capped, whole.	1 cup	123	84	70	1	1
Strawberries: raw, whole berries, capped.	1 cup	149	90	55	1	1
Tangerine, raw, 2⅜-in. diam., size 176, without peel.	1 tangerine	86	87	40	1	Trace
LEGUMES (DRY), NUTS, SEEDS						
Almonds, shelled:						
Chopped (about 130 almonds).	1 cup	130	5	775	24	70
Beans, dry:						
Common cultivars as Great Northern, navy, and others:						
Cooked, drained:						
Great Northern.	1 cup	180	69	210	14	1
Navy.	1 cup	190	69	225	15	1
Red kidney.	1 cup	255	76	230	15	1
Lima, cooked, drained.	1 cup	190	64	260	16	1
Blackeye peas, dry, cooked (with residual cooking liquid).	1 cup	250	80	190	13	1
Brazil nuts, shelled (6–8 large kernels).	1 oz	28	5	185	4	19
Cashew nuts, roasted in oil.	1 cup	140	5	785	24	64
Coconut meat, fresh:						
Piece, about 2 by 2 by ½ in.	1 piece	45	51	155	2	16
Shredded or grated, not pressed down.	1 cup	80	51	275	3	28
Filberts (hazelnuts), chopped (about 80 kernels).	1 cup	115	6	730	14	72
Lentils, whole, cooked.	1 cup	200	72	210	16	Trace
Peanuts, roasted in oil, salted (whole, halves, chopped).	1 cup	144	2	840	37	72
Pecans, chopped or pieces (about 120 large halves).	1 cup	118	3	810	11	84
Walnuts:						
Black:						
Chopped or broken kernels.	1 cup	125	3	785	26	74
Persian or English, chopped (about 60 halves).	1 cup	120	4	780	18	77

Carbohydrate (gm)	Calcium (mg)	Phosphorus (mg)	Iron (mg)	Potassium (mg)	Vitamin A (International Units)	Thiamin (mg)	Riboflavin (mg)	Niacin (mg)	Ascorbic Acid (mg)
22	11	16	0.4	83	30	0.03	0.06	0.1	6
31	16	22	0.6	260	40	0.04	0.08	0.2	8
21	26	12	0.8	226	110	0.14	0.05	0.3	26
8	8	12	0.3	112	160	0.02	0.02	0.3	4
6	3	5	0.1	48	80	0.01	0.01	0.1	1
29	22	34	1.7	298	690	0.04	0.07	0.7	1
112	90	146	5.1	1,106	30	0.16	0.12	0.7	1
17	27	27	1.1	207	160	0.04	0.11	1.1	31
13	31	31	1.5	244	90	0.04	0.10	0.9	88
10	34	15	0.3	108	360	0.05	0.02	0.1	27
25	304	655	6.1	1,005	0	0.31	1.20	4.6	Trace
38	90	266	4.9	749	0	0.25	0.13	1.3	0
40	95	281	5.1	790	0	0.27	0.13	1.3	0
42	74	278	4.6	673	10	0.13	0.10	1.5	—
49	55	293	5.9	1,163	—	0.25	0.11	1.3	—
35	43	238	3.3	573	30	0.40	0.10	1.0	—
3	53	196	1.0	203	Trace	0.27	0.03	0.5	—
41	53	522	5.3	650	140	0.60	0.35	2.5	—
4	6	43	0.8	115	0	0.02	0.01	0.2	1
8	10	76	1.4	205	0	0.04	0.02	0.4	2
19	240	388	3.9	810	—	0.53	—	1.0	Trace
39	50	238	4.2	498	40	0.14	0.12	1.2	0
27	107	577	3.0	971	—	0.46	0.19	24.8	0
17	86	341	2.8	712	150	1.01	0.15	1.1	2
19	Trace	713	7.5	575	380	0.28	0.14	0.9	—
19	119	456	3.7	540	40	0.40	0.16	1.1	2

(continued)

NUTRITIVE VALUE OF THE EDIBLE PARTS OF FOODS

Foods, Approximate Measures, Units, and Weight (Edible Part Unless Indicated Otherwise)	Unit	Weight (gm)	Water (Percentage)	Food Energy (Calories)	Protein (gm)	Fat (gm)
VEGETABLES						
Asparagus, green:						
Cooked, drained:						
Spears, 1⅛-in. diam. at base:						
From raw.	4 spears	60	94	10	1	Trace
From frozen.	4 spears	60	92	15	2	Trace
Beans:						
Lima, immature seeds, frozen, cooked, drained:						
Thick-seeded types (Fordhooks).	1 cup	170	74	170	10	Trace
Thin-seeded types (baby limas).	1 cup	180	69	210	13	Trace
Snap:						
Green:						
Cooked, drained:						
From raw (cuts and French style).	1 cup	125	92	30	2	Trace
From frozen:						
Cuts.	1 cup	135	92	35	2	Trace
French style.	1 cup	130	92	35	2	Trace
Yellow or wax:						
Cooked, drained:						
From raw (cuts and French style).	1 cup	125	93	30	2	Trace
From frozen (cuts).	1 cup	135	92	35	2	Trace
Canned, drained solids (cuts).	1 cup	135	92	30	2	Trace
Beet greens, leaves and stems, cooked, drained.	1 cup	145	94	25	2	Trace
Beets:						
Cooked, drained, peeled:						
Whole beets, 2-in. diam.	2 beets	100	91	30	1	Trace
Canned, drained solids:						
Whole beets, small.	1 cup	160	89	60	2	Trace
Broccoli, cooked, drained:						
From raw:						
Stalk, medium size.	1 stalk	180	91	45	6	1
From frozen:						
Stalk, 4½ to 5 in. long.	1 stalk	30	91	10	1	Trace
Brussels sprouts, cooked, drained:						
From raw, 7–8 sprouts (1¼- to 1½-in. diam.).	1 cup	155	88	55	7	1
From frozen.	1 cup	155	89	50	5	Trace
Cabbage:						
Common varieties:						
Raw:						
Coarsely shredded or sliced.	1 cup	70	92	15	1	Trace
Finely shredded or chopped.	1 cup	90	92	20	1	Trace
Cooked, drained:	1 cup	146	94	30	2	Trace
Red, raw, coarsely shredded or sliced.	1 cup	70	90	20	1	Trace
Savoy, raw, coarsely shredded or sliced.	1 cup	70	92	15	2	Trace
Cabbage, celery (also called pe-tsai or wongbok), raw, 1-in. pieces.	1 cup	75	95	10	1	Trace
Cabbage, white mustard (also called bokchoy or pakchoy), cooked, drained.	1 cup	170	95	25	2	Trace

Carbohydrate (gm)	Calcium (mg)	Phosphorus (mg)	Iron (mg)	Potassium (mg)	Vitamin A (International Units)	Thiamin (mg)	Riboflavin (mg)	Niacin (mg)	Ascorbic Acid (mg)
2	13	30	0.4	110	540	0.10	0.11	0.8	16
2	13	40	0.7	143	470	0.10	0.08	0.7	16
32	34	153	2.9	724	390	0.12	0.09	1.7	29
40	63	227	4.7	709	400	0.16	0.09	2.2	22
7	63	46	0.8	189	680	0.09	0.11	0.6	15
8	54	43	0.9	205	780	0.09	0.12	0.5	7
8	49	39	1.2	177	690	0.08	0.10	0.4	9
6	63	46	0.8	189	290	0.09	0.11	0.6	16
8	47	42	0.9	221	140	0.09	0.11	0.5	8
7	61	34	2.0	128	140	0.04	0.07	0.4	7
5	144	36	2.8	481	7,400	0.10	0.22	0.4	22
7	14	23	0.5	208	20	0.03	0.04	0.3	6
14	30	29	1.1	267	30	0.02	0.05	0.2	5
8	158	112	1.4	481	4,500	0.16	0.36	1.4	162
1	12	17	0.2	66	570	0.02	0.03	0.2	22
10	50	112	1.7	423	810	0.12	0.22	1.2	135
10	33	95	1.2	457	880	0.12	0.16	0.9	126
4	34	20	0.3	163	90	0.04	0.04	0.02	33
5	44	26	0.4	210	120	0.05	0.05	0.3	42
6	64	29	0.4	236	190	0.06	0.06	0.4	48
5	29	25	0.6	188	30	0.06	0.04	0.3	43
3	47	38	0.6	188	140	0.04	0.06	0.2	39
2	32	30	0.5	190	110	0.04	0.03	0.5	19
4	252	56	1.0	364	5,270	0.07	0.14	1.2	26

(continued)

NUTRITIVE VALUE OF THE EDIBLE PARTS OF FOODS

Foods, Approximate Measures, Units, and Weight (Edible Part Unless Indicated Otherwise)	Unit	Weight (gm)	Water (Percentage)	Food Energy (Calories)	Protein (gm)	Fat (gm)
VEGETABLES						
Carrots:						
Raw, without crowns and tips, scraped:						
Whole, 7½ by 1⅛ in., or strips, 2½ to 3 in. long.	1 carrot or 18 strips	72	88	30	1	Trace
Grated.	1 cup	110	88	45	1	Trace
Canned:						
Sliced, drained solids.	1 cup	155	91	45	1	Trace
Cauliflower:						
Raw, chopped.	1 cup	115	91	31	3	Trace
Cooked, drained.	1 cup	125	93	30	3	Trace
Celery, Pascal type, raw:						
Stalk, large outer, 8 by 1½ in., at root end.	1 stalk	40	94	5	Trace	Trace
Collards, cooked, drained:						
From raw (leaves without stems).	1 cup	190	90	65	7	1
From frozen (chopped).	1 cup	170	90	50	5	1
Corn, sweet:						
Cooked, drained:						
From raw, ear 5 by 1¾ in.	1 ear	140	74	70	2	1
From frozen:						
Ear, 5 in. long.	1 ear	229	73	120	4	1
Kernels.	1 cup	165	77	130	5	1
Cucumber slices, ⅛ in. thick (large, 2⅛-in. diam.; small, 1¾-in. diam.):						
With peel.	6 large or 8 small slices	28	95	5	Trace	Trace
Endive, curly (including escarole), raw, small pieces.	1 cup	50	93	10	1	Trace
Kale, cooked, drained:						
From raw (leaves without stems and midribs).	1 cup	110	88	45	5	1
From frozen (leaf style).	1 cup	130	91	40	4	1
Lettuce, raw:						
Butterhead, as Boston types:						
Head, 5-in. diam.	1 head	220	95	25	2	Trace
Leaves.	1 outer or 2 inner or 3 heart leaves	15	95	Trace	Trace	Trace
Crisphead, for example, Iceberg:						
Head, 6-in. diam.	1 head	567	96	70	5	1
Pieces, chopped or shredded.	1 cup	55	96	5	Trace	Trace
Looseleaf (bunching varieties including romaine or cos), chopped or shredded pieces.	1 cup	55	94	10	1	Trace
Muskmelons, raw, orange-fleshed (with rind and seed cavity, 5-in. diam., 2⅓ lb).	½ melon with rind	477	91	80	2	Trace
Honeydew (with rind and seed cavity, 6½-in. diam., 5¼ lb).	½ melon with rind	226	91	50	1	Trace
Mustard greens, without stems and midribs, cooked, drained.	1 cup	140	93	30	3	1
Okra pods, 3 by ⅝ in., cooked.	10 pods	106	91	30	2	Trace

Carbohydrate (gm)	Calcium (mg)	Phosphorus (mg)	Iron (mg)	Potassium (mg)	Vitamin A (International Units)	Thiamin (MG)	Riboflavin (mg)	Niacin (mg)	Ascorbic Acid (mg)
7	27	26	0.5	246	7,930	0.04	0.04	0.4	6
11	41	40	0.8	375	12,100	0.07	0.06	0.7	9
10	47	34	1.1	186	23,250	0.03	0.05	0.6	3
6	29	64	1.3	339	70	0.13	0.12	0.8	90
5	26	53	0.9	258	80	0.11	0.10	0.8	69
2	16	11	0.1	136	110	0.01	0.01	0.1	4
10	357	99	1.5	498	14,820	0.21	0.38	2.3	144
10	299	87	1.7	401	11,560	0.10	0.24	1.0	56
16	2	69	0.5	151	310	0.09	0.08	1.1	7
27	4	121	1.0	291	440	0.18	0.10	2.1	9
31	5	120	1.3	304	580	0.15	0.10	2.5	8
1	7	8	0.3	45	70	0.01	0.01	0.1	3
2	41	27	0.9	147	1,650	0.04	0.07	0.3	5
7	206	64	1.8	243	9,130	0.11	0.20	1.8	102
7	157	62	1.3	251	10,660	0.08	0.20	0.9	49
4	57	42	3.3	430	1,580	0.10	0.10	0.5	13
Trace	5	4	0.3	40	150	0.01	0.01	Trace	1
16	108	118	2.7	943	1,780	0.32	0.32	1.6	32
2	11	12	0.3	96	180	0.03	0.03	0.2	3
2	37	14	0.8	145	1,050	0.03	0.04	0.2	10
20	38	44	1.1	682	9,240	0.11	0.08	1.6	90
11	21	24	0.6	374	60	0.06	0.04	0.9	34
6	193	45	2.5	308	8,120	0.11	0.20	0.8	67
6	98	43	0.5	184	520	0.14	0.19	1.0	21

METRIC CONVERSION TABLE

Into Metric			Out of Metric		
If You Know	Multiply by	To Get	If You Know	Multiply by	To Get

Length

If You Know	Multiply by	To Get	If You Know	Multiply by	To Get
inches	2.54	centimeters	millimeters	0.04	inches
feet	30	centimeters	centimeters	0.4	inches
feet	0.303	meters	meters	3.3	feet
yards	0.91	meters	kilometers	0.62	miles
miles	1.6	kilometers			

Area

If You Know	Multiply by	To Get	If You Know	Multiply by	To Get
sq inches	6.5	sq centimeters	sq centimeters	0.16	sq inches
sq feet	0.09	sq meters	sq meters	1.2	sq yards
sq yards	0.8	sq meters	sq kilometers	0.4	sq miles
sq miles	2.6	sq kilometers	hectares	2.47	acres
acres	0.4	hectares			

Mass (Weight)

If You Know	Multiply by	To Get	If You Know	Multiply by	To Get
ounces	28	grams	grams	0.035	ounces
pounds	0.45	kilograms	kilograms	2.2	pounds
short ton	0.9	metric ton	metric tons	1.1	short tons

Volume

If You Know	Multiply by	To Get	If You Know	Multiply by	To Get
teaspoons	5	milliliters	milliliters	0.03	fluid ounces
tablespoons	15	milliliters	liters	2.1	pints
fluid ounces	30	milliliters	liters	1.06	quarts
cups	0.24	liters	liters	0.26	gallons
pints	0.47	liters	cubic meters	35	cubic feet
quarts	0.95	liters	cubic meters	1.3	cubic yards
gallons	3.8	liters			
cubic feet	0.03	cubic meters			
cubic yards	0.76	cubic meters			

Pressure

If You Know	Multiply by	To Get	If You Know	Multiply by	To Get
$lb/in.^2$	0.069	bars	bars	14.5	$lb/in.^2$
atmospheres	1.013	bars	bars	0.987	atmospheres
atmospheres	1.033	kg/cm^2	kg/cm^2	0.968	atmospheres
$lb/in.^2$	0.07	kg/cm^2	kg/cm^2	14.22	$lb/in.^2$

Rates

If You Know	Multiply by	To Get	If You Know	Multiply by	To Get
lb/acre	1.12	kg/hectare	kg/hectare	0.882	lb/acre
tons/acre	2.24	metric tons/hectare	metric tons/hectare	0.445	tons/acre

TEMPERATURE CONVERSION TABLE*

F	C	F	C	F	C
−26	−32	19	−7	64	18
−24	−31	21	−6	66	19
−22	−30	23	−5	68	20
−20	−29	25	−4	70	21
−18	−28	27	−3	72	22
−17	−27	28	−2	73	23
−15	−26	30	−1	75	24
−13	−25	32	0	77	25
−11	−24	34	1	79	26
−9	−23	36	2	81	27
−8	−22	37	3	82	28
−6	−21	39	4	84	29
−4	−20	41	5	86	30
−2	−19	43	6	88	31
0	−18	45	7	90	32
1	−17	46	8	91	33
3	−16	48	9	93	34
5	−15	50	10	95	35
7	−14	52	11	97	36
9	−13	54	12	99	37
10	−12	55	13	100	38
12	−11	57	14	102	39
14	−10	59	15	104	40
16	−9	61	16	106	41
18	−8	63	17	108	42
				212	100

*Celsius temperatures have been rounded to the nearest whole number.
Fahrenheit to Celsius: Subtract 32 from the Fahrenheit figure, multiply by 5, and divide by 9.
Celsius to Fahrenheit: Multiply the Celsius figure by 9, divide by 5, and add 32.

LIGHT CONVERSION TABLE

Conversion of lux or footcandles (ft-c) from various light sources to watts per sq meter of photosynthetically active radiation (PAR) or to micromols of photosynthetic photon flux (PPF) per sq meter per second

| | To Convert To: | | | |
| | Wm^{-2} PAR Divide: | | $\mu mol\ m^{-2}s^{-1}$ PPF Divide: | |
Light Source	lux	ft-c	lux	ft-c
Daylight[a]	by 247	by 22.9	by 54	by 5.0
Metal halide lamps	326	30.2	71	6.6
Fluorescent, cool-white lamps	340	31.5	74	7.0
Incandescent lamps	250	23.1	50	4.6
Low-pressure sodium lamps	522	48.3	106	9.8
High-pressure mercury lamps[b]	380	35.2	84	7.8

[a]Based on average spectral distribution.
[b]Different types of mercury lamps can vary ±5 percent.

To convert measurements of light intensity in footcandles (ft-c), or illuminance in lux from various light sources into equivalent watts of photosynthetic radiation per square meter (Wm^{-2} PAR), or micromols of photosynthetic photon flux per square meter per second ($\mu mol\ m^{-2}s^{-1}$), divide the reported measurement by the appropriate constant given for the particular light source.

Example: A light intensity of 2,500 ft-c of daylight recorded in the greenhouse may be converted into equivalent Wm^{-2} PAR or $\mu mol\ m^{-2}s^{-1}$ PPF as follows:

$$\frac{2,500}{22.9} = 109\ Wm^{-2}\ PAR \qquad or \qquad \frac{2,500}{5.00} = 500\ \mu mol\ m^{-2}\ s^{-1}\ PPF$$

Terms Used to Describe Visible and Photosynthetic Light Energy

Lux and footcandles (ft-c) are photometric measurements based on the sensitivity of the eye to various wavelengths of electromagnetic radiation. The *lux* is the illuminance resulting from a light energy (luminous) flux of 1 lumen distributed evenly over one square meter (1 lux = 1 lumen m^{-2}). The *footcandle* is the illuminance resulting from a luminous flux of 1 lumen per square foot (1 ft-c = 1 lumen ft^{-2}).

Watts and joules are *radiometric* measurements that express energy per unit of time per unit area without respect to wavelength. Such measurements made in the PAR waveband are termed W PAR.

In electromagnetic terms, the mol is Avogadro's number of photons. Micro-mol per square meter per second ($\mu mol\ m^{-2}\ s^{-1}$) is a flux of 6.022×10^{17} photons intercepted by one square meter over one second. Each of the light sources above has its own specific distribution of wavelengths compared with daylight; each, therefore, requires its own conversion factor. Sometimes Einstein (E) is used in place of mol. The terms are equivalent. However, mol is preferred because E is not a *système international* (SI) unit.

abaxial   Away from the axis or central line; turned toward the base, dorsal.

abscisic acid   A plant hormone involved in abscission, dormancy, stomatal closure, growth inhibition, and other plant responses.

abscission zone   A layer of thin-walled cells extending across the base of a petiole or peduncle, whose breakdown separates the leaf or fruit from the stem and thus causes the leaf or fruit to drop.

acclimatization   The adaptation of an individual plant to a changed climate, or the adjustment of a species or a population to a changed environment, often over several generations.

achene   A simple, dry, one-seeded indehiscent fruit with the seed attached to the ovary wall at one point only. The so-called strawberry and sunflower "seeds" are examples.

acid soil   Soil with a reaction below pH 7; more technically, a soil having a preponderance of hydrogen ions over hydroxyl ions in solution.

action spectrum   A graph indicating the changes in response of an entity to changes in a stimulus.

active transport   Movement of water and solutes into and through a plant by a force that allows a flow through a membrane or against a concentration gradient.

adaptation   The process of change in structure or function of an individual or population caused by environmental changes.

adaxial   Toward the axis or center; turned toward the apex; ventral.

adenosine diphosphate (ADP)   A nucleotide (a nitrogen-based compound) composed of adenine and ribose with two phosphate groups attached. ATP and ADP participate in metabolic reactions (both catabolic and anabolic). These molecules, through the process of being phosphorylated (accepting phosphate groups) or dephosphorylated (losing phosphate groups), transfer energy within the cells to drive metabolic processes.

adenosine triphosphate (ATP)   ATP has three phosphoric groups attached and is the phosphorylated condition of ADP. It conveys energy needed for metabolic reactions, then loses one phosphate group to become ADP (adenosine diphosphate).

adhesion   The molecular attraction between unlike substances such as water and sand particles.

adsorption   The attraction of ions or molecules to the surface of a solid.

adult phase   A growth phase in plants characterized by the ability to flower and reproduce.

adventitious   Refers to structures arising from an unusual place; for example, buds at places other than shoot terminals or leaf axils, or roots growing from stems or leaves.

aeration   The process by which air in the soil is replaced by air from the atmosphere.

aerobic   An environment or condition in which oxygen is not deficient for chemical, physical, or metabolic processes.

agar   A gelatinous substance obtained from certain species of red algae; widely used as a substrate in aseptic cultures.

aggregate   Many primary soil particles held in a single unit as a clod, crumb, block, or prism.

aggregate fruit   Many fruits are attached to a single ovary, for example, blackberry.

agronomy   The art and science of crop production and soil management.

A horizon   The surface layer of varying thickness of a mineral soil having maximum organic matter accumulation, maximum biological activity, and/or eluviation by water of materials such as iron and aluminum oxides and silicate clays.

air-dry   The state of dryness at equilibrium with the moisture content in the surrounding atmosphere.

air layer   An undetached aerial portion of a plant on which roots are caused to develop, commonly as the result of wounding or other stimulation.

air porosity   The proportion of the bulk volume of soil that is filled with air at any given time or under a given condition, such as a specified moisture tension.

albuminous seeds   Seeds that store most of their food in the endosperm, for example, cereal grains.

aleurone   The outer layer of cells surrounding the endosperm of a cereal grain (caryopsis).

alfisol   A fertile tropical soil with a basic argillic horizon.

alkaline soil   Any soil that has a pH greater than 7.

alkalinity   The amount of carbonates in the soil that buffer change in pH.

alkali soil   *See* sodic soil.

alkaloid   Secondary plant compounds that are part of a plant's defense system. They are found in many plant-based medicines, stimulants, and narcotics.

allele   One of a pair or a series of factors that occur at the same locus on homologous chromosomes; one alternative form of a gene.

allelopathy   The excretion of chemicals by plants of some species that are toxic to plants of other species.

alluvial soil   A recently developed soil from deposited soil material that exhibits essentially no horizon development or modifications. Also, soil comprised of sediment that is deposited by flowing water.

alternate   In taxonomy, a leaf arrangement with leaves placed singly at various heights on the stem; i.e., not opposite or whorled.

ambient temperature   Air temperature at a given time and place; not radiant temperature.

amendment   Any substance such as lime, sulfur, gypsum, or an organic material like peat moss, sawdust, or bark used to alter the properties of a soil, generally to improve its physical properties.

amino acids   The fundamental building blocks of proteins. There are twenty common amino acids in living organisms, each having the basic formula NH_2–CHR–COOH.

ammonification   The biochemical process whereby ammoniacal nitrogen is released from nitrogen-containing organic compounds.

ammonium fixation   The incorporation of ammonium ions by soil fractions so that they are relatively insoluble in water and nonexchangeable by the principle of cation exchange.

anaerobic   An environment or condition in which molecular oxygen is deficient for chemical, physical, or biological processes.

anchor root   *See* brace root.

andisol   A highly permeable, glassy soil formed from volcanic ash.

androecium   A group or whorl of stamens

angiosperm   One of a large group of seed-bearing plants in which the female gamete is protected within an enclosed ovary. A flowering plant.

angle of incidence   The angle at which an object intersects a surface, relative to that surface.

anion   A negatively charged particle that, during electrolysis, is attracted to positively charged surfaces.

annual ring   The cylinder of secondary xylem added to a woody plant stem by the cambium in any one year.

annuals   Plants living one year or less. During this time, the plant grows, flowers, produces seeds, and dies.

anther   In a flower, the saclike structure of the stamen in which microspores (pollen grains) are produced; usually borne on a filament.

anthesis   A developmental stage in flowering at which anthers rupture and pollen is shed. A state of full bloom.

anthocyanin   A class of water-soluble pigments that account for many of the red to blue flower, leaf, and fruit colors. Anthocyanins occur in the vacuole of the cell.

antipodal nuclei   The three or more nuclei at the end of the embryo sac opposite the egg nucleus (female gamete). They are produced by mitotic divisions of the megaspore and degenerate following sexual fertilization.

apical dominance   The inhibition of lateral buds on a shoot due to auxins produced by the apical bud.

apical floral bud   A bud forming at the shoot apex that has undergone a transition to produce floral rather than vegetative tissues.

apical meristem   A mass of undifferentiated cells capable of division at the tip of a root or shoot. These cells multiply by division, allowing the plant to grow in depth or height.

apomixis   The asexual (vegetative) production of seedlings in the usual sexual structures of the flower but without the mingling and segregation of chromosomes. Seedling characteristics are the same as those of the maternal parent.

apoplastic   The relatively unimpeded flow of water and ions into the plant through cell walls and intercellular spaces.

aquifer   Natural underground reservoirs of water.

arable land   Land suitable for the production of crops.

argillic   The horizon in soils characterized by an accumulation of clay particles through leaching.

aridisol   Soil formed in areas with low rainfall and little leaching. It can be fertile but may accumulate salts near the surface.

asexual reproduction   The production of a new plant by any vegetative means not involving meiosis and the union of gametes.

assimilation   The transformation of organic and inorganic materials into protoplasm.

authority   The abbreviation of the last name of the person who named the organism. The authority follows the genus and species in scientific nomenclature; for example, the scientific name of marigold is *Tagetes erecta* L., and the L. stands for Linneaus.

autogamy   Pollination within the same flower.

autotroph   Organism that can synthesize its own food from inorganic compounds via a process such as photosynthesis. *See also* heterotroph.

auxins   A class of plant hormones that influences plant growth in several ways, including apical dominance, cell growth, and phototropic response.

available nutrient   That portion of an element or compound in the soil that can be readily absorbed, assimilated, and utilized by growing plants. (*Available* should not be confused with *exchangeable*.)

available water (AW)   The portion of water in a soil that can be readily absorbed by plant roots. The soil moisture held in the soil between field capacity and permanent wilting percentage (available water = FC − PWP).

axil   The angle on the upper side of the union of a branch and main stem or of a leaf and a stem.

axillary bud   A bud formed in the axil of a leaf.

backcross   In breeding, a cross of a hybrid with one of its parents or with a genetically equivalent organism. In genetics, a cross of a hybrid with a homozygous recessive.

bagasse   The dry refuse of sugarcane after the juice has been expressed.

ball and burlap   The digging of a tree and its root ball and wrapping the rootball with damp burlap and wire for shipping and transplanting.

bar   A unit of pressure equal to 1 million dynes/cm^2. Equivalent approximately to 1 atm of pressure.

bare root   Digging dormant trees and removing soil from the roots for shipping and transplanting. The roots must be protected from desiccation until planted.

bark   The tissue of a woody stem or root from the cambium outward.

base pair   The nitrogen bases that pair in the DNA molecule—adenine with thymine, and guanine with cytosine.

berry   A simple fleshy fruit formed from a single ovary; the ovary wall is fleshy and includes one or more carpels and seeds. For example, fruits of the tomato and grape are botanically berries.

B horizon   A soil layer of varying thickness (usually beneath the A horizon) that is characterized by an accumulation of silicate clays, iron and aluminum oxides, and humus, alone or in combination and/or a blocky or prismatic structure. Also known as the *zone of accumulation*.

biennial   A plant that completes its life cycle within two seasons. For most biennial plants, the two seasons are separated by an obligate degree of cold temperature sufficient to initiate flowering and fruit formation, after which the plant dies.

bilayer   The structure of biological membranes formed from two layers of phospholipids.

binomial   In biology, each species is generally indicated by two names; first, the genus to which it belongs, and second, the species name (e.g., *Quercus suber*, cork oak).

biocide A combination of a bactericide and a fungicide used for cut-flower-keeping solutions, or used as a sterilant in horticultural operations.

biodegradable Materials readily decomposed by microorganisms such as bacteria and fungi.

biofuel Fuel that is produced from recently living organic matter.

biological control The use of natural predators or pathogens to control plant pests.

biomass The solid part of living organisms. Sometimes called dry matter.

biodiversity The term used to describe the genetic diversity within and among all species.

biome Large terrestrial ecosystem characterized by specific communities, usually named after the predominant vegetation found there, e.g., temperate forest.

biotechnology Coined by Karl Erecky in 1919, it refers to any product produced from raw materials with the aid of living organisms. For example, monoclonal antibodies (for immunotherapy), enzymes (lactase and rennin), and human insulin.

biotype A subset of a species that has characteristics distinguishing it from others in the same species.

biuret A compound ($H_2NCONHCONH_2 \cdot H_2O$) formed by the thermal decomposition of urea (H_2NCONH_2) that is phytotoxic to many crops. Therefore, biuret-free urea is used for fertilization of crops.

blade The thin and often flat part of a leaf.

bloat Excessive accumulation of gases in the rumen of some animals.

bolting Rapid production of flower stalks in some herbaceous plants after sufficient chilling or a favorable photoperiod.

botany The science of plants, their characteristics, functions, life cycles, and habits.

bourse shoot Shoot arising from the mixed bud on a fruiting spur of apple. Bourse shoots terminate growth by forming floral or vegetative buds.

brace root Also known as *anchor root*. A type of adventitious root that grows from above-ground parts of the stem and serves to support some plants; for example, corn.

bract A modified leaf, from the axil of which arises a flower or an inflorescence.

branched crown A branch arising axially from the strawberry stem. Branched crowns also develop terminal floral buds.

brassinolides Group of steroids that have been recently identified as a class of plant hormones that appears to regulate cell division and elongation.

breeder seed Seed (or vegetative propagating material; e.g., potato) increased by the originating, or sponsoring, plant breeder or institution and used as the source to increase foundation seed.

broadcast seeding Scattering seed or fertilizers uniformly over the soil surface rather than placing in rows.

bryophytes A division of plants that lack a vascular system. Mosses and liverworts are included in this division.

bud A region of meristematic tissue with the potential for development into leaves, shoots, flowers, or combinations; generally protected by modified scale leaves.

budding A form of grafting in which a single vegetative bud is taken from one plant and inserted into stem tissue of another plant so that the two will grow together. The inserted bud develops into a new shoot.

bud scar A scar left on a shoot when the bud or bud scales drop.

bud sport A mutation arising in a bud and producing a genetically different shoot. Includes change due to gene mutation, somatic reduction, chromosome deletion, or polyploidy.

buffering The ability of some soil components to resist a change in soil pH.

bulb A highly compressed underground stem (basal plate) to which numerous storage scales (modified leaves) are attached. Examples are lily, onion, tulip.

bulk density The mass of a known volume (including air space) of soil. The soil volume is determined in place, then dried in an oven to constant weight at 105°C. Bulk density (D_B) = oven dry weight of soil/volume of soil.

bunch type grass Grass that does not spread by rhizomes or stolons.

burn-down herbicide An herbicide that destroys existing weeds in a field before the field is prepared for planting.

by-pass growth Vegetative growth that occurs if the switch to reproductive growth is interrupted. For example, in poinsettias if long days (vegetative photoperiod) follow several short days (reproductive photoperiod).

C₃ cycle The Calvin-Benson cycle of photosynthesis, in which the first products after CO_2 fixation are three-carbon molecules.

C₄ cycle The Hatch-Slack cycle of photosynthesis, in which the first products after CO_2 fixation are four-carbon molecules.

calcareous soil Soil containing sufficient calcium and/or magnesium carbonate to effervesce visibly when treated with cold 0.1 normal (0.1N) hydrochloric acid.

calcium pectate An organic calcium compound found in the middle lamella between plant cells and serving as an intercellular cement.

callus Mass of large, thin-walled parenchyma cells, usually developing as the result of a wound.

calorie Also known as *gram calorie*. Unit for measuring energy, defined as the heat necessary to raise the temperature of 1 g of water from 14.5°C to 15.5°C at standard pressure; 1 kilocalorie (kcal) raises the temperature of 1 kg of water 1°C. Thus, 1 kcal = 1,000 cal.

Calvin cycle The non-light requiring reaction in photosynthesis where carbon fixation occurs. The energy from the light reaction provides the energy for the Calvin cycle.

calyx The collective term for the sepals.

CAM metabolism A water-conserving process found in many arid, high temperature tolerant plants such as cacti and other succulents. CO_2 is taken in during the night when it is cool and stomates can be open. It is stored in the plants as a 4 carbon acid, then released during the day when the sun is out so plants can photosynthesize.

cambium A thin layer of longitudinally dividing cells between the xylem and phloem that gives rise to secondary growth.

capillary water The water held in the capillary or small pores of a soil, usually with a tension greater than 60 cm of water.

capsule   A simple, dry, dehiscent fruit, with two or more carpels.

carbohydrate   Compound of carbon, hydrogen, and oxygen in the ratio of one atom each of carbon and oxygen to two of hydrogen, as in sugar, starch, and cellulose.

carbon cycle   The recycling of carbon through the processes of photosynthesis and respiration.

carbon dioxide compensation point   The level of carbon dioxide available to the leaf for photosynthesis equals the amount of carbon dioxide generated by the plant's respiration. Photosynthesis cannot increase unless the amount of available carbon dioxide is increased by further respiration.

carbon dioxide fixation   The addition of H^+ to CO_2 to yield a chemically stable carbohydrate. The H^+ is contributed by NADPH, the reduced (hydrogen-rich) form of $NADP^+$, produced in the noncyclic phase of the light reactions of photosyn thesis. The H^+ comes originally from the photolysis of water.

carbon-nitrogen ratio   The ratio of the weight of organic carbon to the weight of total nitrogen in a soil or in organic material.

carotene   Yellow plant pigments, precursors of vitamin A. Alpha, beta, and gamma carotenes are converted into vitamin A in the animal body.

carpel   Female reproductive organ of flowering plants. In some plants, one or more carpels unite to form the pistil.

caryopsis   Small, one-seeded, dry fruit with a thin pericarp surrounding and adhering to the seed; the seed (grain) or fruit of grasses.

Casparian strip   A secondary thickening that develops on the radial and end walls of some endodermal cells.

catalyst   Any substance that accelerates a chemical reaction but does not enter into the reaction itself.

cation   A positively charged ion.

cation exchange capacity (CEC)   Also known as *base exchange capacity*. A measure of the total amount of exchangeable cations that a soil can hold; expressed in meq/100 g soil at pH 7.

catkin   A type of inflorescence (a spike) generally bearing either pistillate or staminate flowers. Found on walnuts and willows, for example.

cell   The basic structure and physiological unit of plants and animals.

cell membrane   The membrane that separates the cell wall and the cytoplasm and regulates the flow of material into and out of the cell.

cell plate   The precursor of the cell wall, formed as cytokinesis starts during cell division. It develops in the region of the equatorial plate and arises from membranes in the cytoplasm.

cellulose   A complex carbohydrate composed of long, un-branched beta-glucose molecules, which makes up 40 to 55 percent by weight of the plant cell wall.

cell wall   The outermost, cellulose limit of the plant cell; the barrier that develops between nuclei during mitosis.

center of origin   A geographical area in which a species is thought to have evolved through natural selection from its ancestors.

cereal   A member of the POACEAE family grown primarily for its mature, dry seed.

cereal forage   Cereal crop harvested when immature for hay, silage, green chop, or pasturage.

certified seed   The progeny of foundation, registered, or certified seed, produced and handled to maintain satisfactory genetic identity and purity, and approved and certified by an official certifying agency.

character   The expression of a gene in the phenotype.

chelate   Also known as *sequestering agent*. A large organic molecule that attracts and tightly holds specific cations, like a chemical claw, preventing them from taking part in inorganic reactions but at the same time allowing them to be absorbed and used by plants.

chilling injury   Direct and/or indirect injury to plants or plant parts from exposure to low, but above-freezing temperatures (as high as 9°C or 48°F).

chilling requirement   The duration of exposure to cold needed for a plant to overcome dormancy.

chimera   A plant composed of two or more genetically different tissues. Includes *periclinal chimera*, in which one tissue lies over another as a glove fits a hand; *mericlinal chimera*, where the outer tissue does not completely cover the inner tissue; and *sectorial chimera*, in which the tissues lie side by side.

chisel   A tillage implement with one or more cultivator-type shanks to which are attached knifelike units that shatter or break up hard, compact layers, usually in the subsoil.

chlorophyll   A complex organic molecule that traps light energy for conversion through photosynthesis into chemical energy.

chloroplast   Chlorophyll-containing cytoplasmic body, in which important reactions of sugar or starch synthesis take place during photosynthesis.

chlorosis   A condition in which a plant or a part of a plant is light green or greenish yellow because of poor chlorophyll development or the destruction of chlorophyll resulting from a pathogen or a mineral deficiency.

C horizon   A soil layer beneath the B layer that is relatively little affected by biological activity and pedogenesis and is lacking properties diagnostic of an A or B horizon.

chromosome   A specific, highly organized body in the nucleus of the cell that contains DNA.

circadian rhythm   Rhythmic processes found in many organisms that follow a daily (diurnal) pattern.

class   A group of soils having a definite range in a particular property such as acidity, degree of slope, texture, structure, land-use capability, degree of erosion, or drainage. Also, the taxonomic category that falls between phylum (division) and order.

classification   The systematic arrangement into categories on the basis of characteristics. Broad groupings are made on the basis of general characteristics and subdivisions on the basis of more detailed differences in specific properties.

clay   Soil particles less than 0.002 mm in equivalent diameter. Also, soil material containing more than 40 percent clay, less than 45 percent sand, and less than 40 percent silt.

claypan   A compact, slowly permeable layer of varying thickness and depth in the subsoil having a much higher clay content than the overlying material. Claypans are usually hard when dry, and plastic and sticky when wet.

climacteric   The period in the development of some plant parts involving a series of biochemical changes associated with the natural respiratory rise and autocatalytic ethylene production.

climate   Conditions such as wind, precipitation, and temperature that prevail in a given region.

climax community   The stable community of plants and animals that results from a complex community replacing simpler ones. *See also* succession.

climax vegetation   Fully developed plant community in equilibrium with its environment.

clod   A compact, coherent mass of soil produced artificially, usually by tillage operations, especially when performed on soils either too wet or too dry.

clone   The aggregate of individual organisms originating from one sexually produced individual (or from a mutation) and maintained exclusively by asexual propagation.

clove   One of a group of small bulbs produced, for example, by garlic and shallot plants.

cohesion   Holding together; a force holding a solid or liquid together, owing to attraction between like molecules.

colchicine   An alkaloid, derived from the autumn crocus, used specifically to inhibit the spindle mechanism during cell division and thus cause a doubling of chromosome number.

cold frame   An enclosed, unheated covered frame useful for growing and protecting young plants in early spring. The top is covered with glass or plastic, and sunlight provides heat.

coleoptile   A transitory membrane (first leaf) covering the shoot apex in the seedlings of certain monocots. It protects the plumule as it emerges through the soil.

coleorhiza   Sheath that surrounds the radicle of the grass embryo and through which the young developing root emerges.

collenchyma tissue   Elongated, parenchymatous cells with variously thickened walls, commonly at the acute angles of the cell wall.

colloid   Organic and inorganic matter with very small particle size and a correspondingly large surface area per unit of mass.

community   All the plant populations within a given habitat; usually the populations are considered to be interdependent.

compaction   The compression of soil by the passing of heavy machinery or human and animal traffic.

companion cells   Cells associated with the sieve-tubes in the phloem.

companion crop   A crop sown with another crop and harvested separately. Small-grain cereal crops are often sown with forage crops (grasses or legumes) and harvested in the early summer, allowing the forage crop to continue to grow (e.g., oats sown as a companion crop with red clover).

compensation point   The light intensity at which the rates of photosynthesis and respiration are equal.

complete fertilizer   A fertilizer that contains nitrogen, phosphorus, and potassium.

complete flower   A flower that has pistils, stamens, petals, and sepals, all attached to a receptacle.

compost   A mixture of organic residues and soil that has been piled, moistened, and allowed to decompose biologically. Mineral fertilizers are sometimes added.

compound bud   A bud that contains more than one growing point; usually, one growing point is dominant. The bud may contain floral tissues, vegetative tissues, or both.

compound leaf   A leaf whose blade is divided into a number of distinct leaflets.

cone   The woody, usually elongated seed-bearing organ of a conifer, consisting of a central stem, woody scales and bracts (often not visible), and seeds.

contact herbicide   A chemical that kills plants on contact.

controlled atmosphere (CA) storage   A process of fruit storage wherein the CO_2 level is raised (to 1 to 3 percent) and the O_2 level is lowered (to 2 to 3 percent).

cork cambium   The meristem from which cork develops.

cork tissue   An external, secondary tissue impermeable to water and gases. It is produced by certain kinds of woody plants.

corm   A short, solid, vertical, enlarged underground stem in which food is stored; it contains undeveloped buds (leaf and flower). Examples are crocus, freesia, and gladiolus.

corolla   The collective term for all petals of a flower.

cortex   Primary tissue of a stem or root bounded externally by the epidermis and internally in the stem by the phloem and in the root by the pericycle.

corymb   Inflorescence with outer flowers having longer stems than the inner flowers, resulting in a flat-topped appearance, for example, candytuft.

cotyledons   Leaflike structures at the first node of the seedling stem. In some dicots, cotyledons contain the stored food for the young plant not yet able to photosynthesize its own food. Often referred to as seed leaves.

cover crop   A close-growing crop grown primarily for the purpose of protecting or improving soil between periods of regular crop production or between trees and vines in orchards and vineyards.

critical daylength   The duration of a light period that determines when a photoperiodic plant will flower. *See also* long-day plant; short-day plant.

crop residue   Portion of crop plants remaining after harvest.

crop rotation   Growing crop plants in a different location in a systematic sequence to help control insects and diseases, improve the soil structure and fertility, and decrease erosion.

cross-pollination   The transfer of pollen from a stamen to the stigma of a flower on another plant, except for clones where the plants must be two different clones.

cross section   The surface exposed when a plant stem is cut horizontally and the majority of the cells are cut transversely.

crown   The region at the base of the stem of cereals and forage species from which tillers or branches arise. In woody plants, the root-stem junction. In forestry, the top portions of the tree.

crust   A surface layer on soils, ranging in thickness from a few millimeters to a few centimeters. It is more compact, hard, and brittle when dry than the soil beneath it.

cryptochrome   Light-receiving molecule that mediates light regulation of plant growth and development.

crystal lattice water   Water that is bound inside certain types of soil particles such as clay and is unavailable to plants.

culm   Stem of grasses and bamboos; usually hollow except at the swollen nodes.

cultivar   derived from *cultivated variety*. International term denoting certain cultivated plants that are clearly distinguishable

from others by any characteristic and that, when reproduced (sexually or asexually), retain their distinguishing characteristics. In the United States, *variety* is often considered synonymous with *cultivar*.

cultivation   The growing or tending of crops.

cure   To prepare crops for storage by drying. Dry onions, sweet potatoes, and hay crops are examples. Dehydration of fruits for storage is not considered curing.

cuticle   An impermeable surface layer on the epidermis of plant organs.

cutin   A clear or transparent waxy material on plant surfaces that tends to make the surface waterproof.

cutting   A detached leaf, stem, or root that is encouraged to form new roots and shoots and develop into a new plant.

cutting propagation   The process in which a piece of tissue regenerates the missing part.

cyme   A type of inflorescence that has a broad, more or less flat-topped determinate flower cluster, with the central flower opening first.

cytochrome   A class of several electron-transport proteins serving as carriers in mitochondrial oxidation and in photosynthetic electron transport.

cytokinesis   Division of cytoplasmic constituents at cell division.

cytokinins   A group of plant growth hormones important in the regulation of nucleic acid and protein metabolism and in cell division, organ initiation, and the delay of senescence.

cytology   The study of cells and their components and of the relationship of cell structure to function.

cytoplasm   The living material of the cell, exclusive of the nucleus, consisting of a complex protein matrix or gel. The part of the cell in which essential membranes and cellular organelles are found.

daily light integral (DLI)   The total (accumulated) number of photons received at a defined area during a 24 hour period.

damping off   A pathogenic disorder causing seedlings to die soon after seed germination.

daylength   Number of effective hours of daylight in each twenty-four-hour cycle.

day-neutral plant (DNP)   Plant capable of flowering under either long or short daylengths.

DDT (dichlorodiphenyltrichloroethane)   One of the earliest insecticides of the chlorinated hydrocarbon family. No longer used in some countries because of its persistence in the environment.

deciduous   Refers to trees and shrubs that lose their leaves every fall. Distinguished from evergreens, which retain their leaves throughout the year.

decomposition   Degradation into simpler compounds; rotting or decaying.

defoliant   A chemical or method of treatment that causes only the leaves of a plant to fall off or abscise.

dehiscence   The splitting open at maturity of pods or capsules along definite lines or sutures.

dehulled seed   Seed from which pods, glumes, or other outer covering have been removed, as sometimes with lespedeza and timothy. Also often ambiguously referred to as hulled seed.

denitrification   Biological reduction of nitrate or nitrite to gaseous nitrogen or nitrogen oxides.

deoxyribonucleic acid (DNA)   A molecule composed of repeating subunits of ribose (a sugar), phosphate, and the nitrogenous bases adenine, guanine, cytosine, and thymine. Genes, the fundamental units of inheritance on chromosomes, are sequences of DNA molecules.

dependent variable   The variable that is measured or evaluated as the result of a change in another (independent) variable in an experiment.

desalinization   Removal of salts from saline soil or water.

desiccant   Substance used to accelerate drying of plant tissues.

determinate growth   The flowering of plant species uniformly within certain time limits, allowing most of the fruit to ripen about the same time.

dewpoint   The temperature at which water vapor condenses from the air.

dichogamy   Maturation of male or female flowers at different times, ensuring cross-pollination. Common in maple and walnuts.

dicot   *See* dicotyledons.

dicotyledons   The subclass of flowering plants that have two cotyledons. Also known as *dicots*.

DIF   Average day temperature minus average night temperature. Used in greenhouses to control stem elongation. Positive DIF (day temperature greater than night temperature) promotes stem elongation. Negative DIF inhibits stem elongation.

differentiation   Development from one cell to many cells, together with a modification of the new cells for the performance of particular functions.

diffusion   The movement of molecules, and thus a substance, from a region of higher concentration of those molecules to a region of lower concentration.

dihybrid cross   A cross between organisms differing in two characteristics.

diluent   An essentially inert or nonreacting gas, liquid, or solid used to reduce the concentration of the active ingredient in a formulation.

dioecious   Refers to individual plants having either staminate (male) or pistillate (female) flowers, but not both. Therefore, plants of both sexes must be grown near each other to provide pollen before fruits and seed can be produced. Examples are English holly, asparagus, ginkgo, and date palms.

diploid (2n)   Refers to two sets of chromosomes. Germ cells have one set and are haploid; somatic cells have two sets and are diploid (except for polyploid plants).

dipole   A separation of a positive and negative charges. An object such as water that has positive and negatives sides.

disease   Any change from the state of metabolism necessary for the normal development and functioning of any organism.

disperse   To break up compound particles, such as aggregates, into the individual component particles. To distribute or suspend fine particles, such as clay, in or throughout a dispersion medium.

dissolution   The process by which constituents of parent material dissolve in water or weak, naturally occurring acids and are leached away.

diurnal   Recurring or repeated every day. Going through regular or routine changes daily.

division   A propagation method in which underground stems or roots are cut into pieces and replanted.

dolomite $CaCO_3$—$MgCO_3$; a native limestone source having a significant magnesium content.

domain   Broadest category of taxonomic classification that separates life forms into three categories: prokaryotes, eukaryotes, and archaea bacteria.

dominant gene   The gene (or the expression of the characteristic it influences) that, when present in a hybrid with a contrasting gene, completely dominates in the development of the characteristic. In peaches, for example, white fruit is dominant over yellow.

dormancy   A general term denoting a lack of growth of seeds, buds, bulbs, or tubers due to unfavorable environmental conditions (external dormancy or quiescence) or to factors within the organ itself (internal dormancy or rest).

double fertilization   The process of sexual fertilization in the angiosperms in which one nucleus from the male gametophyte fertilizes the egg nucleus to form the zygote and a second nucleus from the male gametophyte fertilizes two polar nuclei to form endosperm tissue.

drip irrigation   A method of watering plants so that only soil in the plant's immediate vicinity is moistened. Water is supplied from a thin plastic tube at a low rate of flow. Sometimes called trickle irrigation.

drupe   A simple, fleshy fruit derived from a single carpel, usually one-seeded, in which the exocarp is thin, the mesocarp fleshy, and the endocarp hard, for example a peach.

dry land farming   The practice of crop production without irrigation.

dry matter percentage   The percentage of the total fresh plant material left after water is removed. The percentage is determined by weighing a sample of fresh plant material, oven-drying the sample, then reweighing the dried sample. Dry matter percentage equals dry weight divided by fresh weight times 100.

ecology   The study of life in relation to its environment.

ecosystem   A living community and all the factors in its nonliving environment.

ecotype   Genetic variant within a species that is adapted to a particular environment yet remains interfertile with all other members of the species.

edaphic   Pertaining to the influence of the soil on plant growth.

E horizon   The dense layer of soil that forms when ions and organic matter are leached from the A horizon and deposited just above the B horizon.

electrons   Negatively charged particles that form a part of all atoms.

element   A substance that cannot be divided or reduced by any known chemical means to a simpler substance; ninety-two natural elements are known.

eluviation   The removal of soil material in suspension (or in solution) from a layer or layers of a soil. Usually, the loss of materials in solution is described by the term *leaching*.

emasculate   To remove the anthers from a bud or flower before pollen is shed. Emasculation is a normal preliminary step in hybridization to prevent self-pollination.

embryo   A miniature plant within a seed produced as a result of the union of a male and female gamete and resulting in the development of a zygote.

embryo sac   Typically, an eight-nucleate female gametophyte. The embryo sac arises from the megaspore by successive mitotic divisions.

emergy   The amount of solar energy represented in a resource or commodity.

endemic   Species native to a particular environment or locality.

endocarp   Inner layer of the fruit wall (pericarp).

endodermis   In roots, a single layer of cells at the inner edge of the cortex. The endodermis separates the cortical cells from cells of the pericycle.

endogenous   Produced from within. Opposite of *exogenous*.

endophyte   An organism that lives within another organism, often in a symbiotic relationship. Fungal endophytes are associated with many grasses and serve as a deterrent to insects and grazing animals. Endophytes represent a problem in forages because they can be toxic to the animals that graze on the forages.

endoplasmic reticulum (ER)   The lamellar or tubular system of the colorless cytoplasm in a cell.

endosperm   The 3n tissue of angiospermous seeds that develops from sexual fusion of the two polar nuclei of the embryo sac and a male sperm cell. The endosperm provides nutrition for the developing embryo. A food storage tissue.

energy (kinetic)   The capacity to do work. Examples are light, heat, chemical, electrical, or nuclear energy.

entisol   A developing soil without distinct horizons.

entomology   The scientific study of insects.

enzyme   Any of many complex proteins produced in living cells, that, even in very low concentrations, promotes certain chemical reactions but does not enter into the reactions itself.

epicotyl   The upper portion of the embryo axis or seedling, above the cotyledons and below the first true leaves.

epidermis   The outer layer of cells on all parts of the primary plant body: stems, leaves, roots, flowers, fruits, and seeds. It is absent from the root cap and on apical meristems.

epigeal germination   A type of seed germination in dicots in which the cotyledons rise above the soil surface. This occurs in beans, for example.

epigyny   Arrangement of flower parts in which the ovary is embedded in the receptacle so that the other parts appear to arise from the top of the ovary.

epinasty   A twisted or misshapened stem or leaf. Leaves exhibit downward curling of leaf blade due to more rapid growth on the upper side cell than the lower side.

epiphyte   A plant that grows upon another plant yet is not parasitic.

erosion   The wearing away of the surface soil by wind, moving water, or other means.

essential element   Any mineral that is required for proper growth and development of an organism.

Explosion Act (Canada)   see Secure Handling of Ammonia Act (United States of America).

ethephon (2-chloroethyl phosphonic acid)   A compound applied to plants that breaks down after entering the plant to release ethylene gas.

ethylene   A gaseous growth hormone (C_2H_4) regulating various aspects of vegetative growth, fruit ripening, abscission of plant parts, and the senescence of flowers.

etiolation   A condition involving lack of chlorophyll, increased stem elongation, and poor or absent leaf development. It occurs in plants growing under very low light intensity or complete darkness.

eukaryotic cells   Cells that have a nucleus.

eutrophication   A build up of plant nutrients in a body of water that increases plant growth, especially algae, and often results in hypoxia (reduced oxygen) that is needed for aquatic animals to survive. The hypoxia results from the respiration of the densely growing plants, especially at night when they are not photosynthesizing, and decomposition of the plants when they die.

evapotranspiration   The total loss of water by evaporation from the soil surface and by transpiration from plants, from a given area, and during a specified period of time.

evergreen   Trees or shrubs that are never entirely leafless, as in pine or citrus.

evolution   The development of a species, genus, or other larger group of plants or animals over a long time period.

exalbuminous seeds   Seeds that store little or no food in the endosperm but usually in cotyledons, for example, beans.

exchange capacity   The total ionic charge of the adsorption complex active in the adsorption of ions. Also called anion exchange capacity, cation exchange capacity, and base exchange capacity.

exfoliation   The peeling of the outer layers of rocks caused by differential rates of contraction and expansion that in turn are caused by temperature changes.

exocarp   The outermost layer of the fruit wall (pericarp).

exogenous   Produced outside, or originating from or because of external causes. Opposite of *endogenous.*

explant   Living tissue removed from its place in a body and placed in an artificial medium for tissue culture.

F₁   First filial generation in a cross between any two parents.

F₂   Second filial generation, obtained by crossing two members of the F_1 generation, or by self-pollinating plants of the F_1 generation.

facultative   Referring to an organism having the power to live under a variety of conditions; a facultative parasite is either parasitic or saprophytic.

fallow   Cropland left idle for one or more seasons for any number of reasons, such as to accumulate moisture, destroy weeds, and allow the decomposition of crop residue.

family   In plant taxonomy, a group of genera.

fascicle   A bundle of needle-leaves of gymnosperms such as the pines.

fatty acid   Organic compound of carbon, hydrogen, and oxygen that combines with glycerol to make a fat.

fermentation   An anaerobic chemical reaction in foods, such as the production of alcohol from sugar by yeasts.

fertigation   Practice of adding fertilizer to irrigation water.

fertilization   The union of an egg and a sperm (gametes) to form a zygote. Also, application to the soil of needed plant nutrients, such as nitrogen, phosphoric acid, potash, and others.

fertilizer   Any organic or inorganic material of natural or synthetic origin added to a soil to supply elements essential to the growth of plants.

fertilizer injector   A machine that adds a precise amount of fertilizer to water flowing through it.

fibers   Elongated, tapering, thick-walled strengthening cells in various parts of the plant.

fibrous root   Root system characterized by many multidirectional branches.

field capacity   Also known as *field moisture capacity.* The percentage of water remaining in a soil two or three days after having been saturated and after free drainage due to gravity has practically ceased.

filament   Stalk portion of a stamen. *See also* anther; stamen.

fleshy fruit   Any fruit formed from an ovary that has fleshy or pulpy (not dried) walls at maturity. Also, those fruits that include fleshy parts of the perianth, floral tube, or the receptacle.

flocculate   To clump together individual, tiny soil particles, especially fine clay, into small granules. Opposite of *disperse.*

flooding   A method of irrigation by which water is released from field ditches and allowed to spread over the land.

flora   A collective term for all the plant types that grow in a region.

floral incompatibility   A genetic condition in which certain normal male gametes are incapable of functioning in certain pistils.

floricane   A second-year cane (previously a primocane) of raspberries, blackberries, or their hybrids, on which floral buds initiated in the previous season develop, flower, and fruit.

florigen   Term given to a stimulus (suspected of being a protein) that promotes flowering.

flower   Floral leaves grouped together on a stem that, in the angiosperms, are adapted for sexual reproduction.

fodder   Coarse grasses such as corn and sorghum harvested with the seed and leaves and cured for animal feed.

follicle   A simple, dry dehiscent fruit, having one carpel and splitting along one suture. Examples are milkweeds and magnolia.

food chain   The path along which food energy is transferred within a natural plant and animal community (from producers to consumers to decomposers).

forage   Vegetation used as feed for livestock, such as hay, pasture, and silage. The material is fed green or dehydrated.

forcing   A cultural manipulation used to hasten flowering or growing plants outside their natural season.

forb   Flowering herbaceous plants that are not grasses, sedges, or rushes.

fossil   Any impression, natural or impregnated remains, or other trace of an animal or plant of past geological eras that has been preserved in the earth's crust.

foundation seed   Seed stocks increased from breeder seed, and handled to closely maintain the genetic identity and purity of a cultivar.

friable   Generally refers to a soil moisture consistency that crumbles when handled.

frost seeding   *See* oversowing.

fruit   A mature ovary; in some plants, other flower parts are commonly included as part of the fruit, e.g., the hypanthium of the apple flower surrounds the ovary.

fruiting spur   A short, thick, fruit-bearing stem terminated by an apical floral, vegetative, or mixed bud at the end of the growing season, depending on the species. Fruiting spurs usually develop flowers, leaves, and new shoot growth.

fumigation   Control of insects, disease-causing organisms, weeds, or nematodes by gases applied in an enclosed area such as a greenhouse or under plastic laid on the soil.

fungicide   A pesticide chemical used to control plant diseases caused by fungi.

fungus   Plural: **fungi**. A thallus plant unable to photosynthesize its own food (exclusive of bacteria).

furrow   Small V-shaped ditch made for planting seed or for irrigating.

furrow irrigation   A method of irrigation by which the water is applied to row crops in ditches.

gamete   A haploid-generation male sperm cell or a female egg cell capable of developing into an embryo after fusion with a germ cell of the opposite sex.

gametophyte   In a seed plant, the few-celled, haploid generations arising from a meiotic division and giving rise through mitosis to the male or female gametes.

gene   A group of base pairs in the DNA molecule in the chromosome that determines or conditions one or more hereditary characteristics.

genetically modified organism (GMO)   Any organism that is the product of genetic recombination, which includes almost all organisms except those produced clonally or parthenogenically. Sometimes different species, such as plant and bacteria, can exchange genetic material, for example, wheat. Includes, but is not limited to, genetically engineered organisms. *See* genetic engineering.

genetic code   The sequence of nitrogen bases in a DNA molecule that codes for an amino acid or protein. In a broader sense, for example, the full sequence of events from the translation of chromosomal DNA to the final stage of the synthesis of an enzyme.

genetic engineering   The process of moving genetic material (DNA) or genes from their normal location and transferring them to another organism or the same organism in a different combination (genetic recombination) or to a different chromosome. The term *engineering* indicates human intervention.

genetics   The science or study of inheritance.

genotype   The genetic makeup of a nucleus or of an individual.

genus   Plural: **genera**. A group of structurally or phylogenetically related species.

geographic information system (GIS)   A computer system that gathers, stores, and analyzes data relative to position on the surface of the earth.

geotropism   Growth curvature in plants induced by gravity.

germination   Sequence of events in a viable seed starting with imbibition of water that leads to growth of the embryo and development of a seedling.

germplasm   The protoplasm of the sexual reproductive cells containing the units of heredity (chromosomes and genes).

gibberellins   A group of natural growth hormones whose most characteristic effect is to increase the elongation of cells.

glacial till   A product of glacial weathering in which rock particles varying in size from clay to boulders are deposited by the glacier on the land surface as it melts and recedes.

global positioning system (GPS)   The use of satellite triangulation and other technology to locate a position on the earth's surface precisely.

glucose   A simple sugar composed of six carbon, twelve hydrogen, and six oxygen atoms ($C_6H_{12}O_6$).

graft   To place a detached branch (scion) in close cambial contact with a rooted stem (rootstock) so that scion and rootstock unite to form a new plant.

graft incompatibility   The inability of a stable graft to form between two plants.

grain (caryopsis)   A simple, dry, indehiscent fruit with ovary walls fused to the seed. The so-called seed of cereal or grain crops such as corn, wheat, barley, and oats is actually a fruit.

grana   Structures in the chloroplasts usually seen as stacks of parallel lamellae with the aid of an electron microscope.

gravitational water   Water that moves into, through, or out of the soil under the influence of gravity.

grazing land   Land that is suitable for use as pasture for animals though it may not be suitable for growing other crops.

green chop   Forage that is chopped in the field while succulent and green and fed directly to livestock, made into silage, or dehydrated.

green manure   A crop that is plowed under while still green and growing to improve the soil.

groundwater   Water that fills all the unblocked pores of underlying material below the water table, which is the upper limit of saturation.

group   A taxonomic category of cultivated plants at the subspecies level that have the same botanical binomial but have characteristics different enough to warrant naming them for clarity.

growing degree day (GDD)   The measurement of heat accumulation used to predict crop maturation and other phenological events such as stages in an insect life cycle. A growing degree day is a day that has an average temperature 1° above a baseline. Each degree above that baseline is another growing degree day. For example, a single day that as an average temperature 15° above the baseline is considered to be fifteen growing degree days.

growing medium   Soil or soil substitute prepared by combining materials such as peat moss, vermiculite, sand, or composted sawdust. Used for growing potted plants or germinating seeds.

growth   An irreversible increase in cell size and/or cell number. An increase in dry weight, regardless of cause.

growth regulator   A synthetic or natural compound that in low concentrations controls growth responses in plants.

growth retardant   A chemical that selectively interferes with normal hormonal promotion of plant growth but without appreciable toxic effects.

guard cells   Specialized epidermal cells that contain chloroplasts and surround a stoma.

gully erosion A process whereby water accumulates in narrow channels and, over relatively short periods, removes the soil from this narrow area to considerable depths.

guttation Exudation of water in liquid form from plants.

gymnosperm A seed plant with seeds not enclosed by a megasporophyll or pistil.

gynoecium The female part of a flower or pistil formed by one or more carpels and composed of the stigma, style, and ovary.

Haber-Bosch An industrial process that fixes nitrogen into ammonia, much of which is used to make fertilizer.

haploid Having only one complete set of chromosomes; referring to an individual or generation containing such a single set of chromosomes per cell.

hardening off Adapting plants to outdoor conditions by withholding water, lowering the temperature, or lowering the nutrient supply. This conditions plants for survival when transplanted outdoors.

hardpan A hardened soil layer, in the lower A horizon or in the upper B horizon, caused by the cementing of soil particles.

hardscape Nonplant components of a landscape, for example, patios, decks, ponds, and so on.

hardy plants Plants adapted to cold temperatures or other adverse climatic conditions of an area. *Half-hardy* indicates some plants may be able to take local conditions with a certain amount of protection.

harvest index The percentage of a plant's biomass that is incorporated into the harvestable part.

harvest maturity The stage at which the harvestable part of a plant is ready for harvest.

hay Herbage of forage plants, including seed of grasses and legumes, that is harvested and dried for animal feed.

head A type of inflorescence, typical of the composite family, in which the individual flowers are grouped closely together on a receptacle. Example: sunflower. Also, the aerial terminus of a grapevine trunk; in head-pruning systems, it is the position where canes are selected for fruiting.

heaving The partial lifting of plants out of the ground, frequently breaking their roots, as a result of freezing and thawing of the surface soil during the winter.

heavy soil A soil with a high content of the fine separates, particularly clay, or one with a high tractor power requirement and hence difficult to cultivate.

heeling in Temporary storage of bare-rooted trees and shrubs by placing the roots in a trench and covering with soil or sawdust.

heliotropic movement Movement of a plant in response to the diurnal movement of the sun.

helix A spiral form; a term often used in reference to the double spiral of the DNA molecule.

hemicellulose A type of polysaccharide found in cell walls.

herbaceous Refers to plants that do not normally develop woody tissues.

herbage *See* forage.

herbarium A collection of dried and pressed plant specimens cataloged for easy-reference.

herbicide Any chemical used to kill plants; an herbicide may work against a narrow or wide range of plant species.

herbivore Animal that subsists principally or entirely on plants or plant products.

heredity The transmission of morphological and physiological characteristics from parents to their offspring.

hermaphrodite flower A flower having both stamens (male) and pistils (female).

heterosis Also known as *hybrid vigor*. The increased vigor, growth, size, yield, or function of a hybrid progeny over the parents that results from crossing genetically unlike organisms. Also, the increase in vigor or growth of a hybrid progeny in relation to the average of the parents.

heterotroph Organism that depends on the compounds synthesized by autotrophs for nourishment.

heterotrophic Capable of deriving energy for life processes only from the decomposition of organic compounds and incapable of using inorganic compounds as sole sources of energy or for organic synthesis. *Contrast with* autotrophic.

heterozygous Having different genes of a Mendelian pair present in the same cell or organism, for instance, a tall pea plant with genes for both tallness (T) and dwarfness (t).

higher plant parasite Plant that lives on other living plants and takes at least part of its nourishment from the host.

high intensity discharge (HID) A type of lamp that illuminates when an electric current passes through a vapor that contains xenon gas and/or metal halide compounds such as sodium or mercury salts.

hill Raising the soil in a slight mound for planting, or setting plants some distance apart.

hilum The scar on a dicot seed, such as a bean seed, where it was attached to the fruit.

histology The science that deals with the microscopic structure of plant or animal tissues.

homologous chromosomes The two members of a chromosome pair.

homozygous Having similar genes of a Mendelian pair present in the same cell or organism, for instance, a dwarf pea plant with genes for dwarfness (tt) only.

horizon, soil A layer of soil, approximately parallel to the soil surface, with distinct characteristics produced by soil-forming processes.

hormone A chemical substance produced in one part of a plant and used in minute quantities to induce a growth response in another part. For example, auxins are one type of hormone.

horticulture The study of crops that require intense and constant care, from planting through delivery to the consumer.

hotbed A bed of soil enclosed in a low glass or transparent plastic frame and heated with fermenting manure, electric cables, or steam pipes. Used to germinate seeds, root cuttings, and grow other plants for transplanting outside.

hot caps Waxpaper cones, paper sacks, or cardboard boxes with bottoms removed and placed over individual plants in spring for frost and wind protection.

human footprint The impact that human activity has on ecosystems and global resources.

humidity, relative   The ratio of the weight of water vapor in a given quantity of air to the total weight of water vapor that quantity of air is capable of holding at a given temperature, expressed as a percentage.

humus   The more or less stable fraction of the soil organic matter remaining after the major portion of plant and animal residues have decomposed.

hybrid   The offspring of two plants or animals differing in one or more Mendelian characters.

hybridization   The crossing of individuals of unlike genetic constitution. Also, a method of breeding new cultivars that uses crossing to obtain genetic recombinations.

hybrid seeds   Seeds produced by the crossing of parents whose genetic composition is different from each other.

hydration   The addition of molecular water to another compound to form a hydrated material more vulnerable to pulverization.

hydrologic cycle   The movement of water from the atmosphere to the earth and its return to the atmosphere.

hydrolysis   The chemical reaction in which water participates as a reactant and not as a solvent. Usually, the splitting of a molecule to form smaller molecules that incorporate hydrogen and hydroxyl ions, derived from water, in their structures.

hydroponics   Growing plants in aerated water containing all the essential mineral nutrients rather than soil. Also called soilless gardening.

hygroscopic water   Water that surrounds and is tightly held by soil particles, making the water unavailable to plants.

hypha   Plural: **hyphae**. A filament that forms mycelial growth of a fungus.

hypocotyl   Portion of a stem that is located above the root and below the cotyledon.

hypogeal germination   In dicots, a type of seed germination in which the cotyledons remain below the soil surface. Peas are one example.

hypothesis   A proposition or supposition provisionally adopted to explain certain facts. Once proven by ultimate scientific investigation, it becomes a theory or a law.

igneous rock   Rock formed from the cooling and solidification of magma that has not been changed appreciably since its formation.

imbibition   The absorption of liquids or vapors into the ultramicroscopic spaces in materials like cellulose.

immobilization   The conversion of inorganic nitrogen to organic nitrogen.

immune   Free from attack by a given pathogen; not subject to the disease.

imperfect flower   A flower lacking either stamens or pistils.

impervious   Resistant to penetration by fluids or by roots.

inbred line   A pure line usually originating by self-pollination and selection.

inceptisol   A developing soil that is usually wet or moist and shows alterations in horizons.

incompatibility   Failure of a flowering plant to obtain fertilization and seed formation after pollination, usually because of slow pollen tube growth in the stylar tissue. Also, the failure of two graft components (stock and scion) to unite and develop into a successfully growing plant.

incomplete flower   A flower that is missing one or more of the following parts: sepals, petals, stamens, or pistils.

indehiscent fruit   A fruit that does not split open naturally at maturity.

independent variable   The variable in an experiment that is manipulated to evaluate its effect on another (dependent) variable.

indeterminate growth   The flowers of plants are borne on lateral branches, the central stem continues vegetative growth, with blooming continued for a long period. Examples are alfalfa and fuchsia.

indigenous   Produced or living naturally in a specific environment.

indoleacetic acid (IAA)   A natural or synthetic plant growth regulator; an auxin.

inferior ovary   An ovary imbedded in the receptacle, or an ovary whose base lies below the point of attachment of the perianth.

infiltration rate   The maximum velocity at which water can enter the soil under specified conditions, including the presence of an excess of water.

inflorescence   An axis bearing flowers, or a flower cluster (e.g., umbel, spike, panicle).

inheritance   The acquisition of characters or qualities by transmission from parent to offspring.

initiation   *See* differentiation.

inoculate   To induce a disease in a living organism by introducing a pathogen. Also, to treat seeds of leguminous plants with bacteria to induce nitrogen-fixation in the roots.

inorganic compound   A chemical compound that generally is not derived from life processes; compounds that do not contain carbon.

insecticide   Any chemical (organic or inorganic) substance that kills insects.

insolation   The amount of solar energy striking an area of the earth's surface.

integrated plant health management (IPHM)   Using multiple methods such as cultural, biological, and chemical to control pests and diseases in a crop.

integuments   The tissues covering or surrounding the ovule, usually consisting of an inner and outer layer; they subsequently become the seedcoats of the mature ovule.

intercalary growth   A pattern of stem elongation typical of grasses. Elongation proceeds from the lower internodes to the upper internodes through the differentiation of meristematic tissue at the base of each internode.

intercalary meristem   A type of meristem, often found in grasses, that lies across a stem or leaf, resulting in the deposition of new cells between regions of older tissues.

internode   The region of a stem between two successive nodes.

interspecific cross   A cross, natural or intentional, between two species.

in vitro   Latin for *in glass*. Living in test tubes; outside the organism or in an artificial environment.

in vivo   Latin for *in living*. In the living organism.

ions   Atoms, groups of atoms, or compounds that are electrically charged as a result of the loss of electrons (cations) or the gain of electrons (anions).

IPHM   *See* integrated plant health management.

irradiation   In genetics and plant breeding, exposing seed, pollen, or other plant parts to X-rays or other short wavelength (gamma) radiations to increase mutation rates.

irrigation   Applying water to the soil, other than by natural rainfall.

isolation   The prevention of crossing among plant populations because of distance or geographic barriers (geographic isolation).

jasmonate   A compound recently identified as a plant hormone involved in plant response to insect attack.

juvenile phase   A growth phase in plants characterized by the inability to flower and reproduce.

lamina   Blade or expanded part of a leaf.

lateral bud   A bud that grows out from the leaf axil on the side of a stem.

lateral meristem   Meristems (buds) found in the axils of leaves on a stem. They may be dormant until the apical (top) meristem is removed or becomes inactive.

latex   A milky secretion produced by various kinds of plants.

lathhouse   An open structure built of wood lath or plastic screen for protecting plants from excessive sunlight or frost.

layering   A form of vegetative propagation in which an intact branch develops roots as the result of contact with the soil or another rooting medium.

leach   To remove soluble materials from soil or plant tissue with water.

leaf mold   Partially decayed leaves useful for improving soil structure and fertility.

leggy   Weak-stemmed and spindly plants with sparse foliage caused by too much heat, shade, crowding, or overfertilization. *See* etiolation.

legume   Plant member of the family FABACEAE, with the characteristic capability to fix atmospheric nitrogen in nodules on its roots if inoculated with proper bacteria.

lenticel   An opening made up of loosely arranged cells in the periderm that permits passage of gases.

liane   A climbing or twining plant, usually woody.

lifecycle   The stages that an organism goes through during its lifetime.

light compensation point   The level of light intensity where carbon dioxide uptake for daily photosynthesis equals the daily loss of carbon dioxide due to the respiration of a plant.

light quality   The composition of light determined by the wavelengths present and their intensity. For example, visible light is composed of wavelengths of the colors of the rainbow at the appropriate intensity for each color.

light reactions   The reactions of photosynthesis in which light energy is required: the photo (light) activation or excitement of electrons in the chlorophyll molecule, transfer of the electrons, photolysis of water, and associated reactions.

light saturation   Light intensity above which photosynthesis activity does not increase as the intensity increases.

lignin   An organic substance found in secondary cell walls that gives stems strength and hardness. Wood is composed of lignified xylem cells (about 15 to 30 percent by weight).

lime   An agricultural material containing the carbonates, oxides, and/or hydroxides of calcium and/or magnesium. It is used to increase soil pH and to neutralize soil acidity.

line   Group of individuals from a common ancestry. When propagated by seed, it retains its characteristics. A type of cultivar.

liners   Young plants that are grown in a nursery until they are big enough for transplanting.

lipid   Any of a group of fats or fatlike compounds insoluble in water but soluble in certain other solvents.

loam   A textural class for soil with prescribed amounts of sand, silt, and clay.

locus   The fixed position of location of a gene on or in a chromosome.

lodging   A condition in which plants are caused to bend for various reasons at or near the soil surface and fall more or less flat on the ground. Most frequently observed in cereals.

long-day plant   A plant that flowers when the light period is greater (dark period is shorter) than a critical duration.

Lucerne   A name for alfalfa used in Europe, Australia, and other regions.

macronutrient   A chemical element, like nitrogen, phosphorus, and potassium, necessary in large amounts (usually greater than 1 ppm) for the growth of plants.

male sterility   A condition in some plants in which pollen either is not formed or does not function normally, even though the stamens may appear normal.

Mediterranean climate   A climate characterized by cool, wet winters and warm, dry summers.

meiosis   Two successive nuclear divisions, in the course of which the diploid chromosome number is reduced to the haploid and genetic segregation occurs.

Mendel's laws   A set of three laws formulated by Gregor Mendel; each is generally true but there are numerous exceptions. The laws are (1) characteristics exhibit alternative inheritance, being either dominant or recessive; (2) each gamete receives one member of each pair of factors present in a mature individual; and (3) reproductive cells combine at random.

mericlinal chimera   Plant cell mutates and the mutation is maintained in a portion of a cell layer. The mutation appears only in portions of a leaf or stem.

meristem   Undifferentiated tissue whose cells can divide and differentiate to form specialized tissues, such as xylem or phloem.

meristematic tissue   Region of actively dividing and differentiating cells.

mesocarp   Middle layer of the fruit wall (pericarp).

mesophyll   Parenchyma tissue in leaves found between the two epidermal layers.

messenger RNA   Ribonucleic acid produced in the nucleus and capable of carrying parts of the message coded in chromosomal DNA. Messenger RNA moves from the nucleus to the ribosomes, where protein is synthesized in the cytoplasm.

metabolism   The overall physiological activities of an organism.

metamorphic rock A rock that has been greatly altered from its previous condition through the combined action of heat and pressure.

metamorphosis Distinct and often sudden changes in development in organisms, for example changes from egg to larva to pupa to adult insects.

microclimate Atmospheric environmental conditions in the immediate vicinity of the plant, including interchanges of energy, gases, and water between atmosphere and soil.

micronutrient A chemical element necessary in extremely small amounts (less than 1 ppm) for the growth of plants. Examples are boron, chlorine, copper, iron, manganese, and zinc.

micropyle An opening leading from the outer surface of the ovule between the edges of the two integuments inward to the surface of the nucellus.

microspore One of the four haploid spores that originate from the meiotic division of the microspore mother cell in the anther of the flower and that gives rise to the pollen grain.

microspore mother cell Diploid cell in the anther that gives rise, through meiosis, to four haploid microspores.

middle lamella The pectic layer lying between the primary cell walls of adjoining cells.

millimho (mmho) A measure of electrical conductivity, 1 mmho = 0.001 mho. The mho is the reciprocal of an ohm.

mineralization The conversion of organic nitrogen to ammonium, nitrate, and nitrite by microorganisms.

mineral soil A soil whose makeup and physical properties are largely those of mineral matter.

minor element *See* micronutrient.

mist propagation Applying water in mist form to leafy cuttings in the rooting stage to reduce transpiration.

mitochondria Singular: **mitochondrion**. A minute particle in the cytoplasm associated with intracellular respiration.

mitosis A form of nuclear cell division in which chromosomes duplicate and divide to yield two nuclei that are identical with the original nucleus. Usually mitosis includes cellular division (cytokinesis).

mixed bud A bud containing both rudimentary flowers and vegetative shoots.

mode of action The way in which an herbicide or pesticide works to control its target organism.

moisture A basis for representing moisture content of a product. It is calculated from the net weight of water lost by drying, divided by the dried weight of the material, and the answer multiplied by 100 to obtain the percentage.

mol Used in plant science when referring to light and when the waveband stated is 400 to 700 nm (PAR); 1 mol = 6.022×10^{23} photons, or 1 micromol = 6.022×10^{17} photons.

mole (M) Amount of a substance that has a weight in grams numerically equal to the molecular weight of the substance. Also known as *gram-molecular weight*.

molecular biology A field of biology concerned with the interaction of biochemistry and genetics in the life of an organism.

molecule A unit of matter; the smallest portion of an element or a compound that retains chemical identity with the substance in mass. A molecule usually consists of the union of two or more atoms, and some organic molecules contain hundreds of atoms.

monocot *See* monocotyledons.

monocotyledons The subclass of flowering plants that have only a single cotyledon at the first node of the primary stem. Also known as *monocots*.

monoecious A plant with separate male and female flowers on the same plant, such as corn and walnuts.

monsoon Seasonal changes in wind direction and precipitation amount that result in a region receiving heavy rainfall during a specific season and being relatively dry during the rest of the year.

morphology The study or science of the form, structure, and development of plants.

muck Highly decomposed organic material in which the original plant parts are not recognizable.

mulch Any material such as straw, sawdust, leaves, plastic film, and loose soil that is spread on the surface of the soil to protect the soil and plant roots from the effects of rain, soil crusting, freezing, or evaporation.

multiple fruit A cluster of matured fused ovaries produced by separate flowers, for example, pineapple. *See also* aggregate fruit.

mutation A sudden, heritable change appearing in an individual as the result of a change in genes or chromosomes.

mycelium The mass of hyphae forming the body of a fungus.

mycology A branch of botany dealing with the study of fungi.

mycorrhiza The association, usually symbiotic, of fungi with the roots of some seed plants.

nanometer (nm) A unit of length equal to one millionth (10^{-6}) of a millimeter or one millimicron; 1 nm equals 10 angstrom units.

National Organic Program (NOP) The United States Department of Agriculture regulatory program that governs the production of organic food.

naturalized plant A plant introduced from one environment into another in which the plant has become established and more or less adapted to a given region by growing there for many generations.

natural selection Environmental effects in channeling the genetic variation of organisms along certain pathways.

necrosis Death associated with discoloration and dehydration of all or some parts of plant organs.

nectary Flower part that secretes nectar.

nematodes Unsegmented roundworms abundant in many soils; they are important because many species of them attack plants or animals.

net photosynthesis The gain in carbohydrates from photosynthesis after carbohydrate loss through respiration is deducted. *See* total photosynthesis.

neutral soil A soil in which the surface layer is neither acid nor alkaline in reaction.

night-break lighting Low intensity lighting used in the middle of the night to change the photoperiod.

nitrification The conversion of ammonium ions into nitrates through the activities of certain bacteria.

nitrogen assimilation The incorporation of nitrogen into organic cell substances by living organisms.

nitrogen cycle   The cycling of nitrogen from the atmosphere to living organisms and back through fixation, consumption, and decomposition.

nitrogen fixation   The conversion of atmospheric nitrogen (N_2) into oxidized forms that can be assimilated by plants. Certain blue-green algae and bacteria are capable of biochemically fixing nitrogen.

nodes   Enlarged regions of stems that are generally solid where leaves are attached and buds are located. Stems have nodes but roots do not.

no-till planting   Also known as *stubble culture*. A cultural system most often used with annual crops in which the new crop is seeded or planted directly in a field on which the preceding crop plants were cut down or destroyed by a nonselective herbicide rather than being removed or incorporated into the soil, as is common in preparing a seedbed.

nucellus   A tissue originally making up the major part of the young ovule, in which the embryo sac develops.

nucleic acid   An acid found in all nuclei; all known nucleic acids fall into two classes, DNA and RNA.

nucleus   A dense body in the cytoplasm essential for cellular development and reproduction.

nut   A dry, indehiscent, single-seeded fruit with a hard, woody pericarp (shell), such as the walnut and pecan.

nutrient   Element essential to plant growth used in the elaboration of food and tissue.

obligate parasite   An organism that must live as a parasite and cannot otherwise survive.

obligate saprophyte   An organism obliged to live only on non-living animal or plant tissue.

oedema   Intumescence or blister formation on leaves.

opposite   An arrangement of leaves or buds on a stem. They occur in pairs on opposite sides of a single node.

order   The taxonomic category that falls between class and genus.

organ   A part of an animal or plant body adapted by its structure for a particular function.

organelle   A specialized region in a cell, such as mitochondria, that is bound by a membrane.

organic   In chemistry, compounds containing carbon, many of which are associated with living organisms. In, growing systems, the reliance on renewable resources and exclusion or strictly limited use of synthetic fertilizers, synthetic pesticides and plant growth regulators, and genetically modified organisms.

organic base   An organic compound (usually contains nitrogen) that acts as a base. In DNA and RNA they are the molecules that attach to the (deoxy) ribose molecule and determine genetic characteristics.

organic soil   A soil that contains a high percentage (greater than 20 percent) of organic matter.

orthotropic   Upright growth.

osmosis   The diffusion of fluids through a semipermeable or selectively permeable membrane.

ovary   The basal, generally enlarged part of the pistil in which seeds are formed. The ovary, at maturity, is a fruit. It is a characteristic organ of angiospermous plants.

oversowing   The practice of sowing seed over an existing crop. Often done on golf courses by sowing a cool-season turfgrass over a warm-season perennial grass to maintain turfgrass growth during the cool season when the warm-season species is dormant.

ovule   A rudimentary seed, containing, before fertilization, the embryo sac, including an egg cell, all being enclosed in the nucellus and one or two integuments.

oxidation   The reaction of oxygen with soil parent material to form oxides.

oxidation-reduction reaction   A chemical reaction in which one substance is oxidized (loses electrons, or loses hydrogen ions and their associated electrons, or combines with oxygen) and a second substance is reduced (gains electrons, or gains hydrogen ions and their associated electron, or loses oxygen).

oxidative respiration   The chemical decomposition of foods (glucose, fats, and proteins) requiring oxygen as a terminal electron acceptor and yielding carbon dioxide, water, and energy. The energy is commonly stored in ATP.

oxisol   A soil that lacks distinct horizons, contains clay oxides, is highly weathered, and has low fertility.

palatability   Term used to describe how agreeable or attractive feed stuff is to animals or how readily they consume it.

palisade parenchyma cells   The cell layer in leaves immediately below the upper epidermis; it is packed with chloroplasts. Found in dicots but not in monocots.

pallet   Rectangular or square platform, usually wooden, designed for ease of mechanical handling and transportation of material (containers) placed on the platform.

palmate   Arrangement of leaflets of a compound leaf or of the veins in a leaf. Characterized by subunits arising from a common point much as fingers arise from the palm of the hand.

panicle   An inflorescence, common in the grass family, that has a branched central axis. An example is oats.

parasite   An organism obtaining its nutrients from the living body of another plant or animal.

parenchyma tissue   A tissue composed of thin-walled, loosely packed, unspecialized cells.

parent material   Soil component derived from the natural breakdown of rocks by physical and chemical forces. *See also* weathering.

parthenocarpy   Fruit development without sexual fertilization. Such fruits are seedless. Examples are the 'Navel' orange and some fig cultivars.

parthenogenesis   Development of an egg into an embryo without fertilization.

particle size   The effective diameter of a particle measured by sedimentation, sieving, or micrometric methods.

parts per million (ppm)   Weight units of any given substance per 1 million equivalent weight units; the weight units of solute per million weight units of solution (i.e., 1 ppm = 1 mg/1).

pasteurized soil   A soil that has been heated to 82°C (180°F) for at least thirty minutes to kill most of the pathogenic organisms without affecting the saprophytic soil flora.

pasture   Area of domesticated forages, usually improved, on which animals are grazed.

pathogen   An organism that causes disease.

peat Any unconsolidated soil mass of semicarbonized vegetable tissue formed by partial decomposition in water. An example is sphagnum peat moss.

pectin Polysaccharide from the middle lamella of the plant cell wall; jelly-forming substance found in fruit.

pedicel Individual flower stalk of an inflorescence.

peduncle Flower stalk that is borne singly. Also, the main stem of an inflorescence.

percolation The downward movement of water through soil.

perennial A plant that grows more or less indefinitely from year to year and usually produces seed each year.

perfect flower Having both stamens and pistils; a hermaphroditic flower.

perianth The petals and sepals of a flower, collectively.

pericarp The fruit wall, which develops from the ovary wall.

periclinal chimera Mutation in which, mutated plant cells are arranged near the apical dome so that they form a cell layer. The most stable form of chimera.

pericycle The layer of cells immediately inside the endodermis. Branch roots arise from the pericycle.

periderm A corky layer formed by the cork cambium at the surface of organs that are undergoing secondary growth.

permanent wilting point (PWP) The point at which the water potential of the soil is equal to or lower than that of the root, and water cannot move from the soil into the root.

permeable Referring to a membrane, cell, or cell system through which substances may diffuse. Also describes how well water moves through soil.

petal Part of a flower, often brightly colored.

petiole The stalk that attaches a leaf blade to a stem.

pH The negative logarithm of the hydrogen-ion activity of a soil. Also, a measure of acidity or alkalinity, expressed as the negative logarithm (base 10) of the hydrogen-ion concentration. pH 7 is neutral. Values less than this indicate acidity; higher values indicate alkalinity.

phellogen Cork cambium, a cambium layer giving rise externally to cork and, in some plants, internally to phelloderm.

phenology The study of the timing of periodic phenomena such as flowering, growth initiation, or growth cessation, especially as related to seasonal changes in temperature or photoperiod.

phenotype The external physical appearance of an organism.

pheromone Any of a class of hormonal substances secreted by an individual and stimulating a physiological or behavioral response from an individual of the same species.

phloem A tissue through which nutritive and other materials are translocated through the plant. The phloem consists of sieve tube cells, companion cells, phloem parenchyma, and fibers.

phospholipid Lipids that contain a phosphorus molecule. Important in the structure and function of cellular and subcellular membranes

photomorphogenesis Plant shape determined by light quality particularly through relative amounts of red, far-red, and blue light.

photomorphogenic response The reaction of non-shade-adapted plants when they do not receive enough light at the correct red:far-red ratio.

photoperiod That length of day or period of daily illumination required for the normal growth and sexual reproduction of some plants.

photoperiodism Response of a plant to the relative lengths of day and night (light and dark), particularly in respect to flower initiation and bulbing.

photophosphorylation The production of ATP by the addition of a phosphate group to ADP using the energy of light-excited electrons produced in the light reactions of photosynthesis. Photo light, phosphorylation = adding phosphorus.

photosynthesis The process in green plants of converting water and carbon dioxide into sugar with light energy, accompanied by the production of oxygen.

photosynthetic active radiation (PAR) Radiation in the 400–700 nanometer (nm) spectral waveband.

photosynthetic photon flux (PPF) Radiometric irradiance per unit area per unit time at specific wavelengths (400–700 nm waveband for plants). Units micromols $m^{-2}s^{-1}$.

phototropin A blue-light receptor that senses the direction of light and influences phototropic responses.

phototropism A change in the manner of growth of a plant in response to nonuniform illumination. Usually, the response, which is auxin-regulated, is a bending toward the strongest light.

Phylum Plural: **Phyla**. A primary division of the animal and plant kingdom.

physiology The science of the functions and activities of living plants.

phytochrome A reversible protein, red/far-red light-sensing pigment occurring in the cytoplasm of green plants. It is associated with the absorption of light that affects growth, development, and differentiation of a plant, independent of photosynthesis (e.g., in the photoperiodic response). *See* photomorphogenesis.

phytomer *See* tiller.

phytoplasma Round or ovoid prokaryote organisms that lack rigid cell walls and infect plants.

pigments Molecules that are colored by the light they absorb. Some plant pigments are water soluble and are found mainly in the cell vacuole.

pinching The removal of the terminal bud or apical meristematic growth to stimulate branching.

pinnate Leaflets attached to a midrib in a featherlike arrangement.

pistil The seed-bearing organ in the flower, composed of the ovary, the style, and the stigma.

pistillate (female) flower A flower having pistils but no stamens.

pit Channel in plant cell walls that surrounds plasmodesmata.

pith A region in the center of some stems and roots consisting of loosely packed, thin-walled parenchyma cells.

placenta Plural: **placentae**. The tissue within the ovary to which the ovules are attached.

plagiotropic Horizontal growth.

plantae The taxonomic category within the kingdom that includes all plants.

plant development The progression of change during a plant's lifecyle.

plant growth   An irreversible increase in size (biomass) of a plant.

plant growth regulators   Natural and synthetic chemicals that influence plant growth and development.

plant hormone   *See* hormone.

plant pathology   The scientific study of plant diseases and their causal organisms.

plasmalemma   Membrane that surrounds a plant cell.

plasmodesmata   Strand that forms a cytoplasmic connection between two plant cells.

plasmolysis   The separation of the cytoplasm from the cell wall because of the removal of water from the protoplast.

plastids   The cellular organelles in which carbohydrate metabolism is localized.

plow layer   The surface soil layer ordinarily moved in tillage.

plow pan   A hard layer of soil formed by continual plowing at the same depth.

plow-plant   The practice of plowing and planting a crop in one operation, with no additional seedbed preparation.

plumule   The first bud of an embryo or that portion of the young shoot above the cotyledons.

polar nuclei   Two centrally located nuclei in the embryo sac that unite with a second sperm cell in a triple fusion. In certain seeds, the product of this fusion develops into the endosperm.

polar transport   The directed movement within plants of compounds (usually hormones) mostly in one direction; polar transport overcomes the tendency for diffusion in all directions.

pollen   The almost microscopic, yellow bodies that are borne within the anthers of flowers and contain the male generative (sex) cells.

pollenizer   A plant that is a source of pollen.

pollen mother cell   A 2n cell that divides twice (once by meiosis and once by mitosis) to form a tetrad of four pollen grains.

pollen tube   A tubelike structure developed by the tube nucleus in the microspore that helps guide the sperm through the stigma and style of a flower to the embryo sac.

pollination   The transfer of pollen from a stamen (or staminate cone) to a stigma (or ovulate cone).

pollinator   An agent (such as an insect) that pollinates flowers.

polyembryony   The presence of more than one embryo in a developing seed.

polyploidy   A condition in which a plant has somatic (nonsexual) cells with more than 2n chromosomes per nucleus.

polysaccharides   Long-chain molecules composed of units of a sugar; starch and cellulose are examples.

pome   A simple fleshy fruit, the outer portion of which is formed by floral parts that surround the ovary (i.e., apple and pear fruits).

pore size distribution   The volume of the various sizes of pores in a soil. It is expressed as a percentage of the bulk volume (soil plus pore space).

pore space   The open (nonsolid) portion of soil that holds air or water.

porosity   That percentage of the total bulk volume of a soil not occupied by the solid particles.

post-emergence spray   A pesticide or herbicide that is applied after the crop plants have emerged from the soil.

potting mixture   Combination of various ingredients such as soil, peat, sand, perlite, or vermiculite designed for starting seeds or growing plants in containers.

preemergence spray   A pesticide or herbicide that is applied after planting, but before the crop plants emerge from the soil.

pre-plant herbicide   An herbicide applied to a field prior to planting.

primary tissue   A tissue that has differentiated from a primary meristem.

primocane   A vegetative cane of raspberries, blackberries, or their hybrids that arises from crown or rhizome buds. Primocanes develop floral buds in response to physiological or environmental cues but typically do not fruit until their chilling requirements have been met.

primordium   An organ in its earliest stage of development, such as leaf primordium.

profile   A vertical section of the soil through all its horizons and extending into the parent material.

prokaryotic cells   Cells in which DNA is not enclosed in a nucleus, for example, bacteria.

propagation   The process of artificially or naturally increasing plant numbers.

propagule   Any part of a plant used to generate a new plant.

protein   Any of a group of nitrogen-containing compounds that yield amino acids on hydrolysis and have high molecular weights. They are essential parts of living matter and are one of the essential food substances of animals.

protoplasm   The essential, complex living substance of cells on which all vital functions of nutrition, secretion, growth, and reproduction depend.

protoplast   The organized living unit of a single cell.

provenance   The natural origin of a tree or group of trees. In forestry, the term is considered synonymous with geographic origin.

pruning   A method of directing or controlling plant growth and form by removing certain portions of the plant.

pure line   Plants in which all members have descended by self-fertilization from a single homozygous individual.

raceme   An inflorescence in which flowers on pedicels are borne on a single, unbranched main axis.

radial face   The wood surface exposed when a stem is cut along a radius from pith to bark, and the cut parallels the long axis of the majority of the cells.

radicle   The part of the embryonic axis that becomes the primary root. The first part of the embryo to start growth during seed germination.

range   Land and native vegetation that is predominantly grasses, grasslike plants, or shrubs suitable for grazing by animals.

range management   Producing maximum sustained use of range forage without detriment to other resources or uses of land.

raphe   Ridge on seeds, formed by the stalk of the ovule, in those seeds in which the funiculus is sharply bent at the base of the ovule.

ray   A narrow group of cells, usually parenchyma, extending radially in the wood and bark.

reaction *See* pH.

receptacle The enlarged tip of a stem on which a flower is borne.

recessive gene A gene that does not express itself in the presence of the contrasting (dominant) gene.

recombination The mixing of genotypes that results from sexual reproduction.

reduction division A nuclear division in which the chromosomes are reduced from the diploid to the haploid number.

registered seed The progeny of foundation or registered seed produced and handled to maintain satisfactory genetic identity and purity, and approved and certified by an official certifying agency. Registered seed is normally grown for the production of certified seed.

regulatory genes Codes for proteins that regulate structural gene activity. *See also* transcription factors.

relative humidity (RH) The ratio of the amount of water vapor in the air compared to what the air can hold at that temperature. *See also* saturated vapor pressure (SVP).

remobilization The reallocation of nutritional elements from one part of the plant to another. For example, the movement of nitrogen from lower leaves to upper leaves or developing flowers, fruit, and seed.

replication In cell physiology, the production of a second molecule of DNA exactly like the first molecule.

reproduction Sexual reproduction is development of new plants by seeds (except in apomixis). Vegetative reproduction, or vegetative propagation, is reproduction by other than sexually produced seed. Includes grafting, cuttings, layering, and so forth, as well as apomixis.

respiration The oxidation of food by plants and animals to yield energy for cellular activities.

rest period An endogenous physiological condition of viable seeds, buds, or bulbs that prevents growth even in the presence of otherwise favorable environmental conditions. This is referred to by some seed physiologists as *dormancy*.

Rhizobium Genus of bacteria that live symbiotically in the roots of legumes and fix nitrogen that is used by plants.

rhizome An underground stem, usually horizontal and often elongated; distinguished from a root by the presence of nodes and internodes. Capable of producing new shoots.

ribonucleic acid (RNA) A single-strand acid, formed on a DNA template, found in the protoplasm, and controlling cellular chemical activities. Whereas DNA transmits genetic information from one cell generation to the next, ribonucleic acid is an intermediate chemical translating genetic information into action.

ribosome A protoplasmic granule containing ribonucleic acid (RNA) and believed to be the site of protein synthesis.

ribulose bisphosphate carboxylase (RUBISCO) The enzyme in the non-light-requiring reactions of photosynthesis that fixes atmospheric carbon dioxide into a carbohydrate and in the process converts radiant energy to chemical energy.

ripening Chemical and physical changes in a fruit that follow maturation.

rock The material that forms the essential part of the earth's solid crust, including loose incoherent masses such as sand and gravel, as well as solid masses of granite, limestone, and others.

root The descending axis of a plant, usually below ground, serving to anchor the plant and absorb and conduct water and mineral nutrients.

root cap A mass of hard cells covering the tip of a root and protecting it from mechanical injury.

root hair An absorptive unicellular protuberance of the epidermal cells of the root.

rooting media Materials such as peat, sand, perlite, or vermiculite in which the basal ends of cuttings are placed vertically during the development of roots.

root pressure Pressure developed in the root as the result of osmosis and causing bleeding in stem wounds of some plants.

rootstock Also known as *understock* and *stock*. The trunk or root material to which buds or scions are inserted in grafting.

roughage Plant materials that are relatively high in crude fiber and low in digestible nutrients, such as straw.

rumen The stomach in ruminate animals in which cellulose is digested.

ruminant Cud-chewing mammals such as cattle, sheep, goats, and deer, characteristically having a stomach divided into four compartments.

runner *See* stolon.

rural/urban interface The influx of nonagricultural residents and nonfarm land uses into rural areas.

salicylic acid A recently identified plant hormone that plays a role in a plant's response to pathogen attack. Salicylic acid is also the active ingredient in aspirin.

saline soil A nonsodic soil containing sufficient soluble salts to impair plant growth.

samara A dry, indehiscent, simple fruit that has winglike appendages on both sides of the ovary. These appendages help carry the wind-borne fruit.

sand A soil particle between 0.05 and 2.0 mm in diameter.

saprophyte An organism deriving its nutrients from the dead body or the nonliving products of another plant or animal.

saturated vapor pressure (SVP) The greatest pressure that water vapor could exert if the air at a given temperature were saturated with water vapor. *See also* relative humidity (RH).

savanna Grassland having scattered trees, either as individuals or clumps; often a transitional type between true grassland and forest.

scape A strawberry inflorescence.

scarify To scratch, chip, or nick the seed coverings of certain species to enhance the passage of water and gases as an aid to seed germination.

scientific method An approach to a problem that consists of stating the problem, establishing one or more hypotheses as solutions to the problem, testing these hypotheses by experimentation or observation, and accepting or rejecting the hypotheses.

scion A small shoot that is inserted by grafting into a rootstock.

scion-stock interaction The effect of a rootstock on a scion (and vice versa) in which a scion on one kind of rootstock performs differently than it would on its own roots or on a different root-stock.

sclereid A short, thickened cell that provides support or protection in plants. *See also* sclerenchyma tissue.

sclerenchyma tissue Supporting or protective tissue in which the cells have hard lignified walls.

scutellum The rudimentary leaflike structure at the first node of the (embryonic) stem (culm) of a grass plant. The single cotyledon of a monocotyledonous seedling.

secondary mineral nutrients Calcium, magnesium, and sulfur; used in lesser amounts than the macronutrients.

secondary phloem Phloem cells formed by activity of the vascular cambium. Secondary phloem is found in biennials and perennials, but usually not in annuals.

secondary xylem Xylem cells formed by activity of the vascular cambium. The development of the secondary xylem accounts for the so-called annual rings seen in most trees.

sectorial chimera Mutated plant cells are arranged so that they extend through cell layers, thus producing both normal and mutated leaves and stems.

Secure Handling of Ammonia Act The federal law in the United States governing the handling of ammonia. A similar law in Canada is called the Explosives Act.

sedimentary rock Rock formed from material originally deposited as a sediment, then physically or chemically changed by compression and hardening while buried in the earth's crust.

seed The mature ovule of a flowering plant containing an embryo, an endosperm (sometimes), and a seed coat.

seed bank The reserve of seeds in the ground.

seedbed Soil that has been prepared for planting seeds or transplants.

seed coat *See* testa.

seed potatoes Pieces of potato tubers or whole tubers that are planted to produce new plants and subsequent commercial crops.

self-fertile Capable of fertilization and producing viable seed after self-pollination.

self-incompatibility Inability to produce viable seed following self-pollination. The inability is sometimes due to a pollen-borne gene that prevents pollen tube growth on a stigma with the same gene.

self-pollination Transfer of pollen from the stamens to the stigma of either the same flower, other flowers on the same plant, or flowers on other plants of the same clone.

self-sterility Failure to complete fertilization and obtain viable seed after self-pollination.

semiarid Climate in which evaporation exceeds precipitation, a transition zone between a true desert and a humid climate. Usually annual precipitation is between 250 and 500 mm (10 to 20 in.).

senescence A physiological aging process in which tissues in an organism deteriorate and finally die.

sepals The outermost series of floral parts; usually green, leaflike structures at the base of a flower; collectively, they form the calyx.

sessile Used in reference to flowers, florets, leaves, leaflets, or fruits that are attached directly to a shoot and not borne on any type of a stalk.

sexual reproduction Development of new plants by the processes of meiosis and fertilization in the flower to produce a viable embryo in a seed.

shade avoidance response The change in growth in response to overcrowding among plants, most notably an increase in stem length, thinner stems, as well as larger, thinner leaves.

shoot-bearing A term describing a plant that predominantly bears fruit terminally or axially on shoots developed during the current or previous season.

shoot meristem *See* apical meristem.

short-day plants (SDP) Plant that initiates flowers only under short-day (long-night) conditions.

sidedressing Applying fertilizer on a soil surface close enough to a plant that cultivating or watering carries the fertilizer to the plant's roots.

sieve tube Part of phloem tissue formed by the connecting of sieve cells end to end through which nutrients and other soluble substances flow in plants.

silage Forage that is chemically changed and preserved in a succulent condition by partial fermentation in the preparation of food for livestock.

silique A dry, one-seeded, dehiscent (infrequently indehiscent) fruit consisting of two carpels that form a bilocular ovary. Common in the BRASSICACEAE family.

silo A structure for making and storing silage.

silt A soil textural class consisting of particles between 0.05 and 0.002 mm in diameter.

simple fruit A fruit derived from a single pistil.

simple leaf Leaf that is not divided into distinct leaflets. *See also* compound leaf.

sink The deposition of carbon dioxide in the form of carbohydrates and often carbon compounds in plant tissues and organs.

sink tissue Tissue that is actively growing or storing carbohydrates and is importing photosynthates. *See also* source leaf.

sod Top 3 to 7 cm (1 to 3 in.) of soil permeated by and held together with grass roots or grass-legume roots.

sodic soil A soil that contains so much exchangeable sodium (more than 15 percent) that it interferes with the growth of most crop plants. Also, if the total soluble salts is more than 4 mmho per cm^2, the soil is a saline sodic soil; if it is less than 4 mmho per cm^2, the soil is a nonsaline sodic soil.

soil The solid portion of the earth's crust in which plants grow. It is composed of mineral material, air, water, and organic matter both living and dead.

soil air The gaseous phase of the soil; that percentage of the total volume not occupied by solid or liquid.

soil compaction The reduction of soil pore space and the increase in soil bulk density as a result of pressure from sources such as animals, machinery, or humans moving across the surface of the soil.

soil conservation A combination of all management and land use methods that safeguard the soil against depletion or deterioration by natural or by human-induced factors.

soil fertilization *See* fertilization.

soil management The total tillage operations, cropping practices, fertilizing, liming, and other treatments conducted on or applied to a soil for the production of plants.

soil organic matter   The organic fraction of the soil that includes plant and animal residues at various stages of decomposition, cells and tissues of soil organisms, and substances synthesized by the soil population.

soil pasteurization   Treating the soil with heat (usually steam at 60°C to 70°C [140°F to 160°F]) to destroy most harmful pathogens, nematodes, and weed seeds. Less severe temperature treatment than soil sterilization.

soil profile   Distinct layers of soil that develop over time from the weathering of parent material, biological activity, and other factors.

soil salinity   The amount of soluble salts in a soil, expressed as parts per million, millimho/cm, or other convenient ratios.

soil series   The basic unit of soil classification; a subdivision of a family, comprising soils that are essentially alike in all major profile characteristics.

soil solution   The aqueous liquid phase of the soil and its solutes consisting of ions dissociated from the surfaces of the soil particles and of other soluble materials.

soil sterilization   Treating soil by gaseous fumigation, chemicals, or heat (usually steam) at 100°C (212°F) to destroy all living organisms.

soil structure   The arrangement of primary soil particles into secondary particles, units, or peds that act as primary particles. The secondary units are characterized and classified on the basis of size, shape, and degree of distinctness.

soil testing   Chemical analysis of soil to determine its nutrient status, pH, electroconductivity (EC), and other compositional factors.

soil texture   The relative percentages of sand, silt, and clay in a soil.

soil tilth   The physical condition of soil as related to its ease of tillage, fitness as a seedbed, and suitability for plant growth.

soil type   The lowest unit in the natural system of soil classification; a subdivision of a soil series.

solarize   Cause damage to leaves from overexposure to the sun.

soluble salts   Disassociated salts (anions and cations) in the soil that can become toxic to roots when exceeding certain levels.

solute   A substance dissolved in a solvent.

solution   A homogeneous mixture; the molecules of the dissolved substance (the solute) are dispersed among the molecules of the solvent.

solvent   A substance, usually a liquid, that can dissolve other substances (solutes).

somaclonal variation   Existing genetic variation that may not be seen until after plant cells have been through aseptic culture, or the culture may force the change.

somatic tissue   Nonreproductive, vegetative tissue. Tissue developed through mitosis that will not undergo meiosis.

source leaf   Leaf that is producing and exporting more photosynthates than it requires for its own use.

spadix   An inflorescence with sessile flowers on a fleshy stalk, for example, *Spathiphyllum,* or peace lily.

species   A group of similar organisms capable of interbreeding and more or less distinctly different in geographic range and/or morphological characteristics from other species in the same genus.

specific heat   The amount of heat required to raise the temperature of 1 gram of a substance by 1°C.

sperm   A male gamete.

spermatophyte   A seed-bearing plant.

spike   An inflorescence that has a central axis on which sessile flowers are borne. Examples are some grasses, gladioli, and snapdragons.

spiroplasma   Helical prokaryote organisms that lack rigid cell walls and infect plants.

spodosol   A soil that is highly leached, has low fertility, and retains little water.

spongy mesophyll parenchyma cells   The cell layer in a leaf located between the palisade parenchyma and the lower epidermis; these cells have thin cell walls and are loosely packed.

spore   A reproductive cell that develops into a plant without union with other cells.

sprigging   Vegetative propagation by planting stolons or rhizomes (sprigs).

spring ephemerals   Small flowering plants that grow in deciduous forests that start and complete their annual lifecycle in a very short time in the spring before trees leaf out.

spur-bearing   A term describing a plant that predominantly bears fruit on fruiting spurs.

square   An unopened flowerbud in cotton with its accompanying bracts.

stamen   The male reproductive structure of a flower. The stamen produces pollen and is composed of a filament on which is borne an anther.

staminate (male) flower   A flower having stamens but no pistils.

starch   A complex polysaccharide carbohydrate. The form of food commonly stored by plants.

stele   The vascular tissue and closely associated tissues in the axes of plants. The central cylinder of the stem.

stem   The main body of a plant, usually the ascending axis, whether above or below ground in opposition to the descending axis or root. Stems, but not roots, produce nodes and buds.

stigma   In a flower, the portion of the style to which pollen adheres.

stipule   Appendage at the base of a leaf where it attaches to the stem.

stock   *See* rootstock.

stolon   A slender, prostrate above-ground stem. The runners of white clover, strawberry, and Bermuda grass plants are examples of stolons.

stoma (stomate)   Plural: **stomata, stomates,** respectively. A small opening, bordered by guard cells, in the epidermis of leaves and stems, through which gases, including water vapor, pass.

stratification   The practice of exposing imbibed seeds to cool 2°C to 10°C (35°F to 50°F) (sometimes warm) temperatures for a period of time prior to germination and thus break dormancy. This is a standard practice in the germination of seeds of many grass and woody species.

stress   Plant(s) unable to absorb enough water to replace that lost by transpiration. Results may be wilting, cessation of growth, or death of the plant or plant parts.

strig   A currant or gooseberry inflorescence.

strip cropping   The practice of growing crops that require different types of tillage, such as row and sod, in alternate strips along contours or across the prevailing direction of wind.

structural genes   Code for proteins that regulate the structural component of the plant.

style   Slender column of tissue that arises from the top of the ovary in the flower through which the pollen tube grows toward the ovule.

stylet   A mouthpart on some insects and nematodes that allows the organisms to pierce the cells of other organisms.

subapical meristem   Region that produces new cells a few micrometers behind an active shoot or apical meristem.

suberin   A waxy, waterproofing substance in cork tissue.

subirrigation   Application of water from below the soil surface, usually from a ditch or a perforated hose or pipe, or by placing a potted plant on a constantly moist surface.

sublimation   The change of phase of a substance from solid to vapor without going through the liquid stage.

subsoil   That part of the soil below the plow layer.

subsoiling   Breaking of compact subsoils, without inverting them, with a special knifelike instrument (chisel) that is pulled through the soil at depths usually of 30 to 60 cm (12 to 24 in.) and at spacings usually of 60 to 150 cm (2 to 5 ft).

succession   The serial replacement of simple communities by more complex communities. *See also* climax community.

sucrose   Table sugar ($C_{12}H_{22}O_{11}$); a carbohydrate formed by chemically joining a molecule of glucose ($C_6H_{12}O_6$) with a molecule of fructose ($C_6H_{10}O_5$).

summer fallow   The tillage of uncropped land during the summer to control weeds and store moisture in the soil for the growth of a later crop.

sunscald   High temperature injury to plant tissue due to the sun's intense rays warming the trunks of trees during winter, cracking and splitting the bark. It can be prevented by shading or whitewashing tree trunks and larger branches. Sunscald also occurs on unshaded vegetable fruits (tomatoes and melons) or to house-plants exposed to direct sunlight.

superior ovary   An ovary situated above the receptacle; all other floral parts develop below the base of the ovary.

surface tension   The physical property of water in the liquid state that is due to the intermolecular attraction (hydrogen bonding) between the water molecules.

Sustainable Agriculture Research and Education (SARE)   The outreach office the United States Department of Agriculture that *advances farming practices that are profitable, environmentally sound, and good for the community.*

sustainability   The ability to be maintained with little input of nonrenewable resources.

symbiosis   An obligate relationship between two organisms of different species living together in close association for their mutual benefit. An example is the mycelium of a fungus with roots of seed plants.

symplastic   Movement of water into and through the cytoplasm of plant cells to other parts of the plant through the plasmodesmata.

synergids   The two nuclei within the embryo sac at the upper end in the ovule of the flower, which, with the third (the egg), constitute the egg apparatus.

synergism   The mutual interaction of two substances that results in an effect greater than the additional effects of the two when used independently.

systemic   A pesticide material absorbed by plants, making them toxic to feeding insects. Also, pertaining to a disease in which an infection spreads throughout the plant.

systemin   A peptide that has been recently identified as a plant hormone. It is produced in wounded tissue and is transported to remote cells and tissues, where it induces defense genes.

tangential face   The wood surface exposed when a cut is made at right angles to the rays and parallel to the long axis of the majority of cells.

tannin   Broad class of soluble polyphenols with a common property of condensing with protein to form a leatherlike substance that is insoluble in water.

tap root   An elongated, deeply growing primary root.

taxon   Plural: **taxa**. A taxonomic group of plants of any rank such as family, genus, species, and so forth.

taxonomy   The science dealing with describing, naming, and classifying plants and animals.

tendril   A slender coiling modified leaf or stem arising from stems and aiding in their support.

tension   The equivalent negative pressure of water in soil. The attraction with which water is held to soil particles.

terminal bud   A bud at the distal end of a stem.

terrace   A level, usually narrow, plain bordering a river, lake, or the sea. Rivers sometimes are bordered by terraces at different levels. It also refers to a raised, more or less level strip of land usually constructed on a contour and designed to make the land suitable for tillage and to prevent erosion.

testa   Protective tissue derived from the integuments that cover a seed. Also known as *seed coat.*

tetraploid   Having four sets of chromosomes per nucleus.

texture   The relative proportions of the various groups of individual soil grains, of different sizes, in a mass of soil. It refers to the proportions of sand, silt, and clay in a given amount of soil.

thallophytes   A division of plants whose body lacks roots, stems, and leaves (e.g., mushrooms).

thatch   A layer of undecomposed organic material (i.e., roots, leaves, stems) that builds up in turf above the soil and below the green vegetation.

thinning   Removing young plants from a row to provide the remaining plants with more space to develop. Also, the removal of excess numbers of fruits from a tree so the remaining fruits will become larger.

tile drain   Pipe made of burned clay, concrete, or similar material, in short lengths, usually buried at the bottom of a ditch, with open joints to collect and carry excess water from the soil. The newer type polyvinyl tile is made in continuous lengths and is perforated to allow water to enter the pipe for removal from the soil.

tillage   The mechanical manipulation of soil to provide a favorable environment for plant growth.

tiller   A grass stem that arises from an axillary bud.

tilth *See* soil tilth.

tissue A group of cells of similar structure that performs a special function.

tolerance The ability of a plant to be attacked or infested by a pest but show few symptoms.

tonoplast Membrane that surrounds the vacuole in a plant cell.

top-dressing Applying materials such as fertilizer or compost to the soil surface while plants are growing.

top-grafting *See* top-working.

topsoil The upper layer of soil moved in cultivation.

top-working Also known as *top-grafting*. To change the cultivar of a tree by grafting the main scaffold branches.

total photosynthesis The total amount of carbohydrates synthesized from photosynthesis. *See* net photosynthesis.

tracheid An elongated, tapering xylem cell with lignified pitted walls adapted for conduction and support.

training The positioning of limbs, branches, and canes.

transcription factors Proteins that regulate structural gene activity.

transfer RNA *See* ribonucleic acid.

translocation The transfer of food materials or products of metabolism throughout the plant.

transpiration The loss of water vapor through the stomata of leaves.

trifoliate Leaf with three leaflets attached at the same point, for example, clover.

tuber An enlarged, fleshy, underground tip of a stem. The white (Irish) potato produces tubers as food storage organs.

tuberous root An enlarged fleshy, underground root (e.g., sweet potato and dahlia).

turgid Swollen, distended; referring to cells or tissues that are firm because of water uptake.

turgor pressure The pressure within the cell resulting from the absorption of water into the vacuole and the imbibition of water by the protoplasm.

2,4-dichlorophenoxyacetic acid (2,4-D) A selective auxin-type herbicide that kills broad-leaved plants but not grasses.

ultisol A tropical and subtropical soil with acidic, highly weathered layers.

umbel A type of inflorescence in which flowers are borne at the end of stalks that arise like the ribs of an umbrella from one point (e.g., carrot, onion).

unavailable water Water held by the soil so strongly that the root cannot absorb it.

understock *See* rootstock.

understory Those plants that grow under the canopy of other taller plants.

upright A decumbent shoot arising terminally or axially from a cranberry runner that will bear fruit from terminal buds.

vacuole A cavity in the plant's cell bounded by a membrane in which various plant products and by-products are stored.

vapor pressure deficit (VPD) The difference between the amount of moisture actually in the air and the amount the air holds when it is saturated.

variety A subdivision of a species with distinct morphological characters and given a Latin name according to the rules of the International Code of Botanical Nomenclature. A taxonomic variety is known by the first validly published name applied to it so that nomenclature tends to be stable. *See also* cultivar.

vascular bundle A strand of tissue containing primary xylem and primary phloem and frequently enclosed by a bundle sheath of parenchyma or fibers.

vascular cambium A meristem that produces secondary xylem and secondary phloem cells. Vascular cambium is found in biennials and perennials.

vascular plants Plants that have an organized vascular system comprised of xylem and phloem.

vascular system In seed-bearing plants, this system consists of the pericycle, phloem, vascular cambium, xylem, pith rays, and pith.

vector A carrier; for example, an insect that carries pathogenic organisms from plant to plant.

vegetation The plants that cover a region; it is formed of the species that make up the flora of the area.

vegetative growth Referring to asexual (stem, leaf, root) development in plants in contrast to sexual (flower, seed) development.

vein The vascular strand of xylem and phloem in a leaf.

vernalization In reference to flowering, the process by which floral induction in some plants is promoted by exposing the plants to chilling for a certain length of time.

vertisol A soil that has high clay content and poor physical structure. It is prone to shrinking and swelling as moisture content changes.

vessel A series of xylem elements in the stem and root that conduct water and mineral nutrients.

virus A pathogen consisting of a single strand of nucleic acids encapsuled in a protein coat that can replicate only inside a living cell.

visible light Light with wavelengths of 400 to 700 nm, which is the range visible to the human eye.

vitamins Natural organic substances, necessary in small amounts for the normal metabolism of plants and animals.

volatile Evaporating readily or easily dissipated in the form of a vapor.

water potential The difference between the activity of water molecules in pure distilled water at atmospheric pressure and 3°C (standard conditions) and the activity of water molecules in any other system; the activity of these water molecules may be greater (positive) or less (negative) than the activity of the water molecules under standard conditions.

water table The upper surface of groundwater or that level below which the soil is saturated with water.

watts (W) Radiometric units expressing energy per unit of time and unit area independent of wavelength. Irradiance within the photosynthetic active region (PAR) of 400–700 nm is commonly referred to as W PAR.

wavelength The distance between two corresponding points on any two consecutive waves. For visible light, it is minute and is generally measured in nanometers.

weather   The current conditions of temperature, wind, precipitation, etc. in an area.

weathering   All physical and chemical changes produced in rocks, at or near the earth's surface, by atmospheric agents.

weed   A plant not valued for use or beauty. Any plant growing where it is not wanted.

wilting point   Also known as *permanent wilting percentage (PWP)*. The moisture content of soil at which plants wilt and fail to recover even when placed in a humid atmosphere.

windbreak   A planting of trees or shrubs, usually perpendicular or nearly so to the principal wind direction, to protect soil, crops, homesteads, roads, and so on, against the effects of winds.

windrow   Hay, grain, leaves, or other material swept or raked into rows to dry.

winter hardiness   The ability of a plant to tolerate severe winter conditions.

wood   Secondary nonfunctioning xylem in a perennial shrub or tree.

xanthophyll   Yellow carotenoid pigment ($C_{40}H_{56}O_2$) found along with chlorophyll in green plants.

xeriscape   Landscapes that are designed to reduce or eliminate the need for irrigation.

xylem   Specialized cells through which water and minerals move upward from the soil through a plant.

zone of accumulation   *See* B horizon.

zygote   A protoplast resulting from the fusion of gametes (either isogametes or heterogametes). The beginning of a new plant in sexual reproduction.